Forage Evaluation in Ruminant Nutrition

Forage Evaluation in Ruminant Nutrition

Edited by

D.I. Givens

ADAS Nutritional Sciences Research Unit
Alcester Road
Stratford-upon-Avon, UK

E. Owen

Department of Agriculture
University of Reading
Reading, UK

R.F.E. Axford

School of Agricultural and Forest Sciences
University of Wales
Bangor, UK

and

H.M. Omed

School of Agricultural and Forest Sciences
University of Wales
Bangor, UK

CABI *Publishing*

CABI is a trading name of CAB International

CABI Head Office
Nosworthy Way
Wallingford
Oxfordshire OX10 8DE
UK

Tel: +44 (0)1491 832111
Fax: +44 (0)1491 833508
Email: cabi@cabi.org
Web site: www.cabi.org

CABI North American Office
875 Massachusetts Avenue
7th Floor
Cambridge, MA 02139
USA

Tel: +1 617 395 4056
Fax: +1 617 354 6875
Email: cabi-nao@cabi.org

A catalogue record for this book is available from the British Library, London, UK.

Library of Congress Cataloging-in-Publication Data
Forage evaluation in ruminant nutrition / edited by D.J. Givens … [et al.].
p. cm.
Includes bibliographical references.
ISBN 0-85199-344-3
1. Forage plants—Evaluation. 2. Ruminants—Feeding and feeds.
3. Ruminants—Nutrition. I. Givens, D.I. (D. Ian)
SF95.F6725 2000
636.2'0852—dc21

99-052658

ISBN-13: 978-0-85199-344-7
ISBN-10: 0-85199-344-3

First published 2000
Reprinted 2002
Transferred to print on demand 2007

Printed and bound in the UK by CPI Antony Rowe, Eastbourne.

Contents

Contributors

A.T. Adesogan, *Welsh Institute of Rural Studies, University of Wales, Aberystwyth SY23 3AL, UK*

R.F.E. Axford, *School of Agricultural and Forest Sciences, University of Wales, Bangor, Gwynedd LL57 2UW, UK*

A. Bach, *Department of Animal Science, University of Minnesota, St Paul, MN 55108, USA*

N. Ballet, *Aventis Animal Nutrition, 42 Avenue Aristide Briand, BP 100, 92164 Antony Cedex, France*

D.E. Beever, *CEDAR (Centre for Dairy Research), Department of Agriculture, University of Reading, Earley Gate, Reading RG6 6AT, UK*

S. Calsamiglia, *Departament de Patologia i de Producció Animals, Universitat Autónoma de Barcelona, 08193 Bellaterra, Spain*

X.B. Chen, *Rowett Research Institute, Bucksburn, Aberdeen AB21 9SB, UK*

D.J.R. Cherney, *Department of Animal Science, New York State College of Agriculture and Life Sciences, Cornell University, Morrison Hall, Ithaca, NY 14853-4801, USA*

E.R. Deaville, *ADAS Nutritional Sciences Research Unit, Alcester Road, Stratford-upon-Avon CV37 9RQ, UK*

J. Dijkstra, *Animal Nutrition Group, Wageningen Institute of Animal Sciences, Wageningen University, Marijkeweg 40, 6709 PG Wageningen, The Netherlands*

P.C. Flinn, *Agriculture Victoria, Pastoral and Veterinary Institute, Private Bag 105, Hamilton, Victoria 3300, Australia*

J. France, *Department of Agriculture, University of Reading, Earley Gate, Reading RG6 6AT, UK*

M. Gill, *Natural Resources International Ltd, PO Box 258, Chatham, Kent ME4 4PU, UK*

D.I. Givens, *ADAS Nutritional Sciences Research Unit, Alcester Road, Stratford-upon-Avon CV37 9RQ, UK*

J. Hanson, *International Livestock Research Institute, PO Box 5689, Addis Ababa, Ethiopia*

D.S. Himmelsbach, *Richard B. Russell Agricultural Research Center, Agricultural Research Service, PO Box 5677, Athens, GA 30604-5677, USA*

T. Hvelplund, *Danish Institute of Agricultural Sciences, Department of Animal Nutrition and Physiology, Research Centre Foulum, PO Box 50, 8830 Tjele, Denmark*

D.I.H. Jones, *Institute of Grassland and Environmental Research, Plas Gogerddan, Aberystwyth, Ceredigion SY23 3EB, UK*

C. Krueger, *Department of Animal Sciences, University of Wisconsin-Madison, 1675 University Avenue, Room 256, Madison, WI 57306, USA*

S. López, *Department of Animal Production, University of Leon, 24007 Leon, Spain*

D.K. Lovett, *School of Agricultural and Forest Sciences, University of Wales, Bangor, Gwynedd LL57 2UW, UK*

L.R. McDowell, *Department of Animal Science, University of Florida, 125 Animal Science Building, 459 Shealy Drive, Gainesville, FL 32618, USA*

A. MacPherson, *Scottish Agricultural College, Auchincruive, Ayr KA6 5HW, UK*

F.L. Mould, *CEDAR (Centre for Dairy Research), Department of Agriculture, University of Reading, Earley Gate, Reading RG6 6AT, UK*

H.M. Omed, *School of Agricultural and Forest Sciences, University of Wales, Bangor, Gwynedd LL57 2UW, UK*

E.R. Ørskov, *International Feed Resource Unit, Rowett Research Institute, Bucksburn, Aberdeen AB21 9SB, UK*

E. Owen, *Department of Agriculture, University of Reading, Earley Gate, PO Box 236, Reading RG6 6AT, UK*

J.D. Reed, *Department of Animal Sciences, University of Wisconsin-Madison, 1675 University Avenue, Room 256, Madison, WI 57306, USA*

C.K. Reynolds, *Department of Agriculture, University of Reading, Earley Gate, Reading RG6 6AT, UK*

J.C. Robert, *Aventis Animal Nutrition, 42 Avenue Aristide Briand, BP 100, 92164 Antony Cedex, France*

G. Rodriguez, *Department of Animal Sciences, University of Wisconsin-Madison, 1675 University Avenue, Room 256, Madison, WI 57306, USA*

D.L. Romney, *Natural Resources Institute, Chatham Maritime, Kent ME4 4TB, UK*

C. Rymer, *ADAS Nutritional Sciences Research Unit, Alcester Road, Stratford-upon-Avon CV37 9RQ, UK*

M.D. Stern, *Department of Animal Science, University of Minnesota, St Paul, MN 55108, USA*

S. Tamminga, *Wageningen Institute of Animal Sciences, Group Animal Nutrition, Marijkeweg 40, 6709 PG Wageningen, The Netherlands*

M.K. Theodorou, *Institute of Grassland and Environmental Research, Plas Gogerddan, Aberystwyth, Ceredigion SY23 3EB, UK*

G. Valle, *Departamento de Producción Animal, CURLA, Apartado 89, La Ceiba, Honduras*

M.R. Weisbjerg, *Danish Institute of Agricultural Sciences, Department of Animal Nutrition and Physiology, Research Centre Foulum, PO Box 50, 8830 Tjele, Denmark*

R.J. Wilkins, *Institute of Grassland and Environmental Research, North Wyke, Okehampton, Devon EX20 2SB, UK*

B.A. Williams, *Animal Nutrition Group, Wageningen Institute of Animal Sciences (WIAS), Wageningen University, Marijkeweg 40, 6709 PG Wageningen, The Netherlands*

P.E.V. Williams, *Aventis Animal Nutrition, 42 Avenue Aristide Briand, BP 100, 92164 Antony Cedex, France*

Preface

Over 3000 million hectares of the land area of the earth (over 25%) is grazing land and another 4000 million hectares of forest and woodland have some grazing potential. World grasslands support approximately 1500 million cattle equivalents (cattle, buffalo, sheep, goats and camels) and forages provide over 90% of the feed energy consumed by these herbivorous animals. The world's forages therefore indirectly provide a very high proportion of the food for its population. This is achieved without seriously reducing the quantity of food available for direct human consumption.

Although forages generally provide nutrients to animals at lower cost than concentrate feeds, they are inherently variable in nutritive value. This depends on many factors such as forage species, climate, degree of maturity, etc. In many parts of the world, forages are conserved by processes such sun curing and ensiling. These processes can fundamentally change the nutritional characteristics of the original forage, sometimes in unpredictable ways.

Given the importance and variability of forages, it is vital that methods exist that can reliably assess their key nutritional attributes including, crucially, their voluntary intake by animals. In recent years a number of important factors have come into play that are changing the ways in which forage characterization in the laboratory is approached. For instance, in some countries characterization of ruminant feeds in general is rapidly moving away from expressions of energy and protein content to an assessment of the nutrients supplied to the animal both directly and indirectly as a result of microbial activity in the rumen. In addition, in some places there is increasingly powerful public pressure to reduce or stop the use of surgically modified animals in nutritional studies. This may rapidly reduce the use of techniques reliant on rumen fluid and alternatives to these will have to be found. There has been a tremendous upsurge in the use of near infrared reflectance spectroscopy for forage characterization in countries where the expensive technology is available. This emphasizes the need for cheaper but still reliable methods for less well-equipped regions of the world.

In this book the current status of forage evaluation is reviewed and discussed. An attempt has been made to establish the key aspects of forage evaluation given the demands of increasing nutritional complexity and the constraints outlined above. An account is given of the new technologies now available, and consideration given to some of the new nutritional characteristics that may be important in forages of the 21st century.

1 Forages and Their Role in Animal Systems

R.J. Wilkins

Institute of Grassland and Environmental Research, North Wyke, Okehampton, Devon EX20 2SB, UK

This introductory chapter will consider the definition of forage, the types of feed encompassed within this definition and their characteristics, the extent to which forages contribute to animal production in different parts of the world and future prospects.

What are Forages?

The terms forage and its equivalents in French (*fourrage*), Spanish (*foraje*) and Italian (*foraggio*) are commonly used by both scientists and farmers, but there is no widely accepted definition and much variation in the breadth of feeds that may be considered within this term. The definition offered by the authoritative *Oxford English Dictionary* of 'food for horses and cattle' does not correspond with either popular or scientific usage, being too wide in terms of 'food' and too narrow in terms of the animals considered. For the purposes of this chapter, I shall follow the definition given by the Forage and Grazing Terminology Committee (1991), a group endorsed by the International Grassland Congress. Forage is defined as 'edible parts of plants, other than separated grain, that can provide feed for grazing animals or that can be harvested for feeding'. This broad definition is also used by Barnes and Baylor (1995) and includes the classes of feed listed in Table 1.1. The narrower term of 'forage crop' is often used to describe crops, generally annual or biennial, which are grown to be utilized by grazing or harvesting as a whole crop (e.g. maize, sorghum, kale).

Thus a wide range of feeds are included as 'forages', but they generally have substantial contents of cell walls (root crops are an exception) and are suited to utilization by herbivores with their substantial capability for microbial digestion of cell-wall constituents. The composition and nutritive value of forages is extremely variable, both overall and within forage types, as illustrated in Table 1.2. This indicates that different forages can make very different contributions to production

Table 1.1. Feed types included within the definition of forage.

Herbage	Leaves, stems, roots of non-woody species, including sown and permanent grassland and crops that may be grazed or cut.
Hay and silage	
Browse	Buds, leaves and twigs of woody species
Straw	

Table 1.2. Range in nutrient contents of different classes of forages.

	Metabolizable energy ($MJ\ kg^{-1}$ DM)	Crude protein ($g\ kg^{-1}$ DM)
Temperate grasses, hays and silages	7.0–13.0	60–250
Tropical grasses	5.0–11.0	20–200
Maize silage	10.0–12.0	60–120
Cereal straw	5.0–8.0	20–40
Root crops	11.0–14.0	40–130
Kale and rape	9.0–12.0	140–220

systems, varying from feeds not capable of supporting animal maintenance to those with digestibility and energy concentrations as high as in cereal grains.

Forage Production

Grassland

Ruminant production systems throughout the world are based on forages, with grassland feeds being predominant. Food and Agriculture Organization (FAO, 1996) statistics indicate a total area of grassland of some 3500 × 10^6 ha, some 72% of the total agricultural land and 27% of the total land area. Table 1.3 shows large variation between regions in the proportion of land used for permanent grassland, ranging from 33% of total agricultural land in Europe to 89% in Oceania.

This wide extent of permanent grassland is a feature largely of the adaptation and perennial nature of grasses, making grassland either the climax vegetation in much of the world or a vegetation type that may be established following forest clearance and maintained with low levels of management input by grazing and, in some areas, by burning. The current area of permanent grassland is about twice that of natural climax grasslands. Most permanent grassland is on land with substantial limits to arable cropping, because of topographic or climatic factors.

Whilst much permanent grassland is extensively managed with low levels of external inputs, permanent grasslands in Europe and New Zealand may receive high inputs and maintain high stocking rates (see Wilkins, 1995). Green (1982) noted that 40% of grassland over 20 years of age in England and Wales had no

Table 1.3. Permanent grassland in different regions, 1994 (from FAO, 1996).

	Permanent grassland		
	10^6 ha	As % agricultural area	As % total land area
Africa	884	84	30
North and Central America	362	57	17
South America	495	82	28
Oceania	429	89	51
Asia*	1036	67	34
Europe*	92	33	16
Russian Federation	87	40	5
World Total	3385	70	26

* Excluding Russian Federation.

major impediments to cultivation. Much of this grassland will, however, have been reseeded at some time, either following cultivation or by oversowing.

Global figures on the area of 'temporary' grassland (i.e. sown grassland in an arable rotation) are not available, but I estimate that such temporary grassland covers 10% of the total area of arable crops – about 146 × 10^6 ha. Again, there are wide variations between different countries, with Fig. 1.1 showing for a range of European countries that temporary grassland as a percentage of total grassland ranged from 88% in Finland to 3% in Ireland. The high figure in Finland reflects both the small extent of moorland and wetland areas in the country (these are commonly occupied by permanent grassland), widespread production of grass leys in rotation with cereal crops and poor persistence of grassland over winter. Sown or

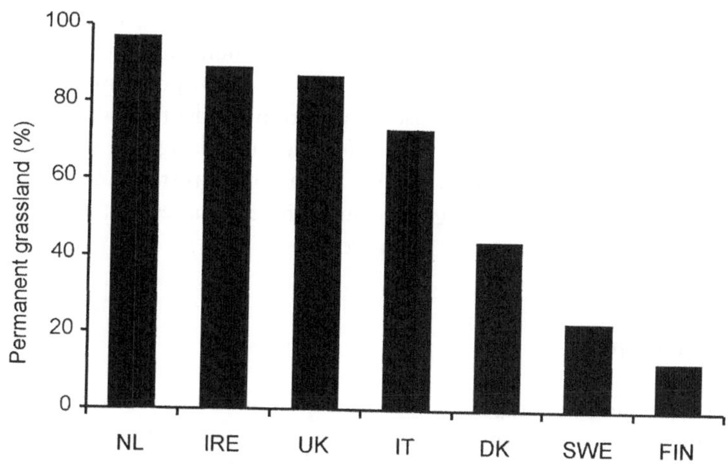

Fig. 1.1. Permanent grassland as a % of total grassland in different European countries (from R.J. Wilkins, unpublished, updated from Lee, 1988). NL, Netherlands; IRE, Ireland; UK, United Kingdom; IT, Italy; DK, Denmark; SWE, Sweden; FIN, Finland.

temporary grassland tends to be managed at a higher level of intensity than permanent grassland, particularly in relation to inputs of fertilizers and other technical chemicals.

The output of dry matter (DM) from sown grassland in Europe is generally in the range of 5–12 t DM ha^{-1} $year^{-1}$, although potential production has been suggested by Leafe (1978) to be some 20 t DM ha^{-1}. Hopkins *et al.* (1990) demonstrated that the production potential of permanent grassland in England and Wales closely approached that of reseeded grassland. Much higher yields may be obtained from tropical grasses, with figures of over 80 t DM ha^{-1} having been recorded in sown swards with adequate water and high fertilizer inputs (Snaydon, 1991). High radiation receipts, linked with the efficient C-4 photosynthetic pathway, contribute to these outputs, which are much higher than those for grain crops, because of continuous crop cover and a high harvest index.

Coupland (1992) reviewed data on the net primary production of above-ground biomass from, mainly, natural grassland ecosystems. The range in values was from 2.4 to 34 t DM ha^{-1}, with the highest values from tropical systems. Applying a figure of 8 t DM ha^{-1} to the world's grassland area of *c.* 3600 $\times$ 10^6 ha gives net primary above-ground production of some 29,800 $\times$ 10^6 t. Coupland (1993), however, estimated that only 7% of this primary production was consumed by domesticated herbivores. Jones *et al.* (1992) drew attention to underestimates in net primary production in much earlier research, because of inadequate allowance for turnover of biomass, and suggested that net primary production may be underestimated by two- to five-fold. Whilst this is highly relevant to arguments about the grassland contribution to carbon sequestration, this extra turnover of biomass would have little effect on quantities harvested by domesticated herbivores. Despite the acknowledged high potential of sown tropical grasses, their impact on world agriculture is relatively small. The resources of water and plant nutrients required to sustain high yields are not readily available in many areas and, particularly in countries with a high human population density, the limited resources of land, water and fertilizers are likely to be directed towards food or cash crops, rather than grassland (and other forage crops) for animal feeds, despite their high production potential.

In addition to producing DM (and energy), permanent and temporary grassland make a major contribution to protein production and supply of mineral elements. Grasses are extremely effective in taking up soil nitrogen and many grasslands include legumes capable of biological nitrogen fixation. Annual yields of crude protein (CP) from grasses and forage legumes are, as for DM, characteristically higher than those for grain crops.

Other forage crops

A wide range of crops in addition to perennial grasses and legumes can be used as forage crops. In temperate conditions, the major forage crops are maize, the leafy brassica crops, such as kale (*Brassica oleracea* var. *acephala*) and rape (*Brassica napus* var. *napus*), and root crops, such as fodder beet (*Beta vulgaris* ssp. *vulgaris*) and swedes (*B. napus* var. *naprobrassica*). Global statistics are not available, but, with the exception of maize, there are few situations in which annual or biennial forage crops provide the basis for ruminant production systems. When used as main crops, they are not only in competition with cash crops and food for direct human consumption, but also with low-cost production of DM and nutrients from perennial grasses and legumes.

Forage crops must have particular features in order to warrant their inclusion in production systems. These relate generally to yield, seasonality of yield, usability or quality, as discussed by Wheeler (1986). Higher yield potential for maize than grass provides a major motivation for growing maize in many areas of North America and Europe. Likewise, high annual yields contribute to the use of sorghum and small-grain cereals (e.g. wheat, barley and oats) as forages. Mitchell (1960) suggested that in New Zealand the potential production of maize and sorghum was 47 t DM ha^{-1}, compared with 29 t DM ha^{-1} for kale and only 13 t DM ha^{-1} for continuously grazed perennial ryegrass pastures, with differences in potential arising from differences in plant canopy characteristics and rooting systems.

Figure 1.2 shows the differences in daily growth rates between perennial ryegrass and the forage crops kale and fodder radish (*Raphanus sativa*) grown in southern England. Resource capture in early season is highest for perennial ryegrass, in mid-season for kale and in the autumn for fodder radish. Fodder radish is a short-season crop that may be sown after harvest of a grain crop, thus illustrating the ability of some forage crops (catch crops) to utilize land during only part of the year within a rotation, providing a contrast to grass in seasonality of feed supply. Winter forage crops also have a role in many areas, particularly when C-3 winter crops with potential to grow at low temperatures can complement summer production based on C-4 grasses or crops. Wheeler (1986) notes that, in the cool temperate areas of northern New South Wales, native pastures dominated by C-4 grasses have to be complemented by C-3 pastures or oats if sheep breeding through the winter is

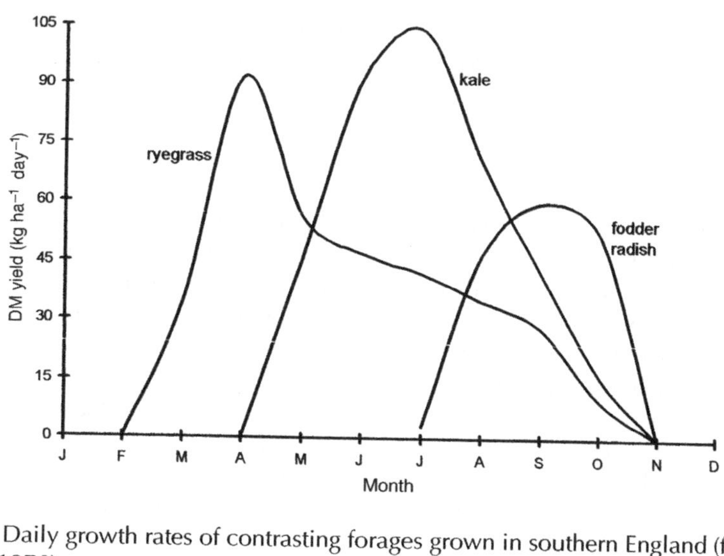

Fig. 1.2. Daily growth rates of contrasting forages grown in southern England (from Wilkins, 1976).

to be successful. Other examples of the impact on system productivity of the inclusion of winter forage crops are given by Dann and Coombe (1987). Oats are commonly used for grazing in the Mediterranean region and in temperate areas of South America and Australia. Berseem (*Trifolium alexandrinum*) is an important winter forage in the Mediterranean basin, the Middle East and India (Knight, 1985; El-Nahrawy *et al.*, 1996; Etman *et al.*, 1998). The high nutritive value of these crops is also relevant, with brassica forages and root crops having organic-matter digestibility up to 90% (Wheeler, 1986). Maraschin and Jacques (1993) discussed the use of winter forage crops in southern Brazil. Natural pastures during the winter period were able to give only 90 kg live-weight gain ha^{-1} $year^{-1}$ from beef cattle. The annual forages oats, ryegrass and rye grown as pure stands increased output to 320–380 kg ha^{-1}, whilst the provision of special-purpose pastures, based on clovers and ryegrass, increased animal production to 600 kg ha^{-1}, associated both with increased stocking rates and improved performance per animal.

When attempts are made to mitigate the effects of seasonality of growth by deferring the grazing of perennial grasses and legumes through to dry or cold periods, there are substantial losses both in quality and in biomass (Wilkins, 1976). These losses may be considerably less with leafy brassica crops and root crops, with their use facilitating grazing over a long season and avoiding the high costs associated with grass conservation as hay or silage. Costs of utilization may also be reduced with maize and other whole-crop cereals grown for silage, because a single cut of high yield may be taken, with lower machinery costs than with perennial grasses and legumes, for which several cuts may be required to achieve the same total yield.

The ability to produce silages with high nutritive value is another reason for the growth of forage crops, particularly maize, which is often of higher digestibility and intake potential than grass silage. High performance of cows fed mixtures of grass silage with maize silage or fodder beet was noted by Phipps *et al.* (1995) and was associated with particularly high levels of intake of the mixed feeds. Winter-grown feeds, such as oats and berseem, are generally of much higher quality than alternatives, such as standing herbage remaining from the previous summer season of growth, as noted by Maraschin and Jacques (1993).

Browse

The consumption of buds, leaves and twigs of woody species, either directly or after harvesting, is an important part of the annual feed supply to ruminants in Mediterranean, tropical and subtropical areas, although good statistics on the magnitude of this feed source are lacking. Browse is characteristically important as a source of feed during dry and winter periods, when either the quantity or quality of available grass is deficient. Dzowela (1993) notes that there are over 200 browse species native to continental Africa which have acceptable nutritional characteristics. Tree leaves may have CP contents up to 250 g kg DM^{-1} and, in addition to their direct contribution to nutrient supply, may increase total DM intake and increase the digestibility of the basal low-CP diet, as discussed by Atta-Krah (1993), with beneficial effects on animal survival and productivity. The yield potential from fodder trees grown alone is high, with Atta-Krah (1993) noting forage yields of up to 40 t DM ha^{-1} for *Leucaena leucocephala* grown in Nigeria with 12-weekly cutting and 0.5 m spacing. More usually, fodder trees are grown at lower density in grassland. Many fodder trees are legumes, with their associated symbiotic nitrogen fixation contributing to increased soil fertility. The importance and utilization of forage tree legumes in tropical agriculture is reviewed by Gutteridge (1994). The contribution of fodder trees to the total diet may be restricted by high contents of tannins and other antimetabolites.

Straw and crop residues

Estimates of the production of fibrous by-products from cereals and other crops are given for 1981 in Table 1.4. It is probable that, with further increases in global crop production, there has been some increase in by-product availability since that time. Kossila (1984) calculated that the total output of by-products could, in theory, supply 84% of the energy and 74% of the CP required by the world's ruminants. The overall contribution is much less than this, because of losses in harvesting, processing and storage and use as fuel and as a raw material. Much straw is also returned to the field, because of poor nutritive value, absence of livestock or shortage of labour. The substantial availability of crop residues in smallholder systems in the tropics was also highlighted by Smith (1993), with the common crop residues listed in Table 1.5. He noted, though, that only a small fraction of the amount available is used strategically. Cereal stems are normally left on the field for *in situ* grazing, but the residues rapidly deteriorate and a large amount is trampled upon and wasted. Further nutrient imbalances characteristic of such residues are normally not corrected by appropriate supplementation. He stresses the need for appropriate harvesting and storage, treatment to improve digestibility and diet supplementation to improve utilization.

Table 1.4. Quantity of total digestible nutrients (TDN) and crude protein (CP) in by-products produced in different regions (from Kossila, 1984).

	TDN (10^6 t) Fibrous by-products from		CP (10^6 t) Fibrous by-products from		By-product/ livestock unit of 500 kg LW (t DM)
	Cereals	Other crops	Cereals	Other crops	
Africa	124.9	56.3	11.7	9.4	2.1
North and Central America	447.1	182.7	44.2	27.7	5.5
South America	100.8	99.9	10.0	13.3	1.7
Asia	542.0	261.3	53.4	44.7	3.0
Europe	173.4	86.7	16.3	15.8	2.3
Oceania	17.2	8.7	1.5	1.0	1.0
USSR	103.4	53.0	9.3	10.4	2.2
World	1508.8	748.6	146.5	122.3	2.8

Table 1.5. Common crop residues in tropical feeding systems (from Smith, 1993).

Crop	Primary product	Field residue	Primary processing residue
Cereals			
Maize	Grain	Stovers	Cob
Rice	Grain	Stubbles	Straw
Sorghum	Grain	Stovers	–
Wheat	Grain	Straw	–
Grain legumes/oil-seeds			
Groundnut	Oil	Haulms	Husk
Cowpea	Grain	Vines	Husk
Pulses	Beans	Vines	–
Roots/tubers			
Cassava	Tubers	Tops	Peels/rejects
Sweet potato	Tubers	Tops	Peels/rejects
Fruits			
Banana/plantain	Fruit	Tops Pseudostems	Peels/rejects
Coconut	Copra	–	Husk
Cocoa	Seeds	–	Pods
Others			
Sugar cane	Cane	Tops	Bagasse

Clearly, crop residues represent an underutilized feed source, although physical or chemical treatment may be necessary in order for fibrous by-products to make a major contribution to the energy requirements of productive livestock (Owen and Jayasuriya, 1989). Greatest reliance on straw occurs generally in areas where the number of ruminant animals is high in relation to the area of productive grassland and in which there is substantial production of cereals for human food. Normally, varieties of cereal will be selected on the basis of efficiency of grain production, but, in some situations, varieties may be used because of enhanced yield or quality of straw. The grazing of crop stubbles is discussed by Dann and Coombe (1987).

Contribution of Forages to Animal Systems

In view of the broad definition adopted for forages and their characteristic substantial contents of cell-wall components, it is not surprising that most ruminant systems are based on forages. Grasses and other forage crops may provide nutrients at low cost, because of the high yields of DM and energy that can be obtained (associated, as noted earlier, both with high light interception through the year and a high harvest index compared with grain crops) and the possibility for *in situ* utilization by grazing. Likewise, straw and browse may be available at low cost. Straw is a by-product from grain production, whilst supply of feed as browse may be associated with tree growth for other products. Fitzhugh *et al.* (1978) estimated that forages provide more than 90% of the feed energy consumed by the herbivorous livestock of the world. It is unlikely that this figure has changed radically in the last 20 years. The contribution of grassland feeds in the dairy cow's ration in the Netherlands, however, decreased from 90% in 1950 to 70% in 1970 and 50% in the late 1980s, associated with a favourable price ratio between animal products and concentrates (Van der Meer and Wedin, 1989). These authors, however, note adverse environmental effects from this intensification in dairy production and suggest a need to reverse this trend.

The major feeds not included within the definition of forages are grain and other seed crops and agroindustrial by-products derived from crop harvesting and processing (e.g. extracted oil-seed meals, bran and molasses, vegetable and fruit waste). Data from Fitzhugh *et al.* (1978) stress the limited use of these feed sources, with grain representing only 7% of the feed energy available in developed regions and a negligible proportion in developing regions (Table 1.6). Fitzhugh *et al.* (1978) and Fitzhugh (1998) draw attention to the discrepancy between the quantity of feed available and that calculated as required by ruminant livestock, with less than half of the feed resources apparently being utilized. The figures in Table 1.6 thus probably overestimate the quantity of forages and crop residues that are consumed by livestock.

Why Other Feeds?

Feeds other than forages, and particularly grain crops, are usually used because forages are not capable of sustaining the required levels of animal production, due to limitations in feeding value. They are widely used in situations in which high product prices (e.g. for milk) encourage systems involving high rates of individual animal production. This will be accentuated if climatic factors lead to relatively

Table 1.6. Feed energy resources available and calculated animal requirements for 1970 (Mcal × 10^9) (based on Fitzhugh *et al.*, 1978).

Source of feed energy	Developed regions	Developing regions
Permanent pasture	2.0	2.8
Forage from non-agricultural land	0.3	0.7
Arable land		
Forages	1.7	1.4
Crop residues	1.4	1.6
Grain	0.4	0.0
Agroindustrial by-products	0.0	0.1
Total available	5.8	6.6
Total required	2.3	3.5

high rates of production from grain as opposed to grassland, such as in the Midwest of the USA or where the feeding value of available forages is low. Characteristically, the digestibility and intake levels of tropical and subtropical grasses are lower than those of temperate grasses (Minson, 1990), with the result that, in order to achieve particular levels of animal performance, higher levels of supplementary feeds are required with these grasses. This contributes, for instance, to the substantial development of grain finishing of cattle in parts of northern Australia. Further factors that may increase the use of non-forage feeds are costs of transport and storage. Transport costs are particularly relevant when the location of animal production is remote from that of feed production. Examples are intensive animal production in Japan, largely dependent on feeds imported into the country, and town dairies in India. In such situations, transport costs will be much lower for grains with high bulk density and high nutrient density than for dried forages of low bulk density, such as straw and hay; fresh forage and silages are perishable and not suited to long-distance transport. Grains may also be used as a drought reserve feed for ruminant livestock, because of their high density and stability and, in some circumstances, the existence of grain reserves for market stabilization and for security of human food supplies.

The supply of processed agroindustrial feeds and vegetable wastes will be limited by the output of the main products and, increasingly, by alternative uses for these by-products. Molasses may be used for fermentation and alcohol production, whilst there is increased interest in using vegetable wastes for bioenergy rather than as a feed. Their role as animal feeds will be determined by their energy and protein value, the location of production in relation to that of the ruminant populations, and transport and storage characteristics.

Reliance on Forages

Some of the factors determining the extent of reliance on different classes of forages have already been mentioned. Table 1.7 illustrates the extent to which grassland and other feeds contribute to energy supply to ruminants in a number of European countries. The contribution from grassland ranges from 97% for Ireland down to 34% for Bulgaria. Countries with extremely high grassland usage are broadly those with high rainfall, giving high grassland yield potential (see Lee, 1988) and often difficult conditions for arable crop production. In contrast, countries with low grassland usage have a combination of low grassland yields and large areas of cereal grains and other arable crops producing an abundant supply of straw and other by-products. In countries with intermediate grassland usage (e.g. France), there is often substantial supply of forage from maize. The rapid increase in milk yields and genetic potential of dairy cows in Western Europe has resulted in a need for increased nutrient density in the complete diet, with increases in the required feeding value from forages and increases in the use of grain and concentrate feeds (Van der Meer and Wedin, 1989).

Animal production in extensive rangeland conditions in America, Africa, Australia and northern Asia is almost entirely dependent on the native grassland vegetation, with animals often being exported to other agroclimatic zones for fin-

Table 1.7. Estimated contribution of grassland, other forages and feeds to ruminant feed composition in selected European countries (%, on dry basis) (from Lee, 1988).

	Grassland	Root crops	Maize silage*	Other annual forages	Straw	Sugar-beet leaves and tops	Other feeds
Ireland	97	1	–	–	1	–	1
UK	83	2	1	–	1	–	13
France	71	3	9	2	–	–	15
Netherlands	54	–	4	–	–	–	42
Denmark	47	8	–	–	4	2	39
Hungary	45	2	4	–	5	2	42
Bulgaria	34	1	6	–	4	2	53

* The use of maize silage will have increased in most European countries over the last decade.

ishing. Low effective rainfall and non-availability of water for irrigation are the major factors that restrict more intensive grassland production or arable cropping. In some areas, however, particularly in South America, low human population pressure and market opportunities for ruminant products have led to extensive production in areas of reasonable climatic potential.

Straw is a particularly important feed resource in areas of Asia and Africa with high human population pressure, as noted earlier. Little cultivable land is available for producing specialist forage crops, so that ruminant animals, often required for draught, are dependent for feed supply on forage from uncultivated land, straw and other crop by-products.

Future Prospects

Forages will provide the basis for most ruminant production systems for the foreseeable future. Extensive production from grassland will continue to be a major form of land use, although management strategies, particularly in North America and Europe, will have to increasingly consider management not only in the context of animal production, but also in relation to effects on landscape and resource sustainability.

There is a major challenge for intensive grassland to maintain its place in production systems, in view of the progress that is being made in improving yields and quality from other forages, such as maize, the high quality required in the diet of high-production animals and the need in some areas to adapt methods of production to restrict environmental pollution, as discussed by Van der Meer and Van der Putten (1995) and Wilkins (1996). However, considerable progress in improving yield potential and nutritive value is now being made by grass breeding (Camlin, 1997) and approaches to restrict pollution risk are being developed (Wilkins, 1996). In some areas, particularly Europe, there will be pressures to reduce intensity of production in order to deliver increased landscape and biodiversity benefits from grassland at the expense of herbage output and quality. The production in both temperate and tropical conditions of improved forage legumes (Cameron *et al.*, 1993; Rhodes and Ortega, 1997) will be of particular importance in relation both to sustaining plant production with low levels of external inputs and to improving feeding value. In order to constrain costs of animal production, attention should continue to be focused on actions to increase the reliance on grazed rather than conserved forage, as discussed by Wilkins (1995). There are undoubtedly situations, though, in some tropical and subtropical countries, in which some increase in use of conserved forages would improve overall production efficiency.

The contribution of specialist forage crops, particularly maize and sorghum, is likely to increase. The large research effort with maize is likely to lead to further improvement in yield potential and stress tolerance with this species. There are opportunities for breeding to enhance the quality of the whole crop or of the straw, but it is disappointing that varieties with enhanced digestibility, associated with the brown-midrib gene, have not yet met with wide commercial success.

The major challenge with straw and many other by-products is to develop ways for increasing feeding value, particularly digestibility and intake potential, in order to increase animal production in straw-based feeding systems and to increase the proportion of straw in the diet. The possibilities for improvement by plant breeding and physical, chemical and microbial processing have been extensively reviewed (e.g. Sundstøl, 1988; Chesson *et al.*, 1995), but, as yet, the global impact has been small. I consider that the incorporation within breeding programmes of criteria in relation to straw quality has high potential for success and should be vigorously pursued. Variation in relevant nutritive characteristics exists and appropriate techniques are available for use in breeding and evaluation.

The increased use of trees and shrubs for forage as browse has considerable potential, particularly in tropical and subtropical regions, as shown by the substantial impact of *L. leucocephala* in northern Australia (Shelton *et al.*, 1991). There is a need for a more concentrated world effort on trees for forage, including both nutritive and agronomic factors.

References

Atta-Krah, A.N. (1993) Trees and shrubs as secondary components of pasture. In: *Proceedings of the XVII International Grassland Congress*, 8–21 February 1993, Palmerston North, New Zealand. New Zealand Grassland Association, Palmerston North, pp. 2045–2052.

Barnes, R.F. and Baylor, J.E. (1995) Forages in a changing world. In: Barnes, R.F., Miller, D.A. and Nelson, C.J. (eds) *Forages*, Vol. 1. *An Introduction to Grassland Agriculture*. Iowa State University Press, Ames, Iowa, pp. 3–13.

Cameron, D.F., Miller, C.P., Edye, L.A. and Miles, J.W. (1993) Advances in research and development with *Stylosanthes* and other tropical pasture legumes. In: *Proceedings of the XVII International Grassland Congress*, 8–21 February 1993, Palmerston North, New Zealand. New Zealand Grassland Association, Palmerston North, pp. 2109–2114.

Camlin, M.S. (1997) Grasses. In: *Occasional Symposium No. 31*. British Grassland Society, Reading, pp. 2–14.

Chesson, A., Forsberg, C.W. and Grenet, E. (1995) Improving digestion of plant cell walls and fibrous feeds. In: Jounet, M., Grenet, E., Farce, M.H., Theriez, M. and Demarquilly, C. (eds) *Recent Developments in the Nutrition of Herbivores*. INRA, Paris, pp. 249–277.

Coupland, R.T. (1992) Approach and generalizations. In: Coupland, R.T. (ed.) *Natural Grasslands. Introduction and Western Hemisphere*. Ecosystems of the World 8A, Elsevier, Amsterdam, pp. 1–6.

Coupland, R.T. (1993) Review. In: Coupland, R.T. (ed.) *Natural Grasslands. Eastern Hemisphere and Résumé*. Ecosystems of the World 8B, Elsevier, Amsterdam, pp. 471–482.

Dann, P.R. and Coombe, J.B. (1987) Utilization of fodder crops and crop residues. In: Wheeler, J.L., Pearson, C.J. and Robards, G.E. (eds) *Temperate Pastures: Their Production, Use and Management*. Australian Wool Corporation/CSIRO, Australia, pp. 517–525.

Dzowela, B.H. (1993) Advances in forage legumes: a sub Sahara African perspective. In: *Proceedings of XVII International Grassland Congress*, 8–21 February 1993, Palmerston North, New Zealand. New Zealand Grassland Association, Palmerston North, pp. 2115–2120.

El-Nahrawy, M.A.Z., Bari, A. and Ramedan, G.A. (1996) Collecting traditional berseem clover varieties in Egypt, and prospects for germplasm enhancement. *Plant Genetic Resources Newsletter* 107, 41–44.

Etman, K.E., Khinizy, A.E., Hathout, M.K., Khafagy, E.A. and Aspila, P. (1998) Effect of feeding berseem as hay or silage with molasses on sheep performance. In: *Proceedings of the 17th General Meeting of the European Grassland Federation*, Debrecen. Organising Committee of the 17th General Meeting of the European Grassland Federation, Debrecen, Hungary, pp. 679–682.

FAO (1996) *FAO Production Yearbook, 1995*. Food and Agriculture Organization of the United Nations, Rome.

Fitzhugh, H.A. (1998) Competition between livestock and mankind for nutrients. In: Waterlow, J.C., Armstrong, D.G., Fowden, L. and Riley, R. (eds) *Feeding a World Population of More than Eight Billion People*. Oxford University Press, New York and Oxford, pp. 223–231.

Fitzhugh, H.A., Hodgson, H.J., Scoville, O.J., Nguyen, T.D. and Byerly, T.C. (1978) *The Role of Ruminants in Support of Man: Winrock Report*. Winrock Foundation, Morrilton, Arkansas.

Forage and Grazing Terminology Committee (1991) *Terminology for Grazing Lands and Grazing Animals*. Pocahontas Press, Blacksburg, Virginia.

Green, J.O. (1982) *A Sample Survey of Grassland in England and Wales, 1970–1972*. Grassland Research Institute, Hurley, Maidenhead, 39 pp.

Gutteridge, R.C. (ed.) (1994) *Forage Tree Legumes in Tropical Agriculture*. CAB International, Wallingford.

Hopkins, A., Gilbey, J., Dibb, C., Bowling, P.J. and Murray, P.J. (1990) Response of permanent and reseeded grassland to fertilizer nitrogen. I Herbage production and herbage quality. *Grass and Forage Science* 45, 43–55.

Jones, M.B., Long, S.P. and Roberts, M.J. (1992) Synthesis and conclusions. In: Long, S.P., Jones, M.B. and Roberts, M.J. (eds) *Primary Productivity of Grass Ecosystems of the Tropics and Sub-tropics*. Chapman and Hall, London, pp. 212–255.

Knight, W.E. (1985) Miscellaneous annual clovers. In: Taylor, N.L. (ed.) *Clover Science and Technology*. American Society of Agronomy, Crop Science Society of America, Soil Science Society of America, Madison, Wisconsin, pp. 547–562.

Kossila, V.L. (1984) Location and potential feed use. In: Sundstøl, F. and Owen, E. (eds) *Straw and Other Fibrous By-products as Feed*. Elsevier, Amsterdam, pp. 4–24.

Leafe, E.L. (1978) Physiological, environmental and management factors of importance to maximum yield of the grass crop. In: Gasser, J.K.R. and Wilkinson, B. (eds) *ARC Symposium Proceedings*. HMSO, London.

Lee, J. (1988) Forages. *Livestock Production Science* 19, 13–46.

Maraschin, G.E. and Jacques, A.V.A. (1993) Grassland opportunities in the subtropical region of South America. In: *Proceedings of the XVII International Grassland Congress*, 8–21 February 1993, Palmerston North, New Zealand. New Zealand Grassland Association, Palmerston North, pp. 1977–1981.

Minson, D.J. (1990) *Forage in Ruminant Nutrition*. Academic Press, San Diego, 483 pp.

Mitchell, K.J. (1960) The structure of pasture in relation to production potential. *Proceedings of the New Zealand Society of Animal Production* 20, 82–92.

Owen, E. and Jayasuriya, M.C.N. (1989) Use of crop residues as animal feeds in developing countries. *Research and Development in Agriculture* 6, 129–138.

Phipps, R.H., Sutton, J.D. and Jones, B.A. (1995) Forage mixtures for dairy cows: the effect on dry matter intake and milk production of incorporating either fermented or urea-treated whole-crop wheat, brewers grain, fodder beet or maize silage with diets based on grass silage. *Animal Science* 61, 491–496.

Rhodes, I. and Ortega, F. (1997) Forage legumes. In: *Occasional Symposium No. 31*. British Grassland Society, Reading, pp. 15–27.

Shelton, H.M., Lowry, J.B., Gutteridge, R.C., Bray, R.A. and Wildin, J.H. (1991) Sustaining productive pastures in the tropics. 7. Tree and shrub legumes in improved pastures. *Tropical Grasslands* 25, 119–128.

Smith, O.B. (1993) Feed resources for intensive smallholder systems in the tropics: the role of crop residues. In: *Proceedings of the XVII International Grassland Congress*, 8–21 February 1993, Palmerston North, New Zealand. New Zealand Grassland Association, Palmerston North, pp. 1969–1976.

Snaydon, R.W. (1991) The productivity of C_3 and C_4 plants: a reassessment. *Functional Ecology* 5, 321–330.

Sundstøl, F. (1988) Improvements of poor quality forages and roughages. In: Ørskov, E.R. (ed.) *World Animal Science B Disciplinary Approach*, 4, *Feed Science*. Elsevier, Amsterdam, pp. 257–277.

Van der Meer, H.G. and Van der Putten, A.H.J. (1995) Reduction of nutrient emissions from ruminant livestock farms. In: *Occasional Symposium No. 29*. British Grassland Society, Reading, pp. 118–134.

Van der Meer, H.G. and Wedin, W.F. (1989) Present and future role of grasslands and fodder crops in temperate countries with special reference to over-production and environment. In: *Proceedings XVI International Grassland Congress*, Nice. The French Grassland Society, Nice, pp. 1711–1718.

Wheeler, J.L. (1986) Complementing grassland with forage crops. In: Morley, F.H.W. (ed.) *World Animal Science B Disciplinary Approach*, *1*, *Grazing Animals*. Elsevier, Amsterdam, pp. 239–260.

Wilkins, R.J. (1976) Limitations to grass for year round ruminant production. In: *Winter Meeting*, British Grassland Society, Reading, pp. 1–10.

Wilkins, R.J. (1995) Optimisation of grass utilization in high rainfall temperate conditions. In: Jounet, M., Grenet, E., Farce, M.H., Theriez, M. and Demarquilly, C. (eds) *Recent Developments in the Nutrition of Herbivores*. INRA, Paris, pp. 363–380.

Wilkins, R.J. (1996) Environmental constraints to grassland systems. In: *Proceedings 16th General Meeting, Grado, Italy*. European Grassland Federation, Gorizia, Italy, pp. 695–703.

2 Forage Evaluation for Efficient Ruminant Livestock Production

D.E. Beever and F.L. Mould

CEDAR (Centre for Dairy Research), Department of Agriculture, University of Reading, Earley Gate, Reading RG6 6AT, UK

Background

Forages represent a diverse range of feedstuffs that make a significant contribution to the overall nutritional economy of meat-, wool- and milk-producing ruminants. However, it is this diversity which presents both opportunities and challenges when attempting to exploit these feeds in ruminant diets. Within the UK, forages are derived from a number of distinct sources. Undoubtedly, the principal forage is grass, which may be consumed *in situ* (i.e. grazing) or after conservation as silage or field-cured hay. The area of forage maize grown in the UK has increased considerably over recent years and, depending upon local climatic conditions, will be harvested and ensiled at dry matter (DM) contents between 250 and 375 g kg^{-1} fresh weight. Forage maize is grown principally for its high energy content and high intake characteristics. In addition, depending upon level of cob formation, starch content in more mature crops can reach 350 g kg^{-1} DM. Such is the popularity of maize that serious consideration is being given to the production of maize grain in more favourable growing areas. However, maize cannot be grown successfully in all areas of the UK, and whole-crop wheat silage may be considered as an alternative. According to preference, the wheat crop can be harvested and ensiled at an immature stage of growth (350 g DM kg^{-1} fresh weight) and offered as a fermented feed. Alternatively, it can be harvested later (550 g DM kg^{-1} fresh weight), when starch levels are higher, but, as such crops are unlikely to undergo fermentation after ensiling, urea is added at harvesting to ensure stability during the storage period (Sutton, 1997).

Cereal straws, principally barley and wheat, are also a source of forage for ruminants, and can form a significant part of the diet for those animals with more modest performance targets (e.g. suckler cows, beef stores). Treatments with either sodium hydroxide or gaseous ammonia are used to upgrade straw quality, although such practices are increasingly questioned on economic grounds. Finally, consideration should be given to the role of legumes. The lucerne area in the UK remains low, for, whilst many farmers appreciate its value, difficulties associated with crop establishment and management, as well as harvesting and storage of the crop, all of which influence both forage yield and quality, remain major limitations to its widespread use. However, an increasing number of farmers are considering the possibility of growing lucerne, especially for inclusion in dairy-cow diets. Again, although the high nutritional value of white clover, especially in grass swards, is recognized, its poor spring growth and lack of persistency under frequent-cutting regimes tend to restrict its use to low-output per hectare and alternative farming systems. Red clover is of relatively minor importance.

From this brief outline of the types of forage available, it follows that the nutritional value of such feeds will vary in terms of their ability to support the maintenance and production requirements of ruminant livestock. This contrasts with many other primary feed sources, including cereal grains and pulses, where variation in nutritional value is relatively small. Principal factors affecting forage nutritive value include forage species, as indicated above, as well as prevailing climatic and management conditions. These include soil type, seasonal variations in the incidence of sunshine and rainfall, fertilizer treatment and stage of growth at harvesting (by animal or machine). For ensiled forages, postharvesting procedures with respect to field wilting, additive application and ensiling conditions are important. Equally, for hays, weather condition, length of the curing period and degree of mechanical handling have a marked influence on the quality of the end material, while drying temperature is important for artificially dried forages. The intention of this chapter is to examine the importance of determining the nutritional value of forages destined for feeding to ruminants, with a comprehensive review of how forage nutritive value is currently assessed. Finally, the chapter will provide some direction as to the way in which forage evaluation should develop, suggesting the potential contribution of newer technologies whilst drawing attention to possible pitfalls. To achieve this overall objective, an initial review of the processes of forage utilization by ruminants will be presented as background to the evaluation of forages as feeds for ruminants, as well as establishing a basis on which current techniques have been developed and future ones may be considered.

Nutrient Digestion and Utilization in Ruminants

Nutrient degradation

The principal reason why non-ruminants and preruminants are unable to utilize significant quantities of forages is that, like other mammals, they do not possess the enzymes capable of breaking down the complex β-linked polymers that form plant cell walls. In ruminants, however, the principal site of digestion in regard to forages is the rumen, where the feed is retained for substantial periods of time and subjected to extensive microbial fermentation under anaerobic conditions (Beever, 1993). Following hydration of the ingested feed, the microbial enzymes break down the complex dietary polysaccharides, namely, cellulose and hemicellulose, which are the principal components of the cell walls. The major end-products are the constitutive sugars, principally glucose derived from cellulose, but both hexoses and pentoses, with variable amounts of uronic acids from hemicellulose, reflecting its more heterogeneous nature. The diet will usually contain other carbohydrates, including water-soluble carbohydrates, such as fructosans, which are found extensively in grasses, and starch, which occurs principally in cereal grains. Whilst these fractions could be potentially digested in the small intestine by enzymes of host-animal origin (Armstrong and Beever, 1969), they too will be subjected to the action of microbial enzymes whilst resident in the rumen.

However, not all ingested carbohydrate is degraded in the rumen, as both the rate and extent of polysaccharide breakdown are influenced by composition of the feed, as well as the nature of the microbial enzymes and accessibility of the dietary polysaccharides to these enzymes. This can be illustrated with feeds containing significant amounts of lignin, indicative of advancing plant maturity. Such forages have impaired rates of cellulose and hemicellulose degradation, for, whilst composition of these fractions will be similar to that in less mature forages of the same species, it is the ingress of lignin into these structures as the plant matures that acts as a major obstacle to their digestion by microbial enzymes (Chesson, 1988). Equally, the nature of the starch in the diet, particularly the relative proportions of amylose and amylopectin, will influence the extent of starch digestion in the rumen. This is best illustrated by the starches in maize grain (and maize silage), which are less rapidly degraded than those in other cereals, including wheat and barley, due to an increased proportion of amylopectin, as well as a more complex protein : lipid matrix found in the endosperm of maize grains. In contrast, there is little evidence that the rate of degradation of water-soluble carbohydrate (WSC), which is primarily associated with cell contents, is influenced by either forage species or stage of maturity, although levels of WSC in the plant decline as forages mature (Beever *et al.*, 1972).

Hexose disposal

Following the initial phase of carbohydrate breakdown, the released constitutive sugars are subjected to one of two metabolic processes, both occurring in the rumen. It is generally assumed that all released sugars undergo microbial fermentation, and, indeed, this is an important route of hexose and pentose utilization. The primary aim of this fermentative pathway is the provision of adenosine triphosphate (ATP), essential for the energy demands of the microbes for maintenance and growth (Issacson *et al.*, 1975). However, the rumen is largely anaerobic and consequently ATP yield per mole of carbohydrate fermented approximates to 10–12% of that achieved under aerobic conditions. Thus, to meet microbial ATP demands, a significant proportion of the available hexose and pentose will be fermented. As such, carbohydrate fermentation comprises a significant component of rumen metabolism (Beever, 1993). It is during this process that significant amounts of carbon dioxide are released, with part used in the removal of ruminally produced hydrogen through the synthesis of methane. This process involves methanogenic bacteria, which proliferate in those environments where excess hydrogen is produced. However, most microbes have some requirement for preformed hexose to support the synthesis of microbial biomass, in particular the synthesis of microbial protein, but also the synthesis of microbial nucleic acids, lipids and polysaccharides (Black *et al.*, 1980/81). Whilst part of the microbial protein component is derived from preformed amino acids, *de novo* synthesis from ammonia and hexose is often the preferred route in overall microbial metabolism (Nolan, 1975) and has been shown to be preferred by amylolytic compared with cellulolytic bacteria. When carbohydrate availability is high (e.g. increased starch feeding), microbial polysaccharide synthesis, as an alternative route for hexose disposal, can be quite significant. There have been no direct attempts to establish the partition of hexose disposal between fermentation and direct incorporation. However, as illustrated by Beever (1993), this can have a significant impact on the net production of fermentation end-products, including the quantity and composition of ruminally derived volatile fatty acids (VFA).

VFA production and utilization

As indicated, the major event associated with the ruminal fermentation of dietary derived monosaccharides is the production of VFA, principally acetate, propionate and butyrate. In most dietary situations, acetic acid is predominant and, with butyric acid, reflects forage-rich diets, in which cellulolytic bacteria proliferate. In contrast, amylolytic bacteria dominate on starch-enriched diets, where increased propionic acid levels are normally observed. This distinction between acetate and butyrate, on one hand, and propionate, on the other, is important with respect to the glucose economy of the animal (MacRae and Lobley, 1982). The overall stoichiometry of carbohydrate utilization by fermentation in the rumen has been established (Baldwin *et al.*, 1977; Murphy *et al.*, 1982) and was recently summarized by Beever (1993), for three contrasting diet types based on high amounts of forage, cereal or molasses. These are presented in Table 2.1, where a number of interesting comparisons are apparent. The most obvious differences relate to the

Table 2.1. The estimated stoichiometric yield of VFA, ATP, carbon dioxide (CO_2) and hydrogen (H_2) (mol mol^{-1} hexose) with the resultant levels of methane (CH_4) production, from the ruminal fermentation of three diets based on different carbohydrate types. Values in parentheses refer to net productions after conversion of available hydrogen to methane.

	Diet types		
Products	High-fibre	High-cereal	Molasses-enriched
Acetate	1.34	0.90	0.94
Propionate	0.45	0.70	0.40
Butyrate	0.11	0.20	0.33
CO_2	1.53 (0.92)	1.30 (0.92)	1.60 (1.06)
H_2	2.44 (0.0)	1.50 (0.0)	2.14 (0.0)
CH_4	(0.61)	(0.38)	(0.54)
ATP	(4.62)	(4.38)	(4.54)

amounts of individual VFA produced, with high acetate levels on the high-fibre diet and relatively high levels of propionate on the high-cereal diet, although acetate was the predominant acid for all diet types. As a consequence of these changes, carbon dioxide yield was reduced on the high-cereal diet, with a net transfer of carbohydrate carbon to VFA carbon of over 78%. Corresponding values for the other two diets were lower (average 74%), representing an increased loss of carbon. The importance of these differences is appreciated when it is recognized that, for average-yielding dairy cows, the amount of carbohydrate fermented in the rumen may approach 50 mol day^{-1}. Associated with these changes were lower yields of hydrogen on the high cereal diet, the difference being most pronounced in comparison with the high-forage diet. The reason for this reduction is related to the utilization of hydrogen in the production of propionate from pyruvate (the initial three-carbon molecule derived from glycolysis). Propionate production can therefore be considered an important sink for the disposal of ruminally derived hydrogen. As a consequence, methane production, which is largely in stoichiometric balance with net hydrogen production, was highest on the fibre-rich diet, with reductions of between 12 and 38% on the molasses-enriched and high-cereal diets, respectively. However, despite such changes, ATP yield mol^{-1} carbohydrate fermented was not significantly affected by diet type, although the highest values were associated with acetate-dominant fermentations.

Following absorption, propionate is utilized extensively by hepatic tissues to support glucose synthesis (Reynolds *et al.*, 1998) and, in most situations, constitutes the primary source of glucose, given that net glucose uptake from the ruminant alimentary tract is generally minimal (Armstrong and Beever, 1969). Indeed, the gut utilizes significant amounts of arterial glucose and, apart from those few instances where small-intestinal starch digestion may be quantitatively important, the gut as a whole is a net utilizer, rather than a net provider, of glucose. In contrast, ruminally derived acetate is largely unchanged by hepatic metabolism and may be augmented by endogenous acetate production in the liver. The posthepatic supply of acetate to peripheral tissues constitutes a major part of the total energy available to the animal and may be either oxidized to produce ATP or used as a substrate in the production of long-chain fatty acids (see following section). While ruminally derived butyrate is quantitatively metabolized to β-OH-butyrate during absorption through the rumen epithelium, in posthepatic tissues it has a similar metabolic fate to that of acetate.

Fatty acid metabolism

Collectively, acetate and β-OH-butyrate make a significant contribution to the synthesis of fatty acids for deposition as triglycerides in adipose tissue or secretion in milk. This comprises chain elongation with two-carbon units up to a maximum of a 16-carbon unit and in part contributes to the preponderance of short- to medium-chain saturated fatty acids in ruminant products. At the same time, preformed fatty acids, either of direct dietary origin or from mobilized body fat (especially in the case of dairy cows), can be used for the provision of ATP by oxidative metabolism or the synthesis of triglycerides. With respect to dietary fatty acids, however, one important distinction between ruminants and non-ruminants is the ruminal hydrogenation of dietary fatty acids. Ingested fats are relatively inert with respect to rumen metabolism, but, after ruminal hydrolysis, a significant proportion of the released fatty acids will be hydrogenated, due to the prevailing hydrogen-rich (i.e. reduced) environment. This constitutes a further route by which a significant amount of ruminally derived hydrogen may be disposed of, a feature not taken into account in the estimates of methane production presented in Table 2.1. Furthermore, as the degree of saturation of fatty acids entering the small intestine of ruminants usually exceeds that of ingested fatty acids (Outen *et al.*, 1975), the degree of saturation of the fatty acids found in ruminant products is generally greater than that found in non-ruminant products. Equally, fatty acids originating from the diet or from mobilized tissue usually comprise longer carbon-chain lengths compared with those synthesized *de novo* from acetate or β-OH-butyrate, with most, although not all, being accounted for as C-18 isomers.

Protein metabolism

One other major metabolic activity in the rumen relates to the utilization of dietary protein to support the synthesis of microbial protein. This has been the subject of considerable research effort, where a series of well-defined studies, conducted principally with sheep, established the quantitative contribution of microbial and undegraded dietary protein to small-intestinal protein supply (Smith and McAllan, 1970; Beever *et al.*, 1974). Following ingestion, dietary protein is subjected to degradation by microbial proteases, the principle end-products being amino acids, together with ammonia, which arises from deamination of the resulting amino acids or the catabolism of non-protein nitrogen (e.g. urea) derived from dietary or endogenous (i.e. saliva) sources. The extent to which dietary protein is degraded in the rumen has been the subject of much debate and, whilst reliable measurements of this process still need to be established, it is accepted that degradation rate is a function of protein structure. With fresh forages, ribulose-1,5-biphosphate carboxylase (Rubisco or fraction 1 protein) is the principal protein and several studies have shown that this protein is highly susceptible to the action of microbial proteases (Mangan, 1982), resulting in a relatively short half-life for the protein and rapid accumulation of ammonia in rumen fluid. Furthermore, at high levels of feed intake, outflow rate from the rumen increases, resulting in an increased proportion of dietary protein surviving intact to the intestines, despite having the potential to be digested in the rumen.

Microbial protein synthesis is dependent upon the availability of a continuous supply of precursors, principally ammonia and amino acids, but the process is energy-consuming and a supply of ATP, provided principally from the fermentation of dietary carbohydrates, is essential. This led to the concept of synchrony of nutrient release in order to ensure maximum rumen microbial protein synthetic rates, as well as optimal utilization of degraded nutrients. However, whilst the theory is well founded, definitive experimentation in support of the hypothesis is still required (Sinclair *et al.*, 1993; Witt *et al.*, 1997). Ultimately, microbial biomass passes to the small intestines, either attached to undegraded feed particles or free-floating within the fluid phase of the digesta. During transit through the small intestine, host enzymes promote significant digestion of the proteins to peptides and amino acids. In turn, these will be absorbed across the intestinal wall, the ultimate contribution directly from peptides being considered to be minimal. As with glucose, gut tissue utilizes significant amounts of individual amino acids, probably from arterial supply, and thus net portal appearance of amino acids is unlikely to equate with small-intestinal loss of amino acids (MacRae and Beever, 1997; Wray-Cahen *et al.*, 1997).

Prediction of Animal Performance

Against this background of nutrient digestion and assimilation, in which dietary nutrients can be substantially modified, due to events principally within the rumen, prediction of animal performance remains relatively complex. Indeed, this task has become increasingly difficult, as the demands placed upon systems have become more prescriptive. When the importance of forage evaluation was first recognized at the turn of the century, the issues were large and apparent. It was relatively easy to establish the superior nutritive value of, for example, grass hays compared with cereal straws and, in the well-quoted studies of Armsby, Kellner and others, establishment of the nutritional advantages of barley grain compared with a range of different forages was quite straightforward. This situation prevailed until the middle of this century, when some of the more subtle differences between forages were appreciated. This led to considerable research effort focusing on three issues: namely, those influencing forage intake, digestibility and utilization. In this respect, a substantial research effort was directed towards consideration of forage digestibility. The resultant data were used effectively in the estimation of animal performance, through prediction of forage intake and the efficiency with which digested nutrients were utilized for maintenance and production. However, it became apparent that measurements of feed or forage digestibility were not sufficient to enable prediction of animal performance with an acceptable degree of precision. This led to in-depth examination of the processes of digestion, with specific focus on events occurring within the rumen. The outcome of this research has been highly rewarding in relation to protein metabolism, where it is now possible to provide more balanced diets to meet the 'protein' requirements of both rumen microbes and the animal. However, there has been no parallel comprehensive evaluation of the dietary carbohydrate fraction, although, with most diets, this will contribute to over 60% of total absorbed nutrients. Equally, only limited attention has been given to an evaluation of dietary lipids, even though these may constitute between 10 and 15% of the total absorbed energy on diets designed for high-producing animals. In partial mitigation, the Cornell net carbohydrate and protein system (CNCPS), as proposed by Sniffen *et al.* (1992), offered some improvement with respect to an appreciation of the importance of difference carbohydrate sources. However, the focus provided by the CNCPS is largely related to consideration of the digestive aspects of different carbohydrate and protein sources and how these influence overall supply of metabolizable energy and protein to the animal. To date, no feed evaluation systems that describe the individual end-products of digestion, other than total amino acids, as included in the plethora of metabolizable protein systems have been developed. This is despite it being well established that the composition of the non-protein part of the total absorbed-energy fraction can have a significant effect on overall efficiency of nutrient utilization by the animal and, in particular, product composition. Such models have only been proposed for research purposes (Baldwin *et al.*, 1977, 1987a; Black *et al.*, 1980/81; France *et al.*, 1982), and, to date, none have been modified for on-farm use.

Whilst progress with respect to protein evaluation, as encompassed in the UK metabolizable protein (MP) scheme and other similar systems, has been considerable, with respect to the prediction of energy utilization, many of the systems proposed over 20 years ago are still in use. The UK metabolizable energy (ME) system, as first outlined by Blaxter (1962), has been extremely valuable in terms of prediction of gross animal events, and the energy unit of this system (MJ ME kg^{-1} diet DM) has been widely adopted within the livestock industry. Many of these data were derived from energy calorimetry studies involving sheep or, in limited instances, beef or dairy cattle, with most of the latter data being derived from US or Dutch studies. Rationalization of such data led to a system of energy evaluation which, at the time of its development, could be supported on conceptual grounds, with many of the proposed relationships being defendable on the basis of best statistical fit of the data. However, acceptance of linearity of response for all functions of energy utilization, at both above- and below-maintenance levels of feeding, has been increasingly questioned as difficult to justify on biological grounds. Equally, some of the distinctions drawn in relation to the efficiency of utilization of ME for different forages, appear to have been based rather more on data availability than on biological grounds, the differences between long and ground and pelleted forages and between spring and autumn forage being cases in point.

At the same time, similar systems were proposed by other countries in Europe, Australia and the USA, often with net energy as the preferred energy unit. However, as systems of animal production become more sophisticated, with greater emphasis on end-product quality to meet the demands of the consumer, such systems have been shown to be lacking in several respects, and it is likely that these will need to be replaced. Indeed, there is already debate as to how such systems should be refined, given the aim of improving the predictability of animal response whilst avoiding unnecessary complication. However, there is no consensus on how alternative systems of feed evaluation should be constructed and, given that most national funding agencies seem to be no longer interested, all future initiatives are likely to come from those actively working in the field. One obvious area of improvement would be systems that predict animal response to deliberate perturbations in nutrient input. Such an approach would be of critical importance in attempting to optimize product output per unit of feed input. Development of this approach would, of necessity, lead to the abandonment of models of feed evaluation based on estimation of the nutrient requirements of the animal and subsequent formulation of a diet to meet such needs. However, in this scenario, the required inputs would be largely unchanged from those currently in use, as it would be the model, rather than the inputs, that would be most affected. Such moves towards the development of nutrient response models are highly supportable, but there is a growing opinion that suitable progress towards reliable prediction of animal performance will only be achieved through recognition of the importance of individual nutrients. Undoubtedly, ME has been a useful nutritional component, but lack of any qualitative description of ME is now a significant limitation. The absence of such information could allow the less informed nutritionist to conclude that wheat and wheat straw would provide similar levels of production (both amount and type) when supplied in equal amounts of ME to the animal. This issue was discussed in a recent Technical Committee on Responses to Nutrients (TCORN) publication (AFRC, 1998), when a number of options to provide a staged development of alternative models were advanced.

Consequently, all aspects of feed evaluation designed to predict animal performance are at an interesting stage of development. On the one hand, there is sufficient information to indicate that current methodologies are inadequate and refinements are required. On the other, however, uncertainty exists with respect to the degree to which systems of feedstuff evaluation will need to change. This ranges from those who advocate minor revisions to existing frameworks to others calling for more radical initiatives, to include recognition of the importance of individual nutrients, as occurs in most non-ruminant feeding systems.

Forage Evaluation: Current Practice

All systems of feedstuff evaluation aim to provide information regarding the capacity of individual feeds to meet the nutritional demands of the animal and, as such, represent some degree of compromise with reality. Undoubtedly, the most precise way to establish the nutritional value of any feedstuff would be to feed it to appropriate animal classes and to observe the level of animal production achieved, but such an approach is neither practical nor justifiable on cost grounds. However, in any system of feed evaluation, it is important to recognize that the ultimate arbitrator of nutritional value will always be the animal, and consideration of this should, at all times, outweigh expediency of laboratory operation in the refinement of current methodologies.

Digestibility and dietary carbohydrates

An excellent example of a systematic approach to the establishment of a feed evaluation system was the development of the first routine *in vitro* system by Tilley and Terry (1963). Recognizing the importance of diet digestibility as an index of nutritional value, these workers established a comprehensive *in vivo* database of forage digestibilities, from which suitable predictive procedures could be developed. To achieve this, a programme of work was undertaken to provide a wide range of forages (both grasses and legumes), which were systematically fed to mature sheep at maintenance levels of feeding to obtain measurements of whole-tract digestibility. Subsequently, these feeds were subjected to an *in vitro* technique, comprised of two stages to simulate ruminal and postruminal digestion, and a series of equations to predict whole-tract digestibility for different diets were proposed. Subsequently, workers in other countries extended this approach to develop suitable prediction equations for tropical forages, and collectively this work formed the basis on which many feedstuffs were evaluated for feeding to ruminant livestock, in both temperate and tropical climates.

Establishment of the technique led to the development of the concept of forage D, or DOMD, value, defined as the content of digestible organic matter in forage DM, and studies by Givens *et al.* (1989, 1990, 1992) and Moss and Givens (1990) led to a series of equations to predict DOMD (%) as indicated below:

Fresh grass	DOMD = 12.0 + 0.857 IVD
High-temperature-dried grass	DOMD = 27.3 + 0.998 IVD
Grass hays	DOMD = 17.2 + 0.710 IVD
Grass silage	DOMD = 10.0 + 0.870 IVD

where IVD (*in vitro* digestibility) is the DOMD estimate obtained by the method of Tilley and Terry (1963).

Estimates of D value have also been widely used in systems of grassland management to optimize the use of grazed grass or the stage of harvesting of grass for ensiling. This concept was also adopted by plant breeders as a screening procedure to examine the nutritional value of new forage cultivars. Equally, DOMD values derived from IVD estimates have been used to predict forage ME contents. A value of 0.15 has been adopted to convert DOMD to ME (i.e. ME (MJ kg^{-1} DM) = 0.15 DOMD%) for fresh and dried grass and hay, and a higher coefficient of 0.16 for high-protein legumes and fermented feeds (see Beever *et al.*, 1999), whilst estimates of forage IVD have also been used to predict forage intake – yet another example of the utility of this index of nutritive value.

With the passage of time, possible alternatives to the *in vitro* system as proposed by Tilley and Terry (1963) have been suggested. As the Tilley and Terry procedure may contain pepsin-insoluble material and microbial residues, Goering and Van Soest (1970) introduced a final washing procedure, using a neutral detergent solution, and termed their end-point estimate the predicted true digestibility. In another development, Jones and Hayward (1975) replaced rumen fluid as the source of microbial enzymes for the ruminal phase of digestion with commercial enzymes. This approach was to be commended, as it offered the possibility of reducing methodological variation attributable to the source of rumen liquor, permitting improved standardization of the method, whilst eliminating the need for surgically modified animals as rumen-fluid donors. However, there are a number of drawbacks to this technique. First, there were doubts as to whether the use of only two enzymes with specific activities was sufficient to provide an accurate measure of digestion, in comparison with the highly complex array of enzymes present in the rumen ecosystem. Secondly, there was a lack of standardization between different batches of commercial enzymes, and consequently the technique has had limited usage. Another modification has been the use of fresh faeces as an alternative microbial inoculum to rumen liquor (El Shaer *et al.*, 1987). Evaluation established that this approach could be used to determine forage digestibility. However, while feedstuffs were ranked in a similar order, as the extent of digestion was generally less than that observed with rumen liquor, alternative calibration formulae were required. It is likely that long-term application of this method will be limited to laboratories where fresh rumen liquor cannot be obtained, a position that prevails in many developing countries.

With the recognition that forage fibre composition was a major determinant of forage quality, with the potential to predict intake and nutritive value, attention turned to developing laboratory methods for forage fibre analysis. Until the mid-1960s, fibre content of feeds, including forages, had been determined largely according to the Weende system of feedstuff evaluation, but there was increasing concern as to the value of crude fibre levels determined by this methodology. By sequential use of neutral- and acid-detergent solutions, the method of Van Soest and Wine (1967) provided, for the first time, a relatively inexpensive and routine method for determining different fibre fractions, namely neutral-detergent fibre (NDF), comprising cellulose, hemicellulose and lignin, and acid-detergent fibre (ADF), comprising cellulose and lignin. A further step involving acetyl bromide was developed to determine lignin content in the ADF fraction. This latter detail has not been widely adopted for routine feedstuff analysis, as it is recognized, on the basis of its relative complexity, that there are no ideal methods for routine analysis of lignin. However, other fibre fractions, especially ADF or modified ADF (MADF), have been used extensively in equations designed to predict forage digestibility, as proposed by Givens *et al.* (1989, 1990, 1992) and Moss and Givens (1990), as illustrated below (as g kg^{-1} DM):

Fresh grass	DOMD = 967 – 0.950 MADF
High-temperature-dried grass	DOMD = 1012 – 1.243 MADF
Grass hays	DOMD = 1023 – 1.222 MADF
Grass silage	DOMD = 996 – 1.04 MADF

The detergent fibre system does, however, have some limitations. Van Soest and Wine (1967) recognized that feeds which have been subjected to heat treatment may contain significant amounts of heat-damaged protein, which, based on its low solubility in detergent solutions, could be a significant contaminant in the isolated ADF fraction. This artefact was particularly evident with distillery by-products and, under such circumstances, it is important to take account of this if realistic estimates of ADF content are to be achieved. However, this potential problem has been turned to advantage, whereby the detergent method has been used to determine acid-detergent-insoluble nitrogen contents (ADIN) of feeds. This provides an estimate of the small-intestinal availability of ruminally undegraded dietary protein, an important component in several of the metabolizable protein systems. A further limitation is the estimation of NDF content in starch-rich feeds. During extension of the methodology from forages to use with other feeds, it was apparent that, under some circumstances, estimated NDF contents were too high and this raised the possibility of starch contamination. The extent of this will depend upon the amount and type of starch present in the original feedstuff, as those starches relatively resistant to chemical or microbial digestion will be more resistant to extraction by detergent solution. Consequently, a pretreatment with commercial amylases to digest the starch was developed, thus facilitating its subsequent removal by the detergent solution. In most forages, such procedures are not required, but, with maize and whole-crop silages, which contain significant amounts of starch, pretreatment of the sample is recommended if reliable estimates of NDF content are to be obtained.

In relation to a more specific evaluation of dietary carbohydrate fractions, one area where uncertainty still remains is the estimation of starch content. With maize or whole-crop wheat silages, the determination of total starch content is essential if appropriate supplements that optimize total starch availability to the animal are to be developed. In the late 1960s, there was considerable effort towards the development of improved methods for determination of total starch content in different feeds. Based on a critical examination of available methods, MacRae and Armstrong (1968) concluded that many were not sufficiently specific and were thus subject to interference from moieties of non-starch origin. This resulted in an enzymatic method being developed by MacRae and Armstrong (1968), using a commercial amyloglucosidase to release the glucose units of starch, followed by determination of the released glucose, using a glucose-specific methodology (glucose oxidase). This methodology became the basis on which several laboratories conducted research to determine the digestion characteristics of different starch-containing feeds in ruminants (MacRae and Armstrong, 1969; Beever *et al.*, 1970), the only potential problem appearing to be some inconsistency in the commercially available amyloglucosidases. It is thus surprising that, over 25 years later, no standard methodology for starch analysis is available in the UK, and the position is remarkably similar in many other countries. Indeed, some laboratories appear to have reverted to techniques that were identified by MacRae and Armstrong as being both unsuitable and unreliable, whilst others have decided not to offer starch determination as a routine analysis. In a recent study, reported by Beever *et al.* (1996), nine laboratories in Europe (including the UK) and the USA were requested to determine the nutrient composition of two contrasting maize silages, including total starch content. Analysis of the resultant data indicated considerable variability, as illustrated in Table 2.2. Further investigation revealed the use of several different techniques, some relying on near-infrared reflectance spectroscopy

Table 2.2. Chemical composition of two maize silages harvested at different stages of maturity and analysed by up to nine different laboratories (units, g kg^{-1} oven-dry matter unless stated; means ± SEM; superscript denotes number of laboratories reporting).

		Low DM	High DM
Dry matter (g kg^{-1} fresh wt)[9]	Mean	276 ± 8.6	335 ± 9.3
	Range	264–290	323–348
Starch [8]	Mean	228 ± 35.6	261 ± 42.4
	Range	165–272	194–311
Crude protein [9]	Mean	101 ± 18.3	101 ± 7.5
	Range	57.0–119	84.0–108
ERDP [3]	Mean	61.3 ± 18.02	68.0 ± 4.16
	Range	27.0–88.0	60.0–74.0
DUP [2]	Mean	23.0 ± 2.00	21.0 ± 4.00
	Range	21.0–25.0	17.0–25.0
ME (MJ kg^{-1} DM) [6]	Mean	11.2 ± 0.55	11.5 ± 0.60
	Range	10.5–12.1	10.7–12.3

SEM, standard error of the mean.

(NIRS) rather than the more conventional chemical/enzymatic approach involving starch extraction and estimation. One US laboratory did not offer starch analysis, on the basis that they were dissatisfied with all of the available methods. Further, in relation to the evaluation of the carbohydrate fraction of forages, it is recognized that some forages may contain appreciable levels of WSC. Under most circumstances, this fraction is extensively and rapidly digested in the rumen, and consequently most analyses are restricted to determination of total WSC content, with no attempt to provide an analysis of the constitutive sugars. Suitable methods for the extraction of WSC exist, although, in estimation of the yield of reducing sugars, it is advisable to select an appropriate sugar standard (e.g. fructose/glucose), whilst ensuring that chemical moieties that could have a significant effect on the final colorimetric reaction are accounted for. When a more detailed analysis of the individual components of the WSC fraction is required, possibly in forage breeding programmes, suitable methods based on gas- or high-pressure liquid chromatography do exist, but these could never be contemplated for routine use.

Protein fractionation

For a significant part of this century, the accepted method for determining forage protein content was by estimation of total nitrogen content, using Kjeldahl methodology. However, this approach did not distinguish between nitrogen of protein and non-protein origin, and thus the resultant fraction was termed 'crude protein'. Attempts to rationalize this into digestible crude protein (DCP) or available protein (AP) were popular for some time, but reversion to crude protein was inevitable when users became dissatisfied with DCP or AP as indices of protein value. With increasing knowledge of dietary protein metabolism, especially in the rumen, Miller (1973) suggested development of a protein-rationing scheme, mirroring an earlier proposal by Burroughs *et al.* (1967) in the USA. Four years after Miller's proposals, the first version of the UK MP system was released, and a revision subsequently accorded some peer recognition (ARC, 1980). Since then a number of further revisions have been published (AFRC, 1984, 1993) and, despite several concerns, the concept of MP is now included in all UK feeding systems.

The main feature of the MP system is the recognition that the nature of the dietary protein will influence both the extent and efficiency of its utilization by rumen microbes. This will then determine the composition of the protein supply to the intestines and therefore the use of dietary protein by the host animal. Thus, the concepts of ruminally degradable and ruminally undegradable protein (RDP and UDP) were established, subsequently refined to effective rumen-degradable protein (ERDP) and digestible undegraded protein (DUP). To support such concepts, it was necessary to develop suitable methodologies to permit quantitative evaluation of the different fractions, and subsequently the *in sacco* technique, as proposed by Mehrez and Ørskov (1978), was recommended. Since then, this methodology has been used extensively to characterize a wide range of feedstuffs for ruminants. The method relies on the incubation of small amounts of test feed retained in artificial-fibre bags within the rumen of experimental animals consuming a feed similar to that being examined. Removal of feed samples at different time intervals and analysis of the residues for total nitrogen content allow the pattern of nitrogen (N) degradation over time to be described. A number of different mathematical models have been developed to analyse these data (Ørskov and McDonald, 1979; McDonald, 1981; France *et al.*, 1993). In summary, three values for the degradation characteristics of the feed can be obtained, namely the rapidly degradable fraction (or *a* fraction), which may be significant in fermented feeds, such as grass silage, the slowly degradable fraction (*b* fraction), which will be digested over time, and the rate of degradation of this fraction (*c* value). Summation of the *a* and *b* fractions provides a measure of total dietary RDP content, whilst protein unaccounted for by these two fractions, $1-(a+b)$, is a measure of dietary UDP content. The quantity of fraction *b* protein degraded is determined by the time spent in the rumen with the feed exposed to microbial fermentation, itself a function of level of feeding and outflow rate. It follows that, at high rates of passage, a significant amount of the *b* fraction is likely to escape rumen degradation. The term ERDP therefore defines the quantity of N available for capture and utilization by the rumen microorganisms for growth and synthesis.

At the same time, it is recognized that the utilization of ERDP by the rumen microbes is influenced by energy availability within the rumen and, in this regard, ruminally digested organic matter (OM) was first proposed as a suitable measure. However, it was subsequently accepted that this was not a sufficiently adequate index of energy supply, and the alternative concept of fermentable ME (FME) was established. Defined as total dietary ME content, less the energy of dietary fermentation products, fats and oils (given that neither contribute significantly to energy supply for microbial metabolism), FME represented a conceptual improvement on previous schemes. However, it suffers from a major weakness, namely that, to date, no values of FME have been obtained from *in vivo* feeding systems. Thus, in a situation where the prediction of ME content may be questionable, the application of relationships between FME and predicted ME, as proposed by Chamberlain *et al.* (1993) and others, which were not based on validated estimates, only serves to add confusion and uncertainty to the ultimate prediction of microbial protein synthesis. Evidence of the potential magnitude of this problem can be obtained from Table 2.2, where further data regarding the two maize silages referred to earlier are presented. Whilst no estimates of FME content were obtained, variations in the estimates of crude protein, ERDP, DUP and ME contents were so large that, if applied to current feeding systems, they would lead to differences in the estimates of MP supply of sufficient magnitude as to make the whole scheme unworkable.

Thus, whilst scientific understanding of the metabolism of dietary proteins by ruminants has advanced considerably over the last two decades, there is still concern regarding the reliability of estimates of MP. More importantly, none of the major components of the scheme, namely the amounts of microbial and undegraded feed protein digested in the intestines, have been systematically validated against *in vivo* data. The absence of this validation clearly makes it difficult to accept or refute the many estimates of MP supply that are currently being reported. In this regard, a major revision to the value used to represent the efficiency of utilization of MP for milk synthesis was recently suggested by Mansbridge *et al.* (1999). Previous studies had suggested that the estimates of MP supply to support a desired level of milk production were too low, and this led to the suggestion that the efficiency value of 0.68, as adopted by the Agricultural and Food Research Council (AFRC, 1993) was too high. In a carefully designed lactation study, Mansbridge *et al.* (1999) varied both protein supply and milk output, and analysis of the data provided an efficiency of protein utilization of 0.625, leading to their conclusion that this lower value should be accepted in all subsequent versions of the UK system. However, whilst Mansbridge *et al.* (1999) tested the system as it is currently being used, they did not provide any validation of their estimates of MP supply. Thus, it is not possible from this study to conclude that the efficiency value is wrong, when the error could equally reside in the computed MP inputs.

Furthermore, it is recognized that the *in sacco* technique (Ørskov *et al.*, 1980) has a number of potential weaknesses. It is difficult to standardize the technique, despite repeated attempts (Madsen and Hvelplund, 1994), this being particularly relevant in relation to the washing of feed residues to remove microbial contamination, whilst there remains concern over the choice of the most appropriate diet for the experimental animals. The technique relies on the use of surgically modified animals and, when the costs and other implications of surgical preparation and animal maintenance are taken into account, the method is relatively expensive and not conducive to routine use. There is also growing concern regarding the possible introduction of animal effects when using mature sheep to determine the digestion kinetics of protein metabolism of feeds for high-yielding dairy cows. Finally, whilst not particularly relevant to the evaluation of feed proteins, the *in sacco* technique has been extended to consider the rumen degradability of other nutrients, including starch and NDF, but often with quite variable and sometimes conflicting results. In this context, de Visser (1993) provided a very convincing data set relating to the degradation characteristics of the starch component of different maize silages, which suggested the presence of increased amounts of ruminally protected starch as the forage crop matured. In contrast, *in vivo* studies from this laboratory undertaken with lactating dairy cows to examine a range of different maturities of maize silage at harvest failed to establish the existence of large amounts of starch entering the small intestine of cattle. In this respect, the use of 'book values' for estimates of the rate of passage of digesta from the rumen in the estimation of the partition of digestion of ruminally degradable nutrients between ruminal and postruminal digestion may be worthy of reconsideration.

In vitro *gas production methodologies*

As indicated, rumen fermentation is associated with the evolution of gas, principally carbon dioxide and methane, with occasional amounts of hydrogen, whilst recently the occurrence of dimethyl sulphide has been reported. On the assumption that the quantity of gas released when feedstuffs are incubated *in vitro* with rumen fluid is closely related to digestibility, and therefore the energy value of the feed to ruminants, Menke *et al.* (1979) proposed a simplified system, based on gas production, to evaluate feedstuffs. In their system, fermentations were conducted in gas-syringes to allow the gas evolved to be retained and measured manually. The approach developed by Wilkins (1974) used a pressure transducer to measure accumulated head-space gases resulting from microbial fermentation. This technique was subsequently exploited by Theodorou *et al.*(1993), who recognized the potential of the methodology to provide estimates of the rate and extent of feedstuff degradation. The technique has now been adopted by a number of laboratories throughout the world, as it undoubtedly offers considerable potential in relation to the throughput of samples, while reducing the reliance on surgically modified animals.

The earlier data were presented simply as cumulative gas-production curves, and this led to difficulties in data interpretation, with many of the more subtle changes between feeds tending to be either masked or lost when mathematical models were fitted. Subsequently, Mould *et al.* (1999) reconsidered this part of the technique in a series of studies designed to examine the effect of a feed additive on the rate and extent of rumen fermentation. Using appropriate corrections (negative controls), they derived profiles for the rate of gas production resembling gas chromatography curves. By using this approach, it became possible not only to 'fingerprint' a feed into its main carbohydrate components but also to provide an estimate of the extent of degradation of each of these components. This greatly increases the value of the technique when attempting to identify small differences between feedstuffs. A further simplification of the methodology was the abandonment of measuring gas volume directly (syringe) for measurement of changes in gas pressure (Mauricio *et al.*, 1999). This, together with a direct data entry system, which reduced operator-associated errors and greatly increased the capacity of the system, has provided a simplified technique for routine use, capable of screening large numbers of substrates.

However, concerns remain over the degree of interpretation being placed upon the data provided by such methodology. There have been few attempts to validate the results with *in vivo* data; indeed, it would be almost impossible to do so. Thus, such data are only comparative, ideal for screening purposes, from which more comprehensive research, possibly involving animals, may be required, before definite recommendations can be made. Gas produced during the ruminal fermentation of feeds represents only a small component of total nutrient flux in a rumen ecosystem. Consequently, changes in gas yield per mole of carbohydrate fermented (see Table 2.1), as well as the partition of degraded carbohydrate between fermentation and microbial incorporation, can have a significant bearing on total gas yield. As Menke *et al.* (1979) concluded, although a high correlation was found between cumulative gas production and OM digestibility, a feedstuff with a low rate of gas production may have a higher *in vivo* digestibility than that suggested from gas production data. Thus, *in vitro* gas production values alone provide little direct information, apart from estimating rate of fermentation. These therefore need to be examined together with simultaneously derived estimates of both rate and extent of degradation to permit estimates of fermentation efficiency to be obtained. It is a matter of concern, therefore, to note that the level to which the technique is now being developed is well beyond its realistic capabilities. Proposed as a first approximation to measurement of *in vivo* ruminal events, it is being overcomplicated by attempts to measure parameters such as protein degradation, ammonia evolution, microbial protein synthesis and the survival of feed protein (Cone and Van Gelder, 1999). Thus, a powerful but, as yet, not fully validated technique is being used to develop further data sets which have not been, and are unlikely ever to be, validated: a very different position from that adopted by Tilley and Terry (1963), as discussed earlier. This is a very questionable basis on which to advance the science of feed evaluation, adding unnecessary complication when the first, and highly laudable, intent was to overcome the complexities of the cow. If current efforts continue, the cow will need to be reconsidered as a seemingly simple experimental tool on which to evaluate feedstuffs for milk production.

Near-infrared reflectance spectroscopy

Since its development, much has been reported about the potential benefits of this technology, particularly for laboratories involved in routine feedstuff analysis, where rapid, comprehensive and relatively inexpensive analyses are claimed (Offer *et al.*, 1998). NIRS is a physical method, which depends on the measurement of light absorption by the surface of a sample using wavelengths in the infrared region of the spectrum (1100–2500 nm). The absorption spectrum produced depends on the chemical bonds within the constituents of the feed sample and it is therefore possible to identify specific regions of the spectrum associated with different chemical entities, such as protein, fibre, starch, etc. To achieve satisfactory estimates of the nutrient content of feeds, however, it is first necessary to calibrate the apparatus against standard reference samples that have been analysed by more routine 'wet chemistry' methodologies. Several workers reported satisfactory estimates for components such as crude protein, and Barber *et al.* (1990) provided useful predictions of *in vivo* OM digestibility for grass silage, compared with more conventional approaches involving *in vitro* techniques or wet chemistry (for more details, see Beever *et al.*, 1999). These studies were subsequently extended by Givens *et al.* (1997) for a wider range of forage types. However, the problem appears to be related to the way in which the technique has been exploited beyond its capabilities, similarly to the gas technique discussed earlier. Prediction of rumen protein degradability by NIRS is never likely to be a simple process, given that most feeds contain several different types of protein, all of which are likely to have different degradation profiles. Furthermore, to use this technique in conjunction with *in sacco* measurements of protein degradability, which have never been validated against *in vivo* observations, is an enormous step of confidence and an approach that should be adopted with caution. A similar situation prevails in relation to the determination of FME contents by NIRS, a procedure used by several laboratories, for which no validation with *in vivo* estimates has been attempted. Consequently, whilst NIRS has much to offer, it should be used only where there is established confidence in the data, and its use to predict *in vivo* events from a questionable set of *in vitro* data is only likely to bring disrepute to a potentially useful technique.

Future Developments in Forage Evaluation

Undoubtedly, feedstuff evaluation is an established component of livestock feeding in the UK, along with most developed countries. It is only through access to reliable information regarding feed quality that rational decisions relating to the feeding of ruminant livestock can be made. This is particularly true with respect to forages, for reasons discussed earlier. However, for the full potential of feedstuff analysis to be realized, several important issues need to be addressed. Some of these will be discussed in this section, along with possible recommendations, accepting that feedstuff evaluation is an evolving science, in which further revisions will be dependent upon the way in which systems of meat and milk production from ruminants develop. In this chapter, as well as in other reviews (e.g. Beever and Cottrill, 1993), concern over the adequacy of present feed rationing schemes, which generate much of the current interest in feed evaluation, has been expressed. Possibly the most crucial issue is the need to consider models that represent animal response, rather than simply meeting animal requirements. Before addressing longer-term aspects, however, it is pertinent to consider how various options may be developed to improve current systems.

Immediate requirements

All feedstuff analysis has a cost, which is always borne by the farmer, even if indirectly through the services of a feed or advisory company. Thus, livestock producers are entitled to question the value of their investment in feedstuff analysis in terms of enhancing the profitability of their business. This would be difficult to test objectively and the best that can be expected is a general perception that their investment is financially rewarding to their business. To achieve this, however, they need to be satisfied with both the analyses they receive and the interpretations placed upon them. Reorganization within the UK Ministry of Agriculture, Fisheries and Food (MAFF) in the late 1980s and privatization of the Agricultural Development and Advisory Service (ADAS) caused considerable fragmentation of the industry in relation to feedstuff analysis and nutritional consultancy. Many more laboratories are now offering such services, backed by an ever-increasing number of consultancy organizations and private consultants. But do all laboratories provide information the farmer wants and that can be used with confidence? Equally, do all laboratories report their data in a format that is easy to comprehend? Undoubtedly, many do, but in too many instances interpretation of the data can be difficult. For estimates of ME content, possibly the most important parameter with respect to forages, several laboratories report more than one value, depending upon the methodology used. An informed consultant may be able to reconcile such issues, but a difference of more than 1 MJ kg^{-1} DM, such as illustrated in Table 2.2, will be difficult for many farmers to ignore, irrespective of which value they believe. Equally, many laboratories report data which are of little or no interest to farmers. It is likely that farmers are already aware of those silages of inferior quality before they receive the silage analysis, which puts in question whether all of the data relating to the volatile components have to be reported, when lactate and acetate levels would probably suffice. The most obvious reason why such data are provided is because they were produced in the laboratory and thus the laboratory felt obliged to report it. All laboratories must be encouraged to be more discriminating in the data they report, as well as paying attention to the presentation of the data.

But there are deeper issues, relating to the plethora of different techniques being used to derive the analytical data. Some of these concerns have already been expressed, the position regarding starch analysis being an excellent example. Currently, there is no agreement over the most appropriate methods to use for all analyses and, whilst most laboratories seek to acquire 'good laboratory practice' status, this does not necessarily guarantee that the data they generate is accurate. Some laboratories have resorted to techniques that were questioned many years ago, whilst others have increased their reliance on indirect methods, such as NIRS. The validity of data produced using NIRS for chemical entities such as crude protein, starch, etc. will never be better than the databases used to establish the calibration curves. There have been many instances where these have been shown to be inadequate, use of grass-silage-derived relationships to evaluate the energy component of whole-crop wheat or maize silages or even fresh grass being suitable examples. But it is the almost uncontrolled use of NIRS to estimate biologically important components, such as ERDP, DUP and FME, which is of greatest concern, with highly questionable databases of *in vitro*-derived information being used with scant regard to the lack of any validation with *in vivo* events. Reference has already been made in Table 2.2 to variation in the estimates of ME content provided by those laboratories involved in the study. Despite this, there was total agreement between laboratories that the later-harvested crop had a higher ME value. This gives little reassurance, however, when *in vivo* determination of the ME contents of the two feeds fed to dairy cows in respiration calorimeters found a higher ME content for the earlier-cut crop. An immediate explanation was the higher gross energy (GE) content of this feed, indicative of an increased extent of fermentation occurring during ensiling of the wetter and more immature crop (Table 2.3). Thereafter, no differences were observed between the two feeds with respect to the partition of GE to faecal, methane or urine energy. Thus, for feeds that have undergone

Table 2.3. Comparison of *in vivo* and laboratory-derived estimates of the metabolizable energy content of two maize silages harvested at different stages of maturity (units, MJ kg^{-1} oven-dry matter unless stated).

	Silage maturity	
	Low DM	High DM
Laboratory estimates*	11.2	11.5
In vivo†		
Maize		
GE content	18.9	17.6
ME content	10.9	10.1
Total mixed ration		
DE / GE (kJ MJ^{-1})	680	676
DE content	12.9	12.2
ME / GE (kJ MJ^{-1})	579	576
ME / DE (kJ MJ^{-1})	850	851

* Mean of six reporting laboratories.
† Not corrected for level of feeding effects.
DE, digestible energy.

different degrees of fermentation, some measurement of GE content to take account of different levels of volatile components may be necessary, if more reliable estimates of ME content are to be obtained.

In summary, the often heard remark that the analytical data were 'unbelievable' is not acceptable and urgent rationalization is needed to provide agreement on appropriate methods, resisting those which have not been validated against *in vivo* observations. However, this can only be achieved through coordination at national level. Additionally, with the level of concern over some of the analyses being provided, it should become a requirement of all laboratories to provide a brief résumé of the methods used, be it NIRS or wet chemistry. This may be of little immediate value to the farmer, but it would provide an important audit trail for the consultant regarding possibly aberrant data. Equally, all laboratories should report only on those parameters where they have sufficient confidence in the methodology, whilst positive encouragement should be given to their committed involvement in schemes designed to examine interlaboratory comparisons. Such procedures are already in place with respect to the analysis of milk constituents, which has major implications with respect to both milk price and quota management. Given that the financial implications of feedstuff evaluation are at least as great, the lack of appropriate standardization is unacceptable.

Longer-term needs

Whilst initiatives similar to those outlined above will provide better support for current systems of feed rationing for ruminants, there is a growing opinion that more radical changes in models of ration formulation are required. This is particularly evident from the current interest being shown in the UK for both the French PDI system and CNCPS. This should not be construed to suggest that such models have been adequately validated or that they necessarily perform better than UK equivalents; rather, they contain certain elements and representations considered to be superior to those in current UK systems of energy and protein rationing. It is not the intention of this chapter to review the strengths and weaknesses of such systems, as others have attempted this elsewhere. The more challenging task is to decide which type of system should be developed. Certain opinion suggests that serious revision of current systems would be adequate, but adoption of this approach would be difficult if one of the agreed challenges was a model of animal response rather than animal requirement. To meet this objective, a new model of energy and protein rationing is required, with an implicit representation of individual nutrients to provide the important characterization of ME and, to a lesser extent, MP that is not evident in present systems. Equally, any new system will require more comprehensive representations of postabsorptive energy and protein metabolism. While these are partially recognized in the rumen element of the current MP system, there is no evidence of nutrient interactions at the tissue level. Such approaches can be found in the models of rumen and whole-animal metabolism proposed by Baldwin and colleagues, the most notable efforts being those of Baldwin *et al.* (1987a, b, c). These models were quite successful in predicting overall nutrient utilization and lactational performance in dairy cows, but were criticized in relation to the required dietary characteristics, which were too comprehensive when compared with those expected from routine feedstuff evaluation. However, in the development of these models, there were no attempts to consider if some of the inputs could be ignored, or at least aggregated, without serious loss of resolution in the models. Thus, it is difficult to provide a definitive list of required feed characteristics, but, in the light of present knowledge, it is possible to suggest some of the most likely candidates.

For those nutrients that the animal ultimately uses for maintenance and production costs, the simplest nutritional representation would be rumen VFA (both lipogenic and gluconeogenic), long-chain fatty acids, glucose and amino acids. Considering these in turn, partition of rumen VFA into lipogenic and gluconeogenic substrates is an obvious necessity, given their contrasting metabolic fates, with further distinction of lipogenic VFA into acetate and butyrate, if only to provide compatibility with *in vivo* data. There is no justification for representing other VFA, such as isobutyrate and valerate, as these are never found in significant amounts in any diets. Propionate is the sole gluconeogenic VFA and, given the crucial role that glucose plays in the overall nutrient economy of the animal, this will need to be represented in all models of nutrient digestion and postabsorptive metabolism. However, as VFA do not occur naturally in most feeds, only with fermented feeds, such as grass silage, will it be necessary to take account of these, acetic acid being the only one of nutritional importance. This will be referred to later in relation to silage analysis.

Long-chain fatty acids, on the other hand, occur in most feedstuffs and, whilst levels are generally low, especially in forages, they may account for up to 15% total ME supply in rations designed for high-producing animals. When augmented with fatty acids from mobilized body tissue in early-lactation cows or underfed animals, free fatty acids become important to the animal's overall energy metabolism. In addition, there are those situations where incomplete fatty acid metabolism can induce nutritional disorders, such as fatty liver syndrome and ketosis. There is also considerable interest, on the grounds of human health, in modifying the fatty acid composition of ruminant products, specifically to increase mono-unsaturated fatty acid levels and the ratio of ω-3 : ω-6 fatty acids. In this respect, both grazed and ensiled grass have been suggested as possible sources of increased ω-3 fatty acid levels. However, unless the market for such products increases dramatically, routine procedures to provide highly comprehensive fatty acid analyses for forages cannot be contemplated. Instead, there may be benefit if an agreed method for determination of total fatty acid content could be established, especially if associated with a rudimentary characterization of the saturated : unsaturated fatty acid ratio. Currently, most laboratories rely on ether extraction, but this method is limited, due to its non-specificity. Other methods, including those based on direct methylation of fatty acids (Outen *et al.*, 1975), may be worth considering, although a final gravimetric determination would be needed to replace existing methodology, which involves relatively time-consuming gas chromatography.

There are few, if any, situations where glucose occurs in significant amounts in the feed, the exception being fresh forages. However, as starch is of increasing interest with respect to the formulation of diets for ruminants, this should be a priority area for analytical technique development. At this point, it would be useful to identify those features of the methodology developed by MacRae and Armstrong (1968) which several laboratories are dissatisfied with, whilst at the same time establishing which of the current methods should be abandoned as being unreliable. Undoubtedly, one important feature of the method of MacRae and Armstrong (1968) was the use of glucose oxidase to determine the released glucose units. Given the high specificity of glucose oxidase, it is unlikely that this is a major area of concern, and thus the problem must be associated with solubilization and hydrolysis of the starch during the initial stages of the method. Starch composition varies quite markedly between different starch sources, particularly the relative proportions of amylose and amylopectin, and it may be desirable to re-examine the initial solubilization procedures as proposed by MacRae and Armstrong (1968) with a wider range of substrates. Thereafter, the main cause of concern must be the specificity, as well as the uniformity, of commercially available amyloglucosidases. This has been recognized as a potential problem in the past, but it was never viewed as being of sufficient severity to cause the technique to be abandoned, especially in favour of much less specific methodologies. Remaining with starch, advances in the protein nutrition of ruminants which led to the concepts of ruminally degradable- and ruminally resistant proteins have recently been extended to starch, leading to nutritional entities such as ruminally fermentable starch (RFS) and ruminally resistant or digestible undegraded starch (DUS). This interest has developed on the unproved hypothesis that starch digested in the small intestines to glucose would be utilized more efficiently by the animal than starch subjected to ruminal fermentation with the resulting hepatic conversion of propionate to glucose. In studies in this laboratory, in which identical amounts of additional starch were provided by infusion into the rumen or the small intestine of dairy cows, posthepatic glucose output showed no effect of treatment, suggesting that the liver exerted considerable control on whole-body glucose status (Reynolds *et al.*, 1998). Equally, the same study failed to establish any difference in overall energy utilization between the different routes of administration. Such data raise doubt regarding the metabolic value of different starch entities such as RFS and DUS. However, interest in such fractions is likely to continue in relation to improving the efficiency of nutrient utilization within the alimentary tract.

Reference was made earlier to the importance of managing energy (carbohydrate) and protein (nitrogen) availability in the rumen in order to optimize microbial synthetic efficiency, and it is recognized that the provision of carbohydrates with different degradation characteristics will be important in this regard. The ability to balance contrasting starch sources, such as wheat and maize, when fed in conjunction with different forages will ensure a more even supply of energy for the microbes, thus avoiding periods of excessive VFA production and substantially reduced rumen pH values. It is known that high starch feeding can significantly depress rumen pH and values approaching 5.0 have recently been reported in this laboratory. On such occasions, overall fibre digestion in the rumen will be reduced, as shown by Beever *et al.* (1989), and ultimately this may lead to a significant reduction in whole-tract digestibility of fibre. Provision of a more continuous supply of degradable carbohydrate will minimize such events, whilst in such situations it may be possible to exploit the ability of the small intestine to digest starch. This would have the effect of reducing overall starch load within the rumen, and research in this laboratory has shown that most dairy cows can successfully digest over 2 kg starch entering the duodenum before overall digestive efficiency is compromised and faecal starch levels increase. To provide suitable methodologies to describe the degradation characteristics of dietary starches, however, two important issues have to be addressed. First, as current *in sacco* methodologies are not suitable, a reliable method that could be used with concentrate feeds that are normally ground prior to pelleting must be developed. Equally, all data on the rate and extent of starch digestion in the rumen obtained with *in vitro* procedures must be validated against *in vivo* data. This would also apply if the methodology were extended to examine aspects of dietary fibre digestion in the rumen.

Remaining with this theme, serious consideration should be given to those procedures currently being used to characterize dietary proteins, and more reliable and repeatable alternatives to the *in sacco* technique should be developed. In this respect, artificial rumen techniques, such as Rusitec (Czerkawski and Breckenridge, 1977), have not been fully exploited and there may be benefit from developing these further, provided, once again, that the data are validated with *in vivo* observations as necessary. Currently, the fractionation of dietary crude protein is limited to that which is or is not of protein origin, with no attempts to identify specific proteins or, more specifically, individual amino acids. It would be difficult to justify major new initiatives in such areas at present, but the issues may need to be reconsidered if environmental pressures on livestock production systems call for tighter controls on nitrogen losses as urine and faeces. When most systems of milk production can claim a milk N output of no more than 300 g kg^{-1} dietary N intake, there appears to be considerable scope in the management of protein in ruminant diets to reduce both dietary input and waste output without compromising milk N output.

Returning finally to the issue of silage analysis, which undoubtedly is the main area of interest with respect to forage evaluation, a number of important suggestions have already been made in this chapter. One possible area worthy of attention is in relation to the determination of true DM content. This will require quantification of the volatile components and, whilst gas-chromatographic techniques are available for determination of most of the major components, such methodologies are not conducive to rapid and routine feedstuff evaluation. This situation is further complicated by the need to quantify lactic acid levels, which, in some silages, can comprise as much as 100 g kg^{-1} of total DM, or over 20% of the energy released in the rumen. Whilst lactate is not utilized to any great extent as an energy source, either directly or after conversion to acetate or propionate by the rumen microbes, it is extensively used in postabsorptive metabolism (Gill *et al.*, 1986). It is also an important component in the estimation of FME, given that this entity, as defined by AFRC (1992), is derived by discounting the ME contribution of fermentation acids and that of oils and fats from the total ME content. However, there is a wider issue relating to the analysis of silage volatiles, as several have been identified as being important in the prediction of silage intake. Finally, until feed rationing schemes move towards a fuller description of individual nutrients, serious consideration should be given to the provision of more reliable estimates of forage ME contents. Currently, these form a major component of all ration formulation schemes and yet, despite such initiatives as 'ME√' (Offer *et al.*, 1996), which is widely used in the UK, there are many highly questionable values being produced by different laboratories. The problem may be due, in part, to the failure to recognize different GE contents, as discussed earlier, but another more fundamental issue may be the predictive relationships that are being used. Many were based on grass silage and appear to have been extended to other forage sources with little regard for the wider implications. Furthermore, the use of sheep to develop such relationships needs to be examined, given the emphasis now placed on the accurate ration formulation of bovine animals and, in particular, high-yielding dairy cows. Recent studies in this laboratory, as reported by Beever *et al.* (1998a, b), showed measured ME intakes of over 310 MJ day^{-1}, equivalent to more than five times the maintenance level of feeding. An error of only 5% in this estimate is equivalent to an under- or overprediction of 3.0 l day^{-1} of milk of standard composition. Equally, errors of the magnitude reported for the two maize silages in Table 2.2 would, if applied to the total diet, equate to an underprediction of milk yield of 8.0 l day^{-1} when comparing the extreme estimates.

References

AFRC (Agricultural and Food Research Council) (1984) *The Nutrient Requirements of Farm Livestock, No. 2, Ruminants. No. 1. Supplement.* CAB International, Farnham Royal.

AFRC (Agricultural and Food Research Council) (1992) Nutritive requirements of ruminant animals: protein. Technical Committee on Responses to Nutrients, Report No. 9. *Nutrition Abstracts and Reviews, Series B* 62, 787–835.

AFRC (Agricultural and Food Research Council) (1993) *Energy and Protein Requirements of Ruminants.* An advisory manual prepared by the AFRC Technical Committee on Responses to Nutrients, CAB International, Wallingford.

AFRC (Agricultural and Food Research Council) (1998) *Responses in the Yield of Milk Constituents to the Intake of Nutrients by Dairy Cows.* AFRC Technical Committee on Responses to Nutrients, Report No. 11, CAB International, Wallingford.

ARC (Agricultural Research Council) (1980) *The Nutrient Requirements of Farm Livestock, No. 2, Ruminants.* Commonwealth Agricultural Bureaux, Farnham Royal.

Armstrong, D.G. and Beever, D.E (1969) Post abomasal digestion of carbohydrates in the adult ruminant. *Proceedings of the Nutrition Society* 28, 121–131.

Baldwin, R.L., Koong, L.J. and Ulyatt, M.J. (1977) A dynamic model of ruminant digestion for evaluation of factors affecting nutritive value. *Agricultural Systems* 2, 255–288.

Baldwin, R.L., France, J. and Gill, E.M. (1987a) Metabolism of the lactating cow. I. Animal elements of a mechanistic model. *Journal of Dairy Research* 54, 77–106.

Baldwin, R.L., Thornley, J.H.M. and Beever, D.E. (1987b) Metabolism of the lactating cow. II. Digestive elements of a mechanistic model. *Journal of Dairy Research* 54, 107–131.

Baldwin, R.L., France, J., Beever, D.E., Gill, E.M. and Thornley, J.H.M. (1987c) Metabolism of the lactating dairy cow. III. Properties of mechanistic models suitable for evaluation of energetic relationships and factors involved in the partition of nutrients. *Journal of Dairy Research* 54, 133–144.

Barber, G.D., Givens, D.I., Kridis, M.S., Offer, N.W. and Murray, I. (1990) Prediction of the organic matter digestibility of grass silage. *Animal Feed Science and Technology* 28, 115–128.

Beever, D.E. (1993) Rumen function. In: Forbes, J.M. and France, J. (eds) *Quantitative Aspects of Ruminant Digestion and Metabolism.* CAB International, Wallingford, pp. 187–215.

Beever, D.E. and Cottrill, B.R. (1993) Protein systems for feeding ruminant livestock: a European assessment. *Journal of Dairy Science* 77, 2031–2043.

Beever, D.E., Coelho da Silva, J.F. and Armstrong, D.G. (1970) The effect of processing maize on its digestion in sheep. *Proceedings of the Nutrition Society* 29, 44A.

Beever, D.E., Coelho da Silva, J.F., Prescott, J.J.D. and Armstrong, D.G. (1972) The effect in sheep of physical form and stage of growth on the sites of digestion of a dried grass. I. Sites of digestion of organic matter, energy and carbohydrate. *British Journal of Nutrition* 28, 347–356.

Beever, D.E., Harrison, D.G., Thomson, D.J., Cammell, S.B. and Osbourn, D.F. (1974) A method for the estimation of dietary and microbial protein in duodenal digesta of ruminants. *British Journal of Nutrition* 32, 99–112.

Beever, D.E., Cammell, S.B. and Edmonds, S. (1996) Inter-laboratory variation in foodstuff evaluation of two contrasting maize samples. *Animal Science* 62, 685A.

Beever, D.E., Sutton, J.D., Thomson, D.J., Napper, D.J. and Gale, D.L. (1998a) Comparison of molassed and unmolassed sugar beet pulp and barley as energy supplements on nutrient digestion and supply in silage fed cows. *Animal Production* 46, 490.

Beever, D.E., Cammell, S.B., Sutton, J.D., Rowe, N. and Prescott, G.E. (1998b) Energy metabolism in high yielding dairy cows. In: *Proceedings of the British Society of Animal Science – 1998*, Scarborough, UK, p. 197.

Beever, D.E., Offer, N. and Gill, M. (1999) The feeding value of grass and grass products. In: Hopkins, A. (ed.) *Grass: Its Production and Utilization*, 3rd edn. Blackwell Science, Oxford.

Black, J.L., Beever, D.E., Faichney, G.J., Howarth, B.R. and Graham, N.M. (1980/81) Simulation of the effects of rumen function on the flow of nutrients from the stomach of sheep. Part I – Description of computer program. *Agricultural Systems* 6, 195–219.

Blaxter, K.L. (1962) *The Energy Metabolism of Ruminants*. Hutchison, London.

Burroughs, W., Trenkle, A. and Vetter, R.L. (1967) A system of protein evaluation for cattle and sheep involving metabolisable protein (amino acids) and urea fermentation potential of feedstuffs. *Veterinary Medicine (Small Animal Clinician)* 69, 713.

Chamberlain, A.T., Seyou, M.K., Chapman, D. and Piotrowski, C. (1993) The fermentable metabolisable energy in grass silage. *Animal Production* 56, 455A.

Chesson, A. (1988) Lignin–polysaccharide complexes of the plant cell wall and their effect on microbial degradation in the rumen. *Animal Feed Science and Technology* 21, 219–228.

Cone, J.W. and Van Gelder, A.H. (1999) Estimation of efficiency of microbial growth in rumen fluid with the gas production technique. In: *Proceedings of the British Society of Animal Science – 1999*, Scarborough, p. 39.

Czerkawski, J.W. and Breckenridge, G. (1977) Design and development of a long-term rumen simulation technique. *British Journal of Nutrition* 38, 371–384.

de Visser, H. (1993) Characterization of carbohydrates in concentrates for dairy cows. In: Garnsworthy, P.C. and Cole, D.J.A. (eds) *Recent Advances in Animal Nutrition – 1993*. Nottingham University Press, Nottingham, pp. 19–38.

El Shaer, H.M., Omed, H.M., Chamberlain, A.G. and Axford, R.F.E. (1987) Use of faecal organisms from sheep for the *in vitro* determination of digestibility. *Journal of Agricultural Science, Cambridge* 109, 257–259.

France, J., Thornley, J.H.M. and Beever, D.E. (1982) A mathematical model of the rumen. *Journal of Agricultural Science, Cambridge* 99, 343–353.

France, J., Dhanoa, M.S., Theodorou, M.K., Lister, S.J., Davies, S.J. and Isac, D. (1993) A model to interpret gas accumulation profiles with *in vitro* degradation of ruminant feeds. *Journal of Theoretical Biology* 163, 99–111.

Gill, M., Siddons, R.C., Beever, D.E. and Rowe, J.B. (1986) Metabolism of lactic acid isomers in the rumen of silage-fed sheep. *British Journal of Nutrition* 55, 399–407.

Givens, D.I., Everington, J.M. and Adamson, A.H. (1989) The digestibility and ME content of grass silage and their prediction from laboratory measurements. *Animal Feed Science and Technology* 24, 27–43.

Givens, D.I., Everington, J.M. and Adamson, A.H. (1990) The nutritive value of spring-grown herbage produced on farms throughout England and Wales over 4 years. III. Prediction of energy values from various laboratory measurements. *Animal Feed Science and Technology* 27, 185–196.

Givens, D.I., Everington, J.M. and Adamson, A.H. (1992) The chemical composition and energy value of high temperature dried grass produced in England. *Animal Feed Science and Technology* 36, 215–218.

Givens, D.I., De Boever, J.L. and Deaville, E.R. (1997) The principles, practices and some future applications of near infrared spectroscopy for predicting the nutritive value of foods for animals and humans. *Nutrition Research Reviews* 10, 83–114.

Goering, H.K. and Van Soest, P.J. (1970) *Forage Fibre Analysis*. Agricultural Handbook No. 379, Agricultural Research Service, USDA, Washington, DC, 20 pp.

Issacson, H.R., Hinds, F.C., Bryant, M.P. and Owens, F.N. (1975) Efficiency of energy utilisation by mixed rumen balance in continuous culture. *Journal of Dairy Science* 58, 1645–1659.

Jones, D.I.H. and Hayward, M.V. (1975) The effect of pepsin pre-treatment of herbage on the prediction of dry matter digestibility from solubility in fungal cellulase solutions. *Journal of the Science of Food and Agriculture* 26, 711–718.

McDonald, I. (1981) A revised model for the estimation of protein degradability in the rumen. *Journal of Agricultural Science, Cambridge* 96, 251–252.

MacRae, J.C. and Armstrong, D.G. (1968) Enzyme method for determination of α-linked glucose polymers in biological materials. *Journal of the Science of Food and Agriculture* 19, 578–581.

MacRae, J.C. and Armstrong, D.G. (1969) Studies on intestinal digestion in sheep. 2. Digestion of some carbohydrate constituents in hay, cereal and hay–cereal rations. *British Journal of Nutrition* 23, 377–387.

MacRae, J.C. and Beever, D.E. (1997) Predicting amino acid supply and utilisation in the lactating ruminant. *Proceedings of the Society of Nutrition Physiology* 6, 15–30.

MacRae, J.C. and Lobley, G. (1982) Some factors which influence thermal energy losses during the metabolism of ruminants. *Livestock Production* 9, 447–456.

Madsen, J. and Hvelplund, T. (1994) Prediction of *in situ* protein degradability in the rumen: results of a European ring test. *Livestock Production Science* 39, 201–212.

Mangan, J.L. (1982) The nitrogenous constituents of fresh forages. In: Thomson, D.J., Beever, D.E. and Gunn, R.G. (eds) *Forage Protein in Ruminant Animal Production*. BSAP Occasional Publication No. 6, BSAP, Thames Ditton, pp. 25–40.

Mansbridge, R., Metcalf, J.A., Oldham, J.D., Blake, J.S. and Newbold, J.R. (1999) The response of net milk protein output to metabolisable protein supply in dairy cows. In: *Proceedings of the British Society of Animal Science – 1999*, Scarborough, UK, p. 19.

Mauricio, R.M., Mould, F.L., Dhanoa, M.S., Owen, E., Channa, K.S. and Theodorou, M.K. (1999) A semi-automated *in vitro* gas production technique for ruminant feedstuff evaluation. *Animal Feed Science and Technology* 79, 321–330.

Mehrez, A.Z. and Ørskov, E.R. (1978) A study of the artificial fibre bag technique for determining the digestibility of feeds in the rumen. *British Journal of Nutrition* 40, 337–345.

Menke, K.H., Raab, L., Salewski, A., Steingass, H., Fritz, D. and Shneider, W. (1979) The estimation of the digestibility and metabolizable energy content of ruminant feedingstuffs from the gas production when they are incubated with rumen liquor *in vitro*. *Journal of Agricultural Science, Cambridge* 93, 217–222.

Miller, E.L. (1973) Evaluation of foods as sources of nitrogen and amino acids. *Proceedings of the Nutrition Society* 32, 79–84.

Moss, A.R. and Givens, D.I. (1990) Chemical composition and *in vitro* digestion to predict digestibility of field-cured and barn-cured grass hays. *Animal Feed Science and Technology* 29, 73–87.

Mould, F.L., Mauricio, R.M., Smith, T. and Owen, E. (1999) The sensitivity of the Reading pressure technique and its ability to describe degradation characteristics of feedstuffs. *Animal Feed Science and Technology* (in press).

Murphy, M.R., Baldwin, R.L. and Koong, L.J. (1982) Estimation of stoichiometric parameters for rumen fermentation of roughage and concentrate diets. *Journal of Animal Science* 55, 411–421.

Nolan, J.V. (1975) Quantitative models of nitrogen metabolism in sheep. In: McDonald, I.W. (ed.) *Digestion and Metabolism in the Ruminant*. University of New England Publishing Unit, Armidale, New South Wales, pp. 416–431.

Offer, N.W., Cottrill, B.R. and Thomas, C. (1996) The relationship between silage evaluation and animal response. In: Jones, D.I.H., Jones, R., Dewhurst, R., Merry, R. and Haigh, P.M. (eds) *Proceedings of the XIth International Conference, held at Aberystwyth, 8–11 September 1996*. Institute for Grassland and Environmental Research, Aberystwyth, pp. 26–38.

Offer, N.W., Percival, D.S., Dewhurst, R.J. and Thomas, C. (1998) Prediction of the voluntary intake potential of grass silage by sheep and dairy cows from laboratory silage measurements. *Animal Science* 66, 357–367.

Ørskov, E.R. and McDonald, I. (1979) The estimation of protein degradability in the rumen from incubation measurements weighted according to rate of passage. *Journal of Agricultural Science, Cambridge* 92, 499–503.

Ørskov, E.R., Hovell, F.D.DeB. and Mould, F.L. (1980) The use of the nylon bag technique for the evaluation of feedstuffs. *Tropical Animal Production* 5, 195–213.

Outen, G.E., Beever, D.E. and Thomson, D.J. (1975) The digestion of the lipids of processed red clover herbage by sheep. *Journal of the Science of Food and Agriculture* 26, 1381–1389.

Reynolds, C.K., Humphries, D.J., Cammell, S.B., Benson, J., Sutton, J.D. and Beever, D.E. (1998) Effects of abomasal wheat starch infusion on splanchnic metabolism and energy balance in dairy cows. In: McCracken, K.J., Unsworth, E.F. and Wylie, A.R.G. (eds) *Energy Metabolism of Farm Animals*. CAB International, Wallingford, pp. 387–390.

Sinclair, L.A., Garnsworthy, P.C., Newbold, J.R. and Buttery, P.J. (1993) Effect of synchronizing the rate of dietary energy and nitrogen release on rumen fermentation and microbial protein synthesis. *Journal of Agricultural Science, Cambridge* 120, 251–263.

Smith, R.H. and McAllan, A.B. (1970) Nucleic acid metabolism in the ruminant. 2. Formation of microbial nucleic acids in the rumen in relation to the digestion of food nitrogen and the fate of dietary nucleic acids. *British Journal of Nutrition* 24, 545–556.

Sniffen, C.J., O'Connor, J.D., Van Soest, P.J., Fox, D.G. and Russell, J.B. (1992) A net carbohydrate and protein system for evaluating cattle diet: II Carbohydrate and protein availability. *Journal of Animal Science* 70, 3562–3577.

Sutton, J.D. (1997) The effect of the replacement of grass silage by increasing proportions of urea-treated whole-crop wheat on food intake and apparent digestibility and milk production in dairy cows. *Animal Science* 65, 343–351.

Theodorou, M.K., Williams, B.A., Dhanoa, M.S., McAllan, A.B. and France, J. (1994) A simple gas production method using a pressure transducer to determine the fermentation kinetics of ruminant feeds. *Animal Feed Science and Technology* 48, 185–197.

Tilley, J.M.A. and Terry, R.A. (1963) A two stage technique for *in vitro* digestion of forage crops. *Journal of the British Grassland Society* 18, 104–111.

Van Soest, P.J. and Wine, R.H. (1967) Use of detergents in the analysis of fibrous feeds. IV.

Contamination of plant cell-wall constituents. *Journal of Association of Analytical Chemists* 50, 50–55.
Wilkins, J.R. (1974) Pressure transducer method for measuring gas production by microorganisms. *Applied Microbiology* 27, 135–140.
Witt, M.W., Sinclair, L.A., Wilkinson, R.G. and Buttery, P.J. (1997) Effects of synchronizing the rate of dietary energy and nitrogen supply to the rumen in diets with two rates of carbohydrate release on growth rate and metabolism in male lambs. In: *Proceedings of the British Society of Animal Science – 1997*, Scarborough, UK, p. 2.
Wray-Cahen, D., Metcalf, J.A., Backwell, F.R.C., Bequette, B.J., Brown, D.S., Sutton, J.D. and Lobley, G.E. (1997) Hepatic response to increased exogenous supply of plasma amino acids by infusion into the mesenteric vein of Holstein-Friesian cows late in gestation. *British Journal of Nutrition* 78, 913–930.

3 Intake of Forages

D.L. Romney[1] and M. Gill[2]

[1]Natural Resources Institute, Chatham Maritime, Kent ME4 4TB, UK; [2]Natural Resources International Ltd, PO Box 258, Chatham, Kent ME4 4PU, UK

Introduction

Forage is a major component of ruminant rations. In temperate regions, systems range from extensive – in which ruminants harvest the forage themselves, to maximize the utilization of the forage – to intensive systems – in which forages are offered with high levels of concentrates, to maximize production per cow. Forages in these systems include rough grazing on hills and uplands, cultivated high-quality grass and legumes, which are grazed or conserved as hay and grass, legumes and maize conserved as silage and straws from cereal crops.

In the tropics, forages tend to be highly variable in quality and quantity, from cultivated forages, such as *Pennisetum purpureum*, through the leaves of fodder trees to poor-quality stovers of maize, sorghum and millet during the dry season. The lack of resources of poor farmers in many of these systems means that concentrate use is limited and, in recent years, there has been increased interest in the use of high-quality legume straws or tree fodders as supplements to poorer-quality forages.

Many authors (e.g. Illius, 1998) have suggested that intake is probably the single most important variable determining animal performance and voluntary intake is generally correlated with the amount of nutrients that can be extracted from a feed, i.e. its digestibility. For forages, digestibility is largely determined by features of the plant, but potential digestibility, and hence potential intake, may not be achieved, due to interactions between feeds or between one or more feeds and the animal itself. To understand, and hence predict, these interactions requires an understanding of the mechanisms that control intake and thus this subject has been studied in detail over the last few decades. Recent reviews of the topic include Gill and Romney (1994), Jung and Allen (1995), Forbes (1995a, b; 1996), Provenza (1995a, b), Allen (1996), Illius and Jessop (1996), Weston (1996) and Wilson and Kennedy (1996), amongst others. The present chapter does not attempt to cover all aspects in such detail, but rather summarizes the information and highlights the aspects that are most important for forage feeds. The chapter starts by considering the main principles of intake control, prior to considering forage factors that determine potential intake. This is followed by summaries of the effects of interactions with animal factors and between dietary components. The chapter ends with sections on prediction and manipulation of intake and priorities for further research.

Principles of Intake Control

In order to understand the principles of intake control, it is important to answer the questions: (i) why do animals eat? and (ii) why do they not eat all the time?

It is generally accepted that animals eat to supply the tissues with the nutrients required to fuel physiological processes of maintenance, growth (including fat deposition in mature animals), milk production and work. However, given the variety of feed components eaten by animals, it is unlikely that the composition of the nutrients supplied will exactly meet the ratio of nutrients required by the animal. Thus, meeting the requirements of one nutrient is likely to result in a deficiency or excess of another. Animals stop eating to limit metabolic or physical discomfort and thus the animal has to decide at what point the disadvantages of the deficiencies or excesses of some nutrients outweigh the advantages of trying to meet the animal's energy requirements, which are thought to be the main intake 'driver' (Emmans, 1997).

In the experimental situation, animals have been shown to be able to make nutritionally wise choices between a range of foods, confirming the theory that animals are monitoring supply and demand of vital nutrients. For example, poultry (Shariatmadari and Forbes, 1993) and pigs (Kyriazakis *et al.*, 1990) were able to select a diet meeting their requirements for protein from foods containing different levels of protein. Similar observations have been made in sheep (Kyriazakis and Oldham, 1993) selecting between diets varying in protein, although it is more likely that diet selection in ruminants is a response to changes in the rumen environment (Cooper *et al.*, 1995), rather than metabolic requirements of the whole animal.

The data considered above all relate to daily intake, while much of the work on factors that stop the animal from eating tend to have been measured at the end of a single meal. The vast literature on experiments to measure intake demonstrate that animals do appear to reach a relatively stable energy intake on a constant diet, but this can be achieved by very different patterns of intake within a day. Early work on factors that limited intake focused on single factors, such as temperature (Brobeck, 1948), blood glucose concentrations (Mayer, 1953) and fat levels in the body (Kennedy, 1953). However, there is plenty of evidence (e.g. Fig. 3.1; Thiago, 1988) that gut fill, for example, varies during the day and is not at a maximum at the end of each meal. Balch and Campling, back in 1962, had concluded that 'food intake is unlikely to be regulated by any single mechanism and that, through the central nervous system, oropharyngeal sensations, gastric contractions and distension, changes in heat production and changes in the levels of circulating metabolites may severally be indicated'. One solution was to envisage integration of different signals within the brain, but argument still remains as to how and where

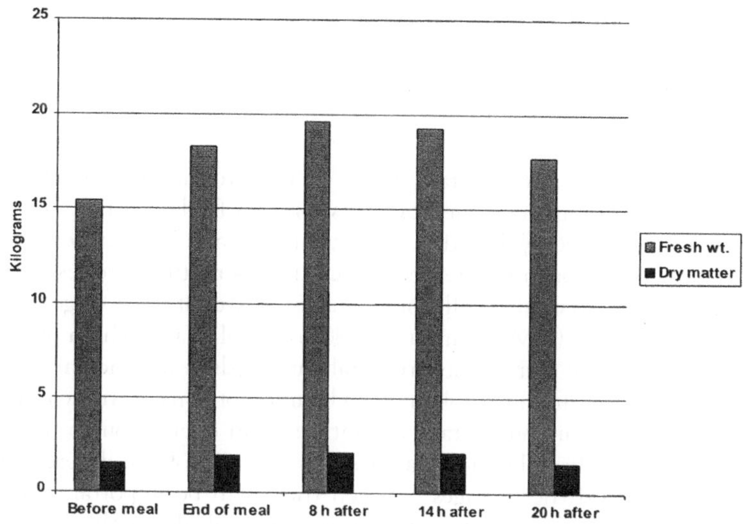

Fig. 3.1. Changes in the amount of fresh and dry digesta in the rumen of steers (liveweight approx. 120 kg) offered hay *ad libitum,* with time of offering a meal (Thiago, 1988).

such integration would take place. Anil *et al.* (1993) and Mbanya *et al.* (1993) reported, consistently, additivity between effects of volatile fatty acids (VFA) infused into the rumen and distension of the rumen, but not for stimuli of two different acids. It may be that it is the ratio of acids relative to the optimal ratio for the metabolism of the animal in question that is important, as discussed by Illius and Jessop (1996), but that ratio may vary within a day and it is not yet clear how an animal could keep count of cumulative metabolism over a 24 h period. This difficulty led Gill and Romney (1994) to hypothesize that factors determining the size of individual meals are related to optimizing nutrient supply, relative to daily metabolism, but this has yet to be confirmed.

It has long been assumed that, for forage diets, it is the 'bulk' of the forages that primarily limits intake, i.e. a combination of the volume and the time during which the undigested food remains in the gastrointestinal tract. Since the original proposal (Lehman, 1941, quoted by Van Soest, 1994) that the presence of undigested feed residues in the gastrointestinal tract limited intake, there has been extensive evidence to support the theory and this has been summarized recently by Allen (1996). Conrad *et al.* (1964) suggested that intake was restricted by gut fill up to a breakpoint in digestibility, beyond which the relationship between intake and digestibility became negative and controlled by the animal's requirements. The breakpoint is thus dependent on the energy requirements of the animal and on the relationship between digestibility and the nature of the constraint exerted at low digestibility for the particular feed under study. A better understanding of this latter relationship should lead to increased accuracy of prediction, particularly of diets with a number of components.

Forage Factors that Affect Intake

'Physical' factors

Plant structure

Physical factors are generally taken to refer to factors that directly influence the initial gut volume occupied by ingestion of a feed and the rate at which that volume is decreased by digestion and onward passage. The content of fibrous cell walls is a major factor in this respect, since these structures are less soluble and take up more space than the cell contents. Forages contain a large proportion of their organic matter (OM) content (35–80%) as cell walls, which provide the structural integrity of the plant. The structural carbohydrates, principally hemicellulose, cellulose and pectins, are broken down by microorganisms in the rumen, which enable the ruminant animal to take advantage of an energy source that is not able to be efficiently utilized by monogastrics. Distribution of the different molecules within the plant and the linkages between them will be important factors affecting the ease with which the microorganisms can break down the cells (Jung and Allen, 1995) and thus the space occupied in the gastrointestinal tract. In addition, physical characteristics of the cell wall or fibre particles themselves, such as tissue origin, shape, buoyancy and specific gravity, affect the rate at which particles are broken down and the ease of passage (Wilson and Kennedy, 1996).

Resistance to comminution (reduction in particle size) is positively related to fibre content; however, relationships between fibre measured using neutral detergent solution (NDF) and dry matter (DM) intake (DMI) are not always consistent. Reid *et al.* (1988) found a significant effect of forage class (C-3 or C-4 grass or legume) on the slope and intercept for regressions of DMI on NDF for both cattle and sheep, indicating that the fill effect of NDF may vary with different forages. This may be explained by the difference in distribution of the different structural polysaccharides. Minson (1990) observed that for groups of forages with similar DM digestibility, fibre content is greater in legumes compared with grasses, temperate compared with tropical grasses and leaf compared with stem. Wilson and Kennedy (1996) suggested that the lower digestibility of tropical grasses compared with temperate grasses and legumes reflects an interlocking and therefore more rigid cell structure. These authors also suggested that the greater digestibility of legumes compared with grasses may reflect leaf length. Grass particles are inherently long and buoyant, with a low functional specific gravity (FSG), and easily entangled, while chewed legume vascular particles are short and chunky with high FSG and therefore likely to escape the rumen quickly. Thus, potential intake is dependent not only on the fibre content, but also on the original structure of the plant and the way in which it breaks down during digestion.

The DM content of feeds may also influence the space occupied within the gut. Prewilting of grass prior to ensiling has consistently been shown to give silages with higher intakes compared with unwilted material from the same sward (Peoples and Gordon, 1989; Teller *et al.*, 1993), sometimes by as much as 44% (Romney *et al.*, 1997). One explanation is that the effectiveness of chewing during eating and hence the rate of particle breakdown was enhanced with the drier material. However, advantages in terms of animal production have been less evident in recent years (Forbes, 1995b), as the control of fermentation in wet silages has improved and grass is cut at an earlier stage of maturity, giving material with higher digestibility, for which the effects of wilting are not as great. Advocacy of wilting continues mainly as a strategy to reduce effluent rather than to improve nutritive value.

Sward structure

In grazing animals, the structure of the sward can restrict intake not only in terms of the space taken up in the gut, but also by limiting the amount of herbage which the animal can actually harvest within a 24 h period. Characteristics of the grazed sward, such as plant density and height, can influence intake through their effect on ease of prehension and thus bite size, which has been shown to be a major factor influencing daily herbage intake (Hodgson *et al.*, 1991). Positive relationships between bite size of sheep and sward height have been observed (Allden and Whittaker, 1970; Penning *et al.*, 1991), while Stobbs (1973) and Chacon and Stobbs (1976) concluded that the sward bulk density and leaf-to-stem ratio were the major factors affecting bite size and intake of cattle. Cushnahan *et al.* (1998) conducted a study on lactating cows fasted for 5 h to enable measurement of potential intake and observed that pregrazing sward height was the major characteristic influencing bite depth and bite volume, but bulk density determined the weight of herbage per bite. Thus, for grazing animals, not only the plant structure, but also the sward structure need to be considered when predicting intake.

'Metabolic' factors

Not all of the differences in intake between forages can be explained by factors that could be anticipated to influence the space occupied in the gastrointestinal tract. For example, it is generally accepted that conservation results in reduced intake (Forbes, 1995b). However, following improvements in silage-making techniques, the low intake of silage compared with hay observed in the past is considered to be less of a problem.

The organic acids and nitrogenous substances, such as amines, produced during the silage-making process and additives used to improve the silage have often been implicated in the reduced intakes observed when fresh grass is ensiled. Ruminal infusions of individual silage constituents or mixtures of them (e.g. Buchanan-Smith and Phillip, 1986) have indicated that no single constituent is responsible for the low intake, but the additive effects of a number of substances may be important.

Interaction between Forage and Animal Factors

Animal size

As discussed earlier, intake is thought to be driven by an animal's requirement for energy and thus, across species, size is the factor most closely correlated with intake. Larger animals consume greater quantities of food; however, the relationship is not isometric but scales allometrically with body mass, and intake is commonly expressed on the basis of metabolic body weight, or live weight $(LW)^{0.75}$. It is not surprising, either, that parameters related to the processing of fibre have also been found to be related to animal size. Welch (1982) found that rumination time per gram of NDF decreased exponentially with LW. Illius and Gordon (1991) showed that time taken to comminute large fibre particles and small-particle retention time scale with $LW^{0.27}$, while the digesta load of gastrointestinal-tract compartments is isometric with LW. These allometric relationships account for much of the intake variation attributable to mass (Illius, 1998) and can be incorporated in models to improve prediction of intake. Illius and Allen (1994) showed that, by scaling large-particle breakdown rate, small-particle passage rate and capacity according to animal size, variation in DMI explained by their model increased from 35 to 61%. However, growing animals appear to be able to consume a greater quantity of feed than mature animals of the same weight and intake appears to be related allometrically to $LW^{0.6}$, as opposed to $LW^{0.75}$, which is generally assumed for mature animals. This difference reflects the effect of metabolic rate and hence physiological status on intake.

Physiological status of the animal

Physiological status affects energy requirements and hence intake. As mentioned earlier, the point at which the relationship between intake and digestibility changes from positive to negative varies with the energy requirements of the animal and appeared to be higher for higher-yielding cows (Conrad *et al.*, 1964). In lactating animals, where nutrient demand is high, the rapid removal of metabolites from the blood may reduce the degree of stimulation of chemoreceptors from the same amount of absorbed nutrients (Forbes, 1995b), or rate of passage may be faster, reducing the bulk effect. Campling (1966) observed that intake is normally higher in lactating compared with dry or pregnant cows. Ingvartsen *et al.* (1992) observed a reduction in intake due to pregnancy, which was consistent with results from earlier literature. Ingvartsen *et al.* observed a decrease at least 12 weeks prepartum and hypothesized that hormonal regulation was implicated, as well as physical limitations resulting from the increased size of the fetus restricting space in the rumen. Evidence to support this hypothesis was provided by Coppock *et al.* (1972) who observed a greater decline in intake in late pregnancy for concentrate compared with forage rations and that intake was markedly reduced in the last 3 months of gestation in heifers fed a complete diet of finely ground roughages and concentrates (Aitken and Preston, 1964).

Grazing animals

Animal factors can also influence grazing intake. Intake of grazing animals is dependent both on intake rate (i.e. bite mass × bite rate) and time spent grazing (Allden and Whittaker, 1970). Although sward structure and herbage mass will affect these parameters, as discussed above, animals are able, within certain limits to alter grazing behaviour, increasing grazing time where herbage is sparse and increasing intake rate. The effect of physiological status on grazing intake has also been shown; for example, when herbage availability is high, increases in intake of up to 460 g OM kg^{-1}, increases in milk yield have been obtained (Stakelum and Dillon, 1991). The effects have been observed on intake rate and/or on grazing time.

Intake rate

Bite mass and bite rate are not independent. Newman *et al.* (1994) point out that, for a given forage requiring a given time of mastication, an increase in bite mass will cause an increase in time of mastication, decreasing bite rate and resulting in intake rates that are similar. The fact that animals are able to alter intake rate is demonstrated by the observed increase in rate in response to fasting (Greenwood and Demment, 1988; Dougherty *et al.*, 1989). This observation also demonstrates that animals rarely graze at their maximum intake rate, although there have been some indications that animals increase intake rate when grazing time is restricted (Romney *et al.*, 1996), once they 'learn' that they will only be allowed to graze for a limited time. Newman *et al.* (1994) suggested that, where cattle increased intake rate through an increase in bite rate, this may reflect a decrease in the degree of mastication. In contrast, sheep increased intake rate by increasing bite mass and apparently have limited scope to alter degree of mastication, either because of plant characteristics, such as shear strength or fibre content, or because of digestive constraints (i.e. sensitivity to particle size).

Time spent grazing

Where sward structure limits bite mass and therefore intake rate, grazing time can be altered to compensate for decreased bite size. However, there appears to be an upper limit to the amount of time a ruminant will spend grazing. Weston (1996) notes observations of grazing times of 13 and 15 h for sheep and cattle, whereas Forbes (1996) suggests that ruminants are unwilling to eat for more than 12 h day^{-1}. Therefore, if bite size falls below a certain limit, animals will not be able to achieve maximum intake capacity. Illius (1998) suggests this occurs as a result of an upper limit to oral processing time, which encompasses prehension, mastication and rumination. However, this limit may in turn be dependent on physiological state, since increasing milk yield was shown to increase grazing time by Journet and Demarquilly (1979).

Experience and learnt aversions

A number of authors have shown that ruminants decrease intake of foods in response to a malaise resulting from ingestion of toxins, such as alkaloids (Thompson and Stuedemann, 1993), condensed tannins (Provenza *et al.*, 1990) and glucosinolates (Duncan and Milne, 1993), and of poor-quality silage (Buchanan-Smith, 1990). The aversion to a particular food increases with severity of the illness (Provenza, 1995a) and decreases with increasing delay between food ingestion and illness caused by that food (e.g. Ralphs and Cheney, 1993). Aversion to feeds diminishes as time passes and the recuperative process counteracts the aversion (Provenza, 1995a), resulting in cyclical intake of nutritious feeds containing toxins, a sharp decrease being followed by a gradual increase. A similar response is also observed in animals consuming grain, which appears to be a result of malaise caused by organic acids produced in starch digestion.

Provenza (1995a) pointed out that, despite the presence of toxic feeds where ruminants are allowed free choice of feeds, there are relatively few cases of toxicosis and that animals are able to limit intake of nutritious plants containing toxins, dependent on the concentration of the toxin. He concluded that animals are able to recognize feeds based on previous experience. Thus, preference for a particular food depends not only on taste, but also on the consequences of that food for the internal environment (Provenza, 1995b).

Interaction between Diet Components

Supplementation can be considered as a means of increasing nutrient supply to animals that are unable to consume sufficient nutrients as forage because of physical fill limitations. Supplementation tends to have a positive effect on overall DMI, but may have positive or negative effects on intake of the basal forage.

Negative effects on DMI

Often there is a substitution effect, where intake of the fibrous feed decreases to an extent that varies with the digestibility of the basal feed, among other factors. In some cases, the effect might be explained according to simple rules of additivity, where fibre is replaced on a 1 : 1 basis; however, this cannot be used as a general explanation. Forbes (1995b) noted that substitution rates for dairy cows can vary between 0.4 and 0.8, with a mean of approximately 0.5. Although some work has shown a linear effect with increasing level of supplementation (Thomas, 1987), other work has shown an increase in substitution rate as level of supplementation increased (e.g. Sutton *et al.*, 1992) and a positive relationship with forage quality (Forbes, 1995b).

Supplements that are high in readily fermentable carbohydrate may have a greater effect on inhibition of fibre intake than more slowly fermentable supplements, through depression of digestibility of the roughage fraction. Rapid fermentation results in an inhibition of cellulolysis, which has been variously attributed to low pH (Terry *et al.*, 1969) or feedback inhibition of key fibre-digesting enzymes (Murphy, 1989).

Positive effects on DMI

In some circumstances, supplementation can be used to increase intake of a poor-quality feed by supplying a limiting nutrient. Wilson and Kennedy (1996) noted that, for forage diets, the rate of microbial fermentation is depressed if ruminal ammonia concentration drops below 50 mg nitrogen (N) l^{-1}.

Meanwhile, Minson (1990) suggested that, for feeds with a crude protein (CP) content of less than 62 g CP kg^{-1} DM, fibre digestion is inhibited and reports a number of trials in which intake of forages have increased by 14–77% following provision of supplementary protein. Where ammonia N concentration limits microbial fermentation, supply of N to the microorganisms increases OM digestion in the rumen, which increases breakdown and rate of passage of a poor quality forage, thereby removing the physical constraint and allowing the animal to consume more feed. Leng (1990) noted that farmers in developing countries recognize the benefit of adding small amounts of green forage to poor-quality roughage diets and cites authors who have demonstrated that supplementation of a straw-based diet with forage of high digestibility boosts digestibility of the basal feed, even when levels of supplementation are low. He suggested that this may be through provision of a highly colonized fibre source to seed bacteria on to the less digestible fibre or through increasing rumen ammonia concentrations above a critical level.

A summary of plant and animal factors affecting forage intake is given in Table 3.1.

Prediction of Intake

Improved accuracy of prediction of intake has been the goal of many scientists in response to the needs of the livestock industry, but, given its multifactorial nature, what degree of accuracy can be achieved? In a review of intake prediction, Poppi (1996) concluded that the purpose for which the prediction equation is to be used should determine the level of accuracy required. The Agricultural and Food Research Council (AFRC, 1991) suggested a target of within 10% of actual intake for the purposes of practical ration formulation. In view of the dominant effect of animal size, referred to above, all prediction equations take account of weight, most relating intake to $LW^{0.75}$, and, for dairy cows, milk yield is often included, although, if energy requirements are driving intake, it should be potential rather than actual yield which is used. In terms of feed components, however, there is less unanimity of approach, and a number of approaches are summarized briefly below and in Tables 3.2 and 3.3.

Chemical composition

The most common chemical components used to predict the intake of forages include a measure of the cell-wall content, with acid- or neutral-detergent fibre fractions (ADF and NDF) (Van Soest, 1965) being the most frequently cited.

In vitro *digestibility*

The relationship between rate of digestion and intake through its effect on rate of passage has resulted in a number of authors using *in vitro* measures of rate of

Table 3.1. A summary of some of the plant and animal factors affecting intake of forages, and circumstances where high or low intakes can be expected.

	High intake	Low intake	Reference
Plant factors			
Fibre content	Low fibre	High fibre	Van Soest (1965)
Forage class	C-3 grasses (temperate), legumes	C-4 grasses (tropical)	Reid *et al.* (1988) Wilson and Kennedy (1996)
DM content	High DM silages	Low DM silages	Peoples and Gordon (1989) Teller *et al.* (1993) Romney *et al.* (1997)
Sward structure	High plant density Optimum sward height for animal species grazed	Low plant density Sward height too low for animal species grazed	Hodgson *et al.* (1991) Penning *et al.* (1991) Cushnahan *et al.* (1998)
Conservation	Good fermentation of silage	Poor fermentation of silage	Forbes (1995b)
Animal factors			
Physiological status	Growing animal Lactating animal	Late pregnancy Mature unproductive animals	Conrad *et al.* (1964) Campling (1966) Ingvartsen *et al.* (1992)
Size	Large animals	Small animals	Illius and Allen (1994) Illius (1998)
Intake rate	Forage characteristics optimize bite mass and minimize mastication	Forage characteristics limit bite mass and increase mastication	Newman *et al.* (1994)
Previous experience	Supplement Licks	Presence of alkaloids, condensed tannins, glucosinolates	Provenza *et al.* (1990) Duncan and Milne (1993) Thompson and Stuedemann (1993)

digestion and *in situ* measures of degradability to predict intake. Parameters used have included gas volumes or DM disappearance at specific times during the fermentation, as well as parameters derived from fitting curves to the gas produced. Blümmel *et al.* (1997) calculated a partitioning factor (PF), reflecting the variation of short chain fatty acid production per unit of substrate degraded and showed this to account for 11% of the variation in DMI. Table 3.2 shows that many authors

Table 3.2. Review of the use of *in vitro* and *in situ* parameters in intake prediction equations.

Authors	Feed type	*In vitro* gas production or *in situ* degradability	Parameters used*	DM intake	R^2
Blümmel and Becker (1997)	54 straws and legume-hay-supplemented straws	*In vitro*	*a, b, c* Whole substrate	g kg^{-1} LW	0.747
			*a*NDF, *b*NDF, *c*NDF NDF fraction	g kg^{-1} LW	0.822
Blümmel *et al.* (1997)	As above	*In vitro*	*a, b, c* and PF	g kg^{-1} LW	0.838
Kibon and Ørskov (1993)	5 browses	*In situ*	$A + B + c$	kg day^{-1}	0.99
Ørskov *et al* (1988)	5 straws	*In situ*	$0.572 + 0.0766(a + b)$ $-0.822 + 0.0748(a + b) + 40.7c$	kg day^{-1}	0.83 0.89
Khazaal *et al.* (1993)	9 legume hays + 1 grass hay	*In vitro* *In situ*	$-47.7 + 4.25a + 2.12b + 444.5c$ $10.3 + 0.53A + 0.70B + 199.4c$	g DM kg^{-1} LW$^{0.75}$ g DM kg^{-1} LW$^{0.75}$	0.630 0.776
Khazaal *et al.* (1995)	10 grass hays	*In vitro* *In situ*	$18.9 - 0.23(a + b) + 687c + 0.11$CP $5.4 + 0.34(a + b) + 342c + 0.119$CP	g DM kg^{-1} LW$^{0.75}$ g DM kg^{-1} LW$^{0.75}$	0.947 0.917
Blümmel and Bullerdieck (1997)	9 legume hays + 10 grass hays	*In situ* *In vitro*	$(a + b)$, c, PF_{48} $(a + b)$, c, PF_{24} $(a + b)$, c, PF_{48}	g kg^{-1} LW$^{0.75}$ g kg^{-1} LW$^{0.75}$ g kg^{-1} LW$^{0.75}$	0.597 0.744 0.597
Ferret *et al.* (1997)	11 maize silages	*In situ*	$-81.21 + 0.21a + 0.22B$	g kg^{-1} LW$^{0.75}$	0.932

*$B = (a + b) - A$ (soluble fraction) = potentially degradable insoluble fraction; *a*, *b* and *c* are constants in the equation $p = a + b(1 - e^{-ct})$, where p = DM degradation or gas production at time t; PF_x is the the partitioning factor reflecting the variation of short-chain fatty acid production per unit of degraded substrate after x h of incubation.

Table 3.3. Some feed parameters used for predicting intake.

	Parameter	Example references
Chemical composition	ADF, NDF	Van Soest (1965)
In vitro digestibility	Gas volumes DM disappearance Rate of digestion	See Table 3.2
Physical structure	Leaf proportion Bulk density Grinding energy Short term intake rate	Minson (1990) Minson (1990) Minson (1990) Romney *et al.* (1998)
NIRS	Spectral	Steen *et al.* (1995)

NIRS, near-infrared reflectance spectroscopy.

have shown high correlations between intake and *in vitro* or *in situ* parameters. However, it should be noted that, in most cases, groups of similar feeds were used. Romney *et al.* (1998) showed that, although high R^2 values between *in vitro* gas production and intake were observed for a range of hays and straws, poor correlations were observed when a wider range of feed types were examined. Discrepancies in predicted intake may relate to the effect of particle size and the physical structure of plant cells, which are not taken into account when ground samples are used. Cherney *et al.* (1988) showed that digestion rates increase as grind size decreases and Allen (1996) suggests that this may result in an overestimation of DMI when *in vitro* or *in situ* digestibility rates of ground samples are used as a predictor. This observation may suggest that some description or measure of physical structure and ease of particle-size breakdown should be used.

Measures of physical structure

Minson (1990) discussed a range of physical measurements that have been related to intake, including leaf proportion, bulk density and grinding energy. An alternative method proposed by Moseley and Manendez (1989) is the use of intake rate measured over 1 min periods. Romney and Gill (1998) have developed the idea further and have considered the use of short-term intake rate (STIR) values as a predictor of *in vivo* parameters, including intake, digestibility and rate of passage. It might be expected that STIR would be closely related to rate of passage and therefore intake, where physical structure and rate of passage were among the principal determinants of intake. These authors found that values were highly correlated with *ad libitum* intake of a range of feeds, when sugar-beet pulp, with a high level of water-soluble carbohydrates, was excluded.

Near-infrared reflectance spectroscopy (NIRS)

Work by Barnes *et al.* (1989) and Barber *et al.* (1990) has resulted in the establishment of standard procedures for the use of near-infrared (NIR) spectra to predict OM digestibility of silage. More recently, Steen *et al.* (1995) have reported R^2 values of 0.86 between NIR spectra and intake of individually fed beef cattle for 136 silages, and Givens and Gill (1998) suggested that NIRS shows potential for application to more complex situations, such as intake responses to forage combinations.

Models

Most of the prediction equations discussed above are based on empirical relationships between feed parameters and potential intake. In most cases, accuracy of prediction depends on the data sets used to develop them (Poppi, 1996) and, normally, they are only reliable when used to predict intake for similar feeds to those used in developing the equations.

Mechanistic models are useful for increasing the understanding of the underlying mechanisms and testing assumptions. If they are able to account for all factors predicting intake then they might be expected to be able to provide much improved predictions. However, they are mainly focused on explaining observed behaviour, rather than predicting what will happen under an unexplored set of circumstances. Few are able to provide useful predictions of intake at present. Even in models, often, the effects of a range of controlling influences are determined and the lowest prediction of intake taken as the correct one (Forbes, 1977; Poppi *et al.*, 1994) and few attempt to embrace the principles of integration. The inputs required are often complex and they remain primarily a research tool.

Manipulation of Intake

Strategies or technologies to alter the intake of animals will depend on the circumstances limiting intake and whether plant or animal factors or a combination of both are the first limiting. An assessment of whether the response in terms of production justifies the cost of improving intake must also be taken into account.

Low-quality roughages

Where low-quality feeds form the major proportion of the diet, intake is constrained by the rate at which the fibre is digested and passes through the gastrointestinal tract. For example, stovers or straws, which are fed particularly during the dry season in the tropics, tend to have a high fibre and low N content, which may limit the rate of fibre digestion. While supplementation with better-quality feeds can increase total intake, as referred to above, treatment of the basal feed itself or the way in which it is fed can also influence intake. Simple physical methods to increase intake include increasing the amount of forage on offer, to allow the animal to select the more palatable or nutritious parts, thereby increasing the quality of the diet consumed as well as intake (e.g. Fernandez-Rivera *et al.*, 1994). Chopping of forage can also have beneficial effects in some cases, but Osafo *et al.* (1997), for example, reported positive effects on intake of chopping sorghum stover in sheep, but not in cattle. Chemical treatment with acid or alkali to break down fibres, with a view to increasing digestibility and therefore intake, has been tried and urea-ammonia treatment has received much interest in the developing world (e.g. Schiere and Ibrahim, 1989). However, use of the technology has been limited, for a variety of reasons (largely economic), except in China, where there has been a recent upsurge in its use (Finlayson, 1993).

Grazing systems

In grazing systems, stocking rate (the number of animals per hectare) can be manipulated to change the intake per animal and sward structure can be managed to maintain a height and density that maximizes intake, as referred to above.

Animal management

The intake of animals also depends on whether they are fed individually or in groups. Experiments to measure intake are often conducted with animals offered feed in individual pens or crates. In practice, farm animals are generally kept in groups and social factors, such as dominance order, can affect average intakes. In many situations, there is insufficient access to feed to enable all animals to have *ad libitum* access at the same time. Intake can be increased by simple husbandry methods to allow all animals adequate access and reduce wastage through soiling.

Manipulation of plant structure through breeding

New techniques of plant breeding mean that the 'ideal' plant can be bred in a much shorter time-scale than previously. The question remains, however, as to what constitutes the 'ideal'. Less fibre may theoretically increase intake, but this may not be ideal if the structure of the plant is changed in such a way as to reduce the efficiency of harvesting.

Examples of manipulating intake are given in Table 3.4.

Future Research

Research on intake is continuing at a number of levels, from metabolic studies, with the objective of identifying metabolites and hormones that initiate or inhibit intake to feeding experiments, to measure the intake of particular feeds under specific conditions. However, neither of these extremes are likely to produce major breakthroughs that will increase our ability either to predict or to manipulate

Table 3.4. Some examples of means of manipulating intake of poor-quality forages.

Manipulation	Mode of action	Reference
Supplement with small amounts of N	Increases rumen digestion and rate of passage where microbial fermentation is limited by N supply	Minson (1990) Leng (1990)
Increase offer rate	Allows animal to select more digestible portions	Fernandez-Rivera *et al.* (1994)
Chopping	Decreases particle size, increasing rate of passage. Can be ineffective if selection becomes constrained	Osafo *et al.* (1997)
Treatment with acid or alkali	Increases fibre digestibility	Schiere and Ibrahim (1989)

forage intake. Increased understanding of forage intake is more likely to arise from studies of factors influencing the rates of digestion and passage and of how nutrient requirements of ruminants and hence 'hunger drive' influence the rate at which an animal is prepared to eat.

At present the emphasis lies very strongly on experiments to study rates of digestion and there is very little research in the UK on factors affecting the passage of digesta down the tract. The explanation for this unbalanced emphasis lies in the increased simplicity of methods that measure rates of digestion. Nevertheless, interpretation of the results is not so simple without a better understanding of rates of passage. There is some evidence (Gill *et al.*, 1999) that fractional rates of passage are not the same for all components that move with the solid fraction and that they are not constant within any 24 h period, and this may have an impact on the pattern in which energy is supplied to the tissues. However, further research is needed to confirm initial findings. Further research on how forages and supplements interact and how mixtures of forages interact in the rumen and beyond would also help to improve ration formulation.

It is now more generally accepted that animals eat to meet their energy requirements, but there is less understanding of how the drive to meet energy requirements initiates and inhibits eating bouts, particularly in the grazing situation. This may be especially important for the development of management systems for feeding the high-genetic-merit dairy cow.

Many other research objectives could be proposed, but, in the current funding climate, proposals must be targeted at the funders' strategic objectives and there is insufficient space to list those here. Suffice it to say that there is a sound base for understanding the control of forage intake, as described above. Extrapolations can be made from existing knowledge to answer questions relating to unconventional feeds or management systems with different goals, although the accuracy of predicting outcomes will be less than for feeds that have already been studied in detail. Future research should aim to generate information that will provide farmers with sufficient information to enable them to make management decisions in a world which, in economic and climatic terms, is changing more rapidly than ever before.

References

AFRC (Agricultural and Food Research Council) (1991) Voluntary intake of cattle. Technical Committee on Responses to Nutrients, Report No. 8. *Nutrition Abstracts and Reviews* 61, 815–823.

Aitken, J.N. and Preston, T.R. (1964) The self feeding of complete milled rations to dairy cattle. *Animal Production* 6, 260A.

Allden, W.G. and Whittaker, McD.I.A. (1970) The determinants of herbage intake by grazing sheep: the interrelationship of factors influencing herbage intake and availability. *Australian Journal of Agricultural Research* 21, 755–766.

Allen, M.S. (1996) Physical constraints on voluntary intake of forages by ruminants. *Journal of Animal Science* 74, 3063–3075.

Anil, M.H., Mbanya, J.N., Symonds, H.W. and Forbes, J.M. (1993) Responses in the voluntary intake of hay or silage by lactating cows to intraruminal infusions of sodium acetate, sodium propionate or rumen distension. *British Journal of Nutrition* 69, 699–712.

Balch, C.C. and Campling, R.C. (1962) Regulation of voluntary food intake in ruminants. *Nutrition Abstracts and Reviews* 32, 669–686.

Barber, G.D., Givens, D.I., Kridis, M.S., Offer, N.W. and Murray, I. (1990) Prediction of the organic matter digestibility of grass silage. *Animal Feed Science and Technology* 28, 115–128.

Barnes, R.J., Dhanoa, M.S. and Lister, S.J. (1989) Standard normal variate transformation and de-trending of near infra-red diffuse reflectance spectra. *Applied Spectroscopy* 43, 772–777.

Blümmel, M. and Becker, K. (1997) The degradability characteristics of fifty four roughages and roughage neutral detergent fibres as described by *in vitro* gas production and their relationship to voluntary feed intake. *British Journal of Nutrition* 77, 757–768.

Blümmel, M. and Bullerdieck, P. (1997) The need to complement *in vitro* gas production measurements with residue determinations from *in sacco* degradabilities to improve the prediction of voluntary intake of hays. *Animal Science* 64, 71–76.

Blümmel, M., Steingass, H. and Becker, K. (1997) The relationship between *in vitro* gas production, *in vitro* microbial biomass yield and ^{15}N incorporation and its implications for the prediction of voluntary feed intake of roughages. *British Journal of Nutrition* 77, 911–921.

Brobeck, J.R. (1948) Food intake as a mechanism of temperature regulation. *Yale Journal of Biology and Medicine* 20, 545–553.

Buchanan-Smith, J.G. (1990) An investigation into palatability as a factor responsible for reduced intake of silage by sheep. *Animal Production* 50, 253–260.

Buchanan-Smith, J.G. and Phillip, L.E. (1986) Food intake in sheep following intraruminal infusion of extracts from lucerne silage with particular reference to organic acids and products of protein degradation. *Journal of Agricultural Science* 106, 611–617.

Campling, R.C. (1966) A preliminary study of the effect of pregnancy and of lactation on the voluntary intake of food by cows. *British Journal of Nutrition* 20, 25–39.

Chacon, E. and Stobbs, T.H. (1976) Influence of progressive defoliation of a grass sward on the eating behaviour of cattle. *Australian Journal of Agricultural Research* 27, 709–725.

Cherney, J.H., Cherney, D.J.R. and Mertens, D.R. (1988) Fiber composition and digestion kinetics in grass stem internodes as influenced by particle size. *Journal of Dairy Science* 71, 2112–2122.

Conrad, H.R., Pratt, A.D. and Hibbs, J.W. (1964) Regulation of feed intake in dairy cows. 1. Change in importance of physical and physiological factors with increasing digestibility. *Journal of Dairy Science* 47, 54–62.

Cooper, S.D.B., Kyriazakis, I. and Nolan, J.V. (1995) Diet selection in sheep: the role of the rumen environment in the selection of a diet from two feeds that differ in their energy density. *British Journal of Nutrition* 64, 39–54.

Coppock, C.E., Noller, C.H., Wolfer, S.A., Callahan, C.J. and Baker, J.S. (1972) Effect of forage–concentrate ratio in complete feeds fed *ad-libitum* on feed intake prepartum and the occurrence of abomasal displacement in dairy cows. *Journal of Dairy Science* 55, 783–789.

Cushnahan, A., McGilloway, D., Laidlaw, A.S., Mayne, C.S. and Kilpatrick, D.J. (1998) The influence of sward surface height and other sward parameters on short term herbage intake rates and grazing behaviour of lactating dairy cows. *Animal Science* (submitted).

Dougherty, C.T., Bradley, N.W., Cornelius, P.L. and Lauriault, L.M. (1989) Short term fasts and the ingestive behaviour of grazing cattle. *Grass and Forage Science* 40, 69–77.

Duncan, A.J. and Milne, J.A. (1993) Effects of oral administration of brassica secondary metabolites, allylcyanide, allyl isothiocyanate and dimethyl disulphide, on the voluntary intake and metabolism of sheep. *British Journal of Nutrition* 70, 631–645.

Emmans, G.C. (1997) A method to predict the food intake of domestic animals from birth to maturity as a function of time. *Journal of Theoretical Biology* 186, 189–199.

Fernandez-Rivera, S., Midou, A. and Marichatou, H. (1994) Effect of food allowance on diet selectivity and intake of pearl millet (*Pennisetum glaucum*) stover leaves by sheep. *Animal Production* 58, 249–256.

Ferret, A., Gasa, J., Plaixats, J., Casanas, F., Bosch, L. and Nuez, F. (1997) Prediction of voluntary intake and digestibility of maize silages given to sheep from morphological and chemical composition, *in vitro* digestibility or rumen degradation characteristics. *Animal Science* 64, 493–502.

Finlayson, P. (1993) Economic aspects of utilising fibrous crop residues for beef production in China. In: Tingshuang, G. (ed.) *Increasing Livestock Production through Utilization of Local Resources*. Bureau of Animal Production and Health, Beijing, pp. 485–493.

Forbes, J.M. (1977) Interrelationships between physical and metabolic control of voluntary food intake in fattening, pregnant and lactating mature sheep: a model. *Animal Production* 24, 90–101.

Forbes, J.M. (1995a) Physical limitation of feed intake in ruminants and its interaction with other factors affecting intake. In: von Engelhardt, W., Leonhard-Marek, S., Breves, G. and Giesecke, D. (eds) *Ruminant Physiology: Digestion Metabolism, Growth and Reproduction. Proceedings of the Eighth International Symposium on Ruminant Physiology.* Ferdinand Enke Verlag, Stuttgart, Germany, pp. 217–232.

Forbes, J.M. (1995b) *Voluntary Food Intake and Diet Selection in Farm Animals*. CAB International, Wallingford, UK.

Forbes, J.M. (1996) Integration of regulatory signals controlling forage intake in ruminants. *Journal of Animal Science* 74, 3029–3035.

Gill, M. and Romney, D. (1994) The relationship between the control of meal size and the control of daily intake in ruminants. *Livestock Production Science* 39, 13–18.

Gill, M., Robinson, P.H. and Kennelly, J.J. (1999) Diurnal patterns in rumen volume and composition of digesta flowing into the duodenum. *Animal Science* 69, 237–249.

Givens, D.I. and Gill, M. (1998) Current and future potential of alternative techniques. In:

Deaville, E.R., Owen, E., Adesogen, A.T., Rymer, C., Huntington, J.A. and Lawrence, T.L.J. (eds) *British Society of Animal Science Occasional Publication No. 22*. British Society of Animal Science, Edinburgh, pp. 161–171.

Greenwood, G.B. and Demment, M.W. (1988) The effect of fasting on short-term cattle grazing behaviour. *Grass and Forage Science* 43, 377–386.

Hodgson, J., Fornes, T.D.A., Armstrong, R.H., Beattie, M.M. and Hunter, E.A. (1991) Comparative studies of the ingestive behaviour and herbage intake of sheep and cattle grazing indigenous hill plant communities. *Journal of Applied Ecology* 28, 205–227.

Illius, A.W. (1998) Advances and retreats in specifying the constraints on intake in grazing ruminants. In: Buchanan-Smith, J.G., Bailey, L. and McCaughey, P. (eds) *Proceedings of XVIII International Grassland Congress*, Vol III. Association Management Centre, Calgary, pp. 39–44.

Illius, A.W. and Allen, M.S. (1994) Assessing forage quality using integrated models of intake and digestion by ruminants. In: Fahey, G.C., Collins, M., Mertens, D.R. and Moser, L.E. (eds) *Forage Quality, Evaluation and Utilization*. American Society of Agronomy, Crop Science Society of America, Soil Science Society of America, Madison, Wisconsin, pp. 869–890.

Illius, A.W. and Gordon, I.J. (1991) Prediction of intake and digestion in ruminants by a model of rumen kinetics integrating animal size and plant characteristics. *Journal of Agricultural Science* 116, 145–157.

Illius, A.W. and Jessop, N.S. (1996) Metabolic constraints on voluntary intake in ruminants. *Journal of Animal Science* 74, 3052–3062.

Ingvartsen, K.L., Andersen, H.R. and Foldager, J. (1992) Effect of sex and pregnancy on feed intake capacity of growing cattle. *Acta Agriculturae Scandinavica, Section A, Animal Science* 42, 40–46.

Journet, M. and Demarquilly, C. (1979) Grazing. In: Broster, W.H. and Swan, H. (eds) *Feeding Strategy for the High Yielding Dairy Cow*. Granada Publishing, London, pp. 295–321.

Jung, H.G. and Allen, M.S. (1995) Characteristics of plant cell walls affecting intake and digestibility of forages by ruminants. *Journal of Animal Science* 73, 2774–2790.

Kennedy, G.C. (1953) The role of depot fat in the hypothalamic control of food intake in the rat. *Proceedings of the Royal Society* B 140, 578–592.

Khazaal, K., Dentinho, M.T., Ribeiro, J.M. and Ørskov, E.R. (1993) A comparison of gas production during incubation with rumen contents *in vitro* and nylon bag degradability as predictors of the apparent digestibility *in vivo* and the voluntary intake of hays. *Animal Science* 57, 105–112.

Khazaal, K., Dentinho, M.T., Ribeiro, J.M. and Ørskov, E.R. (1995) Prediction of apparent digestibility and voluntary intake of hays fed to sheep: comparison between using fibre components, *in vitro* digestibility of characteristics of gas production or nylon bag degradation. *Animal Science* 61, 527–538.

Kibon, A. and Ørskov, E.R. (1993) The use of degradation characteristics of browse plants to predict intake and digestibility by goats. *Animal Production* 57, 247–251.

Kyriazakis, I. and Oldham, J.D. (1993) Diet selection in sheep: the ability of growing lambs to select a diet that meets their crude protein (nitrogen × 6.25) requirements. *British Journal of Nutrition* 69, 617–629.

Kyriazakis, I., Emmans, G.C. and Whittemore, C.T. (1990) Diet selection in pigs: choices made by growing pigs given foods of different protein contents. *Animal Production* 51, 180–191.

Leng, R.A. (1990) Factors affecting the utilisation of 'poor-quality' forages by ruminants particularly under tropical conditions. *Nutrition Research Reviews* 3, 277–303.

Mayer, J. (1953) Glucostatic regulation of food intake. *New England Journal of Medicine* 249, 13–16.

Mbanya, J.N., Anil, M.H. and Forbes, J.M. (1993) The voluntary intake of hay and silage by lactating cows in response to ruminal infusion of acetate or propionate, or both with and without distension of the rumen by a balloon. *British Journal of Nutrition* 69, 713–720.

Minson, D.J. (1990) *Forage in Ruminant Nutrition*. Academic Press, San Diego, California, USA.

Moseley, G. and Manendez, A.A. (1989) Factors affecting the eating rate of forage feeds. In: *XVI International Grassland Congress, Nice, France*. The French Grassland Society, Nice, pp. 789–790.

Murphy, M. (1989) The influence of non-structural carbohydrates on rumen microbes and rumen metabolism in milk producing cows. PhD thesis, Swedish University of Agricultural Sciences, Uppsala.

Newman, J.A., Parsons, A.J. and Penning, P.D. (1994) A note on the behavioural strategies used by grazing animals to alter their intake rate. *Grass and Forage Science* 49, 502–505.

Ørskov, E.R., Reid, G.W. and Kay, M. (1988) Prediction of intake by cattle from degradation characteristics of roughages. *Animal Production* 46, 29–34.

Osafo, E.L.K., Owen, E., Said, A.N., Gill, M. and Sherington, J. (1997) Effects of amount offered and chopping on intake and selection of sorghum stover by Ethiopian sheep and cattle. *Animal Science* 65, 55–62.

Penning, P.D., Parsons, A.J., Orr, R.J. and Treacher, T.T. (1991) Intake and behaviour responses by sheep to changes in sward characteristics under continuous stocking. *Grass and Forage Science* 46, 15–28.

Peoples, A.C. and Gordon, F.J. (1989) The influence of wilting and season of silage harvest and the fat and protein concentration of the supplement on milk production and food utilisation by lactating cattle. *Animal Production* 48, 305–317.

Poppi, D.P. (1996) Predictions of food intake in ruminants from analyses of food composition. *Australian Journal of Agricultural Research* 47, 489–504.

Poppi, D.P., Gill, M. and France, J. (1994) Quantification of theories of intake regulation in growing ruminants. *Journal of Theoretical Biology* 167, 129–145.

Provenza, F.D. (1995a) Post ingestive feedback as an elementary determinant of food preference and intake in ruminants. *Journal of Range Management* 48, 2–17.

Provenza, F.D. (1995b) Role of learning in food preferences of ruminants: Greenhalgh and Reid revisited. In: von Engelhardt, W., Leonhard-Marek, S., Breves, G. and Giesecke, D. (eds) *Ruminant Physiology: Digestion, Metabolism, Growth and Reproduction. Proceedings of the Eighth International Symposium on Ruminant Physiology*. Ferdinand Enke Verlag, Stuttgart, Germany, pp. 233–247.

Provenza, F.D., Burrit, E.A., Clausen, T.P., Bryant, J.P., Reichardt, P.B. and Distel, R.A. (1990) Conditioned flavor aversion: a mechanism for goats to avoid condensed tannins in blackbrush. *American Nature* 136, 810–828.

Ralphs, M.H. and Cheney, C.D. (1993) Influence of cattle age, lithium chloride dose level and food type in the retention of food aversions. *Journal of Animal Science* 71, 373–379.

Reid, R.L, Jung, G.A. and Thayne, W.A. (1988) Relationships between nutritive quality and fibre components of cool season and warm season forages: a retrospective study. *Journal of Animal Science* 66, 1275–1291.

Romney, D.L. and Gill, M. (1998) Measurement of short term intake rate (STIR) to predict *in vivo* parameters in sheep. In: *Proceedings of the British Society of Animal Science, 1998*, Scarborough. British Society of Animal Science, Edinburgh, p. 98.

Romney, D.L., Sendalo, D.S.C., Owen, E., Mtenga, L.A., Penning, P.D., Mayes, R.W. and Hendy, C.R.C. (1996) Effects of tethering management on feed intake and behaviour of Tanzanian goats. *Small Ruminant Research* 19, 113–120.

Romney, D.L., Blunn, V. and Leaver, J.D. (1997) The responses of lactating dairy cows to diets based on grass silage of high or low DM and supplemented with fast and slowly fermentable energy sources. In: *Proceedings of the British Society of Animal Science, 1997*, Scarborough. British Society of Animal Science, Edinburgh, p. 84.

Romney, D.L., Cadario, F.C., Owen, E. and Murray, A.H. (1998) Comparison of parameters from the Theodorou gas production technique using nitrogen-free and nitrogen-rich media as predictors of DM intake and digestibility. In: Deaville, E.R., Owen, E., Adesogen, A.T., Rymer, C., Huntington, J.A. and Lawrence, T.L.J. (eds) *British Society of Animal Science Occasional Publication No. 22*. British Society of Animal Science, Edinburgh, pp. 172–174.

Schiere, J.B. and Ibrahim, M.N.M. (1989) *Feeding of Urea-ammonia Treated Rice Straw.* Pudoc, Wageningen, the Netherlands, 125 pp.

Shariatmadari, F. and Forbes, J.M. (1993) Growth and food intake responses to diets of different protein contents and a choice between diets containing two levels of protein in broiler and layer strains of chickens. *Poultry Science* 34, 959–970.

Stakelum, G. and Dillon, P. (1991) Influence of sward structure and digestibility on the intake and performance of lactating and growing cattle. In: Mayne, C.S. (ed.) *Management Issues for the Grassland Farmer in the 1990s*. Occasional Symposium of the British Grassland Society No. 25. British Grassland Society, Hurley, pp. 30–42.

Steen, R.W.J., Gordon, F.J., Mayne, C.S., Agnew, R.E., Kilpatrick, D.J., Unsworth, E.F., Barnes, R.J., Porter, M.G. and Pippard, C.J. (1995) Prediction of silage intake by cattle. *Animal Science* 60, 514.

Stobbs, T.H. (1973) The effect of plant structure on the intake of tropical pastures I. Variation in the bite size of grazing cattle. *Australian Journal of Agricultural Research* 24, 809–819.

Sutton, J.D., Aston, K., Beever, D.E. and Fisher, W.J. (1992) Body composition and performance of autumn-calving Holstein-Friesian dairy cows during lactation: food intake, milk constituent output and live weight. *Animal Production* 54, 473.

Teller, E., Vanbelle, M. and Kamatali, P. (1993) Chewing behaviour and voluntary grass silage intake by cattle. *Livestock Production Science* 33, 215–227.

Terry, R.A., Tilley, J.M. and Outen, G.A. (1969) Effect of pH on cellulose digestion under *in vitro* conditions. *Journal of the Science of Food and Agriculture* 20, 317–320.

Thiago, L.R.L. de S. (1988) Voluntary intake of forages by ruminants: factors related to eating behaviour and rumen fill. PhD thesis, University of Reading.

Thomas, C. (1987) Factors affecting substitution rates in dairy cows on silage based rations. In: Haresign, L.W. and Cole, D.J.A. (eds) *Recent Advances in Animal Nutrition – 1987*. Butterworth-Heinemann, Oxford, pp. 205–218.

Thompson, F.N. and Stuedemann, J.A. (1993) Pathophysiology of fescue toxicosis. *Agriculture Ecosystems and the Environment* 44, 263–281.

Van Soest, P.J. (1965) Symposium on factors influencing the voluntary intake of herbage by ruminants: voluntary intake in relation to chemical composition and digestibility. *Journal of Animal Science* 24, 834–843.

Van Soest, P.J. (1994) *Nutritional Ecology of the Ruminant*, 2nd edn. Cornell University Press, Ithaca.

Welch, J.G. (1982) Rumination, particle size and passage from the rumen. *Journal of Animal Science* 54, 885–894.

Weston, R.H. (1996) Some aspects of constraint to forage consumption by ruminants. *Australian Journal of Agricultural Research* 47, 175–197.

Wilson, J.R. and Kennedy, P.M. (1996) Plant and animal constraints to voluntary feed intake associated with fibre characteristics and particle breakdown and passage in ruminants. *Australian Journal of Agricultural Research* 47, 199–225.

4 Prediction of Energy Supply in Ruminants, with Emphasis on Forages

S. López,[1] J. Dijkstra[2] and J. France[3]

[1] *Department of Animal Production, University of Leon, 24007 Leon, Spain;* [2] *Animal Nutrition Group, Wageningen Institute of Animal Sciences, Wageningen University, Marijkeweg 40, 6709 PG Wageningen, The Netherlands;* [3] *Department of Agriculture, University of Reading, Earley Gate, Reading RG6 6AT, UK*

Introduction

Energy and protein evaluation systems have been developed virtually independently. The systems use single descriptors of the energy or protein values of feeds for ruminants and assess animal requirements in units of the corresponding descriptor. The use of a single value for each feed, the additivity of the values when several feeds are used in a ration and the factorial approach to balance the estimated requirements of the animal for a target level of production and the nutrient supply from the diet are the main characteristics of the feed evaluation systems (Alderman, 1983; Van der Honing and Steg, 1990), with the aim to be as simple as possible in order to facilitate practical application. However, the simplicity of the systems may be in conflict with accuracy of prediction of animal performance over a wide range of variation of rations, and an appropriate modelling of the underlying physiological processes in the ruminant may be required. This chapter will focus on developments in the energy evaluation of feeds, particularly forages, with attention given to the role of *in sacco* and *in vitro* methods, in combination with mechanistic mathematical modelling, in predicting energy supply to the animal.

Predicting Energy Value and Digestibility in Current Feed Evaluation Systems

Current energy systems for ruminants are based either on metabolizable energy (ME) (CSIRO, 1990; AFRC, 1993) or on net energy (NE) (Van Es, 1978; Vermorel *et al.*, 1987; INRA, 1988; NRC, 1989, 1996). In the Agricultural and Food Research Council (AFRC) and CSIRO systems, a single value of the ME content (called M/D, in MJ ME kg^{-1} dry matter (DM)) is used for each feedstuff.

This value is determined for animals fed at maintenance levels of intake. The Dutch, National Research Council (NRC) and INRA feeding systems use indices of the energy value of feeds relative to a net production output. The determination of the NE value of a feed requires the measurement of the energy balance at two or more levels of intake, and the increment in NE gain resulting from an increment in the intake of the feed is commonly expressed as MJ kg^{-1} DM. However, no feed can be given a single NE value, because the value varies with the purpose for which energy is used by the animal (ME is used with different efficiencies for maintenance, growth and lactation).

The predictive capability of all energy systems relies, largely, upon the energy values assigned to each feed, as the animal requirements can be determined with an acceptable level of accuracy. Energy values (either in ME or in NE) and derived efficiencies of energy utilization have to be determined in balance trials, in which energy inputs (energy intake) and outputs (energy losses in faeces, methane, urine and heat produced) are measured (see Chapter 5). However, energy balance trials are expensive and cannot be conducted for all the forages and major bulk ingredients included in ruminant feeds; thus the energy value must be estimated from other variables.

The determination of feed digestibility is simpler than the determination of energy values in balance trials. Digestibility has been recognized as the main source of variation of the ME content of forages, given the large variability of their digestibility coefficients (values ranging from 0.35 to 0.80) (Minson, 1990; Hvelplund *et al.*, 1995) and the assumed relative constancy of the proportion of energy lost in urine and methane (19% of the digestible energy) (Tyrrell and Moe, 1975; CSIRO, 1990). On the assumption that, for most forages other than silage, the gross energy of the digestible organic matter (OM) is nearly constant, the Ministry of Agriculture, Fisheries and Food (MAFF, 1984) proposed the use of the digestible OM content of feed DM (DOMD or D value) as a useful index of M/D. But digestion trials to measure digestibility *in vivo* are also expensive and are not readily applicable to large numbers of feeds or when small quantities of each forage are available. Thus, the prediction of feed digestibility or energy values from compositional or *in vitro* information has become a necessity in all feeding systems. A predictor of forage nutritive value is a quality-related characteristic of the forage, which can be measured in the laboratory by simple methods. Using analytical results and energy values determined by feeding trials for a number of standard representative feeds, multiple regression equations can be derived statistically and used to predict the energy value of other samples.

A large number of different equations for predicting OM digestibility or energy value of forages from a variety of explanatory variables is provided in the literature (Van Es, 1978; Andrieu *et al.*, 1981; Minson, 1982, 1990; Barber *et al.*, 1984; Andrieu and Demarquilly, 1987; CSIRO, 1990; Givens *et al.*, 1990, 1992; AFRC, 1993; NRC, 1996). In these equations, forage digestibility (Table 4.1) or energy value are predicted from: (i) botanical features (including leafiness, botanical composition, growth or maturity stage), also considering the harvesting and preservation conditions; (ii) chemical composition; (iii) physical properties (density, spectrum for near-infrared reflectance); and (iv) *in vitro* measurements of digestibility (using either buffered rumen fluid or enzymatic solutions) (Minson, 1990; Hvelplund *et al.*, 1995).

Table 4.1. Ranges in correlation coefficients and in residual standard deviations (RSD)* associated with the prediction of *in vivo* digestibility from various predictors.

	Range in correlation coefficient	Range in RSD (%)
Chemical composition		
Crude Protein	+0.44 to +0.79	2.0 to 6.5
NDF	−0.45 to −0.80	2.4 to 5.1
ADF	−0.75 to −0.88	3.6 to 9.0
Lignin	−0.61 to −0.83	4.3
Alternative methods of estimating digestibility		
In vitro digestibility[†]	+0.94	1.6
Enzymatic	+0.94 to +0.99	1.8 to 3.2
In situ	+0.98	1.0 to 3.0
NIR	+0.95	0.6 to 2.7

* Data combined from Minson (1990), Carro *et al.* (1994) and Van Soest (1994).
[†] Tilley and Terry (1963).

Most of these equations are based purely on the statistical relationship between variables and the performance of regression methods facilitated by improved computing facilities, resulting sometimes in equations without biological meaning. One of the consequences of this empirical approach is that the large number of equations available in the literature differ significantly in the predicting variables, in the regression coefficients for the same predictors and in the estimated prediction error. Diverse equations have been obtained for separate classes of feeds and forages, and different laboratories may use different equations for the same combination of predicted and explanatory variables. These empirical prediction equations are a consequence of the specific data set from which they have been derived, so that they are only useful when the situation to be predicted is similar to the original data set. Thus, these equations have a variable degree of unreliability, due to environmental and species variation, interactions among plant species in mixed forages and the difficulty in accounting for basic cause–effect relationships between ruminal digestion and plant composition (Chesson, 1993; Van Soest, 1994). Any error in estimating digestibility is passed along to, and constitutes the largest error in, the final estimation of the energy value of the forage. Despite these criticisms, empirical equations are widely used in feed evaluation systems.

Towards New Approaches in Feed Evaluation

Because of the extensive research on feed evaluation carried out over the last decades, tables of the nutritive value of the forages mostly used in ruminant nutrition are available. These values have been either determined experimentally or predicted using empirical equations with an acceptable degree of precision. Conceptually, though, current evaluation systems have serious limitations, which have been generally recognized (Beever and Oldham, 1986; Gill *et al.*, 1989; Tamminga, 1992; Van Soest, 1994; Reynolds and Beever, 1995; Hanigan *et al.*, 1997).

Variation in energy supply

The first drawback of current systems is when energy content of a feed per unit of mass is used as its energy value, since this accounts for only a part of its potential to contribute to the nutrient supply of the animal and to influence nutrient utilization. Single energy values for a feed quote static conditions, whilst the energy supply from each unit of feed depends on a number of interrelated factors, including site of energy digestion, level of intake and interactions between individual feed substrates.

Site of energy digestion

When considering the major sites of energy digestion in ruminants, the digestive tract can be divided into three sections (Fig. 4.1): the reticulorumen, where digestion is by microbial fermentation; the abomasum and small intestine, where mammalian digestion by host enzymes predominates; and the large intestine, where the caecum is an important site of fermentation (although secondary to the reticulorumen). Microbial fermentation yields volatile fatty acids (VFA), together with

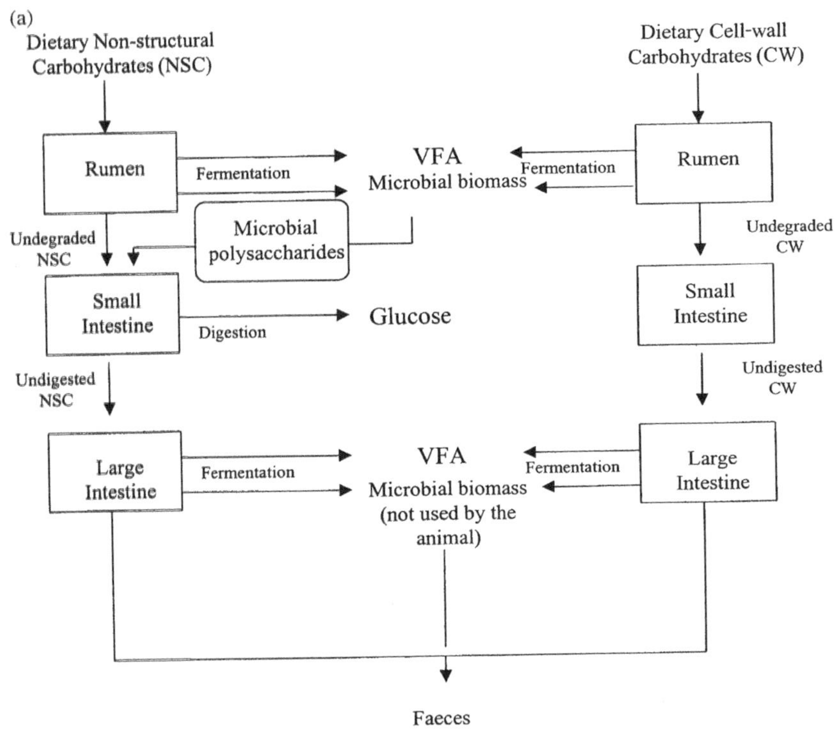

Fig. 4.1 (a) (b and c opposite) Schematic representation of the digestion of (a) dietary carbohydrates, (b) long-chain fatty acids (LCFA) and (c) nitrogenous compounds by ruminants, showing the end-products of the digestion process at each site of the digestive tract. NH_3, ammonia.

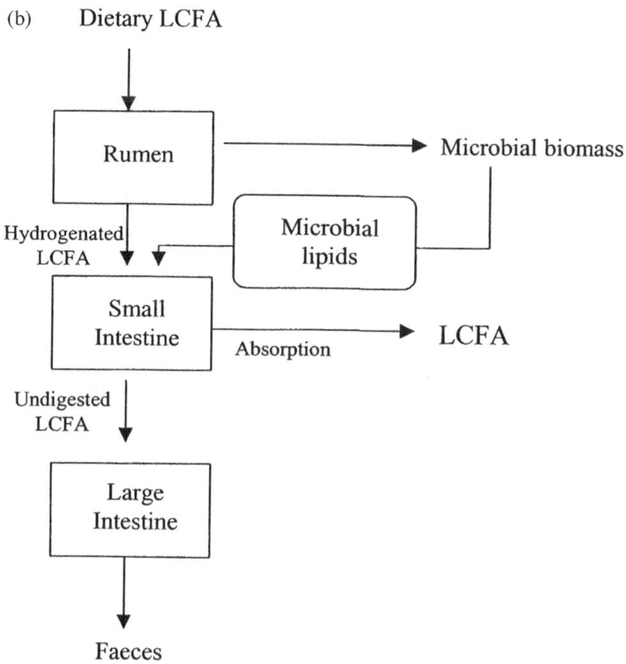
(b)
Dietary LCFA
Rumen
Microbial biomass
Hydrogenated LCFA
Microbial lipids
Small Intestine
Absorption
LCFA
Undigested LCFA
Large Intestine
Faeces

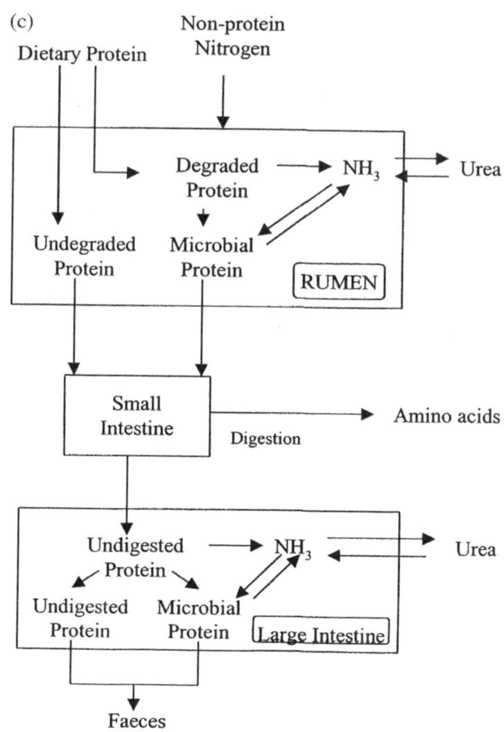
(c)
Dietary Protein
Non-protein Nitrogen
Degraded Protein
NH_3
Urea
Undegraded Protein
Microbial Protein
RUMEN
Small Intestine
Digestion
Amino acids
Undigested Protein
NH_3
Urea
Undigested Protein
Microbial Protein
Large Intestine
Faeces

methane and fermentation heat, which have to be discarded when calculating ME from digestible energy (DE). A carbohydrate may be fermented in the rumen to VFA or digested in the small intestine to glucose, resulting in a different supply of energy from the same initial nutrient (Nocek and Tamminga, 1991) (Table 4.2). An increase in the molar proportion of propionate results in lower fermentation heat and methane losses, although the ratio between the latter remains almost constant (Blaxter, 1989). VFA molar proportions in the rumen from a given feed may differ widely with factors such as microbial growth rate, available substrates, level of feed intake and rumen pH (Dijkstra, 1994). Hence, the fixed factor used in many systems to calculate ME from DE (0.81) is variable (Fig. 4.2).

Level of intake and energy supply

The relationship between energy balance and ME intake is curvilinear, showing a diminishing-returns pattern, though a piecewise linear approximation is generally used (Fig. 4.3). Corrections to feed energy value for the effect of level of intake have been incorporated into some systems (Van Es, 1978; NRC, 1989, 1996; AFRC, 1993), whereas in others it is accepted that the systems have been specifically designed for certain feeding conditions, such as a fixed animal level of production, in which the range of intake levels is not large (INRA, 1988). The regression equations used in many energy systems have been developed from feeding experiments conducted largely with wethers at maintenance levels of feeding, and the validity of corrections for feed intake level for beef and dairy cattle based on sheep data has been questioned. Tyrrell and Moe (1975) reported for mature sheep limited variation of OM digestibility of four contrasting diets based on maize (88 to 92% of

Table 4.2. Differences in the supply of substrates from the gastrointestinal tract associated with changes in the site of digestion of dietary carbohydrates in dairy cows (from Sutton, 1985).

Cereal in the concentrate	Rolled barley	Ground maize
DE intake (MJ day^{-1})	164	158
Proportion of DE digested in the rumen	0.64	0.57
Starch intake (kg day^{-1})	5.75	6.37
Proportion of starch digested		
In whole tract	0.99	0.89
In rumen	0.90	0.67
Nutrient supply (MJ day^{-1})		
Acetate + butyrate	54	55
LCFA	10	14
Amino acids	31	27
Propionate	44	28
Glucose	7	14

LCFA, long-chain fatty acids.

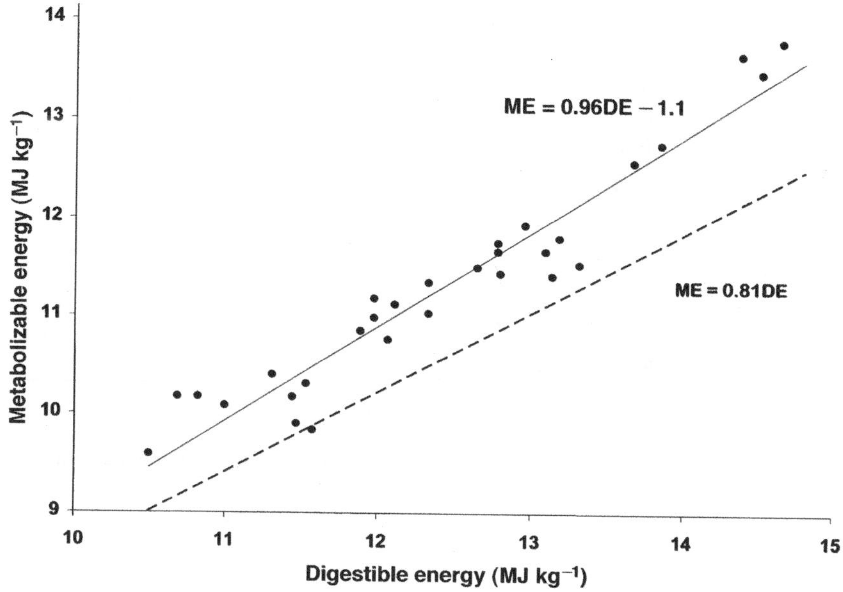

Fig. 4.2. The relation of metabolizable energy (ME) to digestible energy (DE) (adapted from Moe *et al.*, 1972). The commonly assumed factor 0.81 (dashed line) underestimates true ME, which is relatively greater for more digestible feeds.

intake), whilst the same diets fed to cattle resulted in OM digestibilities between 62 and 79% of intake. This clearly indicates the possible errors when extrapolating data on intake levels between species. Next, the use of a fixed correction for intake level is of particular concern in view of interaction between intake level and type of feed (Sutton, 1985). For example, ruminal starch digestion of barley diets decreased with an increase in level of feed intake in dairy cattle (91 and 89% at maintenance and three times maintenance, respectively), but this decrease in ruminal starch digestion was much more pronounced with maize diets (81 and 72% at maintenance and three times maintenance level, respectively) (Sutton, 1980).

Interactions between feed components

Important interactions between feed components exist which cannot be represented by a single energy value (Table 4.3). All the systems fail to consider the associative effects between feeds in a mixed ration, as they use the same energy value for a given feed, whether it is fed alone or in combination with others. However, an adequate combination of ingredients in the ration will result in a significant improvement in the supply of energy to the animal in relation to that obtained by each feed alone, probably due to an enhanced rumen fermentation or a higher efficiency of energy utilization. In contrast, the presence of high quantities of easily degradable soluble sugars and starch in forage or forage supplements will depress fibre degradation in the rumen, owing to a drop in rumen pH due to the accumulation of VFA (Tamminga and Van Vuuren, 1988). Some energy systems endeavour to correct the

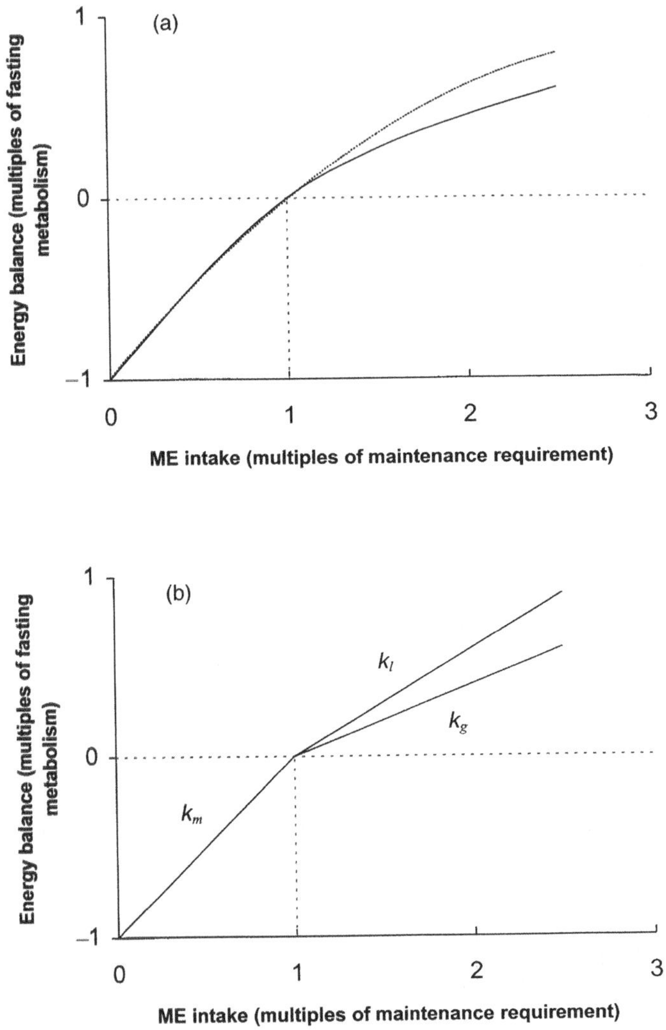

Fig. 4.3. Change in energy balance of the animal in relation to its metabolizable energy (ME) intake. Maintenance is the ME intake that results in zero energy balance. In (a) the dotted line is for a lactating animal whereas the solid line is for a growing animal. (b) A simplified representation of the non-linear relationship using piecewise linear models. The slopes of the lines represent the net efficiency of use of ME for maintenance (k_m), for lactation (k_l) and for growth and fattening (k_g). (Adapted from Blaxter, 1962.)

energy value in the presence of high dietary levels of non-fibre carbohydrates (Van Es, 1978). However, the effect of dietary non-fibre carbohydrates on fibre degradation appears to vary from one feed source to another. For example, replacing maize with barley was shown to have a negative effect on fibre degradability in dairy cattle (DePeters and Taylor, 1985). A possible interaction between carbohydrate

Table 4.3. Associative effects between feeds: influence of forage-to-concentrate ratio on the supply of energy-yielding substrates from the gastrointestinal tract (from Sutton, 1985).

Hay:concentrate	40:60	10:90
DE intake (MJ day^{-1})	157	164
Proportion of DE digested in the rumen	0.67	0.64
Nutrient supply (MJ day^{-1})		
Acetate + butyrate	66	54
LCFA	9	10
Amino acids	28	31
Propionate	24	44
Glucose	3	7

LCFA, long-chain fatty acids.

degradation in the rumen and the availability of rumen-degradable protein and non-protein nitrogen may occur because of the ammonia and branched-chain fatty-acid requirements of fibrolytic microbes. The major supply of ammonia and branched-chain fatty acids is through degradation and fermentation of protein. In dairy cattle, Sutton (1986) estimated an increase of 13 g kg^{-1} in fibre degradation per 10 g kg^{-1} DM increase of crude protein. Hence, the energy value of the feed depends on the whole diet and a single energy value will not necessarily suffice.

Animal requirements versus animal responses

An often recognized weakness of current energy systems relates to the prediction of animal responses (Beever and Oldham, 1986; AFRC, 1991; Reynolds and Beever, 1995; Hanigan *et al.*, 1997). The systems aim to balance the energy requirements of an animal by an appropriate supply of energy in the diet. As such, no attempt is made to consider how animal performance will respond to deliberate changes in the diet. Also, the systems are unable to predict body and milk composition. This is a major drawback in view of the demand for ruminant products of specified compositional standards by industry and consumers, as well as the need to reduce losses to the environment. Without due recognition of the response in production level and product composition to deliberate perturbations in nutrient supply, marginal cost response analysis is not possible.

Composition of the energy-yielding substrates

Energy has been considered as the primary driving force for animal production and dietary protein supply as permissive in allowing the influences of energy supply to be expressed (Thomas, 1990). Therefore, most feeding systems use distinct and independent energy and protein schemes. However, by using single and static descriptors of energy (and protein), the present systems do not consider the composition of the energy- (or protein-) yielding substrates that make up the ME or NE supply (Beever and Oldham, 1986; AFRC, 1991; Reynolds and Beever, 1995). Animals do not metabolize energy (or protein) *per se*, but rather substrates and metabolites derived from the gastrointestinal tract. The availability of these substrates is also influenced by interactions between nutrients at specific sites of digestion and metabolism in different organs and tissues (MacRae *et al.*, 1988). The nature and amount of these available substrates determine how they are used and the proportion of energy they contain that is directed to the various production processes (Table 4.4). For example, Dijkstra *et al.* (1996) evaluated the supplementation of sugar-cane-based diets and simulated the availability of total energy and glucogenic, lipogenic and aminogenic nutrients for milk production (Fig. 4.4). These researchers showed that, in particular feeding situations, dietary energy did not limit milk production, but, rather, specific nutrients (lipogenic or aminogenic) were in short supply, a result that is impossible to predict using current feed evaluation systems.

The Mechanistic Approach

New feeding systems need to be based on the mechanisms that govern the response of animals to nutrients, dealing with quantitative aspects of the digestion and metabolism in the ruminant animal. From the drawbacks of current systems previously described, new mechanistic systems of feed evaluation need to represent the individual nutrients and substrates in organs and tissues, including the major interactions between nutrients and substrates and between organs and tissues. This is where mechanistic mathematical modelling can be applied to represent concepts and mechanisms quantitatively (France and Thornley, 1984; Baldwin, 1995;

Table 4.4. The calorimetric efficiencies of utilization for maintenance and growth of some end-products of ruminant digestion, given to sheep by constant rumen infusion (except values in brackets which are for substrates given by continuous infusion into the abomasum) (from Blaxter, 1962).

Substrate	Efficiency for maintenance	Efficiency for fattening*
Acetate	0.592	0.329
Propionate	0.865	0.563
Butyrate	0.764	0.619
Acid mixture 1†	0.872	0.581
Acid mixture 2‡	0.856	0.318
Glucose	0.940 (1.00)	0.545 (0.715)
Casein	(0.820)	(0.647)

* Increase in energy retained 100 kJ^{-1} of additional ME.
† Mixture 1 contained (mol^{-1} of total VFA): 250 mmol acetate, 450 mmol propionate and 300 mmol butyrate.
‡ Mixture 2 contained (mol^{-1} of total VFA): 750 mmol acetate, 150 mmol propionate and 100 mmol butyrate.

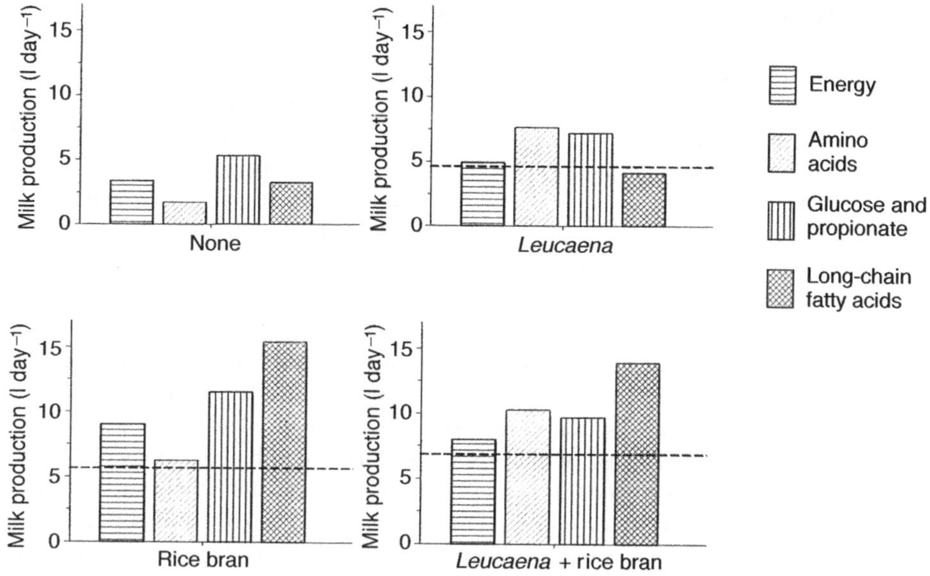

Fig. 4.4. Potential milk production according to simulated availability of energy, amino acids, glucose and propionate and long-chain fatty acids on unsupplemented basal diet (sugar cane and 40 g urea kg^{-1} sugar cane DM), basal diet supplemented with *Leucaena*, basal diet supplemented with rice bran, basal diet supplemented with *Leucaena* and rice bran. Simulations were performed using the model described by Dijkstra *et al.* (1996) and the dashed line represents average milk production observed with Brown-Swiss × Zebu cattle (Alvarez and Preston, 1976; Alvarez *et al.*, 1978).

Dijkstra and France, 1995). These models are constructed by looking at the structure of the system, dividing that structure into its key components and analysing the behaviour of the whole system in terms of its individual components and their interactions with one another.

The accuracy of prediction of animal response using mechanistic models may currently be lower than that achieved by the empirical methods used in practical application; hence, the use of the former models has been constrained mainly to research. However, these research models are very useful in evaluating the adequacy of current knowledge and data, identifying those areas where research efforts should be focused. In contrast with extant models for feed evaluation, new research findings can be easily incorporated and integrated into models constructed mechanistically, which will improve their ability to predict nutrient utilization and availability within the digestive tract (Dijkstra and France, 1995; Hanigan *et al.*, 1997). Therefore, many mechanistic models have been built on the achievements of previously reported models, introducing additional knowledge and detail, based upon newly emerging experimental information and specific failures of previous models to simulate reality (Baldwin, 1995). With this approach, it is expected that, in the future, mechanistic models will yield superior predictions of animal performance and will be more generally applicable than empirical models.

The Roles of *In Situ* and *In Vitro* Methods

For the quantitative description of digestive and metabolic processes, appropriate biological data are required and can be obtained using *in vivo*, *in situ* and *in vitro* methods. Information obtained *in vivo* is the most reliable and should be the reference to evaluate other methods, because it represents the actual animal response to a dietary treatment. However, *in vivo* results are restricted to the experimental conditions under which measurements are carried out. These trials cannot be considered routine in most laboratories and cannot be carried out for all the possible feeding situations found in practice. *In vitro* and *in situ* techniques allow manipulation of parameters defining the state of the animal. These techniques, if properly evaluated against *in vivo* observations, can be appropriate for studying the response of the animal, when one factor is varied and controlled, without the interaction of other related factors that could conceal the main effect. Examples of these techniques include the use of cultures to establish microbial growth yield and efficiency (for a review, see Dijkstra *et al.*, 1998a) and the use of *in vitro* rumen epithelium-cell cultures to study absorption and metabolism of VFA (reviewed by Bergman, 1990). Also, a number of *in vitro* and *in situ* methods have been developed to estimate the digestibility and extent of ruminal degradation of feeds and to study their variation in response to changes in rumen conditions. Thus, *in vitro* and *in situ* techniques may be used to study individual processes, providing information about their nature and sensitivity to various changes. This information is of great importance in the development of mechanistic models. Mechanistic modelling can be used to derive kinetic parameters from data obtained *in vitro* or *in situ*, which can then be incorporated into integrated models to simulate the whole system behaviour. An appropriate model can be a useful tool for linking the data obtained *in vitro* or *in situ* with the processes occurring in organs and tissues *in vivo*.

Predicting events in the rumen

Rumen function involves complex, dynamic interactions between numerous components and processes. Of crucial importance to the supply of energy-yielding nutrients from the ruminant alimentary tract is the presence of a microbial population in the rumen, which serves to degrade nutrients, with the associated production of VFA, and supplies a substantial part of the protein absorbed from the small intestine. Thus, regarding the rumen, major consideration should be given to quantification of the rate and extent of substrate degradation, of the kinetics of microbial growth and synthesis and of the amount and ratio of the different VFA produced and absorbed. Although these aspects will be discussed in separate sections, it is important to realize that there are important interactions between them, which must be represented in whole-rumen models.

Rate and extent of degradation

Kinetic degradation parameters are necessary for the prediction of feed digestibility and thereby the energy available in various circumstances. The amount of substrate degraded in the rumen is the result of the competition between digestion and passage. Several models have been proposed since that of Blaxter *et al.* (1956), in which kinetic parameters for degradation and passage are integrated to estimate the actual extent of degradation of feed in the rumen.

Degradation parameters are usually estimated from degradation profiles (Fig. 4.5) obtained using either the polyester-bag technique or the gas-production technique. In the first case, a time-course disappearance curve for each feed component is obtained *in situ* by measuring the amount of the remaining residue in the bag at several time points (see Chapter 9). Similar gravimetric methods have been employed in which the feed is incubated *in vitro* either with buffered rumen fluid (Chapter 7) or with enzymes (Chapter 8). Disappearance curves are used to evaluate the kinetics of degradation of feeds in the rumen, by assuming that disappearance from the bag equals degradation of feed in the rumen. The gas-production technique (Chapter 10) aims to measure the rate of production of fermentation gases that can be used to predict the rate of feed degradation, assuming that the amount of gas produced reflects the amount of substrate degraded.

To associate disappearance or gas-production curves with digestion in the rumen, models have been developed based on compartmental schemes and assume that the feed component comprises at least two fractions: a potentially degradable fraction S and an undegradable fraction U. Fraction S will be degraded at a fractional rate μ (h^{-1}), after a discrete lag time T (h). The scheme is shown in Fig. 4.6, and the dynamic behaviour of the fractions is described by the differential equations:

$$dS/dt = 0 \qquad 0 \le t < T \qquad (1a)$$

$$= -\mu S \qquad t \ge T \qquad (1b)$$

$$dU/dt = 0 \qquad t \ge 0 \qquad (2)$$

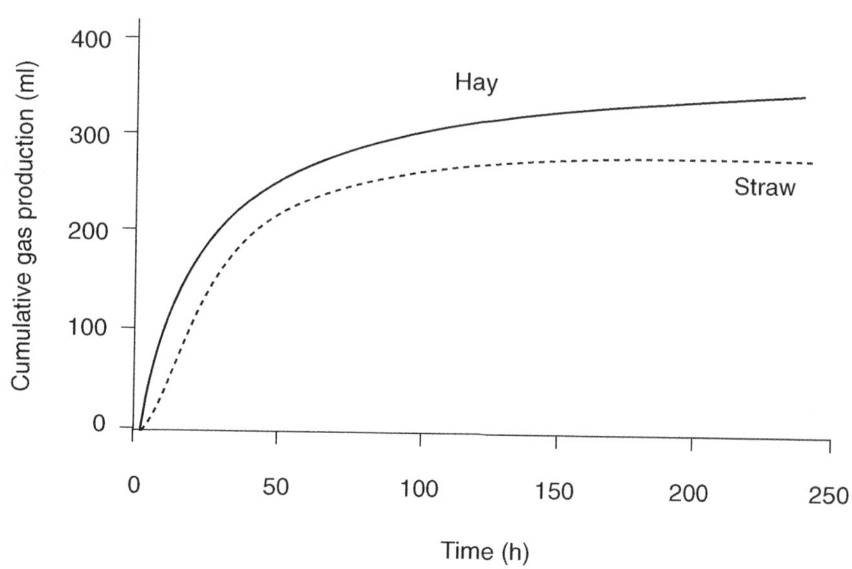

Fig. 4.5. Examples of sigmoidal and non-sigmoidal cumulative gas-production curves *in vitro*.

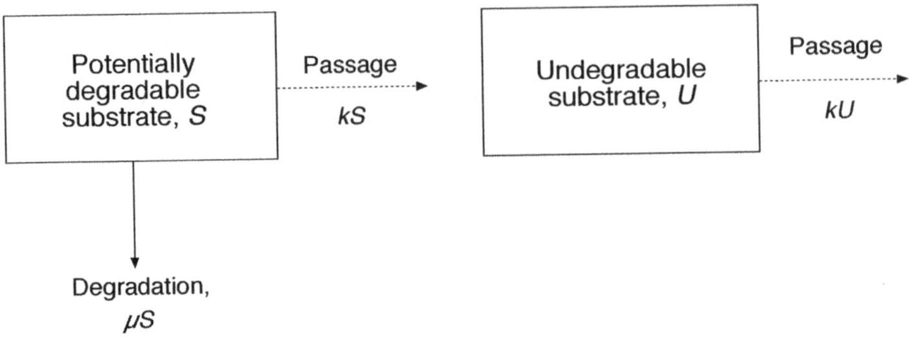

Fig. 4.6. The two-compartment model of ruminal degradation. Deletion of the dashed arrows gives scheme for disappearance during incubation *in vitro* or *in situ.*

Therefore, the parameters to be estimated are the initial size of the fraction S, the size of U, the lag time (T) and the fractional degradation rate (μ).

Estimation of U is critical to the accurate description of degradation kinetics, because the degradation rate, by definition, applies only to the fraction that is potentially degradable, with the assumption that each pool is homogeneous in its kinetic properties. Fraction U of protein and fibre components has been measured by long incubations (from 6 days to several weeks), either *in vitro* or *in sacco*, or estimated from non-linear fitting of degradation profiles. When degradation profiles are obtained by the polyester bag technique, the non-fibre components are assumed to contain a third fraction, called the soluble fraction (wash value, W). W disappears from the bag immediately after incubation begins and is assumed to be degraded instantly in the rumen. This simplification has been adopted because the technique is unable to measure the rate at which the W fraction would actually be degraded. Recently, some degradation rate values for this fraction, albeit very fast, have been reported (Weisbjerg *et al.*, 1998), and these could be incorporated in the models so that the portion of W escaping ruminal degradation can be taken into account. The loss of undegraded particulate matter from the polyester bag leads to overestimation of W and underestimation of the undegradable fraction. Estimation can be significantly improved by measuring the extent of the particle loss from the bag and applying mathematical corrections to the parameter estimates (López *et al.*, 1994a; France *et al.*, 1997). Using *in vitro* techniques allows degradation profiles with many more data points to be obtained, revealing the existence of multiple pools, which would be degraded at different rates. Some models have been reported that include several degradable pools (Robinson *et al.*, 1986). Such models contain a considerable number of parameters, requiring a large number of data points and complicating satisfactory parameter estimation, due to the limitations of the non-linear regression. On the other hand, experimental techniques may result in errors that have an important effect on the fractionation of feed components.

The lag phase of the degradation profiles has been described in terms of either a discrete or a kinetic lag (Van Milgen *et al.*, 1993). The initial lag phase is due in part to the inability of the rumen microbial population and its enzymes to degrade the substrate at a significant rate until microbial growth is sufficient for enzyme production to increase and ultimately to saturate the substrate. Lag may be due to factors other than microbial capacity, such as the rate of hydration of the substrate, microbial attachment to feed particles and nutrient limitations. A discrete lag is not a mechanistic interpretation of the process in the rumen. *In vitro* and *in situ* systems may induce an artificial lag because of the experimental procedures, and this parameter is therefore required in the models representing the system from which the degradation profiles are obtained.

The degradation rate of nutrients in the rumen is a key factor in predicting energy supply to the animal from a given feed, because it can have significant effects on both the ruminal microbes (efficiency of microbial synthesis, changes in rumen environment and effects on relative numbers of microbial species and on cellulolytic activity) and the host (molar composition of VFA produced, VFA absorption rate, amino acid supply postruminally and feed intake). The fractional degradation rate can be considered an intrinsic characteristic of the feed, depending on factors such as the forage chemical composition, the proportion of different plant tissues, as affected by the stage of maturity, surface area and the cell-wall structure. Once feed enters the rumen, the degradation rate may also be affected by factors related to the animal, such as the rate of particle-size reduction by rumination and microbial activity and ruminal conditions (pH, osmotic pressure, mean retention time of the digesta), which have a profound effect on microbial degradative activity. Associative effects of feeds in the diet can become very important. For example, the depressive effect of easily degradable non-fibre carbohydrates on the degradation rate of forage DM is generally recognized (Tamminga and Van Vuuren, 1988).

An essential aspect of estimating the rate of degradation concerns the kinetics assumed for the process. The most commonly used model (Ørskov and McDonald, 1979) assumes first-order kinetics, implying that substrate degraded at any time is proportional to the amount of potentially degradable matter remaining at that time, with constant fractional rate μ (Fig. 4.7), and that only characteristics of the substrate limit degradation. This model has been extensively used, owing to its simplicity, but it is not capable of describing the large diversity of degradation profiles (see Fig. 4.5) which have been observed (Dhanoa *et al.*, 1995) and it cannot represent mechanistically the reciprocal influences of substrate degradation and microbial growth. France *et al.* (2000) postulated that μ may vary with time according to different mathematical functions (Table 4.5). From the various functions used to represent μ, different models can be derived to describe either *in situ* disappearance (López et al., 1999) or *in vitro* gas-production profiles (France *et al.*, 2000) – for example, the generalized Mitscherlich model, the generalized Michaelis–Menten model and standard functions, such as the Gompertz and logistic (Fig. 4.7). Some of these functions are capable of describing both a range of shapes with no inflexion point and a range of sigmoidal shapes in which the inflexion point is variable. On substituting for the function proposed for μ and integrating, equation (1b) yields an equation for the S fraction remaining during the incubation *in situ* or *in vitro* at any time t, which can be expressed in the general form:

$$S = S_0 \times [1 - \Phi(t)] \quad (3)$$

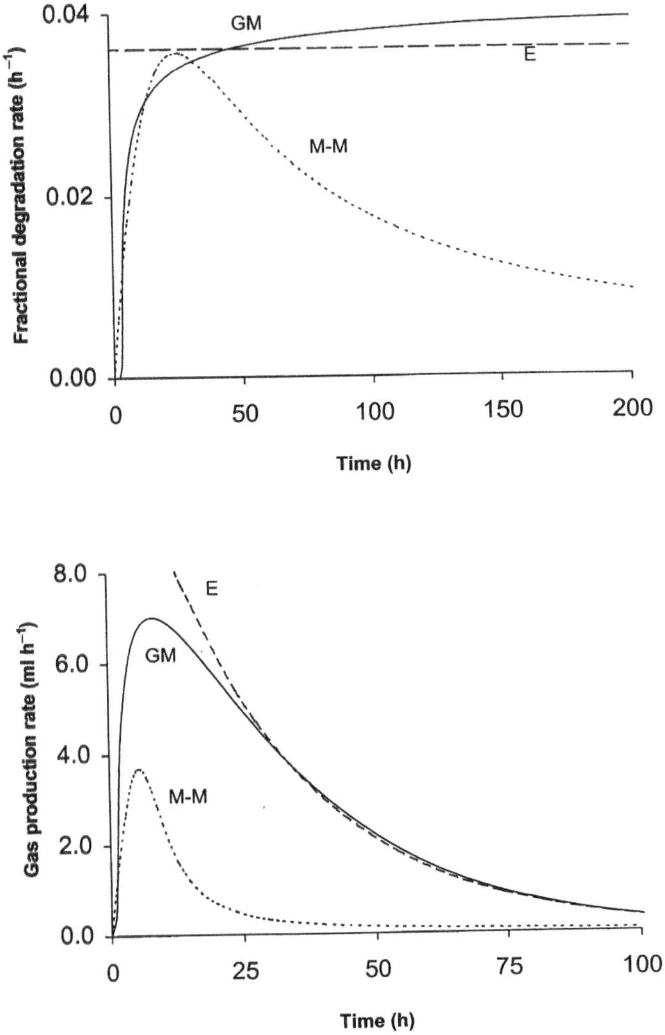

Fig. 4.7. Change in fractional degradation rate and in gas-production rate with time as represented by different mathematical models. E, exponential or simple Mitscherlich; GM, generalized Mitscherlich; M-M generalized Michaelis–Menten.

where S_0 is the zero-time quantity of the S fraction, $\Phi(t)$ is a positive monotonically increasing function with an asymptote at $\Phi(t) = 1$ (Table 4.5) and t is incubation time (h). *In situ* disappearance (D, g g^{-1} incubated) is given by:

$$D = W + S_0 - S = W + S_0 \times \Phi(t) \quad (4)$$

Similarly, gas production profiles observed *in vitro* can be represented by:

$$G = YS_0 \times \Phi(t) \quad (5)$$

Table 4.5. Alternative functions for Φ in the general equations for the *in situ* disappearance curves and the gas-production profiles, with corresponding functions for the fractional degradation rate (μ) of the substrate for each (for the meaning of the constants, which is specific to each model, see France *et al.*, 1990, 2000; López *et al.*, 1999).

	Φ	μ
Generalized Mitscherlich	$1-e^{-c(t-L)-d(\sqrt{t}-\sqrt{L})}$	$c+\frac{d}{2\sqrt{t}}$
Special case of generalized Mitscherlich	$1-e^{-c(t-L)}$	c
Generalized Michaelis–Menten	$\frac{t^c}{t^c+K^c}$	$\frac{ct^{(c-1)}}{t^c+K^c}$
Logistic	$\frac{1-e^{-ct}}{1+Ke^{-ct}}$	$\frac{c}{1+Ke^{-ct}}$
Gompertz	$1-\exp\left[\frac{b}{c}\left(1-e^{ct}\right)\right]$	be^{ct}

where G (ml) denotes total gas accumulation to time t and Y (ml gas g^{-1} degradable DM) is a constant yield factor. For each function, μ could be obtained from equations (1b) and (3) as:

$$\mu = -\frac{1}{S}\frac{dS}{dt} = \frac{1}{(1-\Phi)}\frac{d\Phi}{dt} \tag{6}$$

This function constitutes the mechanistic interpretation of the degradation processes. Rates of degradation and passage can be combined to calculate the extent of degradation of the substrate in the rumen (France *et al.*, 1990, 1993). In the rumen, if S is the amount of potentially degradable substrate remaining which is subjected to both passage and degradation, the rate of disappearance of S is given by (see Fig. 4.6):

$$\frac{dS}{dt} = -kS \qquad t < L \tag{7a}$$

$$= -(k+\mu)S \qquad t \geq L \tag{7b}$$

where k (h^{-1}) is the fractional rate of passage from the rumen and is assumed constant.

To obtain S, the solutions of these differential equations are:

$$S = S_0e^{-kt} \qquad t < L \tag{8a}$$

$$S = S_0 e^{-kt} \times (1 - \Phi) \qquad t \geq L \tag{8b}$$

Using these equations, the extent of degradation in the rumen (E, g degraded g^{-1} ingested) is given by the equations:

$$E = \frac{W + \int_L^\infty \mu S \mathrm{d}t}{W + S_0 + U} = \frac{W + kS_0 \int_L^\infty \Phi e^{-kt} \mathrm{d}t}{W + S_0 + U} \tag{9}$$

for *in situ* disappearance profiles (López *et al.*, 1999) and:

$$E = \frac{\int_L^\infty \mu S \mathrm{d}t}{S_0 + U} = \frac{kS_0 \int_L^\infty \Phi e^{-kt} \mathrm{d}t}{S_0 + U} \tag{10}$$

for *in vitro* gas-production profiles.

These equations provide a general expression for calculating the extent of degradation and are applicable to any model expressed in the form of equations (4) and (5). A number of equations have been proposed in the literature to describe the gas-production curve without considering the quantitative relationship to extent of degradation in the rumen, thus failing to link the *in vitro* technique to animal performance. The major assumption in the gas-production equations is that the rate at which gas is produced is directly proportional to the rate at which substrate is degraded, with constant yield factor Y. However, like any *in vitro* system of microbial activities, the partitioning of ruminally available substrate between fermentation (producing gas) and direct incorporation into microbial biomass may vary, depending on, amongst other things, the size of the microbial inoculum and the balance of energy- and nitrogen-containing substrates (Pirt, 1975). Indeed, Blümmel *et al.* (1997) showed an inverse relationship between gas production and microbial biomass yield. Clearly, during the course of fermentation, the amount of substrate becoming available per unit of microbial mass decreases, and hence (according to the growth and non-growth requirements as described by Pirt (1975)) relatively more substrate is expected to be used for gas production rather than incorporated into biomass, resulting in an increase in the yield of gas in the later phases of incubation. Furthermore, the pattern of fermentation can vary with respect to the yield of individual VFA and, from stoichiometric relationships, this undoubtedly affects Y in time. Whilst Y does not directly appear in the calculation of E (equation (10)), differences in Y during the course of fermentation might well affect the value of E. A further limitation of the gas-production technique is that it merely provides kinetic information on degradation of OM or of cell contents and cell-wall material, but cannot supply detailed information on the flow of aminogenic, glucogenic and lipogenic substrates out of the rumen.

Estimates of the extent of degradation of feeds in the rumen are invariably highly sensitive to the passage-rate values (reviewed by Dijkstra and France, 1996). The removal of digesta from the rumen is also one of the major processes controlling the intake of forages. Kinetics of passage have usually been studied using the faecal-marker excretion technique, enabling calculation of the mean retention time in the rumen and the whole tract. Rate of passage is, in part, an intrinsic characteristic of the forage (Aitchison *et al.*, 1986), but it is also influenced by factors such as particle size (affected by chewing and ruminating), particle functional specific gravity and water kinetics (Poncet, 1991). Main aspects of digesta passage have been identified (Kennedy and Murphy, 1988; Lechner-Doll *et al.*, 1991), but have yet to be quantified. Thus, the representation of outflow of digesta in the models of whole-rumen function is often much less detailed than the representations of substrate degradation and microbial synthesis processes. Some models have been developed to represent the dynamics of particle-size reduction to be applied to passage kinetics (Murphy and Kennedy, 1993), taking into account the particle size distribution in the feed and with a number of pools, each representing a different particle size, with rates of particle size reduction that determine the flow of particles from one pool to another. However, many of the mechanisms of fluid and particle outflow and their interactions with degradation through rumination and microbial activity have not been considered in mechanistic rumen models (for a review, see Dijkstra and France, 1996). Given the important effect of passage on extent of degradation, more quantitative research is required to allow proper integration of passage and degradation and improved prediction of degradation in the rumen.

Microbial synthesis

The partitioning of ruminally available monomers between fermentation (yielding VFA) and direct incorporation into microbial biomass may vary widely. This partitioning will influence the energy supply from the forage and the composition of the energy-yielding substrates. The different approaches to predicting microbial synthesis have been reviewed recently (Dijkstra *et al.*, 1998a). To understand aspects of microbial metabolism, physical analogues of the rumen ecosystem have been employed. The mathematical equations describing these cultures are helpful in discerning the essential parameters of microbial metabolism and in predicting the partitioning of substrate into microbial biomass and fermentation end-products. Most of the physical models for studying ruminal microbial metabolism have been based on the chemostat. The differential equations describing the dynamics of the chemostat are:

$$dX/dt = (\mu_G - D)X \tag{11}$$

$$dS/dt = CF - \mu_G X/Y - DS \tag{12}$$

where X (g) denotes the amount of biomass, S (g) the growth-limiting substrate, μ_G (h^{-1}) the fractional growth rate, Y (g biomass g^{-1} substrate) the growth yield with respect to S, D (h^{-1}) the dilution rate, F (ml h^{-1}) the medium inflow rate and C (mg substrate ml^{-1} medium) the concentration of S in the medium. The substrate consumption by the microbes is assumed to follow enzyme kinetics described by Michaelis and Menten (1913), and therefore the specific growth rate of biomass will be determined by substrate concentration, and can be represented by a rectangular hyperbola (Monod, 1942):

$$\mu_G = \mu_{G,max} /(1 + K/C) \tag{13}$$

where $\mu_{G,max}$ is the maximum value of μ_G (h^{-1}), and K (g substrate ml^{-1}) the saturation constant. K gives a measure of the affinity of the organism for the substrate. The $\mu_{G,max}$ and K values differ widely between microbial species and may be a major determinant of microbial survival in the rumen under conditions of competition (Russell and Baldwin, 1979). These equations show that faster growth rates are achieved when the substrate concentration is increased.

The overall rate of utilization of a substrate can be divided into the rate of usage for non-growth processes (maintenance) and for mass increase (growth). Based on this division, the yield factor in equation (12) is described by the double reciprocal equation of Pirt (1975):

$$1/Y = M/\mu_G + 1/Y_{max} \tag{14}$$

where M (g of S g^{-1} of X h^{-1}) is maintenance coefficient and Y_{max} (g of X g^{-1} of S) is maximum growth yield with respect to S. From equation (14), it is clear that Y decreases when maintenance requirements increase, and Y increases when μ_G or Y_{max} is increased.

Thus, the proportion of substrate fermented to VFA or incorporated into microbial biomass depends on factors that affect growth, maintenance and maximum growth yield (Fig. 4.8). Individual maintenance values can vary widely between species (Russell and Baldwin, 1979) and, since the composition of the rumen microbial population changes with changes in dietary characteristics, this may affect the proportion of the substrate being spent for maintenance. The microbial non-growth requirements are significantly increased when some nutrients other than carbohydrates are limiting growth. In such situations, the potential of microbes to produce energy from catabolic processes is in excess of their potential to use that energy for biosynthetic purposes (referred to as energetic uncoupling; Hespell and Bryant, 1979). Energy requirements per unit of microbial DM synthesized depend on the composition of microbial DM. For example, the biosynthesis of nucleic acids requires far more adenosine triphosphate (ATP) than does that of protein (Stouthamer, 1973). The outflow of microbial biomass to the duodenum is lower than that synthesized within the rumen, due to recycling of microbial biomass within the rumen (Dijkstra *et al.*, 1998b). This recycled biomass will be largely fermented to VFA again, and thus the degree of recycling within the rumen will affect Y as well. In their review, Dijkstra *et al.* (1998a) show that current protein evaluation systems do not take into account key aspects of microbial metabolism and will not predict microbial synthesis satisfactorily. In view of the relation between microbial biomass synthesis and VFA production from degraded substrates, this lack of adequate prediction is of relevance not just to protein nutrition of ruminants but also to energy evaluation, again highlighting the point that protein and energy evaluation systems should not be developed in isolation from each other if they are to cope with the energy and protein interactions within the animal.

VFA production

The VFA produced by fermentation of substrates in the rumen and subsequently absorbed represent the major source of energy, providing at least 50% of the total

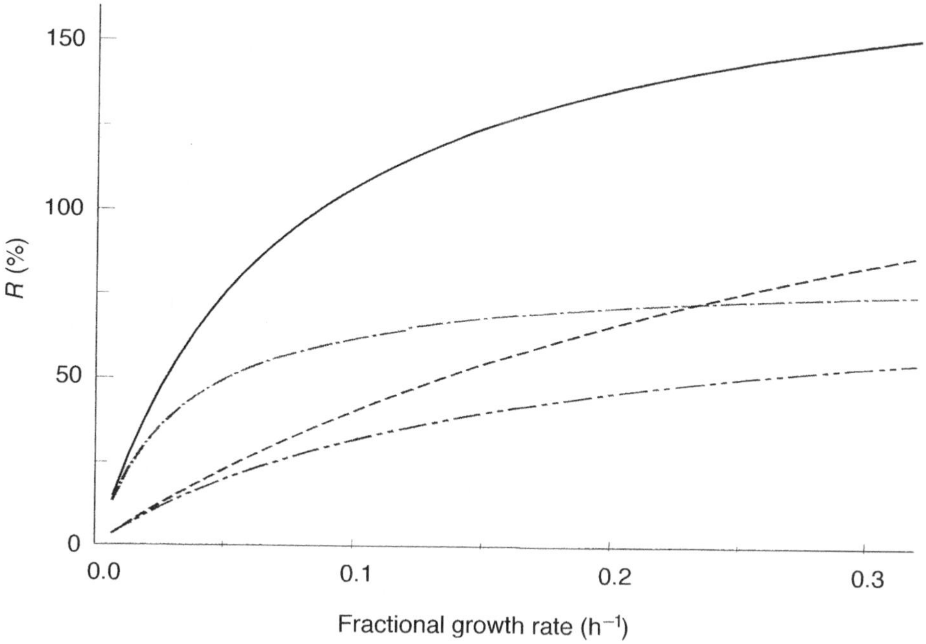

Fig. 4.8. Effect of fractional growth rate (μ_G), microbial maintenance requirement (M; low $M = 0.04$, high $M = 0.20$) and maximum growth yield (Y_{max}) in presence ($Y_{max} = 0.65$) or absence ($Y_{max} = 0.45$) of preformed monomers on the ratio (R) of carbohydrate incorporated into microbial biomass to that fermented to VFA. $M = 0.04$, and $Y_{max} = 0.65$ (——); $M = 0.04$, and $Y_{max} = 0.45$ (–·–); $M = 0.20$, and $Y_{max} = 0.65$ (------); $M = 0.20$, and $Y_{max} = 0.45$ (--——--).

amount of digestible energy (Sutton, 1985). The relative concentrations of the individual acids (mainly acetic, propionic and butyric acids), commonly referred to as the fermentation pattern, are important determinants of energy utilization by ruminants. Methane production (which is an important energy loss) per mole of fermented hexose is variable, depending on the type of VFA produced (less methane if fermentation is directed to the formation of propionate). Within the host animal, propionate is used largely for glucogenesis, being the major source of glucose for the ruminant (Huntington, 1990). In contrast, both acetate and butyrate are used primarily as energy sources, and acetate is also the principal substrate for lipogenesis. This, in addition to the distinct effects of individual VFA on the hormonal balance in the animal, results in a significant effect of the fermentation pattern in the rumen on the partitioning of nutrients between organs and tissues (Bergman, 1990). Thus, diets that stimulate high propionic and low acetic acid concentrations in the rumen are linked to reductions in the milk fat secretion, associated with a repartitioning of nutrients within the body of dairy cattle (Sutton, 1985). Also, the balance between the supply of propionate and that of acetate and butyrate seems to influence the efficiency with which energy is used for productive purposes in dairy cows (MacRae and Lobley, 1982), though the effects of differences in VFA ratios on energy efficiency for growth and growth composition are not consistent (Ørskov and MacLeod, 1990). Therefore, to predict the energy supply from a feed, it is important to estimate accurately the amount of VFA produced and absorbed and the molar proportions of the individual acids.

A number of techniques have been used to measure ruminal VFA production (Dijkstra, 1994). VFA production can be easily measured *in vitro* as the accumulation of VFA when the substrate is incubated in batch or continuous cultures. However, *in vivo* determinations are far more difficult, as VFA do not accumulate in the rumen and either are absorbed or flow out to the omasum. Techniques employing tracers are based on the application of compartmental analysis to interpret isotope dilution data, and some models have been derived based on simple or multiple pool schemes to estimate VFA production (France and Siddons, 1993). Concern exists over the variability and reliability of estimates obtained with ruminants using isotope dilution methods, particularly with dairy cattle (Sutton, 1985). Total VFA production depends upon factors affecting the ruminal digestibility of OM, such as the level of intake or the concentrate proportion of the diet, and also on factors affecting the amount of VFA obtained per unit of fermented OM, such as the fermentation pattern and the efficiency of microbial synthesis (discussed in the previous section). The type of substrate fermented is a major determinant of the type of VFA produced (Murphy *et al*., 1982), but a number of dietary factors, including the rate of fermentation and the level of intake, are responsible for specific interactions of the feed with the microbial population in the rumen and result in significant variations in the fermentation pattern. With forages, the major source of variation affecting the molar proportions of the VFA in the rumen is the OM digestibility, so that feeding more digestible forages is associated with lower proportions of acetate and higher proportions of both propionate and butyrate (Thomas and Rook, 1981). Treatments that change the rate of substrate fermentation may also give rise to marked differences in the fermentation pattern. Reductions in the particle size of forages by grinding are accompanied by less rumination and ruminal cellulolytic bacteria, resulting in a lower acetate-to-propionate ratio in rumen fluid. When starchy feeds are used, higher propionate concentrations at the expense of acetate can be expected, but the fermentation pattern is highly dependent on the source of the starch and its treatment during processing and also on the level of addition and the frequency of feeding of diets rich in sugars (Sutton, 1981).

Based on the well-known specific pathways of VFA production, Baldwin *et al.* (1970) suggested stoichiometric relationships between type of VFA produced and type of substrate fermented. This approach has been adopted in some mechanistic rumen models. Another approach that has been extensively incorporated into rumen models is that of Murphy *et al.* (1982), who provided a series of stoichiometric parameters relating to the yield of individual VFA for five specific dietary substrates. These relationships were not based on the biochemistry of rumen fermentation, but rather on statistical analyses of a large data set of ruminal VFA molar proportions observed for a range of dietary conditions, taking the solution that represented the best estimate on the basis of statistical criteria. Yet current approaches are insufficient in representing the fermentation pathways and in predicting VFA molar proportions accurately (reviewed by Dijkstra, 1994), confirmed by the results of a comparative evaluation of rumen models (Bannink *et al*., 1997). Probably, the mathematical representation of effects other than merely chemical composition of the diet needs to be considered, including dietary factors that induce changes in various parameters of the rumen environment (osmotic pressure, redox potential, pH, turnover rate) (Sutton, 1981). Dijkstra (1994) recommended that factors involved in the maintenance of the redox balance in the rumen (through reduction and reoxidation of nicotinamide adenine dinucleotide (NAD)) should be taken into account. These include ruminal pH, the composition of the substrate, the substrate availability and rate of depolymerization and the microbial species involved in the substrate fermentation. Prins *et al.* (1984) proposed that the pattern of VFA production from a given source and the amount of rumen-fermentable carbohydrate would be affected by the rate of flux of hexose through the microbial cell, which is related to the concentration of bacteria in the rumen (the lower the numbers of bacteria, the higher the flux per cell) and to the rate of fermentation of the substrate. The understanding of the complex relationships between dietary factors, rumen environment parameters and fermentation patterns could significantly improve the prediction of the molar proportions of VFA during fermentation under different dietary conditions.

In addition to VFA production rates, the prediction of VFA availability requires accurate estimates of VFA absorption as well. Mechanisms involved in the absorption of VFA across the rumen wall are not yet fully understood (Carter and Grovum, 1990; Rechkemmer *et al*., 1995). Quantitative information on the absorption rates is also lacking. The rate of VFA appearance in the portal blood has been used to estimate the net VFA absorption into the bloodstream. However, the amounts of VFA appearing in the portal blood are not equal to the amounts produced and absorbed from the rumen, because the former also includes VFA produced and absorbed in the lower tract and because some of the VFA absorbed (in particular, propionate and butyrate) are metabolized by the rumen mucosa during the process of absorption. VFA absorption has also been measured in animals receiving infusions of VFA into the emptied, washed and temporarily isolated rumen. Using this technique, it has been observed that the absorption of VFA through the rumen wall accounts for up to 88% of the infused VFA, that the absorption rates are different for the individual VFA and that absorption can be affected by rumen factors, such as pH, osmotic pressure and VFA concentration (Dijkstra *et al*., 1993; López *et al*., 1994b; López and Hovell, 1996). Further experimental data would be required to characterize the kinetics of VFA production and absorption and of metabolism in the rumen wall, in order to estimate parameters that can be incorporated into mathematical models to account for the dynamic, integrated character of VFA production and absorption processes, allowing for a more accurate prediction of the VFA and energy supply to the animal.

Predicting events postruminally

Small intestine

Nutrients entering the small intestine can be digested by pancreatic, hepatic and enteric enzymes, the end-products being absorbed into the portal bloodstream. The main energy supply to the animal at this level is from starch, protein and long-chain fatty acids (LCFA) (see Fig. 4.1). Some sources of starch are particularly resistant to rumen degradation, and considerable outflow of starch to the duodenum may occur (reviewed by Nocek and Tamminga, 1991). Then, information would be needed on the relative contributions of escaped soluble and insoluble feed starch and of microbial storage starch to the total amount of starch flowing out of the rumen. The amount of starch bypassing the rumen in a variety of circumstances can be predicted using the rumen models of Baldwin *et al.* (1987) and Dijkstra *et al.* (1992). Digestibility coefficients of starch in the small intestine (as proportion of the amount reaching the duodenum) are highly variable, ranging from 0.28 to 0.98 (for a review, see Toullec and Lallès, 1995). Some of the variation is attributed to the variation in amount of starch flowing out of the rumen. In general, as duodenal starch flow increases, the small-intestinal digestibility declines progressively, whereas large-intestinal digestion of starch increases (Beever, 1993). However, the most important factors affecting the site and extent of starch digestion are chemical and structural characteristics associated with the source of starch and the form of processing (Reynolds *et al.*, 1997). The digestion of starch in the small intestine has been considered of special importance for the supply of glucose to the animal. However, the net flux of glucose to the portal vein can only account for a small proportion of the starch digested in the small intestine (Reynolds and Beever, 1995), probably due to the utilization of glucose by the gut epithelial tissue. All these factors affecting the intestinal digestion of undegraded starch must be considered for any prediction of the supply of nutrients from this source. Only a few models have explicitly represented outflow of substrates from the rumen and digestion of substrates postruminally. If represented, the application of fixed intestinal digestion coefficients prevails (Baldwin *et al.*, 1987; Dijkstra *et al.*, 1996).

The absorption of LCFA is almost exclusively from the small intestine. Forages generally contain low amounts of LCFA, though, and hence these do not contribute large amounts of energy to the total energy absorbed from the small intestine. The use of feed supplements rich in LCFA is limited, because of the general adverse effect of LCFA on fibre degradation in the rumen (reviewed by Jenkins, 1993). Microbial LCFA are synthesized in the rumen *de novo* or incorporated from feed LCFA. The synthesis *de novo* may explain why the flow of LCFA into the duodenum exceeds the dietary intake of LCFA on diets containing low or moderate amounts of lipid (Doreau and Chilliard, 1997). Although saturated LCFA are absorbed more slowly than unsaturated LCFA, in general absorption in the small intestine is high, with true digestibility approaching 95–100% (Van Soest, 1994).

Amino acids flowing into the duodenum may form an important contribution to the energy absorbed. Microbial protein is generally highly digestible in the small intestine, with average values of 85% (Storm *et al.*, 1983). However, dietary protein digestibility in the small intestine may be far more variable, especially that of forage protein, and it varies between 50 and 99% of the escaped feed protein entering the small intestine (reviewed by Van Straalen and Tamminga, 1990). As with starch and LCFA, current feed evaluation systems and mechanistic models generally apply fixed intestinal digestion coefficients of protein. The models do vary significantly in predicting the flow of microbial and bypass protein from the rumen, though, thus potentially allowing for considerable variation in the predicted contribution of protein to the energy supply of the ruminant.

Hind-gut

The large intestine may play an important role in the digestion of forages by ruminants. Although up to 95% of the total fermentative activity takes place in the rumen, an active microbial population is also present in the hind-gut (caecum and proximal colon), which may contribute, to a variable extent, to the degradation of carbohydrates and nitrogen compounds (see Fig. 4.1). Digestibility of structural carbohydrates in the large intestine of sheep (as proportion of the amounts entering into the large intestine) ranged from 19 to 49% (Tisserand and Demarquilly, 1995), and substrates absorbed in the caecum–colon may account for 4 to 26% of the total energy absorbed in the whole digestive tract. The proportion of substrates degraded and of digestible energy obtained in the hind-gut in relation to that in the whole digestive tract is greater if the digestibility of the forage is lower, the level of intake is increased, the level of inclusion of concentrates in the diet is increased or with ground and pelletized diets (Demeyer, 1991; Van Soest, 1994). In general, digestion in the hind-gut partly compensates a potential decline in whole-tract digestibility caused by any factor decreasing ruminal digestion of carbohydrates. This also stresses the importance of integrating digestive processes in the rumen and small and large intestines in mechanistic models to obtain approximate estimates of the energy and nutrient supply from the feeds consumed.

Degradation and fermentation processes of substrates in the hind-gut are rather similar to those in the rumen (Demeyer, 1991). After hydrolysis of the substrates, microbial fermentation yields VFA and microbial biomass. Total VFA concentrations in the caecum–colon are in the range of 60 to 90% of those typically observed in the rumen. Some differences between rumen and hind-gut VFA molar proportions have been reported. The molar proportion of acetate in the hind-gut is usually greater and less variable than that observed in the rumen, because in the hind-gut a larger fraction of the available carbohydrates are structural carbohydrates and reductive acetogenesis is not outcompeted by methanogenesis (Demeyer, 1991). Yet many of the research findings provide only qualitative descriptions of the digestive processes in the hind-gut, due to the difficulty of studying digestion and absorption in this compartment. From the available information, it appears that most concepts used for modelling rumen fermentation could be applied to the fermentative processes occurring in the hind-gut as well. Differences between the rumen and the large intestine in the harboured microbial populations are largely related to the shorter retention time of digesta in the hind-gut and to differences in the available substrates for the microorganisms. The retention time of digesta in the hind-gut may vary between 7 and 18 h (Grovum and Hecker, 1973), shorter than retention times in the reticulorumen under similar nutritional conditions. Similar bacterial species to those found in the rumen have been identified, although total counts are considerably smaller (Ulyatt *et al.*, 1975) and relative numbers also differ. No protozoa have been isolated from the ruminant hind-gut, and little is known about any fungal populations at this site. Substrates digested at the caecum–colon are those arriving from the ileum, comprising nutrients that have not been digested in the reticulorumen, abomasum and small intestine and those contained in endogenous secretions and cells. These substrates are more refractory to digestion than those entering other segments of the gut, although their susceptibility to microbial attack may be enhanced following exposure to abomasal and intestinal digestion and by reductions in particle size relative to the feed. Any model to predict energy supply to the animal from the hind-gut has to be integrated in a model of the whole digestive tract, as substrates available for fermentation at this site depend on digestion and absorption in the rumen and small intestine. However, no explicit mathematical representations of hind-gut fermentation in ruminants are available, other than those in whole-animal models, where a fixed digestion coefficient or a fixed proportion of rumen fermentation is applied to predicted flows of nutrients into the hind-gut (Baldwin *et al.*, 1987; Dijkstra *et al.*, 1996).

Concluding Remarks

The prediction of energy supply from forages is primarily based on empirical ME or NE systems. Such predictions do not consider site of nutrient digestion within the ruminant and the major interactions between substrates, and take no regard of the nature of the metabolites that comprise the energy fraction. Moreover, the current systems aim to estimate feed requirements to meet the energy needs of the ruminant, whereas the response of the ruminant to dietary perturbations are of greater interest, along with the animal product composition and excretion of waste material. Data obtained using *in vitro* and *in sacco* methods provide additional information for estimating energy supply that is less expensive to obtain than by using *in vivo* methods. Such information may be of limited value, however, if regarded in isolation. In this chapter, it has been argued that mathematical integration of new and existing data and concepts into mechanistic models is essential in order to utilize the information optimally and to improve the prediction of energy supply to ruminants.

References

AFRC (1991) Technical Committee on Responses to Nutrients, Report No 7. Theory of response to nutrients by farm animals. *Nutrition Abstracts and Reviews, Series B* 61, 683–722.

AFRC (1993) *Energy and Protein Requirements of Ruminants.* CAB International, Wallingford.

Aitchison, E., Gill, M., France, J. and Dhanoa, M.S. (1986) Comparison of methods to describe the kinetics of digestion and passage of fibre in sheep. *Journal of the Science of Food and Agriculture* 37, 1065–1072.

Alderman, G. (1983) Status of the newer European feeding systems. In: Robards, G.E. and Packham, R.G. (eds) *Feed Information and Animal Production.* CAB, Slough, and INFIC, Blacktown, Australia, pp. 305–321.

Alvarez, F.J. and Preston, T.R. (1976) *Leucaena leucocephala* as protein supplement for dual purpose milk and weaned calf production on sugar-cane based ratios. *Tropical Animal Production* 1, 112–118.

Alvarez, F.J., Wilson, A. and Preston, T.R. (1978) *Leucaena leucocephala* as protein supplement for dual purpose milk and weaned calf production on sugar-cane based diets: comparisons with rice polishings. *Tropical Animal Production* 3, 51–55.

Andrieu, J. and Demarquilly, C. (1987) Valeur nutritive des fourrages: tables et prévision. *Bulletin Technique CRZV Theix, INRA* 70, 61–73.

Andrieu, J., Demarquilly, C. and Wegat-Litre, E. (1981) Tables de prévision de la valeur alimentaire des fourrages. In: *Prévision de la valeur nutritive des aliments des ruminants.* INRA Publications, Versailles, pp. 345–580.

Baldwin, R.L. (1995) *Modeling Ruminant Digestion and Metabolism.* Chapman & Hall, London.

Baldwin, R.L., Lucas, H.L. and Cabrera, R. (1970) Energetic relationships in the formation and utilization of fermentation end-products. In: Phillipson, A.T., Annison, E.F., Armstrong, D.G., Balch, C.C., Comline, R.S., Hardy, R.S., Hobson, P.N. and Keynes, R.D. (eds) *Physiology of Digestion and Metabolism in the Ruminant.* Oriel Press, Newcastle upon Tyne, pp. 319–334.

Baldwin, R.L., Thornley, J.H.M. and Beever, D.E. (1987) Metabolism of the lactating cow. II. Digestive elements of a mechanistic model. *Journal of Dairy Research* 54, 107–131.

Bannink, A., de Visser, H. and Van Vuuren, A.M. (1997) Comparison and evaluation of mechanistic rumen models. *British Journal of Nutrition* 78, 563–581.

Barber, W.P.B., Adamson, A.H. and Altman, J.F.B. (1984) New methods of feed evaluation. In: Haresign, W. and Cole, D.J.A. (eds) *Recent Advances in Animal Nutrition – 1984.* Butterworths, London, pp. 161–176.

Beever, D.E. (1993) Rumen function. In: Forbes, J.M. and France, J. (eds) *Quantitative Aspects of Ruminants Digestion and Metabolism.* CAB International, Wallingford, pp. 187–215.

Beever, D.E. and Oldham, J.D. (1986) Nutrient supply and utilization. In: Broster, W.H., Phipps, R.H. and Johnson, C.L. (eds) *Principles and Practice of Feeding Dairy Cows.* AFRC Institute of Food Research, Reading, pp. 45–72.

Bergman, E.N. (1990) Energy contributions of volatile fatty acids from the gastrointestinal tract in various species. *Physiological Reviews* 70, 567–590.

Blaxter, K.L. (1962) *The Energy Metabolism of Ruminants.* Hutchinson, London.

Blaxter, K.L. (1989) *Energy Metabolism in Farm Animals and Man.* Cambridge University Press, Cambridge.

Blaxter, K.L., Graham, N.McC. and Wainman, F.W. (1956) Some observations on the digestibility of food by sheep, and on related problems. *British Journal of Nutrition* 10, 69–91.

Blümmel, M., Steingaß, H. and Becker, K. (1997) The relationship between *in vitro* gas production, *in vitro* microbial biomass yield and ^{15}N incorporation and its implications for the prediction of voluntary feed intake of roughages. *British Journal of Nutrition* 77, 911–921.

Carro, M.D., López, S., Gonzalez, J.S. and Ovejero, F.J. (1994) Comparison of laboratory methods for predicting digestibility of hay in sheep. *Small Ruminant Research* 14, 9–17.

Carter, R.R. and Grovum, W.L. (1990) A review of the physiological significance of hypertonic body fluids on feed intake and ruminal function: salivation, motility and microbes. *Journal of Animal Science* 68, 2811–2832.

Chesson, A. (1993) Mechanistic models of forage cell wall degradation. In: Jung, H.G., Buxton, D.R., Hatfield, R.D. and Ralph, J. (eds) *Forage Cell Wall Structure and Digestibility.* ASA–CSSA–SSSA, Madison, pp. 347–376.

CSIRO (1990) *Feeding Standards for Australian Livestock. Ruminants.* CSIRO Publications, Melbourne.

Demeyer, D.I. (1991) Quantitative aspects of microbial metabolism in the rumen and hindgut. In: Jouany, J.P. (ed.) *Rumen Microbial Metabolism and Ruminant Digestion*. INRA Editions, Paris, pp. 217–237.

DePeters, E.J. and Taylor, S.J. (1985) Effects of feeding corn or barley on composition of milk and diet digestibility. *Journal of Dairy Science* 68, 2027–2032.

Dhanoa, M.S., France, J., Siddons, R.C., López, S. and Buchanan-Smith, J.G. (1995) A non-linear compartmental model to describe forage degradation kinetics during incubation in polyester bags in the rumen. *British Journal of Nutrition* 73, 3–15.

Dijkstra, J. (1994) Production and absorption of volatile fatty acids in the rumen. *Livestock Production Science* 39, 61–69.

Dijkstra, J. and France, J. (1995) Modelling and methodology in animal science. In: Danfær, A. and Lescoat, P. (eds) *Proceedings of the Fourth International Workshop on Modelling Nutrient Utilisation in Farm Animals*. National Institute of Animal Science, Foulum, Denmark, pp. 9–18.

Dijkstra, J. and France, J. (1996) A comparative evaluation of models of whole rumen function. *Annales de Zootechnie* 45 (Suppl.), 175–192.

Dijkstra, J., Neal, H.D.St C., Beever, D.E. and France, J. (1992) Simulation of nutrient digestion, absorption and outflow in the rumen: model description. *Journal of Nutrition* 122, 2239–2256.

Dijkstra, J., Boer, H., Van Bruchem, J., Bruining, M. and Tamminga, S. (1993) Absorption of volatile fatty acids from the rumen of lactating dairy cows as influenced by volatile fatty acid concentration, pH and rumen liquid volume. *British Journal of Nutrition* 69, 385–396.

Dijkstra, J., France, J., Assis, A.G., Neal, H.D.St C., Campos, O.F. and Aroeira, L.J.M. (1996) Simulation of digestion in cattle fed sugar cane: prediction of nutrient supply for milk production with locally available supplements. *Journal of Agricultural Science, Cambridge* 127, 247–260.

Dijkstra, J., France, J. and Davies, D.R. (1998a) Different mathematical approaches to estimating microbial protein supply in ruminants. *Journal of Dairy Science* 81, 3370–3384.

Dijkstra, J., France, J. and Tamminga, S. (1998b) Quantification of the recycling of microbial nitrogen in the rumen using a mechanistic model of rumen fermentation processes. *Journal of Agricultural Science, Cambridge* 130, 81–94.

Doreau, M. and Chilliard, Y. (1997) Digestion and metabolism of dietary fat in farm animals. *British Journal of Nutrition* 78 (Suppl. 1), S15–S35.

France, J. and Siddons, R.C. (1993) Volatile fatty acid production. In: Forbes, J.M. and France, J. (eds) *Quantitative Aspects of Ruminants Digestion and Metabolism*. CAB International, Wallingford, pp. 107–121.

France, J. and Thornley, J.H.M. (1984) *Mathematical Models in Agriculture: A Quantitative Approach to Problems in Agriculture and Related Sciences*. Butterworths, London.

France, J., Thornley, J.H.M., López, S., Siddons, R.C., Dhanoa, M.S., Van Soest, P.J. and Gill, M. (1990) On the two compartment model for estimating the rate and extent of feed degradation in the rumen. *Journal of Theoretical Biology* 146, 269–287.

France, J., Dhanoa, M.S., Theodorou, M.K., Lister, S.J., Davies, D. and Isac, D. (1993) A model to interpret gas accumulation profiles associated with *in vitro* degradation of ruminant feeds. *Journal of Theoretical Biology* 163, 99–111.

France, J., López, S., Dijkstra, J. and Dhanoa, M.S. (1997) Particulate matter loss and the polyester-bag method. *British Journal of Nutrition* 78, 1033–1037.

France, J., Dijkstra, J., Dhanoa, M.S., López, S. and Bannink, A. (2000) Estimating the extent of degradation of ruminant feeds from a description of their gas production profiles observed *in vitro*: derivation of models and other mathematical considerations. *British Journal of Nutrition* 83, 143–150.

Gill, M., Beever, D.E. and France, J. (1989) Biochemical bases needed for the mathematical representation of whole animal metabolism. *Nutrition Research Reviews* 2, 181–200.

Givens, D.I., Everington, J.M. and Adamson, A.H. (1990) The nutritive value of spring-grown herbage produced on farms throughout England and Wales over 4 years. III. The prediction of energy values from various laboratory measurements. *Animal Feed Science and Technology* 27, 185–196.

Givens, D.I., Moss, A.R. and Adamson, A.H. (1992) The chemical composition and energy value of high temperature dried grass produced in England. *Animal Feed Science and Technology* 36, 215–218.

Grovum, W.L. and Hecker, J.F. (1973) Rate of passage of digesta in sheep. 2. The effect of level of food intake on digesta retention times and on water and electrolyte absorption in the large intestine. *British Journal of Nutrition* 30, 221–230.

Hanigan, M.D., Dijkstra, J., Gerrits, W.J.J. and France, J. (1997) Modelling post-absorptive protein and amino acid metabolism in the ruminant. *Proceedings of the Nutrition Society* 56, 631–643.

Hespell, R.B. and Bryant, M.P. (1979) Efficiency of rumen microbial growth: influence of some theoretical and experimental factors on Y_{ATP}. *Journal of Animal Science* 49, 1640–1659.

Huntington, G.B. (1990) Energy metabolism in the digestive tract and liver of cattle: influence of physiological state and nutrition. *Reproduction Nutrition Developpement* 30, 35–47.

Hvelplund, T., Andrieu, J., Weisbjerg, M.R. and Vermorel, M. (1995) Prediction of the energy and protein value of forages for ruminants. In: Journet, M., Grenet, E., Farce, M.H., Theriez, M. and Demarquilly, C. (eds) *Recent Developments in the Nutrition of Herbivores*. INRA Editions, Paris, pp. 205–227.

INRA (1988) *Alimentation des bovins, ovins et caprins*. INRA Editions, Paris.

Jenkins, T.C. (1993) Lipid metabolism in the rumen. *Journal of Dairy Science* 76, 3851–3863.

Kennedy, P.M. and Murphy, M.R. (1988) The nutritional implications of differential passage of particles through the ruminant alimentary tract. *Nutrition Research Reviews* 1, 189–208.

Lechner-Doll, M., Kaske, M. and von Engelhardt, W. (1991) Factors affecting the mean retention time of particles in the forestomachs of ruminants and camelids. In: Tsuda, T., Sasaki, Y. and Kawashima, R. (eds) *Physiological Aspects of Digestion and Metabolism in Ruminants*. Academic Press, Tokyo, pp. 455–482.

López, S. and Hovell, F.D.deB. (1996) Volatile fatty acid absorption and water flux in the rumen of sheep in relation to the volatile fatty acid supply. *Proceedings of the Nutrition Society* 55, 62A.

López, S., France, J. and Dhanoa, M.S. (1994a) A correction for particulate matter loss when applying the polyester-bag method. *British Journal of Nutrition* 71, 135–137.

López, S., Hovell, F.D.deB. and MacLeod, N.A. (1994b) Osmotic pressure, water kinetics and volatile fatty acid absorption in the rumen of sheep sustained by intragastric infusions. *British Journal of Nutrition* 71, 153–168.

López, S., France, J., Dhanoa, M.S., Mould, F. and Dijkstra, J. (1999) Comparison of mathematical models to describe disappearance curves obtained using the polyester bag technique for incubating feeds in the rumen. *Journal of Animal Science* 77, 1875–1888.

MacRae, J.C. and Lobley, G.E. (1982) Some factors which influence thermal energy losses during the metabolism of ruminants. *Livestock Production Science* 9, 447–456.

MacRae, J.C., Buttery, P.J. and Beever, D.E. (1988) Nutrient interactions in the dairy cow. In: Garnsworthy, P.C. (ed.) *Nutrition and Lactation in the Dairy Cow*. Butterworths, London, pp. 55–75.

MAFF (1984) *Energy Allowances and Feeding Systems for Ruminants*. HMSO, London.

Michaelis, L. and Menten, L. (1913) Die Kinetik der Invertinwirkung. *Biochemische Zeitschrift* 49, 334–336.

Minson, D.J. (1982) Effect of chemical composition on feed digestibility and metabolizable energy. *Nutrition Abstracts and Reviews, Series B* 52, 591–615.

Minson, D.J. (1990) *Forage in Ruminant Nutrition.* Academic Press, London.

Moe, P.W., Flatt, W.P. and Tyrrell, H.F. (1972) Net energy value of feeds for lactation. *Journal of Dairy Science* 55, 945–958.

Monod (1942) *Recherches sur la croissance des cultures bactériennes,* 2nd edn. Hermann, Paris.

Murphy, M.R. and Kennedy, P.M. (1993) Particle dynamics. In: Forbes, J.M. and France, J. (eds) *Quantitative Aspects of Ruminants Digestion and Metabolism.* CAB International, Wallingford, pp. 87–105.

Murphy, M.R., Baldwin, R.L. and Koong, L.J. (1982) Estimation of stoichiometric parameters for rumen fermentation of roughage and concentrate diets. *Journal of Animal Science* 55, 411–421.

Nocek, J.E. and Tamminga, S. (1991) Site of digestion of starch in the gastrointestinal tract of dairy cows and its effect on milk yield and composition. *Journal of Dairy Science* 74, 3598–3629.

NRC (1989) *Nutrient Requirements of Dairy Cattle,* 6th edn. National Academic Press, Washington, DC.

NRC (1996) *Nutrient Requirements of Beef Cattle,* 7th edn. National Academic Press, Washington, DC.

Ørskov, E.R. and McDonald, I. (1979) The estimation of protein degradability in the rumen from incubation measurements weighted according to rate of passage. *Journal of Agricultural Science, Cambridge* 92, 499–503.

Ørskov, E.R. and MacLeod, N.A. (1990) Dietary-induced thermogenesis and feed evaluation in ruminants. *Proceedings of the Nutrition Society* 49, 227–237.

Pirt, S.J. (1975) *Principles of Microbe and Cell Cultivation.* Blackwell Scientific Publications, Oxford.

Poncet, C. (1991) The outflow of particles from the reticulo-rumen. In: Jouany, J.P. (ed.) *Rumen Microbial Metabolism and Ruminant Digestion.* INRA Editions, Paris, pp. 297–322.

Prins, R.A., Lankhorse, A. and van Hoven, W. (1984) Gastrointestinal fermentation in herbivores and the extent of plant cell-wall digestion. In: Gilchrist, F.M.C. and Mackie, R.I. (eds) *Herbivore Nutrition in the Subtropics and Tropics.* Science Press, Craighall, South Africa, pp. 408–434.

Rechkemmer, G., Gäbel, G., Diernæs, I., Sehested, J., Møller, P.D. and von Engelhardt, W. (1995) Transport of short chain fatty acid in the forestomach and hindgut. In: von Engelhardt, W., Leonhard-Marek, S., Breves, G. and Giesecke, D. (eds) *Ruminant Physiology: Digestion, Metabolism, Growth and Reproduction.* Enke, Stuttgart, pp. 95–116.

Reynolds, C.K. and Beever, D.E. (1995) Energy requirements and responses: a UK perspective. In: Lawrence, T.J., Gordon, F.J. and Carson, A. (eds) *Breeding and Feeding the High Genetic Merit Dairy Cow.* BSAS Occasional Publication No. 19, Penicuik, pp. 31–41.

Reynolds, C.K., Sutton, J.D. and Beever, D.E. (1997) Effects of feeding starch to dairy cattle on nutrient availability and production. In: Garnsworthy, P.C. and Wiseman, J. (eds) *Recent Advances in Animal Nutrition – 1997.* Butterworths, London, pp. 105–134.

Robinson, P.H., Fadel, J.G. and Tamminga, S. (1986) Evaluation of mathematical models to describe neutral detergent residue in terms of its susceptibility to degradation in the rumen. *Animal Feed Science and Technology* 15, 249–271.

Russell, J.B. and Baldwin, R.L. (1979) Comparison of substrate affinities among several rumen bacteria: a possible determinant of rumen bacterial competition. *Applied and Environmental Microbiology* 37, 531–536.

Storm, E., Brown, D.S. and Ørskov, E.R. (1983) The nutritive value of rumen microorganisms in ruminants. 3. The digestion of microbial amino and nucleic acids in, and losses of endogenous nitrogen from, the small intestine of sheep. *British Journal of Nutrition* 50, 479–485.

Stouthamer, A.H. (1973) A theoretical study on the amount of ATP required for synthesis of microbial cell material. *Antonie van Leeuwenhoek* 39, 545–565.

Sutton, J.D. (1980) Digestion and end-product formation in the rumen from production rations. In: Ruckebusch, Y. and Thivend, P. (eds) *Digestive Physiology and Metabolism in Ruminants.* MTP Press, Lancaster, pp. 271–290.

Sutton, J.D. (1981) Concentrate feeding and milk composition. In: Haresign, W. and Cole, D.J.A. (eds) *Recent Advances in Animal Nutrition – 1981.* Butterworths, London, pp. 35–48.

Sutton, J.D. (1985) Digestion and absorption of energy substrates in the lactating cow. *Journal of Dairy Science* 68, 3376–3393.

Sutton, J.D. (1986) Rumen fermentation and gastro-intestinal absorption: carbohydrates. In: Neimann-Sorensen, A. (ed.) *New Developments and Future Perspectives in Research on Rumen Function.* EEC, Luxemburg, pp. 21–38.

Tamminga, S. (1992) Nutrition management of dairy cows as a contribution to pollution control. *Journal of Dairy Science* 75, 345–357.

Tamminga, S. and Van Vuuren, A.M. (1988) Formation and utilization of end products of lignocellulose degradation in ruminants. *Animal Feed Science and Technology* 21, 141–159.

Thomas, P.C. (1990) Predicting the nutritive value of compound feeds for ruminants. In: Wiseman, J. and Cole, D.J.A. (eds) *Feedstuff Evaluation.* Butterworths, London, pp. 301–317.

Thomas, P.C. and Rook, A.J.F. (1981) Manipulation of rumen fermentation. In: Haresign, W. and Cole, D.J.A. (eds) *Recent Advances in Animal Nutrition – 1981.* Butterworths, London, pp. 157–183.

Tilley, J.M.A. and Terry, R.A. (1963) A two-stage technique for the *in vitro* digestion of forage crops. *Journal of the British Grassland Society* 18, 104–111.

Tisserand, J.L. and Demarquilly, C. (1995) Digestion et absorption dans le gros intestin. In: Jarrige, R., Ruckebusch, Y., Demarquilly, C., Farce, M.H. and Journet, M. (eds) *Nutrition des Ruminants Domestiques.* INRA, Paris, pp. 583–599.

Toullec, R. and Lallès, J.P. (1995) Digestion dans le caillette et l'intestin grêle. In: Jarrige, R., Ruckebusch, Y., Demarquilly, C., Farce, M.H. and Journet, M. (eds) *Nutrition des Ruminants Domestiques.* INRA, Paris, pp. 527–581.

Tyrrell, H.F. and Moe, P.W. (1975) Effect of intake on digestive efficiency. *Journal of Dairy Science* 58, 1151–1163.

Ulyatt, M.J., Dellow, D.W., Reid, C.S.W. and Bauchop, T. (1975) Structure and function of the large intestine of ruminants. In: MacDonald, I.W. and Warner, A.C.I. (eds) *Digestion and Metabolism in the Ruminant.* University of New England Publishing Unit, Armidale, Australia, pp. 119–133.

Van der Honing, Y. and Steg, A. (1990) Comparison of energy evaluation systems of feeds for ruminants. In: Wiseman, J. and Cole, D.J.A. (eds) *Feedstuff Evaluation.* Butterworths, London, pp. 1–19.

Van Es, A.J.H. (1978) Feed evaluation for ruminants. I. The system in use from May 1978 onwards in the Netherlands. *Livestock Production Science* 5, 331–335.

Van Milgen, J., Berger, L.L. and Murphy, M.R. (1993) An integrated dynamic model of feed hydration and digestion, and subsequent bacterial mass accumulation in the rumen. *British Journal of Nutrition* 70, 471–483.

Van Soest, P.J. (1994) *Nutritional Ecology of the Ruminant*, 2nd edn. Cornell University Press, Ithaca.

Van Straalen, W.M. and Tamminga, S. (1990) Protein degradability in ruminant diets. In: Wiseman, J. and Cole, D.J.A. (eds) *Feedstuff Evaluation*. Butterworths, London, pp. 55–72.

Vermorel, M., Coulon, J.B. and Journet, M. (1987) Révision du système des unités fourragères (UF). *Bulletin Technique CRZV Theix, INRA* 70, 9–18.

Weisbjerg, M.R., Hvelplund, T. and Bibby, B.M. (1998) Hydrolysis and fermentation rate of glucose, sucrose and lactose in the rumen. *Acta Agriculturae Scandinavica, Section A, Animal Science* 48, 12–18.

5 Forage Evaluation Using Measurements of Energy Metabolism

C.K. Reynolds

Department of Agriculture, University of Reading, Earley Gate, Reading RG6 6AT, UK

Introduction

Considering the evolutionary adaptations of the ruminant digestive tract, the fundamental importance of forages in ruminant nutrition is self-evident. In wild ruminants and many extensive ruminant production systems, forages often provide the sole source of dietary energy and protein. In these situations, forages are often harvested solely by the animal, but forages may also be mechanically harvested and conserved for use during seasonal changes in forage availability. In more intensive systems seeking maximum production, forages may be fed in conjunction with cereals and other concentrated energy or protein sources, but even in high-yielding dairy cows a minimal level of dietary fibre is required for the maintenance of rumen function and health. In these cases, the effectiveness of by-product fibre sources in maintaining rumen function is often evaluated relative to grass or legume forages (e.g. Clark and Armentano, 1997). In a similar manner, the first feeding standards for cattle developed by Thaer compared the nutritive value of feeds and expressed requirements relative to the fattening value of 'good meadow hay' (Flatt *et al.*, 1972).

Even before the development of Thaer's hay-equivalent system in 1809, ruminant nutritionists and husbandmen sought to develop comprehensive systems for evaluating and expressing the nutritive value of forages and other feeds relative to their ability to meet specific animal requirements. In the intervening years, a number of systems have evolved, but the need for refinement and improvement of current feed evaluation systems is as important today as 200 years ago. Current systems for evaluating the nutritive value of forages for specific production requirements are an improvement over their predecessors, largely on account of collective and integrated experience and technological advances, but they do have limitations. The ideal system of feed evaluation was described by Flatt *et al.* (1969a) as the 'Holy

Grail' of animal nutrition, where, as for religion, a variety of approaches are used around the world, each believed by its users to be the one 'true' system.

Many of the current 'official' feeding systems compare the energetic value of forages based on their relative ability to meet the animal's requirements for net energy (NE) at a standardized level of intake or a specific physiological state. These systems are based on calorimetric measurements of energy balance within animals fed individual feeds or combinations of feeds, although in practice the energetic values of many feeds are estimated using other approaches. This approach has been the subject of much criticism, in part owing to the inability of current approaches to predict production responses in terms of intake, product composition or nutrient partitioning. Energy is not a nutrient *per se*; thus NE approaches represent an amalgamation of the metabolism of a variety of compounds into a single entity, combustible energy, which is measurable at specific points of loss or gain within the animal. This simplification of complex biological reactions is both a strength, in terms of applicability and versatility, and a weakness, in terms of predictability. Many now believe the newer 'religion' of mathematical modelling of nutrient metabolism will ultimately illuminate the path to predictable response, but in application such models will need to be 'fed', using practical criteria for evaluating individual feeds. Although highly desirable and the justification for untold sums of research funding, the one 'true' system capable of predicting production responses to changes in feed composition is as elusive today as the Grail for the crusaders.

This chapter will consider the use of energy-balance measurements for forage evaluation within the context of current feeding standards and research, as well as the limitations and merits of the approach, with an inevitable bias towards the author's background and previous employment.

Historical Perspectives

The work of Thaer and Einhof led to the development of the first 'official' hay-equivalent feeding system and was the basis of the 'Weende' system of feed analysis. This analytical approach was the basis of a 'total digestible nutrients' (TDN) system of feed evaluation and livestock rationing, whose origins were at the Weende Experimental Station in Germany in 1860 (Flatt *et al.*, 1972). The TDN approach that developed was used for rationing dairy cattle in the USA until it was replaced as the 'official' National Research Council (NRC) system for rationing energy in dairy cattle by the NE for lactation (NE_l) system (NRC, 1971). However, the TDN system is still widely used throughout the world. Indeed, because of the difficulties of obtaining measurements of energy balance in dairy cattle, many of the tabular values of ME and NE_l used in the current NRC publication (NRC, 1989) are based on TDN values, often obtained in sheep or steers.

Even as the digestible-nutrient approach was being developed, it was realized that digestible nutrients were not equal in terms of the production they could support and that losses of nutrients other than those in faeces must also be accounted for. In many instances, these differences were due to the efficiency of utilization of metabolizable energy (ME) for NE gain. NE systems for rationing ruminants are based on measurements of energy metabolism and balance, obtained by measuring

ME, by subtracting energy losses as urine (UE) and combustible gases (GE) from digestible energy (DE) and by measuring either heat energy (HE) loss or energy retained as product (RE). The use of calorimeters for measuring HE and NE balance by a number of individuals in Europe and the USA led to the development of two NE energy systems in the early 1900s: Kellner's starch-equivalent system, used widely in Europe, and Armsby's estimated NE system in the USA, which ultimately had limited use as an adjunct to the TDN system. However, both of these systems were based on a limited database of experimental measurements. The limitations of the TDN and starch-equivalent systems and the paucity of their database eventually led to the development of the current NE systems in use today, which were based on an explosion of animal calorimetry studies beginning in the late 1950s. A driving force behind the establishment of a calorimetry facility for dairy cattle at the US Department of Agriculture's Research Center in Beltsville, Maryland, was the scientific debate over the reasons why TDN from concentrates supports higher milk production than TDN from forages (Huffman *et al.*, 1952). Conjointly with the development of large databases on NE metabolism in farm animals, chemical methods of feed evaluation were also developed and refined in an attempt to explain and predict animal responses to changes in diet composition and quality. At the Beltsville laboratory, these efforts led to the simultaneous development of the NE_l system for rationing dairy cows and the neutral-detergent fibre (NDF) system for analytically describing fibre fractions in feeds (Goering *et al.*, 1970). The NDF approach was a leap forward in the evolution of chemical procedures for forage evaluation (Reid, 1994), but the need for enlightened approaches to assessing forage and feed quality and predicting their effects on animal production are as relevant today as for Thaer.

Methodology

Techniques for measuring energy metabolism have been described and critiqued in numerous reviews and books (Flatt, 1969; Blaxter 1971; McLean and Tobin, 1987). As mentioned previously, there are two general approaches to determining NE. One is to measure or estimate HE and calculate RE by subtracting HE from ME and the other is to measure RE, but a measure of ME will also be required to estimate HE. Since the days of Lavoisser's ice calorimeter (McLean and Tobin, 1987), measurements of HE have been obtained using calorimeters that house the experimental subject. Two approaches have typically been used: direct measurement or indirect estimation of HE. As the name implies, direct calorimeters obtain direct measurements of the HE dissipated by the animal, using a variety of technological approaches, such as measuring the difference in the temperature of water entering or leaving a jacket surrounding the chamber, or thermocouples. Indirect approaches include estimation from respiratory exchange or carbon and nitrogen balance. Carbon and nitrogen balance simultaneously estimates RE, but still requires measurement of carbon dioxide and methane losses.

Owing to the historical complexities of direct calorimeters, especially for use with large ruminants, the majority of the calorimeters in use in the later half of the 20th century have been indirect calorimeters. These can be either closed- or open-circuit, but the problem of removing large quantities of moisture and carbon dioxide in closed-circuit systems has meant that most of the calorimeters used for large ruminants are open-circuit. In open-circuit systems, estimates of respiratory exchange are obtained by measuring the difference in gas concentration between incoming and outgoing air and the flow rate of air through the chamber. These measurements are then corrected for effects of temperature, barometric pressure, humidity and the residual volume of the chamber (Flatt, 1969). HE can then be estimated, based on relationships between nutrient oxidation and heat production, with adjustments for the type of nutrient oxidized, based on carbon dioxide production and urinary nitrogen loss. For ruminants, further adjustments are made for incomplete oxidation of carbohydrates, based on methane losses. The constants for these relationships were recommended by a European Association of Animal Production (EAAP) subcommittee on constants and factors (E. Brouwer, K. Blaxter, K. Nehring and W. Wohlbier) and condensed into a single equation, reported by Brouwer (1965).

Whichever approach is used to determine HE, complete energy-balance measurements also require the measurement of DE and UE, using digestion trials, and GE, using respiratory exchange. Gaseous energy losses are mostly represented by methane, but in some cases hydrogen and ethane may also be lost (Flatt, 1969). Digestion trials and measurements of HE may be conducted simultaneously, if the calorimetry system employed is suitable for separation and collection of faeces and urine, or obtained on separate occasions. If this is the case, then effects of level of intake and feed quality must be considered carefully, as changes in intake can have immediate effects on HE, as well as on DE and ME. This is especially a problem in evaluating forages, where day-to-day variations can be substantial, depending on the source, method of harvesting and conservation and experimental management.

Another approach to estimating NE is to use comparative slaughter techniques. To determine RE over a period of time, the body composition of two groups of animals is determined at the beginning and at the end of a prescribed experimental period. DE is determined separately, with GE often estimated. This has obvious limitations for lactating animals, but indirect (specific gravity) estimates of changes in carcass composition are the basis of energy feeding standards for growing beef cattle in the USA (NRC, 1996). However, numerous comparisons between comparative slaughter and respiration calorimetry have found that estimates of HE are higher and thus NE is lower when measurements are obtained by comparative slaughter (Johnson, 1986; Waldo *et al.*, 1990). There are a number of reasons for this discrepancy. One concern with the use of calorimeters is that they require the subject to be confined in a sealed chamber for extended periods, which can elevate HE by causing stress if proper adaptation is not employed. More importantly in the comparison of respiration and comparative slaughter calorimetry, activity is limited when animals are confined to chambers and thus maintenance costs and HE are reduced. Another concern with digestion trials and balance approaches is that errors of measurement are cumulative; thus losses of energy and nitrogen during faecal, urine and scurf collection tend to cause overestimation of their retention in the animal (Martin, 1966). The confinement of animals in chambers also means that traditional respiration-calorimetry approaches cannot be used to measure NE of animals under field or grazing conditions.

Other techniques can be used to estimate respiratory exchange in subjects that do not require total confinement, but at the expense of the precision which can be achieved with a rigorously operated chamber system. Examples include head chambers or face-masks, which would not estimate total GE and also require restriction of activity. Approaches that have been applied to the study of grazing ruminants include tracheal cannulas or isotope-dilution procedures. Tracheal cannulation techniques for measuring energy exchange of ruminants have a history extending over a 100 years (Flatt, 1969). Attempts to use the techniques for measurement of HE in grazing cattle (Flatt *et al.*, 1958) and sheep (Young and Webster, 1963) have been successful, but concerns about the weight of equipment the animal is required to transport and the effects of the cannula on animal health have limited the use of the technique. As for face-masks, the approach does not measure total GE and carbon dioxide loss.

Isotope-dilution procedures can be used to estimate body carbon dioxide production, using labelled carbon dioxide (Corbett *et al.*, 1971) or doubly labelled water (McLean and Tobin, 1987), or body composition (Andrew *et al.*, 1995; Crooker *et al.*, 1998). The doubly labelled water approach is widely used in studies of energy metabolism in 'free-living' humans and appears an attractive option for studies in grazing ruminants, but the agreement between this approach and established calorimetry procedures has not justified the widespread adoption of the procedure. One concern with the use of labelled water dilution in ruminants is the effect of the large and variable gut and milk pools on the dilution profiles obtained (Crooker *et al.*, 1998).

Practical considerations

The limited application of alternative procedures has meant that the evaluation of the energetic value of forages and their effects on animal energy metabolism have largely been obtained using chamber calorimeters and conserved forages. Measurements for fresh herbage have been restricted primarily to zero grazing, which has often required the freezing of large quantities of fresh material to ensure consistent quality over the course of the experiment. However, like other methods of conservation, freezing has effects on forage quality (Minson, 1990). In addition, the cost and difficulties of maintaining calorimeters for large ruminants has meant that measurements from sheep are often used to provide tabular values of the ME or NE value of forages for cattle. However, digestibility and metabolizability of a variety of feeds is higher in sheep than in cattle, due to comparative differences in rumen function (McDonald *et al.*, 1995). Similarly, effects of level of intake are an important consideration, especially when using tabular values from maintenance-fed sheep for rationing lactating dairy cows (Tyrrell and Moe, 1975).

Measurements of HE are not required to estimate ME, and some NE systems assign energy values to feeds in terms of ME and then use constants for the efficiency of ME use for a productive function in rationing the animal to arrive at NE (Alderman and Cottrill, 1995). Thus, in evaluating forages, an ME value is used as the term of reference. Calorimeters are expensive and labour-intensive to operate, but, if calorimeters are not available, then a digestion trial may be conducted and methane losses estimated. The equations derived by Blaxter and Clapperton (1965), based on DE and intake level, are often used for this purpose. However, for dairy cows, a summarization of data by Moe and Tyrrell (1979) found that GE losses were best predicted from amounts of digested soluble residue (the more soluble and readily digested carbohydrates), hemicellulose and cellulose. In a more recent analysis of data from cattle obtained at Beltsville, the equation of Moe and Tyrrell (1979) predicted methane output more closely than any others available (Wilkerson *et al.*, 1995). However, this agreement may reflect the fact that both data sets were obtained in the same laboratory, using similar types of feeds. Another concern is the effect of feeding fats on methane production, which may not be adequately addressed in the prediction equations currently available. There are numerous other approaches for the prediction of ME based on laboratory analyses, which are addressed in other reviews (e.g. Barber *et al.*, 1989; Alderman and Cottrill, 1995). A major concern for the evaluation of conserved forages is the ability to determine volatile components of ensiled material at feeding. Whether fed fresh or conserved, in practice forage quality varies considerably from harvest to harvest, or even day to day; hence the need for rapid, accurate and economical lab-based methods of assessing forage energy value.

In contrast to most concentrates, the energetic value of many forages can be determined by feeding the test forage as the sole diet component. However, this is often not the case in studies with higher-yielding dairy cows. In the case of comparisons or determinations for concentrates (e.g. Andrew *et al.*, 1991) or for forages fed to lactating dairy cows (e.g. Casper *et al.*, 1993), substitution trials are often conducted (Armsby and Fries, 1918; Tyrrell and Moe, 1975). A variety of approaches can be used, but often, in measuring ME or NE values for a specific diet component, all other components of the diet are kept as constant as possible whilst only changing the test component. This can be extremely difficult to achieve if attempts are made to equalize parameters, such as crude or rumen-degradable protein, ME, minerals or fibrous components, such as NDF or acid-detergent fibre, but it represents a creative challenge for the researcher. The associative effects of combining feeds on their digestibility and metabolizability are due to a multitude of factors, such as energy and nitrogen interactions within the rumen, pH, intake and other effects on microbial dynamics. However, these effects must be considered in interpreting substitution trials and applying tabular energy values for individual feeds in diet formulations (Tyrrell and Moe, 1975; Moe, 1981).

Digestibility

Of the losses of dietary energy measured in determining energy balance, the most variable is faecal energy loss; thus DE accounts for the largest proportion of the variation in NE between diets and feeding criteria. In summarizing the results from 543 energy balance trials, Moe *et al.* (1972) found that DE accounted for 86% of the variation in NE measured. Digestibility of forages is determined largely by structural factors, such as the degree of lignification, but is also influenced by factors such as physical processing, level of intake, diet protein concentration and other associative effects within mixed diets (McDonald *et al.*, 1995). All of the factors that determine digestibility, then, have a major impact on the ME the ruminant can derive from a specific feed.

A particular problem in formulating rations for dairy cows is the effect of intake on diet digestibility. A number of studies have shown that increasing intake reduces dry matter (DM) digestibility, but the response varies with the type of diet fed. For example, in heifers fed pelleted diets containing either 75% lucerne hay and 25% concentrates or the inverse proportions, increasing intake from just above maintenance to nearly twice maintenance levels numerically depressed the digestibility of the 75% concentrate diet to a greater extent than the 75% lucerne diet (Reynolds *et al.*, 1991). By increasing rate of particle passage from the rumen, pelleting tends both to lower overall digestibility and to exacerbate effects of intake level on digestibility. However, similar responses have been observed in lactating dairy cows fed lucerne hay-based diets differing in concentrate level (Tyrrell and Moe, 1975). These observations have suggested that increasing the level of cereals in the diet exacerbates depressions in digestibility with increasing intake, but the response is not consistent across all studies (Tyrrell and Moe, 1975). Of the diet components measured, cell-wall components are often most affected. Many of the associative effects of grain feeding on cellulose digestion are known to be associated with, if not mediated by, depressions in rumen pH (Mould *et al.*, 1983). If feeding grain and subsequent effects on rumen acid load reduce cellulolytic activity of the rumen, then the increased rate of passage with increased intake may have a greater impact on cell-wall digestion than in the absence of starch in the diet. Increasing intake reduces methane production to a greater extent as the proportion of concentrate in the diet is increased (Blaxter and Clapperton, 1965), which reflects shifts in the microbial population of the rumen.

One consideration is that studies specifically designed to measure effects of intake level on digestibility often compare the relative digestion of diets in lactating dairy cows at maximal intake and non-lactating animals at maintenance (Tyrrell and Moe, 1975). These comparisons are appropriate, considering that many tabular values for DE, TDN or ME were obtained in steers or sheep, but they tend to magnify the problem relative to changes that may occur across a range of intakes within a group of lactating cows. For example, Bines *et al.* (1988) found that the digestibility of grass hay-based diets was lower in lactating cows than in immature, dry cows, but that, within lactating cows, moderate variations in level of intake had no effect on diet digestibility. This was especially true as level of concentrate in the diet increased. However, elevation of intake did depress digestibility in the non-lactating cows. This suggests either that the response to intake level is curvilinear or that physiological factors other than level of intake *per se* may play a role in the lower digestibility of diets fed to lactating cows compared with dry cows.

It has long been known that there is a positive relationship between dietary crude protein (CP) concentration, DM digestibility and ultimately intake level in ruminants (Schneider and Flatt, 1975). Averaged across a number of studies, the response is roughly 0.01 units of digestibility (1%) for each unit change in diet CP% up to 16%, or higher (Oldham and Smith, 1980). The response appeared to be similar in US and UK studies, although it was suggested that the response might be lower for diets with higher digestibility and CP content. This was more often the case in the UK studies surveyed, where diets tended to have a higher digestibility than US diets because of the use of high-digestibility grass silages (Oldham and Smith, 1980). Increases in DM digestibility with increasing diet CP content can be attributed to the innately high digestibility of protein, or positive effects on microbial fermentation and digestion in the rumen (Tyrrell, 1980). Intake responses may be a result of both increases in digestibility and metabolic effects of improved amino acid supply (Oldham and Smith, 1980). Abomasal infusions of protein are as (or more) effective as dietary protein supplementation in improving grass silage intake (Chamberlain *et al.*, 1989); however, this response might be attributed to increased transfer of urea nitrogen to the rumen. In lactating dairy cows fed maize silage-based diets, feeding urea restored DM digestibility as effectively as an isonitrogenous amount of soybean meal, but only increased DM intake 1.3 kg day^{-1}, compared with an increase of 5.9 kg day^{-1} when soybean meal was fed (Tyrrell, 1980). It is known that urea is less effective than protein as a nitrogen supplement in dairy rations (Clark and Davis, 1980). This suggests either: (i) a protein, peptide or amino acid effect on microbial growth in the rumen; (ii) a metabolic effect of amino acids absorbed from rumen-undegraded feed protein in the small intestine; or (iii) negative effects of excess ammonia in the rumen or tissues such as the liver (Clark and Davis, 1980; Reynolds, 1992). For US dairy rations, the effect of diet CP level on milk yield is greatest when CP content is below 14% but continues to have a diminished effect as CP content increases to as high as 20% (Clark and Davis, 1980). In the UK, the response of DM intake to increasing diet CP content was on average 0.34 kg unit^{-1} increase in CP% across all diets surveyed (Chamberlain *et al.*, 1989).

Retention of DE as ME

Unquestionably, digestibility accounts for the largest variation in ME or NE value of forages. However, forage type and quality can also influence the efficiency of DE use for ME, as well as the use of ME for RE. Methane losses are really digestive losses but are not accounted for when apparent digestion is measured solely by faecal output. Factors affecting methane output are related primarily to the availability of digestible carbohydrate fractions, rumen turnover and microbial population dynamics (Moe and Tyrrell, 1979). In addition to methane, UE losses are the other determinant, albeit a small one, when calculating the amount of DE available as ME in measurements of energy balance. UE losses are determined primarily by urea, which accounts for more than 70% of urine nitrogen (N) (Blaxter and Martin, 1962). However, other metabolites, such as hippurate, also contribute to the energy content of urine, such that the relationship between N and energy content varies with the diet fed (Blaxter *et al.*, 1966). Regardless, factors that increase liver urea production, such as excess rumen-degradable nitrogen, increase UE losses and have a small effect on the ratio of ME/DE. The ratio of ME/DE is normally quite high, especially for diets fed to lactating dairy cows (87%; Flatt *et al.*, 1969b), where GE losses as a proportion of intake energy are lower than for non-lactating animals fed poorer-quality diets at lower intakes (Blaxter and Clapperton, 1965). However, these differences are small, with GE and UE losses combined normally accounting for less than 15–20% of DE.

Efficiency of ME Use for RE

Forages compared with cereals

The lower efficiency of ME retention for forages compared with cereals was clearly demonstrated by Kellner in the development of the starch-equivalent NE feeding standards over 100 years ago (Armsby, 1903; Table 5.1). Oxen fed a supplement of straw or meadow hay had a higher HE increment than when fed an equivalent amount of ME from starch. This was related to the low digestibility of the forages, as alkali treatment of straw dramatically reduced the HE increment when it was fed (O. Kellner as summarized by Armsby, 1903). However, studies by Armsby and Fries (1918) in steer 'J' found a similar NE value for maize (corn) starch and lucerne (alfalfa) hay, which was lower than the value for starch reported by Kellner. This discrepency may have been the result of differences in the basal intakes used in these studies, as the efficiency of utilization of ME for maintenance is greater than the effciency of ME utilization for growth (Blaxter and Graham, 1955). In addition, there were differences in the correction factors used in calculating the results.

These and other conflicting results, often from studies using very limited numbers of observations (e.g. one animal), added confusion to the debate over the reasons for differences between the use of TDN from lucerne compared with maize for milk production. The debate was also clouded by the recent discoveries of a number of vitamins as essential nutrients (Huffman *et al.*, 1952). As mentioned, this debate in part led to the establishment of the Beltsville calorimeters and the conduct of a series of trials comparing the NE value of diets varying in lucerne hay and concentrate (maize, soybean and bone meals) proportion fed to lactating dairy cows. Initial work compared isonitrogenous diets containing 100, 75 or 50% estimated NE from lucerne hay (100, 84 and 63% lucerne hay on a DM basis; Coppock *et al.*, 1964a). They observed clear differences in the ratio of HE/ME across the three diets. When the ME required for maintenance was assumed (131 kcal $kg^{0.75-1}$) and tissue energy losses or gains were adjusted to zero, using efficiency constants, the efficiency of utilization of ME for RE was 54, 61 and 65% for the 100, 84 and 63% hay diets, respectively. Later studies compared similar diets containing 60, 40 or 20% lucerne hay (Flatt *et al.*, 1969) in trials which included the legendary cow Lorna (who achieved a milk yield of 49 kg day^{-1} whilst in the respiration chambers). At higher levels of concentrate inclusion (60% or more), there

Table 5.1. Tissue energy retention (RE) from supplemental ME in oxen fed fodder (from Kellner and Kohler, 1900, as cited by Armsby, 1903).

Supplement	RE/ME
Wheat straw	17.4
Meadow hay	41.4
Starch	61.5
'Extracted' straw	63.0

was no effect of level of forage in the diet on the efficiency of ME use for RE, but clear effects on the partitioning of ME between milk and TE. Thus, in lactating cows, the inclusion of relatively small amounts of concentrate improved the efficiency of ME use for NE_l, but there was little effect of further dilution of the forage component of the diet above minimal levels of concentrate normally fed to dairy cows in practice. For this reason, the effects of the diet forage to concentrate ratio on the efficiency of ME use for NE_l are acknowledged, but are considered too minor to merit inclusion in the NE_l rationing approach (NRC, 1989). As the efficiency of ME use for tissue and milk energy synthesis is similar in lactating dairy cows, the NE_l standards do not attempt to distinguish between the two processes (Moe *et al.*, 1972).

Forage type

In addition, type of forage can also affect the amount of ME lost as HE, and conversely NE (RE) gain. An often-used model in this regard is the comparison of grasses and legumes, where ME from grass is typically used less efficiently than ME from legumes. In growing cattle, the use of ME for RE was lower for cocksfoot grass (orchard-grass) silage than for lucerne silage (Varga *et al.*, 1990; Waldo *et al.*, 1990). Similarly, dairy cows fed diets based on cocksfoot grass silage consumed less DM, had higher HE/ME ratios, produced less milk and had lower TE gains than when fed diets based on lucerne silage (Casper *et al.*, 1993). More recently, a substantial body of data from lactating dairy cows have suggested that feeding grass silage as the sole forage source causes a large elevation of maintenance requirement compared with published values (Yan *et al.*, 1997).

Hypotheses

Solving the riddle of the basis for these differences in the efficiency of ME utilization between forages and concentrates and types of forages has been the 'Rubik's cube' of ruminant nutrition, practically since the discovery that ruminants derive their ME primarily from absorbed volatile fatty acids (VFA). Calorimetric studies suggested that the efficiency of utilization of ME from acetate was lower than for propionate or butyrate (Armstrong and Blaxter, 1957a, b), suggesting that, as feeding forages tends to increase ruminal acetate concentrations, the ratio of absorbed acetate : propionate was responsible for the increase in HE with high-forage compared with high-concentrate diets. It was later suggested that reductions in the efficiency of ME use would be expected when molar acetate proportions in the rumen exceeded 70% of total VFA (Blaxter, 1962). Measurements of ruminal VFA concentration in the studies at Beltsville supported this concept, with a linear negative relationship observed between energetic efficiency and ruminal acetate concentration (Coppock *et al.*, 1964b), but numerous VFA infusion and feeding studies have not (e.g. Ørskov and Allen, 1966; Ørskov *et al.*, 1969). One explanation for the disparity in the results obtained, first proposed by Armstrong (1965), is that the utilization of acetate for fat synthesis requires an adequate supply of reducing power in the form of nicotinamide adenine dinucleotide phosphate (NADPH). The needed reducing equivalents will be more available if ample supplies of absorbed amino acids or glucose are present (Annison and Bryden, 1999). This concept was supported by the finding that the HE increment for acetate infused into the rumen was much lower in dry cows fed concentrate than in those fed lucerne hay alone (Tyrrell *et al.*, 1979).

In addition to variations in the relative proportions of acetate and glucose metabolized, other factors may contribute to differences in the efficiency of ME utilization. In the studies at Beltsville, growing steers fed grass silage had greater gut fill than steers fed lucerne silage (Waldo *et al.*, 1990). In addition, relative to increments in DM intake steers had a much greater increment in oxygen consumption by the portal-drained viscera (PDV) with increasing intake of grass silage compared with the increment for lucerne silage (Huntington *et al.*, 1988). This difference in incremental oxygen use by the PDV accounted for all the difference in incremental body oxygen consumption between the two forages.

In a comparison of high-forage and high-concentrate diets fed at two equalized ME intakes, heifers fed a 75% lucerne diet had greater PDV blood flow and oxygen consumption at equal ME to the 75% concentrate diet (Reynolds *et al.*, 1991). This increase in PDV oxygen consumption accounted for 66% of the difference in body oxygen use between the two diets at an intake near maintenance, and 84% of the difference at an intake near twice maintenance requirements for energy. These studies suggest that a large portion of the difference in ME loss as HE between forages and concentrates and between grasses and legumes is a consequence of increased oxidative metabolism by the PDV. This may be a result of changes in gut fill, the extent of rumination and the work of digestion. In addition, gut mass may be increased, especially the rumen, due to trophic effects of lipogenic VFA, physical abrasion of the epithelium or the work of digestion. Alterations in the profile of metabolites used may alter PDV oxygen consumption as well. In this regard, the PDV is a principal user of acetate, accounting for 50% of total acetate use in sheep fed lucerne at maintenance (Bergman and Wolf, 1971) and as much as 25% of body fat. In the study of Reynolds *et al.* (1991), feeding the high-forage diet increased the ratio of acetate to glucose released by splanchnic tissues (PDV plus liver) twofold compared with the high-concentrate diet (Reynolds *et al.*, 1993).

Excess dietary protein

Many immature, heavily fertilized forages and legumes can be high in rumen-degradable protein. Another consideration for the efficiency of utilization of ME from forages is the negative effect of excess nitrogen on HE and amino acid availability. In a summarization of early calorimetry studies at Beltsville, Tyrrell *et al.* (1970) found that excess protein consumption increased HE and decreased RE in lactating dairy cows. This reduction was equivalent to 30 MJ ME for each g of nitrogen consumed in excess of requirements, and this adjustment was included in the formula used for the calculation of NE_l (Moe *et al.*, 1972). Alternatively, feeding lactating cows protein at a level below requirement reduced RE through effects on DM digestion, as opposed to metabolic effects (Moe and Tyrrell, 1972).

The effect of excess nitrogen intake on HE has been attributed to the energy cost of urea synthesis and excretion, which Martin and Blaxter (1965) estimated to be 3.8 kcal g^{-1} ammonia nitrogen in sheep. This was greater than the theoretical cost of 3.2 kcal, which they attributed to the recycling of urea to the gut and the relatively small cost of renal excretion. However, feeding growing heifers 75% lucerne diets increased the amount of digestible nitrogen compared with a 75% concentrate diet providing equal ME, which markedly increased net PDV absorption of ammonia and liver urea release, but had no effect on liver oxygen consumption (Reynolds *et al.*, 1991). In other work, feeding urea in lucerne hay-based diets caused large (60%) increases in net PDV absorption of ammonia and liver urea release, with little change in the absorption of other nutrients, but had absolutely no effect on liver oxygen consumption (Maltby *et al.*, 1993). These observations may be due, in part, to the fact that, although urea synthesis is described as a cyclic process, it is not isolated from other metabolic processes in the liver and the flow of metabolites through other metabolic pathways will be reduced when urea synthesis is increased. In addition, the cost of urea synthesis in terms of adenosine triphosphate (ATP) is lower on a net basis when the generation of reducing equivalents from fumarate is considered (Reynolds, 1992). The real cost of urea synthesis in terms of ATP and oxygen use is much lower when the cycle is not considered in isolation.

In contrast to the studies just cited, feeding an increased level of soybean meal in a high-concentrate diet increased net PDV absorption of ammonia and liver urea release, as well as liver and body oxygen consumption in growing steers (Reynolds *et al.*, 1992). However, there was also an increase in PDV oxygen use, which was equal to the increase observed for the liver. Together, the PDV and liver accounted for all the increase in body oxygen use, and thus HE, when excess CP was fed. This suggests that there may be specific effects of feeding excess protein, rather than nitrogen *per se*, and that the origin of the increase in HE may not be restricted solely to urea synthesis in the liver. In this study, the increase in liver urea production with increased dietary CP level was associated with an increase in liver removal of methionine, tryptophan and lysine (Reynolds *et al.*, 1995). Effects of increased liver ammonia removal on the concomitant removal of amino acids has been suggested, based on observed changes in α-amino nitrogen (Reynolds, 1992) or amino acid (Parker *et al.*, 1995) metabolism. However, studies in sheep, using isotopic labelling of urea (Lobley *et al.*, 1995, 1996), have not shown conclusive evidence that an increase in ammonia absorption requires an increase in the deamination of amino acids (other than glutamate) to provide the aspartate needed for urea synthesis. If there is an effect of excess ammonia absorption on liver amino acid removal, the response is likely be influenced by the energy and protein status of the animal.

As mentioned previously, abomasal infusions of protein or essential amino acids have improved milk yield and intake in cows fed forages high in rumen-degradable protein (Oldham and Smith, 1980; Chamberlain *et al.*, 1989). However, these responses may reflect effects of excess rumen-degradable protein on the efficiency of microbial protein synthesis in the rumen, rather than effects of excess ammonia absorption on liver removal of absorbed amino acids.

Concluding Remarks

Measurements of energy balance form the basis of most current rationing systems for energy in ruminants and thus are the standard used in judging the effects of forage chemical composition on energy value for production. In this regard, many laboratory procedures for evaluating forage quality are used to predict the digestibility or ME value of forages *in vivo*. Measurements of energy balance are costly and time-consuming to obtain, whilst the resources for the maintenance of a sustained programme of calorimetry are at present limited: hence the need for current, as well as new, innovative, approaches to the evaluation of forage quality and energy value for production.

Current rationing systems for energy and protein fail to adequately predict the response of the animal, both in terms of product composition and nutrient partitioning. However, the systems were never intended to be a replacement for practical wisdom and experience, but a guide and basis for 'rational' decision-making. As discussed, the predictive limitations of current NE systems are, in part, due to the simplification of metabolism to a caloric basis, as well as the simplification of the effects of the animal's metabolism on the recovery of absorbed nutrients in a product to a series of efficiency 'constants'. If the 'nirvana' of a rationing system capable of predicting production response is to be attained, the need for more complex approaches is therefore suggested. This is the logic behind the development of models based on a prediction of the nutrients absorbed from the gut and their subsequent utilization by specific tissues in the animal (AFRC, 1998). If predictive models incorporating the complexities of nutrient metabolism are capable of predicting response, the challenge will be for them to maintain a practical basis at the level of feed-quality evaluation.

References

AFRC (1998) *Response in the Yield of Milk Constituents to the Intake of Nutrients by Dairy Cows*. CAB International, Wallingford, Oxon, UK.

Alderman, G., and Cottrill, B.R. (1995) *Energy and Protein Requirements of Ruminants*. CAB International, Wallingford, Oxon, UK.

Andrew, S.M., Tyrrell, H.F., Reynolds, C.K. and Erdman, R.A. (1991) Net energy value for lactation of a dietary fat, calcium salts of long-chain fatty acids, for cows fed silage-based diets. *Journal of Dairy Science* 74, 2588–2600.

Andrew, S.M., Erdman, R.A. and Waldo, D.R. (1995) Prediction of body composition of dairy cows at three physiological stages from deuterium oxide and urea dilution. *Journal of Dairy Science* 78, 1083–1095.

Annison, E.F. and Bryden, W.L. (1999) Perspectives on ruminant nutrition and metabolism II. Metabolism in ruminant tissues. *Nutrition Research Reviews* 12, 147–177.

Armsby, H.P. (1903) *The Principles of Animal Nutrition*. John Wiley & Sons, New York.

Armsby, H.P. (1917) *The Nutrition of Farm Animals*. Macmillan, New York.

Armsby, H.P. and Fries, J.A. (1918) Net energy values of alfalfa hay and of starch. *Journal of Agricultural Research* 15, 269–286.

Armstrong, D.G. (1965) Carbohydrate metabolism in ruminants and energy supply. In: Douherty, R.W., Allen, R.S., Burroughs, W., Jacobson, N.L. and McGilliard, A.D. (eds) *Physiology of Digestion in the Ruminant*. Butterworths, Washington, DC, pp. 272–288.

Armstrong, D.G. and Blaxter, K.L. (1957a) The heat increment of steam-volatile fatty acids in fasting sheep. *British Journal of Nutrition* 11, 247–271.

Armstrong, D.G. and Blaxter, K.L. (1957b) The heat increment of various mixtures of steam-volatile fatty acids in fasting sheep. *British Journal of Nutrition* 11, 392–408.

Barber, G.D., Offer, N.W. and Givens, D.I. (1989) Predicting the nutritive value of silage. In: Haresign, W. and Cole, D.J.A. (eds) *Recent Advances in Animal Nutrition*. Butterworths, London, pp. 141–158.

Bergman, E.N., and Wolff, J.E. (1971) Metabolism of volatile fatty acids by liver and portal-drained viscera in sheep. *American Journal of Physiology* 221, 586–592.

Bines, J.A., Broster, W.H., Sutton, J.D., Broster, V.J., Napper, D.J., Smith, T. and Siviter, J.W. (1988) Effect of amount consumed and diet composition on the apparent digestibility of feed in cattle and sheep. *Journal of Agricultural Science, Cambridge* 110, 249–259.

Blaxter, K.L. (1962) *The Energy Metabolism of Ruminants*. Hutchison, London.

Blaxter, K.L. (1971) Methods of measuring the energy metabolism of animals and interpretation of results obtained. *Federation Proceedings* 30 (4), 1436–1443.

Blaxter, K.L. and Clapperton, J.L. (1965) Prediction of the amount of methane produced by ruminants. *British Journal of Nutrition* 19, 511–522.

Blaxter, K.L. and Graham, McC.N. (1955) Methods of assessing the energy values of foods for ruminant animals. *Proceedings of the Nutrition Society* 14, 131–139.

Blaxter, K.L. and Martin, A.K. (1962) The utilization of protein as a source of energy in fattening sheep. *British Journal of Nutrition* 16, 397–407.

Blaxter, K.L., Clapperton, J.L. and Martin, A.K. (1966) The heat of combustion of urine of sheep and cattle in relation to its chemical composition and to diet. *British Journal of Nutrition* 20, 449–460.

Brouwer, E. (1965) Report of the sub-committee on constants and factors. In: Blaxter, K.L. (ed.) *Energy Metabolism of Farm Animals*. EAAP Publication No. 11, Academic Press, London, pp. 441–443.

Casper, D.P., Glenn, B.P. and Reynolds, C.K. (1993) Energy metabolism of lactating dairy cows fed two formaldehyde- and formic acid-treated forages with two nonstructural carbohydrate sources. *Journal of Dairy Science* 76 (Suppl. 1), 308.

Chamberlain, D.G., Martin, P.A. and Robertson, S. (1989) Optimizing compound feed use in dairy cows with high intakes of silage. In: Haresign, W. and Cole, D.J.A. (eds) *Recent Advances in Animal Nutrition*. Butterworths, London, pp. 175–194.

Clark, J.H. and Davis, C.L. (1980) Some aspects of feeding high producing dairy cows. *Journal of Dairy Science* 63, 873–885.

Clark, P.W. and Armentano, L.E. (1997) Replacement of alfalfa neutral detergent fiber with a combination of nonforage fiber sources. *Journal of Dairy Science* 80, 675–680.

Coppock, C.E., Flatt, W.P., Moore, L.A. and Stewart, W.E. (1964a) Effect of hay to grain ratio on utilization of metabolizable energy for milk production by dairy cows. *Journal of Dairy Science* 47, 1330–1338.

Coppock, C.E., Flatt, W.P., Moore, L.A. and Stewart, W.E. (1964b) Relationships between end products of rumen fermentation and utilization of metabolizable energy for milk production. *Journal of Dairy Science* 47, 1359–1364.

Corbett, J.L., Farrell, D.J., Leng, R.A., McClymont, G.L. and Young, B.A. (1971) Determination of the energy expenditure of penned and grazing sheep from estimates of carbon dioxide entry rate. *British Journal of Nutrition* 26, 277–291.

Crooker, B.A., Weber, W.J. and Andrew, S.M. (1998) Development and use of deuterium oxide dilution equations to predict body composition of Holstein cows. In: McCracken, K.J., Unsworth, E.F. and Wylie, A.R.G. (eds) *Energy Metabolism of Farm Animals*. CAB International, Wallingford, Oxon, UK, pp. 177–180.

Flatt, W.P. (1969) Methods of calorimetry. (B) Indirect. In: Cuthbertson, D. (ed.) *The International Encyclopedia of Food and Nutrition*. Vol. 17: *Nutrition of Animals of Agricultural Importance*, Part 2. Pergamon Press, Oxford, pp. 1–30.

Flatt, W.P., Waldo, D.R., Sykes, J.F. and Moore, L.A. (1958) A proposed method of indirect calorimetry for energy metabolism studies with large animals under field conditions. In: Thorbeck, G. and Aersoe, H. (eds) *Energy Metabolism of Farm Animals*. EAAP Publication No. 8, Statens Husdyrugsudvalg, Copenhagen, pp. 101–109.

Flatt, W.P., Moe, P.W., Moore, L.A. and Van Soest, P.J. (1969a) Estimation and prediction of the energy value of feeds for ruminants. In: Blaxter, K.L., Kielanowski, J. and Thorbeck, G. (eds) *Energy Metabolism of Farm Animals*. Oriel Press, Newcastle upon Tyne, UK, pp. 59–65.

Flatt, W.P., Moe, P.W., Munson, A.W. and Cooper, T. (1969b) Energy utilization by high producing dairy cows II. Summary of energy balance experiments with lactating Holstein cows. In: Blaxter, K.L., Kielanowski, J. and Thorbeck, G. (eds) *Energy Metabolism of Farm Animals*. Oriel Press, Newcastle upon Tyne, UK, pp. 235–249.

Flatt, W.P., Moe, P.W., Moore, L.A., Breirem, H.C.K. and Ekern, A. (1972) Energy requirements in lactation. In: *Handbuch der Tierernahrung*, Vol. II. Verlag Paul Parey, Hamburg and Berlin, Germany, pp. 341–392.

Goering, H.K, Van Soest, P.J. and Hemken, R.W. (1970) *Forage Fibre Analysis (Apparatus, Reagents, Procedures and Some Applications)*. Agriculture Handbook No. 379, Agricultural Research Service, USDA, Washington, DC, USA.

Huffman, C.F., Duncan, C.W. and Chance, C.M. (1952) Unidentified dietary factors in dairy cattle nutrition II. Further evidence of an unidentified factor(s) in grain needed to balance roughage for milk production. *Journal of Dairy Science* 34, 41.

Huntington, G.B., Varga, G.A., Glenn, B.P. and Waldo, D.R. (1988) Net absorption and oxygen consumption by Holstein steers fed alfalfa or orchardgrass silage at two equalized intakes. *Journal of Animal Science* 66, 1292.

Johnson, D.E. (1986) Fundamentals of whole animal calorimetry: use in monitoring body tissue deposition. *Journal of Animal Science* 63 (Suppl. 2), 111–114.

Kellner, O. (1926) *The Scientific Feeding of Animals*. Duckworth, London.

Lobley, G.E., Connell, A., Lomax, M.A., Brown, D.S., Milne, E., Calder, A.G. and Farringham, D.A.H. (1995) Hepatic detoxification of ammonia in the ovine liver: possible consequences for amino acid catabolism. *British Journal of Nutrition* 75, 217–235.

Lobley, G.E., Weijs, P.J.M., Connell, A., Calder, A.G., Brown, D.S. and Milne, E. (1996) The fate of absorbed and exogenous ammonia as influenced by forage or forage–concentrate diets in growing sheep. *British Journal of Nutrition* 76, 231 – 248.

McDonald, P., Edwards, R.A., Greenhalgh, J.F.D. and Morgan, C.A. (1995) *Animal Nutrition*, 5th edn. Longman Scientific and Technical, Harlow, Essex, UK.

McLean, J.A. and Tobin, G. (1987) *Animal and Human Calorimetry*. Cambridge University Press, Cambridge, UK.

Maltby, S.A., Reynolds, C.K., Lomax, M.A. and Beever, D.E. (1993) The effect of increased absorption of ammonia and arginine on splanchnic metabolism of beef steers. *Animal Production* 56, 462.

Martin, A.K. (1966) Some errors in the determination of nitrogen retention of sheep by nitrogen balance studies. *British Journal of Nutrition* 20, 325–337.

Martin, A.K. and Blaxter, K.L. (1965) The energy cost of urea synthesis in sheep. In: Blaxter, K.L. (ed.) *Energy Metabolism of Farm Animals*. EAAP Publication No. 11, Academic Press, London, pp. 83–91.

Minson, D.J. (1990) *Forage in Ruminant Nutrition*. Academic Press, London.

Moe, P.W. (1981) Energy metabolism of dairy cattle. *Journal of Dairy Science* 64(6), 1120–1139.

Moe, P.W. and Tyrrell, H.F. (1972) Net energy value for lactation of high- and low-protein diets containing corn silage. *Journal of Dairy Science* 55, 318–324.

Moe, P.W. and Tyrrell, H.F. (1979) Methane production in dairy cows. *Journal of Dairy Science* 62, 1583–1586.

Moe, P.W., Flatt, W.P. and Tyrrell, H.F. (1972) Net energy value of feeds for lactation. *Journal of Dairy Science* 55, 945–958.

Mould, F.L., Ørskov, E.R. and Mann, S.O. (1983) Associative effects of mixed feeds. 1. Effects of type and level of supplementation and the influence of the rumen fluid pH on cellulolysis *in vivo* and dry matter digestion of various roughages. *Animal Feed Science and Technology* 10, 15–30.

NRC (1971) *Nutrient Requirements of Dairy Cattle*, 4th revised edn. National Academy Press, Washington, DC.

NRC (1989) *Nutrient Requirements of Dairy Cattle*, 6th edn update. National Academy Press, Washington, DC.

NRC (1996) *Nutrient Requirements of Beef Cattle*, 7th edn. National Academy Press, Washington, DC.

Oldham, J.D. and Smith, T. (1980) Protein–energy interrelationships for growing and for lactating cattle. In: Miller, E.L., Pike, I.H. and Van Es, A.J.H. (eds) *Protein Contribution of Feedstuffs for Ruminants: Application to Feed Formulation*. Butterworth Scientific, London.

Ørskov, E.R. and Allen, D.M. (1966) Utilization of salts of volatile fatty acids by growing sheep. 1. Acetate, propionate and butyrate as sources of energy for young growing lambs. *British Journal of Nutrition* 20, 295–306.

Ørskov, E.R., Flatt, W.P., Moe, P.W. and Munson, A.W. (1969) The influence of ruminal infusion of volatile fatty acids on milk yield and composition and on energy utilization by lactating cows. *British Journal of Nutrition* 23, 443–453.

Parker, D.S., Lomax, M.A., Seal, C.J. and Wilton, J.C. (1995) Metabolic implications of ammonia production in the ruminant. *Proceedings of the Nutrition Society* 54, 549–563.

Reid, R.L. (1994) Milestones in forage research. In: Fahey, G.C. (ed.) *Forage Quality, Evaluation, and Utilization*. American Society of Agronomy, Crop Science Society of America, Soil Science Society of America, Madison, Wisconsin, USA, pp. 1–58.

Reynolds, C.K. (1992) Nitrogen metabolism by ruminant liver. *Journal of Nutrition* 122, 850–854.

Reynolds, C.K., Tyrrell, H.F. and Reynolds, P.J. (1991) Effects of forage-to-concentrate ratio and intake on energy metabolism in growing beef heifers: whole body energy and nitrogen balance and visceral heat production. *Journal of Nutrition* 121, 994–1003.

Reynolds, C.K., Casper, D.P., Harmon, D.L. and Milton, C.T. (1992) Effect of CP and ME intake on visceral nutrient metabolism in beef steers. *Journal of Animal Science* 70 (Suppl. 1), 315.

Reynolds, C.K., Tyrrell, H.F. and Reynolds, P.J. (1993) Effect of diet forage-to-concentrate ratio and intake on net visceral metabolism of VFA in growing beef heifers. *Journal of Dairy Science* 76 (Suppl. 1), 283.

Reynolds, C.K., Harmon, D.L., Prior, R.L., Casper, D.P. and Milton, C.T. (1995) Splanchnic metabolism of amino acids in beef steers fed diets differing in CP content at two ME intakes. *Journal of Animal Science* 73 (Suppl. 1), 270.

Schneider, B.H. and Flatt, W.H. (1975) *The Evaluation of Feeds Through Digestibility Experiments*. University of Georgia Press, Athens, Georgia, USA.

Tyrrell, H.F. (1980) Limits to milk production efficiency by the dairy cow. *Journal of Animal Science* 51, 1441–1447.

Tyrrell, H.F. and Moe, P.W. (1975) Effect of intake on digestive efficiency. *Journal of Dairy Science* 58, 1151–1163.

Tyrrell, H.F., Moe, P.W. and Flatt, W.P. (1970) Influence of excess protein intake on energy metabolism of the dairy cow. In: Schurch, A. and Wenks, C. (eds) *Energy Metabolism of Farm Animals*. EAAP Publication No. 13, Juris Verlag, Zurich, Switzerland, pp. 69–72.

Tyrrell, H.F., Reynolds, P.J. and Moe, P.W. (1979) Effect of diet on partial efficiency of acetate use for body tissue synthesis by mature cattle. *Journal of Animal Science* 48, 598–606.

Varga, G.A., Tyrrell, H.F., Huntington, G.B., Waldo, D.R. and Glenn, B.P. (1990) Utilization of nitrogen and energy by Holstein steers fed formaldehyde- and formic acid–treated alfalfa or orchardgrass silage at two intakes. *Journal of Animal Science* 68, 3780–3791.

Waldo, D.R., Varga, G.A., Huntington, G.B., Glenn, B.P. and Tyrrell, H.F. (1990) Energy components of growth in Holstein steers fed formaldehyde- and formic acid–treated alfalfa or orchardgrass silages at equalized intakes of dry matters. *Journal of Animal Science* 68, 3792–3804.

Wilkerson, V.A., Casper, D.P. and Mertens, D.R. (1995) The prediction of methane production of Holstein cows by several equations. *Journal of Dairy Science* 78, 2402–2414.

Yan, T., Gordon, F.J., Agnew, R.E., Porter, M.G. and Patterson, D.C. (1997) The metabolisable energy requirement for maintenance and the efficiency of utilisation of metabolisable energy for lactation by dairy cows offered grass silage-based diets. *Livestock Production Science* 51, 141–150.

Young, B.A. and Webster, M.E.D. (1963) A technique for the estimation of energy expenditure in sheep. *Australian Journal of Agricultural Research* 14, 867–873.

6 The Measurement of Forage Digestibility *In Vivo*

C. Rymer

ADAS Nutritional Sciences Research Unit, Alcester Road, Stratford-upon-Avon CV37 9RQ, UK

History

The first digestibility trials were conducted at the Weende Experimental Station at the University of Goettingen in Germany, and began before 1860 (Schneider and Flatt, 1975). Since then, a vast number of digestibility trials have been conducted throughout the world and reported in a range of scientific journals. A detailed account of the procedures involved in measuring digestibility *in vivo* was given by Schneider and Flatt (1975), while the special considerations involved in measuring forage digestibility were described by Minson (1981, 1990). The measurement of straw digestibility was described by Cottyn *et al.* (1989) and Givens *et al.* (1989). An overview of the subject was provided by McDonald *et al.* (1995).

Calculation of Digestibility

The proportion of a feed that is not excreted in the faeces is assumed to have been absorbed by the animal, and this is defined as the apparent digestibility of the feed. It is not the true digestibility, as in ruminant animals the methane that arises from the fermentation of carbohydrates is lost as eructation and is not absorbed (McDonald *et al.*, 1995). Also, in addition to undigested feed residues, the faeces contain enzymes and other substances secreted into the gut, which are not reabsorbed. Parts of the gut lining, which slough off as the feed passes through the gut, are also excreted in the faeces (McDonald *et al.*, 1995). The amount of this metabolic material that is unavoidably lost is directly proportional to dry matter (DM) intake, regardless of forage type (Minson, 1990). The amount of metabolic material secreted in the faeces has been observed to range from 0.098 to 0.129 g g^{-1} DM matter intake (Minson, 1990). Determining the true digestibility rather than apparent digestibility may be scientifically more accurate. However, no advantage has been demonstrated from the use of true digestibility in forage evaluation

(Minson, 1990), as the loss of the metabolic secretions is a consequence of feeding the forage and therefore should be accounted for in the assessment of the forage's value to the animal.

A variety of means of estimating the digestibility of feeds have been developed, which are summarized here.

Direct method

To estimate the dry-matter digestibility (DMD) of a diet involves the complete collection over a period of time of all the faeces excreted by an animal once it has been adapted to a diet. The amount of feed dry matter consumed by the animal during this time is also recorded. This is the actual amount of dry matter consumed, and so is the difference between the amount of dry matter offered and the amount of dry matter rejected by the animal. The DMD is then calculated from the equation:

$$\text{DMD} = \frac{\text{DM intake} - \text{Faecal DM excreted}}{\text{DM intake}}$$

The digestibility of other feed fractions may be determined by substituting them for dry matter in the above equation. For forages, the proportion of digestible organic matter (OM) in the dry matter (DOMD or D value) is often calculated, as this gives a measure of the available energy in the forage dry matter. DOMD is calculated using the equation:

$$\text{DOMD} = \frac{\text{OM intake} - \text{Faecal OM excreted}}{\text{DM intake}}$$

Difference method

If a forage is being fed alone, then its digestibility may be determined using the direct method described above. However, it is necessary in many situations to supplement the forage with another feed, as described later ('Use of feed supplements'). This complicates the calculation of digestibility somewhat, and there are two main approaches that have been adopted to cope with this situation. In both approaches, it is assumed that there is no interaction between the feeds in terms of their digestibility.

The first approach is to feed the forage with another 'base feed' of known digestibility (McDonald *et al.*, 1995). The digestibility trial is conducted as normal, and the forage (the test feed) should constitute the major proportion of the diet. The DMD of the forage is then calculated using the equation:

$$\text{DMD of test feed} = \frac{\text{DM intake of test feed} - (\text{Faecal DM} - \text{Faecal DM from excreted base feed})}{\text{DM intake from test feed}}$$

The digestibility of the forage is therefore calculated by difference.

Regression method

The second approach is to feed the forage with a supplement (which may be another forage) at a number of different ratios between forage and supplement (Schneider and Flatt, 1975). The digestibility of each of these different diets is estimated. The diet's digestibility is then regressed with the proportion of the forage in the diet, and then the digestibility of the forage is estimated by extrapolating to zero inclusion of the supplement (1000 g kg^{-1} inclusion of the forage). This approach is illustrated in Fig. 6.1. In this example, the DMDs of four diets with forage inclusion rates of 200, 400, 600 and 800 g kg^{-1} were estimated. They were observed to be 0.72, 0.64, 0.56 and 0.48 respectively. When the inclusion rate of the forage was plotted against the observed digestibility, and the line was extrapolated to 1000 g kg^{-1} inclusion of the forage, it could be estimated that the digestibility of the forage was 0.40.

Givens *et al.* (1989) compared these three methods (direct, difference and regression) as a means of estimating straw digestibility. The largest variability was associated with the difference technique, while the direct method had the smallest variability when straw was fed *ad libitum*.

Indirect method

In some circumstances, particularly with grazing animals, it may not be possible to accurately measure either feed intake or faecal output, or both. The digestibility of the forage may still be estimated, if it contains a component that is known to be completely indigestible. This marker should also be unaltered during its passage through the gut, and it should be completely recovered in the faeces (Omed, 1986). If the concentrations of this component in both the feed and the faeces are

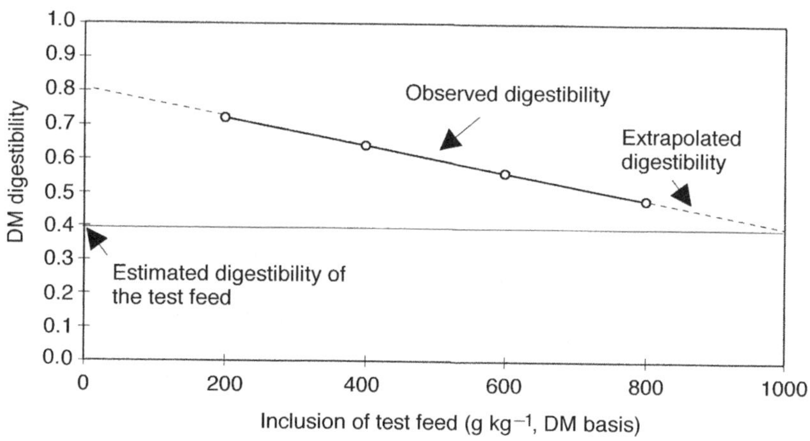

Fig. 6.1. Calculating the dry-matter digestibility of a test feed after feeding it with a supplement at different ratios of test feed : supplement.

then measured, the digestibility of the forage may be predicted. In this method, complete collections of faeces are not needed, so long as representative samples of faeces are collected. Representative samples of the forage that has been consumed are also required for the analysis of the marker.

The equation used to calculate digestibility in this instance assumes that all of the marker that is consumed is excreted. Therefore, at steady state:

$$\text{Amount of marker consumed per day} = \text{Amount of marker excreted per day}$$

$$[\text{Marker}]_{\text{Feed}} \times \text{Feed intake} = [\text{Marker}]_{\text{Faeces}} \times \text{Faecal output}$$

where marker concentrations are expressed in terms of g kg^{-1} DM and feed intakes and faecal outputs are expressed in terms of kg DM day^{-1}. Rearranging this gives:

$$\text{Faecal output} = \frac{[\text{Marker}]_{\text{Feed}} \times \text{Feed intake}}{[\text{Marker}]_{\text{Faeces}}}$$

and since:

$$\text{Digestibility} = \frac{\text{Feed intake} - \text{Faecal output}}{\text{Feed intake}}$$

then:

$$\text{Digestibility} = \frac{1 - [\text{Marker}]_{\text{Feed}} \times \text{Feed intake}}{[\text{Marker}]_{\text{Faeces}} \times \text{Feed intake}}$$

$$= 1 - \frac{[\text{Marker}]_{\text{Feed}}}{[\text{Marker}]_{\text{Faeces}}}$$

The marker may be a naturally occurring constituent of the feed, or it may be a chemical mixed into it. However, it is difficult to mix chemicals with forages, although the animals could be dosed with the marker each day. The disadvantage with this is that it increases the amount of handling that is required and adds to the stress of the animals. Constituents occurring in forages that have been used include lignin, acid-detergent fibre, indigestible acid-detergent fibre (IADF), acid-insoluble ash (AIA) (which is mainly silica) and some naturally occurring *n*-alkanes of long chain length (C-25–C-35). The possibility of using long-chain fatty acids has also been noted (Omed, 1986). The indicator most commonly added to feeds is chromium in the form of chromic oxide, Cr_2O_3, although a number of other markers, including dysprosium and granulated polyamide, have been used as well. The suitability of these markers is discussed later. One problem with external markers is that they can contaminate grazed forage and give biased results in future experiments. This was observed by Sprinkle *et al.* (1995) when chromic oxide was used as a marker.

The indirect, indigestible marker technique is easy to carry out and requires less work than estimating digestibility by the complete collection of faeces. If the concentration of the indicator in the feed is high, then this technique produces reasonably good results (Omed, 1986). However, the technique has limited usefulness in some circumstances, and it should only be used if there is a high possibility of obtaining a representative sample of the forage consumed (Omed, 1986). There are three main sources of error associated with this method. The first is the estimation of the marker's concentration in the forage, if an internal marker is used. This error may arise either from the sampling of the herbage, or from analytical errors. Analytical errors may also be associated with the estimation of the marker concentration in the faeces. It was noted by Newman *et al.* (1998) that analytical errors in estimating the marker concentration in the forage and the faeces may tend to cancel each other out. However, most markers are not completely indigestible, and this third source of error may give rise to very unreliable estimates of digestibility.

Obtaining a representative sample of the forage is difficult, as grazing animals, particularly sheep, can be highly selective in the forage they consume. One approach that can be adopted is to fistulate the animal's oesophagus. A sample of oesophageal extrusa may then be taken and analysed for the indicator to estimate the concentration of the marker in the forage actually consumed.

Prediction of forage digestibility

This method relies on the observation that certain faecal components are related to the digestibility of the diet. If the digestibility of the diet has been measured using the direct method and the faeces from these estimations have been analysed for a particular component, then a regression may be established between the concentration of that component in the faeces and the observed digestibility *in vivo*. In subsequent experiments, spot samples of faeces may be taken (for example, in a grazing experiment), the faeces analysed for the particular component and, from these data, the digestibility of the grazed herbage may be predicted (Omed, 1986). A range of different faecal components have been used, including nitrogen, chromogen, crude fibre and methoxyl. The suitability of these markers will be discussed later.

The technique works better when a local regression is established, using either the same or very similar herbage to that under investigation (Omed, 1986). The general equations that have been established are less precise than expected (Omed, 1986), because herbage factors are a major source of variation.

Conducting a Digestibility Experiment

To obtain an accurate estimate of the digestibility of a forage, it is important that well-established techniques are adopted and also that the objective of the experiment is clearly defined, as this will affect the amount of forage that is fed and what supplements, if any, are included in the diet.

Experimental design

Although it is usually considered that the variation in digestibility between animals is low (Cottyn *et al.*, 1989), the accuracy of the estimation will be improved if some form of crossover design is used. If there is likely to be an effect of time period, then a Latin-square design should be employed. This is particularly true with forages if they are being collected daily and are likely to change in composition during the experiment. If the animals are changing in physiological state during the experiment – for example, if they are lactating or actively growing – then a Latin-square design is also appropriate. However, the physiological changes likely to be encountered in mature wether sheep are small and, if these are the experimental animals, it is questionable what benefit there is in using a Latin square. Although Raymond *et al.* (1954) suggested there may be a decrease in digestive ability during winter, this has not been confirmed. Evans and Potter (1984) observed no effect of time period on the digestibility of cocksfoot grass (orchard-grass) fed to wethers. The advantage of not using a Latin square in this type of experiment is that an extra $(n - 1)$ degrees of freedom are gained, where n is the number of time periods used in the experiment.

The adaptation period

There are two stages to the digestibility trial, the first being the adaptation period and the second the collection period. The adaptation period is designed to ensure that a stable population of rumen microflora has become established, that the animals are eating approximately the same amount of feed daily and at the same time and that the residues being excreted in the faeces arise from the digestion of feed used in the experiment (Omed, 1986).

The variation in daily faecal output is reduced if the animals are fed at the same time each day. The more frequently they are fed, the smaller the variation in faecal output and so, ideally, animals should be fed hourly (Blaxter *et al.*, 1956). However, Raymond *et al.* (1953a) observed that there was little evidence that the frequency of feeding affected the digestibility of the feed, although it might affect the efficiency of utilization of digested feed. Ideally, if animals are only fed twice daily, this should be at 12 h intervals to reduce variation in faecal output (Schneider and Flatt, 1975). This is rarely done, and it is more usual for animals to be fed at the beginning and end of the working day. If this compromise is adopted, it is important that the feeding times are strictly adhered to (Schneider and Flatt, 1975).

The length of time for the animal to be adapted to its diet varies from 4 to 12 days. Normally, it has been shown that 6–8 days are required (Omed, 1986). Chenost and Demarquilly (1982) reported that an adaptation period of 10–14 days was required for the animals to obtain maximum level of intake, but this period could be reduced to 7 days if the animals were familiar with the forage. However, with straw, an adaptation period of 14 days was advocated by Cottyn *et al.* (1989). During this adaptation period, the animals should also be introduced to their digestibility cages if they are being confined (Omed, 1986).

Collection of faeces

Following the adaptation period, there is a collection period, in which the faeces are collected from each experimental animal. During this collection period, representative samples of the feed are also taken and the amount of feed refused is recorded. Representative samples of the refusals are also taken and analysed, so that the amount of each nutrient that was actually consumed may be calculated.

Complete collection of faeces

A complete collection of all the faeces excreted is required if the direct, difference or regression methods are being used. The faeces should not be contaminated with urine, and the collection should be made at the same time each day. The faeces taken each day should be bulked over the whole collection period and a sample taken for analysis. If sheep are used, then there is no need to subsample during the collection period. However, with cows, subsamples can be taken each day (*c.* 10% by weight) and added to the bulked sample (Minson, 1981).

Spot sampling of faeces

Spot samples of faeces are required if the indirect or prediction method is being used. Samples should be taken at least once and preferably twice a day. However, no difference in the estimate of digestibility was observed when samples were taken twice a day as opposed to four times a day when *n*-alkanes were used as a marker (Malossini *et al.*, 1994). If there is diurnal variation in the excretion of the marker (as is the case with chromic oxide), however, more frequent sampling is required.

Spot samples of faeces may be taken from each animal by taking a grab sample from the rectum. This does require that the animals are handled and can cause the animals stress. If they are grazing, then samples of the faeces can be taken from the field. The animals can be dosed with a coloured dye so that the faeces from individual animals can be identified. If this approach is adopted, it is important that the dye selected does not affect the digestibility of the forage or interfere with the marker being used to estimate digestibility.

The length of time of the collection period varies with the nature of the diet, shorter collection periods being required for more uniform diets. Typically, a period of 4–12 days is chosen, but the accuracy of the measurement increases with the length of the sampling period (Blaxter *et al.*, 1956). A collection period of 10 days is necessary for straw (Cottyn *et al.*, 1989).

Apparatus used in digestibility trials

Pens

Animals that are being housed for a digestibility trial need to be penned separately so that they can be fed individually. The arrangement of the pens and the design of the feed boxes need to be such that animals cannot reach the feed in their neighbour's pen. The walls of the feed box also need to be sufficiently high to minimize the amount of feed lost by animals, particularly cows, throwing their feed around. If the feed box is also designed so that the animal's head is restrained while it is feeding, this will reduce the amount of feed that is pulled back into the pen, where it may also be lost. This may mean that the animal needs to be tethered during the digestibility trial. The inside of the feed box should be smooth, so that no feed can lodge in crevices.

Clean, fresh drinking-water should always be available, either via an automatic drinking bowl or by a water bucket that is filled two or three times a day.

Meshes

A metabolism cage that allows the separation of faeces and urine was described by Schneider and Flatt (1975). The animal stands on a mesh floor, which is situated over a screen that is small enough to retain the faeces. The faeces then collect on this screen, while the urine falls through to a sloped or funnelled pan, which directs it into a collection bottle below. The screen on which the faeces collect is removable, so that the faeces may be gathered from it each day. The disadvantage of this type of crate is that the faeces fall over a wide area on the screen below the false floor and become exposed to contamination by urine. Faeces that fall on the floor of the crate may also be trampled and ground by the animal, so that the small particles then pass through both screens (Schneider and Flatt, 1975). This problem may be partly overcome by reducing the size of the cage so that the animal cannot turn around. With male animals, the urine will then tend to fall in the middle of the cage and the faeces around it.

Bags

Instead of using meshes to separate the faeces and urine, bags attached to harnesses that are fitted to the animal may be used. Faeces collect in the bags, which are emptied once or twice a day, or more often if necessary. Bags may be made of canvas, although this is unsuitable if the feed is likely to produce wet faeces (Raymond *et al.*, 1953a). In this case, a rubber or rubberized canvas bag is better, although this will perish with time. Bags may also be lined with plastic bags, which facilitates the sample collection. The bag may be attached to the harness by clips, which are unfastened when the faeces are collected. Alternatively, the bag may be attached to the harness by buckles and a zip-fastener fitted to the bag (if it is made of canvas). Faeces are then collected by simply unzipping the bag and emptying the faeces into a bucket. The harness itself may be made of webbing or leather and consists of two straps, which go round the animal's girth (just behind the shoulders) and just in front of the flank. A strap runs along the animal's back and attaches the bag to these two straps. Smaller straps hold the bag in place by anchoring it to the rear strap across each side of the hindquarters. Care must be taken in fitting harnesses to ensure that they do not rub or chafe. The animals should be adapted to these harnesses for some days before the collection of faeces begins.

A bag of this type is suitable for collecting faeces from wether sheep. A much larger bag is needed for collecting faeces from steers. The advantage of these bags is that they can be used for grazing animals as well, although the presence of the bag may affect their grazing behaviour. With female animals, a plastic tongue needs to be attached above their vulva to direct faeces away from the urine. The faeces may then either be collected in a bag, or in a separate trough if the animals are penned.

Selection of animals

Obviously, the animals used in any digestibility trial should be healthy and, if they have been grazed, they should be wormed. They should be checked before being selected for a digestibility experiment and, in the case of sheep, their feet and teeth in particular should be examined (Givens and Moss, 1994). They should also be used to being handled and, if complete collections of faeces are being made indoors, they should be penned and handled before the beginning of the experiment (Schneider and Flatt, 1975). If they are to wear bags and harnesses, then they need to be adapted to wearing these.

Species

A wide range of species have been used to estimate forage digestibility, and these include golden hamsters, rabbits, goats, sheep and cattle (Minson, 1990). Equids, such as horses, donkeys and mules, have also been used, as have camels. The advantage of the smaller species is that they consume less feed and are therefore less expensive to keep. However, their digestive tract may be very different from the target species. If an estimate of forage digestibility by cattle is required, sheep can be used, and frequently are. Cottyn *et al.* (1989) conducted a number of experiments using maize silages, grass silages and grass hays to investigate whether cattle and sheep differed in their digestive capacity. They concluded that cattle digest poor-quality forages more efficiently, whereas sheep digest high-quality forages more efficiently. They suggested that, for most temperate forages, the difference in digestive capacity between the two species was so small as to be of no practical significance. However, they did suggest that, if sheep were used to estimate forage digestibility, a positive correction should be applied for organic-matter digestibilities (OMDs) below 0.65 and a negative correction for OMDs over 0.80 when predicting the digestibility of the forage by cattle. The size of the applied correction (Y) was:

$$Y \text{ (as a proportion)} = 0.348 - (0.0048 \times \text{Sheep OMD})$$

where Sheep OMD is the observed OMD of the forage by sheep.

A lack of data prevented these authors from providing similar correction factors for other feed fractions, although they did note that protein was usually better digested by sheep than by cows. They suggested that, for grass silage, grass hay and many mixed rations, a mean difference of 0.07 between sheep and cows should be used, as proposed by Van Es (1978). There appears to be no difference in the protein digestibility of maize silage by cows and sheep (Cottyn *et al.*, 1989).

Sex

Male animals are generally preferred, as it is much easier to separate the faeces and urine. However, if the digestibility of forage at high planes of nutrition is required, then high-yielding dairy cows should be selected (Schneider and Flatt, 1975).

Age

The age of the animal appears to have little, if any, effect on DMD. Raymond *et al.* (1954) observed small differences due to age, but acknowledged that they were unlikely to be significant. They did observe some evidence that the efficiency of digestion in sheep increased at the rate of 0.01 unit (when digestibility is expressed as a proportion) per year up to the age of 2 years, with a further, smaller increase between 2 and 3 years. However, Givens and Moss (1994) observed that neither the age nor the body weight of sheep made any significant difference to the digestibility of dried grass, when sheep ranging in age from 3 to 8 years were studied.

Breed

The effect of breed on the DMD of dried grass was investigated by Givens and Moss (1994). They observed a significant (although small) difference between breeds of sheep, with the DMD of dried grass being 0.03 units higher with Cheviot sheep compared with Suffolk × Mule sheep. No significant differences were observed by Ranilla *et al.* (1998) in the DMD and fibre digestibility of good-quality forage at maintentance when fed to either Churra or Merino sheep.

Number

The number of animals that should be used depends on the expected difference between test feeds and on the degree of confidence that is required. Raymond *et al.* (1953a) cited Schneider and Lucas (1950) as stating that the greatest variation in the estimation of digestibility coefficients was between experiment and between experimenter. These authors calculated that the between-animal variation was quite small. However, Raymond *et al.* (1953a) presented a table that provided the number of replicates needed to detect differences between feeds when using a *t* test to analyse differences. These data are presented in Table 6.1.

These numbers could be reduced if a crossover design was used, as the between-animal variation could then be removed. Cottyn *et al.* (1989) noted that, if the feedstuff under investigation was fed alone (and the direct method was used), then three animals were sufficient. However, if it was fed in combination with another feed (and the difference method was used), then four, six or eight animals were required if the test feed constituted 50, 25 or less than 25%, respectively, of the ration.

Table 6.1. The number of replications that are required to obtain a significant result ($P < 0.05$) in a digestibility experiment using Student's *t* test (from Raymond *et al.*, 1953a).

True difference, σ (proportional digestibility)	Number of replications needed to detect σ with a probability of	
	0.8	0.9
0.010	28	38
0.015	14	17
0.020	8	11
0.025	6	7
0.030	4	6
0.040	3	4
0.050	3	3
0.100	2	2

Feed used in the experiment

If complete collections of faeces are being made, it is important that the amount and composition of the feed that is consumed in the experiment are accurately measured. If the indirect method is used, an estimate of the composition of the forage actually consumed is needed, as is an accurate estimate of the concentration of the marker in the consumed forage.

Use of feed supplements

The objective of the digestibility trial is usually to estimate the digestibility of a feed when offered to a healthy animal with a stable rumen. It is therefore important to ensure that the animal's health is maintained throughout the trial and that the diet is balanced. The diet should therefore be supplemented with an adequate mixture of vitamins and minerals. If the OMD of the feed under investigation is below 0.55 or greater than 0.75, then it cannot be fed alone, as to do so would jeopardize the health of the animal (Cottyn *et al.*, 1989). Even if the feed's digestibility is between 0.55 and 0.75, its physical structure may not maintain a healthy rumen function and it should be supplemented. It may also be the case, if high-yielding dairy cows are being used, that a forage may not be able to meet all of the cow's requirements, and supplementation is again required. Experiments investigating straws are particularly difficult, as the forage is so deficient in major nutrients, such as nitrogen (N) and sulphur (S). Cottyn *et al.* (1989) recommended that experimental diets should contain at least 200 g kg^{-1} long forage to prevent an abnormal fermentation, and that they should contain no more than 50 g kg^{-1} DM unsaturated fat and 100 g kg^{-1} DM saturated fat. They further recommended that the diets should contain (on a DM basis) at least 180 g kg^{-1} crude fibre, between 120 and 200 g kg^{-1} crude protein, less than 400 g kg^{-1} starch and less than 200 g kg^{-1} sugars. This means that, for many forages, some form of supplementation is required. The ways in which the digestibility of the supplement may be taken into account when calculating forage digestibility were described above ('Difference method' and 'Regression method').

Level of feed intake

The digestive capacity of animals tends to decrease as the level of feed intake increases. The digestibility of a range of concentrate ingredients was observed to decrease as feed intake was increased from maintenance to twice maintenance (Woods *et al.*, 1999). Givens (1990) noted that OMD decreased by 0.014 units per multiple of maintenance. However, Cottyn *et al.* (1989) concluded that the effect of increased feed intake had little or no effect on the digestibility of long or coarsely chopped forages.

Traditionally, digestibility trials required the feeding of exact quantities of feed over long periods of time. However, Minson (1981) raised the question as to whether, when comparing two forages, the same amount should be fed or whether each should be fed *ad libitum*. Blaxter *et al.* (1956) argued that feeding *ad libitum* necessarily meant that refusals would be left, and the assumption of steady state was invalid if there were refusals. Blaxter *et al.* (1956) therefore concluded that intakes should be limited during digestibility experiments to ensure that no refusals were left. However, Minson (1981) pointed out that feeding close to *ad libitum* levels was a closer approximation to what would occur in the field. He therefore advocated feeding close to *ad libitum*, but did not remove all refusals each day. This 'recycling' of refusals was designed to reduce the amount of selection of forage by the animals. An alternative approach was adopted by Zemmelink (1980), who recommended offering animals four levels of feed rejection. This produced response curves between digestibility and the amount of excess forage offered. From these data, an estimate of the plant's digestibility, depending on the extent of selection the animal was able to practise, could be estimated. Laredo and Minson (1973) also suggested that the leaf and stem could be mechanically separated and then offered separately. However, this is a laborious process that is not easily mechanized. The leaf dust is lost, and the leaf fraction tends to contain leaf sheath and seed heads, which reduce the observed digestibility of the leaf fraction (Minson, 1981).

The object of the experiment must therefore be borne in mind when selecting the appropriate level of feed intake. If the objective is to estimate the digestibility of the whole forage, then feeding should be restricted to maintenance levels, which also has the advantage of being easier to standardize (Cottyn *et al.*, 1989). If the objective is to estimate the digestibility of the forage in a situation where the animal is able to practise a degree of selection, then the amount offered should be closer to *ad libitum*. However, even if this selection is allowed, it is not clear what relationship there is between the forage selection practised by grazing and penned animals (Minson, 1981).

Preparation of feeds

The forage that is offered to animals in a digestibility trial is chopped and mixed to ensure that representative samples of the forage are offered. However, this preparation may in itself affect the digestibility of the forage, although the effect is likely to be very small.

The forage offered to an animal in a digestibility trial may be either fed fresh or harvested before the experiment and preserved by drying or freezing it. It cannot be ensiled to preserve it, unless the object of the experiment is to estimate the digestibility of the ensiled rather than the fresh forage. Alternatively, the animal may be allowed to graze the forage, and the faeces must then be collected from the animal in the field. Each approach has both advantages and disadvantages, and none of them are ideal solutions.

With forages, the upper parts of the plant are more digestible than the lower parts, and animals will therefore tend to select the upper parts. This should be borne in mind when cutting forage for animals in a digestibility trial. It is also a factor that should be considered when sampling pasture that is being grazed by animals in a digestibility trial.

FRESH FORAGE. If fresh forage is to be fed to penned animals, then it needs to be cut each day. This does mean that any change in the composition of the forage during the period of the study may affect the estimation of the forage's digestibility, as the forage will not be of consistent composition throughout the experiment. Since the estimation of digestibility *in vivo* assumes that a steady state has been attained and that feed of the same composition is being fed throughout the experiment, this is a potentially serious flaw in this experimental design. Cutting large quantities of forage can also be difficult when the ground is waterlogged, and it is not possible to compare forages grown in different years or at different times of the year (Minson, 1990).

However, if forage is to be cut daily, then it was recommended by Minson (1981) that forage was cut in the afternoon, and that sufficient forage was cut for both that evening's feed and also the following morning. Cutting in the afternoon avoids contamination of the forage with morning dew. The forage that is to be stored overnight should be kept on dry ice, as the carbon dioxide inhibits further tissue metabolism. Refrigerating the forage in the absence of dry ice is less satisfactory (Minson, 1981). An alternative means of keeping the forage overnight is to freeze it quickly, and then remove it from the freezer the following day and feed it (Raymond *et al.*, 1953b). This does mean that the forage will still be frozen when it is fed, which is likely to increase the amount of feed refused. This is likely to affect the estimated digestibility of the feed. It was pointed out by Minson (1981) that, if neither dry ice nor refrigeration was available, then the forage needed to be cut twice daily.

FREEZING. Harvesting all of the forage that is needed for the experiment beforehand is preferable to cutting fresh forage daily. This is because the composition of the forage will then not change during the experiment. After the forage has been harvested, it should be mixed thoroughly, bagged and frozen. The bags should contain enough forage for 1 day's feed per animal, and should be removed from the freezer the day before they are to be used. The effect of freezing forage has been observed to have negligible effect on the estimation of the DMD of both tropical and temperate forages (Raymond *et al.*, 1953b; Minson, 1990), although the composition of the feed will change to some extent and this has been observed to affect nitrogen digestibility.

DRYING. This method can also be used to preserve forages for use in a digestibility trial. As with freezing, it has the advantage of ensuring that the composition of the forage used in the experiment remains constant. It is important that the forage is dried rapidly, and that the temperature does not exceed 100°C (Minson, 1990). If the temperature rises above 100°C, then the forage will be damaged, because of Maillard reactions. If the temperature is below 100°C, the forage does not dry quickly enough, and this usually results in a decrease in DMD compared with fresh forage (Minson, 1990). It should also be noted that drying changes the physical structure of the forage, and the forage's physical structure can be an important component of its nutritive value.

It was recommended by Minson (1981) that forage should be wilted and then dried for 6 h at 100°C. The forage should then be blown with cold air for 8 h. High-temperature driers should not be used, as these cause a reduction in crude protein digestibility.

FREEZE-DRYING. This is a very good means of preserving forage, as it minimizes the changes in the chemical composition of the forage and removes the need to maintain cold storage facilities. However, freeze-drying is an extremely expensive means of preserving forage and, as with conventional drying, the physical nature of the forage is changed during this process.

GRAZING. Instead of harvesting the forage and then offering it to the animal, the animal can be allowed to graze the material instead. This has the advantage of keeping the animal under more 'normal' conditions, since fresh forage will normally be grazed. However, the disadvantages of feeding cut fresh forages to the animal are also relevant to the grazing situation. The composition of the grazed forage will change during the experiment, and it is not possible to compare forages from different years or different seasons. It is also not possible to measure intake directly, and so either the indirect method of estimating digestibility must be used or some estimate must be made of the amount of forage consumed.

ENSILING. During the ensiling process, there are considerable changes to the chemical composition of the forage. Ensiling cannot therefore be used as a means of preserving forage for use in a digestibility trial if the objective of the experiment is to estimate the digestibility of fresh forage. However, if the digestibility of ensiled forage is to be estimated, then the silage should be mixed well before the experiment begins and stored frozen, as the composition of the silage can change throughout a clamp and its composition will also change with time.

Markers used in the indirect method of estimating digestibility

As was mentioned before, a range of markers has been used to estimate diet digestibility indirectly. The qualities of these markers are described here.

Silicon

Silicon is present as solid silica and monosilicic acid, and the ratio between these two species varies with plant species and plant maturity (Omed, 1986). Silica was first used as an internal marker in 1874, and its determination may now be carried out either colorimetrically or by atomic absorption spectroscopy (Omed, 1986). Recovery rates of silica close to 100% have been reported, and a good agreement between the digestibility predicted using silica as a marker and that observed by a complete collection of faeces has been observed (Duksin and Drozdenko, 1982). However, there is some evidence to suggest that a proportion of the silica is absorbed from the digestive tract and is excreted in the urine or found in the lymph nodes and urinary calculi. It has also been reported by Van Dyne and Lofgreen (1964) that silica can accumulate in the gut, which resulted in a variable recovery rate and a less accurate estimation of digestibility compared with lignin (Omed, 1986). Silica is not widely used as an indicator now, because of its possible absorption from the gut, and the risk of contamination of the herbage with soil which invalidates the estimated intake of silica. There is also some doubt about the accuracy of silica determination (Omed, 1986).

Acid-insoluble ash

Minson (1990) described work using ash insoluble in hydrochloric acid as a marker. Good recoveries of AIA and good agreement between *in vivo* OMD and predicted digestibility were reported. A good prediction of digestibility was also observed in horses by Miraglia *et al.* (1999). However, unsatisfactory results were reported in other studies, with reported recoveries of AIA being 127% (Minson, 1990). The use of AIA was observed to be inferior to acid-detergent fibre methods of predicting DMD (Minson, 1990), although Pereira and de Queiroz (1997) reported a recovery of AIA of 101%. Dos Santos and Petit (1996) noted that, compared with AIA, a controlled-release bolus of chromic oxide gave a much better estimate of faecal output.

Lignin

Lignin is quite widely used as an indicator in digestibility trials and is regarded as a good indicator by many authors (Omed, 1986). However, there are a number of reports which provide evidence that lignin is not completely indigestible. The poor recovery of lignin in horses led Miraglia *et al.* (1999) to suggest that it was an inappropriate marker for measuring digestibility. Mauricio *et al.* (1996) observed that the estimated digestibility of dry matter, crude protein, neutral-detergent fibre (NDF) and gross energy was about 0.05 units lower using lignin as a marker, rather than making a total collection of faeces. Pereira and de Queiroz (1997) observed a recovery of lignin of only 58.3%. Van Soest (1994) outlined a number of factors that could affect the calculated recovery of lignin. These factors were: first, that non-lignin material in the feed could contaminate the lignin in the assay procedure; secondly, that some losses of immature lignin may occur; thirdly, that soluble phenolic matter may form; fourthly, that drying temperatures may affect faeces and herbage samples differently; and, finally, that some finely divided lignin in the faeces may not be recovered in gravimetric methods. It is also difficult to obtain a representative sample of the actual intake (Meijs, 1981). Many workers therefore have little confidence in the use of lignin, although Van Soest (1994) maintained that the technique worked well if the diet dry matter consisted of more than 60 g kg^{-1} DM lignin. It was also noted that, as the lignin content of the diet increased, the standard error of its determination decreased.

Chromogen

Chromogen is a plant pigment that is soluble in 85% aqueous acetone and consists mostly of chlorophylls and their degradation products (Omed, 1986). There have been reports of high (100.5%) recoveries of chromogen in the faeces and a very good agreement between the estimate of digestibility using chromogen and that observed from a complete collection of faeces (Omed, 1986). However, less satisfactory results have also been reported, particularly with autumn and winter forage (Omed, 1986). There are other difficulties associated with this technique as well. Chromogen is unstable in light, and so light must be excluded during the extraction and measuring processes. The use of the technique is also limited to forages with a high chlorophyll concentration, and this restricts its suitability to fresh, green herbage. The technique also requires standardization with indoor digestibility trials, and the wavelength required to measure the concentration of the pigment in the faeces varies with different plant species and stages of growth (Omed, 1986). These drawbacks have resulted in chromogen no longer being used to estimate digestibility (Minson, 1990).

Indigestible acid-detergent fibre

Since a proportion of the cell wall is digestible, the use of lignin as a marker has been questioned. If the potentially digestible fractions of the lignin and cellulose could be removed, the indigestible residue could be used as an indicator. The IADF fraction was studied by Penning and Johnson (1983) as an indicator for estimating digestibility. IADF was estimated by treating the forage and faeces with acid-detergent solution and then treating the residue with cellulase. The remaining fraction was considered to be indigestible cell wall, and Omed (1986) reported that Penning and Johnson (1983) observed that IADF produced satisfactory results over a range of samples. This was also observed by Dove and Coombe (1992). However, Minson (1990) noted that, in another study, only half the marker was recovered in the faeces and in further work the recovery was variable. The technique requires further development and validation to ascertain what causes these variable recoveries, but the method is simple to perform (Omed, 1986).

An alternative means of analysing IADF is to incubate the sample in the rumen for extended periods of time. Huhtanen *et al.* (1994) tried incubating samples for 228 h *in situ* or for 96 h in rumen fluid *in vitro*. The estimation of digestibility that these markers gave was very poor (Huhtanen *et al.*, 1994). However, Fondevila *et al.* (1995) suggested the use of rumen-undegradable NDF or dry matter as an internal marker. They observed that the estimated digestibility with these markers gave a correlation coefficient, with digestibility measured by total collection, of 0.80.

Long-chain n*-alkanes*

The long-chain *n*-alkanes of uneven chain length (C-19–C-35) found in the cuticular wax of forage plants can serve as markers and are increasingly used in the estimation of herbage intake. The recovery of these markers in the faeces increases with increased chain length, although even C-35 has a recovery of only 93% (Mayes *et al.*, 1986). This may limit their usefulness as markers for estimating digestibility, and it was concluded by Newman *et al.* (1998) that estimates of digestibility using *n*-alkanes were unlikely to be reliable. However, Dove and Coombe (1992) observed that the use of them accurately estimated digestibility, in a way that lignin did not. Another advantage with *n*-alkanes is that no cyclical pattern in their excretion has been observed (Mayes *et al.*, 1986). This would reduce the error of estimating faecal alkane concentration by taking only spot samples of faeces. The determination of *n*-alkanes is more accurate than that of lignin, and they are stable, whereas chromogen is not.

Chromic oxide

Chromic oxide is often used as an external marker. If it is added to the ration, it must be mixed in thoroughly, but if it is used in the powdered form it will separate from coarse feeds (Schneider and Flatt, 1975). It will then travel at a different rate through the gut compared with the rest of the diet. It can be mixed with paper pulp and ground to a similar particle size as the coarse feeds, and it will then mix better with the diet. This mixing of the chromic oxide with the feed is not possible in a grazing situation, of course. In this case, the animals need to be dosed with the chromic oxide. This can be achieved with ruminally cannulated animals by inserting chromic oxide preparations into the rumen. J.F.D. Greenhalgh developed a means of impregnating paper with chromic oxide which releases the chromium more slowly into the gut. Alternatively, gelatin capsules containing chromium can be used. Luginbuhl *et al.* (1994) observed that, with a controlled-release bolus containing chromic oxide, a constant faecal excretion of chromium was observed after 8 days. A sharp peak in chromium excretion was then observed about 27 days after administration, and complete excretion was observed after 31 days in seven out of 24 lambs. The accuracy of estimation of digestibility did depend on the diet. It was very poor when a commercial pelleted diet was fed, but very good when lucerne hay was fed (Luginbuhl *et al.*, 1994). The use of a slow-release bolus of chromic oxide resulted in a grab sample taken once providing a reliable ($r = 0.96$, $P < 0.05$) estimate of faecal output (Dos Santos and Petit, 1996). However, a more frequent sampling regime would be advisable.

The recovery of chromic oxide is often not complete (Schneider and Flatt, 1975). If the true recovery rate can be estimated in a separate experiment, then the calculated digestibility can be corrected using this true recovery rate. Another difficulty with chromic oxide is that it exhibits diurnal variation in its excretion pattern (Schneider and Flatt, 1975). This means that random sampling of the faeces may not provide a truly representative sample of the faeces. The diurnal variation is reduced if the chromic oxide is administered more frequently (at least twice a day) or if its release into the gut is continual (by the use of chromic-oxide-impregnated paper or a bolus). It is also important that it is administered for at least 10 days before faecal samples are taken (Schneider and Flatt, 1975). Faeces should also be sampled frequently (at least four times a day) to obtain more representative samples.

When analysing samples for chromium, a calibration curve is required. This should use faecal samples from animals not ingesting chromium. These faecal samples can then be spiked with known amounts of chromium to generate the calibration curve. However, it is essential that these faecal samples are obtained from animals grazing the same forages as those used by the animals dosed with the chromium (Holt, 1993).

Granulated polyamide

Mahler *et al.* (1997) investigated the use of granulated polyamide as a marker. They administered the marker via gelatin capsules every 12 h, and observed that 4 days were needed to obtain a steady state. The marker recovery was 98% and was not affected by diet type or the level of feed intake, unless the diet contained less than 9.26 g N kg^{-1} OM. A good relationship was observed between digestibility calculated using this marker and that observed by making a total collection of faeces. To use this marker, a sampling schedule is needed that takes samples throughout the day (Mahler *et al.*, 1997), but the advantage of this marker is that it is inexpensive and does not require sophisticated analytical techniques.

Markers used in the prediction technique of estimating digestibility

The most commonly used marker in this technique is nitrogen, although a range of other markers have been investigated as well.

Nitrogen

Minson (1990) noted that some of the first equations for predicting DMD from the N concentration in faeces were published by Gallup and Briggs (1948). When regressions were based on a wide range of temperate forages, the residual standard deviation varied between ±0.029 and ±0.057 (Minson, 1990).

However, the regression equations developed from forage cut at different times of the year are very different (Minson, 1990), because of the higher concentration of N in autumn forage and the difference in regressions for leaf and stem fractions. There is some evidence that the application of N as fertilizer may affect the relationship, as the N content of the pasture may be increased without affecting its digestibility (Omed, 1986). In other circumstances, if animals are grazed on sparse pasture, they may ingest large amounts of soil, which causes an increased secretion of metabolic N. High concentrations of faecal N are therefore recorded and so predicted digestibility increases, although the actual digestibility may be depressed (Omed, 1986). The major disadvantage of this technique, however, is that there is no single general equation that is suitable for a wide range of forages under different conditions. Many equations have been reported, but each one is valid only for the particular set of circumstances under which it was produced (Omed, 1986). Some of the variation in these relationships is illustrated in Fig. 6.2. Despite these drawbacks, the technique can produce a reasonably accurate estimation of digestibility, and Omed (1986) noted that it had been observed that 93% of the variation in *in vivo* digestibility could be explained by the concentration of faecal N.

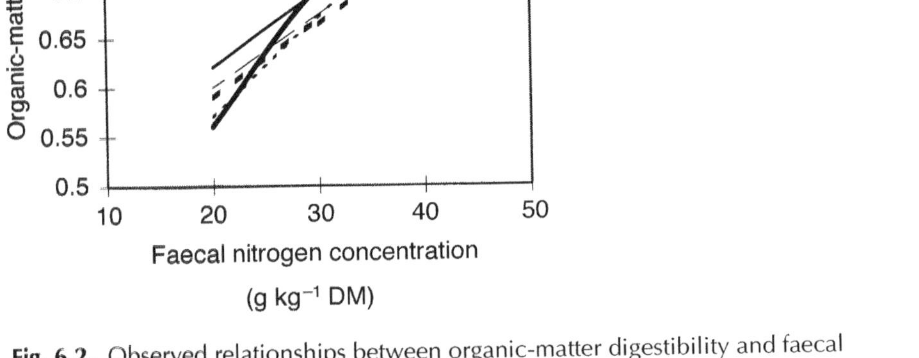

Fig. 6.2. Observed relationships between organic-matter digestibility and faecal nitrogen concentration for temperate forages (from Minson, 1990).

Chromogen

A prediction equation may also be established between faecal chromogen concentration and digestibility. However, with temperate forages in the UK and New Zealand, it was observed that the residual standard deviation of this relationship was greater than regressions based on faecal N (Minson, 1990). The regressions have a seasonal bias similar to that observed with faecal N, and it is now unusual for faecal chromogen to be used to predict digestibility (Minson, 1990).

Fibre

The digestibility of temperate forages is negatively related to the concentration of crude fibre in the faeces. The regression has a residual standard deviation only slightly greater than that of faecal N (Minson, 1990). However, as feed intake is increased, the faecal fibre concentration decreases as more nutrients are excreted and dilute the fibre. The decreased faecal fibre concentration results in a predicted increase in digestibility, although digestibility has in fact decreased. This results in a large bias in the predicted digestibility of grazed forage (Minson, 1990). However, faecal NDF was observed to be well related to the digestibility of both dry matter (r = 0.93) and organic matter (r = 0.87) in the dromedary camel (Abdouli *et al.*, 1992).

Methoxyl

The lignin molecule contains methoxyl groups, and it was noted by Minson (1990) that the faecal methoxyl content was significantly and negatively related to DMD. However, faecal methoxyl concentration does not appear to be used for routine forage evaluation (Minson, 1990).

Preparation of samples for analysis

It is important that samples of feeds and faeces are dried quickly to get an accurate measure of DMD, as slow drying increases the loss of dry matter (Raymond and Harris, 1954). However, if they are dried at too high a temperature, then volatile compounds will be lost, and this will also lead to inaccuracies in the estimation of dry matter. Silages are particularly at risk from loss of volatiles by oven-drying. Drying at too high a temperature may also affect the estimation of fibre fractions in the feed and faeces, and this will affect the estimation of fibre digestibility. Usually, feeds and faeces are oven-dried at 60°C and, following the estimation of dry and organic matter, they may be ground through a 1 mm screen pending further analysis. However, the N and volatile fatty acid content of silage needs to be estimated using a fresh sample, and the gross energy content of forages should be determined using samples of fresh forage that have been chopped under liquid nitrogen.

Accuracy of Estimating Digestibility

The estimation of digestibility *in vivo* is quite accurate, with a standard error of only 0.015 units for most feedstuffs, although this increases to 0.030 for straw (Cottyn *et al.*, 1989).

However, it is important that digestibility trials are conducted with care. For example, if lambs are fed 800 g day^{-1} hay (850 g DM kg^{-1}), and excrete 500 g day^{-1} faeces (400 g DM kg^{-1}), the DMD of hay is 0.706. A 10% error in the weighing out of the feed or in the estimation of its DM content would result in a change in the estimated digestibility to either 0.673 or 0.733. A similar error in the weighing of faeces or its estimation of DM content would change the estimated digestibility to either 0.735 or 0.676. In experiments where the differences between treatments are small, this error would result in no significant difference between treatments being observed.

Relevance of *In Vivo* Measures of Digestibility

The estimation of the DMD of forage does give some measure of the availability of energy and other nutrients, although it ignores the loss of energy in the form of methane. As a means of evaluating forages, it is labour-intensive and time-consuming and requires large quantities of the forage concerned. It also relies on the use of animals. It is therefore not appropriate as a means of routinely evaluating forages. However, it is needed as a technique to validate many of the *in vitro* procedures that have been developed, although it provides no information on the site of digestion, the kinetics of digestion or the supply of nutrients to the animal.

References

Abdouli, H., Khorchani, T. and Nefzaoui, A. (1992) Nutrition of the one-humped camel. 1. Faecal index determination and chromic oxide excretion pattern and recovery. *Animal Feed Science and Technology* 39, 293–301.

Blaxter, K.L., Graham, N.McC. and Wainman, F.W. (1956) Some observations on the digestibility of food by sheep, and on related problems. *British Journal of Nutrition* 10, 69–91.

Chenost, M. and Demarquilly, C. (1982) Measurement of herbage intake by housed animals. In: Leaver, J.D. (ed.) *Herbage Intake Handbook*. British Grassland Society, Hurley, UK, pp. 95–122.

Coop, I.E. and Hill, M.K. (1962) The energy requirements of sheep for maintenance and gain. II. Grazing sheep. *Journal of Agricultural Science* 58, 187–199.

Cottyn, B.G., De Boever, J.L. and Vanacker, J.M. (1989) *In vivo* digestibility measurement of straw. In: Chenost, M. and Reiniger, P. (eds) *Evaluation of Straws in Ruminant Feeding*. Elsevier Applied Science, London, UK, pp. 36–46.

Dos Santos, G.T. and Petit, H.V. (1996) Prediction of total faecal output in sheep fed silage using the Captec chrome controlled-release capsule. *Small Ruminant Research* 20, 223–227.

Dove, H. and Coombe, J.B. (1992) A comparison of methods for estimating supplement intake and diet digestibility in sheep. *Proceedings of the Australian Society of Animal Production* 19, 239–241.

Duksin, Y.U.P. and Drozdenko, N.P. (1982) Estimation of digestibility of nutrients in diets for cows according to the presence of silicon. *Nutrition Abstracts and Reviews* 54, 281.

Evans, E.M. and Potter, J.F. (1984) The reproducibility of *in vivo* estimates of digestibility and voluntary digestible organic matter intake of grass varieties by sheep. *Grass and Forage Science* 39, 101–106.

Fondevila, M., Castrillo, C., Gasa, J. and Guada, J.A. (1995) Rumen-undegradable dry matter and neutral detergent fibre as ratio indicators of digestibility in sheep given cereal straw-based diets. *Journal of Agricultural Science* 125, 145–151.

Gallup, W.D. and Briggs, H.M. (1948) The apparent digestibility of prairie hay of variable protein content, with some observations of faecal nitrogen excretion by steers in relation to their dry matter. *Journal of Animal Science* 7, 110–118.

Givens, D.I. (1990) The effect of growth stage and season on the prediction of the metabolisable energy content of herbage. PhD thesis, University of Newcastle upon Tyne, UK, 244 pp.

Givens, D.I. and Moss, A.R. (1994) Effect of breed, age and bodyweight of sheep on the measurement of apparent digestibility of dried grass. *Animal Feed Science and Technology* 46, 155–162.

Givens, D.I., Dewey, P.J.S., Donaldson, E., Jones, D.I.H. and Adamson, A.H. (1989) Within and between centre variability in the measurement of *in vivo* organic matter digestibility of straw. In: Chenost, M. and Reiniger, P. (eds) *Evaluation of Straws in Ruminant Feeding*. Elsevier Applied Science, London, UK, pp. 62–65.

Holmes, J.C. and Osmon, H.S. (1960) The feed intake of grazing cattle. I. Feed intake of dairy cows on strip and free grazing. *Animal Production* 2, 131–139.

Holt, N.W. (1993) Calibration curves for the determination of low levels of chromium in faeces. *Canadian Journal of Animal Science* 73, 109–115.

Huhtanen, P.K., Kaustell, K. and Jaakola, S. (1994) The use of internal markers to predict total digestibility and duodenal flow of nutrients in cattle given six different diets. *Animal Feed Science and Technology* 48, 211–227.

Lancaster, R.J. (1949) Estimation of digestibility of grazed pasture from faeces nitrogen. *Nature* (London) 163, 330–331.

Laredo, M.A. and Minson, D.J. (1973) The voluntary intake, digestibility and retention time by sheep of leaf and stem fractions of five grasses. *Australian Journal of Agricultural Research* 24, 875–888.

Luginbuhl, J.M., Pond, K.R., Burns, J.C. and Fisher, D.S. (1994) Evaluation of the Captec controlled-release chromic oxide capsule for fecal output determination in sheep. *Journal of Animal Science* 72, 1375–1380.

McDonald, P., Edwards, R.A., Greenhalgh, J.F.D. and Morgan, C.A. (1995) *Animal Nutrition*, 5th edn. Longman Scientific and Technical, Harlow, UK, pp. 221–237.

Mahler, F., Schlecht, E., Sangare, M. and Becker, K. (1997) Granulated polyamide as external marker to estimate total faecal excretion of grazing cattle in extensive management systems. *British Journal of Nutrition* 78, 785–803.

Malossini, F., Bovolenta, S., Piasentier, E. and Valentinotti, M. (1994) Variability of *n*-alkane content in a natural pasture and in faeces of grazing dairy cows. *Animal Feed Science and Technology*, 50, 113–122.

Mauricio, R.M., Goncalves, L.C., Resende, A.C. and Rodriguez, N.M. (1996) Apparent digestibility using total faecal collection, chromic oxide and lignin as markers in equids. *Arquivo Brasileiro de Medicina Veterinaria e Zootecnia* 48, 703–711.

Mayes, R.W., Lamb, C.S. and Colgrove, P.M. (1986) The use of dosed and herbage *n*-alkanes as markers for the determination of herbage intake. *Journal of Agricultural Science*, Cambridge 107, 161–170.

Meijs, J.A.C. (1981) *Herbage Intake by Grazing Dairy Cows*. Agricultural Research Report 909, Centre for Agricultural Publishing and Documentation, Wageningen, the Netherlands.

Minson, D.J. (1981) The measurement of digestibility and voluntary intake of forages with confined animals. In: Wheeler, J.L. and Mochrie, R.D. (eds) *Forage Evaluation: Concepts and Techniques*. Australian Forage and Grassland Council, Melbourne.

Minson, D.J. (1990) *Forage in Ruminant Nutrition*. Academic Press, San Diego, USA, 483 pp.

Minson, D.J. and Kemp, C.D. (1961) Studies in the digestibility of herbage. IX. Herbage and faecal nitrogen as indicators of herbage organic matter digestibility. *Journal of the British Grassland Society* 16, 76–79.

Miraglia, N., Bergero, D., Bassano, B., Tarantola, M. and Ladetto, G. (1999) Studies of apparent digestibility in horses and the use of internal markers. *Livestock Production Science* 60, 21–25.

Newman, J.A., Cribarineto, F. and Jensen, M.J. (1998) The sensitivity of *n*-alkane analysis to measurement error: implications for use in the study of diet composition. *Journal of Agricultural Science* 131, 465–476.

Omed, H.M. (1986) Studies of the relationships between pasture type and quality and the feed intake of grazing sheep. PhD thesis, University College of North Wales, Bangor, UK, 271 pp.

Penning, P.D. and Johnson, R.H. (1983) The use of internal markers to estimate herbage digestibility and intake. II. Indigestible acid detergent fibre. *Journal of Agricultural Science, Cambridge* 100, 133–138.

Pereira, J.C. and de Queiroz, A.C. (1997) Apparent digestibility of elephant grass (*Pennisetum purpureum*, *Schum*) and sugar-cane (*Saccharum officinarum*) in different proportions, fed to horses. *Revisita da Sociedade Brasileira de Zootecnia* 26, 105–110.

Ranilla, M.J., Lopez, S., Giraldez, F.J., Valdes, C. and Carro, M.D. (1998) Comparative digestibility and digesta flow kinetics in two breeds of sheep. *Animal Science* 66, 389–396.

Raymond, W.F. and Harris, C.E. (1954) The laboratory drying of herbage and faeces, and dry matter losses possible during drying. *Journal of the British Grassland Society* 9, 119–130.

Raymond, W.F., Harris, C.E. and Harker (1953a) Studies on the digestibility of herbage. 1. Technique of measurement of digestibility and some observations on factors affecting the accuracy of digestibility data. *Journal of the British Grassland Society* 8, 301–314.

Raymond, W.F., Harris, C.E. and Harker (1953b) Studies on the digestibility of herbage. II. Effect of freezing and cold storage of herbage on its digestibility by sheep. *Journal of the British Grassland Society* 8, 315–320.

Raymond, W.F., Harris, C.E. and Kemp, C.D. (1954) Studies in the digestibility of herbage. V. The variation, with age, of the ability of sheep to digest herbage, with observations on the effect of season on digestive ability. *Journal of the British Grassland Society* 9, 209–220.

Schneider, B.H. and Flatt, W.P. (1975) *The Evaluation of Feeds Through Digestibility Experiments.* University of Georgia Press, Athens, USA, 423 pp.

Schneider, B.H. and Lucas, H.L. (1950) The magnitude of certain sources of variability in digestibility data. *Journal of Animal Science* 9, 504–512.

Sprinkle, J.E., Kress, D.D., Doornbos, D.E., Anderson, D.C., Tess, M.W., Ansotegui, R.P., Olson, B.E. and Roth, N.J. (1995) Chromic oxide contamination of pasture previously used in marker studies. *Journal of Range Management* 48, 194–197.

Van Dyne, G.M. and Lofgreen, G.P. (1964) Comparative digestion of dry annual range forage by cattle and sheep. *Journal of Animal Science* 23, 823–834.

Van Es, A.J.H. (1978) Feed evaluation for ruminants. 1. The systems in use from May 1988 onwards in the Netherlands. *Livestock Production Science* 5, 331–345.

Van Soest, P.J. (1994) *Nutritional Ecology of the Ruminant*, 2nd edn. Cornell University Press, Ithaca, New York, USA, pp. 177–195.

Woods, V.B., Moloney, A.P., Mulligan, F.J., Kenny, M.J. and O'Mara, F.P. (1999) The effect of animal species (cattle or sheep) and level of intake by cattle on *in vivo* digestibility of concentrate ingredients. *Animal Feed Science and Technology* 80, 135–150.

Zemmelink, G. (1980) *Effect of Selective Consumption on Voluntary Intake and Digestibility of Tropical Forages.* Agricultural Research Report 896, Centre for Agricultural Publishing and Documentation, Wageningen, the Netherlands.

7 Faeces as a Source of Microbial Enzymes for Estimating Digestibility

H.M. Omed, D.K. Lovett and R.F.E. Axford

School of Agricultural and Forest Sciences, University of Wales, Bangor, Gwynedd LL57 2UW, UK

The Need for an Alternative to Rumen Fluid

Waentig and Gierisch (1919), cited by Hungate (1966), suggested that *in vivo* digestibility could be predicted from *in vitro* procedures that re-created the conditions of the rumen and abomasum. Of the attempts to achieve this, the most significant was the development of a gravimetric two-stage microbial digestibility assay by Tilley and Terry (1963).

In this method, strained rumen liquor, artificial saliva (McDougall, 1948) and dried milled forage samples are combined in digestion tubes. The tubes are flushed with carbon dioxide (CO_2), closed with Bunsen gas-release valves and incubated for 48 h at 38°C. The samples are agitated and the pH checked at intervals. The acidity is maintained between pH 6.7 and 6.9 to optimize conditions for microbial growth and development. After 48 h of incubation, mercuric chloride ($HgCl_2$) and sodium carbonate (Na_2CO_3) are added to inhibit further microbial activity and aid sedimentation. The supernatant is discarded following filtration or centrifugation. Acid pepsin is added to the residue and the mixture reincubated for 48 h at 38°C. The sample is recentrifuged, the supernatant discarded and the residue washed. This undigested residue is dried to a constant weight. Digestibility is calculated by dividing the weight loss of the sample by its initial weight, after correction for the blank weight of the diluted rumen liquor.

The Tilley and Terry (1963) method has been widely tested and used, and sometimes modified in its detail (e.g. Alexander and McGowan, 1966). Estimates of dry-matter digestibility obtained have shown good correlations with *in vivo* digestibilities in sheep, for a variety of fibrous foods. Marten and Barnes (1980) noted that two-stage *in vitro* fermentations had become universally recognized as methods of choice to predict digestibility of all types of forages, and McLeod and Minson (1980) stated that they were often the standards of excellence against which other procedures were compared. A valuable feature of the method is that *in vitro* estimates not only are closely correlated with *in vivo* values, but also closely approximate to them (Tilley and Terry, 1963; Pigden, 1969). This makes results easy to interpret and minimizes the reliance on standards.

Surprisingly, there are no data available to show if this *in vitro* procedure and sheep digest similar fractions of the forage, or whether the close agreement in overall apparent dry-matter digestibility is purely fortuitous. The Tilley and Terry (1963) first stage does not closely reflect the complexity of ruminant digestion. Bacterial colonies are known to be difficult to maintain unchanged *in vitro* without complex apparatus (Czerkawski, 1976). Fermentation in glass tubes is not regulated by rumination, salivation, absorption of products or fractional outflow rates. The diverse processes of digestion in the hind-gut of the ruminant are not duplicated by second-stage acid-pepsin treatment (Tilley and Terry, 1963). The justification of the method is not the re-creation of ruminant digestive functions, but that it works. The prime function of the rumen liquor is to provide a rich source of fibrolytic organisms, which leaves open the possibility that these might be more readily obtained elsewhere.

Many investigators (e.g. Reid *et al.*, 1964; Milne, 1977) have found that the activity of rumen inocula is affected by the diet of the donor. A more serious practical disadvantage is the need for fistulated donor ruminants to provide the liquor (Abonge, 1988; Omed *et al.*, 1989a). These are subject to restrictive legislation in many countries and are costly to prepare and maintain. Removal of the need for these could make the procedure more widely available. These concerns led to consideration of sheep faeces as an alternative source of fibrolytic organisms.

The use of faecal organisms for the estimation of the digestibility of forages was first explored, using liquor derived from sheep faeces, by Balfe (1985), as an undergraduate project. She used a filtered homogenate of sheep faeces in artificial saliva, instead of rumen liquor, in the modification by Rogers and Whitmore (1966) of the Tilley and Terry (1963) method. Balfe (1985) first showed that filter-paper was digested by faecal liquor, and then obtained a correlation of 0.93 between *in vitro* estimates of dry-matter digestion of three grass samples and one hay sample and their determined *in vivo* dry-matter digestibility. Her method was not convenient for batch operation, as the faecal liquor rapidly blocked the sintered filters, a feature confirmed by El Shaer *et al.* (1987) and Akhter (1994), and the undigested residues were recovered by centrifugation after transfer of the liquor from the fermentation tubes. Later, Williams (1986) demonstrated that increasing the concentration of faeces in faecal liquor increased its activity, and Puw (1986) showed that faeces from horses and cattle also contained activity appropriate for the method.

A procedure using sheep faeces, suitable for batch operation, was subsequently tested by El Shaer *et al.* (1987). The methodology of Balfe (1985) was modified in that fermentation took place in McCartney bottles fitted with wine fermentation locks. Transfer losses were avoided by centrifuging residues within the same bottles, which made it possible to minimize sample size. Omed *et al.* (1989a, b) found that the procedure could be carried out satisfactorily in sealed bottles, dispensing with the fermentation locks. The sample size was reduced to 180 mg and the faecal liquor volume scaled down correspondingly to 18 ml, with the resultant airspace at the top of the McCartney bottles providing sufficient space for gas accumulation. Fermentation gases were released by gently loosening the caps twice during the incubation period.

Despite some conflicting results concerning dietary effect on faecal liquor activity (e.g. El Shaer *et al.*, 1987; Aiple *et al.*, 1992; Akhter, 1994), the impact of this appears to be less than that observed when using rumen liquor. Thus faecal liquor is useful when donor material must be obtained from free-ranging animals.

Rumen and Faecal Fluids as Source of Microbial Enzymes

Physical degradation of ingested forages increases its surface area, resulting in rapid colonization by fibrolytic organisms and rapid solubilization of non-structural carbohydrates (McDonald *et al.*, 1995). Saliva buffers the rumen liquor. This is very important, as cellulolytic bacteria cannot grow at pH values below 6.0 (Stewart, 1977).

The anaerobic environment, fatty acids and rumen fluid rapidly destroy the plasmalemma and much of the cytoplasmic structure of the plant cells (Cheng *et al.*, 1980). Internal solubles leach out, presumably via natural openings, such as stomata, and sites of extensive cell damage (Cheng *et al.*, 1980; Halliday, 1985; McAllister *et al.*, 1994), allowing microbes direct access to the internal structures of the leaf. Bacteria attracted to these sites by the concentration of nutrients (Cheng *et al.*, 1980) bypass the resistant cuticle (Akin, 1979; Bauchop, 1980; McAllister *et al.*, 1990). Large populations of cellulolytic bacteria quickly establish around such openings (Cheng *et al.*, 1980, 1984; Craig *et al.*, 1987). Within 5 min of food ingestion, intimate bonds between substrate particles, bacteria and protozoa have been observed (Bonhomme, 1990). Thus digestion of insoluble feed materials occurs from the inside out (Chesson and Forsberg, 1988; Cheng *et al.*, 1991; McAllister *et al.*, 1994).

Intimate bacterial and forage association is a prerequisite for efficient forage digestion. Between 70 and 80% of all microbial matter in the rumen has been reported to be bonded to forages (Craig *et al.*, 1987); Czerkawski (1986) believed this figure to be nearer 95%. Hemicellulase and cellulase activities are also notably higher in the particulate fraction than in the fluid (Williams and Strachan, 1984), leaving no doubt that particulate-associated microbial populations are responsible for the majority of feed digestion in the rumen. Different internal structures themselves have differing susceptibilities to attachment by rumen microorganisms; high-lignin-content structures, such as vascular and sclerenchymal cells, have a high resistance to attachment, whilst the mesophyll and phloem are easily colonized and rapidly digested (Akin, 1979). Wilson (1990) suggested that the proportion of uncolonizable tissue often dictates the extent to which a forage within the rumen is digested.

Microbes, by attaching themselves to forage particles, not only increase their ability to deliver enzymes but also are able to extend their residence time within the rumen by avoiding passage through the reticulorumen orifice. As a consequence, high population levels can be maintained at low growth rates (Mackie and White, 1990). Four spatially distinct domains coexist and interact within the rumen (Czerkawski, 1986), enabling rapid colonization of ingested forage. These are shown in Table 7.1.

Rumen microbial species differ in their abilities to survive passage through the

Table 7.1. The spatial organization of the rumen microbes according to the compartment model of Czerkawski (1986).

Compartment	Definition	Function
1	Rumen liquor	Transport medium carrying feed particles and microbes. Soluble sugar fermentation
2	Loosely adherent to feed particles	Shuttle compartment between 1 and 3
3	Tightly adherent to feed particles	Breakdown of solid feeds
4	Adherent to rumen wall	O_2 scavenging, urea hydrolysis

O_2, oxygen.

abomasum and small intestines to the large intestine (Pounden *et al.*, 1950; Bergen *et al.*, 1967; Van Soest, 1982; Wallace, 1983). Although isolated microbes are readily killed by acidic environments, the cuticle and other lignin structures of plant particles may insulate embedded microbes from the hostile environment of the abomasum. Bhat *et al.* (1990) found that significant cellulolytic bacteria attachment to substrate occurred down to a pH of 4.0, even though they were unable to grow at a pH lower than 6.0 (Stewart, 1977).

Susceptibility of bacteria to proteolytic digestion ranges between extremes of complete destruction in the abomasum to passage through the entire digestive tract of the host (Baker and Harris, 1947; Bergen *et al.*, 1967; Wallace, 1983; Allison, 1984). Wallace (1983) established that Gram-negative species were more thoroughly digested than Gram-positive ones. Viability was not tested. By combining the findings of Bergen *et al.* (1967), Wallace (1983) and Stewart and Bryant (1988) on the degradability of bacteria exposed to acid pepsin (as shown in Table 7.2), it is possible to speculate that the large intestine will recruit a higher proportion of Gram-positive than Gram-negative bacteria from the rumen. Gram-positive bacteria, however, usually form only a minor component of the rumen microbial population (Hungate, 1966).

It has been suggested from numerical data that *Bacteroides succinogenes*, *Ruminococcus albus*, *Ruminococcus flavefaciens* and *Bacteroides fibrisolvens* were the most important cellulolytic bacteria. *Bacteroides fibrisolvens* is one of the most vigorous digesters of cellulose (Hungate, 1966). *Ruminococcus* and *Bacteroides* species have important roles in cellulose degradation (Cheng *et al.*, 1980). With the exceptions of *R. flavefaciens* and *B. succinogenes*, these have higher than average digestibility values when exposed to an acid-pepsin/pancreatin solution.

While this evidence suggests that most rumen organisms fail to survive into the large intestine, it is likely that the bacterial community there is self-sustaining. The bacterial population of the large intestine and faeces of ruminants includes numerous bacterial species also present in the rumen. These have mostly been identified at the family level and include *Bacteroides*, *Fusobacterium*, *Streptococcus*, *Eubacterium*, *Ruminococcus* and *Lactobacillus* (Allison, 1984). The dwell time permits multiplication, so species depleted by proteolytic activity have a chance to recover, provided niches are available. The hind-gut of the faeces donor needs

Table 7.2. The digestibility of bacteria when exposed to acid pepsin and pancreatin, their substrate and response to Gram stain (data from Bergen *et al.*, 1967; Wallace, 1983; Stewart and Bryant, 1988).

Bacteria	Strain	Substrate	Gram +/-	Digestibility
Bacteroides amylophilus	70	NC	Negative	67.7
Bacteroides fibrisolvens	H10b	C	Negative	84.7
Bacteroides ruminocola	H8a	NC	Negative	73.3
Bacteroides ruminocola	223	NC	Negative	94.0
Bacteroides succinogenes	A3c	C	Negative	62.0
Bacteroides succinogenes	S85	C	Negative	68.6
Bacteroides succinogenes	B21a	C	Negative	67.6
Butyrivibrio fibrisolvens	NOR37	C	Negative	90.0
Butyrivibrio fibrisolvens	B834	C	Negative	91.0
Eubacterium ruminantium	2388	NC	Positive/negative	73.0
Lactobacillus casei	LB17	NC	Positive	58.0
Megasphaera elsdenii	J1	NC	Negative	83.0
Peptostreptococcus elsdenii	H8a	NC	Negative	72.1
Ruminococcus albus	7	C	Positive/negative	80.4
Ruminococcus albus	SY3	C	Negative	96.0
Ruminococcus flavefaciens	B1a	C	Positive	44.2
Ruminococcus flavefaciens	B34b	C	Positive	85.9
Ruminococcus flavefaciens	C1a	C	Positive	63.5
Ruminococcus flavefaciens	C94	C	Positive	80.3
Selenomonas ruminantium	WPL 151/1	NC	Negative	92.0
Selenomonas ruminantium	GA192	NC	Negative	83.4
Staphylococcus	3588/2	NC	Positive	33.0
Streptococcus bovis	no. 26	NC	Positive	19.0
Streptococcus bovis	45S1	NC	Positive	37.0
Streptococcus faecium	SFDC	NC	Positive	48.0
Succinivibrio dextrinosolvens	24	NC	Negative	93.1
Veillonella alcalescens	692	C	Negative	88.0

NC, non-cellulolytic; C, cellulolytic.

slowly fermentable nutrients to stimulate multiplication of cellulolytic digesters (Aiple *et al.*, 1992). An appropriate diet will provide such nutrients for the hindgut, whilst at the same time increasing the residence time of the feedstuff. Both factors should increase the numerical presence of cellulolytic bacteria, which explains the higher *in vitro* digestibility values obtained by Pilling (1997) when diets for donor animals included hay or straw, rather than grass.

Hidayat *et al.* (1993), when investigating digestibility of forages by isolated mixed bacterial populations, established a maximal digestion rate above which the addition of more bacteria or protozoa failed to stimulate further digestion. The same was true when bacteria were added to a 100% protozoa population. This suggests that both populations are highly cellulolytic and are capable of occupying all niches within the rumen environment and presumably also in the large intestine.

Incidences of lower *in vitro* digestibilities obtained using faecal liquor when compared with rumen inocula (Manyuchi *et al.*, 1991; Akhter, 1994; Gonçalves and Borba, 1996) could be due to differences in the microflora present. It is unlikely that lower bacterial numbers in the large intestine are responsible, as the bacterial density there is similar to that in the rumen (10^{10}–10^{11} g^{-1}) (Allison, 1984). This suggests that the difference is due to fewer cellulolytic species being extracted into the liquor. It may be compounded by the supply of nutrients obtained from the rumen fluid. Corbett (1981) found that nutrient deficiencies within test forages were not detected when using rumen inocula *in vitro*, as the liquid contained sufficient trace elements and micronutrients. Faecal fluid may lack some of these, reinforcing the effect of forage deficiencies on digestibility. Few soluble carbohydrates will be available; consequently, non-cellulolytic microbial growth will be inhibited until solubilization occurs under the actions of the cellulolytic microbial population.

The liquor compartment in the rumen is large and can readily be sampled. If the nutritional state of the animal is closely controlled, this fluid should be relatively constant in composition, as should be the activity of its microbial population. Rumination and ruminal mixing continously resuspend particle-bound bacteria in the liquid fraction. There is no equivalent of this compartment in faeces as a source of fibrolytic organisms; the liquor component is too small to be accessible. Fibrolysis in faecal liquor is believed to be due to particle-bound organisms that have been detached and suspended in liquid during the preparation process. Omed *et al.* (1989b) recommended the use of 60 g fresh faeces in each litre of faeces liquor. Thus, each 18 ml aliquot of liquor applied to 180 mg forage is derived from approximately 360 mg faecal dry matter. To seed all the niches in the forage to the level ocurring in the large intestine, the extraction of fibrolytic faecal bacteria must be about 50% efficient. Effective extraction of faeces with artificial saliva is therefore necessary to make an active suspension.

Validation of the Faecal Liquor Method

The only satisfactory proof of this method would be to show that it gives estimates which agree closely and consistently with digestibilities determined *in vivo*. This should be done as a blind test on a range of forage types and on a sufficient scale to allow for the uncertainties in the *in vivo* data, as these too are subject to error (Omed *et al.*, 1989a). Table 7.3 includes data that fall short of these requirements but which are sufficient to justify a properly organized large-scale trial. The table also illustrates a problem that must be addressed before such a trial.

The time course of solubilization of dry matter from forage samples suspended in the rumen follows a well-established curve (Mehrez and Ørskov, 1977). This approaches an asymptote, taken to represent the potential degradability of the sample. Under the *in vitro* conditions of the rumen liquor or faecal liquor methods, fibre degradation follows a similar course (Akhter, 1994; Solangi, 1997). If the liquor is of high activity, degradation of most forages will be substantially completed within the prescribed 48 h, yielding results in which *in vitro* weight loss is close to the *in vivo* apparent digestibility. Unfortunately, it is not easy to gauge if fibre degradation is complete when the prescribed time is reached. Liquor of low activity may need longer than 48 h to achieve the degradation potential. At the cut-off time of 48 h,

Table 7.3. Relationship between *in vitro* dry-matter digestion % (*x*), estimated using faecal liquor from different donor species, and apparent dry-matter digestibility % (*y*), obtained using the specified standard procedure.

Species of faeces donor	Variance accounted for (r^2)	Forage type	Line of best fit	Reference
Ovine	0.93	Grasses and hay	*$y = 7.9 + 0.83x$	Balfe (1985)
	0.98	Variety of feeds	*$y = 1.003x$	El Shaer *et al.* (1987)
	RSD 2.75	Variety of forages	*$y = 0.7 + 1.01x$	Omed *et al.* (1989a)
	0.91	Grasses and straws	*$y = -3.1 + 0.99x$	Faza (1991)
	0.33	Fresh and conserved feeds	*$y = 35.4 + 0.33x$	Borba and Riberiro (1996)
	0.85 (gas)	Tropical feeds	†$y = -20 + 1.26x$	Nsahlai and Umunna (1996)
	0.95	Variety of forages	*$y = 4.0 + 0.94x$	Solangi (1997)
	0.97	Tropical feeds	†$y = 33.8 + 0.55x$	Aboud *et al.* (1999)
Bovine	0.97	Variety of forages	†$y = 17.3 + 0.98x$	Akhter (1994)
(frozen)	0.97	Variety of forages	†$y = 17.2 + 0.97x$	Akhter (1994)
	0.98†	Variety of forages	†$y = 21.4 + 0.83x$	Akhter *et al.* (1995)
	0.88	Variety of forages	*$y = 19.6 + 1.31x$	Akhter *et al.* (1995)
(freeze-dried)	0.83	Variety of forages	*$y = 26.9 + 1.15x$	Akhter *et al.* (1995)
(freeze-dried)	0.92	Variety of forages	*$y = 17.7 + 1.5x$	Akhter *et al.* (1995)
(frozen)	0.80	Variety of forages	*$y = 24.2 + 1.19x$	Akhter *et al.* (1995)
(frozen)	0.77	Variety of forages	*$y = 25.7 + 0.99x$	Akhter *et al.* (1995)
	0.94	Variety of concentrates	†not available	Harris *et al.* (1995)
	0.95	Tropical feeds	†not available	Sileshi *et al.* (1996)
	0.98	Tropical feeds	†$y = -5.8 + 1.04x$	Jones and Barnes (1996)
Equine and donkey	0.90	Tropical forages	†not available	Sileshi *et al.* (1996)
		Tropical feeds	†$y = 29.7 + 0.78x$	Aboud *et al.* (1998)
Goat	0.96	Variety of forages	‡$y = 6.7 + 0.88x$	O'Donovan (1995)
	0.97	and straws	‡$y = 5.0 + 0.86x$§	O'Donovan (1995)
Rabbit	0.96	Variety of forages	‡$y = 11.2 + 0.81x$	O'Donovan (1995)
	0.88	and straws	‡$y = -0.4 + 0.69x$§	O'Donovan (1995)

* *In vivo* apparent digestibility in sheep.
† Rumen liquor *in vitro*.
‡ Ovine faecal liquor *in vitro*.
§ Phosphate buffer used.
RSD, residual standard deviation.

digestion of fibre in some forages may be incomplete and still progressing. No stable, repeatable end-point is reached. Weight losses from standard forages may still show significant correlations with their *in vivo* apparent digestibilities, as indeed they might if merely extracted with water, but the data will reflect variations in the rates of digestion both within and between samples; and forages with highly digestible components may be overvalued. For maximum confidence in the results of a digestibility estimation, the constant a in the following equation should be very small and the coefficient b should approximate to 1, indicating that digestion has reached its natural limits:

In vivo apparent digestibility = $a + b \times$ *in vitro* digestion

Balfe (1985), El Shaer *et al.* (1987) and Omed *et al.* (1989a) all sought to achieve conditions where *in vitro* weight loss equalled *in vivo* digestibility. This particularly involved thorough extraction of the faeces with artificial saliva to suspend the maximum number of the fibrolytic organisms believed to be entrenched in plant debris. It is possible that some later workers have expected the method to be more robust and easily applied than it really is.

To achieve complete digestion when investigating low-digestibility hays and straws, it may be necessary to increase the liquor/sample ratio or to increase the duration of incubation. El Shaer *et al.* (1987) confirmed completion of the fermentation stage after 48 h by noting cessation of gas release from the fermentation locks. Solangi (1997) and Abonge (1998) collected fermentation gases over acidified water in graduated centrifuge tubes. Using the end of gas emission to indicate completion of digestion, Solangi (1997) obtained a close relationship between *in vitro* and *in vivo* digestibilities of a variety of fibrous feeds. Alternatively, routines could be developed that produce potent faecal liquor that rapidly occupies all appropriate niches in the sample (Hidayat *et al.*, 1993) for maximum rate of digestion. Fibrolytic activity could be monitored by including cellulose samples in each batch, as the digestion of pure cellulose is progressive and without asymptote.

Faeces liquor has recently been used in the investigation of the dynamics of gas production (e.g. Altaf *et al.*, 1998). This situation too may be complicated by the variability of faeces liquor preparations. Fibrous carbohydrate is slowly solubilized, at a rate dependent on the activity of the particular liquor preparation, and its conversion becomes rate-limiting on gas production when initial solubles have been used up. Thus, while the total volume of gas should be independent of the quality of the inoculum, the rate of gas production will be characteristic of the particular liquor preparation, as well as of the feedstuff.

Therefore, all potential applications need high-activity faeces liquor.

Factors Affecting the Activity of Faecal Liquor

Source of faeces

Odo (1990), Aiple *et al.* (1992), Akhter (1994) and Aboud *et al.* (1998a) found that gathering naturally voided faecal matter was inconvenient, and collected faeces directly from the anus. Comparisons were not made, but the collection method had no apparent effect.

As ovine faecal output is small (Akhter, 1994) and sheep are not universally available (O'Donovan, 1995; Sileshi *et al.*, 1996), the use of faeces from other species has been tested (Puw, 1986; Akhter, 1994; O'Donovan, 1995; Aboud *et al.*, 1998, 1999; Table 7.3).

Inocula made from bovine faeces have been investigated (Akhter, 1994; Akhter *et al.*, 1994, 1995; Harris *et al.*, 1995; O'Donovan, 1995; Jones and Barnes, 1996; Sileshi *et al.*, 1996); and it has been established that bovine faeces could be successfully utilized in forage evaluation studies. O'Donovan (1995) noted that the percentage of ryegrass digested during 48 h of fermentation with bovine liquor was less than when ovine faeces were used.

Sileshi *et al.* (1996) and Aboud *et al.* (1998, 1999) found a strong relationship between *in vitro* results using rumen liquor and equine faecal liquor, although the extent of overall digestion using equine faeces was lower than when using ruminant faeces. Similarly, O'Donovan (1995) found that faecal liquor made from horse faeces digested 8% to 16% less, respectively, from straw and ryegrass samples than did that made from sheep faeces.

In his study of six different faecal sources, O'Donovan (1995) found that goat and rabbit faeces (in phosphate buffer) produced strong correlations with *in vitro* results using ovine faecal liquor, although total digestibility was lower (Table 7.3). Guinea-pig faeces produced the lowest *in vitro* digestibility of the investigation, possibly because the pellets failed to provide a sufficiently stable internal environment for sufficient numbers of cellulolytic microorganisms to survive. All six faecal sources did, however, rank the forages in the same order of digestibility, suggesting that any of them could be utilized, providing suitable standards were included in the batch.

Where comparison has been made between liquors made from different faecal donor species, sheep faeces have been consistently shown to produce the greatest fibrolytic activity. Accordingly, sheep must be considered to be the preferred faecal source.

Effect of diet

Akhter (1994) found no significant differences between *in vitro* digestibility assays when bovine faecal liquor was prepared from concentrate-, concentrate-and-hay- or hay-fed donors, although the lowest mean results were obtained from the concentrate (barley beef) diets. This agrees with the findings of El Shaer *et al.* (1987), who noted that faeces obtained from sheep fed hay or equal weights of hay and concentrate produced indistinguishable results. Food residues reaching the large intestine are thoroughly leached, leaving mainly residual roughage. Providing sufficient fibrous residue is present in the hind-gut to sustain an active cellulolytic bacterial population, concentrate within the diet would not be expected to influence *in vitro* digestibility results.

Gas-production studies using a variety of feeds have been conducted with liquor from faeces obtained from animals fed different basal diets (Aiple *et al.*, 1992). The results are shown in Table 7.4. Small but significant effects were identified. It was also noted that inocula from hay and concentrate faeces were more

Table 7.4. Effect of diet of the donor sheep on gas production (ml 200 mg^{-1} DM) after 48 h of incubation in 1 : 30 faeces buffer dilution (from Aiple *et al.*, 1992).

Substrate/ diet	Maize : grass pellets : hay 35 : 15 : 50 (%)	Concentrate* : hay 40 : 60 (%)	Hay 1† 100 (%)	Hay 2‡ 100 (%)
Concentrate – standard	65.0^{a}	65.5^{a}	66.1^{ab}	67.4^{b}
Hay – standard	44.3^{a}	46.3^{b}	47.2^{b}	46.5^{b}
Barley	75.6^{a}	77.8^{b}	77.5^{b}	76.7^{ab}
Soybean extract meal	53.6^{a}	56.0^{b}	55.3^{ab}	57.3^{bc}
Hay	50.9^{ac}	49.4^{a}	54.2^{b}	52.3^{c}
Straw	35.4^{a}	34.1^{a}	41.2^{b}	38.6^{c}

* Concentrate, on dry-matter % basis, was wheat (25) : barley (25) : oats (20) : sugar-beet pulp (20) : soybean extract meal (10).
† 8.75 MJ ME kg^{-1} DM.
‡ 7.91 MJ ME kg^{-1} DM.
Means with different subscripts in the same line differ significantly ($P < 0.01$).
ME, metabolizable energy; DM, dry matter.

stable when stored under aerobic conditions than those obtained from donors fed on a pure hay diet.

H.M. Omed (unpublished observation) obtained low digestibilities using faeces from sheep on bare spring pastures. The faeces were greasy and without body, and the resulting liquor was of high colour and high blank weight. All these data suggest that a high fibre dietary component is necessary for donor animals.

Age of faeces

Initial studies on faecal liquor (El Shaer *et al.*, 1987; Omed *et al.*, 1989a, b) used ovine faecal pellets within 1 h of voiding, on the assumption that the pellets would only briefly maintain a suitable environment for cellulolyic bacteria. However, it was noted that faeces voided up to 6 h previously were equally effective for forage digestibility (El Shaer *et al.*, 1987). This confirmed the preliminary results of Balfe (1985), who found no difference in the digestive efficiency of fresh or old ovine faeces when applied to hay. More recently, Pilling (1997) established that ovine faeces maintained their cellulolytic capabilities for up to 24 h when stored either at room temperature or at 5°C, under anaerobic conditions. Storage for 24 h at room temperature did, however, result in a significant decline in activity. Akhter (1994), using bovine faecal liquor, noted a delay of 5 h before digestibility was reduced. In contrast, Aiple *et al.* (1992) found that gas production from hay, but not from concentrates, by faecal liquor was reduced following storage of faeces for 1 h under aerobic conditions at room temperature (Table 7.5), indicating the greater sensitivity of cellulolytic organisms than non-cellulolytic ones to storage conditions. Although total dry-matter disappearance was reduced when preserved faecal material was used, the relationship between faeces liquor digestibility and that of rumen liquor was still maintained (Akhter, 1994). Faecal pellets originating from donor

Table 7.5. Effect of storage of fresh ovine faeces under aerobic conditions and room temperature on gas production after 48 h of incubation (ml 200 mg^{-1} DM) (from Aiple *et al.*, 1992).

		Hours of storage before incubation				
Substrate	Control	1 h	2 h	4 h	12 h	24 h
Concentrate	65.8^{a}	64.9^{a}	63.9^{ab}	64.4^{ab}	61.7^{b}	54.5^{c}
Hay	46.5^{a}	41.0^{b}	39.6^{b}	40.8^{b}	34.7^{c}	33.0^{c}

Means with different superscripts in the same line differ significantly ($P < 0.01$).

sheep fed a mixed diet were observed to provide a more stable environment than those fed roughage only (Aiple *et al.*, 1992).

Most recent investigations have used freshly voided faeces (e.g. Miguel, 1994; Borba and Riberiro, 1996; Buwembo, 1996; Jones and Barnes, 1996). Nsahlai and Umunna (1996) found that faeces voided up to 2 h prior to inoculum preparation produced a significant correlation of *in vitro* faecal liquor digestibility ($r^2 = 0.85$) with *in vitro* rumen liquor, but this was lower than for other investigations using fresh faeces.

Akhter (1994) and Akhter *et al.* (1994, 1995) found that preserved bovine faeces obtained by rectal grab could also be used for digestibility assays. Both freeze-dried and deep-frozen faeces produced inocula with similar digestibility characteristics to those of fresh rumen liquor. The results obtained were highly repeatable.

It can be concluded that the method does not depend on fresh faeces, although these are most effective.

Choice of buffer

El Shaer *et al.* (1987) successfully used artificial saliva (McDougall, 1948), a bicarbonate buffer. Comparable results were obtained by Akhter *et al.* (1995), Jones and Barnes (1996), Sileshi *et al.* (1996) and Aboud *et al.* (1998, 1999) when compared with rumen liquor *in vitro* digestibility. Bicarbonate buffer (McDougall, 1948), the most common choice, when dissolved has a natural pH of 8.23. Thorough gassing with CO_2 is needed to achieve the pH of 6.8 favoured by cellulolytic bacteria (Stewart, 1977). This is a time-consuming procedure. An additional disadvantage associated with this buffer is that direct acidification following completion of the fermentation stage causes extensive frothing. It is therefore necessary to physically separate the residue from the inoculum, prior to second-stage acid digestion.

To overcome these limitations, Menke's medium (Menke *et al.*, 1979) or a modified version of it has been used when investigating gas production (Aiple *et al.*, 1992; Gonçalves and Borba, 1996), or a phosphate buffer (Pigden, 1969) when investigating *in vitro* dry-matter digestibility (Faza, 1991; O'Donovan, 1995; Omed *et al.*, 1998) or for gas-production studies (Morris, 1994; Buwembo, 1996; Solangi, 1997). Gas production is higher when bicarbonate buffer is used than that obtained with phosphate buffer (Morris, 1994; Buwembo, 1996; Solangi, 1997), since bicarbonate buffer reacts with the acid products of fermentation to release CO_2 (Solangi, 1997).

Faza (1991) reported an interaction between buffer type and the bacteria separation method on the *in vitro* digestion of straw. This was greater in bicarbonate buffer when the inoculum was prepared in a food blender than when a pestle and mortar were used. When the bicarbonate buffer was replaced by a phosphate buffer, the opposite was true. O'Donovan (1995) found that sheep and goat faeces were equally effective as agents for *in vitro* digestibility in either a phosphate or a bicarbonate buffer. Rabbit faeces, however, gave lower digestibility when used in a phosphate buffer than in a bicarbonate buffer.

Separation of bacteria

Effective faecal inocula cannot be produced from ovine faeces simply by washing (H.M. Omed, unpublished observation), as the cellulolytic organisms form strong bonds with their substrates. Various techniques have been used to break these bonds (Table 7.6) but few comparisons have been made between these methods.

Solangi (1997) found that the percentage of cellulose samples digested during 48 h incubation was greater when the faeces and artificial saliva were pummelled together in a plastic bag (75%) or ground together in a pestle and mortar (70%) than when shaken with glass beads in a bottle (67%) or homogenized in a food processor (65%). He also found that activity was increased when large sieve sizes (300–800 μm) rather than small sizes (212 μm or doubled muslin) were used, indicating that some bacteria were still particle-bound.

Cattle faeces directly strained through muslin cloth have been used by Akhter (1994) and Sileshi *et al.* (1996). Lower *in vitro* digestibility values, when compared with alternative inocula preparation methods, were obtained, perhaps due to the lack of physical disruption to which the faecal matter was subjected.

Faeces-to-buffer ratio

El Shaer *et al.* (1987) and Omed *et al.* (1989a) used 60 g of faeces 300 ml^{-1} of artificial saliva (McDougall, 1948), although later Omed *et al.* (1989b) found 60 g l^{-1} produced a more active inoculum. Bacterial separation is presumably more efficient when a higher buffer-to-faeces ratio is used.

The effect of faeces dilution on both *in vitro* digestibility systems (Akhter, 1994; Solangi, 1997) and gas production (Aiple *et al.*, 1992; Solangi, 1997) has been studied. Larger concentrations of faeces (up to 120 g l^{-1} of buffer) were found to result in increased *in vitro* digestion and/or rates of gas production (Akhter, 1994; Solangi, 1997). The digestibility was found not to increase directly in proportion to the increase in faeces utilized (Solangi, 1997). Aiple *et al.* (1992) noted that dilutions of dry matter between 1:20 and 1:50 produced similar gas volumes; above or below this, gas production from fibrous substrates was depressed.

A trade-off appears to exist between the mass of faeces used and the potency of

Table 7.6. The effectiveness of different methods of separating the cellulolytic bacteria from the faeces substrate.

Method	Species	Digestibility/ gas production	Correlation coefficient	Reference
Pestle and mortar	Ovine	Digestibility	0.98*	El Shaer *et al.* (1987)
	Ovine	Digestibility	0.88†	Nsahlai and Umunna (1996)
	Ovine	Gas production	0.85†	Nsahlai and Umunna (1996)
Homogenization in food processer	Donkey	Digestibility	0.97†	Aboud *et al.* (1998)
	Ovine		0.97†	
	Ovine	Gas production	0.93*	Aiple *et al.* (1992)
Plastic bag	Rabbit	Digestibility	0.96‡	O'Donovan (1995)
	Goat		0.96‡	
Strained through muslin	Bovine	Digestibility	0.9†	Akhter (1994)
	Bovine	Digestibility	0.982†	Jones and Barnes (1996)
	Bovine Donkey	Digestibility	0.9†	Sileshi *et al.* (1996)

* Correlation with *in vivo* digestibility.
† Correlation with rumen liquor *in vitro*.
‡ Correlation with ovine faecal liquor *in vitro*.

the inoculum. Too small a faeces mass contains insufficient cellulolytic bacteria, while too large a mass of faeces inhibits separation of bacteria (Solangi, 1997).

Addition of a nitrogen supplement

The earliest studies employing faecal liquor (Balfe, 1985; Williams, 1986) did not include a nitrogen supplement. El Shaer *et al.* (1987) found that the inclusion of urea in the buffer improved correlations with *in vivo* digestibility for all assays involving grasses. As lucernes already provided a good correlation, it was assumed that the nitrogen supply was a limiting factor. The addition of urea has now become standard to all buffer solutions utilized in Bangor (e.g. Awemo, 1988; Omed *et al.*, 1989a, b; Faza, 1991; O'Donovan, 1995). Nsahlai and Umunna (1996) and Aboud *et al.* (1998, 1999) replaced urea with ammonium sulphate and found this to be equally successful.

Manyuchi *et al.* (1991) used no nitrogen supplement and obtained low *in vitro* results when digesting tropical grasses with faecal liquor. It was for this reason that Jones and Barnes (1996), whilst digesting tropical browse, added 0.6 ml of a freshly prepared solution of urea (30 g l^{-1}) to every sample and extended the incubation period to 72 h. This produced the best combination when digesting tannin-containing forages, agreeing with earlier work on the rumen inoculum by Drew (1966).

Mixing of the sample and faeces liquor

The liquor must be mixed intimately with the forage sample. Dried and ground forage samples may clump when the inoculum is added, reducing the surface area available for colonization (Miller and Hobbs, 1994). Peel and Schofield (1993) reported a coefficient of variation between gas volumes produced by identical samples of 2.1% for samples stirred with a rod, compared with 4.2% for unstirred samples. A system of agitation during incubation therefore appears necessary.

Samples are normally shaken for 30 s to suspend the solids, after adding faeces liquor and sealing the digestion tubes (e.g. Abonge, 1988; Omed *et al.*, 1989b; Odo, 1990; Miguel, 1994). O'Donovan (1995) shook samples for 30 s. Following this initial suspension, three different forms of agitation have been used: a shaking water-bath (El Shaer *et al.*, 1987; Buwembo, 1996; Solangi, 1997), a rotor contained within an incubator (Faza, 1991; Gonçalves and Borba, 1996) or manual shaking at regular intervals (e.g. Omed *et al.*, 1989b; Odo, 1990; Morris, 1994; O'Donovan, 1995). All aim to ensure that the intimate mixing between inoculum and substrate continues. Only Faza (1991) has made a direct comparison between mixing methods. He found that manual shaking four times daily produced a value for digestibility four percentage units higher for straw than did rotating the samples at 2 r.p.m. inside an incubator. It is possible that periodical suspension allows bacteria to settle and colonize virgin substrate, whereas constant mixing prevents this.

Duration of incubation

The colonization of a fresh forage by microbes is rapid and concentrates around sites of mechanical damage and stomata (Cheng *et al.*, 1980; Halliday, 1985), where leaching of soluble carbohydrates is thought to attract bacteria (Cheng *et al.*, 1980). It is reasonable to assume that, until the internal cells are rehydrated, colonization will be haphazard and slow, leading to a lag in digestion (Spallinger, 1985).

Tilley and Terry (1963) reported that cellulolytic digestion was complete following a 48 h incubation with rumen fluid. El Shaer *et al.* (1987) and Omed *et al.* (1989a, b) showed similarly that 48 h incubations in faecal liquor were sufficient for most forages. However, for highly digestible forages, accurate predictions of digestibility (residual standard deviation (RSD) 0.74) could be obtained with a reduced incubation period of 36 h (Omed *et al.*, 1989b), while 72 h incubations were needed to provide a good correlation with *in vivo* data for straws (El Shaer *et al.*, 1987; Faza, 1991). Solangi (1997) found that at least 96 h incubation was necessary before digestion of straw reached its asymptote. Tannin-containing forages also require extended incubations when using either rumen inocula (Drew, 1966) or faecal fluid (Jones and Barnes, 1996).

Long incubations must be treated with caution as microbial lysis may occur, but variability between duplicate samples may be reduced by extending the incubation time in either rumen inocula (Barnes, 1967) or faecal fluid (Solangi, 1997). The forage being tested may be of unknown quality, preventing prediction of an appropriate incubation period. A possible way of detecting completeness of fibre digestion is to monitor gas evolution from the sample.

Factors Affecting the Second-stage Incubation

In the Tilley and Terry (1963) procedure, the residue remaining after fermentation is further digested in a second incubation, using acid pepsin to solubilize protein and protein-bound materials, before the undigested residue is dried and weighed. This is also necessary in the faecal liquor method. The residue is usually separated from the inoculum before acidification, to avoid frothing of the bicarbonate buffer. However, Alexander and McGowan (1961), using rumen inoculum in a bicarbonate buffer, achieved acidification using a syringe pushed through the cap of the fermentation tube and found this method more convenient and quicker than centrifugation. Abonge (1988), using this procedure with faecal fluid in a bicarbonate buffer, also found it convenient, but noted that great care was needed to minimize frothing.

Centrifugation of faecal liquor at 2000 g for 20 min was recommended by El Shaer *et al.* (1987) and later used by Omed *et al.* (1989a, b). O'Donovan (1995) encountered problems with centrifugation, for, upon completion, some of the forage samples failed to settle, requiring supernatant removal by pipette.

Filtration has been less satisfactory (Tilley and Terry, 1963). Filtration through sintered glass has been reported to be problematic (El Shaer *et al.*, 1987; Ahkter, 1994), as the faecal residue rapidly blocked the filter. Ahkter (1994), however, persevered with the method. Miguel (1994) attempted to utilize filter crucibles when investigating degradability, but consistently obtained low digestibility values. Filtration following secondary digestion with acid pepsin is more rapid and easier (Akhter, 1994).

Phosphate buffer does not need extensive gassing with CO_2 and allows direct acidification without frothing (Marten and Barnes, 1979; Omed *et al.*, 1998).

The acid pepsin treatment used by El Shaer *et al.* (1987) was copied from Tilley and Terry (1963), but it was recommended that the concentration of pepsin be raised from 2 g to 4 g l^{-1}, as 2 g l^{-1} produced lower-than-expected digestibilites. The correlation between *in vitro* estimates and *in vivo* digestibility remained unchanged when the length of incubation with pepsin was reduced to 36 h (Omed *et al.*, 1989b) and later to just 8 h (Solangi, 1997). These results agree with the findings of Pounden *et al.* (1950), Bergen *et al.* (1967) and Wallace (1983) that bacteria which are susceptible to proteolytic agents will be rapidly degraded. Continued solubilization will occur only at a slow rate, for the remaining bacteria population is composed of either very slowly degradable or insoluble bacteria. Short incubations, as reported by Solangi (1997), save time.

An alternative proteolytic agent, biological washing liquid (Persil) at a rate of 0.2–1 ml sample^{-1}, has been tested (Faza, 1991; Morris, 1994; Solangi, 1997; Omed *et al.*, 1998). An advantage of this agent is that acidification is not required, so frothing is avoided. Also, the final separation of the supernatant can be achieved by cautious aspiration following sedimentation, as washing liquid promotes sedimentation. Omed *et al.* (1998) reported an r^2 value of 0.953 for *in vitro* compared with *in vivo* digestibility for a combination of biological washing liquid (to solubilize the protein) and aspiration (to remove the supernatant). This compared with an r^2 of 0.944 for centrifugation followed by acid pepsin when digesting straws for 72 h. The washing liquid increased the sedimentation rate of the sample. Morris (1994) found that the use of biological washing liquid and separation by aspiration increased *in vitro* digestibility. Solangi (1997) found that replacing acid pepsin by biological washing liquid produced digestibilities very close to the known *in vivo* digestibilities for a variety of grasses, legumes and hays after 24 h of second-stage incubation.

Future of the Faecal Liquor Technique

The main strength of the faecal liquor technique is its simplicity, combined with its minimal cost when compared with the methods based upon rumen inocula. It uses freely available waste products from farm animals managed under normal commercial conditions, saving the cost and inconvenience of maintaining fistulated ruminants. The materials needed are intact sheep for faeces, bottles, cheap chemicals, carbon dioxide, a 39°C incubator, a centrifuge and a balance (Omed *et al.*, 1989a). The operator needs few chemist's skills. There is also evidence that the centrifuge and carbon dioxide can be dispensed with (Omed *et al.*, 1998). The other requirements are of insignificant cost, apart from the major items of laboratory equipment. If animal warmth could be used to replace the incubators and the preliminary findings of Omed *et al.* (1998) are confirmed, the irreducible need will be for a sensitive weighing-machine. In view of the potential use of this technique, this needs further investigation.

In technically advanced countries, the faster, cleaner method of near-infrared reflectance spectroscopy (NIRS) is more attractive for estimating digestibility, although faecal liquor may find application in studies on the dynamics of digestion. The faeces liquor procedure would therefore seem to be particularly suitable where potential food sources need to be screened using limited equipment. This type of application would be facilitated if the method could be made consistently direct reading, with digestibility being equal to *in vitro* solubility, as discussed above. This would also give it an advantage over NIRS, which always needs a calibration set of standards.

References

Abonge, G.A. (1988) The use of sheep faecal microorganisms in digestibility studies. MSc thesis, University College of North Wales, Bangor.

Aboud, A.A.O., Owen, E., Chenyambuga, S.W. and Kimambo, A.E. (1998) The use of donkey faeces for *in vitro* evaluation of dry matter digestibility of some tropical forages. In: *Proceedings of Animal Traction Workshop, ILRI, Addis Ababa, 5–9 May 1997.*

Aboud, A.A.O., Chenyambuga, S.W., Owen, E., Kimambo, S.W. and Mnembuka, B.V. (1999) Evaluation of sheep and donkey faeces as sources of inocula for the *in vitro* dry matter digestibility determination in tropical forages. *Sudan Journal of Animal Sciences* (in press).

Aiple, K.P., Steingass, H. and Menke, K.H.O. (1992) Suitability of a buffered faecal suspension as the inoculum in the Hohenheim gas test. *Journal of Animal Physiology and Animal Nutrition* 67, 57–66.

Akhter, S. (1994) Use of cow faeces to provide microorganisms for the *in vitro* digestibility assay of forages. PhD thesis, University of Reading, UK.

Akhter, S., Owen, E., Fall, A., O'Donovan, F. and Theodorou, M.K. (1994) Use of fresh or frozen cow faeces instead of sheep rumen liquor to provide microorganisms for *in vitro* digestibility assays of forages. Presented at British Society of Animal Production Winter Meeting, Paper 100.

Akhter, S., Owen, E., Theodorou, M.K., Tembo, S.L. and Deaville, E.R. (1995) Repeatability of *in vitro* digestibility assays of forages when using fresh, frozen or freeze-dried cow faeces instead of sheep rumen liquor as sources of microorganisms. Presented at British Society of Animal Production Winter Meeting, Paper 110.

Akin, D.E. (1979) Ultra structure of rumen bacteria attachment to forage cell walls. *Applied Environmental Microbiology* 31, 562–568.

Alexander, R.H. and McGowan, M. (1961) A filtration procedure for the *in vitro* determination of herbage digestibility. *Journal of British Grassland Society* 16, 275–277.

Alexander, R.H. and McGowan, M. (1966) The routine determination of *in vitro* digestibility of organic matter in forages – an investigation of the problems associated with continuous large scale operation. *Journal of British Grassland Society* 21, 140–147.

Allison M.J. (1984) Microbiology of the rumen and small and large intestines. In: Swenson, M.J. (ed.) *Duke's Physiology of Domestic Animals*, 10th edn. Cornstock Publishing Associates, Cornell University Press, New York, pp. 340–350.

Altaf, U.R., Mauricio, R., Mould, F.L., Smith, T., Owen, E., Phipps, R.H. and Theodorou, M.K. (1998) Comparison of bovine rumen liquor and bovine faeces as sources of microorganisms for the *in-vitro* gas production technique for assessing silages of maize and maize plant fractions. In: *Proceedings of the British Society for Animal Sciences*. British Society of Animal Sciences, Scarborough, p. 59.

Baker, F. and Harris, S.T. (1947) The role of the microfauna of the alimentary tract of herbivora with special reference to ruminants. 2. Microbial digestion in the rumen (and caecum), with special reference to the decomposition of structural cellulose. *Nutritional Abstracts and Reviews* 17, 3–12.

Balfe, B. (1985) The development of a two-stage technique for the *in vitro* digestion of hay using ovine faeces (instead of rumen liquor) as a source of microorganisms. BSc (Hons) dissertation. University of Wales, Bangor.

Barnes, R.F. (1967) Collaborative *in vitro* rumen fermentation studies on forage substrates. *Journal of Animal Science* 26, 1120–1130.

Bauchop, T. (1980) Scanning electron microscope in study of microbial digestion of plant fragments in the gut. In: Elwood, D.C., Hedger, J.N., Lantham, M.J., Lynch, J.M. and Slater, J.H. (eds) *Contemporary Microbial Ecology*. Academic Press, New York, pp. 101–108.

Bergen, W.G., Purser, D.B. and Cline, J.H. (1967) Enzymatic determination of the protein quality of individual rumen bacteria. *Journal of Nutrition* 92, 357–364.

Bhat, S., Wallace, R.J. and Ørskov, E.R. (1990) Adhesion of cellulolytic ruminal bacteria to barley straw. *Applied and Environmental Microbiology* 56, 2698–2703.

Bonhomme, A. (1990) Rumen ciliates: their metabolism and relationships with bacteria and their hosts. *Animal Feed Science and Technology* 30, 203–266.

Borba, A.E.S. and Riberiro, J.M.C.R. (1996) A comparison of alternative sources of inocula as an *in vitro* digestibility technique. *Annales Zootechnique* 45, 89–95.

Buwembo, F. (1996) *In vitro* digestibility and gas production of selected carbohydrates and forages when incubated in faeces liquor. MSc thesis, University College of North Wales, Bangor.

Cheng, K.J., Fay, J.P., Howarth, R.E. and Costerton, J.W. (1980) Sequence of events in the digestion of fresh legume leaves by rumen bacteria. *Applied Environmental Microbiology* 40, 613–625.

Cheng, K.J., Stewart, C.S., Dinsdale, D. and Costerton, J.W. (1984) Electron microscopy of the bacteria involved in the digestion of plant cell walls. *Animal Feed Science and Technology* 10, 93–120.

Cheng K.J., Forsberg, C.W., Minato, H. and Costerton, J.W. (1991) Microbial ecology and physiology of feed degradation within the rumen. In: Tsuda, T., Sasaki, Y. and Kawashuma, R. (eds) *Physiological Aspects of Digestion and Metabolism in Ruminants*. Academic Press, Toronto, pp. 595–624.

Chesson, A. and Forsberg, C.W. (1988) Polysaccharide degradation by rumen microorganisms. In: Hobson, P.N. (ed.) *The Rumen Microbial Ecosystem*. Elsevier Science Publishing, New York, pp. 251–284.

Corbett, J.L. (1981) Measurement of the forage intake of grazing animals. In: Wheeler, J.L. and Mochrie, R.D. (eds) *Forage Evaluation: Concepts and Techniques*. Commonwealth Scientific and Industrial Research Organisation and America Forage and Grassland Council, East Melbourne, pp. 287–297.

Craig, W.M., Broderick, G.A. and Bicker, D.B. (1987) Quantivation of microorganisms associated with the particulate phase of ruminal ingesta. *Journal of Nutrition* 117, 56–62.

Czerkawski, J.W. (1976) The artificial rumen. *Lab Practice* 25, 15–20.

Czerkawski, J.W. (1986) Degradation of solid feeds in the rumen: spatial distribution of microbial activity and its consequences. In: Milligan, L.P., Grovum, N.L. and Dobson, A. (eds) *Control of Digestion and Metabolism in Ruminants*. Prentice Hall, Oxford, New Jersey, pp. 158–172.

Drew, K.R. (1966) The *in vitro* prediction of herbage digestibility. *Proceedings of the New Zealand Society of Animal Production* 265, 52–70.

El Shaer, H.M., Omed, H.M., Chamberlain, A.G. and Axford, R.F.E. (1987) Use of faecal organisms from sheep for the *in vitro* determination of digestibility. *Journal of Agricultural Science, Cambridge* 109, 257–259.

Faza, A.M. (1991) A study of factors affecting the efficiency and accuracy of a method for the *in vitro* estimation of digestibility of fibrous forages. MSc thesis, University College of Wales, Bangor, UK.

Gonçalves, L.M.B.O. and Borba, A.E.S. (1996) Study of gas producing capacity of three sources of inocula. *Journal of Agricultural Science, Cambridge* 127, 511–515.

Halliday, L.J. (1985) Studies of rumen degradation in relation to the utilisation of forage. PhD thesis, University of North Wales, Bangor, UK.

Harris, D.M., Barlet, A. and Chamberlain, A.T. (1995) The use of dairy cow faeces rather than rumen liquor in the gas pressure transducer technique for assessing digestion kinetics *in vitro*. Presented at British Society of Animal Science, Winter Meeting, Paper 113.

Hidayat, B., Hillman, K., Newbold, C.J. and Stewart, C.S. (1993) The contributions of bacteria and protozoa to ruminal forage fermentation *in vitro*, as determined by microbial gas production. *Animal Feed Science and Technology* 42, 193–208.

Hungate, R.E. (1966) *The Rumen and its Microbes*. Academic Press, New York.

Jones, R.J. and Barnes, P. (1996) *In vitro* digestibility assessment of tropical shrub legumes using rumen fluid or faecal fluid as an inoculum source. *Tropical Grasslands* 30, 374–377.

McAllister, T.A., Cheng, K.J., Rode, L.M. and Forsberg, C.W. (1990) Digestion of barley, maize and wheat by selected species of ruminal bacteria. *Applied and Environmental Microbiology* 56, 3146–3153.

McAllister, T.A., Bae, H.D., Jones, G.A. and Cheng, K.J. (1994) Microbial attachment and feed digestion in the rumen. *Journal of Animal Science* 72, 3004–3018.

McDonald, P., Edwards, R.A., Greenhalgh, J.F.D. and Morgan, C.A. (1995) *Animal Nutrition*. Longman Singapore Publishers, Singapore.

McDougall, E.I. (1948) Studies on ruminant saliva. 1. The composition and output of sheep saliva. *Biochemistry Journal* 43, 99–109.

Mackie, R.I. and White, B.A. (1990) Symposium: Rumen Microbial Ecology and Nutrition: recent advances in rumen microbial ecology and metabolism: potential impact on nutrient output. *Journal of Dairy Science* 73, 2971–2995.

McLeod, M.N. and Minson, D.J. (1969) Sources of variation in the *in vitro* digestibility of tropical grasses. *Journal of the British Grassland Society* 24, 244–249.

Manyuchi, B., Rusike, E. and Chakoma, C. (1991) Comparison of the use of rumen fluid or dung as a source of microbial inoculum for the digestion of forages *in vitro*. *Zimbabwe Journal of Agricultural Research* 29, 17–25.

Marten, G.C. and Barnes, R.F. (1980) Prediction of energy digestibility of forage with *in vitro* rumen fermentation and fungal enzyme systems. In: Balch, C.C. and Gramhan, M. (eds) *Standardisation of Analytical Methodology for Feeds*. Proceedings of International Development Resource Centre, Ottawa, Canada, pp. 61–71.

Mehrez, Z.A. and Ørskov, R.E. (1977) A study of the artificial fibre technique for determining the digestibility of feeds in the rumen. *Journal of Agricultural Science, Cambridge* 88, 645–650.

Menke, K.H., Raab, L., Salewski, A., Steingass, H., Fritz, D. and Schneider, W. (1979) The estimation of the digestibility and metabolisable energy content of ruminant feedstuffs from the gas production when they are incubated with rumen liquor *in vitro*. *Journal of Agricultural Science, Cambridge* 93, 217–222.

Miguel, L.P. (1994) A study of the application of sheep faecal organisms to the estimation of dry matter degradability of forages by ruminants. MSc thesis, University College of Wales, Bangor, UK.

Miller, J.R. and Hobbs, N.T. (1994) Effect of forage hydration on lag time during *in vitro* digestion of meadow hay. *Grass and Forage Science* 49, 107–110.

Milne, J.A. (1977) A comparison of methods of predicting the *in-vivo* digestibility of heather by sheep. *Journal of the British Grassland Society* 32, 141–147.

Morris, C.H. (1994) A study of factors affecting the *in vitro* digestibility of forages by the faecal organisms of sheep. BSc (Hons) dissertation, University of Wales, Bangor, UK.

Nsahlai, I.V. and Umunna, N.N. (1996) Comparison between reconstituted sheep faeces and rumen fluid inocula and between *in vitro* and *in sacco* digestibility methods as predictors of intake and *in vivo* digestibility. *Journal of Agricultural Science, Cambridge* 126, 235–248.

Odo, M.C. (1990) Straw degradability: a comparison of the dacron bag and *in vitro* ovine faecal microorganism technique. MSc thesis, University of Wales, Bangor, UK.

O'Donovan, F. (1995) A comparison of faeces sources and buffers in the application of the faecal liquor technique. MSc thesis, University of Wales, Bangor, UK.

Omed, H.M., Axford, R.F.E., Chamberlain, A.G.C. and Givens, D.I. (1989a) A comparison of three laboratory techniques for estimation of the digestibility of feedstuffs for ruminants. *Journal of Agricultural Science, Cambridge* 113, 35–39.

Omed, H.M., Axford, R.F.E., Chamberlain, A.G.C. and Givens, D.I. (1989b) Development of a method for the *in vitro* estimation of forage in ruminants. In: Phillips, C.J.C. (ed.) *New Techniques in Cattle Production*. Butterworth, London, p. 227.

Omed, H.M., Faza, A., Axford, R.F.E. and Givens, D.I. (1998) A low tech *in-vitro* procedure using faecal liquor for the estimation of digestibility of forages. In: *Proceedings of the British Society for Animal Sciences*. British Society of Animal Sciences, Scarborough, p. 59.

Peel, N.A. and Schofield, P. (1993) Computerised monitoring of gas production to measure forage digestion *in vitro*. *Journal of Dairy Science* 76, 1063–1073.

Pigden, W.J. (1969) Laboratory analysis of herbage used to predict nutritive value. In:

Campbell, J.B. (ed.) *Experimental Methods for Evaluating Herbage*. Publication 1315, Canada Department of Agriculture, Ottawa.

Pilling, D. (1997) The study of the effect of diet types and the age of faeces on the quality of faeces liquor for *in vitro* digestibility. MSc thesis, University of Wales, Bangor, UK.

Pounden, W.D., Ferguson, L.C. and Hibbs, J.W. (1950) The digestion of rumen microorganisms by the host animal. *Journal of Dairy Science* 33, 565–572.

Puw, T. (1986) The development of a two-stage *in vitro* technique for determining the digestibility of forages in ruminants, using equine and bovine faeces as a source of microorganisms. BSc (Hons) dissertation, University of Wales, Bangor, UK.

Reid, R.L., Jung, J.A. and Murray, A.S. (1964) The measurement of nutritive quality of blue grass pasture using *in-vivo* and *in-vitro* techniques. *Journal of Animal Science, Cambridge* 23, 700–710.

Rogers, H.H. and Whitmore, E.T. (1966) A modified method for the *in vitro* determination of herbage digestibility in plant breeding studies. *Journal of the British Grassland Society* 21, 150–152.

Sileshi, Z., Bediye, S. and Owen, E. (1996) Faeces as source of microbial inoculum in *in vitro* digestibility of forages. In: *Proceedings 4th Conference*, ESAP. Ethiopian Society for Animal Production, Addis Ababa, pp. 139–144.

Solangi, A.A. (1997) Studies on the use of faecal organisms in the *in vitro* assessment of forages. PhD thesis, University of Wales, Bangor, UK.

Spallinger, D.E. (1985) The dynamics of forage digestion and passage in the rumen of mule deer and elk. PhD thesis, Washington State University, Pullman, USA.

Stewart, C.S. (1977) Factors affecting the cellulolytic activity of rumen contents. *Applied Environmental Microbiology* 33, 497.

Stewart, C.S. and Bryant, M.P. (1988) The rumen bacteria. In: Hobson, P.N. (ed.) *The Rumen Microbial Ecosystem*. Elsevier Applied Science, London and New York, pp. 21–75.

Tilley, J.M.A. and Terry, R.A. (1963) A two-stage technique for the *in vitro* digestion of forage crops. *Journal of British Grassland Society* 18, 104–111.

Van Soest, P.J. (1982) *Nutritional Ecology of the Ruminant*. O and B Books, Carvallis.

Wallace, R.J. (1983) Digestion of rumen bacteria *in vitro*. *British Journal of Nutrition* 49, 101–108.

Williams, A.G. and Strachan, N.H. (1984) The distribution of polysaccharide degrading enzymes in the bovine rumen digestive system. *Current Microbiology* 10, 215–220.

Williams, K. (1986) The development of a two-stage *in vitro* technique for determining the digestibility of hay in ruminants using ovine faeces as a source of rumen microorganisms. BSc (Hons) thesis, University of Wales, Bangor, UK.

Wilson, J.R. (1990) Influence of plant anatomy on digestion and fibre breakdown. In: Akin, D.E., Ljungdahl, L.G., Wilson, R.J. and Hams, P.J. (eds) *Microbial Opportunities to Improve Lignocellulose Utilisation by Ruminants*. Elsevier Science Publishing, New York, pp. 99–117.

8 Enzyme Techniques for Estimating Digestibility

D.I.H. Jones and M.K. Theodorou

Institute of Grassland and Environmental Research, Plas Gogerddan, Aberystwyth, Ceredigion SY23 3EB, UK

The development and application of laboratory methods for predicting the nutritive quality of forages have been an active and successful area of research during the latter half of this century. An important development has been the introduction of biological methods, using a rumen microbial inoculum as a digestion medium in place of the chemical analysis for constituents such as fibre, cellulose and lignin. Although methods using microbial inocula have been widely and successfully used for a range of different forages, there are inherent problems in their use. Attention has, therefore, been focused more recently on the replacement of microbial inoculum methods by enzyme techniques which use crude cellulase preparations.

The Need for Alternatives to Rumen Fluid

Methods of predicting *in vivo* digestibility using rumen microbial inoculum have become well established, particularly the two-stage *in vitro* technique of Tilley and Terry (1963). It has been shown that their method gives results well related to *in vivo* digestibility for a range of forages and under a variety of growing conditions. There are, however, inherent problems in the use of microorganisms in rumen fluid for forage analysis.

1. The activity of the rumen microorganisms is lost or altered by freeze-drying or other methods of preservation. Fresh rumen fluid must therefore be readily accessible, which entails having fistulated sheep or cattle available as donor animals.
2. The donor animals need to be maintained on a standard feeding regime for long periods to minimize variability in inoculum potency. There is therefore a high and costly labour input involved in the feeding and care of animals.
3. There are analytical difficulties in using rumen fluid, e.g. the need to maintain anaerobic conditions, the high viscosity of digesta precluding filtration, unpleasant smells and the need to observe strict hygiene to avoid possible risk of infection from pathogens.

4. Increasingly, there is concern over the use of surgically modified animals in experimentation, and alternative methodologies need to be used where appropriate.

The main disadvantage of using rumen microbial inoculum in routine analytical procedures, however, is the variability that is found in the potency of the inoculum from week to week. The factors influencing the potency of inoculum have been widely studied, particularly in relation to the Tilley and Terry (1963) technique. Significant variation in the activity of inoculum has been shown to occur between different donor animals and for the same animal on different days (Drew, 1966; Yates and Allden, 1966). Troelsen and Hanel (1966) suggest that the consumption of water by sheep at varying intervals prior to collecting inoculum may explain much of the variation in the potency of inoculum; different methods of processing inoculum have been proposed in attempts to reduce variability (Barnes, 1967). McLeod and Minson (1969) have stressed the need for proper calibration when using microbial inoculum methods and regard the inclusion of appropriate standard forages of known *in vivo* digestibility within each analytical batch to be essential. The inclusion of standards allows correction of the results relative to standard values; in practice, it is often difficult to make a valid correction unless a large number of standard samples are included in each analytical batch.

The use of cellulolytic enzymes as alternatives to rumen fluid is clearly an attractive prospect if their use gives results comparable to those of inoculum methods. The use of enzymes eliminates the need for fistulated animals and anaerobic procedures, thereby simplifying analytical methodology and eliminating the variability in activity between analytical batches.

Characteristics and Activities Required for Enzyme Techniques

The chemical constituents of forages can be divided into those making up the structure of the plant (cell-wall constituents) and those contained within the cell (cell contents). This partition can be conveniently defined analytically by the van Soest (1963a, b) neutral- and acid-detergent procedures. The cell contents are essentially completely digestible *in vivo* and comprise soluble carbohydrates, starch (in some forages), protein, organic acids, lipids and soluble minerals. The cell-wall constituents (essentially cellulose, hemicellulose and lignin) vary in digestibility depending on their polymeric configuration, degree of crystallinity and degree of lignification. Lignin itself is largely indigestible under anaerobic conditions and its presence exerts a profound effect on the digestibility of the cell-wall fractions associated with it. The *in vivo* digestibility of the cellulose and hemicellulose therefore varies from completely digested when unlignified to completely undigested when highly lignified; the cuticle and surface waxes are also indigestible to a large extent.

The enzyme activities sought for analytical procedures to predict *in vivo* digestibility need to reflect the digestive processes in the ruminant. The analysis needs to remove the soluble constituents (cell contents) and to solubilize unlignified and even moderately lignified cell wall to a significant extent. The extent of solubilization does not necessarily need to be as complete as *in vivo* but needs to reflect the effects of lignification in a comparable and correlated way.

Soluble constituents are readily removed in the pretreatments with acid pepsin or neutral detergent, which form part of most of the proposed enzyme methods. The presence of starch in certain forage crops (maize, cereals) needs to be taken into account, since it is not removed by acid pepsin, neutral detergent or incubation with cell-wall-degrading enzymes. An additional treatment with amyloglucosidase or mild acid hydrolysis is therefore required in the analysis of forages containing starch.

The cell-wall-degrading enzyme used must be sufficiently active to achieve the second objective, i.e. the solubilization of a significant amount of the cell wall. Crude 'cellulase' preparations are readily available in liquid or solid form; all appear to hydrolyse both cellulose and hemicelluloses (Table 8.1). There is a wide disparity in the activity of cellulases from different fungal sources (Table 8.2) and it is essential to establish the activity of the cellulase by reference to plant cell-wall material or cellulose paper and not to soluble derivatives, such as carboxymethylcellulose. Crude cellulases from *Trichoderma* species have generally been found to be the most reliable sources of crude cellulase, but the activity of enzymes from different suppliers should always be checked before use.

Sources of Cellulolytic Enzymes

Fungi and other microorganisms producing enzymes able to digest plant fibre are very widely distributed in nature, although those able to degrade crystalline cellulose are less common. The microbial degradation of cellulose and its related polysaccharides in soil is a vital process in the maintenance of the carbon cycle, enabling the decay of vegetation and the recycling of nutrients. Microorganisms in the gut of herbivores enable these animals to survive on diets largely composed of plant fibre; vast areas of natural and improved grasslands are thereby utilized for the production of food.

There is, therefore, a wide choice of potential sources of cellulases for use in forage analysis. Ideally, the enzyme system used to predict *in vivo* digestibility would be derived from rumen microorganisms. Cell-free extracts from disrupted mixed rumen bacteria have been shown to rapidly hydrolyse plant hemicelluloses (Bailey and McRae, 1970); these extracts, however, have little or no activity against purified crystalline cellulose, although they do attack the amorphous polymer to varying degrees. Attempts to extract active cellulase preparations from rumen

Table 8.1. Hydrolysis (%) by *Trichoderma* cellulase of cellulose and hemicellulose isolated from ryegrass (from Jones and Hayward, 1973).

	Hours' incubation			
	3	6	24	48
Cellulose	42	66	98	101
Hemicellulose	39	59	73	77

microorganisms for use in enzyme assays have been largely unsuccessful in comparison with aerobic fungi and gave poor yields. This is undoubtedly related to the anaerobic growth habit of microorganisms in the rumen, which means that the yield of extracellular hydrolases is poor in comparison with that of their aerobic counterparts. Other significant factors in enzyme production include the fact that rumen bacterial cellulases are not extracellular but are often produced in a large complex, the cellulosome, which is tightly associated with the bacterial cell or bound to the substrate that is being degraded. The hydrolytic enzymes produced by the anaerobic gut fungi are an exception, however, and culture filtrates of these organisms do have considerable extracellular hydrolase activity, probably because they have very high specific activities. Nevertheless, the ability to produce significant quantities of gut fungal hydrolases is limited by the fact that these organisms are obligate anaerobes and large-scale industrial fermentation systems do not exist for their exploitation.

In view of the difficulties in obtaining cellulolytic enzymes from rumen sources, attention has been directed towards the aerobic cellulolytic fungi. The cellulase and associated enzyme systems of these fungi are extracellular and are therefore readily recovered from the culture fluid in which the organism is grown. Culture solutions comprise a mixture of salts, nutrients and cellulose at the requisite pH. After a growth period of 7–10 days under aerobic conditions, the culture fluid is removed by filtration and freeze-dried or an acetone powder is prepared. Suitable conditions for the preparation of an active cellulase–hemicellulase from *Trichoderma* species have been described by Reese and Mandels (1963).

Trichoderma cellulases have generally been found to be more active than those of other fungi, when measured using native cellulose or plant cell walls as substrates. Extracts from *Trichoderma* have therefore been extensively studied to determine the nature of the cellulase complex. It is evident that the complex comprises several enzymes necessary for attack of highly orientated forms of cellulose. This includes exocellulase and endocellulases, which attack amorphous regions of the cellulose polymer, producing shorter-chain oligomers, and a cellobiohydrolase, which hydrolyses cellobiose to glucose. All components apparently need to be present for the efficient solubilization of cotton cellulose with complicated synergistic effects existing between the different enzymes. It is clearly essential that cellulases used for forage analysis retain all the components of the cellulase complex. Attempts by suppliers to purify or concentrate crude cellulase extracts may be responsible for the poor activity noted in cellulases from some fungal sources (Table 8.2).

Table 8.2. Activity of cellulases from different fungal sources (from Jones and Hayward, 1975).

	% dry matter solubilized	
	Herbage	Cellulose paper
Trichoderma spp.	57	69
Basidiomycete	48	20
Rhizopus spp.	35	7
Aspergillus niger	45	10
Buffer only	37	5

The Development and Validation of Cellulase-based Techniques

Relationship of solubility in cellulase to in vivo *digestibility*

The earliest attempt to replace rumen inoculum as a digestion medium for the laboratory estimation of forage digestibility appears to be that of Donefer *et al.* (1963). These workers found the solubility of forage in cellulase, or cellulase with pepsin, to be no better related to *in vivo* digestibility than solubility in buffer solution without enzyme; solubility in cellulase was better related to voluntary intake than to digestibility. It should be noted that the cellulase used by Donefer *et al.* appears to have a low activity and that the concentration of cellulase used was much lower than that used by subsequent workers.

The first successful use of cellulase as a replacement for inoculum was reported by Jarrige *et al.* (1970), using a crude cellulase derived from a soil basidiomycete. Jarrige *et al.* found solubility in the cellulase to be highly correlated (r = 0.92) with *in vivo* digestibility for grass and legume forages. Cellulase solubility gave a better prediction of digestibility than the Tilley and Terry (1963) inoculum method and was highly reproducible, with only a third of the standard error of the inoculum method. There was no significant week-to-week variation using cellulase but variability using inoculum was highly significant between weeks. It was shown that the crude cellulase also contained an active hemicellulase and protease.

Jones and Hayward (1973) found the solubility of dried grasses in *Trichoderma* cellulase to be highly correlated (r = 0.92) with *in vivo* digestibility. The crude cellulase used contained 71 g kg^{-1} protein and 640 g kg^{-1} of water-soluble carbohydrate, half in the form of reducing sugars. The enzyme preparation contained active cellulases, hemicellulases and proteases, and completely hydrolysed isolated cellulose and 70% of isolated hemicellulose from ryegrass in 24 h (see Table 8.1); the cellulase removed about half the total protein in the grasses in 24 h.

The amount of dry matter solubilized by the enzyme preparations used by Jarrige *et al.* (1970) and Jones and Hayward (1973) was 15–25 units lower than *in vivo* or *in vitro* (inoculum) values; a regression equation was therefore required to relate cellulase solubility to *in vivo* digestibility. It has been well established that cellulases from different fungal sources or from different suppliers can vary greatly in their capacity to solubilize native cellulose or plant cell walls (Jones and Hayward, 1975; Tinnimit and Thomas, 1976; Rexen, 1977). Cellulases from *Trichoderma* have generally been found to be the most effective and consistent. Some cellulases have no greater solubilizing activity than simple buffer solutions (Table 8.2).

The effectiveness of *Trichoderma* cellulase is confirmed by other studies, showing simple solubility in cellulase to be well related to *in vitro* (inoculum) digestibility of grasses and legumes (Pulli, 1976) and to be applicable to the evaluation of fresh sorghum leaf (Tarumoto and Masaoka, 1978) and treated wood products (Moore *et al.*, 1972).

Other workers have studied the effects of combining cellulase digestion with subsequent treatment with other enzymes, notably proteases. Guggolz *et al.* (1971) found solubility in cellulase followed by pronase to be well related to *in vivo* digestibility for a range of forages, cob residues, straw, etc. Different regression equations were required for the different materials assayed, but, subject to this provision, cellulase solubility was considered a suitable alternative to inoculum methods. *Trichoderma* cellulase was found by McQueen and van Soest (1975) to digest forage cell walls more efficiently than *Aspergillus* cellulase; incubation with acid pepsin following cellulase treatment increased the amount of dry matter solubilized. Cellulase–pepsin solubility was highly correlated (r = 0.87) with *in vivo* digestibility, but different regression equations were required for grasses and legumes. *In vitro* (inoculum) digestibility predicted *in vivo* digestibility more precisely (r = 0.95) and was less influenced by species differences. Similarly, Tinnimit and Thomas (1976) concluded that *in vitro* (inoculum) digestion gave a better prediction of *in vivo* digestibility for tropical grasses and legumes than an enzyme method involving successive incubations with cellulase, amylase and pepsin.

The effect of different pretreatments on cellulase digestion

Incubation with acid pepsin

Incubation with pepsin in 0.1 mol l^{-1} hydrochloric acid prior to cellulase treatment was shown by Jones and Hayward (1975) to increase the amount of dry matter solubilized and to significantly improve the correlation of enzyme solubility with the *in vivo* digestibility of grasses (Table 8.3).

Incubation with pepsin after cellulase only slightly improved the correlation with *in vivo*, while pretreatment with 0.1 mol l^{-1} hydrochloric acid had no effect on the correlation. It was concluded that acid–pepsin treatment improved the accessibility of the cell-wall polysaccharides to the cell-wall-degrading enzymes. Pepsin–cellulase solubility was also found to be highly correlated with *in vitro* (inoculum) digestibility for grasses (r = 0.96) and clover (r = 0.94), with very similar regression equations derived from the different species. The pepsin–cellulase method was found to be highly reproducible, with a coefficient of variation of 1.09% for successive batches of analyses. The technique has been in routine use at the Institute of Grassland and Environmental Research for the past 20 years. The only modifications to the original method described by Jones and Hayward (1975) are: (i) the reduction of incubation time in cellulase from 48 h to 24 h; and (ii) the use of a liquid *Trichoderma* cellulase (Celluclast 1.5 L from Novo Nordisk 4/5, 2880 Bagsvaerd, Denmark). The liquid concentrate is diluted 10 ml l^{-1} with the acetate buffer previously described and 20 ml used 200 mg^{-1} of dried forage. The enzyme concentrate is purchased in 5 l quantities and each batch calibrated against a set of

Table 8.3. The effect of pepsin pretreatment on the relationship of cellulase solubility with *in vivo* digestibility (from Jones and Hayward, 1975).

Method	Correlation coefficient (r)	Residual standard deviation
Cellulase only	0.87	3.74
Pepsin + cellulase	0.93	3.17
In vitro (inoculum)	0.91	3.42

samples of known *in vivo* digestibility. The regression equations used to predict digestibility are then adjusted, if necessary; changes in the equation over many years have been minor and would have little effect on predicted values.

Many workers have shown the pepsin–cellulase technique, as described by Jones and Hayward (1975) or with minor modifications, to predict *in vivo* digestibility with an accuracy comparable to inoculum methods. The method has been successfully applied to a variety of materials, including tropical grasses and legumes (Adegbola and Paladines, 1977), grass silages (Downman and Collins, 1977), hays (Adamson and Terry, 1980), treated straw (Rexen, 1977) and kale cell walls (McCluskey *et al.*, 1984). Goto and Minson (1977) found pepsin–cellulase solubility to be highly correlated ($r = 0.94$) with *in vivo* digestibility for both tropical and temperate grasses over a range of 40–76% *in vivo* digestibility. Using separate regressions for different grass species reduced the standard error of prediction. It was concluded that the enzyme method could accurately predict *in vivo* digestibility, provided samples of known *in vivo* digestibility similar to those being tested were included as standards in each run.

Bugge (1980) showed a simple digestion with cellulase to predict the digestibility of grasses as precisely as the pepsin–cellulase method; predicted digestibility for clover and high-protein feeds (peas, beans, fish-meal), however, was much improved by pepsin pretreatment. Clark and Beard (1977) found pepsin pretreatment to improve correlations of cellulase solubility with *in vivo* digestibility for grain diets and forages; differences in substrate solubility by *Aspergillus* compared with *Trichoderma* cellulases were far less marked if samples were pepsin-treated. A cellulase from *Aspergillus niger* appeared to contain an amylase and was better suited for the analysis of cereal feeds than *Trichoderma* cellulase. Allison and Borzucki (1978) claimed that increasing the acid concentration for the pepsin stage and the incubation temperature to 50°C gave pepsin–cellulase solubility values comparable to *in vitro* digestibility. Jones and Hayward (1982) compared this and other suggested modifications, but could find no improvement over the original procedure.

Terry *et al.* (1978) compared the pepsin–cellulase technique with the two-stage inoculum method (Tilley and Terry, 1963) for predicting *in vivo* digestibility for grasses, lucerne, red clover and sanfoin. They found the enzyme method to be very repeatable and only slightly less accurate than inoculum for predicting the digestibility of grasses (Table 8.4). Enzyme solubility was less accurate than the inoculum method, however, for predicting the digestibility of the legumes. For the inoculum method, a single regression equation was permissible for both grasses and legumes. Separate regressions were required for the grasses and for each legume species when the enzyme technique was used.

Aufrere and Michalet-Doreau (1988) have suggested an interesting variation of the pepsin–cellulase technique to cater for feeds containing starch. The method involves three stages: after the initial acid–pepsin incubation at 40°C, the temperature is raised to 80°C for 30 min to hydrolyse starch; the final stage is an incubation with cellulase after removal of the acid pepsin. The results using this technique were highly correlated ($r = 0.98$) with *in vivo* digestibility for a range of forage, straw and mixed diets, containing varying preparations of concentrates and by-products. A similar procedure with a longer heating time at 80°C has been described by de Boever *et al.* (1988) (see below, 'Hot-acid extraction').

Table 8.4. Relationship between *in vivo* digestibility and (a) *in vitro* digestibility (inoculum) and (b) pepsin–cellulase solubility (derived from Terry *et al.*, 1978).

	Grasses	Lucerne	Red clover	Sainfoin
No. of samples	48	8	8	9
(a) *In vitro* (inoculum)				
r	0.97	0.99	0.89	0.99
RSD	1.46	0.74	2.77	0.70
Slope	1.01	1.03	0.99	0.99
Intercept	0.52	−2.56	3.19	2.00
(b) Pepsin–cellulase				
r	0.96	0.98	0.86	0.99
RSD	1.80	1.49	3.14	0.93
Slope	0.56	1.05	0.77	0.97
Intercept	34.72	−9.25	14.33	2.21

RSD, residual standard deviation.

Neutral-detergent extraction

Neutral-detergent extraction, under reflux or at 100°C, has been widely used as a pretreatment before incubation with cellulase (Abe *et al.*, 1972a, b; Hartley *et al.*, 1974; Jones and Bailey, 1974; Roughan and Holland, 1977; Downman and Collins, 1982).

Jones and Bailey (1974) showed cellulase digestion with neutral-detergent residues from grasses to be highly correlated with the voluntary intake of grasses by sheep. The procedure was considered appropriate for breeding programmes aimed at improving both digestibility and intake. Difficulties were encountered in routine use, however, due to the inhibition of cellulase digestion by traces of lauryl sulphate from the neutral-detergent reagent. Hartley *et al.* (1974) found that the optical density (324 nm) of cellulase extracts of neutral-detergent residues from grasses was highly correlated with *in vivo* digestibility; the ultraviolet (UV) absorption was attributed to the release of carbohydrate esters of phenolic acids by the cellulase.

Roughan and Holland (1977) prepared their own potent crude cellulase from a mutant strain of *Trichoderma*, which readily attacked plant cell wall isolated by neutral-detergent extraction. They proposed a method involving hot neutral detergent, followed by exhaustive digestion with two successive batches of cellulase. Values obtained using this technique were highly correlated ($r = 0.98$) with *in vivo* digestibility of legumes and grasses. Downman and Collins (1982) compared cellulase solubility, using varying concentrations of cellulase, after either pepsin or neutral-detergent pretreatment, with *in vitro* (inoculum) and *in vivo* digestibility for a range of forages, silages, cereals and complete diets. An additional digestion with amyloglucosidase was included in the enzyme methods for materials containing starch. Some of Downman and Collins' (1982) results are summarized in Table 8.5 and show the enzyme techniques to be at least as precise as *in vitro* inoculum for predicting the *in vivo* digestibility of most forages; there was little difference found between the two enzyme methods. The authors point out that the pepsin–cellulase method was easier to manipulate and less time-consuming than the detergent

Table 8.5. Correlation (*r*) of *in vivo* digestibility with cellulase solubility* after pepsin or neutral-detergent pretreatment and with *in vitro* (inoculum) digestibility (derived from Downman and Collins, 1982).

	Pepsin–cellulase	Detergent–cellulase	*In vitro* (inoculum)
Grass silage ($n = 16$)	0.87	0.84	0.81
Maize silage ($n = 16$)	0.72	0.72	0.68
Dried grass ($n = 16$)	0.84	0.70	0.51
Hay ($n = 28$)	0.81	0.88	0.93

* *Trichoderma* cellulase (20 g l^{-1}).

method; an advantage of the latter was that samples could be analysed in fewer working days. It was concluded that both cellulase methods could be regarded as acceptable alternatives to the *in vitro* inoculum method for the routine evaluation of forages. A routine procedure for the estimation of neutral detergent–cellulase digestibility (NCD) has been described by Alderman (1985); Morgan *et al.* (1989) has described a variation of the method using the Fibertec digestion system.

More recently, Downman (1993) has proposed a modified detergent–cellulase method to accommodate the galactomannans found in palm-kernel oil. These polysaccharides are not solubilized by cellulase but can be solubilized by the inclusion of an additional carbohydrase, gamanase, in the buffered cellulase and an increase in digestion time to 40 h. The modified method significantly improved the correlation with *in vivo* metabolizable energy values; the inclusion of gamanase did not affect predicted energy values for feedingstuffs not containing palm-kernel oil. It was concluded that the extra expense of the modified method was justified for compound feedingstuffs containing 10% or more of palm-kernel oil; for samples known to be free of palm oil, the original detergent–cellulase method would remain valid. Downman (1993) points out that new commodities introduced into the feed trade in the future will need to be monitored to decide on the appropriate analytical methodology required. A routine procedure for the determination of neutral detergent cellulase–gamanase solubility (NCGD) has been described by the Ministry of Agriculture, Fisheries and Food (MAFF)-sponsored working party on the prediction of energy values of compound feeds (MAFF, 1993). Appropriate test kits have been prepared for the technique (Biotal Ltd., Cardiff) and the working party specify that kits from this source must be used in support of their prediction equation for the metabolizable energy content of ruminant compound feeds.

McLeod and Minson (1982) compared the accuracy of the detergent and pepsin–cellulase methods for predicting the *in vivo* digestibility of a range of grass and legumes. Both cellulase methods were highly correlated with *in vivo* digestibility values (Table 8.6); pepsin pretreatment appeared to be more precise for grasses and detergent pretreatments for legumes. It was concluded that pretreatment with detergent increased analytical error and that the method failed to produce an overall improvement in the accuracy of predicting *in vivo* digestibility; any advantage gained from a shorter analysis time was offset by an increase in analytical error. Similarly, Coehlo *et al.* (1988) found no advantage of detergent, compared with

Table 8.6. Relation of *in vivo* digestibility to DM solubility in cellulase after either pepsin or neutral-detergent pretreatment (derived from McLeod and Minson, 1982).

	r value	RSD	Analytical error (±)
Pepsin–cellulase			
Grasses (*n* = 50)	0.94	2.6	0.37
Legumes (*n* = 30)	0.92	2.9	0.18
All (*n* = 80)	0.92	2.9	0.29
Detergent–cellulase			
Grasses (*n* = 50)	0.90	3.3	0.90
Legumes (*n* = 30)	0.94	2.7	1.06
All (*n* = 80)	0.89	3.4	0.98

RSD, residual standard deviation.

pepsin, pretreatment for predicting *in vivo* digestibility for lucerne (alfalfa), a range of grasses and ryegrass–lucerne mixtures. The addition of *Aspergillus* hemicellulase to *Trichoderma* cellulase improved the correlation with *in vivo* values for both pepsin and detergent methods. The use of this enzyme combination following pepsin pretreatment predicted *in vivo* digestibility as precisely as the two-stage *in vitro* (inoculum) method studied (r^2 = 0.88).

Givens *et al.* (1990a) compared the accuracy of different laboratory techniques in predicting the *in vivo* digestibility of 173 spring-grown grasses in England and Wales. Samples were taken from commercial farms at different locations and at different growth stages over a 4-year period; the swards varied in age and in the component species and varieties. The relationship of *in vitro* digestibility (inoculum) and fibre content to *in vivo* digestibility was significantly affected by year of harvest, location, predominant variety and age of ley. The detergent–cellulase and pepsin–cellulase methods were less affected by these factors; both, however, were significantly affected by predominant variety. Overall, the enzyme methods were better related to *in vivo* values than either *in vitro* (inoculum) or fibre estimations (Table 8.7); there was a slight advantage in precision indicated for the detergent– compared with the pepsin–cellulase method. It was concluded that the considerable benefits of the cellulase methods, including increased accuracy of prediction and relative freedom from external effects, should ensure that they replace fibre estimations for the routine evaluation of fresh herbage. The same conclusions were shown to be valid for predicting the metabolizable energy content of fresh grass (Givens *et al.*, 1990b).

In a comparison of spring- and autumn-harvested pasture herbage, Givens *et al.* (1993a) found both detergent– and pepsin–cellulase methods to be highly correlated (74.6–76.7% of accountable variance) with the *in vivo* digestibility of spring grass; both enzyme methods were superior to *in vitro* (inoculum) and modified acid-detergent (MAD) fibre methods. Regression equations for predicting *in vivo* digestibility were significantly different, however, from those established in the previous study (Givens *et al.*, 1990a) for all methods except pepsin–cellulase. The difference for detergent–cellulase, though statistically significant, was numerically

Table 8.7. Relationship of *in vivo* digestibility of grasses to enzyme solubility and *in vitro* digestibility (inoculum).

Reference	Method	Accountable variance (%)	SEP
Givens *et al.* (1990a)	Detergent–cellulase	76.6	27.1
	Pepsin–cellulase	75.9	28.8
	In vitro (inoculum)	67.0	33.2
	MAD fibre	66.9	33.3
Givens *et al.* (1993a)	Detergent–cellulase	53.3	29.3
	Pepsin–cellulase	63.4	25.9
	In vitro (inoculum)	40.5	33.1
	MAD fibre	49.4	30.5

SEP, standard error of prediction.

small and would have little effect on predicted values. Both enzyme techniques and *in vitro* (inoculum) were poorly related to the *in vivo* digestibility of autumn pasture (13–20% of accountable variance) and inferior to MAD fibre; the relationships found for the autumn differed significantly in slope and intercept from those for spring herbage for all methods. The best overall (spring and autumn combined) relationship with *in vivo* digestibility was with pepsin–cellulase (Table 8.7); in this study, the relationship for *in vitro* (inoculum) accounted for less of the variance in *in vivo* values than MAD fibre. It was suggested that the poor relationships found for autumn grass could be largely attributable to the small range in *in vivo* values for autumn grass. Other suggested explanations were that environmental factors changed the relationship between plant cell-wall structure and *in vivo* digestibility and changes in species composition, particularly clover, between spring and autumn. It was concluded that cellulase-based techniques, particularly pepsin–cellulase, were more strongly related to digestibility and metabolizable energy than the inoculum or fibre methods, particularly for spring primary growths of grass. The accuracy of prediction, however, could be improved by developing separate regression equations for spring and autumn crops.

When the same enzyme and inoculum techniques were applied to dried samples of 124 clamp silages from farms (Givens *et al.*, 1989), the best single predictor of silage *in vivo* digestibility was the *in vitro* inoculum method (Table 8.8). Of the enzyme methods, pepsin–cellulase was better related to digestibility than detergent–cellulase; all the biological methods were superior to MAD fibre content. The relationships of the laboratory techniques to *in vivo* metabolizable energy were much poorer than for digestibility, the best method (*in vitro* inoculum) only accounting for 24% of the variance. This was increased to 77% by inclusion of a gross energy term in a bivariate equation, and it was concluded that inclusion of this or a related term was essential for predicting silage metabolizable-energy content with acceptable accuracy.

Givens *et al.* (1993b) found the detergent–cellulase method to be the best predictor of *in vivo* digestibility for big-bale silage (Table 8.8). Both enzyme methods, however, gave prediction equations differing significantly in both slope and intercept

Table 8.8. Relationship of the *in vivo* digestibility of clamp and big-bale silages to enzyme and inoculum laboratory techniques.

	Method	Accountable variance (%)	SEP
Clamp silage (Givens *et al.*, 1989)	Detergent–cellulase	58.8	39.7
	Pepsin–cellulase	68.0	32.3
	In vitro (inoculum)	74.1	33.6
Bale silage (Givens *et al.*, 1993b)	Detergent–cellulase	84.0	25.3
	Pepsin–cellulase	75.6	25.7
	In vitro (inoculum)	74.5	32.9

from those previously established for clamp silage; equations for *in vitro* (inoculum) did not differ significantly between clamp and bale. Using the enzyme methods, different prediction equations would therefore be required for clamp and bale silages to predict *in vivo* digestibility with acceptable accuracy. Similarly, Barber *et al.* (1990) found the slope and intercept of the relationship between pepsin–cellulase solubility and *in vivo* digestibility to vary significantly for silages from different centres. Within centres, pepsin–cellulase was as well related to *in vivo* digestibility (r^2 = 0.53 to 0.82) as was *in vitro* (inoculum) digestibility (r^2 = 0.48 to 0.83). The inoculum method, nylon-bag digestion and near-infrared reflectance (NIR) relationships with *in vivo* digestibility were not significantly different between centres. A good relationship was found by Givens *et al.* (1995) between detergent–cellulase solubility and the *in vivo* digestibility of maize silages from different centres; one population of silages, however, gave a regression equation significantly different from that of the others. It was suggested that environmental factors or interactions due to drying may have contributed to the differing equations. In a comparison of field-dried and barn-dried hays, Moss and Givens (1990) found a significant effect of curing on the relationship of detergent–cellulase and *in vitro* (inoculum) with *in vivo* digestibility; relationships for pepsin–cellulase and MAD fibre were not significantly affected by curing and were deemed to be the only methods giving acceptable prediction equations for hay.

Bughrara and Sleeper (1986) found a simple digestion in buffered cellulase to be highly correlated (r = 0.91) with the *in vivo* digestibility of a range of temperate forage species. The correlation with *in vivo* was improved, however, if the forage was pretreated with neutral detergent (r = 0.97) or acid pepsin (r = 0.94) prior to cellulase digestion. They concluded that the one-stage cellulase technique would be adequate as a rapid, low-cost, method for screening in plant-breeding programmes. More precise estimation required pretreatment with detergent or pepsin; both enzyme methods were considered to be more rapid and convenient and less costly methods of predicting *in vivo* digestibility than the two-stage *in vitro* inoculum method of Tilley and Terry (1963). Subsequent work (Bughrara *et al.*, 1989) showed similar results for lucerne. Again, simple digestion in cellulase was well related to *in vivo* digestibility (r = 0.91), but the relationship was improved by pretreatment with detergent (r = 0.94) or acid pepsin (r = 0.96). Similarly, Stakelum *et al.* (1988) found both detergent– and pepsin–cellulase methods to be as precise as *in vitro* inoculum for predicting the *in vivo* digestibility of fresh grass for dairy cows.

Hot-acid extraction

More drastic pretreatments than pepsin or neutral detergent have been proposed in efforts to increase the amount of dry matter solubilized by cellulase to levels comparable with *in vivo* digestion. Kirchgessner and Kellner (1977) found refluxing with 2 mol l^{-1} hydrochloric acid for 30 min, followed by digestion with cellulase and then pepsin, to give values numerically similar to *in vivo* digestion for a range of forages (grasses, legumes and cereal straw). Results obtained with this method were highly correlated (r^2 = 0.85) with *in vivo* digestibility and the method was concluded to be a viable alternative to inoculum methods.

Few comparisons appear to have been made between the acid–reflux method of Kirchgessner and Kellner (1977) and the detergent–cellulase or pepsin–cellulase procedures. Pace *et al.* (1984) found *in vitro* (inoculum) values for grasses and legumes to be highly correlated with pepsin–cellulase solubility (r = 0.96) and the reflux–acid method (r = 0.91). Different regression equations were needed to predict *in vitro* values of grasses and legumes for pepsin–cellulase but not for the acid–reflux procedure. Highly significant week-to-week differences were found between different analytical runs for the inoculum procedure, but not for the enzyme methods. Bughrara *et al.* (1989) found that the reflux–acid method compared favourably with the detergent and pepsin methods when applied to lucerne forages.

De Boever *et al.* (1988) compared the accuracy of different enzyme techniques and *in vitro* inoculum for predicting the *in vivo* digestibility of maize silage, grass silage and grass hays. The enzyme techniques included: (i) the 2 mol l^{-1} acid-pepsin–cellulase procedure; (ii) pepsin–cellulase solubility with varying incubation times for the pepsin and cellulase stages; and (iii) a variant of the pepsin–cellulase method involving three stages, namely: incubation in acid pepsin, heating the acid pepsin solution at 80°C for 45 min to remove starch and finally incubation with cellulase. The detergent–cellulase method was not evaluated, as this method had proved more difficult to manipulate than pepsin–cellulase, resulting in a lower repeatability. The *in vitro* inoculum method was the best predictor of *in vivo* digestibility for grass silage and hay, but the enzyme methods were generally superior for maize silage, particularly methods involving a hot-acid pretreatment (Table 8.9). No one enzyme method was best for all forages, but the pepsin–cellulase-based methods were more consistent than those involving acid–reflux. A similar method to that proposed by De Boever *et al.* (1988), but with a shorter heating time at 80°C, has been suggested by Aufrere and Michalet-Doreau (1988). The method was found to be more precise than the *in vitro* (inoculum) procedure for predicting the *in vivo* digestibility of a diverse range of feeds, including cereals, by-products and dried forages.

Table 8.9. Relationship of the *in vivo* digestibility of silage and hay to enzyme solubility and *in vitro* (inoculum) digestibility (derived from de Boever *et al.*, 1988).

	Correlation coefficient (*r*)		
Procedure	Maize silage	Grass silage	Grass hay
(a) 2 mol l^{-1} HCl (reflux) Cellulase (40°C) Acid pepsin 40°C)	0.89	0.45	0.91
(b) Acid pepsin (40°C) Cellulase (40°C)	0.84	0.77–0.82	0.92–0.93
(c) Acid pepsin (40°C) Acid pepsin (80°C) Cellulase (40°C)	0.88	0.76–0.78	0.90
(d) *In vitro* (inoculum)	0.84	0.90	0.96

(a) Kirchgessner and Kellner (1977); (b) Jones and Hayward (1973); (c) de Boever *et al.* (1988); (d) Tilley and Terry (1963).

Comparative Assessment of Cellulase Techniques, Their Limitations and the Need for Further Work

The experimental work reviewed shows that cellulase-based techniques predict *in vivo* digestibility more precisely than chemical methods, such as MAD fibre, and are at least comparable in accuracy to inoculum digestion methods. Provided an active cellulase (usually from *Trichoderma*) is used, a simple one-stage digestion in cellulase permits a rapid, reliable and low-cost estimate of digestibility, which may be adequate for screening in plant-breeding programmes (Jones and Hayward, 1973; Bugge, 1980).

It has also been shown that the precision of predicting *in vivo* digestibility is significantly improved if the forage is pretreated with acid pepsin, neutral detergent or 2 mol l^{-1} hydrochloric acid prior to the cellulase digestion. The results do not show a clear and consistent advantage for any of the pretreatments, although the drastic 2 mol l^{-1} acid pretreatment has sometimes given poor results (de Boever *et al.*, 1988). This is not surprising, since this treatment would remove hemicelluloses, generally shown in *in vivo* studies to be the least digestible fraction of the cell-wall polysaccharides. This method does, however, have the advantage of removing starch, while the detergent and pepsin methods require an additional amyloglucosidase digestion or acid hydrolysis for its removal.

In nearly all the work reviewed, the relationships between enzyme solubility and *in vivo* digestibility have been examined in one population of forages and predictive equations derived for that population. For a method to have any real value in forage evaluation, it is essential that predictive equations are applicable to other populations, at least of the same or similar forages. Comparisons made by Givens *et al.* (1990a, 1993a) suggest that the pepsin– and detergent–cellulase methods are more robust and less variable than the *in vitro* (inoculum) method of Tilley and

Terry (1963) for predicting the *in vivo* digestibility of fresh grass; the enzyme methods were less affected by factors such as year of harvest, age of crop and location, but were affected by species differences. Equations for predicting *in vivo* digestibility differed significantly between spring and autumn grass for both the enzyme and the inoculum methods. Significantly different predictive equations for *in vivo* digestibility have also been reported for the enzyme methods between clamp and bale silages (Givens *et al.*, 1993b) and between different populations of grass silage (Barber *et al.*, 1990) and maize silage (Givens *et al.*, 1995); predictive equations based on *in vitro* (inoculum) were not significantly affected.

A further complication in deriving general equations for predicting *in vivo* digestibility for pasture crops containing clover is the extent to which enzyme solubility is influenced by species. Jones and Hayward (1975) found no differences in the relationship of *in vitro* (inoculum) digestibility to pepsin–cellulase solubility between grasses and clovers. Terry *et al.* (1978), however, found marked differences in the relationship of pepsin–cellulase solubility to *in vivo* digestibility for arable legumes compared with grasses. Other studies (Goto and Minson, 1977; McLeod and Minson, 1982) have shown the relationship of enzyme solubility to *in vivo* digestibility to be improved if the species are considered separately. There is evidence that species specificity is reduced if the 2 mol l^{-1} acid method of Kirchgessner and Kellner (1977) is used (Pace *et al.*, 1984).

It is, perhaps, not surprising that the relationships between *in vivo* digestibility and enzyme solubility for different crops should prove variable under different environmental and other conditions. It is known that environmental factors can influence both the content and degree of lignification of the cell wall. Enzyme methods solubilize less of the cell wall than *in vivo* digestion and are more influenced by the degree of lignification; it is likely, therefore, that environmental effects on cell-wall structure will influence enzyme solubility differently from *in vivo* digestion. Differences in botanical composition may also be significant, since it appears that the cell walls of legumes are more readily solubilized by cellulase than those of grasses. A further consideration is that *in vivo* digestion can be influenced by a variety of factors that do not affect cellulase solubility, e.g. level of feeding, degree of chop or the nutrient status of the feed. Rees and Minson (1979), for example, found that pepsin–cellulase solubility gave a biased estimate of *in vivo* digestibility in sheep whose digestive efficiency was decreased by inadequate sulphur levels. Additional factors that could influence predictive equations for silages are differential losses of volatiles during sample drying or the influence of the degree and type of fermentation on rumen function and digestibility.

There is clearly a need for further *in vivo* and analytical studies to determine the extent to which predictive equations derived from enzyme solubility apply to different environmental and other conditions. Such studies need to take account of the increased use of grass–clover pastures and the introduction of alternative crops, such as forage maize, kale, red clover and whole-crop cereals, into livestock systems. The current interest in bicropping of protein- and energy-rich crops may also require additional analytical studies on the suitability of different enzymic procedures.

Of the cellulase methods currently in use, the pepsin–cellulase method has been found to be least time-consuming and easiest to manipulate (Downman and

Collins, 1982; de Boever *et al.*, 1988); it also uses inexpensive glassware and chemicals and a higher output per person can be achieved than with methods involving refluxing with detergent or strong acid. Samples containing starch can be accommodated with little extra manipulation if the acid hydrolysis method of de Boever *et al.* (1988) is used. The pepsin–cellulase method has been shown to have a higher analytical accuracy than the detergent technique (McLeod and Minson, 1982). The gamanase modification of the detergent–cellulase procedure (NCGD) was developed by Downman (1993) specifically to cater for polysaccharides in palm-kernel oil not hydrolysed by cellulase. This procedure, however, is being widely used for forages, in addition to concentrates, presumably to avoid a proliferation of methods and to simplify laboratory procedures. It can be argued that, in consequence, an unnecessarily complicated and costly procedure is being applied when simpler methods, such as pepsin–cellulase or detergent–cellulase, would give analytical results for forages of at least equal accuracy and validity and at less cost. A danger in attempting to develop a 'universal' method for all forages and all concentrates is that the method will inevitably become increasingly complicated. Downman (1993) anticipated such a development and emphasized the need to carefully monitor new raw materials being introduced to the feed trade so that appropriate analytical procedures could be devised.

Whatever the merits or otherwise of particular techniques, their usefulness in forage evaluation will ultimately depend on the reliability and consistency of the predictive equations derived for *in vivo* digestibility. Further work is clearly needed to determine the causes of variability in predictive equations and the significance of environmental and other factors.

References

Abe, A., Horii, S. and Kameoka, K. (1972a) [Development and application of cellulase hydrolysis for predicting digestibility of roughage. 2. Influence of pre-treatment of samples on cellulase hydrolysis.] *Japanese Journal of Zootechnical Science* 43, 146–154.

Abe, A., Horii, S. and Kameoka, K. (1972b) [Development and application of cellulase hydrolysis for predicting digestibility of roughage. 3. Application of two step method for predicting dry matter digestibility.] *Japanese Journal of Zootechnical Science* 43, 175–180.

Adamson, A.H. and Terry, R.A. (1980) The relationship between the *in vivo* digestibility of hay and its solubility in pepsin–hydrochloric acid and fungal cellulase solutions. *Journal of the Science of Food and Agriculture* 31, 854–856.

Adegbola, A.A. and Paladines, O. (1977) Prediction of the digestibility of the dry matter of tropical forages from their solubility in fungal cellulase solutions. *Journal of the Science of Food and Agriculture* 28, 775–785.

Alderman, G.A. (1985) Prediction of the energy value of compound feeds. In: Haresign, W., and Cole, J.J.A. (eds) *Recent Advances in Animal Nutrition – 1985*. Butterworths, London, pp. 3–52.

Allison, M. and Borzucki, R. (1978) Cellulase methods for the efficient digestion of grasses and Brassicas. *Journal of the Science of Food and Agriculture* 29, 293–297.

Aufrere, J. and Michalet-Doreau, B. (1988) Comparison of methods for predicting digestibility of feeds. *Animal Feed Science and Technology* 20, 203–218.

Bailey, R.W. and McRae, J. (1970) The hydrolysis by rumen and caecal microbial enzymes of hemicellulose in plant and digesta particles. *Journal of Agricultural Science, Cambridge* 75, 321–324.

Barber, G.D., Givens, D.I., Kridis, M.S., Offer, N.W. and Murray, I. (1990) Prediction of the organic matter digestibility of grass silage. *Animal Feed Science and Technology* 28, 115–128.

Barnes, R.F. (1967) Collaborative *in vitro* rumen fermentation studies on forage substrates. *Journal of Animal Science* 26, 1120–1130.

Bugge, G. (1980) A simplified cellulase digestion technique for predicting dry matter digestibility of grasses. *Zeitschrift Acker- und Pflanzenbau* 149, 206–212.

Bughrara, S.S. and Sleeper, D.A. (1986) Digestion of several temperate forage species by a prepared cellulase solution. *Agronomy Journal* 78, 94–98.

Bughrara, S.S., Sleeper, D.A., Belyea, R.L. and Marten, G.C. (1989) Quality of alfalfa herbage estimated by a prepared cellulase solution and near infrared reflectance spectroscopy. *Canadian Journal of Plant Science* 69, 833–839.

Clark, J. and Beard, J. (1977) Prediction of the digestibility of ruminant feeds from their solubility in enzyme solutions. *Animal Feed Science and Technology* 2, 153–159.

Coelho, M., Hembry, F.C., Barton, F.E. and Saxton, A.M. (1988) A comparison of microbial, enzymatic, chemical and near-infrared reflectance spectroscopy methods in forage evaluation. *Animal Feed Science and Technology* 20, 219–231.

de Boever, J.L., Cottyn, B.G., Buysse, F.X. and Vanacker, J.M. (1988) The use of a cellulase technique to predict digestibility, metabolizable and net energy of forages. *Animal Feed Science and Technology* 19, 247–260.

Donefer, E., Niemann, P.J., Crampton, E.W. and Lloyd, L.E. (1963) Dry matter disappearance by enzyme and aqueous solutions to predict the nutritive value of forages. *Journal of Dairy Science* 46, 965–970.

Downman, M.G. (1993) Modifications to the neutral detergent cellulase digestibility method for the prediction of the metabolizable energy of compound feeds containing palm kernel oil. *Journal of the Science of Food and Agriculture* 61, 327–331.

Downman, M.G. and Collins, F.C. (1977) The prediction of the digestibility of silages using cellulase. *Journal of the Science of Food and Agriculture* 28, 1071–1074.

Downman, M.G. and Collins, F.C. (1982) The use of enzymes to predict the digestibility of animal feeds. *Journal of the Science of Food and Agriculture* 33, 689–696.

Drew, K.R. (1966) The *in vitro* prediction of herbage digestibility. *Proceedings of the New Zealand Society of Animal Production* 26, 52–70.

Givens, D.I., Everington, J.M. and Adamson, A.H. (1989) The digestibility and metabolizable energy content of grass silage and their prediction from laboratory measurements. *Animal Feed Science and Technology* 24, 27–43.

Givens, D.I., Everington, J.M. and Adamson, A.H. (1990a) The nutritive value of spring-grown herbage produced on farms throughout England and Wales over 4 years. II. The prediction of apparent digestibility *in vivo* from various laboratory measurements. *Animal Feed Science and Technology* 27, 173–184.

Givens, D.I., Everington, J.M. and Adamson, A.H. (1990b) The nutritive value of spring-grown herbage produced on farms throughout England and Wales over 4 years. III. The prediction of energy values from various laboratory measurements. *Animal Feed Science and Technology* 27, 185–196.

Givens, D.I., Moss, A.R. and Adamson, A.H. (1993a) Influence of growth-stage and season on the energy value of fresh herbage 2. Relationships between digestibility and metabolizable energy content and various laboratory measurements. *Grass and Forage Science* 48, 175–180.

Givens, D.I., Moss, A.R. and Adamson, A.H. (1993b) Prediction of the digestibility and energy value of grass silage conserved in big bales. *Animal Feed Science and Technology* 41, 297–312.

Givens, D.I., Cottyn, B.G., Dewey, P.J.S. and Steg, A. (1995) A comparison of the neutral detergent cellulase method with other laboratory methods for predicting the digestibility *in vivo* of maize silages from three European countries. *Animal Feed Science and Technology* 54, 55–64.

Goto, I. and Minson, D.J. (1977) Prediction of the dry matter digestibility of tropical grasses using a pepsin–cellulase assay. *Animal Feed Science and Technology* 2, 247–253.

Guggolz, J., Saunders, R.M., Kohler, G.O. and Klopfenstein, T.J. (1971) Enzymatic evaluation of processes for improving agricultural wastes for ruminant feeds. *Journal of Animal Science* 33, 167–170.

Hartley, R.D., Jones, E.C. and Fenlon, J.S. (1974) Prediction of the digestibility of forages by treatment of their cell walls with cellulolytic enzymes. *Journal of the Science of Food and Agriculture* 25, 947–954.

Jarrige, R., Thivend, P. and Demarquilly, C. (1970) Development of cellulolytic enzyme digestion for predicting the nutritive value of forages. In: *Proceedings of the XIth International Grassland Congress*. Surfers Paradise, Australia, pp. 762–766.

Jones, D.I.H. and Bailey, R.W. (1974) Hydrolysis of cell wall carbohydrates of grasses by carbohydrases in relation to voluntary intake by sheep. *Journal of Agricultural Science, Cambridge* 83, 105–111.

Jones, D.I.H. and Hayward, M.V. (1973) A cellulase digestion technique for predicting dry matter digestibility of grasses. *Journal of the Science of Food and Agriculture* 24, 1419–1426.

Jones, D.I.H. and Hayward, M.V. (1975) The effect of pepsin pre-treatment of herbage on the prediction of dry matter digestibility from solubility in fungal enzyme solutions. *Journal of the Science of Food and Agriculture* 26, 711–718.

Jones, D.I.H. and Hayward, M.V. (1982) Estimation of forage digestibility. In: *Annual Report Welsh Plant Breeding Station for 1982*, pp. 126–127.

Kirchgessner, M. and Kellner, R.J. (1977) Estimation of digestibility and net energy of fresh and dried forage by a cellulase method. In: *Proceedings of the 28th Annual Meeting of the European Association for Animal Production*, Brussels, N/2.02, pp. 1–6.

McCluskey, J.G., Allison, M.J., Duncan, M.J. and Jarvis, M.C. (1984) Isolation of anatomically defined cell walls from fodder kale and their contributions to determining the *in vitro* digestibility of the whole plant. *Journal of Agricultural Science, Cambridge* 103, 347–352.

McLeod, M.N. and Minson, D.J. (1969) The use of the *in vitro* technique in the determination of the digestibility of grass/legume mixtures. *Journal of the British Grassland Society* 24, 296–298.

McLeod, M.N. and Minson, D.J. (1982) Accuracy of predicting digestibility by the cellulase technique: the effect of pre-treatment of forage with neutral detergent or acid pepsin. *Animal Feed Science and Technology* 7, 83–92.

McQueen, R. and van Soest, P.J. (1975) Fungal cellulase and hemicellulase prediction of forage digestibility. *Journal of Dairy Science* 58, 1482–1491.

MAFF (1993) *Prediction of the Energy Values of Compound Feedingstuffs for Farm Animals. Summary of the Recommendations of a Working Party Sponsored by the Ministry of Agriculture, Fisheries and Food.* MAFF Publications, London, 17 pp.

Moore, W.E., Effland, M.J. and Millett, M.A. (1972) Hydrolysis of wood and cellulose with cellulolytic enzymes. *Journal of Agricultural Food Chemistry* 20, 1173–1175.

Morgan, D.J., Stekelum, G. and Dwyer, J. (1989) Modified neutral detergent cellulase digestibility for use with the 'Fibertree' system. *Irish Journal of Agricultural Research* 28, 91–92.

Moss, A.R. and Givens, D.I. (1990) Chemical composition and *in vitro* digestion to predict digestibility of field-cured and barn-cured grass hays. *Animal Feed Science and Technology* 31, 125–138.

Pace, V., Barge, M.T., Settineri, D. and Malossini, F. (1984) Comparison of forage digestibility *in vitro* with enzymic solubility. *Animal Feed Science and Technology* 11, 125–136.

Pulli, S. (1976) Cellulase digestion technique compared with the *in vitro* digestibility of forages. *Journal of the Scientific Agricultural Society of Finland* 48, 187–194.

Rees, M.C. and Minson, D.J. (1979) The validity of *in vitro* techniques using rumen fluid or cellulase for predicting changes in the dry matter digestibility of grasses caused by fertiliser calcium, sulphur, phosphorus and nitrogen. *Grass and Forage Science* 34, 19–25.

Reese, E.T. and Mandels, M. (1963) Enzymic hydrolysis of cellulose and its derivatives. In: Whistler, R.L. (ed.) *Methods in Carbohydrate Chemistry*, III, *Cellulose*. Academic Press, New York, pp. 139–143.

Rexen, B. (1977) Enzyme solubility – a method for evaluating the digestibility of alkali-treated straw. *Animal Feed Science and Technology* 2, 205–218.

Roughan, P.G. and Holland, R. (1977) Predicting *in vivo* digestibilities of herbages by exhaustive enzyme hydrolysis of cell walls. *Journal of the Science of Food and Agriculture* 28, 1057–1064.

Stakelum, G., Morgan, D. and Dillon, P. (1988) A comparison of *in vitro* procedures for estimating herbage digestibility. *Irish Journal of Agricultural Research* 27, 104–105.

Tarumoto, I. and Masaoka, Y. (1978) Digestibility of fresh sorghum leaf by simple cellulase digestion method. *Journal of Japanese Society of Grassland Science* 24, 1–9.

Terry, R.A., Mundell, D.C. and Osbourn, D. (1978) Comparison of two *in vitro* procedures using rumen liquor–pepsin or pepsin–cellulase for prediction of forage digestibility. *Journal of the British Grassland Society* 33, 13–18.

Tilley, J.M.A. and Terry, R.A. (1963) A two-stage technique for the *in vitro* digestion of forage crops. *Journal of the British Grassland Society* 18, 104–111.

Tinnimit, P. and Thomas, J.W. (1976) Forage evaluation using various laboratory techniques. *Journal of Animal Science* 43, 1058–1065.

Troelsen, J.E. and Hanel, D.J. (1966) Ruminant digestion *in vitro* as affected by inoculum donor, collection day and fermentation time. *Canadian Journal of Animal Science* 45, 149–156.

van Soest, P.J. (1963a) Use of detergents in the analysis of fibrous feeds. I. Preparation of fibre residues of low nitrogen content. *Journal of the Association of Official Agricultural Chemists* 46, 825–829.

van Soest, P.J. (1963b) Use of detergents in the analysis of fibrous feeds. II. A rapid method for the determination of fiber and lignin. *Journal of the Association of Official Agricultural Chemists* 46, 829–835.

Yates, N.G. and Allden, W.G. (1966) A study of herbage digestibility using an *in vitro* fermentation technique. *Proceedings of the Australian Society of Animal Production* 6, 340–349.

9 The *In Situ* Technique for the Estimation of Forage Degradability in Ruminants

E.R. ØRSKOV

International Feed Resource Unit, Rowett Research Institute, Bucksburn, Aberdeen AB21 9SB, UK

In the title, the technique to be discussed is termed the *in situ* technique. However, this is identical to the *in sacco* technique or the Terylene (Dacron)- or nylon-bag techniques. The procedure is as follows. Samples of dried and milled feed (to pass a 3 mm screen) or wet minced samples are placed in nylon bags (usually 10 × 17 cm). About 2 to 5 g, depending on density, are weighed precisely into each bag. The tied-up bags are incubated in the rumen of sheep or cattle on an appropriate diet by suspending them from a rumen cannula. They are then withdrawn after various intervals of time, washed and dried. Degradability of dry matter, nitrogen, energy, etc., can thus be measured against time. The pore size of the bags, about 40–60 μm, is such that few particles can escape, and yet microorganisms can enter the bag. Thus, fermentation rate inside the bag is similar to that in the rumen. The amount of soluble material in the sample is measured by washing and reweighing an unincubated bag. See also the recent review by Huntington and Givens (1995).

The development of the *in situ* technique was first intended to provide a dynamic assessment of the degradation of protein (Mehrez and Ørskov, 1977; Ørskov and McDonald, 1979). The characteristics of the degradation curve are described by the equation:

$$P = a + b\,(1 - e^{-ct})$$

where P is degradation at time t, a and b are constants and c is the rate constant of b. For protein, the intercept, a, is similar to the soluble fraction (washing loss); b represents the potential degradability.

The *in situ* technique for estimating degradability in the rumen is not new. It was first described by Quin *et al.* (1938), who used silk bags, and it was again employed by McAnally (1942). However, standardization and validation have only occurred recently. Incubation often proceeded for a maximum period of 48 h, and choice of this time was probably influenced by the *in vitro* digestibility method of

Tilley and Terry (1963), who used this interval to reflect *in vivo* retention time for fibre and the solid fractional outflow rate of rumen solids, about 0.025 to 0.030 h^{-1}.

The Use of Dynamic *In Situ* Techniques to Evaluate Forages

When the technique was used for estimating degradation of roughages, it was soon found that the equation $P = a + b\,(1 - e^{-ct})$ had to be reassessed. This was first done by McDonald (1981), who defined the concept of a lag phase.

Figure 9.1 illustrates the degradation of a typical roughage. It can be seen that the intercept a is no longer similar to the solubility; in fact, it is most often negative, as there is a lag-phase period, when no net disappearance of substrate occurs. This is because the microbes take time to adhere to the substrate, so that, at first, there is no net loss in dry matter; in fact, there may be a small increase. While it is possible to describe a separate curve for the soluble fraction, there is always a problem. Increasing mathematical sophistication must be seen in relation to the accuracy required. Is the additional information generated as a result of increased mathematical complexity and is the additional sampling required necessary and cost-effective for describing feed degradation characteristics accurately enough for practical use?

In order to describe the feeding value of roughages, therefore, Ørskov and Ryle (1990) developed another system, in which the soluble fraction of the sample is determined in the laboratory. The method can use the neutral-detergent fibre (NDF) soluble fraction, the loss after washing with buffer solutions or the washing loss from a nylon bag containing substrate that was not incubated in the rumen.

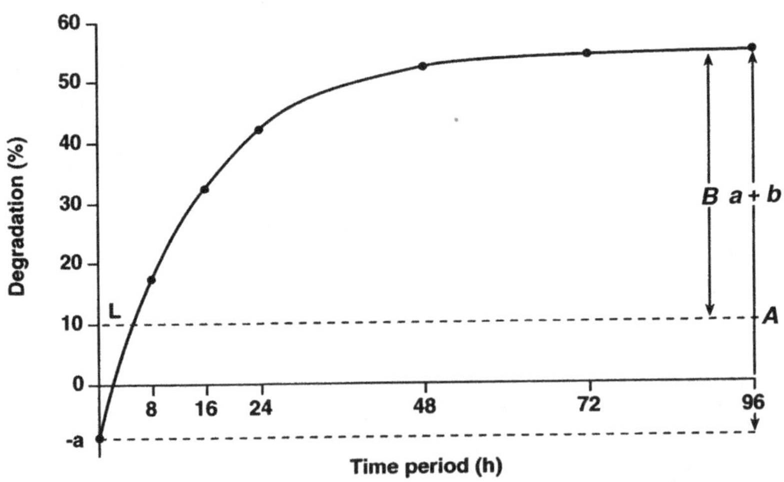

Fig. 9.1. Degradation of a typical roughage diet expressed by the formula $p = a + b\,(1 - e^{-ct})$. Due to a lag phase (L), a is negative. A is the soluble fraction, B the insoluble but potentially fermentable fraction ($B = (a + b) - A$) and c the rate constant of degradation.

Since A represents the soluble fraction, the insoluble but fermentable fraction *B* is worked out as $(a + b) - A$, with *c* as the rate constant (see Fig. 9.1). Particular attention must be given to the choice of incubation times. Ideally, no sample should be taken out of the rumen before the lag phase has ended, normally not before 8 h of incubation. If bags are withdrawn earlier, the curve is distorted and residual standard deviation (RSD) increases. The other rule is that the asymptote must be clearly identified. This is necessary for accurate description of both *B* and *c*. This is achieved by ensuring that the differences in degradability between the last two incubation times is small (less than 10% of the second to last incubation), i.e. 40 and 44%. Sometimes, bags have to be incubated for up to 120 h to describe the asymptote clearly, but usually 72 or 96 h are sufficient.

Some standardization of the microbial fermentation in the animal used for incubation has to be achieved in order to ensure that rumen conditions are optimal for cellulolysis. This is best achieved by feeding the animals on a high-quality roughage diet with about 25 g ntirogen (N) kg^{-1} digestible dry matter (Ørskov, 1992), which means that, if the digestibility is 60%, the diet contains 15 g N or 90–100 g crude protein kg^{-1}, and with rumen pH above 6.2 (Mould and Ørskov, 1984).

Use of the *In Situ* Technique for Evaluation of Roughages

The failure of all feed evaluation systems, including those based on net energy, such as the starch-equivalent systems and the Scandinavian feed-unit systems, or digestible energy, such as the total digestible nutrient (TDN) systems and the metabolizable-energy systems, is that they do not predict intake. While this is not serious as far as concentrate feed is concerned or when roughages are given in restricted amount, it is serious when, as is commonly practised, roughages or forages are grazed or stall-fed *ad libitum*. Ideally, feed evaluations should provide farmers with an exchange rate of one feed compared with another. They should allow planners of livestock production to assess the possible production and predict export or necessity for import of feed, etc. The present systems provide no such guidance and, as a result, farmers who feed mainly or exclusively roughages to their ruminants have no good feed-evaluation system. Digestibility and metabolizability, while positively correlated with feed intake, can generally only account for about a half of the variation in feed intake. While a variety of animal factors have been described which may limit intake of roughages (see Tolkamp and Ketelars, 1994), it does seem that we can advance most readily in practice by giving attention to the physical restriction to intake. This view is supported by the fact that ruminants with a large rumen volume can consume more roughage than animals with a low rumen volume and yet retain it long enough for thorough fermentation, e.g. indigenous vs. Holstein cows in Bangladesh (Mould *et al.*, 1982) and indigenous vs. upgraded sheep (Weyreter *et al.*, 1987).

If we characterize feeds in such a way that potential feed intake is predicted, we may begin to make progress in roughage evaluation. While intake is also influenced by interactions between feed and animal, especially those that affect ruminal retention of fibre, such as reduction of long to small particles, outflow of small particles and, indeed, rumen volume, a feed evaluation system must characterize the feeds themselves, as the animal interaction cannot be determined in each instance. Also, because of the length of time it takes, it would be very expensive and impractical to determine intake and utilization of each feedstuff by feeding trials. Evaluation by chemical determination of feed ingredients, such as by the Weende system for proximate analysis and the more specific system of neutral- and acid-detergent fibre, has been practised for many years. However, these systems suffer from the same problem: they cannot predict feed intake accurately. While the level of acid-detergent fibre in the feed is correlated with feed intake, it accounts for only a half of feed-intake variation (Ørskov *et al.*, 1988). If, for instance, roughages are upgraded by alkali treatment, feed intake is increased, sometimes by as much as 50%, although the fibre content is not decreased or is decreased very little.

In many experiments, the characteristics of feed degradability have been described by the soluble fraction A, the insoluble but degradable fraction *B* and *c*, the speed at which the *B* fraction is degraded. These rate constants have been used in an attempt to develop a system that can predict not only feed nutritive value but consumption as well.

In a first trial (Ørskov *et al.*, 1988), five varieties of cereals were studied whose straw showed a range of different degradation characteristics. These straws were examined in their untreated state and after treatment with ammonia, so that ten different feeds were used. The results are seen in Table 9.1. All the straws were then offered *ad libitum* to groups of beef cattle consuming also 1.5 kg of concentrate daily, with sufficient protein to ensure that fermentation rate of untreated straw was not inhibited by low rumen-ammonia concentrations. The range of feed intakes and daily body-weight gains are also presented. It can be seen that the straws varied in quality, so that there were large differences in growth rate. However, the main purpose was to see if feed intake could be predicted from the multiple regression, using A, *B* and C as explanatory (independent) variables. All of these characteristics have an influence on the volume of ingesta in the rumen. The soluble material, denoted A, arises from cell contents and takes up little space. Generally, this fraction is almost 100% fermentable. The potentially degradable fraction, *B*, takes up space for as long as it remains in the rumen undegraded, and this depends on C. The undegradable fraction, 100 − (A + *B*), of course, takes up space until it is all eliminated by outflow from the rumen.

In order to see if intake and digestibility could be predicted from values for A, *B* and C, multiple regressions were tested on the data (Table 9.2). Here it can be seen that the multiple regressions of using A, *B* and C separately gave the best estimates of intake, because the correlations between A, *B* and C were only weak. This was further corroborated in an experiment with feeding of browse species to goats by Kibon and Ørskov (1993) in Nigeria (Table 9.3). In Tanzania, Shem *et al.* (1995) conducted similar work with steers receiving 17 different feeds from the slopes of Kilimanjaro (Table 9.4), while Khazaal *et al.* (1993), using ten leguminous herbages fed to sheep, showed essentially the same accuracy of prediction of dry-matter intake, digestibility and digestible dry-matter intake (Table 9.5). It is clear that, if digestible dry-matter intake can be predicted accurately, then, given sufficient length of time, growth rate can also be predicted accurately within an experiment.

Table 9.1. The effect of type and variety of straw and of feeding it untreated (U) or treated with ammonia (AM) on growth rate and dry-matter (DM) intake by steers and on *in vivo* digestibility by sheep, together with degradation characteristics (*A*, *B* and *C*).

Type	Variety	Treatment	Growth rate (g day^{-1})	Intake of straw (kg DM day^{-1})	Dry matter digestibility of straw (g kg^{-1})	*A*	*B*	*C*
Winter barley	Gerbel	U	106	3.43	409	12.5	26.3	0.0359
		AM	359	4.70	487	16.0	46.3	0.0257
	Igri	U	126	3.56	412	13.6	29.2	0.0309
		AM	332	4.82	455	15.9	37.3	0.0350
Spring barley	Corgi	U	400	5.16	484	16.0	36.1	0.0481
		AM	608	5.86	596	19.0	47.7	0.0457
	Golden Promise	U	198	4.43	452	15.0	40.5	0.0304
		AM	602	4.93	506	20.1	41.2	0.0377
Winter wheat	Norman	U	273	4.57	343	19.3	29.4	0.0343
		AM	516	5.81	484	24.5	37.5	0.0364
SE			40	0.17	14			

SE, standard error.

Table 9.2. Multiple correlation coefficients (r^2) between factors of the degradation equation and digestibility, dry-matter intake, digestible dry-matter intake and growth rate of steers (from Ørskov and Ryle, 1990).

Factors used in multiple regression analysis	Digestibility	Dry-matter intake	Digestible dry-matter intake	Growth rate
$(A + B)$	0.70	0.83	0.86	0.84
$(A + B) + c$	0.85	0.89	0.96	0.91
$A + B + c$	0.90	0.93	0.96	0.95
Index value	0.74	0.95	0.94	0.96

Table 9.3. Multiple correlation coefficients (r^2) between factors of the exponential equation and digestibility, dry-matter intake, digestible dry-matter intake and growth rate of goats (from Kibon and Ørskov, 1993).

Factors used in multiple regression analysis	Digestibility	Dry-matter intake	Digestible dry-matter intake	Growth rate
$(A + B)$	0.65	0.57	0.15	0.41
$A + B + c$	0.88	0.99	0.92	0.99
Index value	0.75	0.90	0.88	0.81

Table 9.4. Multiple correction coefficients (r) between feed-degradation characteristics and digestibility, dry-matter intake, digestible dry-matter intake and growth rate of steers.

Factors used in multiple regression analysis	Digestibility	Dry-matter intake	Digestible dry-matter intake	Growth rate
$(A + B)$	0.85	0.83	0.84	0.80
$(A + B) + c$	0.95	0.84	0.88	0.90
$A + B + c$	0.98	0.90	0.93	0.93
Index value	0.95	0.90	0.92	0.89

Although A, B and C can all positively influence intake, the intake of diets of similar digestibility can be different. In general, if leguminous herbages are compared with graminaceous herbages of the same digestibility, the intake of leguminous herbages is greater, largely due to a greater C value (Ishida *et al.*, 1996). This implies that digestion of leguminous herbages more rapidly approaches completion in the rumen and therefore is not greatly affected by level of feeding, while differ-

Table 9.5. Multiple correlation coefficients (r^2) between degradation characteristics and leguminous forages and digestible and intake of hay by sheep (from Khazaal *et al.*, 1993).

Factors	Digestibility	Dry-matter intake
$(A + B)$	0.82	0.77
$(A + B) + c$	0.86	0.88
$A + B + c$	0.95	0.88

ences in feeding level and thus rumen fractional outflow can reduce the digestibility of graminaceous herbages (Ishida *et al.*, 1996).

For practical application and advice, it is necessary to generate a simple value, which has been termed feed potential, i.e. the ability of a feed to provide a given level of digestible nutrient intake. This was done in the work of both Ørskov *et al.* (1988) and Shem *et al.* (1995), with quite similar results. Let us assume that intake $= X_1A + X_2B + X_3C$ and simplify by dividing each coefficient by X_1, thus:

$$\left(\frac{X_1}{X_1}\right)A + \left(\frac{X_2}{X_1}\right)B + \left(\frac{X_3}{X_1}\right)C$$

Then the coefficients of A, B and C were 1, 0.4 and 200 respectively from the work referred to. Thus, if a feed – for instance, straw – has values of $A = 12$, $B = 40$ and $C = 0.03$, the index value would be $12 + 16 + 6 = 34$. This index or feed-potential value has no biological meaning, but in the work of Ørskov *et al.* (1988) an index value of about 30 enabled the animals to consume sufficient for energy maintenance. It is likely that different breeds of animals and animals in different physiological states will require feeds of slightly different index values to provide for maintenance, but the ranking of feeds is likely to be similar. Much more work is needed to provide a strong database. It is likely that the coefficients will be similar, since essentially they rely on dietary microbial interactions and only to a limited extent on host animal (see later), except when the rumen environment is less than optimal. The degradation characteristics of a number of roughages are compiled in Table 9.6.

If this concept can be confirmed to be useful, then it may serve as a basis for a new global roughage evaluation system, which could perhaps also be applied to rangeland in different seasons. It may be possible to provide feed-potential inventories in different regions and in different seasons, from which one could match animal potential to feed-resource potential. It may also be possible to estimate the extent to which feed should be imported or upgraded to meet an animal potential greater than the present feed potential can realize and thus help to avoid the many mistaken ways in which animal potential and feed potential have been mismatched. A particular example of this is the introduction of animals with a potential far above the potential of the local feed supply, e.g. the introduction of so-called exotic cattle into areas where only crop residues are available.

Table 9.6. Degradation characteristics of selected roughages expressed as *A*, *B*, and *C* and index values representing a measure of feed potential (from E.R. Ørskov and W. Shand, unpublished).

Forage	*A*	*B*	*C*	Index value
Spring barley straw	10.6	45.3	0.0622	41.1
Barley leaf blade	15.6	70.2	0.0672	57.1
Barley stem	13.5	36.4	0.0406	36.2
Oat straw	11.4	38.2	0.0240	31.4
Oat leaf blade	10.4	50.9	0.0348	37.9
Oat stem	10.7	25.9	0.0196	25.0
Wheat straw	10.9	42.1	0.0369	27.6
Wheat leaf blade	8.9	58.4	0.0526	42.8
Wheat stem	11.2	30.6	0.0274	28.9
Maize stover	10.0	44.3	0.0330	34.3
Maize leaf	19.7	38.0	0.0410	43.1
Maize stem	14.1	37.0	0.0320	35.3
Rape-seed straw	8.6	21.3	0.0697	23.5
Rice straw	17.1	35.9	0.0399	39.5
Rice leaf	14.6	35.0	0.0374	36.1
Rice stem	22.7	32.5	0.0548	46.7
Ryegrass hay	20.9	44.7	0.0485	48.5
Teff (Ethiopia)	20.0	54.3	0.0390	49.6
Leucaena (Ghana)	31.7	52.7	0.0209	56.9
Eicus (Ghana)	11.6	79.9	0.0333	50.2
Lupinus albus straw	13.9	37.5	0.0357	35.9
Vicia bengalensis hay	19.5	34.4	0.0627	51.5

Use of the *In Situ* Technique to Evaluate the Rumen Environment

When this technique is used to evaluate feed, the rumen environment should be optimal, so that the maximum degradability of the feed is expressed. However, the rumen environment is not always optimal; this may be due to acid conditions, to negative interactions between feeds or to deficiencies in the diets, e.g. of N or sulphur (S).

The *in situ* technique can be effectively used for such studies, using a standard feed substrate and varying the rumen environment. A typical example is given in Fig. 9.2. In almost all instances, the rumen environment acts by influencing cellulolysis. From the discussion on degradation characteristics earlier, it is clear that the solubility, A, can scarcely ever by affected by rumen environment. Fermentation of soluble sugars is not affected by rumen pH. It is also logical that *B* is not affected since (A + B) represents the asymptote of potential degradability and, given a sufficient length of time, an asymptote will be reached. The study of rumen environment is centralized in changes in the rate constant, C. In order to make this study more sensitive, it is often an advantage to prepare a large amount of a substrate in which the cellulosic material is highly digestible, such as good-quality hay. This can be made easier by washing away most of the solubles, so that the test material

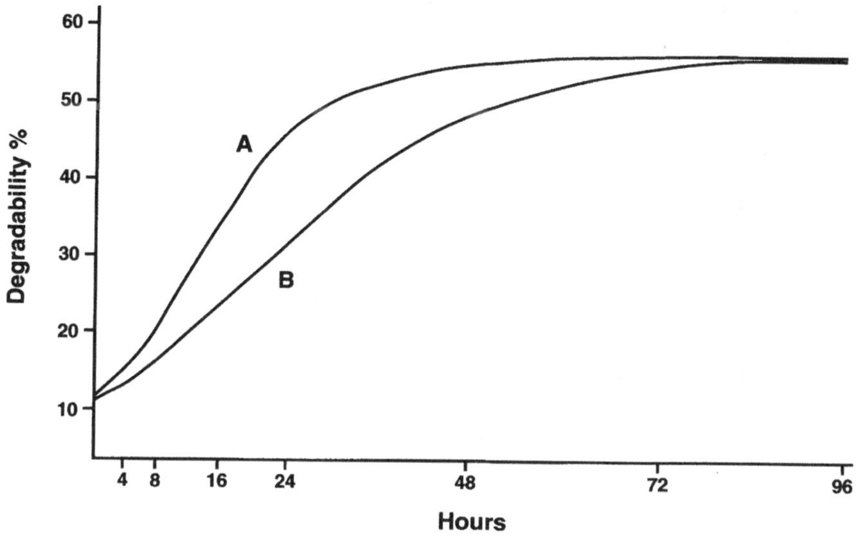

Fig. 9.2. Degradation of a roughage diet in a rumen in which the rumen microbial environment is optimal (A) and suboptimal (B). Note intercept and asymptote are similar.

consists of highly digestible cellulosic roughage. In this way, the feeding regime in question can be studied and the optimal or most cost-effective rumen environment identified. The method was effectively used to identify optimal grain processing for sheep and cattle when grain was used to supplement roughage-based diets (Ørskov and Greenhalgh, 1977). While *in situ* feed evaluation will give similar results for a feedstuff when carried out in different laboratories, since the rumen environment must be optimal, rumen-environment studies can only be made using the feeds, etc., that are to be used locally and these may not be easily transported to a distant laboratory in the quantity needed for feeding.

The method can be used to find the optimal addition of N to a low-N diet and to identify the optimal content of easily fermentable fibre in the diet (Silva and Ørskov, 1988). It is also possible to identify the season at which supplements to pasture-based diets are likely to be cost-effective. Thus, rumen-cannulated cattle or sheep can graze together with the main herd and, every 2 weeks or so, the degradation-rate constant of a standard substrate determined. It would then be possible to identify periods of grazing or seasons in which the degradation-rate constant is reduced. Here, the correct supplement can be used not only as a supplement in its own right, but also to improve intake and digestion of the pasture as well.

Determination of Degradability of Chemical and Botanical Plant Fractions

It is, of course, possible to analyse both botanical and chemical fractions of a feed separately (Ramanzin *et al.*, 1986). One can, for example, determine degradability of NDF in the cell-wall fraction. But, before undertaking such refinements, it is essential to consider the accuracy needed. Are the results going to be used for advice on feeding? If so, it may be difficult to justify more than determination of dry-matter degradability. On the other hand, research to identify varietal differences in fibre degradability may justify such in-depth studies.

Design and Replications

Practical considerations

In earlier studies using the *in situ* technique, it was usual to place all the bags together in the rumen and withdraw them at different intervals of time. However, it is generally more practical to put the bags in at different times and withdraw them all at once. This means that all the bags can be treated and washed together. It is immaterial, on the whole, which system is used. Often, the procedure has to be matched with what is practicable, as it may not be possible to insert bags during the night. Flexibility of incubation times is also possible, since the usual 8, 16, 24 h, etc., are only guidelines; a modified scatter of times is acceptable provided the actual incubation times are entered in the calculations. While the general objective for sample preparation is to imitate chewing, this is obviously not possible and, generally, hammer-milling of dry feed through 2.5–3 mm screens is used. For wet materials, e.g. green forage and silage, a mincing machine is most appropriate, with up to 5 mm screen size. The objective is to avoid small particles leaving the bags as much as possible, and drying silages also leads to loss of volatile materials.

Replications

It is important to recognize that the replication adopted will depend on the purpose or objective of the work. If a feed is evaluated and the results are to be used in diet formulation or to estimate rumen environmental conditions, then the values have to be applicable to the animal population. Mehrez and Ørskov (1977) observed that the greatest source of variation was between-animal variation. The variations within animals and between days within animals were small in comparison with between-animal variation. They came to the conclusion that, for such purposes, three cannulated animals had to be used and that replication within animals and between days was not worthwhile and made little difference to the total variance.

However, if the purpose or objective is to rank the feed potential of forages, which may be the case for plant breeders at early stages of selection for forage or straw quality, then only one animal needs to be used. In fact, perhaps only one sample is required. Here a word of caution. Many research workers use 48 h degradability, as it is close to *in vivo* digestibility. However, if the objective is to rank them, some indication of A, B and C is required. This is best obtained if, for instance, 24 h is used, rather than 48 h. Degradability values at 48 h tend to blur differences in rate of degradation, the C value.

Future Research

With respect to using the *in situ* technique for forage or roughage evaluation, there is a need for a better and more comprehensive database. There will undoubtedly be feeds whose intake cannot be predicted readily from their A, B and C values. Banana pseudostem is one feed that is consumed in smaller amounts than predicted (Shem *et al.*, 1995). This contains 95% water and the fibre is very tough; perhaps in such feeds the breakdown time of long to small particles needs to be taken into account.

There is also a problem relating to small-particle loss, such as starch in maize silage. These particles will appear in the model as soluble particles if they are small enough to pass through the pores of the nylon bag. Many of these aspects have been discussed in some detail by Huntington and Givens (1995), together with possible mathematical solutions. Small starch particles could escape rumen fermentation and be digested in the lower gut (see Ørskov and Ryle, 1992). More production data upon intake of silages, in particular maize silage, are called for.

There is also a need to consider more carefully the measuring of the lag phase. It must be understood that the lag phase refers to net disappearance. There is an increase in dry matter due to microbial biomass adhesion but at the same time some degradation of substrate. In the suggested model, lag phase is not taken into account, because the limited database did not allow improvement in prediction of intake, but with an extended database this may well be important.

Another problem occurs if the feed contains antinutritive factors. Here the *in situ* technique is not ideal, because the very small amount of antinutritive factors in the sample contained in the bag will have little or no effect on the environment of the rumen and so will not affect the sample's degradation characteristics. There are also feeds that cannot be assessed using the nylon-bag technique. Soluble feeds, such as molasses, and small-particle feeds, such as blood meal, single-cell protein, sago meal and cassava meal, cannot be retained in the nylon bag. Here *in vitro* techniques are superior.

Another question is: will feed potential be additive? While an animal can consume twice as much feed with an index value 60 as one with 30, does this mean that animal production will be proportional? Probably not. The total energy cost of eating and rumination may well be similar whether the index value is 30 or 60, and as a result the former provides proportionally more energy for production. This is probably one of the reasons why the metabolizable energy from more concentrated diets is utilized better than that from roughages, as has been known for many years, since proportionally less energy is expended in eating and rumination and standing up to eat.

Since the characteristics of A, B and C are generally a result of microbe–plant interaction, it is quite likely that evaluation systems based upon them can be globally important for assessment of roughage quality, but there is room for further improvement. The association of *B* and C is so firm that perhaps it would be as good to combine them as *B* × C as to treat them separately. There are some roughages that have a very low *B* value and a high C value, in which case the importance of C may be exaggerated.

The *in situ* technique does require use of fistulated animals, which may pose difficulties in some laboratories. The next best technique may be the Menke gas-test technique (Menke and Steingass, 1988), which describes fermentation dynamics by the rate constant of gas production from a feed sample incubated *in vitro* with a rumen inoculum. The rumen inoculum required may be obtained by stomach pump rather than by withdrawal from a rumen fistula or, in some instances, faeces may be used. However, *in vitro* techniques are discussed in detail in another chapter in this book. Blummel and Ørskov (1993) compared results obtained by nylon bag and gas production and found closely similar results. For feeds containing antinutritive factors affecting microbial activity, the gas-production technique is superior, in that the inoculum causes much less dilution than occurs when nylon bags are incubated in the rumen. This can be further refined by the selective removal of particular groups of antinutritive factors (Khazaal *et al.*, 1994; Makkar *et al.*, 1995).

Another intriguing aspect to analyse is whether the feed potential can also predict microbial protein production. The microbial protein is dependent on the extent of fermentation of carbohydrate and also influenced by rate of outflow (see Chen *et al.*, 1992). It is thus possible that both intake and a good estimate of microbial protein supply can be obtained from the concept outlined, but here a new research area must be opened up to provide the evidence for that.

How can chemical analysis of the feed complement results obtained by the *in situ* technique? There is no doubt that determination of N content is essential in order to see if N supplementation is needed to ensure optimal digestion. For some feeds, it may be useful to determine organic-matter rather than dry-matter degradation. Further refinements, such as determination of cell-wall degradation characteristics, may be useful for research purposes but will probably not be valuable for forage evaluation in addition to the characteristics for dry matter, inasmuch as the potential degradability (*B*) of dry matter closely reflects that of the fibre fractions.

For the past 15–20 years, the *in situ* technique has proved to be an extremely useful and robust technique for estimating the feeding value of forages and the quality of the rumen environment. Without that technique, progress would have been much less rapid. In particular, it has been useful in many developing countries, where constancy of electricity supply and consequently of incubation temperature, is difficult to maintain. Without constant temperature, *in vitro* techniques are totally unreliable. The *in situ* technique has no such problems and generally it allows better prediction of feed intake.

Cost-effectiveness

Farmers and the feed industry require reliable information rapidly and, if possible, 'upfront', i.e. before the feed is used. The *in situ* technique can achieve this, whereas measurement of digestibility by feeding trials *in vivo* is slow and laborious and feed evaluation using respiration calorimetry is still more difficult and costly. While the last method has given very valuable information regarding general categories of feedstuffs, it would be prohibitively costly to use routinely for numerous individual batches of feed. When the digestibility of feeds is measured *in vivo*, in the UK, even when sheep are used, the cost is about £1000 per feed; when it is done in respiration chambers, the cost is more like £2000. Estimation of dry-matter degradability using nylon bags costs around £100 per feed. When digestibility or metabolizability is determined *in vivo* at only one feeding level, usually maintenance, only one point on the degradability curve is obtained. The problem, then, is that since the extent to which feeding level influences digestibility depends on the degradation rate (and on retention time in the rumen), intake cannot be predicted accurately from a static measure of *in vivo* digestibility or metabolizability, nor can digestibility account for more than about 50% of variation in intake. The value of estimates *in situ* of degradation characteristics is therefore considerably more than the value of measurements *in vivo* of digestibility or metabolizability at a single feeding level. In relation to *in vivo* measurement, therefore, description of degradation characteristics is very cost-effective. However, as mentioned earlier, gas-production measurements *in vitro* may provide a sufficiently accurate and cheap alternative. It is possible that NIR measurements (spectroscopic measurements in the near-infrared) may also be sufficiently accurate to predict A, *B* and C values of roughages. This would provide a rapid upfront measurement of degradation characteristics, once it is adequately calibrated in relation to *in situ* technique. For assessment of the rumen environment, however, there is, as yet, no adequate cost-effective substitute for the *in situ* technique.

References

Blummel, M. and Ørskov, E.R. (1993) Comparison of *in vitro* gas production and nylon bag degradability of roughages in predicting feed intake in cattle. *Animal Feed Science and Technology* 40, 109–119.

Chen, X.B., Chen, Y.K., Franklin, M.F., Ørskov, E.R. and Shand, W.J. (1992) The effect of feed intake and body weight on purine derivative excretion and microbial protein supply in sheep. *Journal of Animal Science* 70, 1543–1552.

Huntington, J.A. and Givens, D.I. (1995) The *in situ* technique for studying the rumen degradation of feeds: a review of the procedure. *Nutrition Abstract Reviews Series B* 65, 64–93.

Ishida, M., Ando, S. and Oshio, S. (1996) Comparison of digestibility reduction by feed intake increase between lactating cows fed grass-based and alfalfa basal diet. In: *Proceedings of 8th AAAP Animal Science Congress*. Japanese Society of Zootechnical Science, Tokyo, Japan, pp. 164–165.

Khazaal, K., Dentinho, M.T., Riberiro, J.M. and Ørskov, E.R. (1994) A comparison of gas production during incubation with rumen contents *in vitro* and nylon bag degradability as predictors of the apparent digestibility *in vivo* and the voluntary intake of hays. *Animal Production* 57, 105–112.

Kibon, A. and Ørskov, E.R. (1993) The use of degradation characteristics of browse plants to predict intake and digestibility by goats. *Animal Production* 57, 247–251.

McAnally, R.A. (1942) Digestion of straw by ruminants. *Biochemistry Journal* 36, 392–399.

McDonald, I. (1981) A revised model for the estimation of protein degradability in the rumen. *Journal of Agricultural Science, Cambridge* 96, 251–252.

Makkar, H.P.S., Blummel, M. and Becker, K. (1995) Formation of complexes between polyvinyl pyrrolidones or polyethylene glycols and tannins, and their implication in gas-production and true digestibility in *in vitro* techniques. *British Journal of Nutrition* 73(6), 897–913.

Mehrez, A.Z. and Ørskov, E.R. (1977) The use of a Dacron bag technique to determine rate of degradation of protein and energy in the rumen. *Journal of Agricultural Science, Cambridge* 88, 645–650.

Menke, K.H. and Steingass, H. (1988) Estimation of the energetic feed value obtained from chemical analysis and *in vitro* gas production. *Animal Research Development* 28, 7–55.

Mould, F.L. and Ørskov, E.R. (1984) Manipulation of rumen fluid pH and its influence on cellulolysis *in sacco*, dry matter degradation and the rumen microflora of sheep offered either hay or concentrate. *Animal Food Science Technology* 10, 1–14.

Mould, F.L., Saadullah, M., Haque, M., Davis, C., Dolberg, F. and Ørskov, E.R. (1982) Investigation of some of the physiological factors influencing intake and digestion of rice straw by native cattle of Bangladesh. *Tropical Animal Production* 7, 174–181.

Ørskov, E.R. (1992) *Protein Nutrition in Ruminants*, 2nd edn. Academic Press, London and New York.

Ørskov, E.R. and Greenhalgh, J.F.D. (1977) Alkali treatment as a method of processing whole grains for cattle. *Journal of Agricultural Science, Cambridge* 89, 253–255.

Ørskov, E.R. and McDonald, I. (1979) The estimation of protein degradability in the rumen from incubation measurements weighted according to rate of passage. *Journal of Agricultural Science, Cambridge* 92, 499–503.

Ørskov, E.R. and Ryle, M. (1990) *Energy Nutrition in Ruminants*. Elsevier Applied Science, London and New York.

Ørskov, E.R., Reid, G.W. and Kay, M. (1988) Prediction of intake by cattle from degradation characteristics of roughages. *Animal Production* 46(1), 29–34.

Quin, J.I., Van Der Wath, J.G. and Myburgh, S. (1938) Studies on the alimentary tract of Merino sheep in South Africa. 4. Description of experimental technique. *Onderstrepoort Journal of Veterinary Science and Animal Industry* 11, 341–360.

Ramanzin, M., Ørskov, E.R. and Tuah, A.K. (1986) Rumen degradation of straw. 2. Botanical fractions of straw from two barley cultivars. *Animal Production* 43, 271–278.

Shem, M.N., Ørskov, E.R. and Kimambo, A.E. (1995) Prediction of voluntary dry-matter intake, digestible dry-matter intake and growth rate of cattle from the degradation characteristics of tropical foods. *Animal Science* 60, 65–74.

Silva, A.T. and Ørskov, E.R. (1988) Fibre degradation in the rumens of animals receiving hay, untreated or ammonia-treated straw. *Animal Food Science Technology* 19, 277–287.

Tilley, J.M.A. and Terry, R.A. (1963) A two stage technique for the *in vitro* digestion of forage crops. *Journal of the British Grassland Society* 18, 104–111.

Tolkamp, B.J. and Ketelars, J.J.M.H. (1994) Foraging behaviour programs in ruminants. In: Thacker, P.A. (ed.) *Livestock Production in the 21st Century. Priorities on Research Needs*. University of Saskatchewan, Canada, pp. 11–24.

Weyreter, H., Heller, R., Dellow, D., Lechner-Doll, M. and Engelhardt, W.V. (1987) Rumen fluid volume and retention time of digesta in an indigenous and a conventional breed of sheep fed a low quality diet. *Journal of Animal Physiology Animal Nutrition* 58, 89–100.

10 Cumulative Gas-production Techniques for Forage Evaluation

B.A. Williams

Animal Nutrition Group, Wageningen Institute of Animal Sciences (WIAS), Wageningen University, Marijkeweg 40, 6709 PG Wageningen, The Netherlands

Introduction

The quest for a simple, cheap and reliable laboratory method to evaluate the nutritive value of forages for ruminant animals resembles the search for the philosopher's stone during the Middle Ages in Europe. There is an urgent need for such a technique. Not only are cost and animal-welfare considerations making the use of animals less desirable, but also the increasing human population and demand for animal products create a need for new animal feedstuffs and improved varieties of traditional ones. However, like the philosopher's stone, a reliable and accurate technique to predict the nutritive value of all possible feeds is probably impossible to find. The microbial degradation of forages in the rumen is a complex process, involving complex interactions, both between the microorganisms present (bacteria, protozoa and fungi), and between the microbial population and the host (Czerkawski, 1986). To develop such a method, the findings of rumen microbiologists published for the last 40 years provide an important source of information (e.g. Hungate, 1966; Hobson and Wallace, 1982). For example, factors that must be considered when trying to simulate the rumen environment include temperature, pH, buffering capacity, anaerobiosis and nitrogen source.

Existing laboratory techniques concentrate on estimating rumen degradation. The rationale for this is that forage consists largely of carbohydrate in the form of fibre, and this can only be digested by microorganisms in the rumen. However, the effective degradability (the digestibility that can be expected *in vivo*) also depends on the rate of degradation, the potential degradability and the fractional outflow rate from the rumen (Hvelplund *et al.*, 1995). At the moment, there is no *in vitro* technique that can estimate the lattermost aspect of *in vivo* digestibility. Existing techniques measure the potential degradability (Tilley and Terry, 1963) or both the degradability and rate of degradation (*in situ*, and cumulative gas production).

The *in situ* technique has been widely used for the last 50 years. Whatever it is called (*in sacco*, nylon bag, Dacron bag, cloth bag), it requires the incubation of substrates in bags within the rumen of fistulated animals. These bags are removed at regular intervals, the weight loss is plotted (usually as dry matter (DM)) and a curve is fitted to the resultant data points (see Chapter 12). Van Soest (1994) has listed the advantages and disadvantages of this kind of technique.

While by no means a new concept, the measurement of gas production to evaluate forages has come to prominence in the last 10 years. There are several methodologies in current use, and the differences and similarities between them will be described, along with current research into the fundamentals of what the technique can predict and how best to carry it out. The general stoichiometry of gas related to volatile fatty acid (VFA) production is described for the rumen fermentation process. The use to which the technique is being put and potential future developments are then discussed in relation to feed evaluation.

A History of Gas-production Techniques for Forage Evaluation

Early techniques

Gas (carbon dioxide (CO_2) and methane (CH_4)) is a waste product of fermentation. In 1943, Quin incubated feedstuffs in gas-tight flasks and measured gas production manometrically. By the 1960s, the measurement of gas using liquid displacement systems was suggested as a technique to evaluate ruminant feedstuffs (Johnson, 1963; Trei *et al.*, 1970) and this is still used today (Waghorn and Stafford, 1993).

In order to simplify measurement of gas by manometry, O'Hara *et al.* (1974) used a technique described by El-Shazly and Hungate (1965), whereby gas production was measured by use of a syringe attached to a flask in which the substrate was fermented. They found characteristic patterns of change in rates of gas production according to the substrate used.

The Hohenheim gas test

The group at Hohenheim University (Germany) were the first to conduct the fermentation process within large glass syringes, so that the gas produced pushed the plunger outwards, and the total volume produced was recorded after 24 h (Menke and Ehrensvard, 1974; Fig. 10.1). This evaluation system combines the total gas volume after 24 h with the concentration of crude protein (CP), crude fat, crude fibre and ash of the feed to predict metabolizable energy (ME). In Germany, the Hohenheim *in vitro* gas test is now widely used for the estimation of *in vivo* digestibility and ME for ruminants (Menke and Steingass, 1988).

Recently, Blümmel and Ørskov (1993) have adapted the Hohenheim gas test by recording the change in gas production in the glass syringes at regular time intervals, so that a measure of the kinetics of fermentation is obtained, as well as the end-point value.

Fig. 10.1. Equipment for the Hohenheim gas test (photo courtesy of Klaus Aiple and Klaus Becker, University of Hohenheim).

Manual transducer technique

A new method, using a pressure transducer to measure head-space gases in sealed serum bottles, was first described in 1991 by Theodorou *et al.* from the Institute of Grassland and Environmental Research (IGER) in the UK (Fig. 10.2). The pressure and volume were recorded at regular intervals: approximately every 3–4 h for the first 24 h, and then less frequently to the end of fermentation at about 120 or 144 h. The head-space gas is removed after each reading. The cumulative gas profile represents the kinetics of the fermentation process of individual feedstuffs. An example of such a profile comparing wheat straw and ryegrass is shown in Fig. 10.3.

Automated systems

The earliest reported automated device was developed by Beuvink *et al.* (1992) at the Institute for Animal Science and Health, Lelystad, the Netherlands. This system was based on the weight of fluid displaced by fermentation gas for 24 individual bottles. The change in weight was recorded and a calculation made to convert the weight to gas volume, which was then registered by a data logger.

Pell and Schofield (1993), from Cornell University in the USA, first described the use of computerized pressure sensors to monitor gas production. They investigated

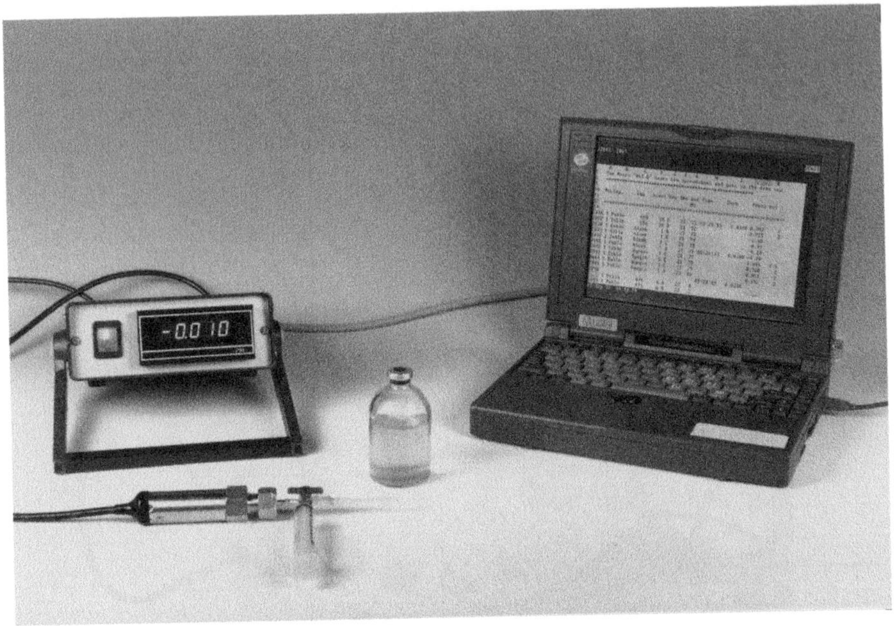

Fig. 10.2. The pressure transducer, complete with serum bottle, connectors and computer, as used for the manual transducer technique.

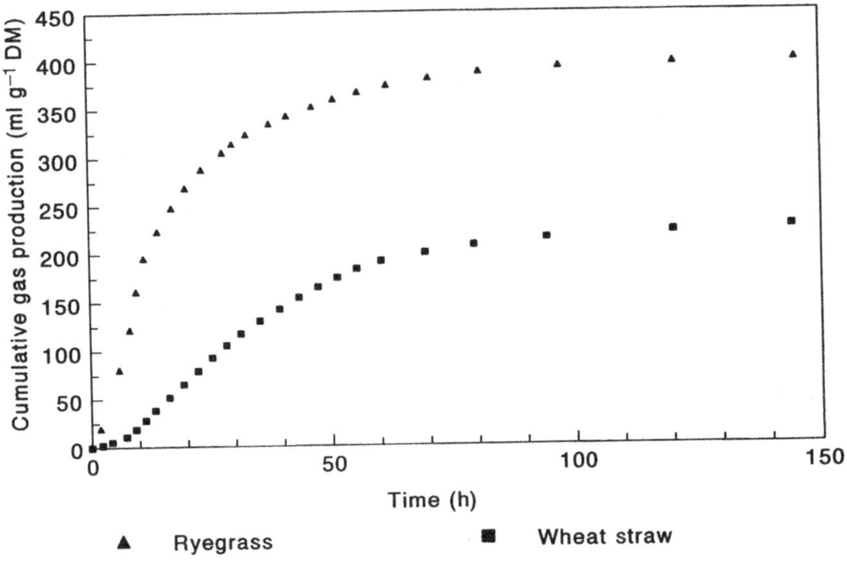

Fig. 10.3. Example of cumulative gas-production profiles generated using the manual transducer technique, showing the differences between the fermentation of wheat straw and ryegrass.

the use of several variables, including sample size, inoculum size, vessel size and type of pressure sensor, and reported the ranges within which gas production could be accurately measured. For the Cornell technique, gas is not vented during the fermentation but the head space of the bottles is much larger than for the IGER technique and the sample much smaller (Fig. 10.4).

An automated version of the manual transducer technique (called the automated pressure evaluation system (APES)), developed at IGER, was first reported by Davies *et al.* (1995). The system allowed for 50 gas-tight culture bottles, each of which was fitted with a pressure sensor and solenoid valve linked to a computer, for continuous monitoring of all bottles (Fig. 10.5). During fermentation, each solenoid valve opens to release the accumulated gas when the pressure sensor registers a pre-set gas pressure. The number of vents and time between each vent for each bottle are recorded automatically by a computer and are then used to plot cumulative gas profiles. This system is now also in place at Wageningen Agricultural University, and was used to produce the profiles for wheat straw and ryegrass shown in Fig. 10.6. The large number of data points produce very detailed fermentation profiles.

Cone *et al.* (1996) described what was called 'a new fully automated time-related gas-production apparatus', developed at ID-DLO, Lelystad, in the Netherlands. This apparatus uses a combination of pressure transducer and an

Fig. 10.4. Equipment used for the automated gas system at Cornell University (photo courtesy of Alice Pell, Cornell University).

Fig. 10.5. The IGER automated pressure evaluation system at Wageningen University, The Netherlands.

electric valve for every bottle. Each valve opening represented a fixed amount of gas and the fermentation kinetics were determined by recording the time of each valve opening on a data logger. A diagrammatic representation of the apparatus was published as part of the methods paper.

A General Description of the Method

Broadly speaking, the general principles of what the different techniques measure are the same. The technique is similar to other *in vitro* digestibility procedures which use milled substrates, an anaerobic medium and an inoculum of a mixed microbial population from the rumen. The pre-weighed substrate is suspended in a medium, the mixture warmed to ~39°C and a freshly collected sample of rumen fluid added as inoculum. From that moment, the production of gas resulting from fermentation is recorded, either at the end of the fermentation or at regular time intervals for those methods that measure fermentation kinetics. A cumulative gas profile is plotted from the results at the end of fermentation. This curve records the

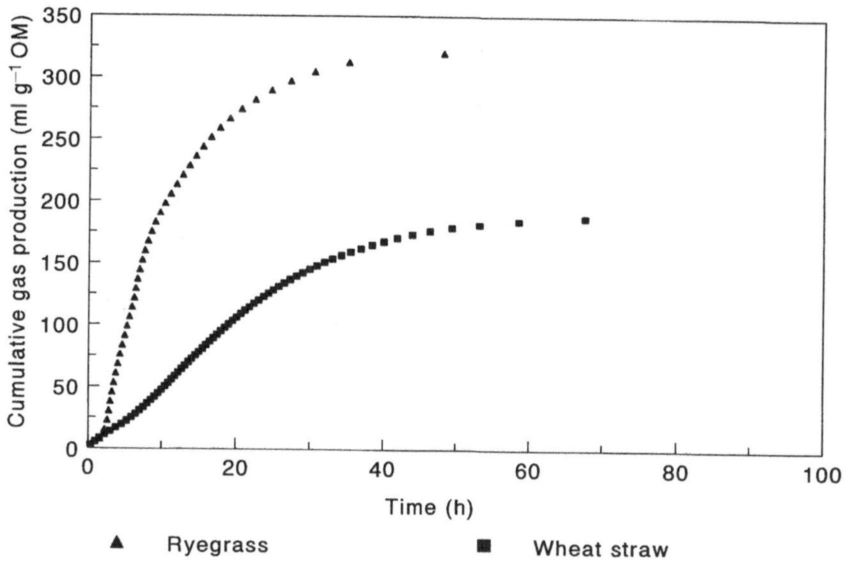

Fig. 10.6. Example of cumulative gas-production profiles generated using the automated pressure evaluation system, showing the differences between the fermentation kinetics of wheat straw and ryegrass (from Pablo Chilibroste, unpublished data). OM, organic matter.

production of a waste product by the microorganisms, and so reflects the kinetics of microbial activity within the bottle.

Sample preparation

As with all laboratory techniques, the preparation of a forage sample is fraught with many factors to be considered, which have been dealt with exhaustively for other *in vitro* methods (Goering and Van Soest, 1970; Huntington and Givens, 1995). It is generally recognized that a smaller particle size increases the surface area available for microbial degradation. Consequently, for routine gas-production analysis most authors mill the test substrates through a 1 mm screen (Menke *et al.*, 1979; Pell and Schofield, 1993; Theodorou *et al.*, 1994), in keeping with other *in vitro* methods (Tilley and Terry, 1963; Goering and Van Soest, 1970).

The method and temperature of drying a substrate were reviewed for the *in situ* technique by Huntington and Givens (1995). Most groups either freeze-dry samples (which, in the case of silages, can lead to the loss of volatile components) or oven-dry at a low temperature, either 60°C (Pell and Schofield, 1993) or 70°C (Beauvink *et al.*, 1992). Cone *et al.* (1994) reported a comparison of different drying conditions on the fermentation kinetics of grass and maize stems, and concluded that freeze-drying was better than oven-drying (at 30, 50, 70 and 105°C).

Media

The living rumen provides an ideal environment for its anaerobic microorganisms in terms of pH, temperature, buffering capacity and nitrogen in the form of urea, while the microorganisms behave symbiotically by providing each other with micronutrients, such as branched-chain fatty acids and vitamins. There are also likely to be other unknown factors present in the rumen milieu, which cannot yet be replicated *in vitro*. In the past, quite a number of media used to characterize rumen microorganisms overcame this problem by addition of a large proportion of sterile rumen fluid to ensure the presence of all possible factors for microbial growth.

All the media in current use have the following in common: bicarbonate and phosphate buffers, a reducing agent, a source of nitrogen, various minerals and resazurin as an indicator of redox potential. In all cases, CO_2 is used during the preparation of the medium to ensure a low redox potential at the time of inoculation. A low redox potential is very important for the fermentation of very fibrous forages, or even for those rich in starch. Leedle and Hespell (1983) showed that the absence of strict anaerobiosis resulted in partial losses of cellulolytic and amylolytic bacterial groups from rumen fluid. Therefore a low redox potential is very important, not only for the fermentation of fibrous forages, but also for those rich in starch. Grant and Mertens (1992) recommended that all *in vitro* systems used to assess fibre digestion kinetics must use continuous CO_2 gassing and reducing agents to promote both the rapid initiation and the progression of fermentation. Minimizing the limitations of the *in vitro* system means that differences between substrates will be more readily detected.

A modified version of medium B (Williams *et al.*, 1995b) is used most commonly in Wageningen; it is a complex semi-defined medium, which, in principle, should provide most of the rumen microorganisms with all of their requirements (Lowe *et al.*, 1985), except for energy, which is provided by the test substrate. This medium, while suitable for a wide range of the rumen microorganisms, comprises several component solutions and so is time-consuming and complicated to use for routine analyses. Not surprisingly, several other media have been developed, most of which seem to originate either from the medium of Menke and Ehrensvard (1974) or from that of Goering and Van Soest (1970).

Research continues to develop media that are appropriate for different situations. The composition and strength of the buffer are still a focus of interest (Pell and Schofield, 1993; Nsahlai *et al.*, 1995). Measurement of pH at the end of fermentation is to be recommended to ensure that sufficient buffering capacity was available for novel or concentrate substrates.

Inoculum

Conditions for the keeping of donor animals, their feeding, and the timing of rumen fluid collection, so as to obtain the most consistent result between runs for *in vitro* techniques was reviewed for *in vitro* techniques by Alexander and McGowan (1966). It was recommended that a sample should be taken after overnight fasting from at least three animals, all of which were fed the same diet of medium-quality hay with a small amount of concentrate.

Animal species

Sheep (usually wethers) are most often used as a routine rumen fluid donor for *in vitro* techniques (Beuvink *et al.*, 1992; Khazaal *et al.*, 1994; Theodorou *et al.*, 1994; Bonsi *et al.*, 1995; Williams *et al.*, 1995b). They are chosen mostly for their size, which makes them cheaper to feed and house for an extended period. However, cattle are also used as inoculum donor (Grant and Mertens, 1992; Pell and Schofield, 1993; Pienaar, 1994; Makkar *et al.*, 1995; Sileshi *et al.*, 1996). Differences in the activity of microorganisms from rumens of different species or of the same species but different diets mean that all descriptions of *in vitro* gas-production evaluations should describe the conditions of the donor animal (species and breed, diet, time of collection in relation to feeding, etc.).

Time of collection

Rumen fluid taken after overnight fasting is less active than that taken 2 h after feeding, but is more consistent in its composition and activity. As a rule, therefore, it has been recommended (Menke and Steingass, 1988) that rumen fluid be collected before feeding, which is the routine followed by many groups (Blümmel and Ørskov, 1993; Pell and Schofield, 1993; Theodorou *et al.*, 1994; Williams *et al.*, 1995b).

Preparation of inoculum

Another important consideration is that of homogenization of the rumen fluid before inoculation. It has been shown that blending increases the number of adherent bacteria (which are most often the cellulolytic species) in the inoculum, but it can also increase the number of small feed particles present (Pell and Schofield, 1993), which can lead to higher gas production in the blank bottles. The blending procedure can also increase the risk of oxygen exposure if the stream of CO_2 is not forceful enough.

Proportion of inoculum to medium

The proportion of inoculum to medium has been investigated by Pell and Schofield (1993), who recommended an inoculum of 20% of a total volume of 10 ml. However, the amount of added inoculum varies widely between groups. Hidayat *et al.* (1993) found increased rates of gas production from forages, in response to increased bacterial density.

What does the Technique Measure?

In vitro gas production is related solely to fermentation in the rumen, rather than to digestibility in the whole digestive tract, which also includes enzymatic digestion, absorption and hind-gut fermentation. Rumen fermentation is mainly the result of the activity of the bacteria, protozoa and fungi that inhabit it. Some species from each of these groups possess cellulolytic and hemicellulolytic activity (Hungate, 1966; Orpin 1983/4; Williams and Coleman, 1988), though the relative contribution of each to fermentation is still not completely understood.

Stoichiometry of rumen fermentation

Energy for the growth of rumen microorganisms is derived from the fermentation of carbohydrates, particularly starch and cell walls, which are first broken down into hexoses and pentoses, from which the main end-products are acetate, propionate and butyrate (VFA), CO_2 and CH_4 (Wolin, 1975), but which can also supply essential carbon skeletons for the synthesis of microbial biomass (Beever, 1993b). The VFA are absorbed across the rumen mucosa and provide energy for the mucosal tissue, as well as for the host animal. The gases produced from VFA production and from the bicarbonate buffer in the rumen are eructated to the atmosphere. It is this gas that is being measured *in vitro*. The stoichiometric reactions of fermentation of cell-wall hexoses were described by Hungate (1966):

$$1 \text{ mol Hexose} + 2H_2O \rightarrow 2 \text{ Acetate} + 2CO_2 + 4H_2 \quad (1)$$

$$1 \text{ mol Hexose} + 2H_2 \rightarrow 2 \text{ Propionate} + 2H_2O \quad (2)$$

$$1 \text{ mol Hexose} \rightarrow 1 \text{ Butyrate} + 2CO_2 + 2H_2 \quad (3)$$

$$CO_2 + 4H_2 \rightarrow CH_4 + 2H_2O \quad (4)$$

Formation of propionic acid is the only reaction that requires the uptake of hydrogen (H_2) (Van Soest, 1994). Any remaining H_2 is usually converted in the rumen to methane by methanogenic bacteria. Also, the production of propionate does not involve the generation of CO_2. Therefore, to compare gas data between different forages, it is also useful to examine differences in the acetate:propionate ratio at the end of fermentation (Stefanon *et al.*, 1996; Groot *et al.*, 1998).

The complete pathway of VFA production for the hexose equivalents from high-forage diets was summarized by Beever (1993b) as:

$$1 \text{ mol Hexose} \rightarrow 1.34 \text{ mol acetate} + 0.45 \text{ mol propionate} + 0.11 \text{ mol butyrate} + 0.61 \text{ mol methane} + 4.62 \text{ mol adenosine triphosphate (ATP)}$$

All techniques, except for the Hohenheim gas test, rely on the measurement of pressure in a receptacle of fixed volume. The empirical gas laws of Boyle and Gay-Lussac, from which the volume of gas can be calculated are:

$$PV = nRT \quad (5)$$

where:

P = pressure of gas (atm)
V = volume of gas (l)
n = number of moles of gas
R = the molar gas constant at 0°C and 760 mm Hg
T = temperature (°K)

In most studies, it is assumed that the pressure of gas is at 1 atm, so 1 mmol of gas occupies 25.62 ml at 39°C (Weast, 1968). However, if laboratories are located in areas where the atmospheric pressure is not 1 atm, an adjustment must be made accordingly.

Microbial biomass production

It has been shown that an inverse relationship can exist between gas volume (or VFA) and microbial biomass production (Naga and Harmeyer, 1975; Leng, 1993). Krishnamoorthy *et al.* (1991) indicated that the type of carbohydrate and its rate of fermentation could influence total synthesis of microbial protein per unit volume of gas produced. On the one hand, it has been suggested that the efficiency of microbial growth can vary depending on feeding conditions (Leng, 1993). Yet it has also been pointed out that, if there are no deficiencies of microbial growth factors, both the source of carbohydrate (whether sugar, starch or fibre) and the flow rate of liquor from the rumen have little effect on microbial growth yield. The methodology used to measure cumulative gas production can easily be adapted to investigate the effect of microbial growth factors and sources of carbohydrates under controlled conditions, which could help to answer some of these questions.

Little work has been done to examine the relation between gas production and microbial growth, though Blümmel *et al.* (1997b) have done some work in that direction, using the Hohenheim system. They demonstrated that some substrates with proportionally high gas volumes had comparatively low biomass yields. Gas-volume measurements were complemented by a determination of the truly degraded substrate *in vitro* to avoid selection of substrates with proportionally high VFA production and low microbial biomass yield (Blümmel *et al.*, 1997a), which is relevant for forages with a low nitrogen content. It has been pointed out by Leng (1993) that one of the factors affecting microbial growth efficiency is the relative amounts of carbohydrate and protein that are fermented. For example, a high protein-to-carbohydrate ratio could lead to a relatively low microbial protein-to-VFA ratio. These relationships could be examined more closely using cumulative gas-production techniques, particularly if other parameters, such as VFA, ammonia (NH_3) and microbial protein production are also measured, using a medium that is known to contain all possible microbial growth factors, or, contrarily, is deficient in one or more of them.

Data Processing and Interpretation of Results

Conversion of raw data to cumulative volumes

Gas volumes recorded at each time interval are added together to give a cumulative gas profile. In the case of the measurement of head-space pressure and volume, a correction is made for irregularities in head-space volume between bottles, by a regression of pressure against volume for each bottle. A correction is also made to cumulative volumes for the starting DM/organic matter (OM) weight of the substrate. Ideally, if remaining substrates are not required for further analysis, OM data should be used, since ash makes no significant contribution to gas or VFA production and can vary significantly between substrates. The yield of gas or VFA (Y_G or Y_{VFA}: the amount of gas or VFA produced per gram material disappeared) can also be calculated (as DM or OM) if the substrate losses at the end of fermentation are measured.

Correction for blanks

A series of blank bottles containing medium and inoculum but no substrate is routinely included in each run. The total gas recorded for the blanks is subtracted from the total gas produced from the test substrates to obtain the total gas derived from fermentation of the substrate. However, the blank profile is not subtracted from the substrate profiles prior to curve fitting, particularly in the case of the multiphasic curves, as the phases present in the blank may not be relevant to what occurs in the presence of substrate (Fig. 10.7).

Curve fitting

Fitting cumulative gas profiles to an appropriate equation allows one to summarize the kinetic information therein. Several equations have been developed for this purpose (Beuvink and Kogut, 1993; France *et al.*, 1993; Groot *et al.*, 1996).

The choice of equation to fit gas profiles must take into account not only mathematical considerations (such as the statistical fit) but also the biological meaning of the parameters. Increasingly frequent readings for gas production lead to very detailed profiles, and considerations of mathematical fit (particularly very low Durbin–Watson values) have led to the development of a multiphasic model (Groot *et al.*, 1996), which was based on the multiphasic growth model of Koops (1986). The potential value of this model is that it separates the test substrate into the more rapidly and more slowly fermentable fractions (Oostdam *et al.*, 1995; Groot *et al.*, 1998).

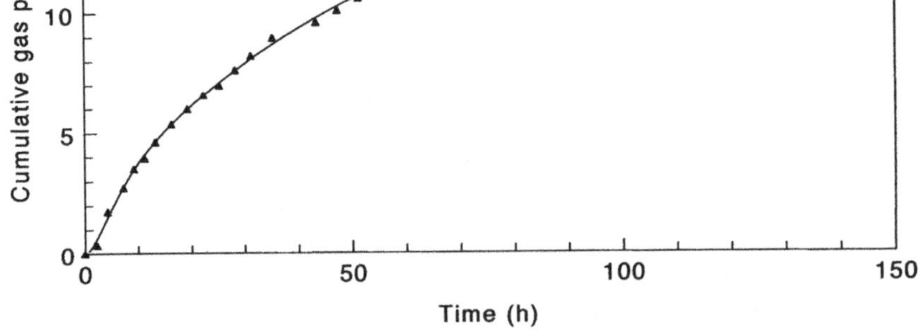

Fig. 10.7. Cumulative gas-production profile of a blank bottle, containing medium B and sheep rumen fluid as the inoculum.

France *et al.* (2000) have compared the different equations available in mathematical terms, while Dhanoa *et al.* (2000) have compared the same equations, using a large series of data sets from widely different forages, and have analysed the results from the point of view of the use of the equations in practice.

Applications of the Technique

A wide range of forages has been investigated using cumulative gas-production techniques, of which a selection is shown in Table 10.1. The largest proportion of materials are those which are fed as a source of energy, i.e. carbohydrates. It is important to realize that a nitrogen-rich medium should provide ample nitrogen for microorganisms, so a proteinaceous feed fermented in a nitrogen-rich medium would actually assess the potential energy, not the protein, content of that feed.

Grass, hay and silage

Aspects of grass, hay and silage quality that have been investigated include the detailed differences between substrates due to plant maturity (Stefanon *et al.*, 1996; Williams *et al.*, 1999) and conditions of growth. Gas-production kinetics are also used to rank different species or cultivars (Piva *et al.*, 1988).

Table 10.1. Cumulative gas production characteristics after 144 h of fermentation of a range of substrates inoculated with sheep rumen fluid.

Substrate	DMCV* (ml g^{-1} DM)	$t_{\frac{1}{2}}$ (h)	t_{RM}† (h)	RM† (ml h^{-1})	DM loss (%)
Glucose	392	19.7	7.5	0.032	100
Starch	425	23.6	22.2	0.037	99.3
Crop residues					
Sorghum stem	231	44.4	18.7	0.014	68.4
Ammoniated wheat straw	232	34.0	31.0	0.025	60.3
Wheat	223	36.2	30.7	0.021	52.3
Red clover	207	22.6	18.8	0.034	79.6
Ryegrass hay‡	352	23.2	14.4	0.028	52.6
Silage					
Grass§	283	17.4	14.0	0.050	67.9
Maize‖	258	22.1	24.2	0.060	67.6

* DMCV is the measured total gas production at 144 h g^{-1} dry matter of the substrate.
† RM (fractional rate of substrate digestion) and t_{RM} (time at which the fractional rate occurs) were both calculated according to the monophasic version of the model reported in Groot *et al.* (1996).
‡ From Groot *et al.* (1998).
§ From Calabrò and Williams (1997).
‖ From Williams *et al.* (1995b).

Crop residues

Cumulative gas production has been used to determine differences in fermentability between different chemical (urea: Prasad *et al.*, 1994) and physical (steam: Castro *et al.*, 1994; steam explosion: Williams *et al.*, 1995b; extrusion: Williams *et al.*, 1997) pretreatments of straw. The fermentabilities of crop residues of different cultivars (Hermanto *et al.*, 1991) or those grown under differing environmental conditions (Williams *et al.*, 1996) have also been examined. The technique is also increasingly being used to investigate the effect of addition of supplements on the fermentability of straws (Prasad *et al.*, 1994).

Browses

The technique has been used to rank different browse trees as potential feed supplements for poorly degradable roughages (Topps, 1992). Many browses contain antinutritive factors (ANFs), as well as significant amounts of protein, which complicates *in vitro* assessment of their nutritive value. The effect of such ANFs on gas production and on DM and protein degradability depends on an array of factors, which includes their molecular structure and reactivity (Hagerman *et al.*, 1992; Kaitho *et al.*, 1998). Newbold *et al.* (1997) reported that some ANFs could be quite selective in their effects on certain microorganisms, while others only affect the host animal.

Some phenolic compounds from browses have been shown to be negatively related to gas production (Siaw *et al.*, 1993; Khazaal *et al.*, 1994). The latter compared gas production with the *in situ* technique and concluded that gas production was more efficient for the assessment of feeds containing ANFs. They suggested that, while physical binding of polyphenols to substrate (protein or carbohydrate) could be detected in a nylon bag incubated in a large environment (i.e. rumen), other effects, such as toxicity to microbes or binding to their enzymes, would be diluted and would therefore be more difficult to detect.

Given that there are several possible mechanisms by which browse ANFs could affect microbial activity, the technique has been used to pinpoint which mechanism is responsible for changes to fermentation. Several variations, combining different inocula (Nsahlai *et al.*, 1995), the use of tannin-binding agents (Makkar *et al.*, 1995; Wood and Plumb, 1995) or supplementation of crop residues, have been used to determine the exact effect of specific ANFs.

Feed components

Protein (legumes, browses and grasses) and starch (maize and whole-wheat silages) are often important components of forages, and their fermentation can provide insight into the fermentation of the whole forage.

Starch

Trei *et al.* (1970) reported that gas production was highly correlated with *in vitro* DM disappearance, total VFA production and starch digestion. More recently,

Opatpatanakit *et al.* (1994) found that gas production from fermentation of several starch-based substrates proved to be an accurate index of VFA production and change in pH.

Protein

Raab *et al.* (1983) described a method for the determination of protein degradation based on measurements of NH_3 concentration and gas production. NH_3 liberated during incubation is partly used for microbial protein synthesis, while the production of CO_2 and CH_4 is a measure of the energy available for protein synthesis. The ratio gas production : incorporation of NH_3 nitrogen was estimated by addition of starch to the substrate.

Until now, gas production has most commonly been used to assess protein (particularly browses) as a supplement to determine whether supplementation with a concentrate or browse legume could enhance the fermentation of poor-quality crop residues, such as teff and wheat straw (Prasad *et al.*, 1994; Bonsi *et al.*, 1995). Investigations have started in Wageningen to determine whether the cumulative gas-production technique can be used as an indicator of the relationship between carbohydrate and protein on *in vitro* fermentation. For example, the form of nitrogen supply to microorganisms can affect the total gas production of carbohydrate, as shown in Fig. 10.8 (Debersaques *et al.*, 1998).

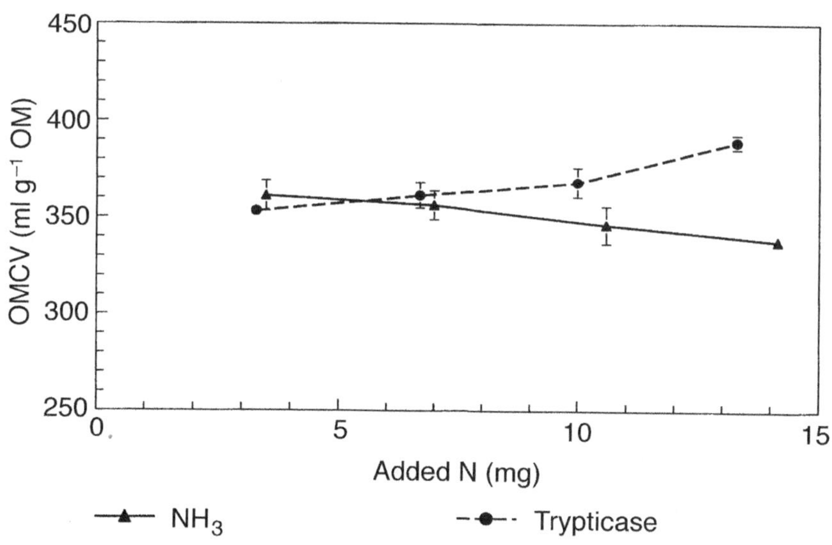

Fig. 10.8. Total gas production from citrus pulp as affected by increasing amounts of nitrogen originating either from trypticase or from ammonia (from Debersaques *et al.*, 1998). OMCV, organic matter cumulative volume.

Relation of Fermentation Data to *In Vivo* Parameters

All *in vitro* techniques have been developed to provide a cheaper, easier and faster technique to predict the availability of certain nutrients for an animal. Comparing *in vitro* techniques with each other is pointless, unless they are both being related to the same set of *in vivo* data and one can determine which technique is a better predictor of the *in vivo* parameters.

Voluntary intake of roughages is considered to be an essential factor in quality assessment (Minson, 1990). The relation between gas production and feed intake is interesting, as being of immediate practical relevance. However, a correlation between the *in vitro* relationship of a given microbial population and a specific substrate cannot take into account important elements of animal behaviour, including appetite, preference and the unmeasurable (until now) factor of taste. Uden and Van Soest (1984) noted that because there was a balance between digestion and passage in the rumen *in vivo*, *in vitro* batch systems could be expected to give higher digestibilities than *in vivo* values. However, substrates that are not readily digested in the rumen but are readily digested elsewhere (such as herbage protein) could have higher *in vivo* digestibilities than *in vitro* (Prasad *et al.*, 1994).

Several groups have compared *in vivo* digestibility and/or intake with results of *in vitro* gas production. Results of 400 digestibility trials and corresponding total gas production were used to calculate the regression equations between *in vivo* digestibility and total gas, and results of 300 sets of data were then used to check these equations (Menke and Steingass, 1988). There was a close correlation between gas production and *in vivo* digestibility, with the exception of succulent feeds. They concluded that the inclusion of CP content for concentrate feeds and CP and ash content for succulent feeds improved the accuracy of prediction.

Blümmel and Ørskov (1993) correlated total gas production from ten straws with DM intake (r = 0.88) and digestible DM intake (r = 0.93) of the straw and growth rate (r = 0.95) of steers. In another experiment, Blümmel and Becker (1997) reported that 75% of the variation in intake of whole roughage was accounted for by *in vitro* gas production after 8 h of incubation, while the average gas production from neutral-detergent fibre (NDF) measured from 24–96 h accounted for 81% of the variation in intake.

However, given that cumulative gas production can provide kinetic fermentation data, it seems a pity to compare it with end-point values, such as *in vivo* intake, or even apparent or true digestibility, though this is the comparison made most commonly at the moment, as these are the parameters most commonly measured. Models to predict intake usually include parameters related to feedstuff characteristics, particularly to carbohydrate degradability (Fisher, 1996), and, from this point of view, a kinetic parameter related to fermentability could prove a valuble contribution to the model as a whole, rather as the rate of degradation measured using the *in situ* techniques is used now (Forbes, 1993).

Advantages and Limitations of the Technique

Cumulative gas-production techniques have several major advantages in terms of animal welfare, sample size and cost and, most importantly, in the fact that they describe the kinetics of microbial activity in response to a given substrate.

Animal welfare

While the use of animals is far less than is required for *in vivo* digestibility trials, gas-production techniques still require animals as a source of rumen fluid. To eliminate this need for fistulated animals, some investigations have been carried out to determine whether or not faeces could be used as an alternative inoculum (Aiple, 1993). A comparison between rumen fluid and faeces from cows fed a silage diet (Williams *et al.*, 1995a) showed that the faecal inoculum resulted in significantly lower total gas, DM loss and VFA production for all substrates (wheat straw, rye grass, soybean hulls, wheat, maize and soybean meal). Consequently, it is not yet clear whether faeces would be an ideal replacement for rumen fluid for this method.

Sample size

The amount of material required to assess the fermentation kinetics of substrates ranges from 0.1 to 1 g. More is required than is actually fermented, in order to ensure a representative sample, but these are still small amounts compared with other techniques, making it suitable for materials available in limited quantities (e.g. from plant-breeding trials, plants grown under special conditions, etc.).

Costs

The costs of laboratory tests are usually much lower than those required to maintain animals. Whether the manual or automated systems are cheaper to run depends, to a large extent, on the cost of labour in the country where the laboratory is located. The initial equipment required can be quite expensive for the automated systems, but, as the hours for measuring are significantly reduced, it is a matter of balancing the capital costs of the equipment with the labour costs.

Information obtained

Cumulative gas production measures the kinetics of fermentation (i.e. microbial activity), rather than a reduction in particle size (Williams *et al.*, 1997). Reduction of particle size is an indication of substrate degradation, but does not necessarily indicate the availability of material to microorganisms. However, gas production remains an indicator of what occurs in the rumen, rather than in the whole animal.

Limitations

The limitations of the gas-production technique rest, for the most part, with the expectations of those using the technique. If someone hopes that the technique can be used to predict all aspects of animal health and behaviour, he/she will be disappointed. However, it does measure the kinetics of fermentation of a given substrate with a given microbial population, thereby giving a reasonable simulation of what occurs in the rumen, and, as such, will be a useful tool for evaluation of ruminant feedstuffs.

One limitation to the use of the technique for forage evaluation is the lack of uniformity in methodology. This makes it difficult to compare results from different groups. An urgent task facing those who are using the technique is to remedy this situation. It should be possible to standardize the technique in terms of the kind of medium being used, the timing of inoculum collection and the diet of the donor animals. Steps have been made in this direction. Two ring tests were organized in 1994 and 1996 (Agricultural Development and Advisory Service (ADAS) – Jim Huntington and Caroline Rymer, personal communication). Preliminary results from the first test, in which all groups used their own equipment and methodology, were reported at an informal meeting held at the University of Reading in 1995, while results of the second test, in which the medium, inoculum collection and, to some extent, animal diet were standardized, have also been reported (Rymer, 1997).

The use of standard substrates, as pioneered for the Hohenheim gas test, may need to be considered in order to overcome the inevitable differences between inocula, not only between groups, but also between runs. Other important details which still need to be investigated include: medium (e.g. an ideal buffer concentration, presence of micronutrients and anaerobiosis) and treatment of inoculum to maximize cellulolytic organism activity; the use of blanks; sample preparation in terms of method and duration of drying and grinding; and, very importantly, the relationship between gas, VFA and microbial production within the *in vitro* system. Inevitably, compromises will have to be made between what would be ideal in terms of the metabolism and growth requirements of microorganisms and what is practical on a day-to-day basis. Some flexibility within the system will need to be maintained to allow for the requirements of different research questions, such as the potential ATP and/or biomass yield of forages.

Future Use and Avenues for Further Research

The need to evaluate the nutritive characteristics of ruminant feeds must increase in importance in coming years. However, the feed-evaluation systems which exist now are inadequate (Beever, 1993a) and are in a state of flux, as new methodologies are developed and new discoveries made concerning animal and microbial physiology. Determination of the fermentation kinetics of substrates for their routine evaluation is likely to be an important facet of evaluation systems of the future, particularly as the dynamics of nutrient utilization are taken into account.

Synchronization of energy and nitrogen supply to the rumen is considered to be an important approach to improve the efficiency of rumen fermentation (Rooke and Armstrong, 1989). To apply this principle in ruminant feed formulation, one needs sufficient data on rates of protein degradation and carbohydrate fermentation by rumen microbes. It is evident that energy supplements differ in their rate and extent of ruminal fermentation (Sutton, 1985; Krishnamoorthy *et al.*, 1991) and cumulative gas-production techniques could be used to generate such information *in vitro*.

Cumulative gas production is being used in combination with alkane analysis of forages as a measure of grazing intake in dairy cows (Sip van Wieren and Pablo Chilibroste, personal communication). This methodology could be extended further to investigate the value of feed consumed by wild animals, by analysing the alkanes present in their faeces and then carrying out a fermentation study of the plants that are then known to have been eaten.

In addition to routine screening and feed evaluation, the technique is also being used to study more fundamental aspects of gastrointestinal tract (GIT) fermentation, particularly in relation to plant cell walls and other fermentable carbohydrates. Valuable information is being obtained, particularly when used in combination with other techniques already available. For example, van Laar *et al.* (1999) have recently used cumulative gas production in conjunction with cell-wall sugar analysis at various time intervals to evaluate differences in the fermentation of different anatomical components of soybean cell walls (Fig. 10.9).

Given the flexibility of the technique in terms of medium chosen and inoculum used, it is also being adapted to investigate more fundamental questions relating the source of substrate to the amount of microbial biomass produced. For this kind of research, it is essential to remember that one is not only examining the test substrate, but also microbial behaviour in terms of that test substrate and the medium being used. For example, when investigating the relation between biomass and gas production, it must be realized that the microbes must have access to all possible micronutrients they may require, or the results from an *in vitro* test will bear little resemblance to what occurs *in vivo*.

Progress will continue to be made in terms of the technical details of the methodology. However, it is clear that once these details have been refined, the technique can be used to measure the dynamic interaction that exists between the substrate and microbes for fermentation in the rumen or, indeed, the rest of the gastrointestinal tract. Results of fermentation kinetics are likely to change our perceptions of both feed evaluation and the kinetics of microbial metabolism.

Acknowledgements

Walter Gerrits, Kate Lessels and Seerp Tamminga are thanked for their critical evaluation of the manuscript.

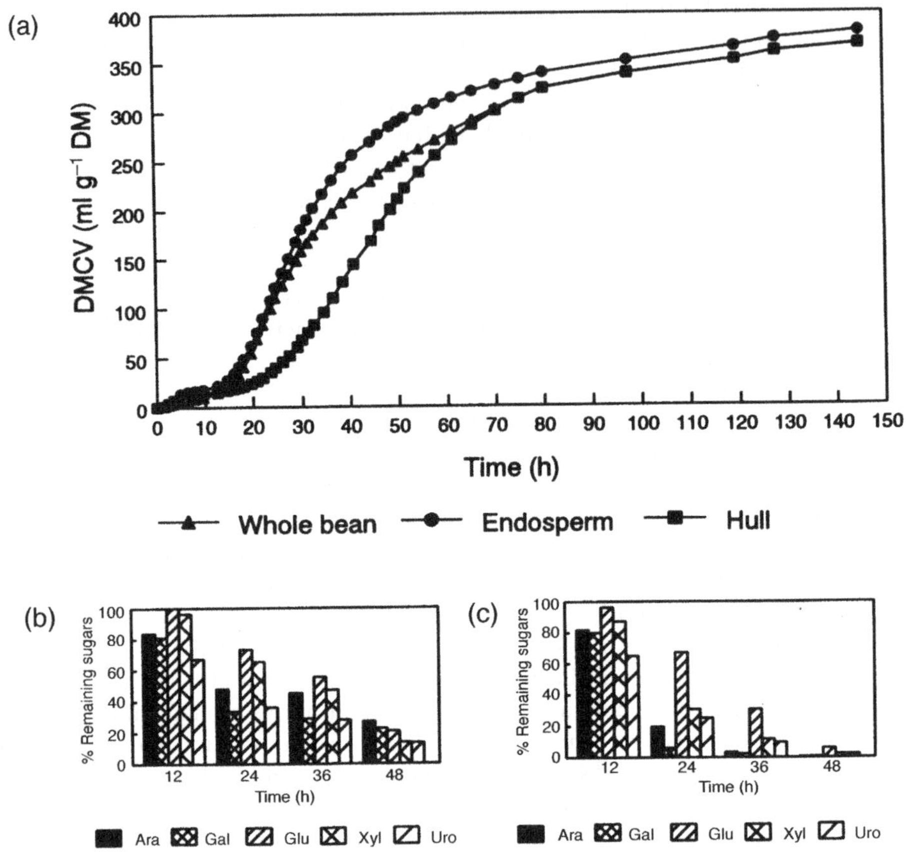

Fig. 10.9. Cumulative gas-production profiles of soybean hull and endosperm (a), and the progressive loss of cell-wall sugars: (b) hull, (c) endosperm (van Laar *et al.*, 1999). DMCV, dry matter cumulative volume; Ara, arabinose; Gal, galactose; Glu, glucose; Xyl, xylose; Uro, uronic acid.

References

Aiple, K.-P. (1993) *Vergleichende Untersuchungen mit Pansensaft und Kot als Inokulum im Hohenheimer Futterwerttest.* Dissertation, University of Hohenheim, Stuttgart, Germany.

Alexander, R.H. and McGowan, M. (1966) The routine determination of *in vitro* digestibility of organic matter in forages – an investigation of the problems associated with continuous large-scale operation. *Journal of the British Grassland Society* 21, 140–147.

Beever, D.E. (1993a) Characterisation of forages: appraisal of current practice and future opportunities. In: Garnsworthy, P.C. and Cole, D.J.A. (eds) *Recent Advances in Animal Nutrition.* Nottingham University Press, Loughborough, pp. 3–17.

Beever, D.E. (1993b) Rumen function. In: Forbes, J.M. and France, J. (eds) *Quantitative Aspects of Ruminant Digestion and Metabolism.* CAB International, Wallingford, UK, pp. 187–215.

Beuvink, J.M.W. and Kogut, J. (1993) Modelling gas production kinetics of grass silages incubated with buffered ruminal fluid. *Journal of Animal Science* 71, 1041–1046.

Beuvink, J.M.W., Spoelstra, S.F. and Hogendorp, R.J. (1992) An automated method for measuring time-course of gas production of feedstuffs incubated with buffered rumen fluid. *Netherlands Journal of Agricultural Science* 40, 401–407.

Blümmel, M. and Becker, K. (1997) The degradability characteristics of fifty-four roughages and roughage neutral detergent fibres as described by *in vitro* gas production and their relationship to voluntary feed intake. *British Journal of Nutrition* 77, 757–768.

Blümmel, M. and Ørskov, E.R. (1993) Comparison of *in vitro* gas production and nylon bag degradability of roughages in predicting food intake in cattle. *Animal Feed Science and Technology* 40, 109–119.

Blümmel, M., Makkar, H.P.S. and Becker, K. (1997a) *In vitro* gas production: a technique revisited. *Journal of Animal Physiology and Animal Nutrition* 77, 24–34.

Blümmel M., Steingaß, H. and Becker, K. (1997b) The relationship between *in vitro* gas production, *in vitro* microbial biomass yield and ^{15}N incorporation and its implications for the prediction of voluntary feed intake of roughages. *British Journal of Nutrition* 77, 911–921.

Bonsi, M.L.K., Osuji, P.O. and Tuah, A.K. (1995) Effect of supplementing teff straw with different levels of *Leucaena* or *Sesbania* leaves on the degradabilities of teff straw, *Sesbania*, *Leucaena*, tagasaste and veronia and on certain rumen and blood metabolites in Ethiopian Menz sheep. *Animal Feed Science and Technology* 52, 101–129.

Calabrò, S. and Williams, B.A. (1997) Differences between fermentative abilities of rumen fluid from buffalo and dairy cows to measure cumulative gas production. In: In Vitro *Techniques for Measuring Nutrient Supply to Ruminants*. BSAS Occasional Meeting held in Reading, 8–10 July. Abstract.

Castro, F.B., Hotten, P.M. and Orskov, E.R. (1994) Inhibition of rumen microbes by compounds formed in the steam treatment of wheat straw. *Bioresource Technology* 50, 25–30.

Cone, J.W., van Gelder, A.H. and Marvin, H.J.P. (1994) *The Effect of Different Drying Conditions on Chemical Composition, Physical Properties, Organic Matter Digestibility and Fermentation Kinetics of Grass and Maize Stems* [in Dutch]. Report, Instituut voor Veehouderij en Diergezondheid, Dienst Landbouwkundig Onderzoek No. 266, Lelystad, the Netherlands, 60 pp.

Cone, J.W., van Gelder, A.H, Visscher, G.J.W. and Oudshoorn, L. (1996) Influence of rumen fluid and substrate concentration on fermentation kinetics measured with a fully automated time related gas production apparatus. *Animal Feed Science and Technology* 61, 113–128.

Czerkawski, J.W. (1986) *An Introduction to Rumen Studies*. Pergamon Press, Oxford.

Davies, D.R., Theodorou, M.K., Baughan, J., Brooks, A.E. and Newbold, J.R. (1995) An automated pressure evaluation system (APES) for determining the fermentation characteristics of ruminant feeds. *Annales de Zootechnie* 44 (Suppl. 1), 36.

Debersaques, F., Williams, B.A. and Tamminga, S. (1998) Effect of nitrogen source on fermentation characteristics of citrus pulp. In: *Proceedings of the BSAS Winter Meeting held in Scarborough, 23–25 March*, Abstract 174.

Dhanoa, M.S., Lopez, S., Dijkstra, J., Davies, D.R., Sanderson, R., Williams, B.A., Sileshi, Z. and France, J. (2000) Estimating the extent of degradation of ruminant feeds from a description of their gas production profiles observed *in vitro*: comparison of models. *British Journal of Nutrition* 83, 131–142.

El-Shazly, K., and Hungate, R.E. (1965) Fermentation capacity as a measure of net growth of rumen microorganisms. *Applied Microbiology* 13, 62–69.

Fisher, D.S. (1996) Modeling ruminant feed intake with protein, chemostatic, and distension feedbacks. *Journal of Animal Science* 74, 3076–3081.

Forbes J.M. (1993) Voluntary feed intake. In: Forbes, J.M. and France, J. (eds) *Quantitative Aspects of Ruminant Digestion and Metabolism*. CAB International, Wallingford, UK, pp. 479–494.

France, J., Dhanoa, M.S., Theodorou, M.K., Lister, S.J., Davies, D.R. and Isac, D. (1993) A model to interpret gas accumulation profiles associated with *in vitro* degradation of ruminant feeds. *Journal of Theoretical Biology* 163, 99–111.

France, J., Dijkstra, J., Dhanoa, M.S., Lopez, S. and Bannink, A. (2000) Estimating the extent of degradation of ruminant feeds *in vivo* from a description of their gas production profiles observed *in vitro*. 1. Derivation of models and other mathematical considerations. *British Journal of Nutrition* 83, 143–150.

Goering, H.K. and Van Soest, P.J. (1970) *Forage Fiber Analysis (Apparatus, Reagents, Procedures and Some Applications)*. Agricultural Handbook 379, ARS, USDA, Washington, DC.

Grant, R.J. and Mertens, D.R. (1992) Impact of *in vitro* fermentation techniques upon kinetics of fibre digestion. *Journal of Dairy Science* 75, 1263–1272.

Groot, J.C.J., Cone, J.W., Williams, B.A., Debersaques, F.M.A. and Lantinga, E.A. (1996) Multiphasic analysis of gas production kinetics for *in vitro* fermentation of ruminant feeds. *Animal Feed Science and Technology* 64, 77–89.

Groot, J., Williams, B.A., Oostdam, A., Boer, H. and Tamminga, S. (1998) The use of cumulative gas and VFA production to predict *in vitro* fermentation kinetics of Italian ryegrass leaf cell walls and contents at various time intervals. *British Journal of Nutrition* 79, 519–525.

Hagerman, A.E., Robbins, C.T., Weerasuriya, Y., Wilson, T.C. and McArthur, C. (1992) Tannin chemistry in relation to digestion. *Journal of Range Management* 45, 57–62.

Hermanto, Soebarinoto, Ørskov, E.R. and van Bruchem, J. (1991) *In vitro* gas production as a predictor of digestibility and voluntary intake of rice straw. In: Ibrahim, M.N.M., de Jong, R., van Bruchem, J. and Purnomo, H. (eds) *Livestock and Feed Development in the Tropics*. Brawijaya University, Malang, pp. 239–244.

Hidayat, Hillman, K., Newbold, C.J. and Stewart, C.S. (1993) The contributions of bacteria and protozoa to ruminal forage fermentation *in vitro*, as determined by microbial gas production. *Animal Feed Science and Technology* 42, 193–208.

Hobson, P.N. and Wallace, R.J. (1982) Microbial ecology and activities in the rumen. Parts 1 and 2. *CRC Critical Reviews in Microbiology* 9, pp. 165–225, 253–319.

Hungate, R. (1966) *The Rumen and Its Microbes*. Academic Press, New York.

Huntington, J.A. and Givens, D.I. (1995) The *in situ* technique for studying the rumen degradation of feeds: a review of the procedure. *Nutrition Abstracts and Reviews* 65, 63–93.

Hvelplund, T., Andrieu, J., Weisbjerg, M.R. and Vermorel, M. (1995) Prediction of the energy and protein value of forages for ruminants. In: Journet, M., Grent, E., Farce, M.-H., Thériez, M. and Demarquilly, C. (eds) *Recent Developments in the Nutrition of Herbivores. Proceedings of the IVth International Symposium on the Nutrition of Herbivores*. INRA Editions, Paris, pp. 205–227.

Johnson, R.R. (1963) *In vitro* rumen fermentation techniques. *Journal of Animal Science* 22, 792–800.

Kaitho, R.J., Nsahlai, I.V., Williams, B.A., Umunna, N.N., Tamminga, S. and van Bruchem, J. (1998) Relationships between preference, rumen degradability, gas production and chemical composition of browses. *Agroforestry Systems* 39, 129–144.

Khazaal, K., Boza, J. and Ørskov, E.R. (1994) Assessment of phenolics-related antinutritive effects in Mediterranean browse: a comparison between the use of the *in vitro* gas production technique with or without insoluble polyvinylpolypyrrolidone or nylon bag. *Animal Feed Science and Technology* 49, 133–149.

Koops, W.J. (1986) Multiphasic growth curve analysis. *Growth* 50, 169–177.

Krishnamoorthy, U., Soller, H., Steingass, H. and Menke, K.H. (1991) A comparative study on rumen fermentation of energy supplements *in vitro*. *Journal of Animal Physiology and Animal Nutrition* 65, 28–35.

Leedle, J.A.Z. and Hespell, R.B. (1983) Brief incubations of mixed ruminal bacteria: effects of anaerobiosis and sources of nitrogen and carbon. *Journal of Dairy Science* 66, 1003–1014.

Leng, R.A. (1993) Quantitative ruminant nutrition – a green science. *Australian Journal of Agricultural Research* 44, 363–380.

Lowe, S.E., Theodorou, M.K., Trinci, A.P.J. and Hespell, R.B. (1985) Growth of anaerobic rumen fungi on defined and semi-defined media lacking rumen fluid. *Journal of General Microbiology* 131, 2225–2229.

Makkar, H.P.S., Blümmel, M. and Becker, K. (1995) Formation of complexes between polyvinyl pyrrolidones or polyethylene glycols and tannins, and their implications in gas production and true digestibility in *in vitro* techniques. *British Journal of Nutrition* 73, 897–913.

Menke, K.H. and Ehrensvard, U. (1974) Die Bestimmung der Stoffbilanz und die berechnung der stoichiometrie und der energiebilanz bei der vergarung von pansensaft *in vitro*. *Zeitschrift Tierphysiologie Tierernährung und Futtermittelkunde* 34, 24–45.

Menke, K.H. and Steingass, H. (1988) Estimation of the energetic feed value obtained from chemical analysis and *in vitro* gas production using rumen fluid. *Animal Research and Development* 28, 7–55.

Menke, K.H., Raab, L., Salewski, A., Steingass, H., Fritz, D. and Schneider, W. (1979) The estimation of the digestibility and metabolizable energy content of ruminant feeding stuffs from the gas production when they are incubated with rumen liquor *in vitro*. *Journal of Agricultural Science (Cambridge)* 93, 217–222.

Minson, D.J. (1990) *Forage in Ruminant Nutrition*. Academic Press, San Diego.

Naga, M.A. and Harmeyer, J.H. (1975) Gas and volatile fatty acid production at different rates of rumen microbial protein synthesis *in vitro*. *Journal of Animal Science* 40, 374–379.

Newbold, C.J., El Hassan, S.M., Wang, J., Ortega, M.E. and Wallace, R.J. (1997) Influence of foliage from African multipurpose trees on activity of rumen protozoa and bacteria. *British Journal of Nutrition* 78, 237–249.

Nsahlai, I.V., Umunna, N.N. and Negassa, D. (1995) The effect of multipurpose tree digesta on *in vitro* gas production from napier grass or neutral detergent fibre. *Journal of the Science of Food and Agriculture* 69, 519–528.

O'Hara, M., Hozumi, S. and Ohki, K. (1974) Studies on the mode of gas production in artificial rumen and its application to the evaluation of feedstuffs IV. On the role of bicarbonate buffer for the gas production. *Japanese Journal of Zootechnical Science* 45, 1–7.

Oostdam, A., Williams, B.A. and Groot, J. (1995) Het effect van veroudering op de fermentatie van gehele bladeren alsmede celwand- en celinhoudfracties bij Italiaans raaigras, volgens de cumulatieve gas produktie methode. In: *20ste Nederlandstalige Voedingonderzoeders Dag*. Catholic University of Leuven, Belgium, pp. 48–49.

Opatpatanakit, Y., Kellaway, R.C., Lean, I.J., Annison, G. and Kirby, A. (1994) Microbial fermentation of cereal grains *in vitro*. *Australian Journal of Agricultural Research* 45, 1247–1263.

Orpin, C. (1983/84) The role of ciliate protozoa and fungi in the rumen digestion of plant cell walls. *Animal Feed Science and Technology* 10, 121–143.

Pell, A.N. and Schofield, P. (1993) Computerized monitoring of gas production to measure forage digestion *in vitro*. *Journal of Dairy Science* 76, 1063–1073.

Pienaar, J.P. (1994) Theoretical and applied aspects of voluntary feed intake by ruminants, with special reference to the kinetics of rumen digestion. PhD thesis, University of Natal, Pietermaritzburg, South Africa, pp. 75–94.

Piva, G., Santi, E., Belladonna, S. and Curto, O. (1988) Kinetics of *in vitro* fermentation of forages. *Journal of Alimentation and Nutrition of Herbivores* 28, 101–102.

Prasad, C.S., Wood, C.D. and Sampath, K.T. (1994) Use of *in vitro* gas production to evaluate rumen fermentation of untreated and urea treated finger millet straw (*Eleusine coracana*) supplemented with different levels of concentrate. *Journal of the Science of Food and Agriculture* 65, 457–464.

Quin, J.I. (1943) Studies on the alimentary tract of merino sheep in South Africa 7. Fermentation in the forestomachs of sheep. *Onderspoor Journal of Veterinary Science and Animal Industry* 2, 91–117.

Raab, L., Cafantaris, B., Jilg, T. and Menke, K.H. (1983) Rumen protein degradation and biosynthesis. 1. A new method for determination of protein degradation in rumen fluid *in vitro*. *British Journal of Nutrition* 50, 569–582.

Rooke, J.A. and Armstrong, D.G. (1989) The importance of the form of nitrogen on microbial protein synthesis in the rumen of cattle receiving grass silage and continuous intro-rumen infusions of sucrose. *British Journal of Nutrition* 61, 113–121.

Rymer, C. (1997) A study of the inter-laboratory variation in results obtained using the gas production technique. In: In Vitro *Techniques for Measuring Nutrient Supply to Ruminants*. BSAS Occasional Meeting held in Reading, 8–10 July, Abstract.

Rymer, C. (1998) A study of the inter-laboratory variation in results obtained using the gas production technique. In: Deaville, E.R., Owen, E., Adesogan, A.T., Rymer, C., Huntingdon, J.A. and Lawrence, T.L.J. (eds) In Vitro *Techniques for Measuring Nutrient Supply to Ruminants*. BSAS Occasional Publication, No. 22. British Society of Animal Science, Penicuik, pp. 278–281.

Siaw, D.E.K.A., Osuji, P.O. and Nsahlai, I.V. (1993) Evaluation of multipurpose tree germplasm: the use of gas production and rumen degradation characteristics. *Journal of Agricultural Science, Cambridge* 120, 319–330.

Sileshi, Z., Owen, E., Dhanoa, M.S. and Theodorou, M.K. (1996) Prediction of *in situ* rumen dry matter disappearance of Ethiopian forages from an *in vitro* gas production technique using a pressure transducer, chemical analyses or *in vitro* digestibility. *Animal Feed Science and Technology* 61, 73–87.

Stefanon, B., Pell, A.N. and Schofield, P. (1996) Effect of maturity on digestion kinetics of water-soluble and water-insoluble fractions of alfalfa and brome hay. *Journal of Animal Science* 74, 1104–1115.

Sutton, J.D. (1985) Digestion and absorption of energy substrates in the lactating cow. *Journal of Dairy Science* 68, 3376–3393.

Theodorou, M.K., Williams, B.A., Dhanoa, M.S. and McAllan, A.B. (1991) A new laboratory procedure for estimating kinetic parameters associated with the digestibility of forages. In: *Proceedings of the Conference on Plant Cell Wall Digestion*. US Dairy Forage Research Center, Madison, Wisconsin, p. 20.

Theodorou, M.K., Williams, B.A., Dhanoa, M.S., McAllan, A.D.B. and France, J. (1994) A simple gas production method using a pressure transducer to determine the fermentation kinetics of ruminant feeds. *Animal Feed Science and Technology* 48, 185–197.

Tilley, J.M.A. and Terry, R.A. (1963) A two stage technique for the *in vitro* digestion of forage crops. *Journal of the British Grassland Society* 18, 104–111.

Topps, J.H. (1992) Potential composition and use of legume shrubs and trees as fodders for livestock in the tropics. *Journal of Agricultural Science, Cambridge* 118, 1–8.

Trei, J., Hale, W.H. and Theurer, B. (1970) Effect of grain processing on *in vitro* gas production. *Journal of Animal Science* 30, 825–831.

Uden, P. and Van Soest, P.J. (1984) Investigations of the *in situ* bag technique and a comparison of the fermentation in heifers, sheep, ponies and rabbits. *Journal of Animal Science* 58, 213–221.

van Laar, H., Tamminga, S., Williams, B.A., Verstegen, M.W.A. and Engels, F.M. (1999) Fermentation characteristics of cell walls sugars from soya bean meal, and from separated endosperm and hulls of soya beans. *Animal Feed Science and Technology* 7, 179–193.

Van Soest, P.J. (1994) *Nutritional Ecology of the Rumen*, 2nd edn. Cornell University Press, Ithaca, New York.

Waghorn, G.C. and Stafford, K.J. (1993) Gas production and nitrogen digestion by rumen microbes from deer and sheep. *New Zealand Journal of Agricultural Research* 4, 493–497.

Weast, R.C. (1968) *CRC Handbook of Chemisty and Physics*, 49th edn. Chemical Rubber Co., Cleveland, USA.

Williams, A.G. and Coleman, G.S. (1988) The rumen protozoa. In: Hobson, P.N. (ed.) *The Rumen Microbial Ecosystem*. Elsevier Applied Science, London, pp. 77–128.

Williams, B.A., Bhatia, S.K., Boer, H. and Tamminga, S. (1995a) A preliminary study using the cumulative gas production technique to compare the kinetics of different fermentations by use of standard substrates. *Annales de Zootechnie* 44 (Suppl. 1), 35.

Williams, B.A., van der Poel, A., Boer, H. and Tamminga, S. (1995b) The use of cumulative gas production to determine the effect of steam explosion on the fermentability of two substrates with different cell wall quality. *Journal of the Science of Food and Agriculture* 69, 33–39.

Williams, B.A., Chuzaemi, S., Soebarinoto, van Bruchem, J., Boer, H. and Tamminga, S. (1996) A comparison of ten rice straw varieties grown at two different altitudes during a wet and dry season, using the *in vitro* cumulative gas production technique. *Animal Feed Science and Technology* 57, 183–194.

Williams, B.A., van der Poel, A., Boer, H. and Tamminga, S. (1997) The effect of extrusion conditions on the fermentability of wheat straw and corn silage. *Journal of the Science of Food and Agriculture* 74, 117–124.

Wolin, M.J. (1975) Interactions between the bacterial species of the rumen. In: McDonald, I.W. and Warner, A.C. (eds) *Digestion and Metabolism in the Ruminant*. University of New England Publishing Unit, Armidale, Australia, pp. 134–148.

Wood, C.D. and Plumb, V.E. (1995) Evaluation of assays for phenolic compounds on the basis of *in vitro* gas production by rumen microorganisms. *Animal Feed Science and Technology* 56, 195–206.

11 Animal-based Techniques for the Estimation of Protein Value of Forages

S. TAMMINGA[1] AND X.B. CHEN[2]

[1]Wageningen Institute of Animal Sciences, Group Animal Nutrition, Marijkeweg 40, 6709 PG Wageningen, The Netherlands; [2]Rowett Research Institute, Bucksburn, Aberdeen AB21 9SB, UK

Introduction

Forages are important components of diets for ruminant farm animals. Not only do forages contribute to the energy supply of ruminants, but they also provide a significant amount of protein. In diets for dairy cows, good-quality grass can easily contribute 50% of the energy and up to 60% of their protein requirement. Proteins in feed are important sources of amino acids, the building blocks of many biologically active compounds in animals, of which proteins in muscles, connective tissues and enzymes are quantitatively the most important. The protein value of feeds for ruminants contains two important components: feed protein escaping microbial degradation and microbial protein. Protein degradation and microbial protein synthesis occur in the forestomachs and both sources of protein are subsequently hydrolysed in and absorbed from the small intestine. However, both processes are quite variable and their combined result determines the protein supply to the small intestine. This supply broadly depends on the combination of rate of degradation, rate of microbial growth and rate of outflow of feed particles and microbes. Therefore, a proper evaluation of the protein value of forages requires techniques that take these dynamics into account. Various *in vivo* techniques have been developed and this chapter gives a critical review of such techniques.

In Vivo Estimation of Protein Degradation and Microbial Protein Synthesis

The most widely applied way of measuring protein degradation and microbial protein synthesis *in vivo* is in animals equipped with a re-entrant or a T-piece cannula in the proximal duodenum. Total protein passing into the proximal duodenum is measured and sampled and the proportion of microbial protein is determined by using an appropriate marker.

Various surgical techniques for providing a ruminant with intestinal cannulae have been developed and discussed (McGilliard, 1982). The use of cannulated animals has a number of disadvantages. First of all, it is an invasive and very costly technique. Measuring intestinal digesta flow is also tedious (MacRae, 1975; Faichney, 1993; Reynolds *et al.*, 1994; Poncet *et al.*, 1995). Although digesta can be sampled reliably, flow rates may become affected after opening the cannulae for various reasons, such as changes in pressure (Sutton and Oldham, 1977; Faichney, 1993). Therefore, the use of markers is also recommended with re-entrant cannulae; another reason is because the measuring periods are usually too short to compensate for the often observed initial flow reduction after opening the cannulae. Use of markers, however, is a prerequisite when T-piece cannulae are used (Faichney, 1975, 1993).

Further limitations of measurements in cannulated animals are due to the difficulty in accurately measuring microbial protein (Hvelplund *et al.*, 1995). Various microbial markers have been developed. Examples were summarized and discussed (Harrison and McAllan, 1980; Tamminga, 1982; Broderick and Merchen, 1992), and these include nucleic acids (RNA, DNA), total purines, amino acids (2,6-diaminopimelic acid (DAPA), D-alanine) and radioactive as well as stable isotopes (sulphur-35 (^{35}S), phosphorus-32 (^{32}P), nitrogen-15 (^{15}N). Based on the assumption that feed protein or endogenous protein do not contain DAPA or RNA, the proportion of microbial N (MN) in the non-ammonia-N (NAN) entering the small intestine is estimated from:

MN/NAN = (DAPA/NAN in digesta)/(DAPA/N in microbes)
MN/NAN = (RNA/NAN in digesta)/(RNA/N in microbes)

As pointed out by Harrison and McAllan (1980) and Stern *et al.* (1994), the results obtained using different markers seldom agree. For reference material, bacteria isolated from strained rumen fluid by differential centrifugation are often used, but these may not represent the daily flow of bacteria out of the rumen and arriving at the proximal duodenum. The ratio between bacteria associated with the fluid (FAB) and those associated with feed particles (PAB) is subject to considerable variation. The proportion of bacteria associated with feed particles may vary between 50 and 80% (Owens and Goetsch, 1986; Oosting, 1993). Chemical composition differs between FAB and PAB (Merry and McAllan, 1983) and differences exist in outflow rate between rumen fluid and feed particles and, as a result, also between FAB and PAB. Besides, diurnal variation in composition of FAB and PAB is often not taken into account, a problem that can be overcome to some extent by taking samples more frequently and spread over a period of 24 h. As an alternative, bacteria have sometimes been isolated from duodenal digesta (Yang and Poncet, 1993). Estimating the contribution of protozoa is even more cumbersome (Harrison and McAllan, 1980; Schelling *et al.*, 1982). Currently, the two most commonly used microbial markers to differentiate between microbial and dietary protein are DAPA and purines. However, DAPA is absent in protozoa and some bacteria and is subject to extensive metabolism. In their extensive review on markers for quantifying microbial protein synthesis in the rumen, Broderick and Merchen (1992) concluded that a simultaneous determination of total purines and the use of the ^{15}N method has the best chance of getting reliable results. On the other hand, Stern *et al.* (1994) concluded that 'because no absolute standard exists, it is impossible to state which marker method, microbial isolation procedure or site of microbial isolation provides the most accurate estimate of microbial N flow into the proximal duodenum'. The choice of method may also be determined by the availability of analytical equipment. Extremes are the analytical procedures for purines, which can easily be done in a spectrophotometer, and the use of ^{15}N, which depends on having access to quite costly isotope-ratio mass spectroscopy (IRMS) equipment.

Depending on the site of cannulation, the mix of undegraded feed protein and microbial protein may be diluted by a variable amount of protein originating from the animal itself and giving no net contribution to the postruminal protein supply. Methods for estimating endogenous contribution have been developed, but more frequently in non-ruminants than in ruminants (Tamminga *et al.*, 1995; Nyachoti *et al.*, 1997). A widely used method in pigs is that of the isotope dilution technique, often with ^{15}N leucine as precursor. In ruminants, ^{15}N-labelled feeds have been used (Van Bruchem *et al.*, 1997), but corrections for ^{15}N contamination of microbial protein passing along the small intestine are necessary.

Furthermore, when ruminal protein degradation is based on the appearance of protein in the small intestine, this measurement only gives the final outcome and it remains uncertain whether differences result from differences in rate of degradation or differences in rumen retention time or both. Yet another disadvantage of the use of cannulated animals is that, when mixed diets are fed, as is usually the case with dairy cows, no digestive information is obtained on the individual dietary ingredients. Again, the conclusion of Stern *et al.* (1994) was that:

> *in vivo* estimates of protein degradation are expensive, labour intensive, time consuming, and subject to error, associated with use of digesta flow rate markers, microbial markers and animals variation. Therefore, alternative procedures for measuring ruminal degradation of dietary protein are needed.

This has led to the development of alternative methods and in many countries the *in situ* method, in which nylon bags filled with the ingredient to be investigated are incubated in the rumen for various lengths of time, has been adopted as the reference method. This method also has its limitations, which will be discussed in the next chapter. An alternative method, which as yet has received little attention, is the rumen evacuation method (Robinson *et al.*, 1987). This method, when applied carefully at different times after feeding and combined with measuring the rate of passage out of the rumen (the reciprocal of rumen retention time), which is possible with the application of markers, can give useful information on the dynamics of rumen degradation and synthesis. It would make it possible to quantify PAB and FAB and, when appropriate outflow rates for fluid and particles are applied, give an estimate of microbial protein supply and its variation. An additional advantage would be that animals equipped with a rumen cannula only would suffice.

In Vivo Estimation of Postruminal Availability

Postruminal availability of protein can be measured *in vivo* as the difference between protein entering the proximal duodenum and protein passing out of the terminal ileum. A complicating factor is that two sets of cannulae are needed, one in the proximal duodenum and one in the terminal ileum. Using re-entrant cannulae in dairy cattle certainly creates severe problems (Reynolds *et al.*, 1994). A summary of the apparent fractional absorption of NAN and amino acids is given in Table 11.1 (NRC, 1985).

As already discussed, a complicating factor is the presence of a variable amount of endogenous protein at both sites. Besides, what enters the duodenum is a mixture of undegraded feed protein and microbial protein, which can be differentiated in the proximal duodenum, but can no longer be differentiated at the terminal ileum. Solutions have been found in infusion studies, either on top of the normal passage (Schwarting and Kaufmann, 1979) or in animals kept on intragastric infusion (Hvelplund *et al.*, 1994). The techniques have been applied both with isolated rumen bacteria (Storm *et al.*, 1983) and with feedstuffs (Hvelplund, 1985). Both protein sources – feedstuffs in particular – contain non-protein dry matter (DM), which may induce the release of extra endogenous protein; this may explain why the results of infusion studies usually show a slightly lower intestinal digestion than the method based on mobile nylon bags. Using the technique of intragastric infusion offers a possibility to correct for endogenous contamination (Hvelplund *et al.*, 1994), with the restriction that the endogenous secretions of animals fed with intragastric infusion may not represent a practical situation.

An alternative approach is to insert small nylon bags, containing a small amount of the protein source to be investigated, into the proximal duodenum and to recover them either from the distal ileum or from the faeces. This mobile nylon-bag technique is now a widely applied method and further details on its suitability will be discussed in the next chapter.

Measuring the net appearance of absorbed amino acids in the portal blood has also been recommended as a means of estimating postruminal availability of protein. It should, however, be realized that the gut (wall) itself uses considerable and variable amounts of amino acids (MacRae *et al.*, 1996), with an apparent preference for non-essential amino acids.

Table 11.1. Summary of apparent fractional absorption of non-ammonia nitrogen and amino acids from the small intestine in ruminants (NRC, 1985).

	n	mean	SD	CV
Non-ammonia nitrogen				
Lactating cattle	12	0.65	0.04	0.07
Non-lactating cattle	17	0.66	0.04	0.06
Sheep	29	0.64	0.06	0.09
All	58	0.65	0.05	0.08
Amino acids				
Lactating cattle	21	0.69	0.05	0.08
Non-lactating cattle	11	0.62	0.06	0.10
Sheep	22	0.70	0.06	0.09
All	54	0.68	0.06	0.10

SD, standard deviation; CV, coefficient of variation.

Use of Flow Markers

As already mentioned, the use of intestinal cannulae requires the use of markers. Quite a variety of different markers have been applied. One category follows the passage of fluid, examples being polyethylene glycol (PEG) and various elements in a complex with ethylenediamino-tetra-acetic acid (EDTA), notably cobalt (Co)-EDTA and chromium (Cr)-EDTA. Examples of markers representing the passage of particles are chromic oxide (Cr_2O_3), Cr-mordanted neutral-detergent fibre (NDF) (Cr-NDF) and a wide variety of rare-earth elements, such as cerium (Ce), dysprosium (Dy), lanthanum (La), ruthenium (Ru), often applied as Ru-phenanthroline, and ytterbium (Yb). Rare-earth elements adsorb strongly to feed particles, and so are thought to represent the behaviour of feed particles. Migration between particles may cause problems, however (Faichney, 1993). Chromium oxide (Cr_2O_5) and titanium(IV) oxide (TiO_2) have also been widely used as markers, but, because of their high density, they may not represent either of the two fractions. A third category are the internal markers, such as lignin, lignin fractions (cellulose), indigestible acid-detergent fibre (IADF) and rumen-indegradable neutral-detergent fibre (RINDF) (Tamminga *et al.*, 1989a, b) and *n*-alkanes (Mayes *et al.*, 1995). A thus far under-explored type of internal marker is that based on carbon-13 (^{13}C), which seems applicable either based on the difference in natural abundance between C-3 and C-4 plant species (Südekum *et al.*, 1995) or based on artificially enriched plants (parts) grown in a carbon dioxide ($^{13}CO_2$)-rich environment (Svejcar *et al.*, 1993).

The most straightforward method is to infuse the marker continuously into the rumen at a constant rate. Intestinal flows can than be estimated by dividing the marker dose rate by the mean marker concentration in the digesta. When using T-piece cannulae, sampling of digesta may become problematic and non-representative. This seems particularly the case when the cannula is placed immediately behind the pylorus (P.H. Robinson and M.W. Bosch, personal communication). A solution may then be the double-marker method, a combination of a solute marker and a particle-associated marker, from which a reconstitution factor can be calculated (Faichny, 1993).

When a steady state has been achieved and is maintained by continuous infusion of a solute marker (*S*) and a particle-associated marker (*P*) and their concentrations are expressed as fractions of the daily dose per unit of digesta or its phases, the following relationship is valid:

$$R = (P_{DG} - Z \cdot S_{DG})/(Z \cdot S_{FP} - P_{FP})$$

Where R = the reconstitution factor, i.e. the number of units of fluid phase (FP) that must be added to (or removed from) one unit of digesta (DG) to obtain true digesta (TD), and Z = the marker concentration ratio, P/S, in TD. The concentration (C_{TD}) and digesta flow (F_{TD}) can than be calculated from:

$$C_{TD} = (C_{DG} + R \cdot C_{FP})/(1 + R)$$

and:

$$F_{TD} = 1/S_{TD}$$

However, this does not solve the problem of non-representative sampling within the particulate phase.

Markers can also be used to estimate the ruminal outflow of fluid or particles. In this case, a pulse dose of the marker is given into the rumen and fractional passage rates are calculated either from its dilution in rumen fluid or from its faecal excretion pattern with time, based on the assumption that fractional passage rates out of the rumen follow first-order kinetics (Grovum and Williams, 1973):

$$C_t = C_0 \cdot e^{-kt}$$

Where C_t and C_0 represent marker concentrations at times t and zero, and k is the fractional outflow rate.

A still unresolved question is whether external markers are superior to internal markers or vice versa. Till now, most researchers would argue it to be an advantage for a marker not to take part in the digestive process. However, inert markers may behave differently from feed particles, because they do not posses the right functional density and are not subjected to particle-size reduction or entrapment in the mat of fibrous material normally present in the rumen. Recent findings (Tamminga, 1993; Van Straalen, 1995) would suggest that passage out of the rumen of a feed particle to some extent depends on its degree of degradation. As a result, a relationship between degradation and passage cannot be ruled out. Under these circumstances, internal markers, particularly those which take part in the digestive process, could prove to be superior to external and inert markers. The use of ^{13}C-enriched feed components could possibly solve this problem.

Use of Purine-derivative Excretion as Index of Microbial Protein Supply

The *in vivo* methods discussed so far in this chapter all require cannulation of animals. With an increasing concern for animal welfare, there is a greater need to develop non-invasive or less invasive methods. The technique for the estimation of microbial protein supply using urinary excretion of purine derivatives (PD) as a parameter is a recent development in this line. Unlike the commonly used methods based on digesta flow measurement, this technique does not require any surgical preparation. Since only urine collection is needed, an estimation of microbial protein supply can be incorporated into N balance and digestibility trials without much additional labour input.

Urinary PD comprise allantoin, uric acid, xanthine and hypoxanthine collectively. All four components are found in urine of sheep, goats, red deer and llamas, but only allantoin and uric acid are found in urine of cattle and buffaloes. These derive mainly from the degradation of absorbed microbial purines (exogenous source) and a smaller fraction from the turnover of tissue nucleic acids (endogenous source). The endogenous PD excretion has been determined in several species, including sheep, cattle, goats, llamas and buffaloes. It has been repeatedly demonstrated in sheep and cattle that the daily PD excretion is directly related to the uptake of exogenous (i.e. intestinal) purines (Chen *et al.*, 1990b; Verbic *et al.*, 1990; Balcells *et al.*, 1991). Therefore, the excretion of PD can provide a quantitative estimate of the intestinal flow of microbial protein if the purine:protein ratio in mixed rumen microbes is constant. A practical example of the application of this technique is given in Chen *et al.* (1992b). In a number of studies (Oosting *et al.*, 1995; Pérez *et al.*, 1996, 1997; Samaniego *et al.*, 1997), it was shown that estimates of microbial protein in sheep based on this technique were in close agreement with direct measurements using microbial markers (^{15}N, amino acid profile, DAPA and RNA). A numerical comparison is given in Table 11.2.

In using this technique, however, one should be aware of its limitations and uncertainties. The technique is based on two assumptions. The first assumption is that all the nucleic acid purines entering the small intestine are of microbial origin. Considering that residual dietary purines after microbial degradation may be negligible for most diets and that the quantity is difficult to determine in intact animals, at this stage this assumption may have to be accepted. However, special precaution needs to be taken when a large amount of fishmeal is fed. The second assumption is that the purine:protein ratio in mixed rumen microbes is a constant. Whether a dietary regime can drastically alter the value of this ratio is not known. There is not yet sufficient information to provide an answer, but earlier results (Schelling *et al.*, 1982) do not indicate a large variation.

Marked species differences in purine metabolism have been demonstrated in several species including sheep, cattle, buffaloes and llamas (Chen *et al.*, 1990a, 1996a; Bakker *et al.*, 1996; for a review, see Stangassinger *et al.*, 1995). It is important to note that different equations will need to be used with different animal species for the calculation of microbial N supply based on PD excretion. Equations have been established so far for European sheep (Chen *et al.*, 1990b) and cattle (Verbic *et al.*, 1990), the parameters of which have been confirmed by studies in separate laboratories (Balcells *et al.*, 1991; Beckers and Thewis, 1994).

Table 11.2. Duodenal-flow microbial N in sheep estimated with purine bases and other methods.

Duodenual flow	Bacterial marker	Flow of bacterial amino acid N (exp. 1) or total N (exp. 2)(g day^{-1})			
Experiment 1 (Oosting *et al.*, 1995)					
Diet:		UWS	AWS	AWSC	AWSP
MN1	DAPA	6.5	8.0	10.0	11.5
MN2	AA profile	6.8	8.4	11.7	11.0
MN3	Purine bases	6.7	10.5	11.3	12.0
MN2	Purine bases	4.8	7.2	12.0	11.7

Diets: UWS, untreated wheat straw; AWS, ammonia-treated wheat straw; AWSC, ammonia-treated wheat straw supplemented with casein; AWSP, ammonia-treated wheat straw supplemented with potato protein.

Experiment 2 (Pérez *et al.*, 1996)					
Diet:		0	220	400	550
MN1	^{15}N	6.2	9.0	14.0	14.4
MN2	Purine bases	4.8	7.2	12.0	11.7

Diets: lucerne (550 g day^{-1}) supplemented with 0, 220, 400 and 550 g of barley.

An equation derived from the sheep experiment of Chen *et al.* (1990b) is:

$$Y = (0.150\ W^{0.75}\ e^{-0.25X}) + 0.84X$$

and one derived from the cattle experiment of Verbic *et al.* (1990) is:

$$Y = (0.385\ W^{0.75}) + 0.85X$$

where Y is urinary excretion of total PD (mmol day^{-1}), $W^{0.75}$ the metabolic body weight and X the absorbed exogenous purine (mmol day^{-1}). The equation for sheep may also be applied to goats. However, there are strong indications that the equations developed using European animals may not work with animals in the tropics. Animals in the tropics appeared to excrete a smaller proportion of the plasma PD in the urine than those in temperate areas (Liang *et al.*, 1994; Chen *et al.*, 1996a). Recent data suggest that this proportion is rather consistent within each animal individual and is not affected by dietary regime, such as feed intake (P. Prasitkusol, unpublished). It is unlikely that one single equation will be applicable universally. In the future, specific equations will need to be established and validated for the target animal groups to which the method is applied. This may seem an insurmountable task, but methodologies are now being developed to make it easier to establish the equation (Chen *et al.*, 1997; IAEA, 1997). The accuracy of the results obtained based on the purine technique has to be assessed more extensively in the future. At this stage, one should not take the estimates of microbial N based on PD excretion as absolute values, but as values for comparative purposes.

Measurements of PD excretion have proved useful in many studies on microbial protein production from forage feeding. This technique has been used in studies to examine: (i) how the source and level of N in the diet affected microbial protein supply (Chen *et al.*, 1992a; Puchala and Kulaseck, 1992; Balcells *et al.*, 1993a; Giraldez *et al.*, 1993; Susmel *et al.*, 1994a, b); (ii) the efficacy of supplementing straw or silage with carbohydrates or concentrates (Chen *et al.*, 1992a; Balcells *et al.*, 1993b; Chamberlain *et al.*, 1993; Dewhurst *et al.*, 1994; Gomes *et al.*, 1994; Moorby *et al.*, 1994); (iii) the effect of phosphorus supply on microbial protein synthesis (Gunn and Ternouth, 1994; Scott *et al.*, 1994); and (iv) the effect of rumen outflow rate on microbial protein flow (Chen *et al.*, 1992b; Dewhurst and Webster, 1992).

Research has also been carried out to simplify the technique further for application in the field, where total urine collection is impractical. The results indicate that it is possible to use the averaged 'PD-to-creatinine' ratio in several spot urine samples collected within a day as an index of microbial protein supply (Chen *et al.*, 1995, 1996b), but the sensitivity is, of course, lower than that based on total urine collection. Spot testing may be of value only to indicate large differences. Therefore, where facilities allow, total urine collection is recommended. Recent studies indicated that the ratio of purine N to total N (called the 'purine nitrogen index') in spot urine samples could potentially be used to identify situations where rumen-degradable N (RDN) is inefficiently used for the synthesis of microbial protein (Chen and Jayasuriya, 1997). Similarly, it has been suggested that milk allantoin could be used as an indicator of microbial protein production in dairy cows. Its potential use may be limited, because concentrations in milk are much lower than in urine and milk allantoin as a proportion of the total excretion appears to vary with milk production (Stangassinger *et al.*, 1995).

Some technical details of the method are given in the laboratory manual prepared by Chen and Gomez (1992) and the International Atomic Energy Agency (IAEA) (1997). The chemical analysis of purine derivatives can be carried out with any of the following instruments: high-performance liquid chromatographs (HPLC), autoanalysers and spectrophotometers.

Application to Forages

Figure 11.1 illustrates the procedures that are currently used or available for the *in vivo* evaluation of the protein values of forages. The need to do so may arise, for example, in order to evaluate the nutritive value of new varieties of pasture or to compare the different methods of preservation or different ways of feeding the forage. Typically, the forages in question need to be fed as the sole or main component of the diet. Feeding trials should be conducted to measure the voluntary feed intake and the flow of dietary and microbial protein to the small intestine kg^{-1} digestible organic matter (OM) intake. There are two options, depending on the extent of surgical intervention to be imposed on the animals. One is to use animals fitted with a rumen cannula and a duodenal catheter to measure the flow of dietary and microbial protein into the small intestine. Microbial and flow markers need to be used. Protein availability can be measured if a cannula is also fitted at the ter-minal ileum. This approach involves painstaking labour input and few animals can be used at one time. Most earlier work for measuring the protein value of forages was done in this way (see review by Thomson and Beever, 1980).

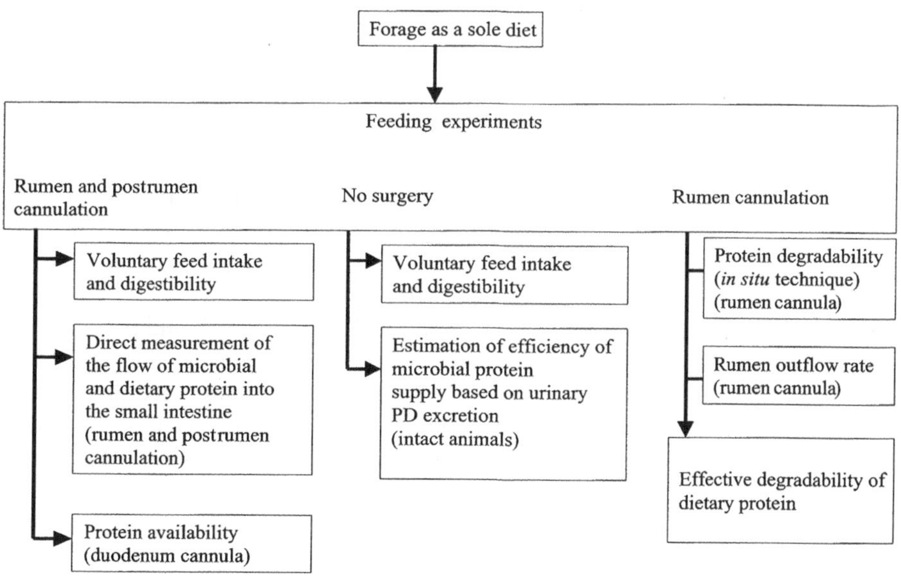

Fig. 11.1. Commonly used procedures for the estimation of the protein value of forages.

The alternative approach is to use an indirect estimation of microbial protein flow based on urinary PD excretion, which can be carried out on intact animals. The undegradable dietary protein contribution can be estimated based on protein degradability, using the *in situ* technique, and the rumen digesta outflow rate can be estimated in separate animals fitted with rumen cannula. As was discussed before, the latter approach does not give an absolute value for the flow of microbial protein and dietary protein and is thus more suitable for comparison. The simplicity of this approach outweighs its limitation when the object of the study is to compare two or more forages. There have been an increasing number of studies reported in which this approach is used. For instance, Verbic (1997) compared different methods of grass forage preservation (fresh-cut silage, wilted silage, highly wilted silage, formic acid-treated silage and hay). He reported that the hay and highly wilted silage had higher contents of rumen-escape protein kg^{-1} DM and a greater amount of microbial protein supply (estimated based on PD excretion) kg^{-1} digestible OM intake. Formic acid-treated silage also had a higher protein value than untreated silage. Verbic *et al.* (1997) used the same approach to compare maize silages made from dent- or flint-type hybrids. Silage made from the flint-type hybrid supported higher microbial protein production than that from the dent-type hybrid. O'Kiely *et al.* (1994) showed that addition of dry beet pulp to the ensiling of whole-crop fodder beet (*Beta vulgaris*) not only acted as an effluent absorbent but also improved the microbial protein supply from the silage. The intake of tropical grasses is often limited by their low N content and supplementation with *Leucaena leucocephala* improves intake and animal productivity. In coastal Kenya, where *Leucaena* production was affected by a psyllid pest, Abdulrazak *et al.* (1996, 1997) showed that *Gliricidia sepium* could be used as an alternative to *Leucaena* as a protein supplement to Napier grass or *Zea mays* stover.

The combined use of different methods often has added value in terms of extra information. From the *in situ* measurements, Verbic (1996) found that the rates of degradation of OM and protein were more synchronized in hay than in silage, and this may in part explain why highly wilted silage and hay have a greater efficiency of microbial protein production than fresh-cut silage. In using the *in situ* degradation measurements, most people would only carry out the measurements for DM or protein alone, but, with forages fed as the sole or main component of the diet, it pays to determine the degradation of both. The ratio of the content of RDN to that of degradable organic matter (DOMR) – ideally 32 g RDN kg^{-1} DOMR, according to the Agricultural Research Council (ARC, 1984) – should provide an indication of whether the RDN content of the forage is deficient or sufficient. Moreover, the amounts of RDN or OM released at different times after feeding can be calculated from the derivative of the degradation curves, and the variation in the amount of RDN released with time relative to that of OM provides an index of the synchronization of the N and energy supply. The term 'synchrony index' (Sinclair *et al.*, 1993) was proposed in this context. Although its practical significance has yet to be explored (Chamberlain and Choung, 1995), a measurement of the synchronization index may shed light on why forages vary in their protein value (i.e. in the efficiency of microbial protein production).

Further Research Needed

The aims of methods developed to measure the protein value of forages are accuracy, simplicity, cost-effectiveness and minimum surgical intervention (particularly to avoid postrumen cannulation).

In direct measurement of intestinal flow of microbial protein and dietary protein, chemical analysis is not a problem. The key element that affects the accuracy of the measurement is the determination of digesta flow. Although a lot of work has been done in this area and an increasing number of liquid and solid markers are being used, there is still a lack of an ideal marker that can represent all particle sizes/densities. The development of a simple and accurate method for digesta flow measurement is still required.

Direct measurement as an approach will probably not be developed much further, due to its need for surgical intervention. Its application will remain, however, for some time to come, particularly when absolute protein value measurements are needed.

It is to be expected that future developments will be in non-invasive methods (e.g. using PD for estimation of microbial protein supply). Further work on the technique based on PD excretion in urine includes the following areas: (i) the results need to be validated by simultaneous comparison with two other methods; and (ii) due to species differences in purine metabolism, it is most likely that specific 'calibration curves' will need to be constructed for each target animal.

Cost Implications

Table 11.3 summarizes the cost elements of the techniques covered in this chapter. These include invasiveness, labour and chemical analysis. Most of the techniques require the use of surgically modified animals. Not only is such surgery (including facilities, skills and staff involved) costly, but the care and maintenance of such animals, particularly animals with postrumen cannulation, is also costly. Animals may also lose their carcass value. In addition, measuring periods can be very labour-intensive. Besides, the invasiveness of surgical techniques meets increasing resistance from animal-welfare groups and the general public.

The major consideration for the cost of chemical analysis is the requirement for capital equipment. Where flow measurement using metal-based markers is needed, the basic requirement is often an atomic-absorption spectrometer. Determination of microbial markers in digesta samples requires an amino acid analyser (DAPA), ultraviolet (UV) spectrophotometer (RNA), liquid scintillation counter (^{35}S or ^{32}P) or mass spectrometer (^{15}N), depending on the choice of marker. RNA is commonly used, because its analysis does not need expensive equipment. Apart from the requirement for facilities to collect total urine, this technique for the indirect estimation of microbial protein supply based on urinary PD excretion appears to be the simplest and least expensive. The basic equipment required for the chemical analysis is a UV spectrophotometer, although chromatographic equipment can also be used. Apart from tracers, the reagents used for chemical analysis in most techniques are relatively inexpensive on a 'per sample' basis.

Table 11.3. Cost elements of various *in vivo* techniques for the estimation of protein value of forages.

	Cannulae	Flow measurement	Microbial marker	Others	Main cost elements	Equipment required for chemical analysis
Microbial protein						
Direct determination	Rumen, duodenum	Yes	RNA, DAPA, ^{15}N ^{32}P, ^{35}S		Surgery, labour, chemical analysis, loss of carcass value	DAPA: amino acid analyser/HPLC RNA: spectrophotometer ^{15}N: mass spectrometer ^{32}P and ^{35}S: liquid scintillation counter Flow markers: atomic absorption spectrometer
Indirect estimation (urinary marker)		No		Complete urine collection	Facility to collect total urine, chemical analysis	Spectrophotometer or HPLC
Undegradable protein						
Flow of undegradable protein	Duodenum	Yes			Surgery, chemical analysis	Flow markers: atomic absorption spectrometer Kjeldahl N
In situ (nylon bag)	Rumen	No			Surgery, labour	Kjeldahl N
Rumen-emptying	Rumen	Yes			Surgery, labour	Kjeldahl N
Postruminal availability						
Difference between duodenum and ileum	Proximal duodenum, terminal ileum	Yes			Invasiveness, labour, chemical analysis	Flow markers: atomic absorption spectrometer Kjeldahl N
Infusion into abomasum	Rumen, Abomasum,			Intragastric infusion	Surgery, labour	
Mobile nylon bag	Duodenum	No			Surgery, labour	Kjeldahl N

References

Abdulrazak, S.A., Muinga, R.W., Thorpe, W. and Ørskov, E.R. (1996) The effects of supplementation with *Gliricidia sepium* or *Leucaena leucocephala* forage on intake, digestion and live-weight gains of *Bos taurus* × *Bos indicus* steers offered napier grass. *Animal Science* 63, 381–388.

Abdulrazak, S.A., Muinga, R.W., Thorpe, W. and Ørskov, E.R. (1997) Supplementation with *Gliricidia sepium* or *Leucaeca leucocephala* on intake, digestibility, rumen fermentation and live weight of crossbred steers offered *Zea mays* stover. *Livestock Production Science* 49, 53–62.

ARC (Agricultural Research Council) (1984) *The Nutrient Requirements of Ruminant Livestock, Supplement No 1*. Commonwealth Agricultural Bureaux, Slough, UK.

Bakker, M.L., Chen, X.B., Kyle, D.J., Ørskov, E.R. and Bourke, D.A. (1996) Urinary and plasma purine derivatives in fed and fasted llamas (*Lama glama* and *L. guanacoe*). *Comparative Biochemistry and Physiology (B)* 113, 367–374.

Balcells, J., Guada, J.A., Castrillo, C. and Gasa, J. (1991) Urinary excretion of allantoin and allantoin precursors by sheep after different rates of purine infusion into the duodenum. *Journal of Agricultural Science, Cambridge* 116, 309–317.

Balcells, J., Guada, J.A., Castrillo, C. and Gasa, J. (1993a) Rumen digestion and urinary excretion of purine derivatives in response to urea supplementation of sodium-treated straw fed to sheep. *British Journal of Nutrition* 69, 721–732.

Balcells, J., Fondevila, M., Guada, J.A., Castrillo, C. and Surra, J.C.E. (1993b) Urinary excretions of purine derivatives and nitrogen in sheep given straw supplemented with different sources of carbohydrates. *Animal Production* 57, 287–292.

Beckers, Y. and Thewis, A. (1994) Excretion of purine derivatives in urine of Belgian Blue bulls following duodenal infusion of purines from Torula yeast. *Proceedings of the Society of Nutrition Physiology* 3, 235.

Broderick, G.A. and Merchen, N.R. (1992) Markers for quantifying microbial protein synthesis in the rumen. *Journal of Dairy Science* 75, 2618–2632.

Chamberlain, D.G. and Choung, J.-J. (1995) The importance of rate of ruminal fermentation of energy sources in diets for ruminants. In: Garnsworthy, P.C. and Cole, D.J.A. (eds) *Recent Advances in Animal Nutrition 1995*. Nottingham University Press, Nottingham, pp. 3–28.

Chamberlain, D.G., Robertson, S. and Choung, J.-J. (1993) Sugars versus starch as supplements to grass silage: effects on ruminal fermentation and the supply of microbial protein to the small intestine, estimated from the urinary excretion of purine derivatives, in sheep. *Journal of the Science of Food and Agriculture* 63, 189–194.

Chen, X.B. and Gomes, M.J. (1992) *Estimation of Microbial Protein Supply to Sheep and Cattle Based on Urinary Excretion of Purine Derivatives – an Overview of the Technical Details*. Occasional Publication of the International Feed Resources Unit, Rowett Research Institute. Aberdeen, UK.

Chen, X.B. and Jayasuriya, M.C.N. (1997) Purine nitrogen index, a new parameter for rapid feed evaluation in ruminants: *in vitro* techniques for measuring nutrient supply to ruminants. In: *BSAS International Symposium, 8–10 July, 1997, University of Reading, Whiteknights, Reading, UK: Programme and Summaries*. p. 26.

Chen, X.B., Ørskov, E.R. and Hovell, F.D.D. (1990a) Excretion of purine derivatives by ruminants: endogenous excretion, differences between cattle and sheep. *British Journal of Nutrition* 63, 121–129.

Chen, X.B., Hovell, F.D.DeB., Ørskov, E.R. and Brown, D.S. (1990b) Excretion of purine derivatives by ruminants: effect of exogenous nucleic acid supply on purine derivative excretion by sheep. *British Journal of Nutrition* 63, 131–142.

Chen, X.B., Abdulrazak, S.A., Shand, W.J. and Ørskov, E.R. (1992a) The effect of supplementing straw with barley or unmolassed sugar beet pulp on microbial protein supply in sheep estimated from urinary purine derivative excretion. *Animal Production* 55, 413–417.

Chen, X.B., Chen, Y.K., Franklin, M.F., Ørskov, E.R. and Shand, W.J. (1992b) The effect of feed intake and body weight on purine derivative excretion and microbial protein supply in sheep. *Journal of Animal Science* 70, 1534–1542.

Chen, X.B., Mejia, A.T., Kyle, D.J. and Ørskov, E.R. (1995) Evaluation of the use of purine derivative:creatinine ratio in spot urine and plasma samples as an index of microbial protein supply in ruminants: studies in sheep. *Journal of Agricultural Science, Cambridge* 125, 137–143.

Chen, X.B., Samaraweera, L., Kyle, D.J., Ørskov, E.R. and Abeygunawardene, H. (1996a) Urinary excretion of purine derivatives and tissue xanthine oxidase activity in buffaloes, with special reference to differences between buffaloes and *Bos taurus* cattle. *British Journal of Nutrition* 75, 397–407.

Chen, X.B., Susmel, P., Stefanon, B. and Ørskov, E.R. (1996b) On the use of purine derivatives in spot urine, plasma and milk samples as indicators of microbial protein supply in sheep and cattle. In: Numes, A.F., Portugal, A.V., Costa, J.P. and Ribiero, J.R. (eds) *Protein Metabolism and Nutrition*. EAAP publication no. 81, Estacao Zootecnica Nacional, Santarem, Portugal, pp. 325–329.

Chen, X.B., Calder, A.G., Prasitkusol, P., Kyle, D.J. and Jayasuriya, M.C.N. (1997) Determination of ^{15}N isotopic enrichment and concentrations of allantoin and uric acid in urine by gas chromatography/mass spectrometry. *Journal of Mass Spectrometry* (in press).

Dewhurst, R.J. and Webster, A.J.F. (1992) A note on the effect of plane of nutrition on fractional outflow rates from the rumen and urinary allantoin excretion by wether sheep. *Animal Production* 54, 445–448.

Dewhurst, R.J., Offer, N.W. and Thomas, C. (1994) Urinary purine derivative excretion and milk protein production by lactating dairy cows. *Proceedings of Society of Nutrition and Physiology* 3, 236.

Faichney, G.J. (1975) The use of markers to partition digestion within the gastrointestinal tract of ruminants. In: McDonald, I.W. and Warner, A.C.I. (eds) *Digestion and Metabolism in the Ruminant.* University of New England Publishing Unit, Armidale, New South Wales, pp. 261–276.

Faichney, G.J. (1993) Digesta flow. In: Forbes, J.M. and France, J. (eds) *Quantitative Aspects of Ruminant Digestion and Metabolism*. CAB International, Wallingford, UK, pp. 53–86.

Giraldez, F.J., Peláez., R., Zorita, E. and Valdés, C. (1993) Purine derivatives excretion in sheep urine in relation to metabolizable energy and rumen degradable protein intake. *Animal Production* 56, 466A.

Gomes, M.J., Hovell, F.D.DeB., Chen, X.B., Fickremariam, D. and Nengomasha, E.M. (1994) The effect of starch supplementation of straw on microbial protein supply in sheep. *Animal Feed Science and Technology* 49, 277–286.

Grovum, W.L. and Williams, V.J. (1973) Rate of passage of digesta in sheep. *British Journal of Nutrition* 30, 313–329.

Gunn, K.J. and Ternouth, J.H. (1994) The urinary excretion of purine derivatives in sheep offered nitrogen and phosphorus deficient diets. *Proceedings of Australian Society of Animal Production* 20, 444A.

Harrison, D.G. and McAllan, A.B. (1980) Factors affecting microbial growth yields in the reticulo-rumen. In: Ruckebusch, Y. and Thivend, P. (eds) *Digestive Physiology and Metabolism in Ruminants*. MTP Press, Lancaster, UK, pp. 205–226.

Hvelplund, T. (1985) Digestibility of rumen microbial protein estimated in the small intestine of sheep and by *in sacco* procedure. *Acta Agriculturae Scandinavica Supplement* 25, 132–144.

Hvelplund, T., Hovell, F.D.DeB., Orskov, E.R. and Kyle, D.J. (1994) True intestinal digestibility of protein estimated with sheep on intragastric infusion and with the mobile nylon bag technique. *Proceedings Society of Nutrition Physiology* 3, 64.

Hvelplund, T., Andrieu, J., Weisbjerg, M.R. and Vermorel, M. (1995) Prediction of the energy and protein value of forages for ruminants. In: Journet, M., Grenet, E., Farce, M.-H., Theriez, M. and Demarquilly, C. (eds) *Recent Developments in the Nutrition of Herbivores*. INRA, Paris, pp. 205–227.

IAEA (1997) *Estimation of Rumen Microbial Protein Production from Purine Derivatives in Urine. A Laboratory Manual for the FAO/IAEA Co-ordinated Research Programme on Development, Standardization and Validation of Nuclear Based Technologies for Measuring Microbial Protein Supply in Ruminant Livestock for Improving Productivity*. IAEA, Vienna.

Liang J.B., Matsumoto, M. and Young, B.A. (1994) Purine derivative excretion and ruminal microbial yield in Malaysian cattle and swamp buffalo. *Animal Feed Science and Technology* 47, 189–199.

McGilliard, A.D. (1982) Surgical techniques – advances and cautions. In: Owens, F.N. (ed.) *Protein Requirements for Cattle: Symposium*. Oklahoma State University, pp. 31–36.

MacRae, J.C. (1975) The use of re-entrant cannulae to partition digestive function within the gastrointestinal tract of ruminants. In: McDonald, I.W. and Warner, A.C.I. (eds) *Digestion and Metabolism in the Ruminant*. University of New England Publishing Unit, Armidale, New South Wales, pp. 261–276.

MacRae, J.C., Backwell, F.R.C., Bequette, B.J. and Lobley, G.E. (1996) Protein metabolism in specific organs. In: Nunes, A.F., Portugal, A.V., Costa, J.P. and Ribeiro, J.R. (eds) *Protein Metabolism and Nutrition*. Estacao Zootecnica Nacional, Santarem, Portugal, pp. 297–308.

Mayes, R.W., Dove, H., Chen, X.B. and Guada, J.A. (1995) Advances in the use of faecal and urinary markers for measuring diet composition, herbage intake and nutrient utilisation in herbivores. In: Journet, M., Grenet, E., Farce, M.-H., Theriez, M. and Demarquilly, C. (eds) *Recent Developments in the Nutrition of Herbivores*. INRA, Paris, pp. 381–406.

Merry, R.J. and McAllan, A.B. (1983) A comparison of the chemical composition of mixed bacteria harvested from the liquid and solid fractions of rumen digesta. *British Journal of Nutrition* 50, 701–709.

Moorby, J.M., Dewhurst, R.J., McKendrick, E.J. and Roberts, D.J. (1994) Urinary purine derivative excretion and milk protein production by lactatating dairy cows. *Proceedings of Society of Nutrition and Physiology* 3, 237.

NRC (1985) *Ruminant Nitrogen Usage*. National Research Council, National Academy Press, Washington.

Nyachoti, C.M., De Lange, C.F.M., McBride, B.W. and Schulze, H. (1997) Significance of endogenous gut nitrogen losses in the nutrition of growing pigs: a review. *Canadian Journal of Animal Science* 77, 149–163.

O'Kiely, P., Moloney, A.P. and Chen, X.B. (1994) Growth and rumen fermentation in steers offered whole-crop fodder beet silage. *Irish Journal of Agricultural and Food Research* 33, 93A.

Oosting, S. (1993) Wheat straw as ruminant feed. PhD thesis, Wageningen Agricultural University.

Oosting, S.J., Van Bruchem, J. and Chen, X.B. (1995) Intake, digestion and small-intestinal protein availability in sheep in relation to ammoniation of wheat-straw with or without protein supplementation. *British Journal of Nutrition* 74, 347–368.

Owens, F.N. and Goetsch, A.L. (1986) Digesta passage and microbial protein synthesis. In:

Milligan, L.P., Grovum, W.L. and Dobson, A. (eds) *Control of Digestion and Metabolism in Ruminants*. Prentice Hall, Englewood Cliffs, USA, pp. 196–227.

Pérez, J.F., Balcells, J., Guada, J.A. and Castrillo, C. (1996) Determination of rumen microbial-N production in sheep: a comparison of urinary purine excretion with methods using ^{15}N and purine bases as markers of microbial-N entering the duodenum. *British Journal of Nutrition* 75, 699–709.

Pérez, J.F., Balcells, J., Guada, J.A. and Castrillo, C. (1997) Rumen microbial production estimated either from urinary purine derivative excretion or from direct measurement of ^{15}N and purine bases as microbial markers: effect of protein source and rumen bacteria isolate. *Animal Science* 65, 225–236.

Poncet, C., Michalet-Doreau, B., McAllister, T. and Rémond, D. (1995) Dietary compounds escaping rumen digestion. In: Journet, M., Grenet, E., Farce, M-H., Theriez, M. and Demarquilly, C. (eds) *Recent Developments in the Nutrition of Herbivores*. INRA, Paris, pp. 167–204.

Puchala, R. and Kulaseck, G.W. (1992) Estimation of microbial protein flow from the rumen of sheep using microbial nucleic acid and urinary excretion of purine derivatives. *Canadian Journal of Animal Science* 72, 821–830.

Reynolds, C.K., Harmon, D.L. and Cecava, M.J. (1994) Absorption and delivery of nutrients for milk protein synthesis by portal drained viscera. *Journal of Dairy Science* 77, 2787–2808.

Robinson, P.H., Tamminga, S. and Van Vuuren, A.M. (1987) Influence of declining level of feed intake and varying the proportion of starch in the concentrate on rumen ingesta quantity, composition and kinetics of ingesta turnover in dairy cows. *Livestock Production Science* 17, 37–62.

Samaniego, M.A., Chen, X.B., Hovell, F.DeB., MacRae, J.C. and Mayes, R.W. (1997) Comparison of methods for the measurement of microbial protein supply in sheep. *Proceeding of the British Society of Animal Science* 1997, 125A.

Schelling, G.T., Koenig, S.E. and Jackson, T.C. (1982) Nucleic acids and purine and pyrimidine bases as markers for protein synthesis in the rumen. In: Owens, F.N. (ed.) *Protein Requirements for Cattle: Symposium*. Oklahoma State University, pp. 1–9.

Schwarting, G. and Kaufmann, W. (1979) Die Verdaulichkeit des Proteins beim Wiederkauer. *Zeitschrift für Tierphysiologie, Tierernährung und Futtermittelkunde* 40, 6–18.

Scott, D., Robins, S.P., Nicol, P., Chen, X.B. and Buchan, W. (1994) The effects of low phosphate intake on calcium and phosphorus retention, plasma osteocalcin levels and microbial protein synthesis in the lamb. *Experimental Physiology* 79, 183–187.

Sinclair, L.A., Garnsworthy, P.C., Newbold, R.J. and Buttery, P.J. (1993) Effect of synchronizing the rate of dietary energy and nitrogen release on rumen fermentation and microbial protein synthesis in sheep. *Journal of Agricultural Science* 120, 251–263.

Stangassinger, M., Chen, X.B., Lindberg, J.E., and Giesecke, D. (1995) Metabolism of purines in relation to microbial production. In: Von Engelhardt, W., Leonhard-Marek, S., Breves, G. and Giesecke, D. (eds) *Ruminant Physiology: Digestion, Metabolism, Growth and Reproduction*. Ferdinand Enke Verlag, Stuttgart, pp. 387–406.

Stern, M.D., Varga, G.A., Clark, J.H., Firkins, J.L., Huber, J.T. and Palmquist, D.L. (1994) Evaluation of chemical and physical properties of feeds that affect protein metabolism in the rumen. *Journal of Dairy Science* 77, 2769–2786.

Storm, E., Brown, D.S. and Ørskov, E.R. (1983) The nutritive value of rumen microorganisms in ruminants. 3. The digestion of microbial amino acids and nucleic acids in and losses of endogenous nitrogen from the small intestine. *British Journal of Nutrition* 50, 479–485.

Südekum, K.-H., Ziggers, W., Roos, N., Sick, H., Tamminga, S., and Stangassinger, M. (1995) Estimating the passage of digesta in steers and wethers using the ratio ^{13}C to ^{12}C and Titanium(IV)-oxide. *Isotopes Environment Health Studies* 31, 219–227.

Susmel, P., Spanghero, M., Stefanon, B., Mills, C.R. and Plazzotta, E. (1994a) Digestibility and allantoin excretion in cows fed diets differing in nitrogen content. *Livestock Production Science* 39, 97–99.

Susmel, P., Stefanon, B., Plazzotta, E., Spanghero, M. and Mills, C.R. (1994b) The effect of energy and protein intake on the excretion of purine derivatives. *Journal of Agricultural Science, Cambridge* 123, 257–265.

Sutton, J.D. and Oldham, J.D. (1977) Feed evaluation by measurement of sites of digestion in cannulated ruminants. *Proceeding Nutrition Society* 36, 203–209.

Svejcar, T.J., Judkins, M.B. and Boutton, T.V. (1993) Technical note: labeling of forages with ^{13}C for nutrition and metabolism studies. *Journal of Animal Science* 71, 1320–1325.

Tamminga, S. (1982) Protein metabolism in ruminants with special reference to forage protein utilisation. In: Griffith, T.W. and Maquire, M.F. (eds) *Forage Protein Conservation and Utilisation*. CEC, Luxemburg, pp. 83–98.

Tamminga, S. (1993) Influence of feeding management on ruminant fiber digestibility. In: Hatfield, R.D., Jung, H.-G., Ralph, J., Buxton, D.R., Mertens, D.R. and Weimer, P.J. (eds) *Forage Cell Wall Structure and Digestibility*. American Society of Agronomy, Madison, USA, pp. 571–602.

Tamminga, S., Robinson, P.H., Meijs, S. and Boer, H. (1989a) Feed components as internal markers in digestion studies with dairy cows. *Animal Feed Science and Technology* 27, 49–57.

Tamminga, S., Robinson, P.H., Vogt, M. and Boer, H. (1989b) Rumen ingesta kinetics of cell wall components and nitrogen in dairy cows. *Animal Feed Science and Technology* 27, 89–98.

Tamminga, S., Schulze, H., Van Bruchem, J. and Huisman, J. (1995) The nutritional significance of endogenous N losses along the gastro-intestinal tract of farm animals: a review. *Archives Animal Nutrition* 48, 9–22.

Thomson, D.J. and Beever, D.E. (1980) The effect of conservation and processing on the digestion of forage by ruminants. In: Ruckebusch, Y. and Thivend, P. (eds) *Digestive Physiology and Metabolism in Ruminants*. MTP Press, Lancaster, UK, pp. 291–308.

Van Bruchem, J., Voigt, J., Lammers-Wienhoven, T.S.C.W., Schönhusen, U., Ketelaars, J.J.M.H. and Tamminga, S. (1997) Secretion and re-absorption of endogenous protein along the small intestine of sheep: estimates derived from ^{15}N dilution of plasma non-protein-N. *British Journal of Nutrition* 77, 273–286.

Van Straalen, W.M. (1995) Modelling of nitrogen flow and excretion in dairy cows. PhD thesis, Wageningen Agricultural University, Wageningen, the Netherlands.

Verbic, J. (1996) Degradation and synthesis of protein in the rumen in different preserved forages from grassland. PhD thesis. Biotehniška fak., Oddelek za zootehniko, Ljubljana, Slovenia.

Verbic, J. (1997) The effect of preservation method on the protein value of grassland forage. *Alpenlandisches Expertenforum Grundfutterqualitat und Grundfutterbewertung, Gumpenstein, 21–22 January 1997*. Irdning, Bundesanstalt für alpenlandische Landwirtschaft, Gumpenstein, pp. 81–86.

Verbic, J., Chen, X.B., Macleod, N.A. and Ørskov, E.R. (1990) Excretion of purine derivatives by ruminants: effect of microbial nucleic acid infusion on purine derivative excretion by steers. *Journal of Agricultural Science, Cambridge* 114, 243–248.

Verbic, J., Babnik, D., Znidarsic-Pongrac, V. and Resnik, M. (1997) Ruminal starch digestion and microbial protein supply in sheep given maize silage made from dent (*Zea mays indentata*) or flint (*Zea mays indurata*) type hybrid. In: Van Arendonk, J.A.M. (ed.) *Book of Abstracts of the 48th Annual Meeting of the European Association for Animal Production*. Wageningen Press, Wageningen, p. 116.

Yang, W.Z. and Poncet, C. (1993) Measurement of microbial nitrogen flow in the duodenum of sheep. *Annales de Zootechnie* 42, 167.

12 *In Situ* Techniques for the Estimation of Protein Degradability and Postrumen Availability

T. HVELPLUND AND M.R. WEISBJERG

Danish Institute of Agricultural Sciences, Department of Animal Nutrition and Physiology, Research Centre Foulum, PO Box 50, 8830 Tjele, Denmark

Introduction

Obtaining the protein value of a feed for ruminants, expressed as amino acids truly absorbed from the small intestine, which is now common according to the new protein evaluation systems, involves measurements of several characteristics of the feed (Verité *et al.*, 1987; AFRC, 1992; Tamminga *et al.*, 1994; Madsen *et al.*, 1995). These characteristics are related to the fate of the protein in the feed during its passage through the digestive tract and to the energy-yielding processes in the rumen (mainly fermentation of carbohydrates) which drive microbial protein synthesis in the rumen. Factors related to the feed protein are degradability of the feed protein in the rumen, content of amino acids in the undegraded crude protein and, finally, small-intestinal digestibility of the amino acids in the undegraded dietary protein. Microbial protein synthesis in the rumen is related to the energy made available during fermentation in the rumen, but also of importance is the content of amino acids in the microbial protein produced in the rumen and the digestibility of microbial amino acids in the small intestine.

To determine the contribution of amino acids truly absorbed due to consumption of a forage, it would be necessary to quantify the factors mentioned above and develop easy and reliable methods for estimating values for those factors, which are variable among feeds and also within feeds. This chapter is restricted to *in situ* methods used to estimate protein degradability in the rumen and intestinal digestibility of undegraded dietary protein.

Rumen Protein Degradation

Feed protein is, to a large extent, degraded in the rumen. Rumen protein degradability is therefore one of the most important qualitative factors determining the protein value of a feed, as the protein degradability determines the supply to the rumen microbes of both ammonia and branched-chain fatty acids, as well as the supply to the small intestine of rumen-undegraded feed protein as a potential source for amino acid absorption.

Rumen protein degradation follows the scheme: protein → oligopeptides → dipeptides → amino acids → ammonia (Wallace, 1995; Cotta and Russell, 1996). The degradation of amino acids to ammonia is an intracellular process, but the degradation of protein to amino acids is an extracellular, but cell-associated, process (Cotta and Russell, 1996). According to Wallace (1995), the rate-limiting step in protein degradation is the degradation of oligopeptides to dipeptides from the N terminal.

It is well known from many studies that protein degradability can vary substantially both between and within feed groups. Therefore, estimation of protein degradability is of major importance in feed evaluation. As *in vivo* estimation of rumen protein degradability is problematic (see Chapter 11), other methods have been applied. Since the introduction of the new protein evaluation systems, the *in situ* (nylon bag, *in sacco*) method has been the most widely used method, and has commonly been used also as reference method for validation of other laboratory methods.

The technique to estimate rumen fermentation by incubation of small feed samples in fibre bags in the rumen was first used by Quin *et al.* (1938), who used silk bags to study degradation in the rumen of sheep. However, it was the introduction by Ørskov and McDonald (1979) of mathematical tools to weight the obtained data with a fractional rumen outflow rate and thereby transform the degradability values to the so-called effective protein degradability (EPD) which led to today's extensive use of the *in situ* method. Beside protein degradability, the method has been used to examine rumen degradability of dry matter (DM), organic matter, neutral-detergent fibre (NDF) (Stensig *et al.*, 1997) and starch (Tamminga *et al.*, 1990). Recently, the *in situ* method has been thoroughly reviewed by Huntington and Givens (1995); therefore, this chapter will mainly deal with those aspects that have major importance for forages.

Description of the in situ *method*

The principle in the *in situ* method is incubation of small feed samples in the rumen in fibre bags. The bags have pores small enough to retain the feed sample, but large enough to allow bacteria to enter the bags. Due to the small amounts of feed sample incubated, the feed under examination will not affect the rumen fermentation, and it is assumed that the conditions within the bag are similar to the conditions in the surrounding rumen content.

It has been attempted to standardize the *in situ* method, and the following method has been agreed upon for concentrate feeds by a European Economic Community (EEC)–European Association of Animal Production (EAAP) seminar (Oldham, 1987), as described in detail by Madsen and Hvelplund (1994). However, the standard is also commonly used for forages.

Bag specification
Bags should be made of polyester (Dacron) or nylon, with a pore size of 30–50 μm. Sample size should be 10–15 mg of material cm^{-2} of bag surface, and the ratio of width to length of the bags should be between 1:1 to 1:2.5.

Sample preparation
Samples for the incubation in the bags should normally be air-dry. For some materials that would otherwise be difficult to process, low temperature (less than 60°C) or freeze-drying is acceptable before milling through a 1.5–2.5 mm screen. After milling, there should be no fine sieving/screening to remove fine particles.

Animals and feeding
Animals should be cattle, sheep or goats, fed according to maintenance level or slightly above, with feeding twice per day and with a minimum interval between meals of 8 h. The diet should contain grass or legume hay plus concentrates, in the ratio two parts forage and one part concentrate on a DM basis, with a minimum of 13% crude protein in total ration DM. The concentrates used should contain as wide a mix of ingredients as possible, with a minimum of three sources of protein, and ideally the feedstuff under test should be included and high-starch concentrates should be avoided.

Replications and procedure
A minimum of three animals are to be used in each estimation, with incubation time 0, 2, 4, 8, 16, 24 and 48 h. Bags are to be inserted at either feeding time.

Washing procedure
For washing, cold water should be used and automatic machine washing is preferable for standardization, with a washing time of 10–15 min.

Compared with concentrates, it may be necessary to modify sample pretreatment, incubation times and treatment of the residues when forages are examined. Most forages need drying before milling. Here, it is very important to avoid heat damage of the protein, and therefore freeze-drying is recommended (López *et al.*, 1995). Compared with the original fresh forages, drying and milling can be argued to result in changes, which may affect the measured degradability. Therefore freezing and chopping or homogenizing the frozen material is used as an alternative (Huntington and Givens, 1997). Compared with dried and ground samples, the frozen and chopped samples will probably reflect the original fresh samples better. However, handling and weighing out of representative samples are easier with dried and ground samples.

With low protein and high fibre content, microbial colonization of the bag residues means that a significant proportion of the nitrogen (N) in the residues may be of microbial origin. This and other problems that are of special interest when evaluating forages will be discussed below.

Data presentation

The most detailed data presentation is to give the degradabilities measured at each incubation time, which can be used to draw the degradation profile (see Fig. 12.2).

For many feeds, first-order kinetics (equation 1) can describe the degradation profile satisfactorily (Ørskov and McDonald, 1979):

$$\text{Deg}(t) = a + b(1 - e^{-ct}) \tag{1}$$

where:

Deg(t) = fraction of protein degraded at time t of incubation
a = fraction of immediately degradable (soluble) protein
b = fraction of not soluble, but degradable protein
c = the fractional rate of degradation of fraction b (h^{-1})

The degradation profile describes the degradation of protein retained in the rumen. However, some protein will escape rumen degradation. Traditionally, the escape of protein from the rumen has been assumed to follow first-order mass action kinetics, according to equation (2) (Ørskov and McDonald, 1979):

$$R(t) = e^{-kt} \tag{2}$$

where:

$R(t)$ = fraction of protein retained in the rumen at time t of incubation
k = fractional outflow rate from the rumen (h^{-1})

The EPD can then be calculated according to equation (3) (Ørskov and McDonald, 1979):

$$\text{EPD} = a + b[c/(c + k)] \tag{3}$$

The EPD, which is the predicted *in vivo* rumen protein degradability, can also be calculated by the summation method, proposed by Kristensen *et al.* (1982), as shown in equation (4):

$$EPD = N_0 + \sum_{i=0}^{n}\left[\left(N_{t_{i+1}} - N_{t_i}\right) * \frac{e^{-kt_i} + e^{-kt_{i+1}}}{2}\right] \tag{4}$$

where:

N_0 = fraction of soluble protein determined by washing
N_{t_i} = fraction of degraded protein at incubation time t_i
i = incubation number, where $n + 1$ indicates the total number of the considered intervals and k and t are explained above

Equation (4) is a method that needs no assumptions regarding the degradation profile to follow a special model. Furthermore, this method requires only a pocket calculator to calculate EPD. However, with very few incubation times, the method may result in an underestimation of EPD, and also, if the potential degradability is not reached within the longest incubation time and this is combined with a low fractional outflow rate, it can result in an underestimation of EPD.

The assumption that the particle outflow from the rumen follows first-order mass action kinetics is questionable, especially for forages. Huhtanen *et al.* (1993) showed a selective retention (two-compartment model with an escapable and non-escapable pool) of particles from barley, barley fibre, rape-seed meal and soybean meal, which heavily affected calculated EPD. As the passage from the first to the second (escapable) pool is slower for forage particles than for concentrates (Huhtanen and Kukkonen, 1995; Stensig *et al.*, 1997), the assumption about simple first-order passage kinetics is probably more erroneous for forages than for concentrates. If this is true and two-compartment passage models are applied for calculation of EPD, this may result in a decrease in the protein value of forages compared with concentrates.

In vivo *validation*

In vivo estimation of rumen protein degradability requires a separation of the duodenal protein flow into undegraded feed protein (including peptides and amino acids of feed origin) and the rest, which consists of microbial and endogenous protein. Microbial protein can be estimated using different markers, but endogenous protein flow is difficult to estimate; therefore, *in vivo* estimation of the flow of undegraded feed protein is problematic. Furthermore, most concentrates and also many forages must be fed in combination with other feeds, and therefore only the flow of undegraded feed protein from the whole diet can be estimated, and not undegraded feed protein from the individual feeds. As the rumen protein degradation is a dynamic process, the effective protein degradation is highly dependent on the passage rate, as discussed above, and will therefore depend on, for example, feeding level. When validating nylon-bag data, it is therefore important also to judge the passage rate, used for calculation of the effective degradability, based on nylon-bag data in relation to the expected passage rate at the feeding level in the *in vivo* experiments used for validation.

Due to these difficulties only a few *in vivo* validations of the nylon-bag method have been performed. Madsen and Hvelplund (1985) validated the nylon-bag method on *in vivo* data, where microbial synthesis was estimated using diaminopimelic acid (DAPA) as a marker and endogenous non-ammonia nitrogen (NAN) was estimated as 0.27% of duodenal DM flow. Values for individual feeds were obtained using the difference approach and correcting for extramicrobial and endogenous protein when additional feeds (protein-rich concentrates, beet pulp and grass pellets) were added to a basal ration. The results are shown in Fig. 12.1. Figure 12.1 shows a reasonable general relationship between *in vivo* and nylon-bag data, but also reveals that there are severe differences between the two methods for some of the samples.

Vanzant *et al.* (1996) compared *in situ* with *in vivo* degradability for lucerne (alfalfa) and prairie hay, as shown in Table 12.1. Table 12.1 shows a reasonable agreement between *in vivo* and *in situ* protein degradability. However, with three different assumptions for endogenous N supply to the duodenal N flow, Vanzant *et al.* (1996) calculated *in vivo* lucerne protein degradabilities from 79 to 90% and prairie-hay protein degradabilities from 41 to 72%. These large variabilities in *in*

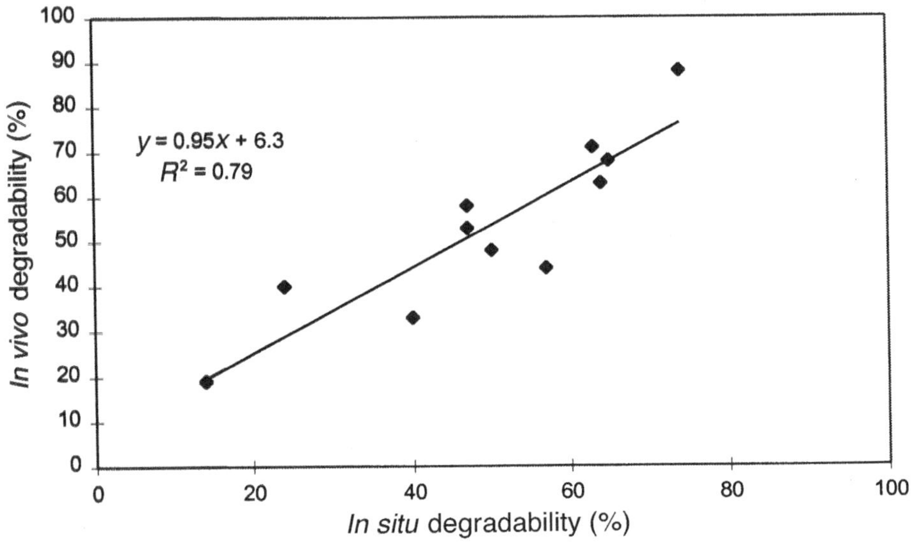

Fig. 12.1. The relation between *in vivo* and *in situ* protein degradability in protein-rich concentrates, beet pulp and grass pellets. *In situ* degradability is calculated using a fractional outflow rate of 8% h^{-1} (redrawn from Madsen and Hvelplund, 1985).

vivo estimates depending on endogenous flow estimates clearly show the problems faced in estimating *in vivo* degradability, and these problems will increase in severity with decreasing proportion of undegraded feed protein in the duodenal flow.

Compared with the extensive use of nylon-bag data, the validation on *in vivo* values has been very scarce and dubious, due to the lack of proper estimates of duodenal flow of endogenous N.

Loss of small particles from the bags

As a result of milling the samples, small particles are produced that can escape through the pores in the bags without any further degradation. According to equations (1), (3) and (4), this loss will be regarded as soluble and immediately degraded, which is probably not true.

Experience from our laboratory with concentrate feeds showed that effective protein degradabilities measured with the *in situ* method were not additive. Compounded feedstuffs often gave higher EPD values than expected from the EPD values obtained with the ingredients. The lack of additivity could be explained by a higher particle loss from the bags for the compounded feeds compared with the ingredients. This is probably due to extra milling and particle-size reduction during the mixing and pelleting process.

The extent of small-particle loss can be estimated as the difference between the loss from the nylon bags when they are only washed and the solubility measured on filter-paper. The method used to measure water solubility in our laboratory

Table 12.1. Comparison of *in vivo* with *in situ* effective protein degradability (%) (from Vanzant *et al.*, 1996).

	Lucerne	Prairie hay
In vivo	83.4	55.5
*In situ**	91.5	58.3
In situ†	87.2	57.2

* Incubated in the same cows as used for *in vivo* data.
† Incubated in steers fed brome hay *ad libitum*.

is as follows: 0.5 g of sample (same milling as in *in situ* study) is weighed out in a beaker, 40 ml tap water is added and the sample is soaked for 1 h at room temperature (≈ 20°C). Then the material is transferred to an N-free filter-paper with retention value 2 and washed four times with 40 ml of water. The residual N is determined and the water-soluble N is calculated as the difference between the original amount of N and the N in the residue.

Assuming the particle loss is degraded similarly to the particles remaining in the bag, corrections can be made for the loss. Individual degradability values, *a*, *b* and *c* values and values for EPD can be corrected.

Individual degradability values (Weisbjerg *et al.*, 1990):

$$\mathrm{DEG_{cor}}(t_i) = \mathrm{DEG}(t_i) - P[1 - ((\mathrm{DEG}(t_i) - (P + \mathrm{SOL}))/(1 - (P + \mathrm{SOL})))] \quad (5)$$

a, *b* and *c* values (Weisbjerg *et al.*, 1990):

$$a_{cor} = a - P \quad (6)$$

$$b_{cor} = b + P[b/(1 - (P + \mathrm{SOL}))] \quad (7)$$

$$c_{cor} = c \text{ (as this is the assumption)} \quad (8)$$

EPD values (Madsen *et al.*, 1995):

$$\mathrm{EPD_{cor}} = \mathrm{SOL} + [((1 - \mathrm{SOL})/(1 - (P + \mathrm{SOL}))) \times (\mathrm{EPD} - (P + \mathrm{SOL}))] \quad (9)$$

where:

$\mathrm{DEG_{cor}}(t_i)$ = corrected degradability at incubation time t_i
$\mathrm{DEG}(t_i)$ = measured degradability at incubation time t_i
P = loss of small particles
SOL = water solubility
$\mathrm{EPD_{cor}}$ = corrected effective protein degradation
EPD = measured effective protein degradation

The effect of correction for particle loss on the degradation profile is shown in Fig. 12.2. The *a* value (intercept) will be reduced by a value that corresponds to the estimated particle loss and, with increasing incubation time, the corrected curve will approach the measured curve, as shown in Fig. 12.2. The particle loss can be large, as shown in Table 12.2. The assumption, used in equations (5)–(9), that the

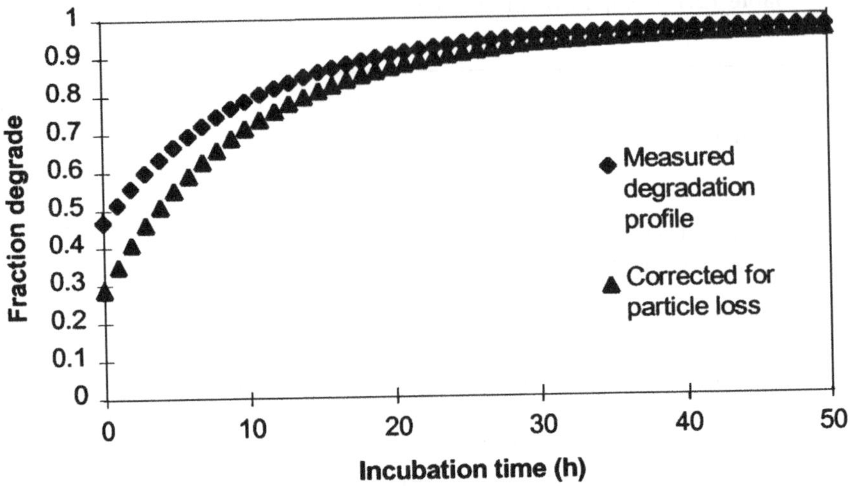

Fig. 12.2. Degradation profile for protein in ryegrass (values from Table 12.4).

particle loss would be degraded similarly to the fraction remaining in the bag after washing is certainly not true for some feeds, e.g. oat grain, where the particle loss is very high (from 42 to 67% of total N in our laboratory) and the remaining protein is, to a large extent, hull-associated protein. In such situations, correction for particle loss will result in severe underestimation of EPD, as the assumption used for correction will not be true.

As shown in Table 12.4, whole crop also gives large particle losses and, as whole crop is also a feed with a heterogeneous composition (straw and grain), similar problems to those with oat grain can be expected. In our laboratory, we examined how correction for particle loss affected whole crop of barley, field beans, peas and two samples of wheat in an experiment, where water solubility and *in situ* degradability were determined both on the whole crop and on samples that had been split up into the straw part and grain part. The results are shown in Table 12.2. It is obvious from Table 12.2 that the particle loss from whole crop is higher for the grain fraction than for the straw fraction. However, except for barley whole crop, the corrected EPD (EPD II) values obtained for the whole crop correspond reasonably well to the corrected EPD calculated from the grain and straw fraction. Therefore, based on these results, corrections for particle loss seem reasonable also for whole-crop cereals.

Reproducibility

One of the major problems with the *in situ* method is the poor reproducibility. In a ring test with participation from 23 laboratories representing 17 different countries, very poor reproducibility was found (Madsen and Hvelplund, 1994). For soybean meal, cottonseed meal, coconut meal, fish-meal and barley, effective degradabilities

Table 12.2. The effect of correction for particle loss on whole-crop cereals and parts of whole-crop cereals.

	Proportion of N in grain and straw	% N in DM	EPD I*	EPD II†	N lost with particles‡
Barley					
Grain	0.380	1.44	86.5	75.4	42.1
Straw	0.620	1.18	68.6	59.5	19.6
Whole crop		1.18	73.3	58.5	32.0
Calculated§		1.28	75.4	65.5	28.2
Wheat sample 1					
Grain	0.314	1.72	87.1	78.6	40.4
Straw	0.686	1.08	62.2	45.6	25.7
Whole crop		1.30	71.8	55.6	29.0
Calculated§		1.28	70.0	56.0	30.3
Wheat sample 2					
Grain	0.584	1.55	92.9	83.9	45.1
Straw	0.416	0.71	69.7	51.5	24.2
Whole crop		1.04	83.5	68.0	31.5
Calculated§		1.20	83.2	70.4	36.4
Field bean					
Grain	0.123	3.56	80.1	62.9	40.5
Straw	0.877	2.72	79.5	69.0	21.6
Whole crop		2.76	80.1	69.5	23.7
Calculated§		2.82	79.6	68.2	23.9
Peas					
Grain	0.288	4.06	91.3	85.7	38.1
Straw	0.712	3.15	74.0	69.0	17.2
Whole crop		3.41	76.3	69.5	22.1
Calculated§		3.41	79.0	73.8	23.2

* Effective protein degradability calculated at 8% passage rate without correction for particle loss.
† Effective protein degradability calculated at 5% passage rate with correction for particle loss.
‡ % of total N.
§ Values for grain and straw are weighted according to the proportion they make up of whole-crop N.

(mean ± standard deviation) of 63% ± 11; 51% ± 9; 47% ± 8; 23% ± 6 and 72% ± 8, respectively, were found. Standard deviations of this magnitude means that the obtained values cannot be considered as absolute values and that one should be very careful when values from different laboratories are mixed. However, the ranking of the feeds within the laboratory was reasonably consistent.

In a ring test between the Nordic countries (unpublished), where standardization of the *in situ* method has been attempted for many years, the reproducibility was better for barley, rape-seed meal and artificial dried grass (mean ± standard deviation) – 68% ± 7; 64% ± 4 and 42% ± 4, respectively – but these differences between laboratories were still too large if values were to be adopted from other laboratories.

A major part of the difference between laboratories seems to be due to differences in the washing procedure (Madsen and Hvelplund, 1994), and the difference between laboratories can be extremely high even with a high degree of standardization, as shown in a Nordic ring test. Some of the poor reproducibility can be explained by the repeatability. Data from our laboratory showed the following means and standard deviations for EPD when measurements were repeated five times for barley, rape-seed meal and artificial dried grass: 65% ± 3, 64% ± 3 and 45% ± 2, respectively. The variation within the laboratory is mainly due to variation between periods and, to a smaller degree, variation between cows and between bags within cow and period (van der Koelen *et al.*, 1992). Especially the N disappearance after just washing showed a substantial variation between periods, supporting the theory that variation in disappearance after washing (0 h incubation) is one of the major sources for variation in EPD within and between laboratories.

Microbial contamination

The principle in the *in situ* method is that microbes should enter the bag and degrade the feed in a similar way to that in the rumen. Therefore, the samples in the bag will be colonized by microbes. The normal washing procedure after incubation removes degraded feed and a large part of the microbes, but some microbes will be so attached to the residues that they will not be removed. This contamination of the residues with microbes will have only a minor influence on DM degradation, but, due to the high protein content in microbes, protein degradation can be underestimated, especially in high-fibre and low-protein feeds (Varvikko and Lindberg, 1985). As many forages are high in fibre and some are low in protein, microbial contamination of the residues can severely affect the EPD estimation of forages.

Different methods have been used to overcome this problem. Chilling, sonication and stomaching have all been used for removal of bacteria, and stomaching has been shown to be the most effective method (Michalet-Doreau and Ould-Bah, 1992). T. Hvelplund and J. Lindberg (unpublished) compared protein degradability estimated with four different methods using ^{15}N-labelled feeds. The four estimates were true degradability obtained as ^{15}N degradability, degradability based on residual N after machine washing, degradability based on residual N after the residues had been treated in a stomacher and degradability based on residual N after NDF boiling of the residues, as shown in Table 12.3.

It can be seen from Table 12.3 that machine washing as the only treatment resulted in an underestimation of the protein degradability, and the underestimation became more pronounced with increasing maturity of the crop. The use of the stomacher resulted in an overestimation of rumen protein degradability for feeds like whole-crop cereals and underestimation for feeds like straw and grass. These results indicate that the stomaching process does not remove all microbes, but increases the particle loss in feeds like whole crops. Therefore, the stomaching method is not perfect, but is clearly superior to no treatment.

In our laboratory, we have used the stomacher (individual residues are treated in the stomacher for 5 min) for all forage samples for the last 10 years. However,

Table 12.3. Effective protein degradability in different crops estimated with the *in situ* technique using four different methods (N = normal procedure, S = treatment of the residue in a stomacher, ^{15}N = ^{15}N degradability, NDF-N = N residue in the bag estimated as NDF-N) and calculated according to Kristensen *et al.* (1982), using a passage rate of 8% h^{-1} (from T. Hvelplund and J. Lindberg, unpublished).

Feed	Phase of vegetation (days after germination)	Stage of maturity	N	S	^{15}N	NDF-N
Pea whole crop	56	Full blooming	0.79	0.80	0.78	0.93
Pea straw I	77	Start of filling	0.72	0.77	0.75	0.86
Pea straw II	90	Filled pea shell	0.61	0.72	0.68	0.83
Pea straw III	104	Matured	0.48	0.56	0.50	0.72
Barley whole crop	56	Start of heading	0.43	0.60	0.53	0.72
Barley straw I	77	Milky stage	0.20	0.40	0.47	0.66
Barley straw II	90	Yellow stage	−0.12	0.20	0.33	0.54
Barley straw III	104	Matured	−0.19	0.14	0.31	0.52
Mixture A*	104	Matured	0.66	0.75	0.66	0.90
Mixture B*	104	Matured	0.45	0.50	0.48	0.77
Barley straw IV		Matured	−0.01	0.13	0.48	0.63
Ryegrass hay	51	4–5 leaves	0.65	0.66	0.70	0.84

* Composition of mixture: barley straw + pea straw + pea seeds; A, original; B, NaOH-treated. NaOH, sodium hydroxide.

use of the stomacher is a very time-consuming process. Therefore, Michalet-Doreau and Ould-Bah (1989) derived equations (equations 10 and 11) that make it possible to correct EPD calculated at 6% h^{-1} passage rate for the microbial contamination from information about the crude-protein content of the feed or, better, information about both crude-protein and NDF content:

$$\Delta EPD = 20.2 - 0.674(\%\text{ crude protein in DM}) \qquad (10)$$

$$\Delta EPD = 6.4 - 0.353(\%\text{ crude protein in DM}) + 0.170(\%\text{ NDF in DM}) \qquad (11)$$

where ΔEPD is the value which should be added to the measured (uncorrected) EPD.

Escape of soluble protein

According to equations (3) and (4), soluble protein is assumed to be immediately degraded in the rumen. For forages like silage, with a protein solubility as high as 84%, as shown for wheat whole-crop silage in Table 12.4, the fraction is close to the estimated EPD, even if corrections are made for particle loss. The fractional outflow rate of fluid from the rumen is high (≈ 12–15% h^{-1}), compared with particle outflow, which means that the degradation rate of the soluble protein should be very high if no soluble proteins are to escape, and this indicates that the present method results in an overestimation of rumen protein degradability for feeds high in soluble protein. Aaes and Kristensen (1997) measured fluid passage rate and the

Table 12.4. Degradation characteristics for different forages, measured *in situ*.

		% N in DM		*a*		*b*		*c*			% N lost with particles‡	
Feed	OBS*	Mean	SD	Mean	SD	Mean	SD	Mean	SD	EPD†	Mean	SD
Beets												
Fodder beets	10	1.14	0.12	0.759	0.023	0.198	0.037	0.080	0.035	0.88	11.5	2.6
Grass												
Clover grass	78	3.38	0.79	0.428	0.069	0.543	0.073	0.115	0.022	0.81	18.2	4.0
Italian ryegrass	1	2.83	–	0.466	–	0.497	–	0.129	–	0.82	24.0	–
Cocksfoot	21	2.80	0.84	0.446	0.116	0.539	0.144	0.095	0.034	0.80	15.9	4.5
Ryegrass	19	3.37	0.92	0.467	0.088	0.509	0.090	0.098	0.036	0.80	18.0	4.3
Legumes												
Lucerne	22	3.27	0.46	0.577	0.064	0.362	0.058	0.103	0.028	0.82	–	–
Galega	4	2.83	0.13	0.355	0.029	0.577	0.047	0.074	0.019	0.70	16.4	2.8
White clover	4	4.20	0.81	0.582	0.094	0.408	0.589	0.066	0.025	0.81	–	–
Red clover	11	3.54	0.73	0.272	0.150	0.692	0.149	0.099	0.018	0.73	13.7	3.4
Field bean	3	2.68	0.28	0.653	0.107	0.263	0.118	0.082	0.015	0.82	28.1	4.5
Vetches	1	3.40	–	0.610	–	0.322	–	0.168	–	0.86	15.1	–
Peas	18	3.11	0.33	0.671	0.080	0.248	0.058	0.070	0.028	0.82	28.3	14.2
Whole crop												
Barley	11	1.34	0.11	0.564	0.121	0.369	0.235	0.074	0.073	0.78	24.6	6.6
Oat	1	1.16	–	0.691	–	0.173	–	0.195	–	0.83	47.4	–
Triticale	12	1.53	0.59	0.555	0.080	0.310	0.110	0.145	0.093	0.79	16.5	5.7
Wheat	49	1.32	0.30	0.629	0.083	0.257	0.085	0.079	0.103	0.79	26.3	10.1
Rye	18	1.71	0.69	0.637	0.077	0.256	0.098	0.138	0.076	0.82	26.4	5.3
Grass silage												
Clover grass	38	2.61	0.65	0.758	0.057	0.191	0.053	0.097	0.025	0.88	9.8	3.6
Cocksfoot	3	2.87	0.72	0.648	0.051	0.293	0.024	0.100	0.016	0.84	12.8	1.8
Ryegrass	2	2.12	0.06	0.749	0.010	0.200	0.016	0.095	0.008	0.88	19.8	1.6
Legume silage												
Field beans	2	2.78	0.11	0.633	0.019	0.298	0.012	0.082	0.013	0.82	29.9	2.8
Vetches	4	3.05	0.60	0.723	0.042	0.189	0.058	0.093	0.014	0.85	–	–
Peas	5	2.85	0.35	0.795	0.120	0.157	0.105	0.107	0.032	0.90	16.9	4.1
Whole crop silage												
Barley	10	1.72	0.15	0.823	0.066	0.127	0.061	0.025	0.015	0.87	19.7	2.4
Wheat	28	1.39	0.25	0.839	0.057	0.075	0.040	0.081	0.120	0.89	19.8	6.0
Maize	5	1.71	0.13	0.798	0.020	0.147	0.042	0.019	0.006	0.84	28.3	5.2
Hay												
Clover grass	5	2.32	0.37	0.400	0.097	0.554	0.098	0.059	0.016	0.70	10.1	0.4

* Number of observations.

† EPD calculated from the mean *a*, *b* and *c* values, using a passage rate of 5% h^{-1}.

‡ % of total N. Number of observations for particle loss are in most cases less than the number of observations used for the calculation of degradation characteristics. Samples were milled through a 1.5 mm screen.

SD, standard deviation.

content of soluble amino acids in rumen fluid of cows fed either fresh clover, fresh ryegrass or grass silage and found, especially for grass silage, a significant contribution of rumen-soluble amino acid N to the duodenal amino acid N flow. As the content of soluble amino acid N was highest for silage and the DAPA content in the soluble amino acid N fraction was very low, this indicates that most of the soluble amino acid N was originating from the feed and not from endogenous or microbial origin. This is in agreement with several studies that show that the degradation rate of soluble proteins and amino acids in the rumen is far from infinite (Mangan, 1972; Volden *et al.*, 1998). This implies that proper estimation of rumen protein degradation, at least for feeds with a high proportion of the N in soluble amino acid N, as in silage, will necessitate that the soluble *a* fraction is weighted, with the ratio between rate of degradation of soluble protein ($k_{d-fluid}$) and the rate of passage of fluid ($k_{p-fluid}$) as shown in equation (12):

$$EPD = [a(k_{d-fluid}/(k_{d-fluid} + k_{p-fluid})] + [b(c/(c + k))] \qquad (12)$$

where EPD, *a*, *b*, *c* and *k* are as explained under equation (3).

Fluid passage rate has been determined in many experiments and it would be easy to find appropriate values for calculation of EPD. Rate of degradation of soluble proteins has rarely been determined, as it is difficult to trace the soluble proteins, as some end up in ammonia and some in microbial protein. Mangan (1972) estimated the half-life of soluble proteins. For casein it was 20 min and for ovalbumin the half-life was 180 min. Assuming first-order kinetics, this results in rates of degradation of 210% h^{-1} and 23% h^{-1} for casein and ovalbumin, respectively, and assuming a fluid passage rate of 13% h^{-1} this will result in rumen degradabilities of 94 and 64%, respectively. Peltekova and Broderick (1996) estimated *in vivo* degradation rates for casein (48.6% h^{-1}) and for soluble proteins from lucerne hay and lucerne silage (27.3 and 28.5% h^{-1}, respectively), but the proportions of soluble true protein were small in both lucerne hay and silage. However, inclusion of the soluble fraction in the potentially escapable pool also raises the question as to how far the protein should be degraded to be regarded as degraded. From a nutritional point of view, it is the amino acid supply of feed origin to the intestine that is interesting, and amino acids will be lost from the feed amino acid pool when they are either degraded to ammonia or used for microbial amino acid synthesis.

Values for forage protein degradation

In Table 12.4, values for protein degradation obtained in our laboratory for forages are shown. Large differences have been observed between feeds, but the most striking difference is the difference between fresh and ensiled feeds. The ensiling process results in a large increase in the water-soluble fraction of the protein, and thereby in the *a* value. The *b* value is reduced similarly to the increase in *a*, but the changes result in increased EPD values. The values in Table 12.4 show EPD values for silage of clover plus grass, ryegrass, pea whole crop, barley whole crop, and wheat whole crop, which are 7 to 10 percentage-units higher than for fresh feeds.

Crude-protein versus amino acid degradation

The measured EPD values are normally used directly for prediction of amino acid protein degradability, although they are based on crude protein (total N × 6.25) degradability. In forages, non-amino acid N generally makes up a bigger part of total N than in concentrates, which can make it more problematic to use total N values as predictors for amino acid N. In a study with 14 different concentrate ingredients, Weisbjerg *et al.* (1996) found, as a mean, no differences in crude protein and amino acid protein effective degradability, although small differences were found for some of the feeds. However, water solubility was considerably higher for crude protein than for amino acid protein. In a study with eight fresh and ensiled forages, Skiba *et al.* (1996) found a higher solubility/degradability at short incubation times for crude protein compared with amino acid protein, a difference which was so pronounced that EPD values were also considerably higher for crude protein compared with amino acid protein, as shown in Table 12.5.

The lower degradability of amino acids compared with crude protein is in accordance with results obtained by Susmel *et al.* (1989) for ensiled lucerne. These results indicate that, for forages, the use of crude-protein degradabilities results in an overestimation of amino acid protein degradation. However, in most protein evaluation systems, this is probably more than counteracted in an overestimated assumption with respect to the amino acid N proportion of total N in forages, especially in silages.

Intestinal Digestibility of Undegraded Feed Protein

Estimates of the true digestibility of protein and amino acids in the small intestine, from experiments in which undegraded protein was infused into the intestine of ruminants, are scarce. Applying regression techniques, Tas *et al.* (1981) reported a value of 0.82 for the true digestibility of undegraded amino acids passing to the duodenum of sheep fed a diet of grass and concentrates. Hvelplund (1985)

Table 12.5. Effective rumen degradability of crude protein (N) and total amino acids (AA) from different roughages determined *in situ* (from Skiba *et al.*, 1996).

	Effective degradability*		Effective degradability,* corrected for particle loss	
	N	AA	N	AA
Clover grass silage	0.885	0.858	0.818	0.792
Clover grass I	0.769	0.739	0.597	0.581
Ryegrass	0.762	0.722	0.648	0.606
Clover grass II	0.763	0.722	0.631	0.614
Winter wheat whole-crop silage	0.899	0.858	0.795	0.712
Pea whole-crop silage	0.857	0.838	0.702	0.635
Artificial dried grass	0.597	0.502	0.506	0.414
Galega	0.700	0.663	0.568	0.523

* 5% passage rate.

measured the increments in protein and amino acids passing the terminal ileum when different proteins, which had previously been exposed to some degradation in the rumen, were infused into the abomasum of sheep. This experiment showed that the protein and amino acids in different undegraded proteins were digested to a variable degree in the small intestine. Thus, protein and amino acid digestibility of undegraded feed protein varies, especially if the protein sources are treated to protect the protein against degradation. Protection may influence digestibility negatively if the protection is overdone, as shown by Kaufmann and Lüpping (1982) and Hvelplund (1985). Because of the variable digestibility of undegraded feed protein, there is a need to assess the digestibility of undegraded dietary protein for all types of feeds. However, *in vivo* experiments and the use of infusion techniques for the estimation of postruminal digestion of undegraded dietary protein can only be applied to a limited extent, as they are costly and time-consuming. Therefore, alternative methods have been evaluated.

The mobile nylon-bag method

The mobile-bag method for estimation of intestinal protein digestibility was introduced for pigs by Sauer *et al.* (1983). Since then, the method has also been used to study intestinal digestibility in ruminants and has been introduced in many different laboratories (Hvelplund, 1985; Rae and Smithard, 1985; Rooke, 1985; Voigt *et al.*, 1985; DeBoer *et al.*, 1987). Intestinal digestibility, using the mobile-bag technique, is obtained as the disappearance of protein from the bags during passage through the intestine. Many factors may influence the disappearance using this procedure to predict digestibility, including bag-cloth characteristics, pepsin–hydrochloric acid (HCl) pretreatment of the sample and place of recovery.

Pore size of the bag cloth

An appropriate pore size of bag cloth should allow free inflow of digestive enzymes into the bag and free outflow from the bags of nutrients originating from digested feed, without losing feed particles that have not been degraded. Pore sizes of bag material ranging from 9 to 53 μm have been reported (Vanhatalo, 1995) and comparative data from different sources indicate that the difference in the disappearance obtained due to bag pore size tends to be rather small and thus of minor importance. However, the pore size of the bags is a compromise between an efficient flushing of the bags with intestinal contents during the passage through the intestine and avoiding losses of undegraded feed particles from the bags. Based on this, a bag pore size of approximately 10–15 μm can be recommended and seems to be a good compromise with respect to the points mentioned above, provided, however, that the free surface area of the bag cloth is large enough (> 5%), as pointed out by Vanhatalo (1995).

Pepsin–HCl treatment

Exposure of feed residues to pepsin–HCl treatment is a normal process during passage through the abomasum. In the mobile-bag procedure, however, pretreatment with pepsin-HCl is not always included, as the necessity for this is doubted, mainly because there are acidic conditions in the proximal part of the ruminant small intestine. This extends the activity of pepsin also to the section where the mobile bags are inserted. In some situations, the influence of pretreatment is very pronounced, as observed for formaldehyde-treated soybean meal, where the pH in the abomasal content used for the pretreatment varied from 2.21 to 3.22, which resulted in an intestinal disappearance varying from 0.75 to 0.54 (Hvelplund, 1985). As a consequence, pretreatment with pepsin–HCl at a pH of 2.4 is recommended, as this pH value was found to represent the average conditions in the abomasum of cattle, based on measurements of abomasal content from a donor cow and values obtained from the literature (Hvelplund, 1985).

Place of recovery: ileum vs. faeces

In the modern protein evaluation systems, the small-intestinal digestibility is an important parameter. This implies that the digestibility between duodenum and ileum should be measured. However, recovery of bags from the ileum cannulae is difficult, compared with recovery from the faeces. Place of recovery may influence the estimated disappearance of protein from the mobile bag due to the activity in the large intestine. An increase in disappearance from ileum to faeces is obtained if further breakdown of protein takes place in the hind-gut. If available substrate for microbial digestion is present in the bag residue, the possibility for microbial growth also exists and, if the microbes are strongly adhering to the residues, this will lead to an increase in protein content and, thereby, decreased protein disappearance during passage through the hind-gut. Comparative studies, reviewed by Vanhatalo (1995), have shown that collecting bags from faeces instead of from the ileum usually only causes a slight increase in protein disappearance from the bags with plant-based concentrates, nearly no impact on protein feeds of animal origin and a variable response on different forages, which is reflected in the amount of substrate left in the bags for microbial fermentation.

In a study including different forages labelled with ^{15}N, the influence of place of recovery on both true (^{15}N) and apparent (Kjeldahl-N) disappearance was estimated (Jarosz *et al.*, 1994). The results from this study (Table 12.6) showed that the disappearance of Kjeldahl N underestimated true N disappearance at both recovery sites. Kjeldahl-N disappearance measured using bags recovered from the faeces generally agreed with estimates of ^{15}N disappearance measured using bags recovered from the ileum, except for very mature high-fibre feeds. This leads to the conclusion that bags recovered from the faeces can be used to predict true small-intestinal disappearance.

Table 12.6. Disappearance of Kjeldahl nitrogen (N_K) and ^{15}N from mobile bags recovered at the terminal ileum (TI) or from the faeces (F), and statistical significance of place of recovery on disappearance, as well as significance between apparent (N_K) and true (^{15}N) nitrogen digestibility. Each value is the mean of three observations (from Jarosz *et al.*, 1994).

	Disappearance				Test probability (*P*)			
	N_K		^{15}N		TI vs. F		N_K vs. ^{15}N	
Feedstuffs*	TI ($g\ kg^{-1}$)	F ($g\ kg^{-1}$)	TI ($g\ kg^{-1}$)	F ($g\ kg^{-1}$)	N_K	^{15}N	TI	F
Pea whole crop	946	970	955	978	0.06	0.008	0.2	0.4
Pea straw I	911	936	913	950	0.003	0.001	0.7	0.02
Pea straw II	833	870	858	895	0.02	0.5	0.02	0.02
Pea straw III	636	641	804	846	0.8	0.03	0.0003	0.0001
Barley whole crop	885	896	930	949	0.2	0.06	0.004	0.001
Barley straw II	669	626	735	765	0.3	0.2	0.01	0.02
Barley straw III	565	470	646	677	0.003	0.4	0.04	0.0003
Mixture A†	730	775	809	859	0.009	0.02	0.003	0.001
Mixture B†	907	917	941	957	0.3	0.1	0.03	0.002
Ryegrass hay	852	873	908	922	0.2	0.06	0.04	0.002

* For stage of maturity of the crop, see Table 12.3.
† Composition of mixture: barley straw + pea straw + pea seeds; A, original; B, NaOH-treated.
NaOH, sodium hydroxide.

Description of the mobile-bag procedure

No attempts have been made to standardize the mobile-bag procedure between laboratories and, to our knowledge, no ring test has been performed. The procedure mentioned below is the one that is used at Research Centre Foulum to estimate total digestibility of protein, from which the digestibility of rumen-undegraded protein can be calculated.

Bag specification and sample preparation

Nylon material with a pore size of 11 μm and with a 6% open bag area is used for preparation of bags measuring 6 cm × 6 cm. The feed is milled through a 1.5 mm screen before weighing. Sample size is *c.* 1 g of material per bag. If the feed is bulky (e.g. some roughages), the sample size is reduced to 0.8 g. The bags are closed by heat-sealing.

Incubation procedure

For each feed, six bags are prepared. The bags are incubated in the rumen for 16 h, with two bags in each of three cows. Incubation for 16 h is used for all feeds at present, as this incubation time is believed to reflect the influence of rumen metabolism on the feed before passage to the intestine, but this needs further clarification. After incubation, the bags are washed in a washing-machine, using cold water (2 × 22 l). After this, the bags are soaked for 1 h in a 0.004 mol l^{-1} HCl solution at pH 2.4. Then, the bags are incubated for 2 h in a pepsin–HCl solution at a pH of 2.4 (100 mg pepsin 1:10,000 l^{-1} of 0.004 mol l^{-1} HCl solution) in a shaking water-bath at 40°C. The bags are then inserted into the small intestine through the duodenal cannula in three cows, with two bags per feed in each cow. A maximum of 12 bags are introduced into the duodenum per cow per day. After passage through the intestine, the bags are recovered from the faeces and washed in a washing-machine, using cold water (2 × 22 l), and, finally, the residues are quantitatively transferred to N-free filter-paper and analysed for nitrogen.

In vivo *validation: relation between disappearance from bags and true digestibility*

Disappearance of protein from the mobile bag during passage through the intestine is assumed to be related to the true digestibility of protein in the intestine, and this assumption has been tested in a limited number of experiments, where the true digestibility was estimated in infusion experiments or with the ^{15}N dilution method (Hvelplund, 1985; Varvikko and Vanhatalo, 1990; Todorov and Girginov, 1991; Hvelplund *et al.*, 1994; Vanhatalo and Varvikko, 1995). Good agreement was obtained in the experiment by Todorov and Girginov (1991) and, as illustrated in Fig. 12.3, in the experiment by Hvelplund *et al.* (1994), whereas the agreement was less satisfactory for the other experiments. The reason for this may be ascribed to different factors with respect to the bag method used and the method used for obtaining *in vivo* values as discussed by Vanhatalo (1995). Further research is needed to clarify the relationship between disappearance from the bags and true intestinal digestibility, but, until this has been tested further on a wide range of

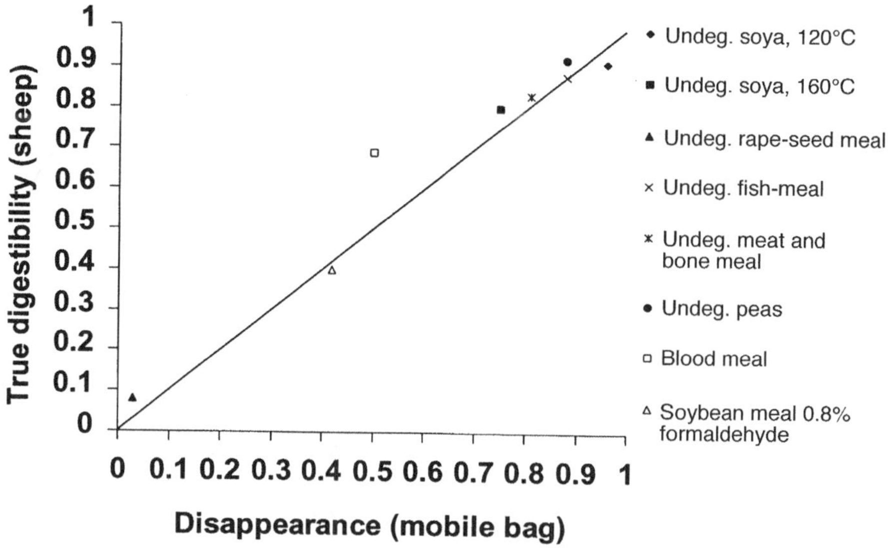

Fig. 12.3. The relationship between the disappearance of protein from the nylon bags and the true digestibility of protein in the small intestine of sheep. —, $x = y$ (from Hvelplund *et al.*, 1994).

feeds, it is reasonable to treat the results from the mobile bag as direct estimates of the true intestinal digestibility.

Practical use of mobile-bag results

The value needed in the protein evaluation systems is the digestibility of rumen-undegraded feed protein at the actual rumen degradability. This digestibility can be predicted from total digestibility of the original feed protein and information on the actual rumen degradability according to equation (13) (Hvelplund *et al.*, 1992), as illustrated in Fig. 12.4:

$$TD = (UDN - TU)/UDN \qquad (13)$$

where:

TD = true digestibility in the small intestine of undegraded feed protein
UDN = fraction of undegraded dietary protein
TU = fraction of true indigestible protein in the feed

The equation assumes that the fraction of true digestible protein in the feed, which is estimated from experiments with the mobile bag, is equivalent to maximum rumen degradability. The equation has been found valid for a variety of different concentrates and roughages, where total digestibility has been estimated as the disappearance from mobile bags passing through the intestine. The practical implication

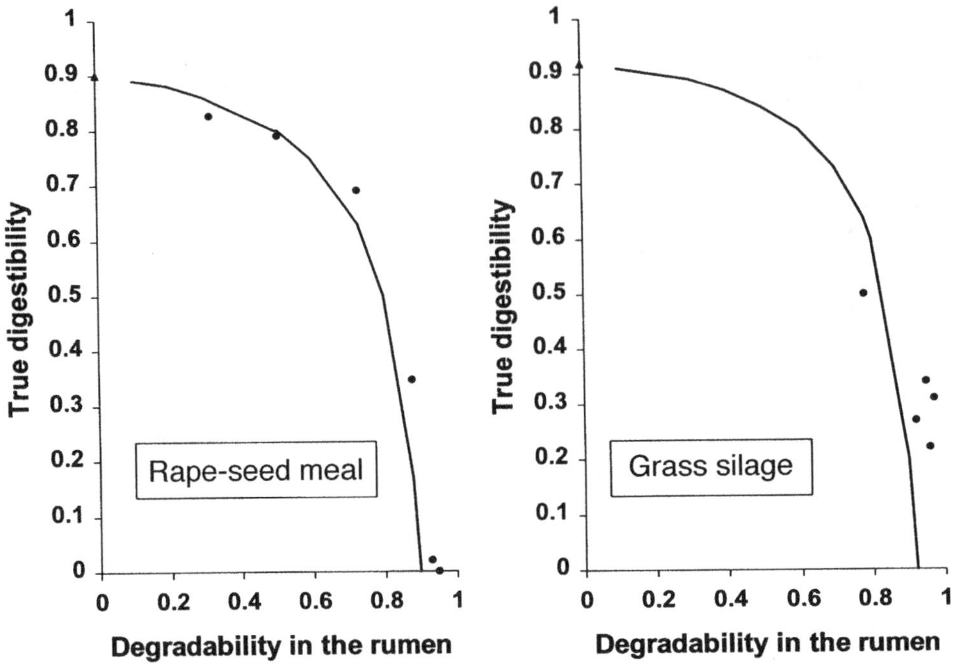

Fig. 12.4. The relationship between intestinal digestibility of undegraded feed protein and rumen protein degradability on two different feeds. ▲ True digestibility of intact feed protein used to estimate the digestibility (curve) from equation 13. ● True digestibility observed on samples of undegraded dietary protein (from Hvelplund *et al.*, 1992).

of this is that the intestinal digestibility of undegraded dietary protein cannot be considered as a constant value for a particular feed, but will vary according to the actual degradability in the rumen. If rumen degradability is assumed to vary according to feeding level, as recommended in the British protein evaluation system (AFRC, 1992), this also means that the same feed will have different intestinal digestibility of the undegraded dietary protein. This is illustrated for different roughages in Table 12.7, where the intestinal digestibility of undegraded dietary protein is calculated at two different rumen degradabilities, based on a passage rate of 2% h^{-1} or 5% h^{-1}.

The assumption that total digestibility can be estimated as small-intestinal disappearance from the mobile bags can be questioned. Mgheni *et al.* (1994) showed that, for tropical roughages, the potential degradability in the rumen was substantially higher than the digestibility of original protein in the small intestine measured with the mobile-bag technique. This is probably due to the content of tannins and to a high proportion of cell-wall-bound nitrogen in tropical roughages, which cannot be solubilized by intestinal enzymes but may be digested in the rumen, due to the combined action of proteolytic and cellulolytic microbial enzymes. This is also true for temperate roughages at a late stage of maturity, as

Table 12.7. Total digestibility of original protein (TDIG), degradability calculated at a passage rate of 2 or 5% h^{-1} (DEG) and the digestibility of the undegraded dietary protein at these degradabilities (DIG).

		Passage rate (2% h^{-1})		Passage rate (5% h^{-1})	
	TDIG	DEG	DIG	DEG	DIG
Clover grass silage	0.937	0.922	0.192	0.885	0.452
Clover grass I	0.937	0.865	0.533	0.769	0.727
Ryegrass	0.952	0.869	0.664	0.762	0.798
Clover grass II	0.963	0.875	0.704	0.763	0.844
Winter wheat whole-crop silage	0.906	0.906	0.000	0.899	0.069
Pea whole-crop silage	0.951	0.907	0.473	0.857	0.657
Artificial dried grass	0.797	0.722	0.269	0.597	0.496
Galega	0.830	0.782	0.220	0.700	0.433

shown by Vanhatalo *et al.* (1996). Similar problems may occur with other feeds, such as grains, in which protein is protected by starch, which limits the access of intestinal enzymes (Volden and Harstad, 1995), or with some protein-rich feeds, such as rape-seed meal, with a relatively high fibre content (Vanhatalo and Varvikko, 1995). Therefore, for estimation of total digestibility, it is recommended to include preincubation of the mobile bags in the rumen and, for temperate feeds, 16 h seems reasonable.

Crude-protein versus amino acid digestibility

The mobile bag is usually used to measure crude-protein digestibility, but the values are also assumed to represent amino acid digestibilities, as amino acid digestibility is the value required in the modern protein evaluation systems. In a recent study by Weisbjerg *et al.* (1996), which included 15 different concentrates, it was shown that total amino acid digestibilities were similar to crude-protein digestibilities. Large differences in amino acid digestibilities were observed among different concentrates, with the highest for protamyl (0.98) and the lowest for cottonseed cake (0.84). With respect to individual amino acids, glutamic acid and arginine had higher and threonine and glycine lower digestibilities, compared with total amino acids, whereas the remainder of the individual amino acids were similar to total amino acid digestibility.

The mobile nylon-bag method has also been used to evaluate the digestibility of protein and amino acids in different roughages. In one study, it was found that digestibility of amino acids was higher than for protein (Skorko-Sajko *et al.*, 1994), whereas no differences were observed in the study by Skiba *et al.* (1996), as shown in Table 12.8. Both studies, however, showed that big differences exist in the digestibility of both protein and amino acids among the different roughages.

Table 12.8. Total digestibility of original protein (N), total amino acids (AA), lysine (LYS) and methionine (MET) in different forages (from Skiba *et al.*, 1996).

	N	AA	LYS	MET
Clover grass silage	0.937	0.939	0.942	0.932
Clover grass I	0.937	0.939	0.951	0.946
Ryegrass	0.952	0.944	0.961	0.939
Clover grass II	0.963	0.959	0.970	0.961
Winter wheat whole-crop silage	0.906	0.890	0.857	0.882
Pea whole-crop silage	0.951	0.946	0.953	0.927
Artificial dried grass	0.797	0.798	0.806	0.749
Galega	0.830	0.832	0.827	0.863

Values for total digestibility of original protein in forages

Table 12.9 shows values obtained in our laboratory for total digestibility of original protein in different forages. The mean digestibility varies between 0.86 and 0.94, except for straw, where the digestibility is 0.65. Within each feed, there is a considerable difference between maximum and minimum values, which is mainly a reflection of stage of maturity of the crop (mowing time), but there is also some influence of cut (first or later cuts).

Table 12.9. Total digestibility of original protein in different forages.

Feed	Number of samples	Mean digestibility	Maximum digestibility	Minimum digestibility	SD
Alfalfa	27	0.922	0.954	0.866	0.027
Clover grass	20	0.897	0.911	0.715	0.068
Ryegrass	32	0.889	0.958	0.751	0.053
Cocksfoot	20	0.884	0.941	0.753	0.044
White clover	6	0.918	0.932	0.885	0.019
Red clover	13	0.939	0.973	0.886	0.028
Lupin whole crop	15	0.934	0.967	0.908	0.014
Lupin whole-crop silage	7	0.925	0.960	0.900	0.025
Horsebean whole crop	4	0.864	0.897	0.839	0.026
Triticale whole crop	12	0.915	0.967	0.879	0.030
Barley whole crop	14	0.863	0.931	0.769	0.056
Wheat whole crop	59	0.897	0.956	0.834	0.031
Rye whole crop	32	0.931	0.966	0.884	0.027
Pea whole crop	16	0.906	0.945	0.857	0.025
Grass silage	9	0.919	0.933	0.897	0.011
Wheat straw	2	0.652	0.669	0.634	0.026
Barley straw	1	0.642	–	–	–

SD, standard deviation.

Future Developments

In situ techniques for the estimation of protein degradability in the rumen and intestinal digestibility of rumen undegraded dietary protein require access to cannulated animals. *In situ* methods are therefore not acceptable as routine methods in feed evaluation, and laboratory methods need to be developed. A lot of effort has been devoted to this field in recent years, but universal methods to predict degradability, which can be used on a broad spectrum of feeds, have not yet been developed. The reason for this can probably be ascribed to a number of factors. An *in vitro* method may give satisfactory results against the *in situ* method in one laboratory on a limited number of samples, but fail in another laboratory. The reason for this could be the low reproducibility of the *in situ* method, as shown in the European ring test (Madsen and Hvelplund, 1994). To proceed in developing, testing and implementing laboratory methods, including the *in situ* method, it is necessary to have feed samples available where 'true' *in vivo* degradability has been estimated at a standardized feeding level, as degradability is a dynamic process that is dependent on a balance between the rate of degradation and the rate of outflow from the rumen. However, it is difficult to estimate 'true' *in vivo* degradabilities, as this requires a separation of duodenal protein flow in undegraded feed protein, microbial protein and endogenous protein. Microbial protein can be estimated with reasonable precision, using different microbial markers, but further research is needed to obtain more trustworthy estimates of endogenous protein flow, if undegraded feed protein is to be estimated by difference.

The values used in protein evaluation systems are effective protein degradabilities and, to obtain these values, the nylon-bag results have to be weighted with a passage rate. The traditional approach, using first-order kinetics, seems to be insufficient to describe the passage behaviour of feed particles from the rumen, especially for forages. As the assumptions about passage are of major importance both for the actual values obtained for effective degradabilities and for the ranking of feeds, further research in passage kinetics is necessary.

Intestinal digestibility of rumen-undegraded protein has only been investigated in a few *in vivo* experiments, and the mobile bag has commonly been accepted as giving values that can be regarded as equal to estimates of *in vivo* digestibility. However, at present, the number of comparisons between *in vivo* estimates and the mobile-bag method are scarce and have not been fully conclusive, and further comparisons are necessary to accept the mobile-bag method as a general reference method.

References

Aaes, O. and Kristensen, T. (1997) Proteinværdi af græs. In: *Intern Rapport nr. 88, Statens Husdyrbrugsforsøg*, pp. 30–45.

AFRC (1992) Nutritive requirements of ruminant animals: protein. AFRC Technical Committee on Responses to Nutrients: Report no. 9. *Nutrition Abstracts and Reviews (Series B)* 62, 787–835.

Cotta, M.A. and Russell, J.B. (1996) Digestion of nitrogen in the rumen: a model for metabolism of nitrogen compounds in gastrointestinal environments. In: Mackie, R.I. and White, B.A. (eds) *Gastrointestinal Microbiology*. Vol. 1. *Gastrointestinal Ecosystems and Fermentations*. Chapman and Hall Microbiology Series. International Thomson Publishing, New York, pp. 380–423.

DeBoer, G., Murphy, J.J. and Kennelly, J.J. (1987) Mobile nylon bag for estimating intestinal availability of rumen undegradable protein. *Journal of Dairy Science* 70, 977–982.

Huhtanen, P. and Kukkonen, U. (1995) Comparison of methods, markers, sampling sites and models for estimating digesta passage kinetics in cattle fed at two levels of intake. *Animal Feed Science and Technology* 52, 141–158.

Huhtanen, P., Jaakola, S., Kukkonen, U. and Nousianen, J. (1993) Influence of passage model in estimating ruminal protein degradability. *Journal of Animal Science* 71 (Suppl. 1), 263.

Huntington, J.A. and Givens, D.I. (1995) The *in situ* technique for studying the rumen degradation of feeds: a review of the procedure. *Nutritional Abstracts and Reviews (Series B)* 65, 63–93.

Huntington, J.A. and Givens, D.I. (1997) Studies on *in situ* degradation of feeds in the rumen 3. The effect of freezing forages before and after rumen incubation. *Animal Feed Science and Technology* 68, 131–138.

Hvelplund, T. (1985) Digestibility of rumen microbial protein and undegraded dietary protein estimated in the small intestine of sheep or by *in sacco* procedure. *Acta Agricultura Scandinavica* Suppl. 25, 132–144.

Hvelplund, T., Weisbjerg, M.R. and Andersen, L.S. (1992) Estimation of the true digestibility of rumen undegraded dietary protein in the small intestine of ruminants by the mobile bag technique. *Acta Agricultura Scandinavica, Section A, Animal Science* 42, 34–39.

Hvelplund, T., Hovell, F.D.DeB., Ørskov, E.R. and Kyle, D.J. (1994) True intestinal digestibility of protein estimated with sheep on intragastric infusion and with the mobile nylon bag technique. *Proceedings of the Society of Nutrition Physiology* 3, 64.

Jarosz, L., Hvelplund, T., Weisbjerg, M.R. and Jensen, B.B. (1994) True digestibility of protein in the small intestine and the hind gut of cows measured with the mobile bag technique using ^{15}N labelled roughage. *Acta Agricultura Scandinavica, Section A, Animal Science* 44, 146–151.

Kaufmann, W. and Lüpping. W. (1982) Protected proteins and amino acids for ruminants. In: Miller, E.L., Pike, I.H. and Van Es, A.J.H. (eds) *Protein Contribution of Feedstuffs for Ruminants: Application to Feed Formulation*. Butterworth Scientific, London, pp. 36–75.

Kristensen, E.S., Møller, P.D. and Hvelplund, T. (1982) Estimation of the effective protein degradability in the rumen of cows using the nylon bag technique combined with the outflow rate. *Acta Agricultura Scandinavica* 32, 123–127.

López, S., Howell, F.D.DeB., Manyuchi, B. and Smart, R.I. (1995) Comparison of sample preparation methods for the determination of the rumen degradation characteristics of fresh and ensiled forages by the nylon bag technique. *Animal Science* 60, 439–450.

Madsen, J. and Hvelplund, T. (1985) Protein degradation in the rumen. *Acta Agricultura Scandinavica* Suppl. 25, 103–124.

Madsen, J. and Hvelplund, T. (1994) Prediction of *in situ* protein degradability in the rumen: results of a European ringtest. *Livestock Production Science* 39, 201–212.

Madsen, J., Hvelplund, T., Weisbjerg, M.R., Bertilsson, J., Olsson, I., Spörndly, R., Harstad, O.M., Volden, H., Tuori, M., Varvikko, T., Huhtanen, P. and Olafsson, B.L. (1995) The AAT/PBV protein evaluation system for ruminants: a revision. *Norwegian Journal of Agricultural Science* Suppl. 19, 1–37.

Mangan, J.L. (1972) Quantitative studies on nitrogen metabolism in the bovine rumen: the rate of proteolysis of casein and ovalbumin and the release and metabolism of free amino acids. *British Journal of Nutrition* 27, 261–283.

Mgheni, D.M., Hvelplund, T. and Weisbjerg, M.R. (1994) Intestinal digestibility of rumen undegraded dietary protein from tropical roughages estimated by the mobile bag technique. *Acta Agricultura Scandinavica, Section A, Animal Science* 44, 230–235.

Michalet-Doreau, B. and Ould-Bah, M.Y. (1989) Estimation of the extent of bacterial contamination in bag residues and its influence on *in sacco* measurements of forage nitrogen degradation in the rumen. In: *Proceedings XVI International Grassland Congress, Nice, France*, pp. 909–910.

Michalet-Doreau, B. and Ould-Bah, M.Y. (1992) *In vitro* and *in sacco* methods for the estimation of dietary nitrogen degradability in the rumen: a review. *Animal Feed Science and Technology* 40, 57–86.

Oldham, J.D. (1987) *Towards a European Standard Method for Assessment of Protein Degradability*. Circular in connection with EAAP–EEC workshop, Brussels, 4 pp.

Ørskov, E.R. and McDonald, I. (1979) The estimation of protein degradability in the rumen from incubation measurements weighted according to rate of passage. *Journal of Agricultural Science, Cambridge* 92, 499–503.

Peltekova, V.D. and Broderick, G.A. (1996) *In vitro* ruminal degradation and synthesis of protein on fractions extracted from alfalfa hay and silage. *Journal of Dairy Science* 79, 612–619.

Quin, J.I., van der Wath, J.G. and Myburgh, S. (1938) Studies on the alimentary tract of Merino sheep in South Africa. IV. Description of experimental technique. *Onderstepoort Journal of Veterinary Science and Animal Industry* 11, 341–360.

Rae, R.C. and Smithard, R.R. (1985) Estimation of true nitrogen digestibility in cattle by a modified nylon bag technique. *Proceedings of the Nutrition Society* 44, 116A.

Rooke, J.A. (1985) The nutritive value of feed proteins and feed protein residues resistant to degradation by rumen microorganisms. *Journal of Science of Food and Agriculture* 36, 629–637.

Sauer, W.C., Jørgensen, H. and Berzins, R. (1983) A modified nylon bag technique for determining apparent digestibilities of protein in feedstuffs for pigs. *Canadian Journal of Animal Science* 63, 233–237.

Skiba, B., Weisbjerg, M.R. and Hvelplund, T. (1996) Rumen and total intestinal tract digestibility of protein and amino acids from different roughages, determined *in situ*. *Journal of Animal and Feed Sciences* 5, 347–363.

Skorko-Sajko, H., Hvelplund, T. and Weisbjerg, M.R. (1994) Rumen degradation and intestinal digestibility of amino acids in different roughages estimated by nylon bag techniques. *Journal of Animal and Feed Sciences* 3, 1–10.

Stensig, T., Weisbjerg, M.R. and Hvelplund, T. (1997) Evaluation of different methods for the determination of digestion and passage rates of fibre in the rumen of dairy cows. *Acta Agricultura Scandinavica, Section A, Animal Science* 48, 141–154.

Susmel, P., Stefanon. B., Mills, C.R. and Candido, M. (1989) Change in amino acid composition of different protein sources after rumen incubation. *Animal Production* 49, 375–383.

Tamminga, S., Van Vuuren, A.M., Van der Koelen, C.J., Ketelaar, R.S. and Van der Togt, P.L. (1990) Ruminal behaviour of structural carbohydrates, non-structural carbohydrates and crude protein from concentrate ingredients in dairy cows. *Netherlands Journal of Agricultural Science* 38, 513–526.

Tamminga, S., Van Straalen, W.M., Subnel, A.P.J., Meijer, R.G.M., Steg, A., Wever, C.J.G. and Blok, M.G. (1994) The Dutch protein evaluation system: the DVE/OEB system. *Livestock Production Science* 40, 139–155.

Tas, M.V., Evans, R.A. and Axford, R.F.E. (1981) The digestibility of amino acids in the small intestine of the sheep. *British Journal of Nutrition* 45, 167–174

Todorov, N.A. and Girginov, D.G. (1991) Comparison of the infusion method, mobile bag procedure and *in vitro* method for determination of the true protein digestibility in small intestine of cattle. In: Eggum, B.O., Boisen, S., Børsting, C., Danfær, A. and Hvelplund, T. (eds) *Protein Metabolism and Nutrition. Proceedings of the 6th International Symposium on Protein Metabolism and Nutrition*, Vol. 2. National Institute of Animal Science, Herning, Denmark, pp. 80–82.

van der Koelen, C.J., Goedhart, P.W., van Vuuren, A.M. and Savoini, G. (1992) Sources of variation of the *in situ* nylon bag technique. *Animal Feed Science and Technology* 38, 35–42.

Vanhatalo, A. (1995) Assessment of intestinal feed nitrogen digestibility in ruminants by the mobile-bag method. Dissertation, Agricultural Research Centre of Finland, Institute of Animal Production, 124 pp.

Vanhatalo, A. and Varvikko, T. (1995) Effect of rumen degradation on intestinal digestion of nitrogen of ^{15}N-labelled rapeseed meal and straw measured by the mobile-bag method in cows. *Journal of Agricultural Science, Cambridge* 125, 253–261.

Vanhatalo, A., Dakowski, P. and Huhtanen, P. (1996) Effect of stage of growth and duration of rumen incubation time on intestinal digestibility of rumen-undegradable nitrogen of grass by mobile-bag method in cows. *Acta Agricultura Scandinavica, Section A, Animal Science* 46, 1–10.

Vanzant, E.S., Cochran, R.C., Titgemeyer, E.C., Stafford, S.D., Olson, K.C., Johnson, D.E. and St Jean, G. (1996) *In vivo* and *in situ* measurements of forage protein degradation in beef cattle. *Journal of Animal Science* 74, 2773–2784.

Varvikko, T. and Lindberg, J. (1985) Estimation of microbial nitrogen in nylon-bag residues by feed ^{15}N dilution. *British Journal of Nutrition* 54, 473–481.

Varvikko, T. and Vanhatalo, A. (1990) The effect of differing types of cloth and of contamination by non-feed nitrogen on intestinal digestion estimates using porous synthetic-fibre bags in a cow. *British Journal of Nutrition* 63, 221–229.

Verité, R., Michalet-Doreau, B., Chapoutot, P., Peyreau, J.L. and Poncet, C. (1987) Révision du système des protéines digestibles dans l'intestin (PDI). In: *Bulletin Technique No. 70*. CRZV de Theix Ceyrat, France, pp. 19–34.

Voigt, J., Piatkowski, B., Engelmann, H. and Rudolph, E. (1985) Measurement of the postruminal digestibility of crude protein by the bag technique in cows. *Archiv für Tierernährung* 35, 555–562.

Volden, H. and Harstad, O.M. (1995) Effect of rumen incubation on the true indigestibility of feed protein in the digestive tract determined by nylon bag techniques. *Acta Agricultura Scandinavica, Section A, Animal Science* 45, 106–115.

Volden, H., Velle, W., Harstad, O.M., Aulie, A. and Sjaastad, O. (1998) Apparent ruminal degradation and rumen escape of lysine, methionine and threonine administered intraruminally in mixtures to high yielding cows. *Journal of Animal Science* 76, 1232–1240.

Wallace, R.J. (1995) The proteolytic systems of ruminal microorganisms. In: *Proceedings of the Symposium on Rumen Microbiology, 16–17 September 1995, Clermont-Ferrand, France*, pp. 20–29.

Weisbjerg, M.R., Bhargava, P.K., Hvelplund, T. and Madsen, J. (1990) *Anvendelse af Nedbrydningsprofiler i Fodermiddelvurderingen*. Beretning, 679, Statens Husdyrbrugsforsøg, Foulum, Denmark, 33 pp.

Weisbjerg, M.R., Hvelplund, T., Hellberg, S., Olsson, S. and Sanne, S. (1996) Effective rumen degradability and intestinal digestibility of individual amino acids in different concentrates determined *in situ*. *Animal Feed Science and Technology* 62, 179–188.

13 Enzymatic and Microbial-cell Preparation Techniques for Predicting Rumen Degradation and Postruminal Availability of Protein

S. Calsamiglia,[1] M.D. Stern[2] and A. Bach[2]

[1]Departament de Patologia i de Producció Animals, Universitat Autónoma de Barcelona, 08193 Bellaterra, Spain; [2]Department of Animal Science, Univeristy of Minnesota, St Paul, MN 55108, USA

Need for an Alternative Technique to *In Situ*

Current feeding systems recognize the need to select feeds based on the supply of ruminal degradable and undegradable protein (Madsen, 1985; INRA, 1989; NRC, 1989; ARC, 1984). Several systems have incorporated dynamic aspects of nutrient degradation in the rumen (Sniffen *et al.*, 1992), estimates of amino acid supply (Madsen, 1985; INRA, 1989; O'Connor *et al.*, 1993) and intestinal protein digestion (Madsen, 1985; INRA, 1989; Sniffen *et al.*, 1992) into their models. Most systems have used *in vivo* and *in situ* data to develop feed tables. Determination of protein degradation in the rumen using these techniques is expensive, labour-intensive and time-consuming for industry and field application. The advantages and disadvantages of these techniques have been discussed in previous chapters. Although the *in situ* procedure has often been used as a reference method to compare degradation values, ring-test studies have shown that there is substantial variation among and within laboratories on estimates of protein degradation obtained with this technique (Madsen and Hvelplund, 1994; Wilkerson *et al.*, 1995). Most of this variation depends on diet and individual animals, and it is difficult to control. Two additional problems are encountered when using the *in situ* technique for measuring protein degradation in forages: (i) forages may have a high proportion of water-soluble materials, which the technique will erroneously consider degradable; and (ii) the effect of microbial contamination may be more important in forages, because of high fibre and low protein content. To take full practical advantage of the new feeding systems, it is necessary to develop techniques that allow fast, affordable, reliable and accurate measurements of ruminal degradation and intestinal digestion of proteins. These techniques need to be validated with true estimates of protein degradability and digestibility, but these measurements are difficult to obtain *in vivo* or *in situ* (Stern *et al.*, 1997). In the absence of true estimates with which to validate *in vitro* techniques, the use of values obtained with the *in situ* technique provides the best available data. However, the limitations of this technique should be carefully evaluated when using these values to validate other techniques.

Enzymes and Activities Needed

Proteins are linear chains of amino acids, linked together by peptide bonds. Protein degradation and utilization in the rumen require the action of different proteases, peptidases and deaminases. The rate and extent to which protein degradation occurs will depend on the type and concentration of enzymes and on the number of susceptible peptide bonds and their accessibility. The secondary and tertiary structure of proteins will influence the accessibility of proteases to specific peptide bonds, thereby affecting the rate and extent of ruminal protein degradation (Wallace and Kopecny, 1983). Because the number of different bonds within a single protein is large, the synergistic action of different proteases is necessary for complete protein degradation.

Several studies revealed that the predominant protease type produced by ruminal bacteria is a cysteine protease, while serine protease (0–41%), metalloprotease (9–30%) and aspartic protease (2–15%) are generally present at lower levels (Brock *et al.*, 1982; Kopecny and Wallace, 1982; Prins *et al.*, 1983). Several authors (Brock *et al.*, 1982; Prins *et al.*, 1983; Wallace and Kopecny, 1983) reported that trypsin-like activity was the predominant proteolytic activity in the rumen. Lower chymotrypsin-like activity has also been identified. Peptide breakdown is accomplished mainly by the activity of a dipeptidyl peptidase, while leucine and alanine aminopeptidase and carboxypeptidase have been found at lower concentrations (Wallace and McKain, 1991; McKain *et al.*, 1992; Depardon *et al.*, 1996). The complexity of the rumen and the difficulty of simulating its proteolytic activity suggest that the role of quantitatively unimportant microorganisms or enzymes may play a strategic role in protein degradation. For example, Wallace (1985) demonstrated that the combined activity of two ruminal bacteria resulted in a protein degradation larger than the one obtained from the sum of both activities individually (Fig. 13.1). Therefore, it appears that a combination of enzymes with the same activities and specificities as those of the rumen environment would be required to fully simulate ruminal degradation of proteins. However, the relative importance of each protease or peptidase changes with type of diet (Nuget and Mangan, 1981) and animal (Wallace and Cotta, 1988), creating further difficulties in understanding and modelling ruminal protein degradation.

In addition to the proteolytic activity needed for protein degradation, other non-proteolytic enzymes may be required. Most proteins interact with other nutrients within feeds. Assoumani *et al.* (1992) demonstrated that starch interferes with protein degradation, and the addition of amylase increased total protein degradation of cereal grains by between 6 and 20 percentage units. Debroas and Blanchart (1993) found that neutral-detergent fibre (NDF) protein was degraded by proteolytic

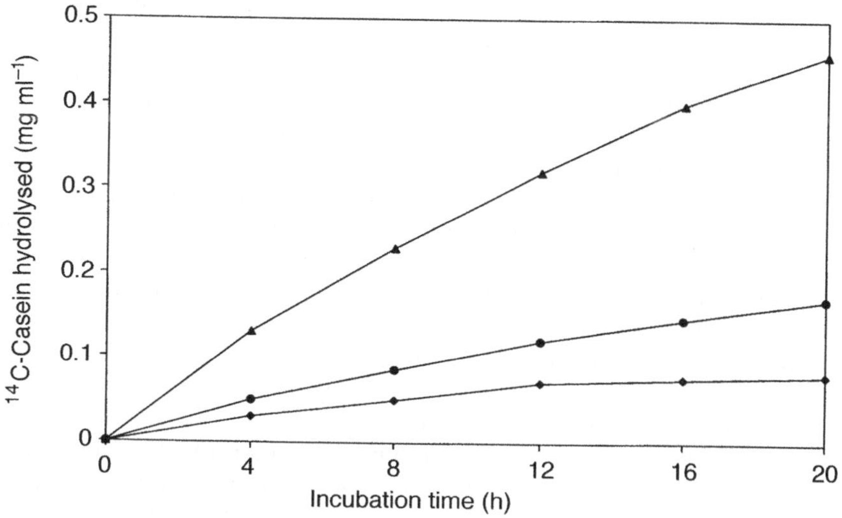

Fig. 13.1. Effect of pure cultures of *Selenomonas ruminantium* (◆), *Streptococcus bovis* (●) or a coculture of both (▲) on hydrolysis of carbon-14 (^{14}C)-casein (adapted from Wallace, 1985).

bacteria only after microbial depolymerization of cellulose began. Kohn and Allen (1995b) also reported an increase in protein degradation (from 42.4 to 53.1%) when cellulases were added to an *in vitro* proteolytic technique, and suggested that fibre interference may be particularly important for protein degradation during long-term incubation of forages. Therefore, it appears that protein degradation in the rumen requires the presence of several proteolytic and non-proteolytic enzymes. *In vitro* systems should provide these same activities to accurately simulate ruminal degradation of proteins.

Cell-free Enzymes versus Microbial-cell Preparations

Ruminal proteolysis is the result of the activity of an array of enzymes synthesized by bacteria, protozoa and fungi. Approximately 70–80% of ruminal microorganisms are attached to undigested feed particles in the rumen (Craig *et al.*, 1987), and 30–50% of these have proteolytic activity (Prins *et al.*, 1983). The role of protozoa in ruminal protein degradation is mostly related to the degradation of feed particles and bacterial proteins (Wallace and Cotta, 1988), but their overall effect on protein degradation is still controversial. The significance of the proteolytic activity of fungi has not been clearly established, but it is likely that they play an important role in the degradation of plant cell-wall proteins (Wallace and Munro, 1986). Development of an *in vitro* method should attempt to simulate the activities and interactions of these microorganisms. Several approaches have been used, from incubations with commercially available enzymes of non-ruminal origin to complex simulations of ruminal fermentation (Stern *et al.*, 1997).

There has been considerable interest in the development of techniques using commercial enzymes to predict protein degradation in the rumen (Table 13.1). These techniques should be cheaper, easy to standardize and completely independent from the animal. Krishnamoorthy *et al.* (1983) used a protease from *Streptomyces griseus*, because its endo- and exopeptidase activities were similar to those found on major proteolytic ruminal bacteria. Ruminal proteolysis *in vitro* was simulated by choosing an enzyme concentration (0.066 units ml^{-1}) that would provide a proteolytic activity similar to that of whole ruminal fluid. Feed samples were subjected to proteolysis for 18 or 48 h to resemble mean ruminal retention times for grain and roughages, respectively. Undegradable nitrogen estimated *in vitro* was moderately correlated (r^2 = 0.61) with values estimated *in vivo*, and suggested that the technique could be used for comparing ruminal protein degradation of feedstuffs on a relative basis. Other research (Aufrère and Cartailler, 1988; Terramoccia *et al.*, 1992; Susmel *et al.*, 1993) also reported good correlations between *S. griseus* protease and *in situ* estimates of protein degradation (Table 13.1). The most comprehensive validation study of the *S. griseus* protease *in vitro* technique was conducted by Aufrère *et al.* (1991). Ninety-seven concentrate feeds were incubated for 1 and 24 h with the *S. griseus* protease, and estimates of ruminal protein degradation were highly correlated to *in situ* results (r^2 = 0.89). In contrast, other authors reported moderate to poor correlations (from r^2 = 0.21 to r^2 = 0.67) between *S. griseus* protease and *in situ* estimates of protein degradation (Table 13.1), suggesting that this enzyme may not be specific for simulation of ruminal degradation across a wide variety of feeds. Fahmy *et al.* (1991) studied the pattern of protein degradation by *S. griseus* protease activity, using gel electrophoresis, and reported that the type of proteins degraded and resulting end-products were different *in situ* and *in vitro*, suggesting that the enzyme did not have the same specificity as that found in the rumen. Therefore, it appears that the relationship between *S. griseus* protease and *in situ* protein degradation is an empirical association, rather than a solid simulation of ruminal proteolytic activity. Few efforts have been made to validate this technique for forages. Aufrère *et al.* (1989) indicated that the use of the crude-protein content and estimates of *S. griseus* protease degradation could be used for predicting protein degradation of hay (r^2 = 0.79; n = 21).

Another commercial enzyme used for *in vitro* determination of protein degradation is a neutral protease extracted from *Bacillus subtilis*. Protein degradation by *B. subtilis* protease resulted in high correlations with ruminal degradation estimated *in situ* (Roe *et al.*, 1991, r^2 = 0.78; Assoumani *et al.*, 1992, r^2 = 0.92). However, these tests were conducted with few samples and the technique has not been tested with forages. A different neutral fungal protease extracted from *Aspergillus orizyae* (Poos-Floyd *et al.*, 1985) has also been tested in a small number of concentrate samples with positive results (Table 13.1).

Ficin is a plant protease with a wide spectrum of activity extracted from *Ficus glabatra*. Poos-Floyd *et al.* (1985) reported that the highest correlation between enzymatic and ruminal protein degradation *in vivo* was obtained after 1–4 hours digestion with ficin (r^2 = 0.76), but the relationship decreased as incubation time increased. Aufrère and Cartailler (1988) also observed a high correlation (r^2 = 0.94) between protein degradation obtained with 24 h of *in situ* versus *in vitro* ficin incubations, although variation was lower using the protease from *S. griseus*.

Table 13.1. Summary of validation studies conducted using commercially available enzymes.

Enzyme	Samples*	Comments	r^2 †	SE‡	Reference
S. griseus	12C/2F	pH = 8/18–48 h	0.61		Krishnamoorthy *et al.*, 1983
S. griseus	9C	1 h	0.67		Poos-Floyd *et al.*, 1985
S. griseus	12C	pH = 8/1–24 h	0.90	7.0	Aufrère and Cartailler, 1988
S. griseus	21F	pH = 8/1–24 h	0.79	3.9	Aufrère *et al.*, 1989
S. griseus	3C2F	pH = 8	0.07	21.0	Kopecny *et al.*, 1989
S. griseus	96C	pH = 8/1–24 h	0.89	5.8	Aufrère *et al.*, 1991
S. griseus	6C	12 h	0.21	88.0	Roe *et al.*, 1991
S. griseus	8C	pH = 6.5	0.91	8.1	Assoumani *et al.*, 1992
S. griseus	8C		0.77	9.6	Terramoccia *et al.*, 1992
S. griseus	10C	pH = 6.5	0.72		Susmel *et al.*, 1993
S. griseus	13FC	pH = 8	0.39	9.8	Tománková and Kopecny, 1995
Ficin	9C	4 h	0.76		Poos-Floyd *et al.*, 1985
Ficin	42 FC	4 h	0.04		Lindquist *et al.*, 1989
Ficin	4F	4 h	0.20		Lindquist *et al.*, 1989
Ficin	6C	12–24 h	0.01	4.2	Roe *et al.*, 1991
Ficin	23C/38F	pH = 7/2–3 h	0.92		Kosmala *et al.*, 1996
Bromelain	9C	1 h	0.72		Poos-Floyd *et al.*, 1985
Bromelain	13FC	pH = 7.2	0.73	6.5	Tománková and Kopecny, 1995
Bromelain	40C	pH = 7.2	0.70	7.1	Tománková and Kopecny, 1995
Bromelain	28C	pH = 7.2	0.19	10.2	Tománková and Kopecny, 1995
Bromelain	41F	pH = 7.2	0.53	5.8	Tománková and Kopecny, 1995
Bromelain	68F	pH = 7.2	0.55	7.3	Tománková and Kopecny, 1995
Bromelain	9F	pH = 7.2	0.77	4.5	Tománková and Kopecny, 1995
Papain	9C	1–4 h	0.72		Poos-Floyd *et al.*, 1985
Papain	3C2F	pH = 6	0.81	4.5	Kopecny *et al.*, 1989
Papain	13FC	pH = 6	0.59	8.1	Tománková and Kopecny, 1995
B. subtilis	6C	12h	0.78	30	Roe *et al.*, 1991
B. subtilis	8C	pH = 6.5	0.92	7.6	Assoumani *et al.*, 1992
A. orizyae	9C	4–8–24 h	0.90		Poos-Floyd *et al.*, 1985
Alkalase	3C2F	pH = 9	0.35	13.6	Kopecny *et al.*, 1989
Fromase	3C2F	pH = 6	0.14	6.5	Kopecny *et al.*, 1989
Chymosin	3C2F		0.00	21.0	Kopecny *et al.*, 1989
Trypsin	3C2F		0.00	14.6	Kopecny *et al.*, 1989
Trypsin and fromase	3C2F		0.09	8.2	Kopecny *et al.*, 1989

* Type of feed: C = concentrate; F = forage; the value indicates the number of samples used in each study.
† Coefficient of determination between *in vitro* and *in vivo* or *in situ*.
‡ Standard error of estimate.
S. griseus, Streptomyces griseus; B. subtilis, Bacillus subtilis; A. orizyae, Aspergillus orizyae.

Kosmala *et al.* (1996) showed that the incubation of feed samples with 9 or 18 units of ficin 100 mg^{-1} of sample crude protein from concentrates and forages, respectively, resulted in an overall $r^2 = 0.92$ ($n = 61$) compared with *in situ*. When the regression was performed only with forages ($n = 38$), the coefficient of determination was slightly lower ($r^2 = 0.85$), probably due to the potential interaction with fibre components of forages. These results disagree with observations by Lindquist *et al.* (1989) and Roe *et al.* (1991), as shown in Table 13.1.

Other enzymes tested, including fromase, alkalase, chymosin, or trypsin (Kopecny *et al.*, 1989), pepsin, pancreatin and protease type XIV (Siddons *et al.*, 1985), have not yielded positive results (Table 13.1). Luchini *et al.* (1996a) used a mixture of enzymes (*S. griseus* protease, chymotrypsin and proteinase K; or trypsin, carboxypeptidase B, chymotrypsin and carboxypeptidase A) in an attempt to provide similar proteolytic activity and specificity to those found in the rumen. However, these mixtures failed to simulate the proteolytic activity of strained ruminal fluid.

Two other plant proteases, bromelain and papain, have also been tested to estimate protein degradation in the rumen. Bromelain yielded moderate correlations (Table 13.1), ranging from $r^2 = 0.55$ for forages (Tománková and Kopecny, 1995) to $r^2 = 0.72$ for concentrates (Poos-Floyd *et al.*, 1985). Higher correlations were obtained by Kopecny *et al.* (1989) using papain in five different feeds ($r^2 = 0.81$). Poos-Floyd *et al.* (1985) also obtained moderate correlations ($r^2 = 0.72$) between *in vivo* estimates of protein degradation and papain, although ficin and neutral fungal protease had slightly better correlations. However, these results are in conflict with those reported by Tománková and Kopecny (1995).

Other factors, in addition to proteolytic enzymes, may affect protein degradation of forages in the rumen. For example, Tománková and Kopecny (1995) showed that the addition of α-amylase to a bromelain solution improved the coefficient of determination between *in situ* and bromelain digestion from 0.26 to 0.81. The positive effect of amylases on protein degradation has also been reported by other authors (Aufrère and Cartailler, 1988; Assoumani *et al.*, 1992). It appears that the use of α-amylase was necessary when studying protein degradation of feeds with a starch content of 230 g kg^{-1} or greater (Assoumani *et al.*, 1992). Although it has not been tested in forages, this observation may be of particular interest when measuring protein degradation of maize silage, where starch may account for as much as 400 g kg^{-1} of the dry matter (DM). Furthermore, the addition of cellulolytic enzymes resulted in an increase in protein degradation from 424 to 512 g kg^{-1} after 16 h incubation with ruminal enzymes (Kohn and Allen, 1995b). Similar results were obtained by Abdelgadir *et al.* (1996) when forages were pretreated with 8000 units of cellulase before undergoing an *in vitro* digestion with *S. griseus* protease. On the other hand, factors related to the incubation medium may also play an important role in *in vitro* protein degradation. Protein conformation and solubility, and therefore degradability, depend on the pH of the medium. Assoumani *et al.* (1992) indicated that, as pH was reduced from 8 to 6.5, correlation between *S. griseus* protease and *in situ* protein degradation increased from $r^2 = 0.68$ to $r^2 = 0.91$. The presence of reducing factors also affects *in vitro* protein degradation (Kohn and Allen, 1995b). These authors reported that, as the environment was more reduced, more protein was degraded (Fig. 13.2).

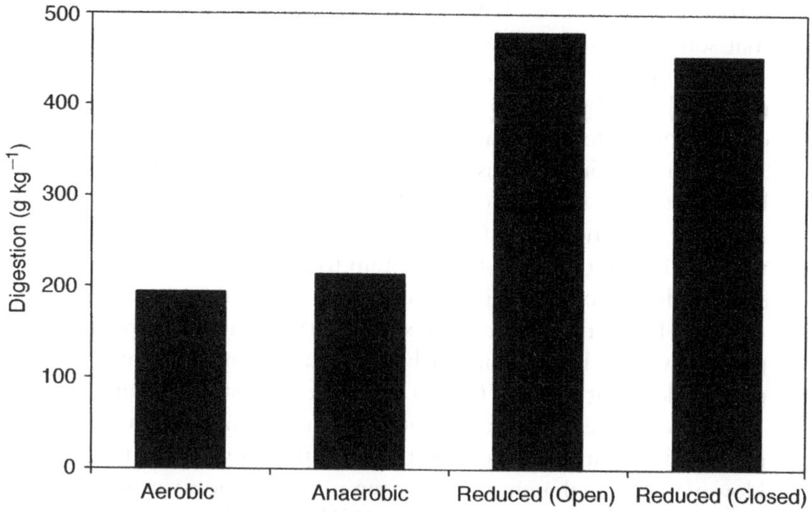

Fig. 13.2. Effect of the addition of a reducing agent to an *in vitro* incubation medium on protein degradation of soybean meal after 16 h digestion with enzymes extracted from the rumen (adapted from Kohn and Allen, 1995b).

It is apparent from data available on *in vitro* protein degradation that the use of non-ruminal enzymes and incubation conditions different from those of the rumen are unlikely to successfully mimic the complex interactions occurring during protein degradation by ruminal microbes. At best, these methods provide an empirical relationship with estimates of ruminal degradation of proteins. Because of these limitations, caution is required when using results obtained with these techniques (Mahadevan *et al.*, 1987; Roe *et al.*, 1991; Luchini *et al.*, 1996a).

A more physiological approach would be to use proteolytic enzymes isolated from the ruminal contents. Most protease and peptidase activities in ruminal contents are associated with the outer side of bacterial cell walls, and can be liberated by treatment with detergents or gentle shaking (Brock *et al.*, 1982; Kopecny and Wallace, 1982; Prins *et al.*, 1983). Susmel *et al.* (1993) used cell-free strained rumen fluid as inoculum and reported a moderate correlation (r^2 = 0.69) between *in situ* and *in vitro* estimates of protein degradation. Mahadevan *et al.* (1987) proposed an extraction method based on release of enzymes after acetone, butanol and chilled-water rinses. These extractions allowed the recovery of approximately 30–35% of the proteolytic activity of strained ruminal fluid. Enzymes could be stored at −20°C for at least a year without losing proteolytic activity. Kohn and Allen (1995a) slightly modified this method and increased recovery of proteolytic activity in ruminal fluid up to 62%. The proteolytic activity did not decrease with time of incubation, and the addition of new enzyme after 8 h of digestion did not increase estimates of protein degradation at 16 h (Kohn and Allen, 1995b). These observations suggest that enzymatic activity was not limiting, and that long-term incubation periods are possible. When this enzyme preparation was evaluated on feed samples (Kohn and Allen, 1995b), protein degradation was lower compared with values from the National Research Council (NRC, 1989). It was suggested that some enzymes may be selectively eliminated or inactivated by the extraction method, or that some enzymes may function differently when associated with the microbial membrane. However, the technique successfully showed relative differences between forages, stage of maturity and methods of forage preservation (Kohn and Allen, 1995c). The use of enzymes extracted from the rumen may provide the basis for further research on a technique that may be low in cost, fast and reliable. However, it is necessary to carefully observe methods of enzyme isolation, determine optimal ratio of substrate protein to proteolytic activity and provide all necessary conditions of the incubation medium to simulate protein degradation.

It is likely that the complex interactions involved in ruminal degradation of protein will be best simulated by live ruminal microorganisms. Initial studies on protein degradation used *in vitro* incubations of strained ruminal fluid (Little *et al.*, 1963). Feeds were incubated with the inoculum under anaerobic conditions and protein degradation was calculated from the ammonia nitrogen that was released. However, the final ammonia nitrogen concentration was the result of both protein degradation and ammonia nitrogen (NH_3-N) utilization by ruminal microbes, with the latter being greatly influenced by the amount and nature of carbohydrates fermented. For example, Broderick (1982) showed that readily fermentable carbohydrates, such as maize and sorghum grains, resulted in negative NH_3-N release because microbial uptake was greater than NH_3-N released.

To prevent nitrogen utilization by ruminal microorganisms, Broderick (1987) developed an *in vitro* inhibitor procedure (IIV) for estimating ruminal protein degradation by using chloramphenicol and hydrazine sulphate to inhibit microbial protein synthesis and amino acid and ammonia nitrogen utilization by ruminal microbes. Because neither of these inhibitors depresses proteolytic activity, the IIV method might be expected to yield useful estimates of protein degradation kinetics. A limited substrate incubation was used to guarantee the first-order kinetic reaction required for calculations of rates of degradation. Results demonstrated that soluble proteins have different rates of degradation and are not completely degraded. Using the IIV technique, Broderick *et al.* (1988) reported that rates of protein degradation of 18 protein supplements ranged from 0.103 to 0.813 h^{-1}, yielding estimates of protein degradation ranging from 730 to 940 g kg^{-1}. Rates of degradation estimated with the IIV technique were slightly higher, but were highly correlated with *in situ* estimates (r^2 = 0.92, n = 7). Broderick *et al.* (1988) suggested that the lower rates of degradation observed *in situ* could be partially explained by: (i) microbial growth inside the bags, which may reduce apparent rates of degradation; (ii) the microbial population inside the bag may be different compared with the rumen environment; and (iii) the efflux of material from the bag may leave a less degradable material inside the bag, but the soluble fraction may not be all degradable. As a result, the *in situ* method may overestimate degradation of the soluble fraction and underestimate rate of degradation. In contrast, the relatively short incubation time in the IIV (a maximum of 4 h) may overestimate rates of degradation of feeds containing a large pool of rapidly degraded proteins. The IIV technique has been successful in identifying differences in protein degradation of concentrates (Broderick *et al.*, 1988) and forages (Broderick and Buxton, 1991; Broderick and Albrecht, 1997). Recent data by Broderick and Albrecht (1997) provided information on rates of degradation of forage protein obtained *in vitro* that agree with those reported *in vivo* (Table 13.2).

However, the IIV technique has some limitations: (i) the accumulation of ammonia and amino acids may result in end-product inhibition of enzymes; (ii) microbial activity decreases with time due to the difficulty of maintaining anaerobic conditions and pH; (iii) growth inhibitors are expected to eventually interfere with enzyme synthesis and microbial growth, affecting protein degradation; (iv) degradation rates determined for feeds with high levels of ammonia and free amino acids (as in forage silages) are less accurate because breakdown of more slowly degraded residual proteins must be computed from the appearance of additional ammonia and amino acids in the presence of high background nitrogen; and (v) quantification of degradation rates from the slope of slowly degraded proteins (rates less than 0.01 h^{-1}) is less accurate.

Broderick and Clayton (1992) used alternative mathematical models based on the Michaelis–Menten approach to overcome some of the problems associated with limited substrate degradation. Results using this new approach were obtained more rapidly (2 h) and were more consistent with *in vivo* estimates of ruminal protein degradation. Dialysed ruminal fluid has also been used to reduce background nitrogen contamination and increase the accuracy of estimated rates of degradation (Broderick and Albrecht, 1997).

Neutze *et al.* (1993) modified the inhibitor *in vitro* method by estimating the degradable protein fraction as trichloracetic acid (TCA)-soluble nitrogen. Estimates of the soluble and insoluble potentially degradable fractions and rates of degradation were different from those obtained *in situ*, but estimated protein degradation was strongly correlated to *in situ* values ($r^2 = 0.75$). The coefficients of

Table 13.2. Estimates of the rate and extent of degradation of the potentially degradable pool of protein at two rates of passage (adapted from Broderick and Albrecht, 1997).

	n	B_0*	Rate of degradation† (h^{-1})	Escape protein‡	
				0.06 h^{-1}	0.03 h^{-1}
Alfalfa	2	94.3	0.219	20.3	11.4
White clover	1	94.6	0.269	17.3	9.5
Red clover	2	95.1	0.140	28.5	16.8
Kura clover	1	96.3	0.220	20.6	11.6
Cicer milkvetch	2	94.4	0.213	20.8	11.7
Crownvetch	1	93.2	0.219	20.0	11.2
Canada milkvetch	1	94.7	0.252	18.2	10.1
Birdsfoot trefoil	10	93.9	0.197	21.9	12.4
Grassland Maku	2	95.4	0.037	59.0	42.7
Sainfoin	6	95.2	0.027	65.9	50.4
Lespedeza	12	96.2	0.009	83.2	73.2

* Potentially degradable protein fractions.
† Estimates using the inhibitory *in vitro* technique (Broderick, 1987).
‡ Estimated as $B_0 \times (k_p/(k_p + k_d))$, at passage rates ($k_p$) of 0.06 and 0.03 h^{-1}.

variation of the degradation parameters were smaller *in vitro* than *in situ*, which indicates that the *in vitro* system was more accurate.

To avoid the use of inhibitors of nitrogen utilization by microbial cells, Hristov and Broderick (1994) developed a system using $^{15}NH_3$ to quantify microbial nitrogen utilization. The advantage of this method is that it allows for normal microbial growth and reduces the potential for end-product inhibition. Degradation rates were computed as appearance of NH_3-N plus net microbial protein synthesis. Results showed similar trends to the limited-substrate IIV method, but average degradation was about 28% greater in the uninhibited system, suggesting that the inhibition of microbial growth in the IIV technique results in a reduction of proteolysis. Estimates of protein degradation of seven protein supplements were consistent with previously reported values. This method may be of particular interest in the analysis of samples with high non-protein nitrogen content as percentage of total protein, such as legume and grass silages, although the technique still needs to be tested on these types of samples. Peltekova and Broderick (1996) used this modified procedure to estimate the rate and extent of protein degradation in lucerne (alfalfa) hay and silage, using a two-compartment model. Estimates of ruminal protein degradation were comparable with NRC (1989) values. An *in vitro* method based on use of $^{15}NH_3$ to estimate microbial nitrogen utilization may serve as a more theoretically sound procedure with which to compare other *in vitro* tests that estimate protein degradability.

Raab *et al.* (1983) used a different approach to determine nitrogen utilization by bacteria. Their method consisted of adding graded amounts of fermentable carbohydrate to an *in vitro* fermentation system with the protein source to be tested. Gas production and NH_3-N concentrations were measured up to 24 h. Extrapolation of NH_3-N release to zero gas production was assumed to correct for NH_3-N and amino acid utilization for microbial growth, because zero gas production represents no incorporation of nitrogen into microbial protein. The technique assumes that gas production is a reflection of microbial growth and that this relationship is linear over the conditions of the experiment. Estimates of protein degradation were similar to those obtained *in vivo*. However, a total of 15 incubation tubes were required per sample tested, which may impose a limit on its routine application.

Nuget and Mangan (1981) used an *in vitro* fermentation system to test if rates of proteolysis could be measured *in vitro*. Rates of degradation of fraction I leaf protein compared well with previously reported estimates obtained *in vivo* and *in vitro*. Results using this *in vitro* system showed an initial rate of proteolysis that was approximately 82% of that *in vivo*, and only slowed significantly after 2 h incubations. When the fraction I leaf protein was labelled with carbon-14 (^{14}C), the system allowed for a precise study of the kinetics of protein degradation.

All batch-culture systems require the use of a microbial inoculum and incubation conditions similar to those of the ruminal environment. Because about 75% of the proteolytic activity was recovered in the particulate fraction of the rumen contents (Brock *et al.*, 1982), it is necessary to obtain ruminal samples with solid-phase-associated bacteria. Chilling and blending of whole rumen contents will suffice to recover a large amount of its proteolytic activity (Craig *et al.*, 1984). Estimates of protein degradation may be affected by the donor animal, time of ruminal fluid collection and type of diet (Wallace and Cotta, 1988). However, Furchtenicht and Broderick (1987) indicated that changes in proteolytic activity of the inoculum were largely due to its microbial numbers and that the variation in proteolytic activity among inocula obtained from cows fed different diets could be reduced by standardizing the inoculum to a constant microbial weight. Luchini *et al.* (1996b) tested the possibility of using preserved ruminal microorganisms as inoculum. The use of frozen microbes in the IIV technique of Broderick (1987) resulted in lower proteolytic activity, but feeds were ranked in the same order as with fresh ruminal fluid. This approach may be promising, but needs further validation. Other factors, such as pH and anaerobic and reducing conditions, also affect microbial protein degradation and should be carefully controlled to allow for optimal microbial fermentation (Grant and Mertens, 1992). Available information suggests that the refinement of *in vitro* ruminal-fluid incubations may provide an appropriate method for measuring ruminal protein degradation. However, the need for ruminally cannulated animals still creates a problem for routinely using *in vitro* ruminal-fluid incubations by industry.

There have been very few attempts to use a combination of pure microbial cultures to simulate ruminal fermentation. If *in vitro* rumen microbial fermentation is able to provide reasonable estimates of protein degradation, it is possible that a combination of major proteolytic, peptidolytic, amylolytic and cellulolytic bacteria may provide the basis for a standardized *in vitro* technique based on ruminal microbial activities. At the present time, only the research conducted by Laycock *et al.* (1985) has attempted this approach. A mixture of *Butyribivrio* sp. and *Streptococcus bovis* was used to predict protein degradation of soybean meal. However, after 24 h incubation, estimates of protein degradation were about half of those obtained *in situ*. It is possible that a combination of other microorganisms may improve estimates of protein degradation, but this hypothesis needs to be examined further.

The most physiological approach to determining protein degradation *in vitro* would be to design a system that carefully simulates ruminal fermentation. Various continuous-culture fermentation systems have been designed to simulate the ruminal environment, enabling the study of ruminal microbial ecology and digestion of nutrients (Hoover *et al.*, 1976; Czerkawski and Breckenridge, 1977; Teather and Sauer, 1988; Fuchigami *et al.*, 1989; Miettinen and Setälä, 1989). Advantages of these systems compared with *in vivo* measurements include reduction in cost, time and variation among experimental units. Also, there are no complications from endogenous sources, and digesta flow markers are not required, because passage rates are regulated and measured directly. However, similar to *in vivo* measurements, reliable techniques are required for isolation of microbial cells and for differentiation of effluent digesta into microbial and dietary N fractions. The most commonly used continuous-culture systems are the single-flow, Rusitec (Czerkawski and Breckenridge, 1977) and the dual-flow continuous culture (Hoover *et al.*, 1976). The Rusitec system has a single outflow, and residence time in the rumen is simulated by placing feedstuffs into nylon bags inside the reaction vessel for 48 h. Recently, Prevot *et al.* (1994) evaluated the Rusitec and concluded that the system cannot reproduce the *in vivo* fermentation because ciliated protozoal populations, and probably many bacterial species, are eliminated from the fermenter. The dual-flow continuous-culture system developed by Hoover *et al.* (1976) has been successful in maintaining ruminal fermentation parameters and microbial populations similar to those found in the rumen. Hannah *et al.* (1986) and Mansfield *et al.* (1995) compared ruminal fermentation patterns and nutrient flow between *in vivo* and the dual-flow continuous-culture fermenters. The continuous-culture system was able to obtain digestibility values for organic matter, total non-structural carbohydrate, fibre and protein within physiological ranges, although some digestibility measurements were significantly different compared with *in vivo* data. Some of the differences in fermentation parameters observed between the dual-flow continuous-culture fermenters and the rumen can be attributed to lack of absorptive capacity and defaunation *in vitro*. However, it should be noted that interpretation of results *in vitro* and *in vivo* were similar for 80% of the individual parameters evaluated, supporting the dual-flow continuous-culture system as an excellent model for studying ruminal microbial fermentation (Mansfield *et al.*, 1995). Although the protozoal population was much lower in fermenters, Mansfield *et al.* (1995) found small differences between *in vivo* and *in vitro* systems in total and cellulolytic counts and no differences in amylolytic, proteolytic and fungal counts. High protozoal numbers have been accomplished by intermittent mixing and compartmentalization (Teather and Sauer, 1988; Fuchigami *et al.*, 1989). These systems are excellent tools for research and modelling of ruminal fermentation, but are elaborate and expensive, require inoculation with ruminal digesta and may not be suitable for routine analysis of microbial digestion for individual feed ingredients.

Can Enzymes Estimate Rate of Degradation?

Rates of energy and protein degradation affect energy and protein utilization by ruminal microorganisms and the animal. Feeds are composed of a mixture of proteins that may have different degradation patterns. Pitchard and Van Soest (1977) proposed that there are different pools of proteins with different rates of degradation within each feed. Overall degradation in the rumen is a function of pool sizes and rates of degradation and passage of the potentially degradable pools.

Determination of rates of degradation requires that the digestion process follows first-order kinetics. Therefore, *in vitro* procedures should be designed to contain an excess of enzyme (Krishnamoorthy *et al.*, 1983). Krishnamoorthy *et al.* (1983) and Aufrère and Cartailler (1988) observed that rates of degradation with the *S. griseus* protease were higher than those obtained by the *in situ* method. Similarly, Kosmala *et al.* (1996) and Coblentz *et al.* (1997) observed that degradation rates of feed proteins were faster with ficin than those estimated *in situ*. Rates of degradation estimated with ficin were inconsistent with those obtained *in situ*, suggesting that ficin was not suitable for measuring dynamics of protein degradation in the rumen (Kosmala *et al.*, 1996). Roe *et al.* (1991) compared the dynamics of protein degradation *in situ* and with *S. griseus*, ficin and neutral proteases. However, none of these *in vitro* tests resulted in degradation curves that had consistent relationships with those generated by the *in situ* technique. Inconsistencies in rates of degradation were also found when untreated and chemically treated soybean-meal protein fractions were digested *in vitro* by bacterial proteases, pepsin or trypsin (Lynch *et al.*, 1988). Several authors suggested that adjusting the protease concentration to provide a similar proteolytic activity to that found in the rumen could provide better estimates of rates of protein degradation (Krishnamoorthy *et al.*, 1983; Kosmala *et al.*, 1996; Coblentz *et al.*, 1997). Mahadevan *et al.* (1987) tested rates of degradation of insoluble feed protein, using enzymes extracted from the rumen and an *S. griseus* protease. Results revealed considerable differences in degradation rates between the two techniques. However, the ruminal enzymes were more consistent with reported values. It was suggested that the use of non-ruminal enzymes may not be adequate for the determination of rates of protein degradation. Kohn and Allen (1995b) also obtained more reasonable estimates of rates of degradation of soybean meal and lucerne hay proteins (0.15 and 0.06 h^{-1}, respectively) using 2 h incubations with ruminal enzyme extract. At the present time, it appears that the IIV technique of Broderick (1987), with recent modifications (Hristov and Broderick, 1994; Broderick and Albrecht, 1997), may be the most adequate *in vitro* method for determining rates of protein degradation. The use of enzymes extracted from the ruminal fluid may also provide reasonable estimates of rates of degradation, but the technique needs further testing.

The question that needs to be answered is how accurate should the measurements of rates of protein degradation be. From a scientific standpoint, accuracy should always be an objective. For field application, accuracy will depend on its impact on estimates of effective ruminal degradation. Figure 13.3 demonstrates the effect of different rates of passage and degradation on changes in estimates of effective ruminal degradation. A feed protein with 20% soluble protein, 60% potentially degradable fraction and 20% undegradable fraction was used for the model. Degradation was calculated using the model of Ørskov and McDonald (1979).

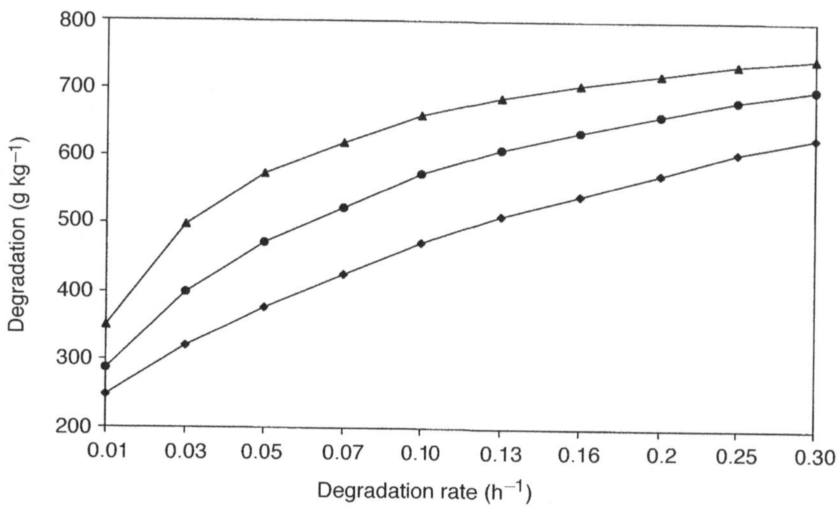

Fig. 13.3. Effect of rate of degradation on estimated effective degradation at different passage rates: forage-particles pool (k_p = 0.03 h^{-1}, ▲); concentrate-particles pool (k_p = 0.06 h^{-1}, ●); and liquid pool (k_p = 0.12 h^{-1}, ◆).

From Fig. 13.3, it can be concluded that: (i) for all rates of passage, the importance of an accurate estimate of degradation rate is greater at low degradation rates; and (ii) rate of degradation is more important than rate of passage at low degradation rates, but the opposite occurs at high degradation rates. Therefore, small errors in the estimation of rates of degradation in the insoluble fraction of forages (low passage rates) may have a great impact on estimates of degradability within a range of rates of degradation between 0 and 0.15 h^{-1}. For example, a change from 0.03 to 0.05 h^{-1} in degradation rate of an insoluble forage protein with a rate of passage of 0.03 h^{-1} results in an increase in degradation from 500 to 575 g kg^{-1} (an increase of 75 g kg^{-1}). In contrast, the same increment from 0.20 to 0.23 h^{-1} only affects estimates of degradation from 722 to 731 g kg^{-1} (an increase of 0.9 percentage units). In contrast, soluble protein flows with the liquid fraction and rates of passage are higher (i.e. 0.12 h^{-1}). Under these conditions, an error in the estimation of rates of degradation will have a smaller effect on estimated degradability, although this effect will influence estimates of degradation over a wider range of degradation rates (Fig. 13.3). Unfortunately, *in vitro* techniques are less accurate as degradation rate decreases. It is important to keep these observations in mind when attempting to improve *in vitro* techniques for industry application.

Intestinal Digestion

Protein supply from forages represents a high proportion of total protein intake. In preserved or heat-damaged forages, the amount of undegradable protein reaching the small intestine may be considerable. The large differences in intestinal digestion of undegraded protein from forages (from 554 g kg^{-1} to 860 g kg^{-1}; Frydrych, 1992) justify the need to incorporate these estimates into new feeding systems. *In vivo* and *in situ* techniques to measure intestinal digestion of proteins and their advantages and disadvantages have been discussed elsewhere (Stern *et al.*, 1997).

The Association of Official Analytical Chemists (AOAC, 1984) approved a standardized pepsin digestion procedure for estimating unavailable nitrogen. Pepsin-insoluble nitrogen (PIN) was strongly correlated to total-tract unavailable nitrogen in forages (Goering *et al.*, 1972; Shelford *et al.*, 1980). In contrast, Britton *et al.* (1986) and Zinn and Owens (1982) indicated that PIN was a poor predictor of total-tract crude protein (CP) digestibility. Zinn and Owens (1982) reported that about 200 g kg^{-1} of the nitrogen leaving the ileum was soluble in acid–pepsin solution, suggesting that PIN may underestimate unavailable protein of feeds. In addition, total-tract digestion does not provide information on site of protein digestion in the gastrointestinal tract.

Because of the limited spectrum of specificity of any single enzyme, it is unlikely that single enzymatic techniques will be able to predict protein digestion in the small intestine. The most commonly used enzymatic technique was developed by Akeson and Stahmann (1964). The system was designed to simulate abomasal (pepsin) and intestinal (pancreatin) digestion. Results were highly correlated (n = 12; r^2 = 0.98) with rat growth studies. Antoniewicz *et al.* (1992) compared results from the mobile-bag technique with pepsin–pancreatin digestion in four protein supplements treated with five different doses of formaldehyde. Results obtained with the two techniques were strongly correlated ($r^2 = 0.90$). Van Straalen *et al.* (1993) also reported a strong correlation ($n = 28$; $r^2 = 0.83$) between the mobile-bag technique and a pepsin–pancreatin digestion test.

More recently, Calsamiglia and Stern (1995) developed a three-step procedure, modified from that of Akeson and Stahmann (1964), to estimate intestinal digestion of proteins in ruminants. After 16 h of ruminal preincubation, samples were incubated for 1 h in a hydrochloric acid (HCl)–pepsin solution, followed by a 24 h phosphate buffer–pancreatin digestion. Results of the pancreatin digestion step were highly correlated ($r^2 = 0.83$) with 34 duodenal samples from which small-intestinal protein digestion was determined *in vivo* (Fig. 13.4). The technique has been succesfully applied to numerous feeds (Yoon *et al.*, 1994; Howie *et al.*, 1996; Calsamiglia and Stern, 1997; Calsamiglia *et al.*, 1997).

Further Research Needed

All *in vitro* techniques currently available for protein quality evaluation have limitations (Stern *et al.*, 1997). Efforts should continue in the search of a scientifically sound technique that allows for further understanding of protein degradation and utilization by ruminal microbes and the ruminant animal. Aspects of protein digestion that should be studied in the near future include the following.

1. The search for a fast, inexpensive and reliable *in vitro* technique should continue. It is likely that current techniques may be sufficient for specific purposes (e.g. industry quality control). However, few techniques have been tested and validated

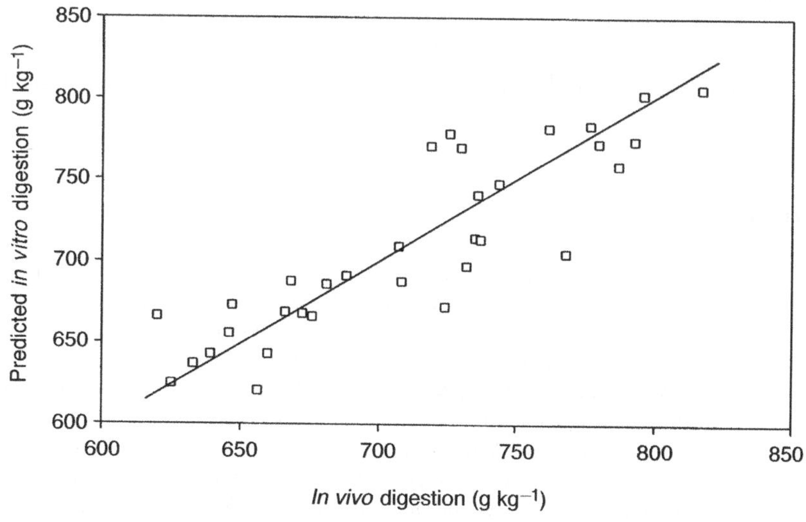

Fig. 13.4. Observed *in vivo* intestinal digestion of protein versus predicted value, using the three-step *in vitro* procedure ($n = 34$; $r^2 = 0.83$) (from Calsamiglia and Stern, 1995).

for forages. Recent research suggests that enzymatic techniques should carefully consider the use of combinations of proteolytic, peptidolytic and non-proteolytic enzymes (amylases and cellulases) in a medium where conditions (pH, redox potential, etc.) are maintained similar to those of the ruminal environment.
2. New techniques should be developed to determine pool sizes and their rates of degradation and ruminal passage. The application of pool sizes and their respective rates of degradation and passage may allow for modelling of the dynamics of protein availability in the rumen and true protein passage to the small intestine.
3. Feeding systems are incorporating daily requirements of amino acids. None of the *in vitro* techniques currently available has addressed the issue of amino acid profile of ruminally undegraded and intestinally digested protein. Therefore, efforts should also concentrate on the development of *in vitro* techniques capable of determining the amino acid profile of the undegraded protein in the rumen and of the digestible protein in the small intestine.

Cost Implications

One of the major advantages of *in vitro* techniques is the reduction in cost and labour compared with *in vivo* and *in situ* procedures. Cost of commercially available enzymes and reagents per analysis is generally very low and many samples may be processed at the same time, making these techniques suitable for industry application. The need for ruminal fluid in several of the techniques discussed is still a problem for their routine application in the field. However, when compared with other *in vivo* or *in situ* techniques, these *in vitro* techniques are still generally faster and cheaper. The cost of determining rates of degradation is higher than measuring extent of degradation, because more measurements are required. Some techniques (e.g. those of Raab *et al.*, 1983; Hristov and Broderick, 1994) are more complex and expensive, but may have the potential of yielding results of rates of degradation and effective degradability more in accordance with the physiological mechanisms found *in vivo*.

References

Abdelgadir, I.E.O., Cochran, R.C., Titgemeyer, E.C. and Vanzant, E.S. (1996) Comparison of ruminal protein degradability of alfalfa and prairie hay determined *in vivo* and *in vitro* via a commercial protease in the presence or absence of cellulase or driselase. *Journal of Animal Science* 74 (Suppl. 1), 79 (Abstract).

Akeson, W.R. and Stahmann, M.A. (1964) A pepsin pancreatin digest index of protein quality evaluation. *Journal of Nutrition* 83, 257–261.

Antoniewicz, A.M., van Vuuren, A.M., van der Koelen, C.J. and Kosmala, I. (1992) Intestinal digestibility of rumen undegraded protein of formaldehyde treated feedstuffs measured by mobile bag and *in vitro* technique. *Animal Feed Science and Technology* 39, 111–124.

AOAC (1984) *Official Methods of Analysis*, 14th edn. AOAC, Washington, DC.

ARC (1984) *The Nutrient Requirements of Ruminant Livestock, Supplement No. 1.* Commonwealth Agicultural Bureau, Slough, England.

Assoumani, M.B., Vedeau, F., Jacquot, L. and Sniffen, C.J. (1992) Refinement of an enzymatic method for estimating the theoretical degradability of proteins in feedstuffs for ruminants. *Animal Feed Science and Technology* 39, 357–368.

Aufrère, J. and Cartailler, D. (1988) Mise au point d'une méthode de laboratoire de prévision de la dégradabilité des protéines alimentaires des aliments concentrés dans le rumen. *Annales de Zootechnie* 37, 255–270.

Aufrère, J., Michalet-Doreau, B., Graviou, D. and Vérité, R. (1989) Predicting *in sacco* degradability of hay protein by chemical or enzymatic methods. In: *XVI International Grassland Congress, Nice, France.* Association Française pour la Production Fourragere, France, pp. 887–888.

Aufrère, J., Graviou, D., Demarquilly, C., Vérité, R., Michalet-Doreau, B. and Chapoutot, P. (1991) Predicting *in situ* degradability of feed proteins in the rumen by two laboratory methods (solubility and enzymatic degradation). *Animal Feed Science and Technology* 33, 97–116.

Britton, R.A., Klopfenstein, T.J., Cleale, R., Goedeken, F. and Wilkerson, V. (1986) Methods of estimating heat damage in protein sources. Proc. Dist. Gr. Conf., *Cincinnati, Ohio* 41, 67–75.

Brock, F.M., Forsberg, C.W. and Buchanan-Smith, J.G. (1982). Proteolytic activity of rumen microorganisms and effects of proteinase inhibitors. *Applied Environmental Microbiology* 44, 561–569.

Broderick, G.A. (1982) Estimation of protein degradation using *in situ* and *in vitro* methods. In: *International Symposium Proceedings: Protein Requirements for Cattle.* Miscellaneous Publication No. 109, Oklahoma State University, Stillwater, p. 72.

Broderick, G.A. (1987) Determination of protein degradation rates using a rumen *in vitro* system containing inhibitors of microbial nitrogen metabolism. *British Journal of Nutrition* 58, 463–475.

Broderick, G.A. and Albrecht, K.A. (1997) Ruminal *in vitro* degradation of protein in tannin-free and tannin containing forage legume species. *Crop Science* 37, 1884–1891.

Broderick, G.A. and Buxton, D.R. (1991) Genetic variation in alfalfa for ruminal protein degradability. *Canadian Journal of Plant Science* 71, 755–760.

Broderick, G.A. and Clayton, M.K. (1992). Rumen protein degradation rates estimated by non-linear regression analysis of Michaelis–Menten *in vitro* data. *British Journal of Nutrition* 67, 27–42.

Broderick, G.A., Wallace, R.J., Ørskov, E.R. and Hansen, L. (1988) Comparison of estimate of ruminal protein degradation by *in vitro* and *in situ* methods. *Journal of Animal Science* 66, 1739–1745.

Calsamiglia, S. and Stern, M.D. (1995) A three-step *in vitro* procedure for estimating intestinal digestion of protein in ruminants. *Journal of Animal Science* 73, 1459–1465.

Calsamiglia, S. and Stern, M.D. (1997) Development, validation and application of a three step *in vitro* procedure for estimating intestinal digestion of protein in ruminants. In: *International Symposium:* In vitro *Techniques for Measuring Nutrient Supply to Ruminants, 8–10 July 1997, University of Reading, Reading, UK.* British Society of Animal Science, Penicuik, p. 20.

Calsamiglia, S., Ferret, A., Sensada, M., Miquel, N. and Doblas, A. (1997) Ruminal degradation and intestinal digestion of protein supplements commonly fed to ruminants. *Journal of Dairy Science* 80 (Suppl. 1), 399 (Abstract).

Coblentz, W.K., Fritz, J.O., Cochran, R.C., Rooney, W.L. and Bolsen, K.K. (1997) Protein degradation in response to spontaneous heating in alfalfa hay by *in situ* and ficin methods. *Journal of Dairy Science* 80, 700–713.

Craig, W.M., Hong, B.J., Broderick, G.A. and Bula, R.J. (1984) *In vitro* inoculum enriched with particle-associated microorganisms for determining rates of fibre digestion and protein degradation. *Journal of Dairy Science* 67, 2902–2909.

Craig, W.M., Broderick, G.A. and Ricker, D.B. (1987) Quantitation of microorganisms associated with particulate phase of ruminal digesta. *Journal of Nutrition* 117, 56–62.

Czerkawski, J.W. and Breckenridge, G. (1977) Design and development of a long-term rumen simulation technique (Rusitec). *British Journal of Nutrition* 38, 371–389.

Debroas, D. and Blanchart, G. (1993) Interactions between proteolytic and cellulolytic rumen bacteria during hydrolysis of plant cell wall protein. *Reproduction and Nutrition Developments* 33, 283–288.

Depardon, N., Debroas, D. and Blanchart, G. (1996) Breakdown of peptides from a casein hydrolysates by rumen bacteria: simultaneous study of enzyme activities and physiological parameters. *Reproduction and Nutrition Developments* 36, 457–466.

Fahmy, W.G., Aufrère, J., Graviou, D., Demarquilly, C. and El-Shazly, K. (1991) Comparison between the mechanism of protein degradation of two cereals by enzymatic and *in situ* methods, using gel electrophoresis. *Animal Feed Science and Technology* 35, 115–130.

Frydrych, Z. (1992) Intestinal digestibility of rumen undegraded protein of various feeds as estimated by the mobile bag technique. *Animal Feed Science and Technology* 37, 161–172.

Fuchigami, M., Senshu, T. and Horiguchi, M. (1989) A simple continuous culture system for rumen microbial digestion study and effects of defaunation and dilution rates. *Journal of Dairy Science* 73, 3070–3078.

Furchtenicht, J.E. and Broderick, G.A. (1987). Effect of inoculum preparation and dietary energy on microbial numbers and rumen protein degradation activity. *Journal of Dairy Science* 70, 1404–1410.

Goering, H.K., Gordon, C.H., Hemken, R.W., Waldo, D.R., van Soest, P.J. and Smith, L.W. (1972) Analytical estimates of nitrogen digestibility in heat damaged forages. *Journal of Dairy Science* 55, 1275–1280.

Grant, R.J. and Mertens, D.R. (1992) Impact of *in vitro* fermentation techniques upon kinetics of fibre digestion. *Journal of Dairy Science* 75, 1263–1272.

Hannah, S.M., Stern, M.D. and Ehle, F.L. (1986) Evaluation of a dual flow continuous culture system for estimating bacterial fermentation *in vivo* of mixed diets containing various soya bean products. *Animal Feed Science and Technology* 16, 51–62.

Hoover, W.H., Crooker, B.A. and Sniffen, C.J. (1976) Effects of differential solid–liquid removal rates on protozoal numbers in continuous culture of rumen contents. *Journal of Animal Science* 43, 528–534.

Howie, S.A., Calsamiglia, S. and Stern, M.D. (1996) Variation in ruminal degradation and intestinal digestion of animal byproduct proteins. *Animal Feed Science and Technology* 63, 1–7.

Hristov, A. and Broderick, G.A. (1994) *In vitro* determination of ruminal protein degradability using [15N]-ammonia to correct for microbial nitrogen uptake. *Journal of Animal Science* 72, 1344–1354.

INRA (1989) *Ruminant Nutrition: Recommended Allowances and Feed Tables*, edited by R. Jarrige. Libbey Eurotex, Paris.

Kohn, R.A. and Allen, M.S. (1995a) Enrichment of proteolytic activity relative to nitrogen in preparations from the rumen for *in vitro* studies. *Animal Feed Science and Technology* 52, 1–14.

Kohn, R.A. and Allen, M.S. (1995b) *In vitro* protein degradation of feeds using concentrated enzymes extracted from rumen contents. *Animal Feed Science and Technology* 52, 15–28.

Kohn, R.A. and Allen, M.S. (1995c) Effect of plant maturity and preservation method on *in vitro* protein degradation of forages. *Journal of Dairy Science* 78, 1544–1551.

Kopecny, J. and Wallace, R.J. (1982) Cellular location and some properties of proteolytic enzymes of rumen bacteria. *Applied Environmental Microbiology* 43, 1026–1033.

Kopecny, J., Vencil, B., Kyselova, J. and Bcezina, P. (1989) Determination of rumen degradable protein with enzymes. *Archives of Animal Nutrition* 39, 635–645.

Kosmala, I., Antoniewicz, A., De Boever, J., Hvelplund, T. and Kowalczyk, J. (1996) Use of enzymatic solubility with ficin (EC 3.4.22.3) to predict *in situ* feed protein degradability. *Animal Feed Science and Technology* 59, 245–254.

Krishnamoorthy, U., Sniffen, C.J., Stern, M.D. and Van Soest, P.J. (1983) Evaluation of a mathematical model of rumen digestion and an *in vitro* simulation of rumen proteolysis to estimate the rumen-undegraded nitrogen content of feedstuffs. *British Journal of Nutrition* 50, 555–568.

Laycock, K.A., Hazlewood, G.P. and Miller, E.L. (1985) Potential use of proteolytic rumen bacteria for assessing feed protein degradability *in vitro*. *Proceedings of Nutrition Society* 44, 54 (Abstract).

Lindquist, K.J., Crooker, B.A., Illg, D.J. and Stern, M.D. (1989) Evaluation of ficin assay for estimating ruminal degradation of dietary protein. *Journal of Animal Science* 67 (Suppl. 1), 496 (Abstract).

Little, C.O., Borroughs, W. and Woods, W. (1963) Nutritional significance of soluble nitrogen in dietary proteins for ruminants. *Journal of Animal Science* 22, 358–363.

Luchini, N.D., Broderick, G.A. and Combs, D.K. (1996a) Characterization of the proteolytic activity of commercial proteases and strained ruminal fluid. *Journal of Animal Science* 74, 685–692.

Luchini, N.D., Broderick, G.A. and Combs, D.K. (1996b) Preservation of ruminal microorganisms for *in vitro* determination of ruminal protein degradation. *Journal of Animal Science* 74, 1134–1143.

Lynch, G.L., Van der Aar, P.J., Berger, L.L., Fahey, G.C. and Merchen, N.R. (1988) Proteolysis of alcohol-treated soybean meal proteins by *Bacteroides ruminocola*, *Bacteroides amilophilus*, pepsin, trypsin, and in the rumen of steers. *Journal of Dairy Science* 71, 2416–2427.

McKain, N., Wallace, R.J. and Watt, N.D. (1992) Selective isolation of bacteria with dipeptidyl aminopeptidase type I activity from the sheep rumen. *FEMS Microbiology Letters* 95, 169–174.

Madsen, J. (1985) The basis for the proposed Nordic Protein Evaluation System for ruminants: the AAT–PBV system. *Acta Agricultura Scandinavica* Suppl. 25, 9–20.

Madsen, J. and Hvelplund, T. (1994) Prediction of *in situ* protein degradability in the rumen: results of a European ringtest. *Livestock Production Science* 39, 201–212.

Mahadevan, S., Sauer, F.D. and Erfle, J.D. (1987) Preparation of protease from mixed rumen microorganisms and its use for the *in vitro* determination of the degradability of true protein in feedstuffs. *Canadian Journal of Animal Science* 67, 55–64.

Mansfield, H.R., Endres, M.I. and Stern, M.D. (1995) Comparison of microbial fermentation in the rumen of dairy cows and dual flow continuous culture. *Animal Feed Science and Technology* 55, 47–66.

Miettinen, H. and Setälä, J. (1989) Design and development of a continuous culture system for studying rumen fermentation. *Journal of Agricultural Science, Finland* 61, 463–473.

Neutze, S.A., Smith, R.L. and Forbes, W.A. (1993) Application of an inhibitor *in vitro* method for estimating rumen degradation of feed protein. *Animal Feed Science and Technology* 40, 251–265.

NRC (1989) *Nutrient Requirements of Dairy Cattle*, 6th revised edn. National Academy of Science, Washington, DC.

Nuget, J.H.A. and Mangan, J.L. (1981) Characteristics of the rumen proteolysis of fraction I (18S) leaf protein from lucerne (*Medicago sativa* L.). *British Journal of Nutrition* 46, 39–58.

O'Connor, J.D., Sniffen, C.J., Fox, D.G. and Chalupa, W. (1993) A net carbohydrate and protein system for evaluating cattle diets. IV. Predicting amino acid adequacy. *Journal of Animal Science* 71, 1298–1311.

Ørskov, E.R. and McDonald, I. (1979) The estimation of protein degradability in the rumen from incubation measurements weighted according to rate of passage. *Journal of Agricultural Science, Cambridge* 92, 499–503.

Peltekova, V.D. and Broderick, G.A. (1996) *In vitro* ruminal degradation and synthesis of protein on fractions extracted from alfalfa hay and silage. *Journal of Dairy Science* 79, 612–619.

Pitchard, G. and Van Soest, P.J. (1977) Protein solubility of ruminant feeds. In: *Proceedings of the Cornell Nutrition Conference, November 1977, Syracuse, New York.* Cornell University, Ithaca, New York, pp. 91–98.

Poos-Floyd, M., Klopfenstein, T. and Britton, R.A. (1985) Evaluation of laboratory techniques for predicting ruminal protein degradation. *Journal of Dairy Science* 68, 829–839.

Prevot, S., Senaud, J., Bohatier, J. and Prensier, G. (1994) Variation in the composition of the ruminal bacterial microflora during the adaptation phase in an artificial fermenter (Rusitec). *Zoological Science* 11, 871–882.

Prins, R.A., van Rheenen, D.L. and van Klooster, A.T. (1983) Characterization of microbial proteolytic enzymes in the rumen. *Antonie van Leeuwenhoek* 49, 585–595.

Raab, L., Cafantaris, B., Jilg, T. and Menke, K.H. (1983) Rumen protein degradation and biosynthesis. 1. A new method for determination of protein degradation in rumen fluid *in vitro*. *British Journal of Nutrition* 50, 569–582.

Roe, M.B., Chase, L.E. and Sniffen, C.J. (1991) Comparison of *in vitro* techniques to the *in situ* technique for estimation of ruminal degradation of protein. *Journal of Dairy Science* 74, 1632–1640.

Shelford, J.A., Cathcart, E.B., France, R.T. and Dobbs, J.L. (1980) A survey of composition and heat damage in dehydrated alfalfa products over the growing season from two B.C. processing plants. *Canadian Journal of Animal Science* 60, 197–200.

Siddons, R.C., Paradine, J., Gale, D.L. and Evans, R.T. (1985) Estimation of the degradability of dietary protein in the sheep rumen by *in vivo* and *in vitro* procedures. *British Journal of Nutrition* 54, 545–561.

Sniffen, C.J., O'Connor, J.D., Van Soest, P.J., Fox, D.G. and Russell, J.B. (1992) A net carbohydrate and protein system for evaluating cattle diets: II. Carbohydrate and protein availability. *Journal of Animal Science* 70, 3562–3577.

Stern, M.D., Bach, A. and Calsamiglia, S. (1997) Alternative techniques for measuring nutrient digestion in ruminants. *Journal of Animal Science* 75, 2256–2276.

Susmel, P., Mills, C.R., Colitti, M. and Stefanon, B. (1993) *In vitro* solubility and degradability of nitrogen in concentrate feeds. *Animal Feed Science and Technology* 42, 1–13.

Teather, R.M. and Sauer, F.D. (1988) A naturally compartmented rumen simulation system for the continuous culture of rumen bacteria and protozoa. *Journal of Dairy Science* 71, 666–673.

Terramoccia, S., Puppo, S., Rizzi, L. and Martillotti, F. (1992) Comparison between *in sacco* and *in vitro* protein rumen degradability. *Annales de Zootechnie* 41, 20.

Tománková, O. and Kopecny, J. (1995) Prediction of feed protein degradation in the rumen with bromelain. *Animal Feed Science and Technology* 53, 71–80.

van Straalen, W.M., Dooper, F.M.H., Antoniewicz, A.M., Kosmala, I. and van Vuuren, A.M. (1993) Intestinal digestibility in dairy cows of protein from grass and clover measured with mobile bag and other methods. *Journal of Dairy Science* 76, 2970–2981.

Wallace, R.J. (1985) Synergism between different species of proteolytic rumen bacteria. *Current Microbiology* 12, 59–64.

Wallace, R.J. and Cotta, M.A. (1988) Metabolism of nitrogen-containing compounds. In: Hobson, P.N. (ed.) *The Rumen Microbial Ecosystem*. Elsevier Applied Science, London, pp. 217–249.

Wallace, R.J. and Kopecny, J. (1983) Breakdown of diazotized proteins and synthetic substrates by rumen bacterial proteases. *Applied Environmental Microbiology* 45, 212–217.

Wallace, R.J. and McKain, N. (1991) A survey of peptidase activity in rumen bacteria. *Journal of General Microbiology* 137, 2259–2264.

Wallace, R.J. and Munro, C.A. (1986) Influence of the rumen anaerobic fungus *Neocalimastix frontalis* on the proteolytic activity of a defined mixture of rumen bacteria growing on a solid substrate. *Letters in Applied Microbiology* 3, 23–26.

Wilkerson, V.A., Klopfenstein, T.J. and Stroup, W.W. (1995) A collaborative study of *in situ* forage protein degradation. *Journal of Animal Science* 73, 583–588.

Yoon, I.K., Calsamiglia, S., Crooker, B.A. and Stern, M.D. (1994) Estimation of absorbable fish meal protein for ruminants. *Journal of Animal Science* 72 (Suppl. 1), 238 (Abstract).

Zinn, R.A. and Owens, F.N. (1982) Predicting net uptake of non-ammonia nitrogen from the small intestine. In: Owens, F.N. (ed.) *Protein Requirements of Cattle: Proceedings of an International Symposium*. MP-109, Oklahoma State University, Stillwater, p. 133.

14 Characterization of Forages by Chemical Analysis

D.J.R. Cherney

Department of Animal Science, New York State College of Agriculture and Life Sciences, Cornell University, Morrison Hall, Ithaca, NY 14853-4801, USA

What is Worth Measuring?

The best evaluator of forage quality is animal performance. Intake, digestibility and efficiency of utilization are characteristics of forages that determine animal performance, with variation in intake accounting for 60–90% of the variation in digestible energy (Mertens, 1994). It would therefore be desirable to measure those forage characteristics that relate most closely to intake and digestibility. Chemical fractions that have been associated with intake and digestibility include fibre, lignin and protein (Cherney and Mertens, 1998). A routine system of chemical characterization of forages for ruminants, then, must include determination of fibre, lignin and protein and critical to these is an accurate dry matter (DM) determination. Some characterizations, such as the determination of water-soluble carbohydrates (WSC), may be of particular importance in silage research. Others, such as tannin assays, starch and soluble fibres, may be of particular importance for some types of forages and some feeding situations. The worth of any particular assay, other than for fibre, lignin, and protein, is thus dependent on the researcher's objectives and the likely economic impact of not performing the assay. If an anti-quality factor limits production, it may become the most important factor to analyse for.

What can it Tell us about Forages?

Chemical analyses can provide valuable information about the actual chemical constituents influencing digestion, unlike *in vitro* methods (Van Soest, 1994). Component chemical analyses, often expensive, can provide important biochemical information, leading to a better understanding of the factors that may limit animal performance. Chemical characterization methods cannot give a direct estimate of nutritive value, but rather rely on statistical association to measure digestibility and intake. Using these statistical associations, characterization of forage fibre, lignin, protein and other chemical components are used increasingly to predict animal performance.

Chemical analyses, along with the use of models, are increasingly used to predict performance. Forage chemical analyses can be used to identify factors in forages that may be limiting animal performance (Minson, 1981).

Chemical Characterization Methods

Dry-matter concentration

While not considered a chemical characterization *per se* by many, DM determination is included here because accurate determination of forage DM is essential for successful characterization of any other chemical constituent. Analytical results must be reported on a total DM basis for meaningful comparisons among feeds. Additionally, the accurate determination of DM is critical for balancing the diets of ruminants. Clancy *et al.* (1977) demonstrated that small errors in DM determination were amplified in the calculation of quality parameters, such that significant differences in quality resulted solely from the method of DM determination. Fresh forage or samples preserved as hay have several options for DM determination, while samples collected after ensiling must account for volatile compounds in ensiled forages. Options for determination of DM directly or by chemical means will be briefly discussed here (Table 14.1).

Oven-drying is commonly used for DM determination. At a drying temperature of 100°C, mechanically trapped water is evaporated, leaving chemically bound water (Brusewitz *et al.*, 1993). The amount of water removed from the sample is directly related to drying temperature, although a small amount of chemically bound water remains in samples dried at temperatures as high as 135°C. Officially approved oven-drying methods, according to the Association of Official Analytical Chemists (AOAC), include drying at 100°C under a vacuum of 1.3×10^4 Pa for 5 h (AOAC-934.01, 1990), drying samples in a forced-air oven at 135°C for 2 h (AOAC-930.15, 1990), or drying at 105°C for 16 h (AOAC-967.03, 1990). An officially approved method in the UK involves drying in a forced-air oven at 100°C for 18 h (MAFF-1, 1986). Further analyses should not be attempted on the dried sample, as drying at 100°C will alter some forage compounds making them less available (Van Soest, 1994).

A two-step procedure is generally used in the USA. Wet samples are dried in a forced-air oven at 55 or 60°C overnight to determine a partial DM value. Samples are then ground and can be used for the second step in total DM determination or for other analyses. The second step of the DM procedure usually involves one of the AOAC procedures.

Fermented feeds contain volatile components, which may be lost during the drying process. Direct laboratory measurement of water content of an as-received silage sample can reduce or eliminate DM determination errors due to loss of volatile components. Toluene distillation is one method for the direct measurement of water in a forage sample. Its use is warranted only for fermented samples,

Table 14.1. Comparison of laboratory methods for determining forage moisture contents (adapted from Brusewitz *et al.*, 1993).

Method	Operating principle	Moisture range	Accuracy*
Hot-air oven	Drying	0–1000	Good for dried forage, poor for silage
Hot-air oven	Drying with correction for volatiles	0–1000	Good
Toluene distillation	Boiling	0–900	Good
Toluene distillation	Boiling with correction	0–900	Very good for silages
Saponification	Chemical binding of water	0–1000	Very good
Karl Fischer	Titration	0–1000	Excellent
Gas chromatography	Physical separation of water from other extracted chemical	0–1000	Excellent
Near-infrared Reflectance Spectroscopy	Electromagnetic properties	0–400	Very good

* Precision for all listed procedures is ±10 g kg^{-1}.

due to the hazards associated with toluene (NFTA-2.3.1, 1993). The small amount of water not removed from the sample has been assumed to offset the volume of volatile components in the distilled-water phase (NFTA-2.3.1, 1993). A more accurate method is to measure the volatile components in the water phase in order to adjust the volume of water for these components (Dewar and McDonald, 1961; AOAC-925.04, 1990). Most studies have concluded that uncorrected toluene distillation results in higher DM content than oven-drying at a range of temperatures (Larsen and Jones, 1973).

The Karl Fischer titration method (Fischer, 1935) is based on the reaction of water with iodine in the presence of pyridine and methanol. This method also involves hazardous chemicals, and the relatively small sample size analysed will increase potential sampling error. Karl Fischer titration for DM determination has been considered too time-consuming for routine analyses (Hood *et al.*, 1971; Robertson and Windham, 1983), although others considered it a simple, rapid procedure for silage samples (Galletti and Piccaglia, 1988; Kaiser *et al.*, 1995).

Gas-chromatography procedures have been developed to measure the water content of forage (Fenton *et al.*, 1981; Burdick and McHan, 1982). Water and alcohol are separated using a gas chromatograph and the ratio of water to alcohol in unknown samples compared with the standards allows determination of sample water content. Accuracy and repeatability are excellent.

Because they can contain significant quantities of volatile compounds, fermented feeds present special problems. Several methods of correcting silage DM content for loss of volatile components are used. Corrections to oven-dried silage DM have been made by analysing samples for fermentation products (Fatianoff and Gouet, 1969; Dulphy *et al.*, 1975). Galletti and Piccaglia (1988) did not find significant differences between Karl Fischer titration values and oven-dried values, after oven-dried silage values were corrected for fermentation products. Equations have been proposed for estimating a toluene DM determination from oven-dried grass-silage samples (Haigh and Hopkins, 1977; Barber *et al.*, 1984; Haigh 1995a, b, c). Such equations will work only if the exact oven-drying procedure is used. Loss of DM during drying is a function of the type and concentration of volatiles present, as well as oven temperature (Minson and Lancaster, 1963).

Kaiser *et al.* (1995) compared eight DM determination methods for silages. In general, silage × method interaction was highly significant, due to greater differences among methods at low silage DM compared with high silage DM. Kaiser *et al.* (1995) concluded that Karl Fischer DM values can be estimated in feed testing laboratories with sufficient accuracy from oven DM values and prediction equations.

Fibre determination

The proximate analysis system (Weende method) has been used for almost 150 years. Components of proximate analysis include crude protein, ether extract, crude fibre, ash and, by difference, nitrogen (N)-free extract (Fisher *et al.*, 1995). While this method is simple, repeatable and relatively inexpensive, there are several problems associated with the method that caution against its use. Ether removes waxes and some other compounds not considered digestible. The crude-protein determination assumes that all N is contained in protein with an N content of 16%, which is incorrect. Crude fibre does not provide an accurate representation of the least digestible fibrous part of the feed in many cases (Sollenberger and Cherney, 1995). Inversely, nitrogen-free extracts are not always a reasonable estimate of the highly digestible carbohydrate fraction. Crude-fibre digestibility exceeds nitrogen-free extract digestibility in about 30% of feedstuffs (Van Soest, 1994), with the largest error occurring in tropical grasses and straws. Perhaps the most serious error is the assumption that crude fibre contains all dietary cellulose, hemicellulose and lignin (Fig. 14.1). For these reasons, crude fibre and nitrogen-free extract are not and should not be routinely used. Crude protein has long been the standard for evaluating the protein value of forages and is still acceptable for many conditions (Vérité, 1980).

The primary standard for fibre evaluation of forages in the USA is the Van Soest system of fibre analysis (Marten, 1981). The Van Soest system separates carbohydrates into fractions based on nutritional availability (Sollenberger and Cherney, 1995). This separation results in a more reasonable estimate of the structural carbohydrates than does crude fibre (Fig. 14.1) and allows for prediction of other indices of forage quality, such as digestibility (Van Soest, 1994). The Van Soest system does not fractionate feeds into chemically pure fractions. It must be remembered, however, that what is important is the nutritional uniformity of the fraction in question.

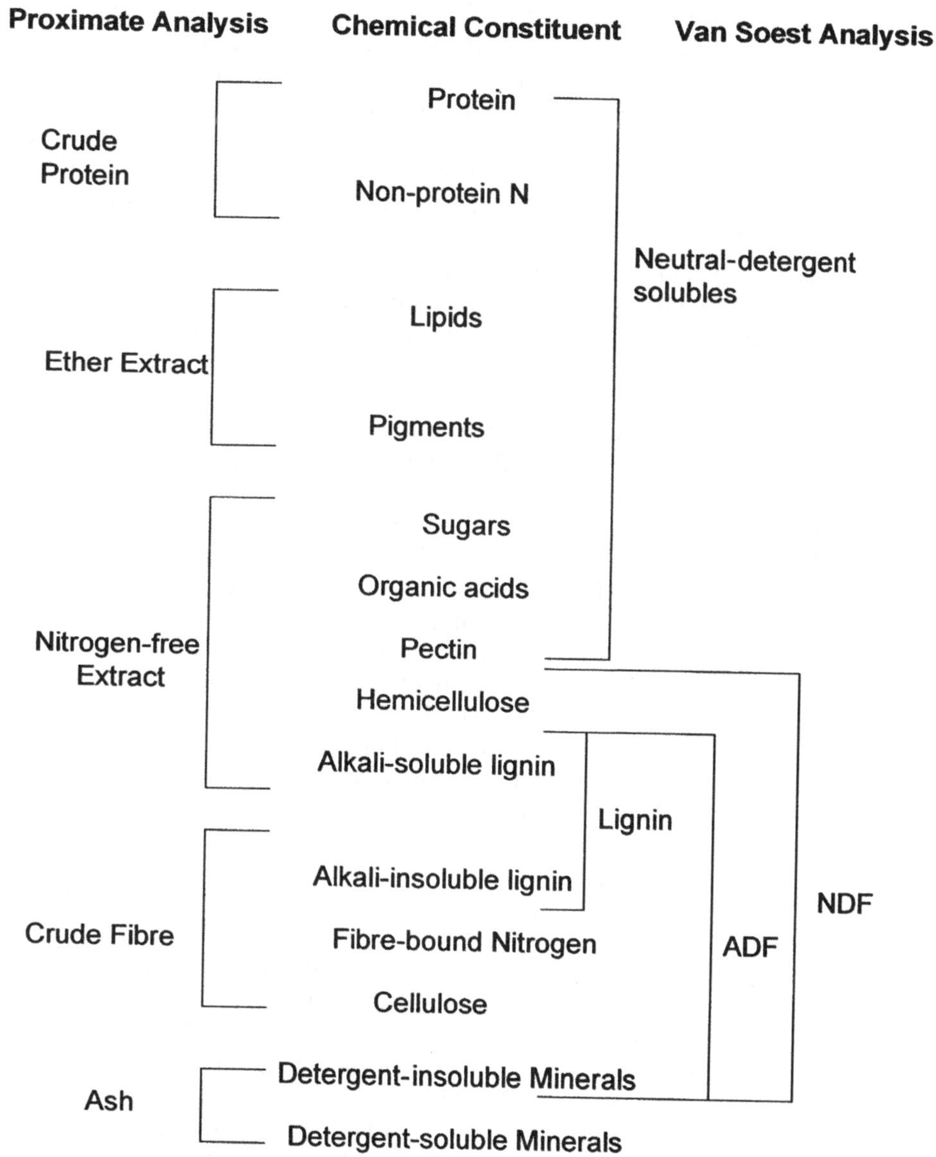

Fig. 14.1. Contrast of Weende system and Van Soest system of carbohydrate analysis (adapted from Fisher *et al.*, 1995). ADF, acid-detergent fibre; NDF, neutral-detergent fibre.

The Van Soest system of fibre analysis has remained relatively unchanged over the years, although a number of modifications have been proposed for special samples (Mascarenhas-Ferreira *et al.*, 1983). Current recommended procedures for neutral-detergent fibre (NDF) (Van Soest *et al.*, 1991) recommend elimination of sodium sulphite, discussed further below, and decalin from the original procedure

(Goering and Van Soest, 1970) and replacement of 2-ethoxyethanol with triethylene glycol. Other research has suggested that 2-ethoxyethanol is not necessary for the determination of NDF in forages and should be eliminated rather than replaced (Cherney *et al.*, 1989). Triethylene glycol may be used where excessive foaming is a problem during filtration. The use of amylase to aid in the removal of starch is also recommended (Van Soest *et al.*, 1991).

Starch contamination of NDF, particularly in those forages containing grain, can lead to overestimates of the NDF value (Robertson and Van Soest, 1977). Insufficient removal of starch from a sample often leads to difficulties with filtration, resulting in elevated NDF values (Cherney *et al.*, 1989). Generally, the use of heat-stable α-amylase is recommended (Van Soest *et al.*, 1991). Pretreatment with urea to remove starch may be necessary in forages with large concentrations of starch (Van Soest *et al.*, 1991).

Sodium sulphite, originally used to reduce nitrogenous contamination in fibre analyses, was removed from the system because it was thought to solubilize some lignin (Van Soest *et al.*, 1991). Recently, the inclusion of sodium sulphite in the NDF solution of high-protein forage samples was re-evaluated (Hintz *et al.*, 1996). Addition of 0.5 g sodium sulphite decreased nitrogen concentration in fibre and lignin values and reduced within-sample variance. Use of sodium sulphite in high-protein forages is therefore recommended.

Recently, a new modification, using the ANKOM fibre analyzer (ANKOM Technology, Fairport, New York), has been developed, which uses preweighed fibre filter bags (Komarek *et al.*, 1994). Fibre values obtained using this method compared to conventional systems were similar (Komarek *et al.*, 1994). This system may be advantageous for silage samples or samples with high soil contamination, which are routinely difficult to filter, as the ANKOM system does not require filtration. This system also offers advantages of reduced initial cost and reduced labour over the conventional system.

Another modification of the NDF system worthy of mention is the micro-NDF method (Pell and Schofield, 1993). There are numerous instances when the amount of sample available may be too small for conventional analysis. In the micro-NDF method, samples in the range of 10–50 mg are placed in 50 ml serum bottles with neutral-detergent solution and autoclaved for 60 min at 105°C. Results compare favourably with standard Van Soest fibre methodology (Van Soest *et al.*, 1991; Pell and Schofield, 1993).

Of the fibre fractions, NDF relates best to feed intake, representing the total insoluble fibre matrix (Van Soest, 1994). This makes NDF the most important fibre fraction to consider. The intended use of acid-detergent fibre (ADF) is as a preparative residue for the determination of cellulose, lignin, Maillard products, silica, acid-insoluble ash and acid-detergent-insoluble N (ADIN). Despite the statistical association in some cases, there is no valid theoretical basis to link ADF to digestibility (Van Soest *et al.*, 1991).

Modified ADF, slightly higher in acid concentration (0.5 mol l^{-1}) and longer in boiling time (MAFF-30, 1986) than Van Soest ADF, is common for fibre analysis in the European Union (EU). The original study for the method (Clancy and Wilson, 1966) reported that prolonged boiling with acid of higher concentration reduced bound nitrogen and improved the relationship between fibre and digestibility. The use of modified ADF as a means of assaying for heat damage and unavailable protein is precluded, because this system includes oven-drying at 95°C as a preliminary step (Van Soest, 1994).

Lignin

Lignin and associated phenolics are most often identified as the chemical constituents in cell walls most limiting digestion (Van Soest, 1994; Buxton *et al.*, 1996). There are a number of procedures used to determine lignin in forage samples (Table 14.2). Insolubility of lignin in 72% sulphuric acid is the basis of most procedures (Barnes, 1973). Measurement of lignin in forage species is complicated by the presence of proteins, chemically bonded cinnamic acids (Iiyama and Wallis, 1990), Maillard products and other non-lignin substances insoluble in 72% sulphuric acid (Van Soest and Robertson, 1985). Klason lignin (72% sulphuric acid method), suitable for most woody species, leads to an overestimation of lignin, because protein coprecipitates with lignin (Norman and Jenkins, 1934). Most methods to determine lignin in forages, therefore, involve preliminary removal of protein, either by enzymes (Crampton and Maynard, 1938) or by pretreatment with acid-detergent solution (Van Soest, 1973).

Barnes (1973) suggested that problems associated with the use of lignin concentration and digestibility could be summarized as: (i) complex and unknown structure of lignin; (ii) lack of a standardized lignin procedure; and (iii) the inherent variation in lignin content among forages. Some 25 years later, the definition or concept of what constitutes lignin is still subject to debate (Van Soest, 1993) and there is, as yet, no universally accepted lignin standard. As to the third point, Van Soest (1993) suggested that a very large part of the diversity between species in the relationship between lignin content and DM digestibility is due to interaction with NDF content. If lignin is expressed on a fibre basis, lignin's relationship to digestibility appears to be reasonably uniform across legumes and grasses (Van Soest, 1993). It would thus appear logical to report lignin on a fibre basis, or to include fibre values when reporting lignin.

Giger (1985) performed a thorough review of current lignin methodology and concluded that a pure analytical lignin fraction, without any lignin loss or contamination by proteins, carbohydrates, cutins or tannins, is not possible with current methods. Methodological insufficiencies may be due to the complexity and diversity of the 'lignin fraction' (Barnes, 1973; Giger, 1985; Van Soest, 1994). Van Soest (1994) suggested that lignin assays be evaluated with the following criteria: (i) lignin recovery in digestion balances; (ii) degree of correlation between lignin and digestibility; (iii) low nitrogen content of lignin preparation; and (iv) recovery of phenolic matter. While this may not be a 'true definition' of lignin (Van Soest, 1994), it is certainly an operational definition. Lignin's value is in its relationship to digestibility or indigestibility. None of the methods listed in Table 14.2 or the more extensive list of Giger (1985) is clearly better based on the four criteria listed above. Some of the newer procedures, such as pyrolysis mass spectrometry, may be beneficial in determining the true structure of lignin. At this time, however, the equipment may be too expensive for use in a routine evaluation scheme. For any

Table 14.2. Methods for the determination of lignin.

Method	Comment	Reference
Quantitative		
Gravimetric		
Klason lignin	Most severe limitation is protein contamination	Giger, 1985
Acid-detergent lignin (ADL)	May contain cutin and other non-lignin carbohydrates	Van Soest, 1973
Difference after lignin removal		
Permanganate lignin	Yields higher values than ADL; cutin excluded, but may contain polyphenolic substances	Van Soest and Wine, 1968
Chlorite	Values may be higher than permanganate	Collings *et al.*, 1978
Triethylene glycol–HCl	Yields values similar to permanganate lignin; lignin preserved	Edwards, 1973
Absorbance	Most of these methods require calibration with gravimetric method	
Acetyl bromide	Protein and non-lignin fractions may contribute to absorbance	Morrison, 1972
Modified acetyl bromide	Non-lignin fractions cause little interference	Iiyama and Wallis, 1990
Bjorkman's ball-milling	Low lignin yields, carbohydrate contamination; used for structural studies	Lam *et al.*, 1990
Saponification	Only applicable to monocotyledonous plants containing ester bridges	Lau and Van Soest, 1981
Calorimetric	Heats of combustion vary due to lignin structure, cannot be used across genera	Schramm and Bergner, 1969
Pyrolysis mass spectrometry	Equipment expensive, must have definition of lignin; can distinguish differences between different lignin chemical methods	Morrison and Mulder, 1995
Non-quantitative	Used for anatomical studies	
Phloroglucinol–HCl		Akin *et al.*, 1990
Chlorine–sulphite		Akin *et al.*, 1990
Coupling with diazonium compounds		Akin *et al.*, 1990
UV fluorescence		Hartley *et al.*, 1990

HCl, hydrochloric acid; UV, ultraviolet.

new lignin method to be routinely used in evaluation, it would have to improve on the relationship with indigestibility, as well as be competitive with respect to time and effort *vis-à-vis* currently used methods (Van Soest, 1994).

Starch

In situations where forages containing high starch levels are fed, specific analysis for starch may be useful. Assays for the reliable extraction and hydrolysis of starch have been developed (Faichney and White, 1983). Accuracy of starch analysis, however, is dependent on the specificity of enzymes used in the analysis and on complete hydrolysis of starch to glucose (Hall, 1997). Starch digestibility is probably as critical as starch quantity in determining its value in a forage sample, and can have a large impact on animal performance (Allen *et al.*, 1997). Methods that evaluate starch digestibility quickly and reliably in the laboratory have not yet been developed, so starch digestibility is not currently routinely determined (Allen *et al.*, 1997).

Water-soluble carbohydrates

WSC concentration in pre-ensiled forage allows assessment of the available supply of energy for lactic acid production during ensiling, and therefore is an important measurement for studies involving these forages. Insufficient WSC will inhibit fermentation, while excess WSC provides substrate for undesirable organisms during storage and feed-out. Forage monosaccharides, glucose and fructose, the oligosaccharides, sucrose, melibiose, raffinose and stachyose, and fructosans can be extracted with cold water (McDonald and Henderson, 1964). Cold-water extractions of legumes and subtropical and tropical grasses do not yield starch, because amylose is partially soluble in hot water, while amylopectin is not soluble in water (Smith, 1981). Procedures that solubilize starch by enzyme or dilute-acid methods, as well as the sugars, are referred to as total non-structural carbohydrate methods (Smith, 1981).

If not inactivated, forage enzyme systems will quickly modify WSC concentration. Rapid freezing, followed by freeze-drying, can minimize modifications to WSC concentration, although freeze-drying does not inactivate enzymes. Rapid oven-drying will also minimize changes in WSC. Enzymes have been inactivated with hot water or ethanol (Kerepesi *et al.*, 1996), but cold-water extraction is recommended. Hot water may solubilize fibre components (McDonald and Henderson, 1964). Protein solubility and other interferences can be minimized by extracting sugars with alcohol solutions (Wiseman *et al.*, 1960).

Spectrophotometric methods used in the UK, involving the use of anthrone, are described elsewhere (Yemm and Willis, 1954; Thomas, 1977; MAFF-14, 1986). Total WSC concentration may also be determined using a spectrophotometric method based on the reaction of phenol–sulphuric acid with carbohydrate (Dubois *et al.*, 1956).

Other carbohydrates

Analysis for sugar components may yield valuable basic information. Methods used will depend on the carbohydrates present and the degree of fractionation required (Moore and Hatfield, 1994). Traditionally, schemes for fractionating carbohydrates were long and laborious (Van Soest, 1994). The development of automated high-performance liquid chromatography (HPLC) and gas-liquid chromatography (GLC) equipment has eliminated this problem, but other problems remain. General schemes cannot be applied without knowing the peculiarities of the forage involved (Van Soest, 1994).

Soluble carbohydrates are approximately 98% digestible (Van Soest, 1994), but the diversity of carbohydrate fractions included prevents their treatment as a uniform nutritional entity. One fraction, soluble fibre, which includes non-strarch polysaccharides, such as pectic polysaccharides, is of particular interest. Unlike sugars and starches, with which they may be coextracted, soluble fibre is not digested by mammalian enzymes. Compared with starch and sugars, which tend to ferment to lactic acid and result in decreases in ruminal pH, soluble-fibre fermentation tends not to produce lactic acid (Hall *et al.*, 1997). The practical implications of this for ruminant nutrition, particularly high-producing dairy cows, justifies the separation of soluble carbohydrate into a fibre and non-fibre fraction. Colorimetric or chromatographic methods to extract fibre have problems with incomplete extractions, partial degradation of carbohydrates or interference by other substrates (Hall, 1997). Recently, a gravimetric method, using the neutral-detergent solution of the Van Soest method, has been proposed to estimate the neutral-detergent-soluble fibre fraction in forages. This method is relatively simple and inexpensive, suggesting its possible use in routine evaluation schemes for forages containing significant quantities of pectic substances.

Nitrogen fractions

Total N content is usually determined by macro- or micro-Kjeldahl methods. Samples are digested in sulphuric acid, with a copper sulphate or copper–selenium (MAFF-48, 1986; AOAC-984.13, 1990) catalyst. Nitrogen is converted to ammonia, distilled and titrated. A variety of other macro- and micro-Kjeldahl methods are available, based on similar principles. Total N content can also be determined effectively by combustion in elemental analyser instruments (Emteryd *et al.*, 1990). These instruments have become more affordable, and there are fewer environmental concerns related to chemical use and disposal than with traditional Kjeldahl methods. Because the N content of forage crude protein is not always 16%, it is suggested that N content be used with the correction factor for crude protein.

Total N may be only 60–80% true protein in fresh and ensiled forages, with the remainder being non-protein N and unavailable N (Van Soest, 1994). Simple crude-protein concentration determination is inadequate to describe protein quality, particularly in high-protein fresh or ensiled forages. Ideally, any method of assessing protein quality should be able to describe the degree to which a protein contributes to bacterial crude protein and undegraded intake protein (Broderick,

1994). Models such as the Cornell net carbohydrate and protein system (CNCPS) model (Fox *et al.*, 1992) have developed the need for even more comprehensive fractionation of forage proteins (Licitra *et al.*, 1996).

Protein solubility has long been used as a technique for determining degradation characteristics of proteins (Broderick, 1994). An ideal method for determining soluble protein would be a single laboratory analysis, and there has been much research into this methodology. Solubility and extraction methods for protein fractions are subject to many of the limitations of carbohydrate solubility methods (Van Soest, 1994). Unlike carbohydrate fractionation, however, in which the Van Soest method is standard in the USA for evaluating carbohydrate fractions in forages, there is no such standardization with protein fractions. This sometimes leads to difficulties in interpreting reported values for nitrogen fractions.

Sniffen *et al.* (1992) recently proposed the use of a series of chemical fractionations to identify five N components in crude protein. It provides a rational basis for the fractionation of protein, based on nutritional availability, much like the carbohydrate fractionation system of Van Soest (1994). A suggested methodology for these N components is reported by Licitra *et al.* (1996). This suggested methodology is outlined in Table 14.3.

Non-protein nitrogen is traditionally the protein passing into the filtrate after precipitation with a protein-specific reagent (Licitra *et al.*, 1996). Trichloroacetic acid (Krishnamoorthy *et al.*, 1982) and tungstic acid (Pichard and Van Soest, 1977) are two commonly referenced methods. These methods differ in the size of the amino acids included in the precipitate. Other methods are also available (Licitra *et al.*, 1996). The choice of the method will depend on the objective of the experiment.

Table 14.3. Suggested partitioning of nitrogen and protein fractions in forages (adapted from Licitra *et al.*, 1996).

Fraction	Estimation or definition	Enzymatic degradation	Classification*
Non-protein N	Not precipitable	Not applicable	A
True protein	Precipitate with tungstic acid		
True soluble protein	Buffer soluble but precipitable (TP-IP)	Fast	B_1
Insoluble protein	Buffer insoluble	Variable	B_2
Neutral-detergent soluble protein	IP-NDIP		
ND-insoluble protein, but soluble in AD	Protein insoluble in ND, but soluble in AD	Variable to slow	B_3
Insoluble in acid detergent	Includes heat-damaged protein and N associated with lignin	Indigestible	C

* From Pichard and Van Soest, 1977; Van Soest, 1994.
TP, true protein; IP, insoluble protein; ND, neutral detergent; AD, acid detergent; NDIP, neutral-detergent-insoluble protein.

A borate phosphate buffer (Krishnamoorthy *et al.*, 1982), which ensures pH stabilization, is recommended to estimate soluble true protein as the difference between nitrogen and non-protein nitrogen (Licitra *et al.*, 1996). Other fractions (Table 14.3) are based on Kjeldahl analysis of acid-detergent and neutral-detergent residues. The ADIN fraction is assumed to be indigestible (Licitra *et al.*, 1996). This fraction is also an indication of heat damage in silages (Van Soest, 1994). The neutral-detergent-insoluble protein fraction is that protein fraction which is not soluble in neutral-detergent protein (Licitra *et al.*, 1996).

Tannins

Tannins may modify the digestibility of dietary protein and structural carbohydrates in forages used for ruminants (Mueller-Harvey and McAllan, 1992; see also Chapter 20, this volume). This may be considered to be an anti-quality component or a desirable attribute, depending on the type of tannin and its concentration in forage (Broderick, 1995; Reed, 1995). High concentrations (50–100 g kg^{-1} DM) are generally considered toxic to ruminal microorganisms, while lower concentrations reduce the soluble non-protein nitrogen content of silage and improve efficiency of protein utilization (Albrecht and Muck, 1991).

Numerous tannin extraction and quantification procedures have been developed (Table 14.4). Most procedures are specific for condensed tannins. The majority have some difficulty associated with their use, either because the procedure is not specific or because the procedure is long and laborious. Additionally, toxic effects of tannins are strongly influenced by an interaction between animal species, animal adaptation and type of tannin, so that it may be difficult to predict (Robbins *et al.*, 1991). Because of these factors, no one procedure has emerged as the prevalent method in tannin analysis. An improved method, based on precipitation with trivalent ytterbium (Giner-Chavez *et al.*, 1997), exhibits promise for more routine quantification of tannins in forages.

Tannins are not prevalent in grasses or most temperate legumes. Routine evaluation for tannins for these forages is not necessary. Tannins are prevalent, however, among dicotyledonous forbs, shrubs and tree leaves (Haslam, 1979). For animals consuming diets with these types of forages, performance may be adversely affected. In these situations, energy or nitrogen content may not limit performance, but rather this anti-quality aspect of tannins. In these cases, it becomes important to routinely analyse for tannins. Other anti-quality forage components should be handled similarly (see Chapter 20).

Vitamins and minerals

Minson (1981) stated that a complete analysis of the elements would be his first priority where animal production was low. Forages clearly provide an important source of vitamin A and E and the minerals essential for growth (Fisher *et al.*, 1995). There are no simple procedures that can be routinely used for vitamin analysis in forages (Fisher *et al.*, 1995). Forages can be prepared for mineral analysis

Table 14.4. Methods for isolating and quantifying tannins.

Method	Comment	Reference
Extraction		
Absolute methanol		Terrill *et al.*, 1990
Aqueous methanol		Terrill *et al.*, 1990
Aqueous acetone		Terrill *et al.*, 1990
Aqueous acetone with ascorbic acid and diethyl ether	Ascorbic acid minimizes oxidation of tannins, but can interfere with subsequent analysis Some high molecular weight condensed tannins not soluble	Terrill *et al.*, 1992; Reed, 1995; Giner-Chavez *et al.*, 1997
Hot methanol		Reed, 1995
Isolation methods		
Sephadex LH-20	Commonly used method	Makkar and Becker, 1994
Trivalent ytterbium	Newer method yields results similar to Sephadex LH-20	Reed *et al.*, 1985; Giner-Chavez *et al.*, 1997
Binding procedures	Non-specific isolation	
Protein		Makkar *et al.*, 1988
Polyvinylpyrrolidine (PVP)		Makkar *et al.*, 1993
Polyethylene glycol (PEG)		Silanikove *et al.*, 1996
Quantification methods		
Gravimetric	Crude estimate of tannins	
Trivalent ytterbium	Does not work well for low levels of tannins	Reed, 1995
PVP	Not specific for condensed tannins	Makkar *et al.*, 1993
Colorimetric	All require standards; internal standards required	
Redox	Subject to numerous interferences	Hagerman and Butler, 1989
Vanillin–HCl	Commonly used, condensed tannins and flavan-3-ols specific	Broadhurst and Jones, 1978; Reed, 1995
Acid butonol	Best assay for condensed tannins; very specific	Hagerman and Butler, 1989
Protein binding	Less precise than colorimetric, may relate to biological effects	
Radial diffusion	Depends on protein structure and reaction conditions	Hagerman and Butler, 1989
Radioactivity		
^{14}C PEG	Pre-extraction of tannins is not required	Silanikove *et al.*, 1996

HCl, hydrochloric acid; ^{14}C, carbon-14.

by dry ashing at 550°C or wet digestion with nitric acid (AOAC-968.08, 1990). Mineral solutions are then analysed by atomic absorption spectrophotometry or flame photometry for individual elements or by inductively coupled plasma emission spectrophotometry for multiple elements. These procedures are labour-intensive and, unless a mineral deficiency or toxicity is expected, would not warrant routine analysis except for the macrominerals.

Chemical Analyses versus *In Vitro* Techniques

Many of the chemical characterization assays are designed to estimate *in vivo* digestibility. Numerous studies have concluded, however, that chemical methods are usually less highly correlated with *in vivo* methods of measuring forage quality than microbial and enzymatic methods (Barnes, 1973; Marten, 1981; Van Soest, 1994). The inability of a single chemical analysis or solubility index to accurately predict an *in vivo* parameter is due to the complexity of biological systems (Barnes, 1973). Legumes and grasses may have the same *in vivo* digestibility but be considerably different in chemical composition. Conversely, forages may be similar in chemical composition (e.g. NDF) but considerably different in digestibility (Cherney *et al.*, 1990). It is likely, therefore, that *in vitro* and/or *in situ* assays will continue to be used (Van Soest, 1994).

In vitro assays are not without difficulty. Standardization of techniques can be difficult and many factors will affect results (Marten, 1981; Cherney *et al.*, 1993). Soluble protein is often determined by direct *in vivo* measurement, although results obtained using this methodology have not always been successful and are difficult for commercial application (Broderick, 1994). Chemical characterization of forages is easier, quicker and less expensive than conducting *in vitro*, enzymatic or *in situ* assays (Weiss, 1994). Intensifying environmental and welfare concerns will make maintaining animals for research or commercial analyses increasingly difficult.

Microbial methods are sensitive to most intrinsic factors that limit *in vivo* digestion of a forage and can yield information concerning rates of digestion (Van Soest, 1994). Microbial methods, however, provide little information regarding the actual chemical constituents influencing digestion. Information from both of these types of analyses is likely to continue to be needed.

Future

The use of modelling to predict animal performance will continue to increase. Many of these models will rely heavily on chemical characterization, because of the speed, repeatability and generally low cost of chemical characterization. Fibre, lignin and protein assays will continue to be important, due to their strong association with factors affecting animal performance. Other assays, such as soluble-fibre and starch analyses, will become increasingly important as our level of understanding of the factors affecting animal performance increases. Automation and refinement of equipment will make chemical characterization faster and more repeatable.

There are numerous methods available for fibre, lignin, protein and other chemical components. It is likely that new methods will continue to be developed. As with lignin, it is imperative that any new method be judged on its ability to improve on the relationship with intake or indigestibility, as well as be competitive with respect to time and effort *vis-à-vis* currently used methods. In addition, because there may be numerous assays to determine the same component, such as lignin, it is the responsibility of the researcher to know the limitations of their methods and what they actually are analysing (e.g. permanganate lignin versus acid-detergent lignin). Researchers should also take special care to report the exact methodology that they are using. With the advent of new methodologies and a better understanding of factors limiting animal performance, the chemical characterization of forages will continue to be a worthwhile field of endeavour.

References

Akin, D.E., Hartley, R.D., Morrison, I.M., Morrison, W.H. and Himmelsbach, D.S. (1990) Diazonium compounds localize grass cell wall phenolics: relation to wall digestibility. *Crop Science* 30, 985–989.

Albrecht, K.A. and Muck, R.E. (1991) Proteolysis in ensiled forage legumes that vary in tannin concentration. *Crop Science* 31, 464–469.

Allen, M.S., Oba, M. and Choi, B.R. (1997) Nutritionist's perspective on corn hybrids for silage. In: *Silage: Field to Feedbunk.* NRAES-99, Northeast Regional Agricultural Engineering Service, Ithaca, New York, pp. 25–36.

AOAC-925.04 (1990) Moisture in animal feed. By distillation with toluene. In: *Official Methods of Analysis of the Association of Official Analytical Chemists*, 15th edn. AOAC, Arlington, Virginia, p. 69.

AOAC-930.15 (1990) Moisture in animal feed. Drying at 135°. In: *Official Methods of Analysis of the Association of Official Analytical Chemists*, 15th edn. AOAC, Arlington, Virginia, p. 69.

AOAC-934.01 (1990) Moisture in animal feed. Drying *in vacuo* at 95–100°. In: *Official Methods of Analysis of the Association of Official Analytical Chemists*, 15th edn. AOAC, Arlington, Virginia, p. 69.

AOAC-967.03 (1990) Moisture in peat. In: *Official Methods of Analysis of the Association of Official Analytical Chemists*, 15th edn. AOAC, Arlington, Virginia, p. 37.

AOAC-968.08 (1990) Minerals in animal feed. Atomic absorption spectrophotometric method. In: *Official Methods of Analysis of the Association of Official Analytical Chemists*, 15th edn. AOAC, Arlington, Virginia, p. 84.

AOAC-984.13 (1990) Protein (crude) in animal feed. Copper catalyst Kjeldahl method. In: *Official Methods of Analysis of the Association of Official Analytical Chemists*, 15th edn. AOAC, Arlington, Virginia, p. 74.

Barber, W.P., Adamson, A.H. and Altman, J.F.B. (1984) New methods of forage evaluation. In: Haresign, W. and Cole, D.J.A. (eds) *Recent Advances in Animal Nutrition – 1984.* Butterworths, London, UK, pp. 161–176.

Barnes, R.F. (1973) Laboratory methods of evaluating feeding value of herbage. In: Butler, G.W. and Bailey, R.W. (eds) *Chemistry and Biochemistry of Herbage.* Academic Press, London, pp. 179–214.

Broadhurst, R.B. and Jones, W.T. (1978) Analysis of condensed tannins using acidified vanillin. *Journal of the Science of Food and Agriculture* 29, 783–794.

Broderick, G.A. (1994) Quantifying forage protein quality. In: Fahey, G.C., Collins, M., Mertens, D.R. and Moser, L.E. (eds) *Forage Quality, Evaluation, and Utilization.* ASA, CSSA and SSSA, Madison, Wisconsin, pp. 200–228.

Broderick, G.A. (1995) Desirable characteristics of forage legumes for improving protein utilization in ruminants. *Journal of Animal Science* 73, 2760–2773.

Brusewitz, G.H., Chase, L.E., Collins, M., Delwiche, S.R., Garthe, J., Muck, R.E. and Pitt, R.E. (1993) *Forage Moisture Determination*. NRAES-59, Northeast Regional Agricultural Engineering Service, Ithaca, New York, 29 pp.

Burdick, D. and McHan, F. (1982) Rapid gas chromatographic procedure for measuring moisture in silage. *Journal of Dairy Science* 65, 2396–2401.

Buxton, D.R., Mertens, D.R. and Fisher, D.S. (1996) Forage quality and ruminant utilization. In: Moser, L.E., Buxton, D.R. and Casler, M.D. (eds) *Cool-season Forage Grasses*. American Society of Agronomy, Madison,Wisconsin, pp. 229–266.

Cherney, D.J.R. and Mertens, D.R. (1998) Modelling grass utilization for dairy cows. In: Cherney, J.H. and Cherney, D.J.R. (eds) *Grass for Dairy Cattle*. CAB International, Wallingford, Oxon, UK, pp. 351–371.

Cherney, D.J.R., Patterson, J.A. and Cherney, J.H. (1989) Use of 2-ethoxyethanol and α-amylase in the neutral detergent fiber method of feed analysis. *Journal of Dairy Science* 72, 3079–3084.

Cherney, D.J.R., Mertens, D.R. and Moore, J.E. (1990) Intake and digestibility by sheep as influenced by forage morphology at three levels of offer. *Journal of Animal Science* 68, 4387–4399.

Cherney, D.J.R., Siciliano-Jones, J. and Pell, A.N. (1993) Technical note: forage *in vitro* dry matter digestibility as influenced by fiber source in donor cow diet. *Journal of Animal Science* 71, 1335–1338.

Clancy, M.J. and Wilson, R.K. (1966) Development and application of a new chemical method for predicting the digestibility and intake of herbage samples. In: *Proceedings of the 10th International Grasslands Congress, Helsinki.* Valtioneuvoston Kirjapaino, Helsinki, pp. 445–452.

Clancy, M.P.J., Wangsness, P.J. and Baumgardt, B.R. (1977) Effects of moisture determination method on estimates of digestibilities and intakes of conserved alfalfa. *Journal of Dairy Science* 6, 216–223.

Collings, G.F., Yokoyama, M.T. and Bergen, W.T. (1978) Lignin as determined by oxydation with sodium chlorite and a comparison with permanganate lignin. *Journal of Dairy Science* 61, 1156–1180.

Crampton, E.W. and Maynard, L.A. (1938) The relation of cellulose and lignin content to animal feeds. *Journal of Nutrition* 15, 383–395.

Dewar, W.A. and McDonald, P. (1961) Determination of DM in silage by distillation with toluene. *Journal of the Science of Food and Agriculture* 12, 790–795.

Dubois, M., Gilles, K.A., Hamilton, J.K., Rebers, P.A. and Smith, F. (1956) Colorimetric method for determination of sugars and related substrates. *Analytical Chemistry* 28, 350–356.

Dulphy, J.P., Demarquilly, C. and Henry, M. (1975) Determination of silage DM after oven-drying and losses of volatile components. *Annales de Zootechnie* 24, 743–756.

Edwards, C.S. (1973) Determination of lignin and cellulose in forages by extraction with triethylene glycol. *Journal of the Science of Food and Agriculture* 24, 381–388.

Emteryd, O., Olsson, L. and Wallmark, H. (1990) Regeneration of copper-II oxide with hydrogen gas in the reduction column of an elemental analyzer. *Microchemistry Journal* 42, 336–338.

Faichney, G.J. and White, W.T. (1983) *Methods for the Analysis of Feeds Eaten by Ruminants*. CSIRO, Melbourne, Australia, 36 pp.

Fatianoff, N. and Gouet, J.J. (1969) Relation permitting to correct rapidly and with accuracy the DM of oven dried silages. *Annales de Zootechnie* 18, 407–418.

Fenton, T.W., Fenton, M. and Mathison, G.W. (1981) Determination of water in fermented feeds by gas chromatography. *Canadian Journal of Animal Science* 61, 827–831.

Fischer, K. (1935) Neues Verfahren zur Massanalytischen bestimmung des Wassergehaltes von Flussigkeiten and festen Korpern. *Angewandte Chemie* 48, 394–396.

Fisher, D.S., Burns, J.C. and Moore, J.E. (1995) The nutritive evaluation of forage. In: Barnes, R.F., Miller, D.A. and Nelson, C.J. (eds) *Forages: Vol. I. An Introduction to Grassland Agriculture*. Iowa State University Press, Ames, pp. 105–115.

Fox, D.G., Sniffen, C.J., O'Connor, J.D., Russell, J.B. and Van Soest, P.J. (1992) A net carbohydrate and protein system for evaluating cattle diets: III. Cattle requirements and diet adequacy. *Journal of Animal Science* 70, 3578–3596.

Galletti, G.C. and Piccaglia, R. (1988) Water determination in silages by Karl Fischer titration. *Journal of the Science of Food and Agriculture* 43, 1–7.

Giger, S. (1985) Revue sur les méthodes de dosage de la lignine utilisées en alimentation animale. *Annales de Zootechnie* 34, 85–122.

Giner-Chavez, B.I., Van Soest, P.J., Robertson, J.B., Lascano, C., Reed, J.D. and Pell, A.N. (1997) A method for isolating condensed tannins from crude plant extracts with trivalent ytterbium. *Journal of the Science of Food and Agriculture* 74, 359–368.

Goering, H.K. and Van Soest, P.J. (1970) *Forage Fiber Analyses (Apparatus, Reagents, Procedures, and Some Applications)*. Agricultural Handbook 379, ARS, USDA, Washington, DC.

Hagerman, A.E. and Butler, L.G. (1989) Choosing appropriate methods and standards for assaying tannin. *Journal Chemical Ecology* 15, 1795–1810.

Haigh, P.M. (1995a) A note on the relationship between oven and toluene determined DM concentrations in bunker-made grass silages. *Irish Journal of Agricultural and Food Research* 34, 63–67.

Haigh, P.M. (1995b) A note on the relationship between oven and toluene determined DM concentrations in big-bale grass silages. *Irish Journal of Agricultural and Food Research* 34, 189–191.

Haigh, P.M. (1995c) A note on the relationship between oven and toluene determined DM concentrations in maize silages. *Irish Journal of Agricultural and Food Research* 34, 193–195.

Haigh, P.M. and Hopkins, J.R. (1977) Relationship between oven and toluene DM in grass silage. *Journal of the Science of Food and Agriculture* 28, 477–480.

Hall, M.B. (1997) Interpreting feed analyses: uses, abuses and artifacts. In: *8th Annual Florida Ruminant Nutrition Symposium, 16–17 January, 1997, Gainesville, Florida*. University of Florida, Gainesville, pp. 161–169.

Hall, M.B., Lewis, B.A., Van Soest, P.J. and Chase, L.E. (1997) A simple method for estimation of neutral detergent-soluble fibre. *Journal of the Science of Food and Agriculture* 74, 441–449.

Hartley, R.D., Akin, D.E., Himmelsbach, D.S. and Beach, D.C. (1990) Microspectrometry of bermudagrass (*Cynodon dactylon*) cell walls in relation to ligninfication and wall biodegradability. *Journal of the Science of Food and Agriculture* 50, 179–189.

Haslam, E. (1979) Vegetable tannins. In: Swain, T., Harbone, J.B. and Van Sumere, F. (eds) *Biochemistry of Plant Phenolics*. Recent Advances in Phytochemistry, Plenum Press, New York, pp. 475–523.

Hintz, R.W., Mertens, D.R. and Albrecht, K.A. (1996) Effects of sodium sulfite on recovery composition of detergent fiber and lignin. *Journal of the Association of Official Analytical Chemists International* 79, 16–22.

Hood, R.L., Allen, C.E., Goodrich, R.D. and Meiske, J.C. (1971) A rapid method for the direct chemical determination of water in fermented feeds. *Journal of Animal Science* 33, 1310–1314.

Iiyama, K. and Wallis, A.F.A. (1990) Determination of lignin in herbaceous plants by an improved acetyl bromide procedure. *Journal of the Science of Food and Agriculture* 51, 145–161.

Kaiser, A.G., Mailer, R.J. and Vonarx, M.M. (1995) A comparison of Karl Fischer titration with alternative methods for the analysis of silage DM content. *Journal of the Science of Food and Agriculture* 69, 51–59.

Kerepesi, I., Toth, M. and Boross, L. (1996) Water-soluble carbohydrates in dried plant. *Journal of Agricultural and Food Chemistry* 44, 3235–3239.

Komarek, A.R., Robertson, J.B. and Van Soest, P.J. (1994) A comparison of methods for determining ADF using the filter bag technique versus conventional filtration. *Journal of Dairy Science* 77 (Suppl. 1), 114.

Krishnamoorthy, U., Muscato, T.V., Sniffen, C.J. and Van Soest, P.J. (1982) Nitrogen fractions in selected feedstuffs. *Journal of Dairy Science* 65, 217–255.

Lam, T.B.T., Iiyama, K. and Stone, B.A. (1990) Lignin in wheat internodes. Part 2. Alkaline nitrobenzene oxidation by wheat straw lignin and its fractions. *Journal of the Science of Food and Agriculture* 51, 493–506.

Larsen, R.E. and Jones, G.M. (1973) Effects of different DM determination methods on chemical composition and *in vitro* digestibility of silages. *Canadian Journal of Animal Science* 53, 753–760.

Lau, M.M. and Van Soest, P.J. (1981) Titratable groups and soluble phenolics as indicators of the digestibility of chemically treated roughages. *Animal Feed Science and Technology* 6, 123–131.

Licitra, G., Hernandez, T.M. and Van Soest, P.J. (1996) Standardization of procedures for nitrogen fractionation of ruminant feeds. *Animal Feed Science and Technology* 57, 347–358.

McDonald, P. and Henderson, A.R. (1964) Determination of water-soluble carbohydrates in grass. *Journal of the Science of Food and Agriculture* 15, 395–398.

MAFF-1 (1986) Method 1. Preparation of samples of plant material. In: *The Analysis of Agricultural Materials. A Manual of the Analytical Methods used by the Agricultural Development and Advisory Service.* Reference Book 427, Ministry of Agriculture, Fisheries and Food, Her Majesty's Stationery Office, London, pp. 1–5.

MAFF-14 (1986) Method 14. Carbohydrates, soluble, in herbage. In: *The Analysis of Agricultural Materials. A Manual of the Analytical Methods used by the Agricultural Development and Advisory Service.* Reference Book 427, Ministry of Agriculture, Fisheries and Food, Her Majesty's Stationery Office, London, pp. 43–45.

MAFF-30 (1986) Method 30. Fibre, modified acid-detergent, in plant material. In: *The Analysis of Agricultural Materials. A Manual of the Analytical Methods used by the Agricultural Development and Advisory Service.* Reference Book 427, Ministry of Agriculture, Fisheries and Food, Her Majesty's Stationery Office, London, pp. 93–97.

MAFF-48 (1986) Method 48. Nitrogen in plant material. In: *The Analysis of Agricultural Materials. A Manual of the Analytical Methods used by the Agricultural Development and Advisory Service.* Reference Book 427, Ministry of Agriculture, Fisheries and Food, Her Majesty's Stationery Office, London, pp. 148–149.

Makkar, H.P.S. and Becker, K. (1994) Isolation of tannins from leaves of some trees and shrubs and their properties. *Journal of Agricultural and Food Chemistry* 42, 731–734.

Makkar, H.P.S., Dawra, R.K. and Singh, B. (1988) Determination of both tannin and protein in a tannin–protein complex. *Journal of Agricultural and Food Chemistry* 36, 523–525.

Makkar, H.P.S., Blummel, M., Borowy, N.K. and Becker, K. (1993) Gravimetric determination of tannins and their correlations with chemical and protein precipitation methods. *Journal of the Science of Food and Agriculture* 61, 161–165.

Marten, G.C. (1981) Chemical, *in vitro*, and nylon bag procedures for evaluating forage in the USA. In: Wheeler, J.L. and Mochrie, R.D. (eds) *Forage Evaluation: Concepts and Techniques.* Griffen Press, Netley, South Australia, pp. 39–55.

Mascarenhas-Ferreira, A., Kerstens, J. and Gast, C.H. (1983) The study of several modifications of the neutral detergent fibre procedure. *Animal Feed Science and Technology* 9, 19–28.

Mertens, D.R. (1994) Regulation of forage intake. In: Fahey, G.C., Collins, M., Mertens, D.R. and Moser, L.E. (eds) *Forage Quality, Evaluation, and Utilization.* ASA, CSSA and SSSA, Madison, Wisconsin, pp. 450–493.

Minson, D.J. (1981) An Australian view of laboratory techniques for forage evaluation. In: Wheeler, J.L. and Mochrie, R.D. (eds) *Forage Evaluation: Concepts and Techniques.* Griffen Press, Netley, South Australia, pp. 57–71.

Minson, D.J. and Lancaster, R.J. (1963) The effect of oven temperature on the error in estimating the DM content of silage. *New Zealand Journal of Agricultural Research* 6, 140–146.

Moore, K.J. and Hatfield, R.D. (1994) Carbohydrates and forage quality. In: Fahey, G.C., Collins, M., Mertens, D.R. and Moser, L.E. (eds) *Forage Quality, Evaluation, and Utilization.* ASA, CSSA and SSSA, Madison, Wisconsin, pp. 229–280.

Morrison, I.M. (1972) Improvements in the acetyl bromide technique to determine lignin and digestibility and its application to legumes. *Journal of the Science of Food and Agriculture* 23, 1463–1469.

Morrison, I.M. and Mulder, M.M. (1995) Pyrolysis mass spectrometry for the characterisation of straw and chemically treated straw. *Phytochemical Analysis* 6, 74–84.

Mueller-Harvey, I. and McAllan, A.B. (1992) Tannins: their biochemistry and nutritional properties. *Advances in Plant Cell Biochemistry and Biotechnology* 1, 151–217.

NFTA-2.3.1 (1993) Method 2.3.1. Toluene determination of moisture. In: Undersander, D., Mertens, D.R. and Thiex, N. (eds) *Forage Analysis Procedures.* National Forage Testing Association, Omaha, Nebraska, pp. 49–51.

Norman, A.G. and Jenkins, S.H. (1934) The determination of lignin. II. Errors introduced by the presence of proteins. *Biochemistry Journal* 28, 2160–2168.

Pell, A.N. and Schofield, P. (1993) Computerized monitoring of gas production to measure forage digestion *in vitro*. *Journal of Dairy Science* 76, 1063–1073.

Pichard, G. and Van Soest, P.J. (1977) Protein solubility of ruminant feeds. In: *Proceedings of the Cornell Nutrition Conference.* Department of Animal Science, Cornell University, Ithaca, New York, pp. 91–98.

Reed, J.D. (1995) Nutritional toxicology of tannins and related polyphenols in forage legumes. *Journal of Animal Science* 73, 1516–1528.

Reed, J.D., Horvath, P.J., Allen, M.S. and Van Soest, P.J. (1985) Gravimetric determination of soluble phenolics including tannins from leaves by precipitation with trivalent ytterbium. *Journal of the Science of Food and Agriculture* 36, 255–261.

Robbins, C.T., Hagerman, A.E., Austin, P.J., McArthur, C. and Hanley, T.A. (1991) Variation in mammalian physiological responses to a condensed tannin and its ecological implications. *Journal of Mammalogy* 72, 480–486.

Robertson, J.A. and Windham, W.R. (1983) Oil and water analysis of sunflower seed by near infrared reflectance spectroscopy. *Journal of the American Oil Chemistry Society* 61, 543–547.

Robertson, J.B. and Van Soest, P.J. (1977) Dietary fibre estimation in concentrate feedstuffs. *Journal of Animal Science* 45 (Suppl. 1), 254.

Schramm, S. and Bergner, H. (1969) Studies on a new method for the indirect calorimetric determination of lignin in foodstuffs. II. Evidence for the general applicability and reproducibility of the method by experiments with various foodstuffs. *Archiv für Tierernäehrung* 19, 281–288 (in German).

Silanikove, N., Shinder, D., Gilboa, N., Eyal, M. and Nitsan, Z. (1996) Binding of poly(ethylene glycol) to samples of forage plants as an assay of tannins and their negative effects on ruminal degradation. *Journal of Agricultural and Food Chemistry* 44, 3230–3234.

Smith, D. (1981) *Removing and Analyzing Nonstructural Carbohydrates from Plant Tissue.* Research Report R2107, University of Wisconsin College of Agricultural and Life Sciences, Madison, Wisconsin.

Sniffen, C.J., O'Connor, J.D., Van Soest, P.J., Fox, D.G. and Russell, J.B. (1992) A net carbohydrate and protein system for evaluating cattle diets: II. Carbohydrate and protein availability. *Journal of Animal Science* 70, 3562–3577.

Sollenberger, L.E. and Cherney, D.J.R. (1995) Evaluating forage production and quality. In: Barnes, R.F., Miller, D.A. and Nelson, C.J. (eds) *Forages*, Vol. II: *The Science of Grassland Agriculture*, 5th edn. Iowa State University Press, Ames, pp. 97–110.

Terrill, T.H., Windham, W.R., Evans, J.J. and Hoveland, C.S. (1990) Condensed tannin concentration in *Sericea lespedeza* as influenced by preservation method. *Crop Science* 30, 219–224.

Terrill, T.H., Rowan, A.M., Douglas, G.B. and Barry, T.N. (1992) Determination of extractable and bound condensed tannin concentrations in forage plants, protein concentrate meals and cereal grains. *Journal of the Science of Food and Agriculture* 58, 321–329.

Thomas, T.A. (1977) An automated procedure for the determination of soluble carbohydrates in herbage. *Journal of the Science of Food and Agriculture* 28, 639–642.

Van Soest, P.J. (1973) Collaborative study on acid-detergent fibre and lignin. *Journal of Official Analytical Chemists* 56, 781–784.

Van Soest, P.J. (1993) Cell wall matrix interactions and degradation – session synopsis. In: Jung, H.G., Buxton, D.R., Hatfield, R.D. and Ralph, J. (eds) *Forage Cell Wall Structure and Digestibility*. ASA, CSSA and SSSA, Madison, Wisconsin, pp. 377–395.

Van Soest, P.J. (1994) *Nutritional Ecology of the Ruminant.* Cornell University Press, Ithaca, New York, 476 pp.

Van Soest, P.J. and Robertson, J.B. (1985) *Analysis of Forages and Fibrous Feeds*. Laboratory Manual for Animal Science 613, Cornell University, Ithaca, New York, 202 pp.

Van Soest, P.J. and Wine, R.H. (1968) Determination of lignin and cellulose in ADF with permanganate. *Journal of Official Analytical Chemists* 51, 780–785.

Van Soest, P.J., Robertson, J.B. and Lewis, B.A. (1991) Methods for dietary fiber, neutral detergent fiber, and nonstarch polysaccharides in relation to animal nutrition. *Journal of Dairy Science* 74, 3583–3597.

Vérité, R. (1980) Appreciation of the nitrogen value of feeds for ruminants. In: *Standardization of Analytical Methodology for Feeds*. International Development Research Centre, Ottawa, Canada, pp. 87–96.

Weiss, W.P. (1994) Estimation of digestibility of forages by laboratory methods. In: Fahey, G.C., Collins, M., Mertens, D.R. and Moser, L.E. (eds) *Forage Quality, Evaluation, and Utilization*. ASA, CSSA and SSSA, Madison, Wisconsin, pp. 644–681.

Wiseman, H.G., Mallack, J.C. and Jacobson, W.C. (1960) Determination of sugar in silages and forages. *Journal of Agricultural and Food Chemistry* 8, 78–80.

Yemm, E.W. and Willis, A.J. (1954) The estimation of carbohydrates in plant extracts by anthrone. *Journal of Biochemistry* 57, 508–514.

15 Near-infrared (NIR) Spectroscopy: an Alternative Approach for the Estimation of Forage Quality and Voluntary Intake

E.R. Deaville[1] and P.C. Flinn[2]

[1]*ADAS Nutritional Sciences Research Unit, Alcester Road, Stratford-upon-Avon CV37 9RQ, UK;* [2]*Agriculture Victoria, Pastoral and Veterinary Institute, Private Bag 105, Hamilton, Victoria 3300, Australia*

Introduction

The demand for objective estimates of forage quality has increased significantly in recent years for several reasons. The intensive feeding of housed animals in Europe and North America requires constant monitoring of diet composition to ensure consistent levels of milk or meat production. In Australia, where grazed pasture is the foundation of the major livestock industries, the growth and quality patterns of pastures vary widely. Hence real-time decisions have to be made by producers on whether or when to purchase and utilize feed supplements (Flinn and Downes, 1996). Selection or breeding for nutritive value is now considered an important factor by many forage-plant breeders, with the need often to test a very large number of lines.

The widely used but empirical techniques to measure fibre content of forages are labour-intensive and hence expensive, even with modern digestion equipment. The highly regarded *in vitro* (Tilley and Terry, 1963) and enzyme-based techniques for estimating digestibility of forages require approximately one week to complete.

The physical, non-destructive technique of near-infrared (NIR) spectroscopy has been described as 'the most practicable and exciting analytical technique to hit the agricultural and food industries since Johann Kjeldahl introduced the Kjeldahl test' (Williams and Norris, 1987). Since its introduction in the 1960s for measuring moisture in grains (Norris and Hart, 1965), the number of analytical applications of NIR in quality control has expanded dramatically. The technique is also 'clean and green', in that no chemicals are used and there are no health and safety problems.

NIR spectroscopy represents a radical departure from conventional analytical methods, in that the entire sample of a forage is characterized in terms of its absorption properties in the NIR region, rather than separate subsamples being treated with various chemicals to isolate specific components. This forces the analyst to abandon his/her traditional narrow focus on one sample and one analyte at a time and to take a broader view of the relationship between components within the sample and between the sample and the population from which it comes.

The main disadvantages of NIR spectroscopy are the high initial cost of the instrumentation and its reliance on chemometrics (Norris, 1989a), leading to its traditional but unfair reputation as a 'black box', due to a lack of understanding of the way in which NIR spectra relate to composition.

This chapter reviews the principles and practice of NIR spectroscopy in the assessment of forage quality, and outlines the major steps required for successful use of the technique. Other reviews (see Givens *et al.*, 1997) have also extended into the use of NIR spectroscopy for predicting the nutritive value of foods for animals and humans.

NIR Spectra: What are we Really Measuring?

Spectroscopy literally means looking at light and is based on the interaction of electromagnetic radiation with the matter to be analysed. The theory of NIR spectrosocopy is covered in depth in the books of Williams and Norris (1987) and Osborne *et al.* (1993).

The NIR region of the electromagnetic spectrum lies between the visible and infrared regions, and is usually defined by the wavelength range 700–3000 nanometres (nm) (Norris, 1989b). However, most analytical use of NIR is between 1100 and 2500 nm. NIR spectra are essentially plots of reciprocal $\log_{10}$ reflectance ($\log 1/R$) versus wavelength. The spectra appear as smooth, rolling lines with few well-defined features (Murray, 1988), but they consist of many overlapped bands, since the reflectance spectrum of a forage sample is the summation of the spectra of its major chemical components. The challenge is to extract useful information on nutritive value of forages from the reflectance data.

Prominent absorptions in NIR spectra of forages (Fig. 15.1) include water, with two bands at 1940 nm (combination band) and 1450 nm (first overtone band); aliphatic carbon (C)-hydrogen (H) bands (lipids) at 2310, 1725, 1400 and 1210 nm; and oxygen (O)-H bands (carbohydrates) around 2100 and 1600 nm (Murray, 1983). The nitrogen (N)-H absorptions due to amide structures present in protein occur predominantly at 2180 and 2055 nm (Wetzel, 1983), but these are often masked by the broad O-H band at 2100 nm. It is the shape of the spectral line, or the rate of change in slope with wavelength, that conveys chemical information (Murray and Williams, 1987). Hence first- or second-derivative plots of $\log 1/R$ are useful, as they can resolve overlapping bands into component absorptions, which may appear as shoulders on larger peaks, as well as minimizing the effect of particle size (Shenk and Westerhaus, 1994). The second derivative is particularly useful, as it produces a negative peak that almost coincides with the wavelength position where the $\log 1/R$ peak occurs (although frequently not apparent from the $\log 1/R$ spectra) (Fig. 15.2).

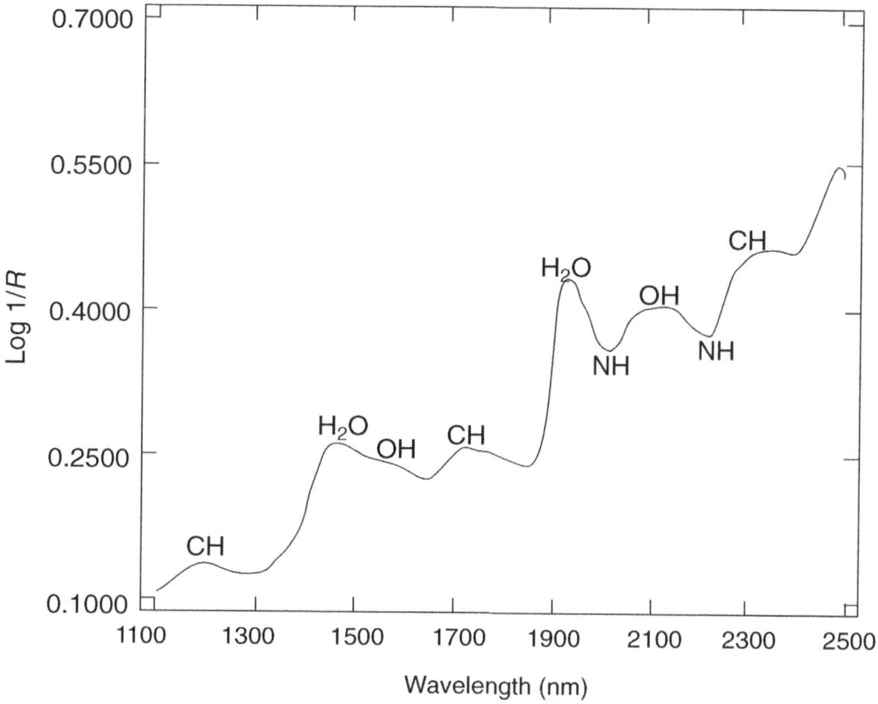

Fig. 15.1. NIR log 1/*R* spectrum of hay sample showing locations of absorptions due to major functional groups.

NIR Instruments and Sample Presentation Methods

There are many instrument types available for NIR analysis, and a detailed comparison is beyond the scope of this chapter (see Osborne *et al.*, 1993). However, the three major types of devices for wavelength selection are monochromators, filters and diode arrays. Monochromators scan the whole wavelength range by using a grating or prism and they provide maximum versatility. Tilting-filter instruments have three or six tilting filters, which, in effect, allow segments of spectra to be scanned. Fixed-filter instruments measure log 1/*R* at discrete wavelengths only, but are quite satisfactory for applications where the constituents are well defined and the appropriate wavelengths well known. Diode-array instruments use no moving parts for wavelength selection and have the advantage of being able to scan at high speed.

There are basically two alternative procedures for presenting a forage sample to an NIR instrument. Small sample cups are used for finely ground, uniform materials and large rectangular cells are used to pack unground and often high-moisture heterogeneous sample (Shenk and Westerhaus, 1994). Until recently, most NIR forage analysis has been done using dried, ground samples and the recommended procedures for drying (if required), grinding, mixing and packing are well established

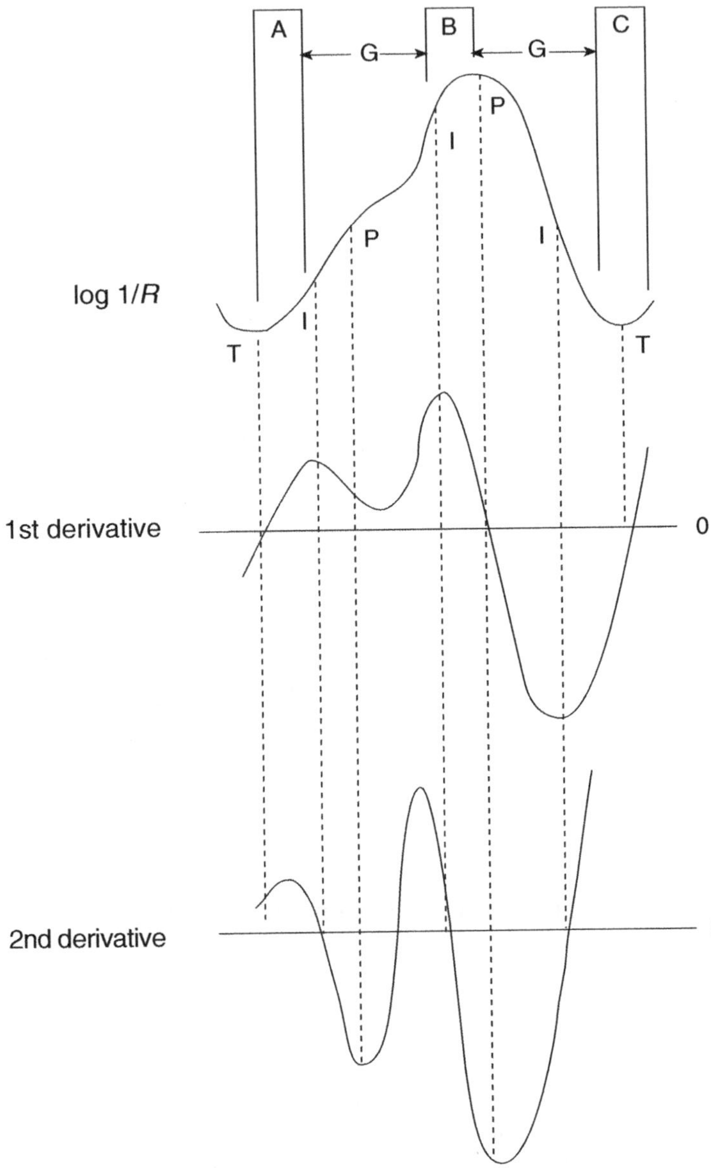

Fig. 15.2. The relationship between log 1/*R* and first- and second-derivative log 1/*R* spectra. The first derivative is computed from the difference (A − B), whilst the second derivative is computed from (A − B) − (B − C), or A − 2B − C (after Murray, 1988). A, B, C, segments; G, gap; T, trough; P, peak; I, point of maximum inflection.

(Shenk and Westerhaus, 1994). However, there are advantages in analysing forages in their natural (undried), unground state, due to reduced sample preparation time and, in the case of fermented forages, avoidance of analytical errors due to losses of volatile compounds by oven-drying.

NIR Data Processing and Calibration

Calibration is the mathematical process required to relate NIR optical measurements to the desired constituent or property used to define the nutritional quality of forages. Calibration is the key to successful use of the NIR technique, and many different approaches have been proposed. There are a number of major steps which are essential to obtain a satisfactory working calibration, i.e. sample selection, acquisition of spectra and reference data, pretreatment of spectral data, derivation of the regression model and validation of the model.

Sample population structuring

Selection of the appropriate samples is critical for a successful calibration. Dardenne (1996) confirmed that extended time spent on the development of appropriate calibration sample sets is rewarded by more stable and accurate calibrations. A decision must first be made on the intended use of the calibration, e.g. if the purpose is to analyse ryegrass samples, the calibration population should consist only of ryegrass. Conversely, if many different types of hay are to be analysed, all types must be represented in the population. The calibration population must include all sources of variation likely to be found in future unknown samples of similar material, such as chemical, physical and botanical characteristics (Windham *et al.*, 1989; Shenk and Westerhaus, 1994). The traditional method of sample selection involved either structured or random sampling from a population, using information already available on chemical or botanical composition (Abrams, 1989). A more convenient procedure is to choose representative samples from a large population on the basis of their spectral characteristics, thus reducing the number of samples requiring reference data. Software is available for this purpose, with one approach using algorithms known as CENTER and SELECT, where spectral boundaries for a product are first established using principal-component scores and then samples are selected from defined neighbourhoods from within the population (Shenk and Westerhaus, 1991).

There is no easy answer to the question of how many samples are needed for calibration. For agricultural products, calibrations based on fewer than 50 samples are seldom satisfactory. Windham *et al.* (1989) concluded that 50–100 samples were adequate for narrow-based or 'closed' populations, whereas 150 or more samples were necessary for broad-based or 'open' populations. With recent advances in software and computer power, it is now possible to establish calibration populations or 'libraries' containing hundreds or even thousands of samples, which are necessary for multi-product forage calibrations (Shenk and Westerhaus, 1993).

Quality of reference analyses

The quality of reference-method analysis has a crucial effect on the accuracy of NIR calibrations. Poor calibration accuracy has often been incorrectly blamed on NIR instruments, when, in most cases, the laboratory reference method and other associated factors were at fault (Williams, 1987). Before assessing the accuracy of an NIR calibration, the error associated with the reference method should be known, but this is often ignored.

Pretreatment of spectral data

Because NIR spectra are affected by particle size, light scatter and path-length variation, pretreatment of the spectral data can often (but not always) improve calibration accuracy. Commonly used corrections include standard normal variate–detrending (SNV-D) (Barnes *et al.*, 1989), multiplicative scatter correction (MSC) (Martens and Naes, 1989) and derivatization. Detrending removes the linear and quadratic curvature of each spectrum, SNV scales each spectrum to have a standard deviation of 1.0 and MSC expands or contracts each spectrum and shifts it up or down to look most like the target spectrum (usually the mean of a file of spectra). First- or second-derivative mathematical treatments are most frequently employed.

Calibration methods

Multiple regression techniques relating NIR absorbance values (X variables) at selected wavelengths to reference values (Y variables) have the longest history of use in NIR calibration, and two common variations are step-up and stepwise regression. Step-up regression selects the X variable most highly correlated with Y and then progressively adds other X variables with the next highest partial correlations. This process is stopped when no additional improvement in correlation is found, by which time 'overfitting' occurs, which is detected by testing (validating) the model on an independent set of samples. Stepwise regression is similar, except that, when a new X variable is added, all variables are re-evaluated before continuing, until no further changes result in improvement to the overall fit (Shenk *et al.*, 1981). Errors are usually lower than for step-up regression, as more combinations of X variables are evaluated.

Alternative calibration methods include principal-component regression (PCR) (Cowe and McNicol, 1985), partial least-squares regression (PLS) (Martens and Naes, 1989) and artificial neural networks (Naes *et al.*, 1993). In PCR, the original spectral data are compressed into a small number of independent components, or scores, which are used in the regression equation instead of the spectra themselves. This overcomes the problem of intercorrelation between spectral data points. Typically, a model containing between five and ten principal components gives satisfactory predictions. PLS is similar to PCR, except that PLS uses the laboratory reference data (Y) as well as the spectral data (X) in computing the factors used in regression. This often results in better predictions. A variation of this procedure, called modified PLS (MPLS), is frequently used in forage calibrations and involves standardizing the variables in the regression after each factor is calculated (Shenk and Westerhaus, 1991).

There is no 'best' calibration technique to use in all situations. MPLS has been shown to be superior to multiple linear regression (MLR) (Shenk and Westerhaus, 1991; Windham and Flinn, 1992), but this is not always the case. Time is usually better spent on obtaining high-quality reference data for calibration than on trying endless combinations of data treatments and regression techniques.

Validation: how do you know the calibration is working?

Different approaches to validation of an NIR calibration equation may be needed, depending on the way in which the equation is used. The traditional procedure using MLR is to randomly select samples from the calibration population, omit them from the calibration process and use them as a validation set to test the calibration equation. Error is measured as standard error of prediction (SEP). This test is satisfactory if working with a closed population but is inadequate for open populations, such as in routine testing of widely diverse forage samples. In this situation, a separate, independent set of samples should be used to test the robustness of the equation before it is implemented. Windham *et al.* (1989) have recommended a protocol for this purpose.

An alternative now widely used in PCR or PLS/MPLS regression is cross-validation, where the calibration set is split into several groups and calibration progressively performed on each group until every sample has been used for both calibration and validation. The validation errors are combined into a standard error of cross-validation (SECV) (Shenk and Westerhaus, 1994). It can thus be argued that both calibration and validation are able to be done in one process.

The performance of a calibration equation on sets of validation samples will depend on the degree to which all sources of variation in the validation samples are encompassed in the calibration set. The equation must be monitored over time to ensure that its accuracy is maintained. Also, random samples during routine analysis should be set aside, analysed by the reference method and subjected to a monitoring test such as that recommended by Shenk *et al.* (1989).

Can a calibration equation be too broad?

It is now widely accepted that broad-based NIR calibrations for forage analysis are more useful than those based on separate species. Whilst the latter can be slightly more accurate, they lack robustness and involve more work and expense (Flinn, 1990; Smith and Flinn, 1991). Studies have also found that multi-forage calibrations (e.g. comprising hay, fresh forage, haylage and silage) can be nearly as accurate as those for single products, provided enough samples are present in the calibration population (Shenk and Westerhaus, 1993).

An alternative approach is to derive a local calibration for a new sample set, using samples in a broad-based calibration population which best match the spectra of the new samples and, if necessary, expanding this local calibration with a few of the new samples (Shenk *et al.*, 1993). The local calibrations in this study were found to be more accurate than the broad-based calibrations.

A disadvantage of this procedure is the increased number of equations required in routine analysis, increasing the complexity of the analytical system. Another development involves computing a specific calibration equation for each sample analysed, based on selected similar samples within a large database. Consistent reduction in SEP was found using this technique for forage data sets (Sinnaeve *et al.*, 1994), but it was considered too slow for real-time analysis.

Evaluation of Forages by NIR Spectroscopy: a Brief History

Measurement of forage quality by NIR was first reported by Norris *et al.* (1976). In what has become a classic paper, the authors found standard errors of 0.95, 3.1, 2.5, 2.1 and 3.5%, respectively, for crude protein (CP), neutral-detergent fibre (NDF), acid-detergent fibre (ADF), lignin and dry-matter digestibility (DMD) *in vitro*, measured in a diverse range of forage samples. This paper stimulated wide interest and a great many investigations soon began across the world to extend and refine the use of NIR spectroscopy in forage analysis. A significant achievement was the adoption of NIR as an official method by the Association of Official Analytical Chemists for the measurement of CP and ADF in forages (Barton and Windham, 1988).

The successful use of NIR to measure the major organic components of forages has been demonstrated by many workers, and was reviewed extensively by Murray (1993). As well as protein and fibre fractions, water-soluble carbohydrates have been measured successfully by NIR in perennial ryegress (Parnell and White, 1983) and total non-structural carbohydrates in lucerne roots (Brink and Marten, 1986) and tropical grasses (Brown *et al.*, 1987).

Introduction of alternative sampling devices for NIR instruments has allowed the analysis of fresh (undried) and unmillled forages (e.g. silage and fresh grass) by NIR spectroscopy (Shenk, 1992). Although errors are often slightly higher for components such as CP, ADF, NDF and DMD *in vitro* than on dried, ground samples, this is balanced by the ability to scan much larger samples, the avoidance of compositional losses and changes due to oven-drying, and a major reduction in analysis time and cost, due to no sample preparation being necessary (Sinnaeve *et al.*, 1994; de la Roza *et al.*, 1996; Kennedy *et al.*, 1996). Park *et al.* (1996) have also reported that NIR spectroscopy can be used for the prediction of fermentation characteristics and biological parameters on undried silages.

A major limitation in the use of NIR spectroscopy for forage analysis is the inadequacy of many conventional laboratory reference methods. The problem was described succinctly by Murray (1988) when he stated that 'we are using 19th century chemistry to calibrate 20th century technology'.

The Role of NIR Spectroscopy for Forage Evaluation

Predicting the digestibility in vivo *of forages*

The use of NIR spectroscopy to predict digestibility *in vivo* has been widely reported by several research groups, including work on mixed forages by Lindgren

(1983) and on grasses by Robert *et al.* (1986). In most cases, NIR spectroscopy provided a better prediction of digestibility than traditional laboratory procedures. In the UK, the most important development was reported by Barber *et al.* (1990). For grass silage, they showed for the first time that NIR spectroscopy was able to predict the organic-matter digestibility (OMD) *in vivo* more accurately than a range of commonly used laboratory methods. This work used a stepwise MLR of eight first- and second-derivatized spectral terms. A summary of the results obtained after validation of the various relationships is shown in Table 15.1.

The work of Barber *et al.* (1990) highlighted the fact that most traditional relationships that had been developed for predicting digestibility from various laboratory methods were, strictly speaking, calibrations and had never been exposed to the validation procedure necessary for NIR spectroscopy calibrations. Recently, De Boever *et al.* (1996) have also shown for grass silages made in Belgium that NIR spectroscopy provided the best calibration relationship for OMD *in vivo* with a variance accounted-for value (R^2) of 0.79, compared with values of 0.68, 0.64 and 0.53 for cellulase digestibility, rumen-fluid digestibility and acid-detergent lignin determinations, respectively.

NIR calibrations have also been successfully developed for prediction of the digestibility and metabolizable energy (ME) content of other forages including grasses and cereal straws. Givens *et al.* (1992b) showed that the metabolizable energy (ME) content of mixed-growth grasses could be predicted by a five-term NIR calibration equation and that this technique was equal to the best traditional laboratory method (pepsin–cellulase). One of the main advantages of the five-term NIR spectroscopy equation was that it did not discriminate between spring, summer and autumn grasses or between grasses harvested in different years.

Gordon *et al.* (1998) have recently reported on an extensive study that compared a range of sample preparation methods and NIR spectral model types, using fresh grass silage, for predicting OMD *in vivo* and voluntary intake. They compared a total of eight sample preparation/NIR scanning methods involving three extents of comminution, two liquid extracts and scanning via either an external probe (1100–2200 nm) or an internal cell (1100–2500 nm). In addition, all log $1/R$ spectral data were studied using three modelling techniques, together with a range of data transformations. A summary of the main findings is presented in Table 15.2.

Table 15.1. Comparison of NIR spectroscopy with several laboratory methods for predicting the organic matter digestibility *in vivo* of grass silage (after Barber *et al.*, 1990).

	Validation statistics			
Method	R^2	SEP	Slope	Bias
NIR – 8-term	0.76	2.6	0.93	−0.79
In vitro OMD	0.64	3.6	0.89	−1.85
Pepsin–cellulase	0.40	4.7	0.71	2.33
Acetyl bromide lignin	0.14	5.3	0.48	1.18
Modified acid-detergent fibre	0.20	5.1	0.52	−0.59

SEP, standard error of prediction (%).

Table 15.2. Analysis of variance of standard error of prediction data for comparisons of sample preparation methods and spectral model type for predicting organic-matter digestibility *in vivo* and voluntary intake of fresh (undried) grass silage (after Gordon *et al.*, 1998).

	Voluntary intake (g kg^{-1} W$^{0.75}$)	OMD *in vivo* (g kg^{-1} OM)
Spectral model type (means across sample method)		
MPLS	6.57	38
PCR	6.68	41
MLR	7.37	40
SEM	0.123	0.9
Significance	$P < 0.05$	$P < 0.06$
Silage sample preparation method (means across spectral model type)		
Scanned by external NIR probe		
Intact, lightly packed in open tray	7.88	41
Intact, lightly packed in tube	5.97	36
Coarsely chopped, lightly packed in tube	6.86	34
Finely milled, lightly packed in tube	6.69	33
Scanned by internal NIR cups		
Coarsely chopped, lightly packed in NIR cup	6.73	29
Finely milled, lightly packed in NIR cup	6.14	26
Expressed liquor, 5 drops on viscous sample cup	6.82	54
Extracted eluent, 5 drops on viscous sample cup	8.37	62
SEM	0.349	1.4
Significance	$P < 0.01$	$P < 0.001$

MPLS, modified partial least squares; MLR, multiple linear regression; PCR, principal-components regression; SEM, standard error of the mean.

For predicting OMD *in vivo*, they showed that the fine-cup method of preparation was the best, although, for voluntary intake, it came second to the intact tube. However, the authors highlight the fact that, when compared on the basis of SECV (valid for MPLS and PCR models), both OMD *in vivo* and intake predictions were always best using the fine-sample cup method. Overall, these findings support the overall concept that enhanced comminution of forages prior to NIR analysis is an important element in improving the accuracy of predictions of both OMD *in vivo* and intake.

Prediction of voluntary intake of forages

In many ruminant production systems, energy intake from forage is more influenced by voluntary dry-matter (DM) intake than by the energy concentration of the forage DM. Accordingly, there have been many attempts to predict intake from various attributes describing forage composition (for a review, see Minson, 1982), although there are often many non-forage factors that may influence voluntary intake, including the amount of supplementary feed given and animal and environmental factors.

It was proposed by the Agricultural and Food Research Council (AFRC, 1991) that NIR spectroscopy could be one of the most promising areas for predicting the voluntary intake of grass silage, since the spectra contain information on the entire chemistry of the feed. In fact, NIR spectroscopy has been successfully used to predict the intake of mainly non-fermented forages. The early work of Norris *et al.* (1976) and the more recent studies of Ward *et al.* (1982) and Flinn *et al.* (1992) are summarized in Table 15.3. The work of Flinn *et al.* (1992) was unusual in that it involved the use of an MPLS regression, where the NIR spectral data of faeces were related to forage intake, intake having been originally calculated from the *n*-alkane dosing procedure of Mayes *et al.* (1986).

Recent studies by Park *et al.* (1997) and Steen *et al.* (1998) have shown that NIR spectroscopy can also accurately predict the voluntary intake of grass silage by cattle. Park *et al.* (1997) measured the DM intakes of 136 grass silages, using 192 individually fed beef cattle (range 45 to 113 g DM kg^{-1} $W^{0.75}$), and showed that an MPLS spectra relationship could achieve a prediction accuracy (SECV) of ±5.05 g DM kg^{-1} $W^{0.75}$ (R^2 cross-validation 0.78). Steen *et al.* (1998) showed NIR spectroscopy to be superior to some 35 other measurements of silage quality, digestibility *in vivo* and rumen DM degradability *in situ*. Of the other measurements, the next best relationships with intake were obtained with electrometric titration (R^2 0.53), digestibility *in vivo* (R^2 0.30) and DM degradability *in situ* (R^2 0.28).

The fact that NIR spectroscopy seems to offer considerable potential for predicting voluntary intake of forages suggests that such studies should be extended to examine such effects as concentrate supplementation and the impact of feeding forage mixtures.

Table 15.3. Prediction of voluntary dry-matter intake of forages by NIR spectroscopy.

Forage type	Number of measurements	Intake (g kg^{-1} $LW^{0.75}$ day^{-1})		NIR prediction		Reference
		Species	Range	R^2	SEP	
Mixed	76	Sheep	39.7–114.3	0.62	7.8	Norris *et al.* (1976)
Grazed pasture	21	Cattle	52.6–112.3	0.72	9.6	Ward *et al.* (1982)
Grazed pasture	80	Sheep	430–1458*	0.80	140	Flinn *et al.* (1992)

* Organic-matter intake $animal^{-1}$ day^{-1}; LW, live-weight.
SEP Standard error of prediction.

Prediction of new aspects of energy value of forages

It is now becoming clear that the unifying concept of digestibility or energy value is inadequate and, in particular, it is now recognized (AFRC, 1992) that it is important to distinguish between the energy in the forages that is available for rumen microbial growth (fermentable energy) and that which is directly available to the host animal. The importance of estimating fermentable energy is because it represents the portion of dietary energy available for microbial protein synthesis, which in many situations represents at least 0.7 of the total α-amino nitrogen available to the host animal.

Of particular importance for ensiled forages is the fact that they may contain substantial amounts of lactic acid and volatile fatty acids (VFA), which provide little or no fermentable energy (AFRC, 1992; Deaville and Givens, 1998b). Thus, an important future role of NIR spectroscopy would be to predict the concentrations of the individual or total fermentation acids (see Abrams *et al.*, 1988; Sinnaeve *et al.*, 1994). An alternative approach to the problem has been reported by Snyman and Joubert (1992), who showed that fermentation acids in sorghum silages could be adequately predicted by NIR spectroscopy if the acids were first converted to their corresponding salts before microwave-drying. Recently, Deaville and Givens (1996a) (Table 15.4) showed that an MPLS model of the NIR spectra obtained from 800 fresh (undried) silages could be used to predict lactic acid and VFA.

The prediction of silage fermentation acids represents an approach to estimating fermentable energy content by difference. An alternative approach would be a direct prediction of fermentability. The gas-production technique, originally described by Menke *et al.* (1979) and modified by, for example, Cone *et al.* (1996), is a method that has the potential to measure fermentability more directly, including fermentation rate. Recent integrated studies on forages, using both the gas-production technique and NIR spectroscopy, by Herrero *et al.* (1996) showed that it was possible to derive reasonable NIR predictive relationships for cumulative gas volumes produced between 6 and 48 h of incubation, although rates of gas production could not be predicted.

Table 15.4. Estimate of fermentation acids in silages by NIR spectroscopy (Deaville and Givens, 1996a).

	Calibration		Validation	
Fermentation acid	R^2	SEC	R^2	SECV
Lactic acid	0.73	4.24	0.72	4.37
Acetic acid	0.69	1.72	0.66	1.80
Propionic acid	0.59	0.47	0.56	0.49
n-Butyric acid	0.87	0.82	0.86	0.87

SEC, standard error of calibration (g kg^{-1} fresh weight); SECV, standard error of cross-validation (g kg^{-1} fresh weight).

For most forages, the cell-wall fraction represents the major fermentable energy-yielding substrate. However, the rumen degradability/fermentability of cell walls is very variable. Generally, there appear to be relatively few estimates of the extent and rate of degradation of forages, although some *in vitro* estimates of cell-wall digestion dynamics of grass silages have been presented by Foulkes (1993) and Deaville and Givens (1996b). The latter authors also showed that NIR spectroscopy gives considerable scope for predicting potential and effective cell-wall degradability, although the rate of degradation and the lag phase could not be predicted (Table 15.5). As proposed by Herrero *et al.* (1996), the failure of NIR spectroscopy to predict dynamic values may relate to the exponential nature of the models currently used, thereby altering the distribution and range of the parameters and thus preventing a good fit by the NIR calibration approaches, which are all multivariate and linear.

Prediction of new aspects of protein value of forages

Many reports in the literature show the use of NIR spectroscopy to predict the CP content of forages and other feeds. However, for ruminant animals, protein quality is more accurately described by estimating three aspects of protein digestion in the rumen: the immediately soluble fraction, the insoluble but degradable fraction and the rate of degradation of the degradable fraction (AFRC, 1992). These measurements are normally carried out using the *in situ* technique, which not only requires the use of specially modified animals, but is also slow and expensive.

For 19 previously frozen fresh grasses, Waters and Givens (1992) demonstrated that it was possible to derive NIR calibrations for the nitrogen degradation values derived *in situ*, although overall the calibrations were relatively weak and the shortage of samples prevented any validation. Better results were obtained when the population was split into primary growths and regrowths, although the population sizes were very small.

Since a very high but variable proportion (0.45–0.75) of the nitrogen present in grass silage is water-soluble, and hence soluble in the rumen, the ability to predict nitrogen solubility by NIR spectroscopy would be a significant improvement over simply predicting total nitrogen or CP concentration. Abrams *et al.* (1988)

Table 15.5. Prediction of *in vitro* cell-wall degradability characteristics of grass silage ($n = 113$) by NIR spectroscopy (Deaville and Givens, 1996b).

Degradability parameter	Mean	Calibration		Cross-validation	
		R^2	SEC	R^2	SECV
Potential degradability (%)	73.3	0.73	3.76	0.66	4.27
Effective degradability (%)	53.8	0.83	2.42	0.70	3.21
Rate of degradation (h^{-1})	0.07	0.50	0.01	0.36	0.01
Lag phase (h)	2.49	0.32	0.81	0.17	0.89

SEC, standard error of calibration; SECV, standard error of cross-validation.

showed that with various fresh (undried) silages (DM content 394 ± 137 g kg^{-1}), the solubility and insolubility of nitrogen in buffered sodium chloride solution could be estimated by a six-term multiple stepwise regression (MSR) calibration of second-order derivatized log $1/R$ spectral data (Table 15.6). Recent work (E.R. Deaville and D.I. Givens, unpublished) with fresh (undried) grass silage showed that an MPLS calibration had an ability to predict the soluble nitrogen and ammonia nitrogen concentrations.

Use of NIR Spectra to Monitor Forage Digestion

Most of the initial NIR calibrations were derived by MLR without any inherent knowledge about which regions of the NIR spectrum contained the most information in relation to forage digestibility. Reviews of the literature (Clark and Lamb, 1991; Deaville and Baker, 1993) relating to the prediction of forage digestibility by NIR spectroscopy indicates that several wavelength regions regularly appear as being the most important. Table 15.7 presents a brief summary of some of the reports.

Table 15.6. Estimation of nitrogenous constituents in silage by NIR spectroscopy (after Abrams *et al.*, 1988).

Nitrogen fraction	Mean value (g kg^{-1} fresh weight)	SD	Calibration statistics R^2	SEC
Total nitrogen	9.55	4.49	0.98	0.70
Soluble nitrogen	3.48	1.99	0.92	0.58
Insoluble nitrogen	5.97	2.70	0.98	0.35

SD, standard deviation; SEC, standard error of calibration.

Table 15.7. Primary and secondary NIR spectroscopy wavelength segments used to predict forage digestibility.

Forage	Predicted term	Wavelength (nm) Primary	Secondary	Reference
Mixed	DMD *in vivo*	2266	1662	Norris *et al.* (1976)
Grass	ME	1668	1364	Lindgren (1983)
Grass silage	Lignin	1658	2286	Murray *et al.* (1987)
Hays	DMD *in vivo*	1326	2266	Coelho *et al.* (1988)
Hays/grass	ME	1666	1690	Lindgren (1988)
Grass silage	OMD *in vivo*	2266	1662	Kridis (1989)
Grass	ME	1658	2280	Givens *et al.* (1992b)
Cereal straw	OMD *in vitro*	1662	ns	Guzmán *et al.* (1996)

DMD, Dry-matter digestibility; ns, not stated.

Table 15.7 shows that the spectral regions around 1650–1670 and 2260–2280 nm consistently appear as the most important regions associated with digestibility. Interestingly, the wavelength segments associated with gas production (Herrero *et al.*, 1996) indicated that the primary wavelength is in the same region for digestibility, located between 1664 and 1696 nm. Subsequently, by sequentially subtracting spectra associated with the residues of *in situ* bag incubations in the rumen, Givens *et al.* (1992a) demonstrated that, in cereal straws, the 1650 and 2254 nm regions related to the indigestible fractions of the forage. Deaville *et al.* (1992) went on to show that these regions related to the most indigestible fractions of the plant cell wall and Deaville and Givens (1998a) have shown that these regions are also important in relation to the rumen degradation of grass silage, maize silage and fresh grass. Other findings (Russell *et al.*, 1989; Deaville and Givens, 1998a) indicate that these regions relate to lignin bonding and this gives further confidence that regions of the NIR spectrum have biological meaning. Further aspects of this have also been discussed by Deaville and Baker (1993).

Minerals and Antinutritional Factors

Mineral analysis of forages by NIR appears unlikely, as minerals do not absorb energy in the NIR region. However, correlations between minerals and other organic components allow reasonable NIR calibrations to be obtained in some cases. In a study involving a wide range of minerals in various forages, Clark *et al.* (1987) concluded that accurate NIR analysis was limited to calcium (Ca), phosphorus (P), potassium (K) and magnesium (Mg), which were found to have coefficients of variation (CV) ranging from 10 to 20%. The CV values for other minerals generally exceeded 20%. Smith *et al.* (1991) obtained a calibration for Mg in perennial ryegrass with a CV of 16% and concluded that NIR could be a useful tool for preliminary screening of ryegrass lines for Mg in a breeding programme. Shenk and Westerhaus (1994) concluded that mineral analysis by NIR can be useful if the alternative is to assume an average value from a feed-composition table.

NIR has also been used to screen various forages for antinutritional factors/secondary plant products. These are normally present in small quantities but may be detectable if they affect NIR spectra, again by means of secondary correlations. Examples include total alkaloids and condensed and total phenolics.

Conclusions

NIR spectroscopy has revolutionized the analysis and nutritional characterization of forages. This is extremely important since, worldwide, human food production from ruminant animals is highly dependent on efficient utilization of a diverse range of forage sources. The developments in NIR spectroscopy have allowed for the rapid evaluation of forages thus allowing for rapid, strategic and economic decisions to be made regarding appropriate supplementation or forage upgrading. Current evaluation systems for describing the nutritional characteristics of forages and other feeds for ruminants are undergoing very substantial changes in many parts of the world, with greater emphasis being placed on estimating the nutrient supply to the animal rather than the all-embracing values of energy and protein. Such approaches will undoubtedly require more complex methods of evaluation and it seems likely that they will only be put into widespread practice if they can be predicted by NIR spectroscopy.

References

Abrams, S.M. (1989) Calibration – populations. In: Marten, G.C., Shenk, J.S. and Barton, F.E., II (eds) *Near Infrared Reflectance Spectroscopy (NIRS): Analysis of Forage Quality.* USDA Agriculture Handbook No. 643, Washington, DC, p. 37.

Abrams, S.M., Shenk, J.S. and Harpster, H.W. (1988) Potential of near infrared reflectance spectroscopy for analysis of silage composition. *Journal of Dairy Science* 71, 1955–1959.

AFRC (1991) AFRC Technical Committee on Responses to Nutrients, Report Number 8. Voluntary intake of cattle. *Nutrition Abstracts and Reviews (Series B)* 61, 815–823.

AFRC (1992) AFRC Technical Committee on Responses to Nutrients, Report No. 9. Nutritional requirements of ruminant animals: protein. *Nutrition Abstracts and Reviews (Series B)* 62, 787–835.

Barber, G.D., Givens, D.I., Kridis, M.S., Offer, N.W. and Murray, I. (1990) Prediction of the organic matter digestibility of grass silage. *Animal Feed Science and Technology* 28, 115–128.

Barnes, R.J., Dhanoa, M.S. and Lister, S.J. (1989) Standard normal variate transformation and detrending of near infrared diffuse reflectance spectra. *Applied Spectroscopy* 43, 772–777.

Barton, F.E., II and Windham, W.R. (1988) Determination of acid detergent fibre and crude protein in forages by near infrared reflectance spectroscopy: collaborative study. *Journal of the Association of Official Analytical Chemists* 71, 1162–1167.

Brink, G.E. and Marten, G.C. (1986) Analysis of alfalfa root carbohydrate concentration by near infrared reflectance spectroscopy. *Crop Science* 26, 159–161.

Brown, W.F., Piacitelli, C.K. and Mislevy, P. (1987) Near infrared reflectance analysis of total nonstructural carbohydrate concentration in tropical grasses. *Crop Science* 27, 786–788.

Clark, D.H. and Lamb, R.C. (1991) Near infrared reflectance spectroscopy: a survey of wavelength selection to determine dry matter digestibility. *Journal of Dairy Science* 74, 2200–2205.

Clark, D.H., Mayland, H.F. and Lamb, R.C. (1987) Mineral analysis of forages with near infrared reflectance spectroscopy. *Agronomy Journal* 79, 485–490.

Coelho, M., Hembry, F.G., Barton, F.E. and Saxton, A.M. (1988) A comparison of microbial, enzymatic, chemical and near infrared reflectance spectroscopy methods in forage evaluation. *Animal Feed Science and Technology* 20, 219–231.

Cone, J.W., van Gelder, A.H., Visscher, G.J.W. and Oudschoorn, L. (1996) Influence of rumen fluid and substrate concentration on fermentation kinetics measured with a fully automated time related gas production apparatus. *Animal Feed Science and Technology* 61, 113–128.

Cowe, I.A. and McNicol, J.W. (1985) The use of principal components in the analysis of near infrared spectra. *Applied Spectroscopy* 39, 257–266.

Dardenne, P. (1996) Stability of NIR spectroscopy equations. *NIR News* 7, 8–9.

Deaville, E.R. and Baker, C.W. (1993) Spectral interpretations in relation to feed characterisation. In: *Proceedings of ADAS Drayton Feed Evaluation Unit Conference, NIRS Developments in Agriculture and Food, 12 October 1993.* ADAS, Oxford, 20 pp.

Deaville, E.R. and Givens, D.I. (1996a) Prediction of the non-fermentable energy fraction of grass silage using near infrared reflectance spectroscopy. *Animal Science* 60, 681 (Abstract).

Deaville, E.R. and Givens, D.I. (1996b) The prediction of the *in vitro* cell wall degradability characteristics of grass silage by near infrared reflectance spectroscopy. *Animal Science* 60, 681 (Abstract).

Deaville, E.R. and Givens, D.I. (1998a) Regions of normalised near infrared reflectance difference spectra related to the rumen degradation of fresh grass, grass silage and maize silage. *Animal Feed Science and Technology* 72, 41–51.

Deaville, E.R. and Givens, D.I. (1998b) Investigation of direct gas production from silage fermentation acids. In: *Proceedings of the British Society of Animal Science 1998*. British Society of Animal Science, Pericuik, p. 64.

Deaville, E.R., Givens, D.I. and Baker, C.W. (1992) Use of normalised NIR difference spectra to identify regions related to digestion of cell walls in straws. In: Hildrum, K.L., Isaksson, T., Naes, T. and Tandberg, A. (eds) *Near Infrared Spectroscopy*. Ellis Horwood, Chichester, pp. 219–222.

De Boever, J.L., Cottyn, B.G., De Brabander, D.L., Vanacker, J.M. and Boucqué, C.V. (1996) Prediction of the feeding value of grass silages by chemical parameters, *in vitro* digestibility and near-infrared reflectance spectroscopy. *Animal Feed Science and Technology* 60, 103–115.

de la Roza, B., Martinez, A., Modrono, S. and Santos, B. (1996) Determination of the quality of fresh silages by near infrared reflectance spectroscopy. In: Davies, A.M.C. and Williams, P.C. (eds) *Near Infrared Spectroscopy: The Future Waves*. NIR Publications, Charlton, UK, pp. 537–541.

Flinn, P.C. (1990) Analysis of ruminant feeds in southern Australia using near infrared reflectance (NIR) spectroscopy. MSc thesis, University of Aberdeen, UK.

Flinn, P.C. and Downes, G.J. (1996) The importance of near infrared spectroscopy in deciding appropriate feeding strategies for Australian livestock. In: Davies, A.M.C. and Williams, P.C. (eds) *Near Infrared Spectroscopy: The Future Waves*. NIR Publications, Charlton, UK, pp. 512–518.

Flinn, P.C., Windham, W.R. and Dove, H. (1992) Pasture intake by grazed sheep estimated using natural and dosed *n*-alkanes – a place for NIR? In: Hildrum, K.L., Isaksson, T., Naes, T. and Tandberg, A. (eds) *Near Infrared Spectroscopy*. Ellis Horwood, Chicester, pp. 175–178.

Foulkes, G.R. (1993) Cell wall digestion kinetics. MPhil thesis, University College of Wales, Aberystwyth, 110 pp.

Givens, D.I., Baker, C.W. and Zamime, B. (1992a) Regions of normalised near infrared reflectance difference spectra related to the rumen digestion of straws. *Animal Feed Science and Technology* 36, 1–12.

Givens, D.I., Moss, A.R., Baker, C.W. and Adamson, A.H. (1992b) Influence of growth type and season on the prediction of the metabolisable energy content of herbage by near infrared reflectance spectroscopy. *Animal Feed Science and Technology* 37, 281–296.

Givens, D.I., De Boever, J.L. and Deaville, E.R. (1997) The principles, practice and future applications of near infrared spectroscopy for predicting the nutritive value of foods for animals and humans. *Nutrition Research Reviews* 10, 83–114.

Gordon, F.J., Cooper, K.M., Park, R.S. and Steen, R.W.J. (1998) The prediction of intake potential and organic matter digestibility of grass silage by near infrared spectroscopy of undried samples. *Animal Feed Science and Technology* 70, 339–351.

Guzmán, J.L., Garrido, A., Gómez-Cabrera, A., Guerrero, J.E. and Murray, I. (1996) Use of NIRS to estimate the improvement in digestibility after ammonia treatment of cereal straws. *Animal Feed Science and Technology* 57, 149–157.

Herrero, M., Murray, I., Fawcett, R.H. and Dent, J.B. (1996) Prediction of the *in vitro* gas production and chemical composition of kikuyu grass by near-infrared reflectance spectroscopy. *Animal Feed Science and Technology* 60, 51–67.

Kennedy, C.A., Shelford, J.A. and Williams, P.C. (1996) Near infrared spectroscopic analysis of intact grass silage and fresh grass for dry matter, crude protein and acid detergent fibre. In: Davies, A.M.C. and Williams, P.C. (eds) *Near Infrared Spectroscopy: The Future Waves*. NIR Publications, Charlton, UK, pp. 524–530.

Kridis, M.S.F. (1989) Energy evaluation of grass silage. PhD Thesis, University of Glasgow and West of Scotland College, Ayr, 302 pp.

Lindgren, E. (1983) The use of near infrared reflectance spectroscopy for predicting energy value of grass. *Swedish Journal of Agricultural Research* 13, 229–233.

Lindgren, E. (1988) Prediction of energy value and protein content of forages by near infrared reflectance spectroscopy. *Swedish Journal of Agricultural Research* 18, 21–26.

Martens, H. and Naes, T. (1989) *Multivariate Calibration*. John Wiley, New York, USA.

Mayes, R.W., Lamb, C.S. and Colgrove, P.M. (1986) The use of dosed and herbage *n*-alkanes as markers for the determination of herbage intake. *Journal of Agricultural Science, Cambridge* 107, 161–170.

Menke, K.H., Raab, L., Salewski, A., Steingass, H., Fritz, D. and Schneider, W. (1979) The estimation of digestibility and metabolisable energy content of ruminant feedstuffs from the gas production when they are incubated with rumen liquor *in vitro*. *Journal of Agricultural Science, Cambridge* 193, 217–222.

Minson, D.J. (1982) Effects of chemical and physical composition of herbage eaten upon intake. In: Hacker, J.B. (ed.) *Nutritional Limits to Animal Production for Pastures*. Commonwealth Agricultural Bureaux, Farnham Royal, pp. 167–182.

Murray, I. (1983) Fundamentals of NIR analysis of feedstuffs. In: *NIR Analysis: How Near Infrared Reflects Composition*. Chemistry Division, School of Agriculture, Aberdeen, Scotland, 24 pp.

Murray, I. (1988) Aspects of the interpretation of NIR spectra. In: Creaser, C.S. and Davies, A.M.C. (eds) *Analytical Applications of Spectroscopy*. Royal Society of Chemistry, London, UK, pp. 9–20.

Murray, I. (1993) Forage analysis by near infrared reflectance spectroscopy. In: Davies, A., Baker, R.D., Grant, S.A. and Laidlaw, A.S. (eds) *Sward Management Handbook*. British Grassland Society, Reading, UK, pp. 285–312.

Murray, I. and Williams, P.C. (1979) Chemical principles of near infrared technology. In: Williams, P.C. and Norris, K. (eds) *Near Infrared technology in the Agricultural and Food Industries*. pp. 17–34.

Murray, I., Kridis, M.S.F., Barker, G.D., Givens, D.I. and Offer, N.W. (1987) *Grass Silage in* vivo *Digestibility Prediction by NIR*. CEC Monograph, Brussels.

Naes, T., Kvaal, K., Isaksson, T. and Miller, C. (1993) Artificial neural networks in multivariate calibration. *Journal of Near Infrared Spectroscopy* 1, 1–12.

Norris, K.H. (1989a) Introduction – definition of NIRS analysis. In: Marten, G.C., Shenk, J.S. and Barton, F.E., II (eds) *Near Infrared Reflectance Spectroscopy (NIRS): Analysis of Forage Quality*. USDA Agriculture Handbook No. 643, Washington, DC, p. 6.

Norris, K.H. (1989b) Equipment – NIRS instrumentation. In: Marten, G.C., Shenk, J.S. and Barton, F.E., II (eds) *Near Infrared Reflectance Spectroscopy (NIRS): Analysis of Forage Quality*. USDA Agriculture Handbook No. 643, Washington, DC, pp. 12–17.

Norris, K.H. and Hart, J.R. (1965) Direct spectrophotometric determination of moisture content of grain and seeds. In: Wexler, A. (ed.) *Principles and Methods of Measuring Moisture in Liquids and Solids*, Vol. 4. Reinhold, New York, USA, pp. 19–25.

Norris, K.H., Barnes, R.F., Moore, J.E. and Shenk, J.S. (1976) Predicting forage quality by infrared reflectance spectroscopy. *Journal of Animal Science* 43, 889–897.

Osborne, B.G., Fearn, T. and Hindle, P.H. (1993) *Practical NIR Spectroscopy with Applications in Food and Beverage Analysis*. Longman Scientific and Technical, Harlow, 227 pp.

Park, R.S., Gordon, F.J. and Steen, R.W.J. (1996) The use of near infrared reflectance spectroscopy (NIRS) on undried silages to predict a range of chemical and biological parameters. In: Jones, D.I.H., Jones, R., Dewhurst, R., Merry, R. and Haigh, P.M. (eds) *Proceedings of the XI International Silage Conference*. IGER, Aberystwyth, Wales, pp. 66–67.

Park, R.S., Gordon, F.J., Agnew, R.E., Barnes, R.J. and Steen, R.W.J. (1997) The use of near infrared reflectance spectroscopy on dried samples to predict biological parameters of grass silage. *Animal Feed Science and Technology* 74, 2664–2673.

Parnell, A. and White, J. (1983) The use of near infrared reflectance analysis in predicting the digestibility and the protein and water soluble carbohydrate contents of perennial ryegrass. *Journal of the National Institute of Agricultural Botany* 16, 221–227.

Robert, P., Bertrand, D. and Demarquilly, C. (1986) Prediction of forage digestibility by principal component analysis of near infrared reflectance spectra. *Animal Feed Science and Technology* 16, 215–224.

Russell, J.D., Murray, I. and Frazer, A.R. (1989) Near and mid-infrared studies of the cell wall structure of cereal straw in relation to its rumen degradability. In: Chesson, A. and Ørskov, E.R. (eds) *Physico-chemical Characterisation of Plant Residues for Industrial and Feed Use*. Elsevier Applied Science, London, pp. 13–24.

Shenk, J.S. (1992) NIRS analysis of natural agricultural products. In: Hildrum, K.I., Isaksson, T., Naes, T. and Tandberg, A. (eds) *Near Infrared Spectroscopy: Bridging the Gap between Data Analysis and NIR Applications*. Ellis Horwood, Chichester, UK, pp. 235–240.

Shenk, J.S. and Westerhaus, M.O. (1991) Population definition, sample selection and calibration procedures for near infrared reflectance spectroscopy. *Crop Science* 31, 469–474.

Shenk, J.S. and Westerhaus, M.O. (1993) Near infrared reflectance analysis with single and multiproduct calibrations. *Crop Science* 33, 582–584.

Shenk, J.S. and Westerhaus, M.O. (1994) The application of near infrared reflectance spectroscopy (NIRS) to forage analysis. In: Fahey, G.C., Jr, Collins, M., Mertens, D.R. and Moser, L.E. (eds) *Forage Quality, Evaluation, and Utilization*. American Society of Agronomy, Madison, Wisconsin, USA, pp. 406–449.

Shenk, J.S., Landa, I., Hoover, M.R. and Westerhaus, M.O. (1981) Description and evaluation of a near infrared reflectance spectro-computer for forage and grain analysis. *Crop Science* 21, 355–358.

Shenk, J.S., Westerhaus, M.O. and Abrams, S.M. (1989) Supplement 2. Protocol for NIRS calibration: monitoring analysis results and recalibration. In: Marten, G.C., Shenk, J.S. and Barton, F.E., II (eds) *Near Infrared Reflectance Spectroscopy (NIRS): Analysis of Forage Quality*. USDA Agriculture Handbook No. 643, Washington, DC, pp. 104–110.

Shenk, J.S., Fales, S.L. and Westerhaus, M.O. (1993) Using near infrared reflectance product library files to improve prediction accuracy and reduce calibration costs. *Crop Science* 33, 578–581.

Sinnaeve, G., Dardenne, P. and Agneessens, R. (1994) Global or local? A choice for NIR calibrations in analyses of forage quality. *Journal of Near Infrared Spectroscopy* 2, 163–176.

Smith, K.F. and Flinn, P.C. (1991) Monitoring the performance of a broadbased calibration for measuring the nutritive value of two independent populations of pasture using near infrared reflectance (NIR) spectroscopy. *Australian Journal of Experimental Agriculture* 31, 205–210.

Smith, K.F., Willis, S.E. and Flinn, P.C. (1991) Measurement of the magnesium concentration in perennial ryegrass (*Lolium perenne*) using near infrared reflectance spectroscopy. *Australian Journal of Agricultural Research* 42, 1399–1404.

Snyman, L.D. and Joubert, H.W. (1992) Near infrared reflectance analysis of the fermentation characteristics of silage prepared by chemical treatment to prevent volatilisation of fermentation end products. *Animal Feed Science and Technology* 37, 47–58.

Steen, R.W.J., Gordon, F.J., Dawson, L.E.R., Park, R.S., Mayne, C.S., Agnew, R.E., Kilpatrick, D.J. and Porter, M.J. (1998) Factors affecting the intake of grass silage by cattle and prediction of silage intake. *Animal Science* 66, 115–127.

Tilley, J.M.A. and Terry, R.A. (1963) A two-stage technique for the *in vitro* digestion of forage crops. *Journal of the British Grassland Society* 18, 104–111.

Ward, R.G., Smith, G.S., Wallace, J.D., Urquhart, N.S. and Shenk, J.S. (1982) Estimates of intake and quality of grazed range forage by near infrared reflectance spectroscopy. *Journal of Animal Science* 54, 399–402.

Waters, C.J. and Givens, D.I. (1992) Nitrogen degradability of fresh herbage: effect of maturity and growth type, and prediction from chemical composition and by near infrared reflectance spectroscopy. *Animal Feed Science and Technology* 38, 335–349.

Wetzel, D.L. (1983) Near infrared reflectance analysis – sleeper among spectroscopic techniques. *Analytical Chemistry* 55, 1165A–1176A.

Williams, P.C. (1987) Variables affecting near infrared reflectance spectroscopic analysis. In: Williams, P.C. and Norris, K.H. (eds) *Near Infrared Technology in the Agricultural and Food Industries*. American Association of Cereal Chemists, St Paul, USA, pp. 143–167.

Williams, P.C. and Norris, K.H. (1987) *Near Infrared Technology in the Agricultural and Food Industries*. American Association of Cereal Chemists, St Paul, USA, 330 pp.

Windham, W.R. and Flinn, P.C. (1992) Comparison of MLR and PLS regression in NIR analysis of quality components in diverse feedstuff populations. In: Hildrum, K.I., Isaksson, T., Naes, T. and Tandberg, A. (eds) *Near Infrared Spectroscopy: Bridging the Gap between Data Analysis and NIR Applications*. Ellis Horwood, Chichester, UK, pp. 459–464.

Windham, W.R., Mertens, D.R. and Barton, F.E., II (1989) Supplement 1. Protocol for NIRS calibration: sample selection and equation development and validation. In: Marten, G.C., Shenk, J.S. and Barton, F.E., II (eds) *Near Infrared Reflectance Spectroscopy (NIRS): Analysis of Forage Quality*. USDA Agriculture Handbook No. 643, Washington, DC, pp. 96–103.

16 NMR and Other Physicochemical Techniques for Forage Assessment

D.S. HIMMELSBACH

Richard B. Russell Agricultural Research Center, Agricultural Research Service, P.O Box 5677, Athens, GA 30604-5677, USA

Introduction

The use of high-resolution (HR) spectroscopic techniques (e.g. high-field nuclear magnetic resonance (NMR), mid-infrared (MIR), Raman spectroscopy and pyrolysis mass spectrometry (PyMS)) are finding increased usage in forage assessment. These techniques, as opposed to low-resolution (LR) spectroscopic techniques (e.g. near-infrared (NIR) and ultraviolet (UV) spectroscopy), generally provide more specific and detailed information of a primary nature. The trade-off for the gain in resolution is often the sacrifice of time, increased cost and the requirement to use more skilled operators. However, the employment of HR techniques is frequently necessary when the LR techniques fail to provide the desired information or require verification.

UV spectroscopy can be an excellent primary source of information, if the information sought is related to a single or isolated chromophore. However, forages tend to contain a multitude of chromophores, making such analyses difficult. NIR spectroscopy is generally not a source of direct information but an indirect or secondary technique, requiring independent calibration by a primary technique, using chemometric approaches (Kowalski, 1977; Martens *et al.*, 1991). The primary technique is normally some form of gravimetric analysis for moisture, protein, lignin or acid-detergent fibre (ADF) (AOAC, 1997). Although this approach is sufficient and cost-efficient for many needs, the dependence on proximate analyses limits the effectiveness of NIR spectroscopy (Preston *et al.*, 1997).

To date, the focus of the HR techniques has been directed towards providing a route to primary molecular structures. If the HR spectroscopic results can be linked to the LR techniques or be made more cost-effective themselves, the accessibility to more reliable results should be greatly enhanced. The use of chemometrics with any of the techniques can also increase their utility (Bro *et al.*, 1997). Since anatomy and chemical composition have combinatory effects on the digestibility of forages, additional benefits can be realized by the integration of any of these spectroscopic with microscopic techniques.

NMR Spectroscopy

Both solution and solid-state NMR have been found to be powerful tools for examination of forages and their residues. True solution NMR studies are limited to extractable components that are soluble in suitable deuterated NMR solvents. If the isolated material is pure, a wide array of single and multidimensional experiments may be employed to unequivocally assign the molecular structure and solution confirmation of a molecule (Bruch, 1996). Solid-state NMR is well suited for examining the intact structure of forage materials, without requiring extraction. It is especially suited for the *in situ* study of forages and their residues. However, it is more limited than solution NMR in the variety of experiments that can be performed and the line widths of signals are broader (i.e. show lower apparent resolution).

New experiments, which lie somewhere in the realm between the solution and solid methods, are now in vogue. These are experiments in which all or part of the sample is made mobile by hydration. The carbon-13 (^{13}C) version of this experiment permits separation of component spectra based on an increased separation of relaxation times, due to selective hydration (Newman, 1992). The proton (^{1}H) version of this experiment is characterized by relatively narrow line widths and permits greater use of multidimensional experiments (Gil *et al.*, 1997). These experiments show potential application to studies on forages, but have not yet been employed for that purpose.

Solution NMR spectroscopy

Solution NMR spectroscopy continues to benefit from the development of superconducting magnets with increased field strengths. Commercially available systems, with field strengths of 18.5 tesla (T) (800 MHz for protons), are now being delivered. This is a significant increase in field strength over the previously available highest fields of 11.5 and 14 T (500 and 600 MHz). This means the possibility of even greater signal dispersion and sensitivity and the ability to unravel even more complex polymeric structures, such as those habitually encountered in plant materials used as forages.

Examples of the effective use of the currently available solution NMR technology are provided in studies directed at determination of the specific involvement of hydroxycinnamic ester moieties in restricting the availability of cell-wall polysaccharides of forage grasses to ruminant digestion. Several hydroxycinnamic oligosaccharide esters have been isolated and their structures have been subjected to detailed characterization by solution NMR. Figure 16.1 shows examples of such compounds that are currently known. These compounds have been isolated from bagasse (Kato *et al.*, 1983, 1987, 1990), barley straw (Mueller-Harvey *et al.*, 1986), bamboo (Ishii and Hiroi, 1990a, b; Ishii *et al.*, 1990; Ishii, 1996), coastal Bermuda grass (Hartley *et al.*, 1990; Himmelsbach and Hartley, 1993; Himmelsbach *et al.*, 1994) and maize (Kato and Nevins, 1985).

The characterization of compound 3b in Fig. 16.1 provides a good example of how solution NMR can be used for unequivocal structural assignment of these

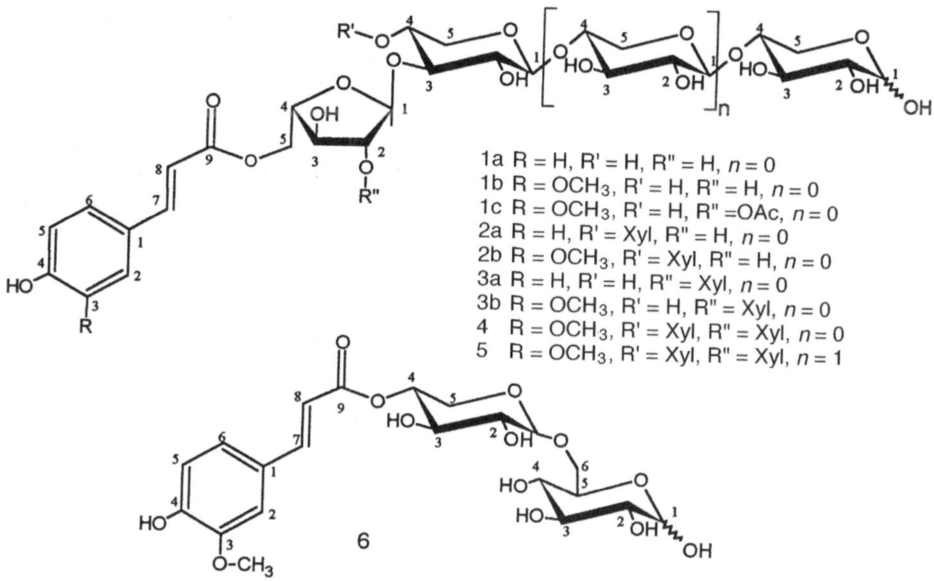

Fig. 16.1. Hydroxycinnamic oligosaccharide esters isolated from grass cell walls. Primary esters: (1a) *O*-[5-*O*-((*E*)-*p*-coumaryl)-α-L-arabinofuranosyl]-(1→3)-*O*-β-D-xylopyranosyl-(1→4)-D-xylopyranose; (1b) *O*-[5-*O*-((*E*)-*p*-feruloyl)-α-L-arabinofuranosyl]-(1→3)-*O*-β-D-xylopyranosyl-(1→4)-D-xylopyranose; (1c) *O*-[5-*O*-((*E*)-*p*-feruloyl)(2-*O*-acetyl)-α-L-arabinofuranosyl]-(1→3)-*O*-β-D-xylopyranosyl-(1→4)-D-xylo pyranose; (2a) *O*-β-D-xylopyranosyl-(1→4)-*O*-[5-*O*-((*E*)-*p*-coumaryl)-α-L-arabinofuranosyl]-(1→3)-*O*-β-D-xylopyranosyl-(1→4)-D-xylopyranose; (2b) *O*-β-D-xylopyranosyl-(1→4)-*O*-[(5-*O*-((*E*)-*p*-feruloyl)-α-L-arabinofuranosyl]-(1→3)-*O*-β-D-xylopyranosyl-(1→4)-D-xylopyranose; (3a) *O*-[5-*O*-((*E*)-*p*-coumaryl) (*O*-β-D-xylopyranosyl)- (1→2)-α-L-arabinofuranosyl]-(1→3)-*O*-β-D-xylopyranosyl-(1→4)-D-xylopyranose; (3b) *O*-[5-*O*-((*E*)-feruloyl)(*O*-β-D-xylopyranosyl)-(1→2) -α-L-arabinofuranosyl]-(1→3)-*O*-β-D-xylopyranosyl-(1→4)-D-xylopyranose; (4) *O*-β-D-xylopyranosyl-(1→4)-*O*-[5-*O*-((*E*)-feruloyl)(*O*-β-D-xylopyranosyl)-(1→2)-α-L-arabinofura nosyl]-(1→3)-*O*-β-D-xylopyranosyl-(1→4)-D-xylopyranose; (5) *O*-[5-*O*-((*E*)-feruloyl)(*O*-β-D-xylopyranosyl)-(1→2)-α-L-arabinofuranosyl-(1→3)]-*O*-β-D-xylopyranosyl-(1→4)-*O*-β-D-xylopyranosyl-(1→4)—D-xylopyranose. (6) Secondary ester: *O*-[4-*O*-((*E*)-feruloyl)-β-D-xylopyranosyl]-(1→6)-D-glucopyranose.

compounds. The process is begun by first assigning the proton NMR spectrum. This assignment is straightforward for the phenolic acid portion of the molecule, but requires a more sophisticated approach for the carbohydrate portion, due to the many overlapping signals. The assignment of the carbohydrate portion of the proton spectrum can be accomplished most easily by use of the one-dimensional (1-D) analogue of the 2-D {^{1}H, ^{1}H} homonuclear Hartmann-Hahn (HOHAHA) or total correlation spectroscopy (TOCSY) experiment (Davis and Bax, 1985). This involves the selective irradiation of a known position, such as the anomeric proton of each residue. This generates edited subspectra of each of the monosaccharide components. Then the simple one-bond proton–proton scalar coupling system is worked through to assign the proton signal for each unit. Knowing the proton assignments, the corresponding ^{13}C spectrum can be assigned by employing an experiment such as the inverse detected (^{1}H) one-bond {^{1}H, ^{13}C} heteronuclear multi-quantum correlation (HMQC) experiment (Bax and Subramanian, 1986). The results of this type of experiment are shown in Fig. 16.2.

After both the ^{1}H and ^{13}C spectra have been assigned for the individual subunits, the linkages between them can be found by employing an experiment such as the inverse-detected long-range {^{1}H, ^{13}C} heteronuclear multiple-bond connectivity (HMBC) experiment (Bax and Summers, 1986; Bax and Marion, 1988). The circled cross-peaks, shown in Fig. 16.3, establish the linkages between the subunits

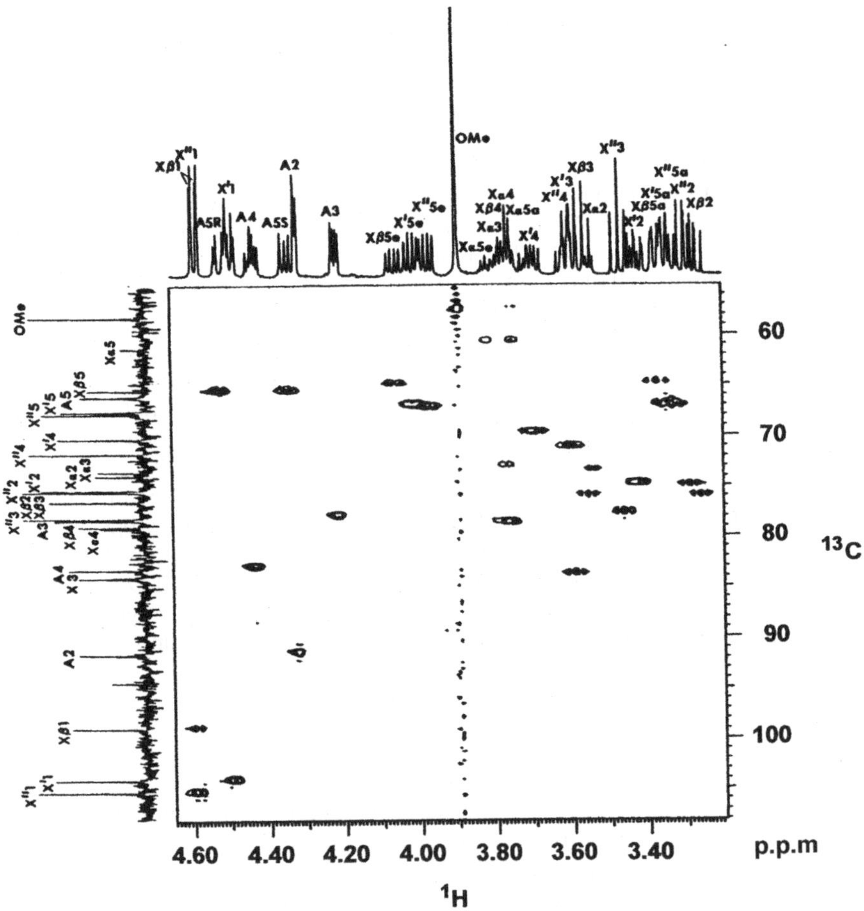

Fig. 16.2. Results of the inverse-detected one-bond HMQC experiment conducted on 5 mg of compound 3b (Fig. 16.1) obtained at 500 MHz for ^{1}H and 125 MHz for ^{13}C in a 20% d_6-acetone/D_2O solution at 25°C.

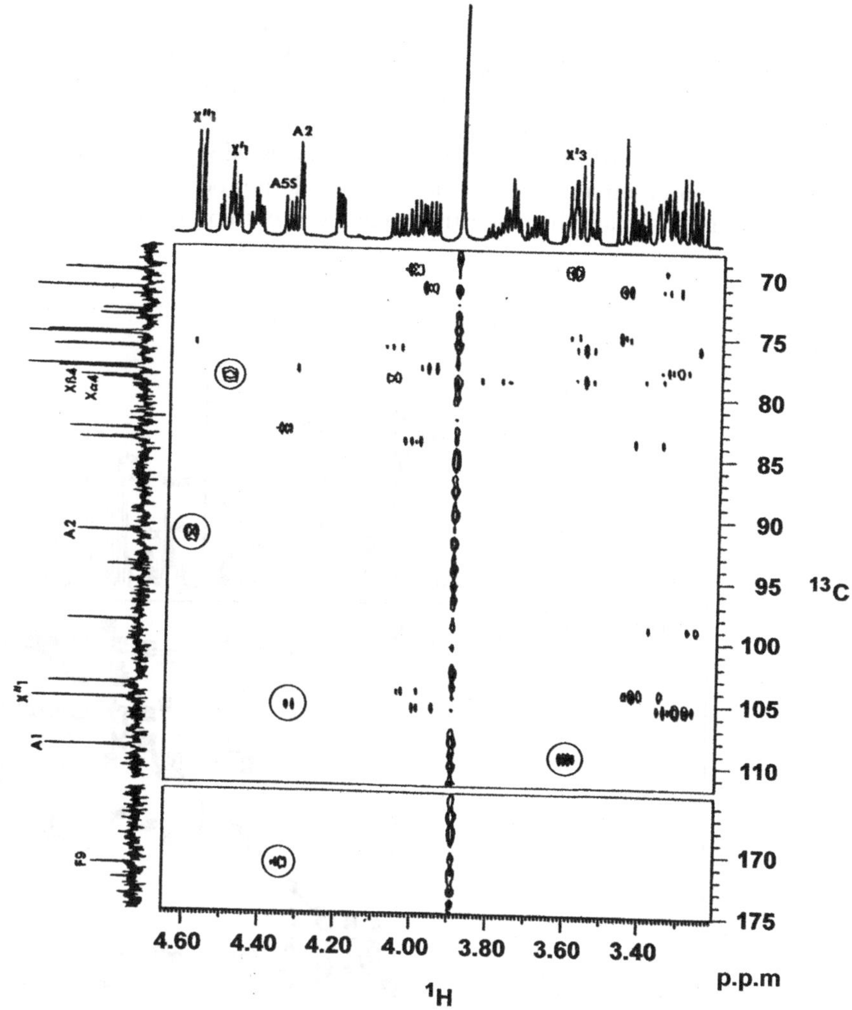

Fig. 16.3. Results of the inverse-detected long-range HMBC experiment conducted on compound 3b (Fig. 16.1) obtained at 500 MHz for ^{1}H and 125 MHz for ^{13}C in 20% d_6-acetone/D_2O solution at 25°C.

of 3b (Fig. 16.1). It shows that the ferulic acid is linked through an oxygen attached to the carbonyl carbon at F9 to the carbon at the 5 position (A5S proton shown) of the arabinofuranoside. This arabinofuranoside, in turn, is linked to two different xylopyranoside units at the 1 (A1) and 2 (A2) positions. At the 1 position, it is linked to the 3 position (X′3) of a xylopyranoside, which is further linked at its 1 position (X′1) to the 4 position of the reducing terminal xylopyranose, which displays signals for both α and β anomers (Xα4, Xβ4).

The actual solution conformation of the molecule in the NMR tube can be established by another 2-D experiment, the phase-sensitive {^1H,^1H} rotating-frame Overhauser enhancement spectroscopy (ROESY) experiment (Bax and Davis, 1985). The ROESY results are shown (Fig. 16.4b), compared with a {^1H,^1H} double-quantum filtered correlation spectroscopy (DFQ-COSY) experiment (Fig. 16.4a) (Aue *et al.*, 1976). The circled peaks in the ROESY contour map indicate the through-space interunit interactions. These do not occur in the DFQ-COSY. The stronger the signals, the closer the nuclei lie to each other, in space.

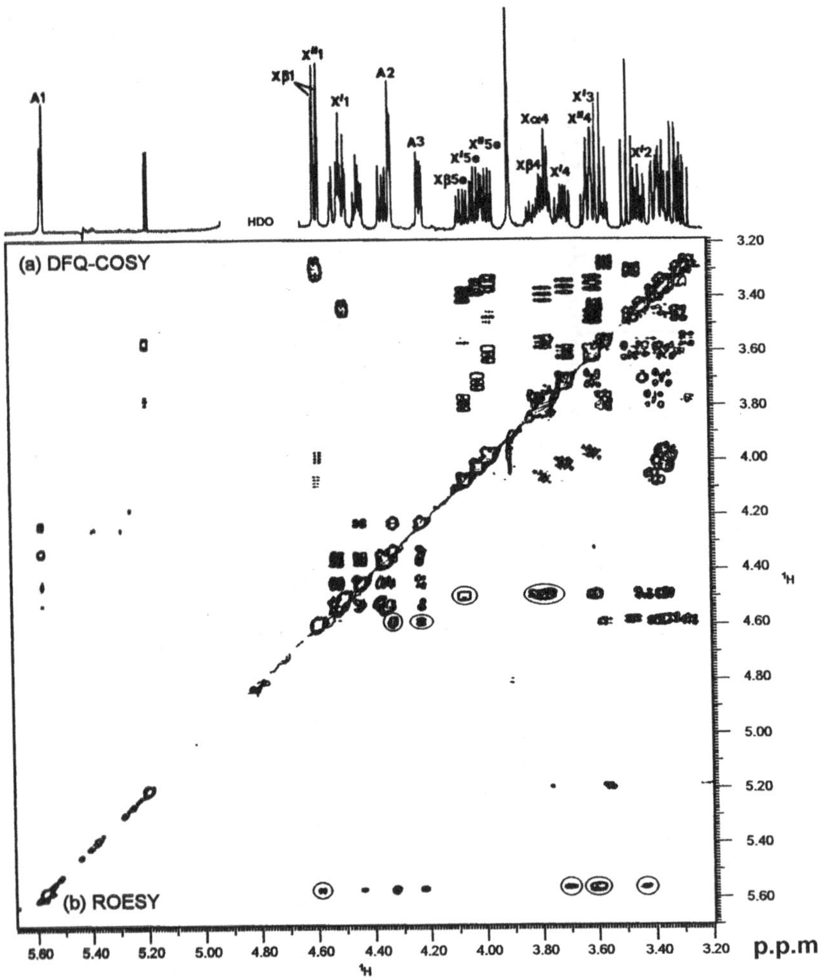

Fig. 16.4. {^1H,^1H} correlation plots of 3b (Fig. 16.1): (a) DFQ-COSY and (b) ROESY experiments. Both were obtained at 500 MHz for ^1H. The circles in the ROESY indicate the through-space interunit interactions.

In addition, the angle of rotation of the hydrocinnamic acid in relation to the arabinofuranoside can be estimated by analysing the three-bond coupling constants ($^3J_{HH}$) made between the proton attached to the A4 carbon and each of the two protons attached to the A5 carbon (Hoffmann *et al.*, 1992). The final conformation can then be expressed as a population of rotamers. In this case, the rotamer population was *trans* = 0.52, gauche$^-$ = 0.38 and gauche$^+$ = 0.10. These energy-minimized structures are shown in Fig. 16.5. Although these conformations will be fixed in the solid-state matrix of the cell walls of forage and not free to rotate, as in solution, this does give some idea of what the preferred (minimum-energy) conformation might be. The solid-state conformation may be critical to how these compounds are incorporated into the rest of the cell wall.

Solution NMR has also been used to provide insight as to how the hydroxycinnamic acids linked to hemicellulosic polysaccharides are incorporated into the cell wall. It has been shown that [2+2] cyclo-addition dimers can be formed in a solid-state photochemical reaction, yielding α-truxillic acids from compounds 1a and 1b (Fig. 16.1) (Hartley *et al.*, 1990). Proton NMR, along with mass spectrometry (MS), has been used to verify that the hydroxycinnamic acids all prefer to couple with a head-to-tail orientation towards each other (Morrison *et al.*, 1992). In the case of the ferulate polysaccharide esters, the 8 position of the hydroxycinnamic

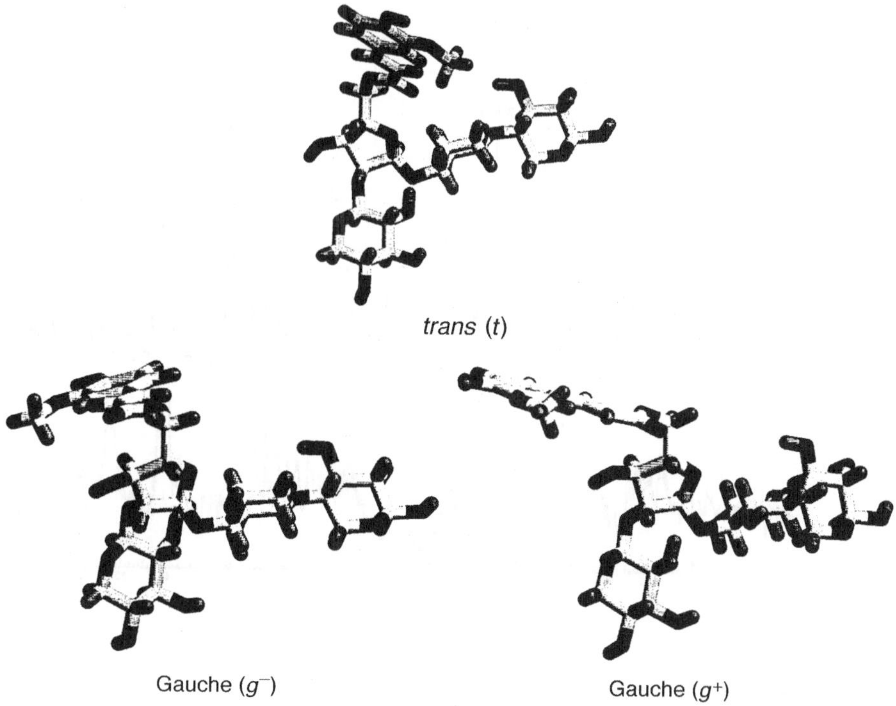

Fig. 16.5. The three minimum energy-staggered rotamers of 3b (Fig. 16.1) that describe its solution conformation in a 20% d_6-acetone/D_2O solution at room temperature.

acid has been shown (by the HMBC experiment) to couple to the β (8) position of both guaiacyl and syringyl alcohol monomers (Ralph *et al.*, 1995). The ferulate esters are also known to form dehydrodiferulates (Harris *et al.*, 1980) and possibly link to core lignin by ether bonds (Jung and Ralph, 1990), although this has not been verified by NMR spectroscopy.

Solution ^{13}C NMR has additionally been used to compare isolated forage lignins and show that warm-season grass forages differ from cool-season in that they incorporate *p*-coumaryl units (Himmelsbach and Barton, 1980). The solution ^{13}C NMR spectra of two representative spectra of these two classes of forages are shown in Fig. 16.6. The incorporation of *p*-coumaryl units in the warm-season forage (Fig. 16.6a) is indicated by the signals labelled '2,6H' and '3,5H'. It has been shown, from solution ^{13}C NMR spectra of bagasse, that treatment of lignin isolated from warm-season grasses with alkali releases most of the *p*-coumaryl units (Fernandez, 1990). This indicates that most of the *p*-coumaryl units exist as acids esterified to lignin and the remainder are esterified to arabinoxylans. This esterification to lignin has been confirmed, using the HMBC experiment, in isolated maize lignin, as compared with synthetic materials. Further, it has been revealed by this experiment that *p*-coumaric acid is extensively incorporated into the α and γ positions of

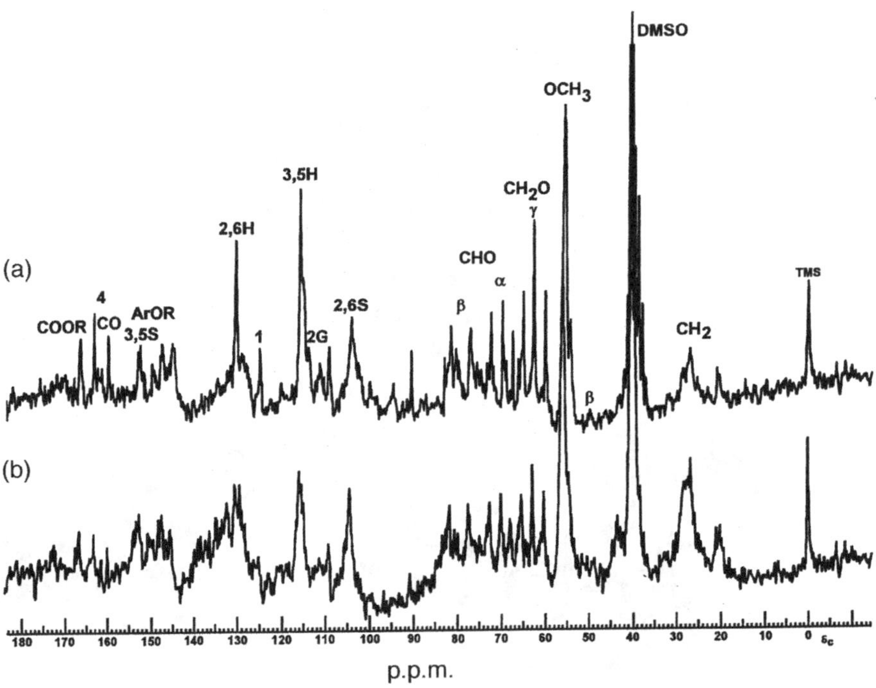

Fig. 16.6. Solution ^{13}C NMR spectra at 25 MHz of (a) Coastal Bermuda grass (warm-season grass) and (b) Kentucky-31 tall fescue (cool-season grass) lignins in d_6-DMSO and referenced to internal TMS. H, 4-hydroxycinnamyl (*p*-coumaryl); G, guaiacyl; S, syringyl; α, β and γ, 7, 8, and 9 positions, respectively, of aliphatic three-carbon side-chains off aromatic rings of lignin precursors.

other lignin subunits (Ralph *et al.*, 1994). It has been suggested (based on this NMR evidence) that *p*-coumaric acid initially occurs esterified to arabinoxylans (see Fig. 16.1) during primary wall development, but is later incorporated by esterification into lignin.

The use of 2-D NMR experiments, for the unambiguous assignment of large polymeric structures, like lignin, has been rather limited because of the poor resolution in the proton dimension. This problem has been somewhat overcome by the use of 3-D solution NMR spectroscopy. An example of this is provided by the use of the 3-D HMQC-HOHAHA experiment, which separates the ^{1}H spectrum into two dimensions and the ^{13}C spectrum into the third dimension. This has been effectively employed to verify that the signal at 50 p.p.m. in the ^{13}C NMR spectrum in an acetylated ^{13}C-enriched lignin is that of the β carbon in a β-1 linkage (Kilpeläinen *et al.*, 1994). An efficient protocol is to assign as many as possible of the signals by 2-D NMR spectroscopy and handle those that are still ambiguous by applying 3-D NMR spectroscopy. The inability of solution NMR spectroscopy to accommodate anything but isolatable and readily soluble materials, such as these, is quickly realized and other methods are required to study larger, more complex materials or intact structures.

Solid-state NMR Spectroscopy

Solid-state NMR spectroscopy can overcome many of the limitations of solution NMR spectroscopy and is ideally suited to the study of intact forages or their residues. Unlike solution NMR spectroscopy, a higher field is not always better. This is especially true for carbonaceous materials, if one is only interested in obtaining isotropic chemical-shift information. Higher fields generate spinning side bands, which must often be spun out at higher speeds. Achieving higher spinning speeds has practical limits and can cause selective loss of some signals. The most popular technique for obtaining solid-state ^{13}C spectra has been that of cross-polarization/magic-angle spinning (CP/MAS) (Schaefer and Stejskal, 1979).

Figure 16.7 shows the use of ^{13}C CP/MAS NMR spectroscopy in an attempt to discern the gross differences between a warm- and a cool-season grass. The only readily apparent difference, at this level, is the presence of additional signals in the region of 25–40 p.p.m. and increased carbonyl intensity at ~175 p.p.m. in the cool-season grass fescue (Fig. 16.7b) over that of the warm-season grass (Fig. 16.7a). This would appear to indicate that there is more protein present in the cool-season grass.

Modifications of the basic solid-state ^{13}C CP/MAS NMR experiment can often be more effectively used to study forages. An example of such a modification is the interrupted decoupling experiment. This experiment, also called the delayed-decoupling experiment, is employed to selectively observe non-protonated carbons (Opella and Fry, 1979). This has been used to detect differences in normal versus brown midrib (*bmr*) mutant lines in sorghum (Akin *et al.*, 1986) and pearl millet (Morrison *et al.*, 1993). The mutant lines are characterized by the brown coloration in the mid-vein of their leaves or stems and are generally more digestible by ruminants than the normal lines. By employing the interrupted decoupling experiment,

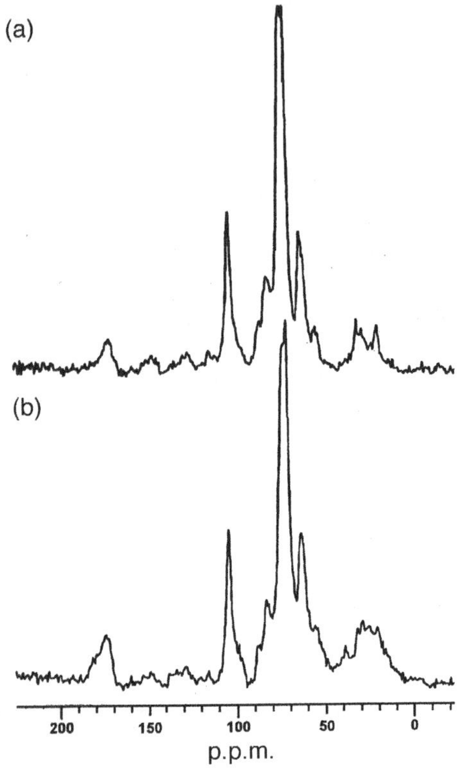

Fig. 16.7. Solid-state ^{13}C CP/MAS NMR spectra at 50 MHz of: (a) Coastal Bermuda grass (warm-season grass) and (b) Kentucky-31 tall fescue (cool-season grass). Spectra were externally referenced to the aromatic signal in hexamethylbenzene, set at 132.3 p.p.m.

a change in lignin structure can be directly detected in the intact forage. Thus, the isolation of the lignin is not required in order to obtain the desired information. A comparison of the results of using the basic CP/MAS and interrupted decoupling experiments is shown for pearl millet in Fig. 16.8. The loss of the signal intensity at 154 p.p.m., due to decrease of syringyl units, can barely be detected in going from the normal to the mutant (Fig. 16.8a and b, respectively) using the standard CP/MAS experiment. However, using the interrupted decoupling experiment, this difference is readily apparent (Fig. 16.8c and d). It also shows that the experiment is not effective when molecular motion averages ^{1}H-^{13}C interactions of proton-bearing carbons. Thus, signals also appear for some protonated carbons, the aliphatic CH_3 in lipids (~22 p.p.m.) and the aromatic OCH_3 (~56 p.p.m.) in lignin. Fortunately, these signals do not interfere with others and the aromatic OCH_3 actually helps in providing confirming information for the decrease in syringyl units.

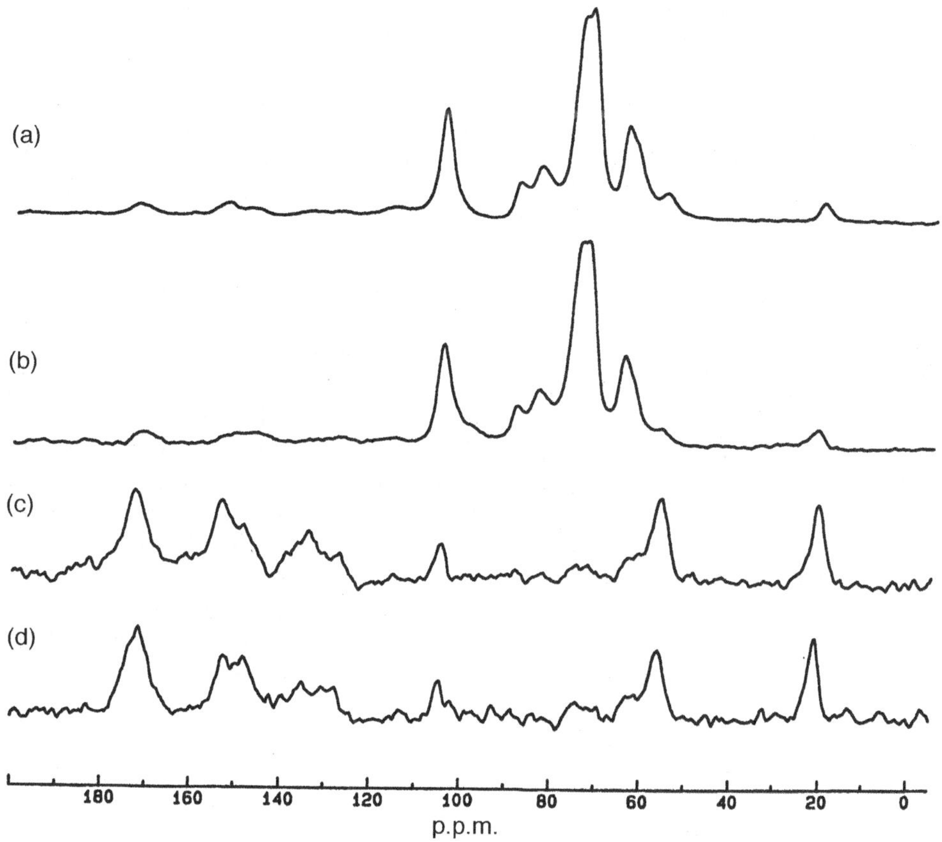

Fig. 16.8. Solid-state ^{13}C NMR spectra at 75 MHz of pearl millet rind tissue: (a) CP/MAS of normal line; (b) CP/MAS of mutant line; (c) CP/MAS with interrupted decoupling of normal line; (d) CP/MAS with interrupted decoupling of mutant line.

Solid-state ^{13}C CP/MAS NMR spectroscopy was used, in the same study (Morrison *et al.*, 1993), to show that treatment with 1 mol l^{-1} sodium hydroxide (NaOH) removed almost all of the phenolics from the *bmr* mutant rind tissue but was not effective in doing so in the normal tissue. It required treatment with 4 mol l^{-1} NaOH to remove phenolics from the normal rind tissue of pearl millet. This indicated that the phenolics in the mutant were ester-linked, rather than ether-linked as in the normal line. This suggests that lignin polymerization had not occurred to a large extent in the mutant and this could account for the greater digestibility of *bmr* mutants by ruminants.

One of the most effective uses of solid-state ^{13}C CP/MAS NMR spectroscopy for the study of forages is the investigation of the nature of the indigestible residues that remain after rumen microbial digestion. It has been used to characterize the residues from both *in vitro* (Akin *et al.*, 1993b) and *in sacco* digestion (Himmelsbach *et al.*, 1988).

Figure 16.9 shows the spectra of intact residues from a legume (lucerne (alfalfa)) and a cool-season grass (cocksfoot (orchardgrass)) after 3 × 72 h *in vitro* digestions by rumen bacteria. Thus, this represented the material remaining after all potentially bacterially degradable material had been removed. The spectra were compared by setting the total integrated area under the trace to 100 and determining the contribution of the areas of each of six specified regions to the total. This provides a semi-quantitative means by which to compare the relative chemical composition of the residues.

The solid-state ^{13}C CP/MAS NMR spectra of residues resulting from neutral-detergent treatment of mixed silage containing 280 g kg^{-1} dry matter (DM) crude fibre after *in sacco* digestion are shown in Fig. 16.10. The samples had originally been subjected to digestion in the rumen of a cannulated steer and removed after 0 (dipped), 48 and 336 h time periods. These time periods represent no digestion, typical digestion and completed digestion. From these spectra, it can be readily seen that carbohydrate is digested over time by the decrease in the signals *c.* 75 and 105 p.p.m. in going from 0 to 336 h. Furthermore, there appears to be no preference for amorphous over crystalline carbohydrates, due to the nearly equal loss of signals at 84 and 89 p.p.m., respectively. Lignin and lipids appear not to be digested. In fact, there is a slight increase in aromatic signals (115–165 p.p.m. region) and in hydrocarbon CH signals (20–30 p.p.m.), which are indicative of these components. There appears to be a loss of one type of carbonyl (CO), as indicated *c.* 175 p.p.m. This may represent the loss of digestible protein.

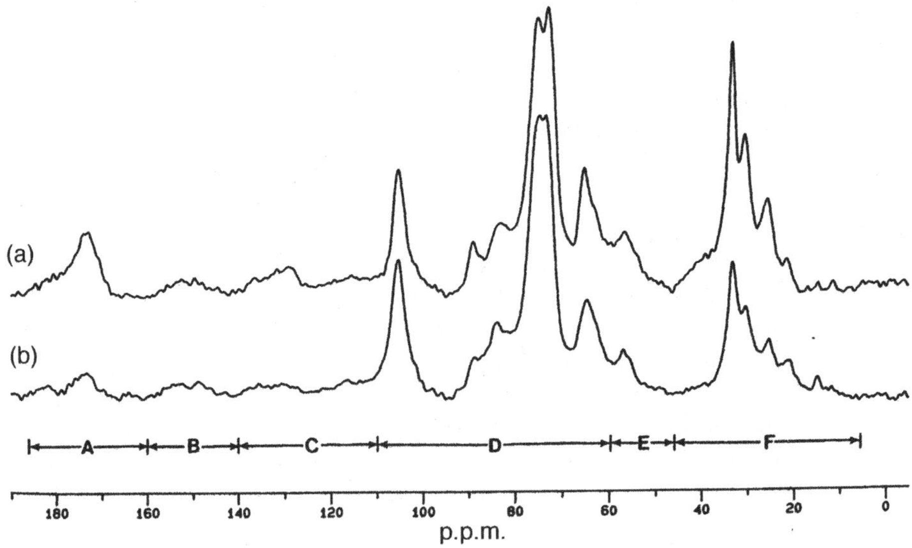

Fig. 16.9. Solid-state ^{13}C CP/MAS NMR spectra at 75 MHz for digested residues of: (a) lucerne and (b) cocksfoot grass. Region A, carbonyls; B, oxygenated aromatics (phenolics); C, non-oxygenated aromatics (phenolics); D, carbohydrates; E, methoxyl; and F, lipids.

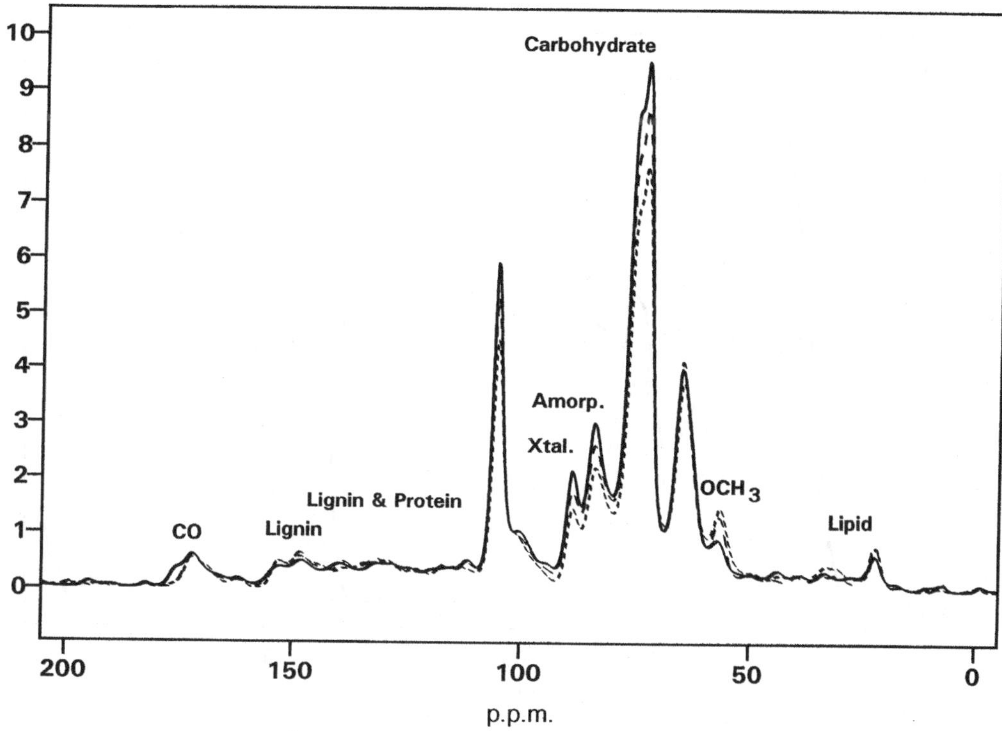

Fig. 16.10. Solid-state ^{13}C CP/MAS NMR spectra at 75 MHz of neutral detergent residues (NDRs) from *in sacco* digestion of a 28% crude-fibre mixed silage after: —— 0 h (dipped in digesta), ---- 48 h and -------- 336 h digestions.

Other Spectroscopic Techniques

Other spectroscopic techniques have been used in attempts to obtain similar information to that obtained by NMR spectroscopy. Diffuse-reflectance infrared Fourier-transform spectroscopy (DRIFTS), in the MIR region, performed on the same samples as were used in the study on *in sacco* residues (Himmelsbach *et al.*, 1988) are shown in Fig. 16.11. Comparison of the MIR spectra reveal, even more clearly, the digestion of carbohydrate. This is shown by the decrease in O-H stretch bands (~3000–3600 cm^{-1}) and C-O mixed-mode heavy-atom vibrations (~900–1200 cm^{-1}). Like the NMR spectra, the MIR spectra show that there is no digestion of lignin or lipids. Again, the indications for lignin or phenolics (1265 and 1505 cm^{-1} bands) actually show a slight increase as digestion proceeds. One indicator for lipids, the C-H stretch at 2850 cm^{-1}, is overshadowed by the loss of similar bands due to carbohydrate, but a sharpening of the shoulder at this position is evident. The other indicator for lipids is the carbonyl (CO) band at 1750 cm^{-1}. It shows the most dramatic change, indicating that lipids are not digested.

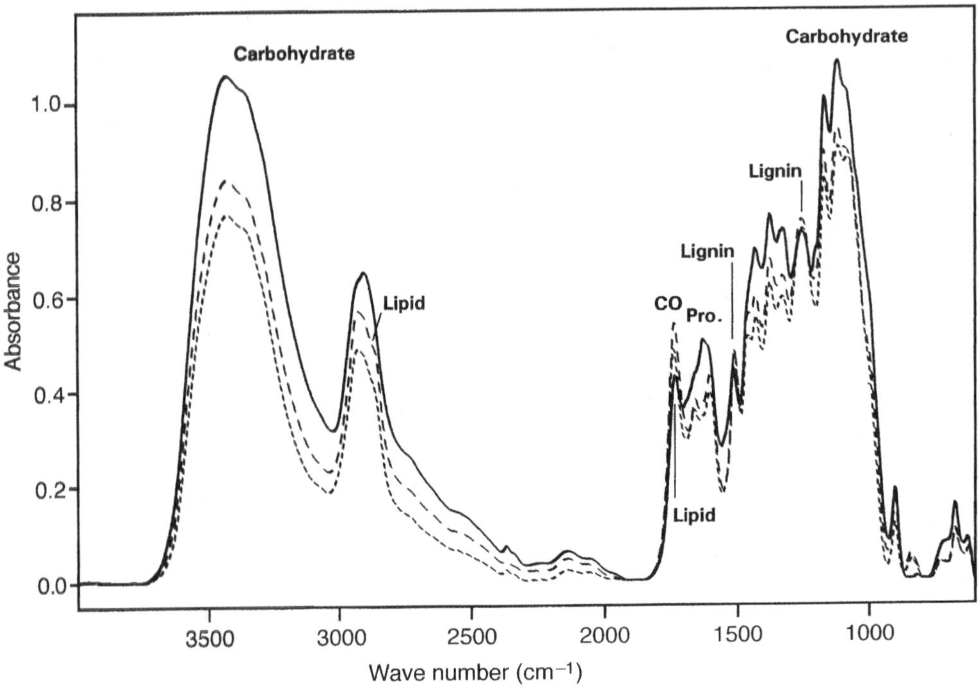

Fig. 16.11. DRIFTS MIR spectra of NDRs from *in sacco* digestion of a 28% crude-fibre mixed silage after: ——— 0 h (dipped in digesta), ----48 h and --------- 336 h digestions. All spectra are baseline-corrected.

The Fourier-transform (FT) Raman spectra of these same samples, taken using an NIR laser source, are shown in Fig. 16.12. The baselines in these spectra have been corrected to remove fluorescence effects. The loss of carbohydrate is not as evident in the Raman spectra as in the MIR or NMR spectra. However, the bands in the 900–1200 cm^{-1} region do decrease in intensity as digestion proceeds. The biggest difference in the Raman spectra is the response to aromatic components. This is to be expected, since the Raman active vibrations must be accompanied by a change in polarizability (Colthup *et al.*, 1990) and aromatic compounds contain highly polarizable bonds. Small differences in the phenolic (lignin or hydroxycinnamic acid) content of the residue, barely detectable in the solid-state NMR and MIR spectra, are revealed as large differences in the Raman spectra at ~1600 cm^{-1}. Lipids are only detectable by a sharpening of the shoulder for C-H stretch bands at 2850 cm^{-1}, as in the MIR. The carbonyl band gives only a small response in the Raman, compared with the MIR. The bands in the ~1200–1500 cm^{-1} region cannot be unequivocally assigned to any single component, but could be associated with an initial loss in protein with digestion at 48 h and a reassociation of microbial protein by 336 h of digestion.

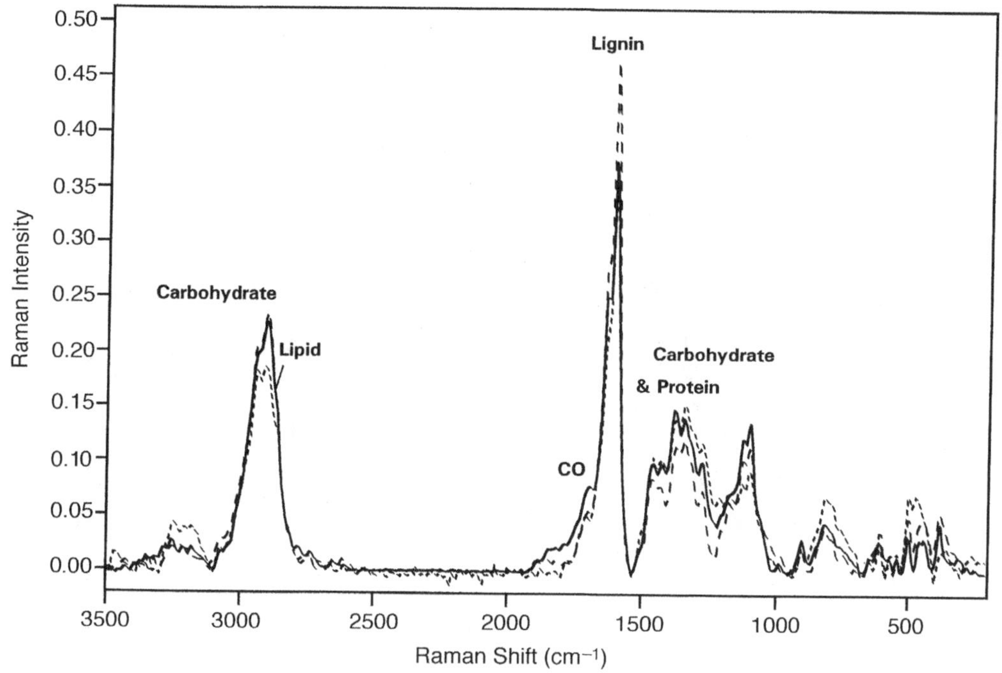

Fig. 16.12. NIR-FT-Raman spectra of NDRs from *in sacco* digestion of a 28% crude-fibre mixed silage after: ——— 0 h (dipped in digesta), ---- 48 h and --------- 336 h digestions. All spectra baseline-corrected, to remove fluorescence effects.

Diffuse-reflectance NIR spectra of the same samples are shown in Fig. 16.13. Here the base lines are corrected to remove the slope, which is mainly due to the influence of an intense O-H stretch band that occurs in the MIR region and somewhat due to particle-size effects. These are normally corrected with the use of derivatives, but linear baseline removal was done in the same manner as for the other spectral methods for purposes of comparison. Even though NIR spectroscopy is an LR technique, the same basic differences in the intact samples can be detected. The loss of carbohydrates is best observed in the band centred around 2100 nm. The response for lignin or phenolics, at 2260 nm appears to be on a par with that observed in the MIR spectra. The response to aromatics is greatest in the Raman spectra. The responses for lipid, at 1725, 2300 and 2340 nm, are at about the level observed in the Raman and NMR spectra, which is not as good as that observed with the MIR (carbonyl band). Changes in protein are not observable in the NIR spectra of these samples; however, the locations where these bands should appear are shown in Fig. 16.13.

Pyrolysis mass spectrometry (PyMS) has also been employed effectively to examine compositional differences in forages. When low-energy (16 eV) electron impact (EI) is used, phenolic components can be examined. When even softer ionization of pyrolysis products is used, such as chemical ionization (CI),

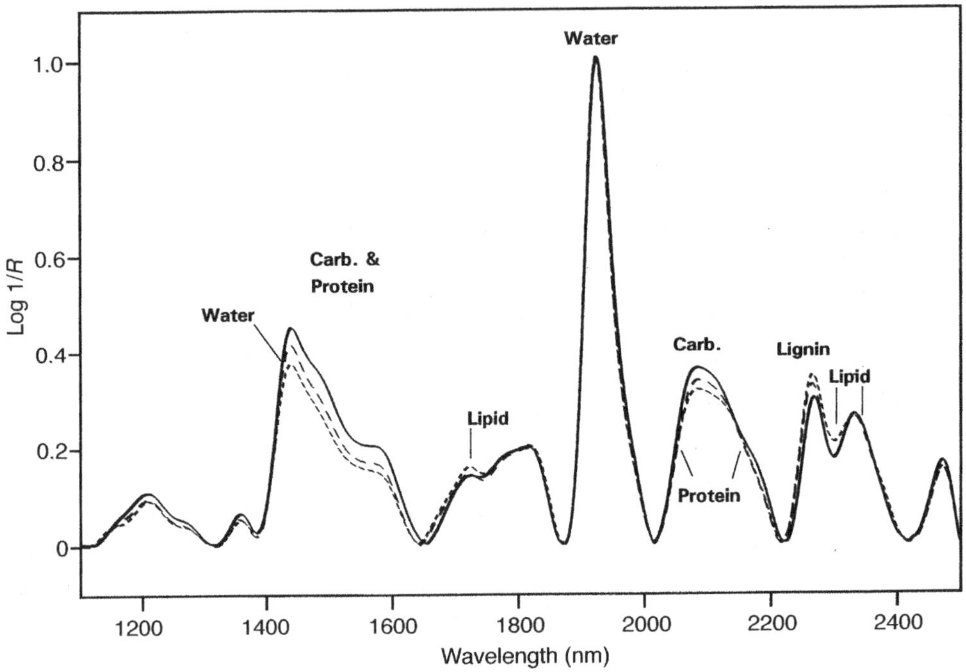

Fig. 16.13. NIR spectra of NDRs from *in sacco* digestion of a 28% crude-fibre mixed silage after: ——— 0 h (dipped in digesta), – – – – 48 h and -------- 336 h digestions. All spectra baseline-corrected and normalized around the water band.

carbohydrates can be emphasized. An example of the use of PyMS is provided in the examination of Coastal Bermuda grass and Kentucky-31 tall fescue, using EI PyMS (Fig. 16.14; Morrison *et al.*, 1991). Specific markers were found for *p*-coumaric and ferulic acid esters at 120 and 150 mass-to-charge ratio (m/z), respectively. The marker for *p*-coumaric acid showed greater abundance in Coastal Bermuda grass. This is in agreement with the NMR spectroscopic results on lignin already discussed. In addition, PyMS results revealed that (after sequential ozone and base treatment) all traces of phenolics could be removed from the Kentucky-31 tall fescue residue but not from the Coastal Bermuda-grass residue. This suggested that all cell-wall carbohydrates were more available in Kentucky-31 tall fescue and could explain the greater rumen digestibility of this forage after treatment with ozone.

Microspectroscopy and Imaging

Morphological, anatomical and chemical factors all potentially contribute to limiting ruminant digestibility of plants (Akin and Chesson, 1989). Thus, being able to study chemical factors within anatomical compartments at specific morphological stages is a highly desirable goal.

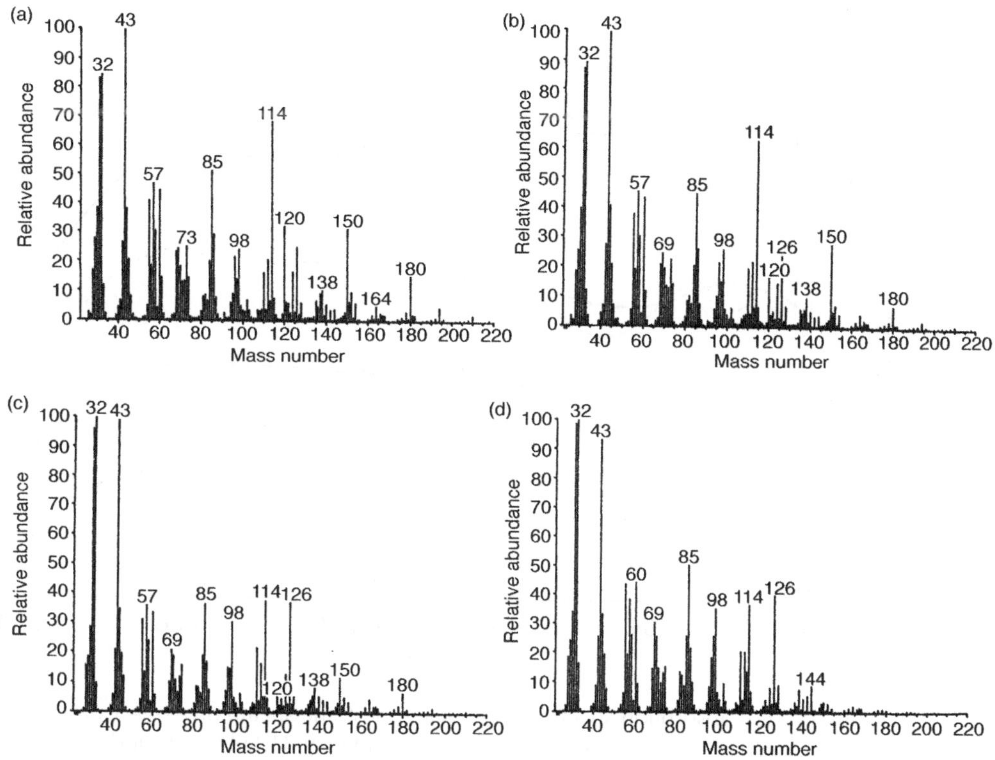

Fig. 16.14. PyMS, electron impact (16 eV) fingerprint of Coastal Bermuda grass and Kentucky-31: (a) and (b) cell walls, respectively, and (c) and (d) chemically treated residues, respectively.

UV spectroscopy has been employed in conjunction with microscopy, called UV microspectroscopy or microspectrophotometry, in an attempt to attain this goal. By restricting the spectra to specific cell types, the interference between compounds that give similar spectral responses can also be reduced. One of the better examples of the use of UV microspectroscopy is its use in the detection of phenolic components in specific cells of forages that show different degradabilities (Akin and Rigsby, 1992). Figure 16.15 shows the UV spectra of the mestome sheath of leaf blades, which is mostly degraded in legumes, partially degraded in cool-season grasses and not degraded in warm-season grasses by rumen microorganisms after 7 days. The important bands in the UV spectra are the absorptions for aromatics, which occur at ~280 nm for condensed lignin or lignin-like compounds and at ~320 nm for those with extended conjugation (e.g. *p*-coumaryl and feruloyl esters). This suggests that, when only condensed lignin or lignin-like structures are present, digestibility is not particularly hindered in this tissue, as in the case of lucerne. When esters are present, digestion is restricted, as with the cool-season grasses (cocksfoot and tall fescue). When both are present to a large extent, then digestion is blocked, as with the warm-season grass Coastal Bermuda grass.

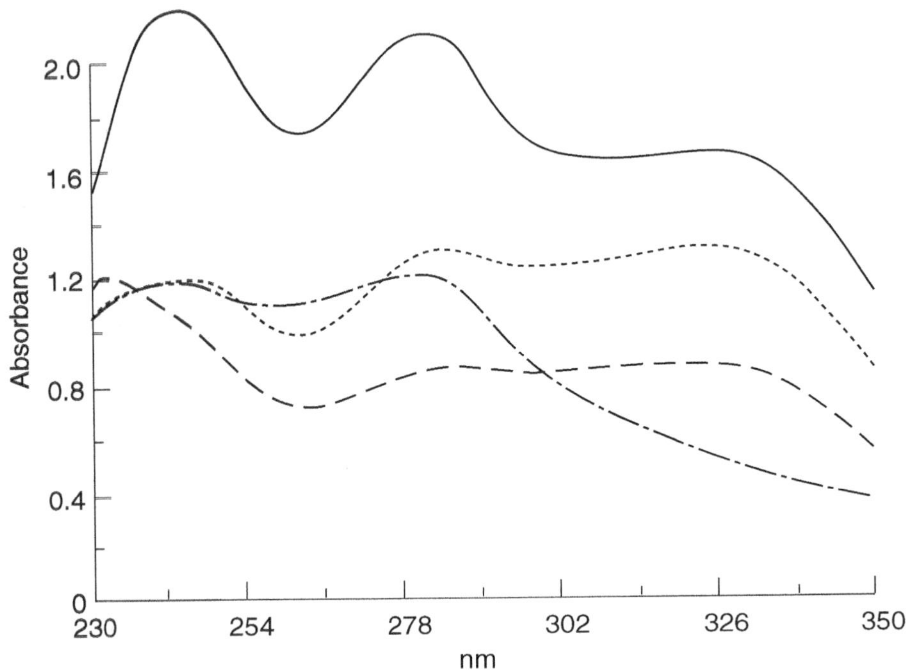

Fig. 16.15. UV absorption spectra of leaf blade mestome sheath of: —— Coastal Bermuda grass, ------- cocksfoot grass, — — — Kentucky-31 tall fescue and — - — lucerne.

Figure 16.16 shows the UV spectra of the parenchyma-bundle sheath, which is mostly degraded in cool-season grasses and only partially degraded in warm-season grasses. In lucerne, this tissue is totally degraded and thus is not shown. From these spectra it appears that the parenchyma-bundle sheath is essentially non-lignified and the restriction to digestion is provided by the presence of phenolic esters.

The concept of the use of microspectroscopy in the MIR region is also extremely useful. This technique has been around since the 1940s but experienced a resurgence in the 1980s with the development of the FT infrared (Messerschmidt and Harthcock, 1988). The state of the art of this technique has been recently reviewed (Reffner, 1998). An example of the use of MIR microspectroscopy on forages is again provided by pearl millet (Akin *et al.*, 1993a). It has been used to compare the level of phenolics found in the vascular bundle and pith of normal and *bmr* mutant tissue. Figure 16.17 shows that the band for lignin (~1510 cm^{-1}) is less intense in the vascular bundle of the mutant than in the normal line and that the pith showed an even lower intensity than the rind tissue. This was shown to be consistent with the digestibility of these tissues.

An example of the use of MIR imaging on forages is provided by an investigation of a leaf vascular bundle of Coastal Bermuda grass (Wetzel *et al.*, 1998). In this approach, a synchrotron light source was used to enhance the signal/noise ratio and a spatial resolution of ~6 μm was obtained. Here the phenolics, lipid and protein of

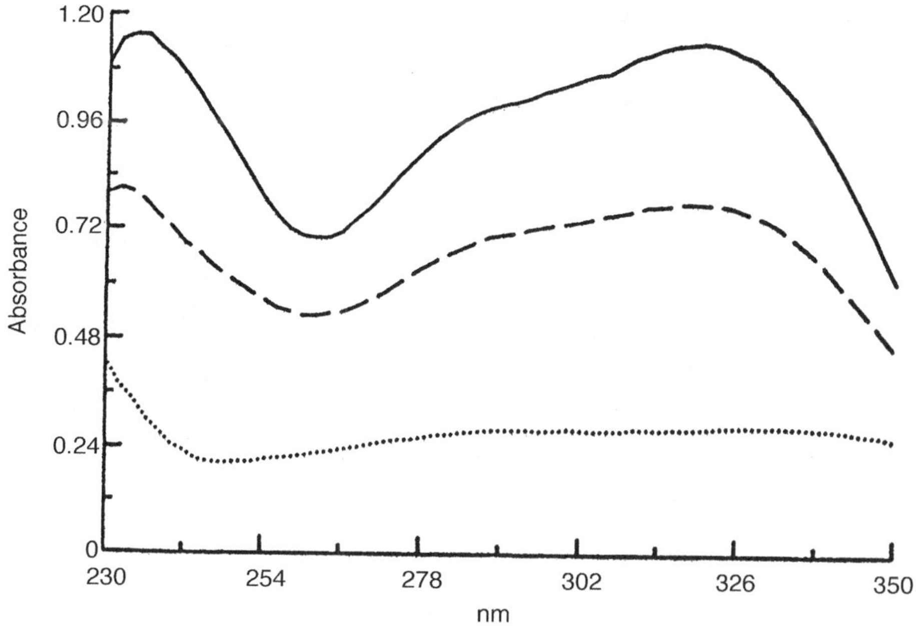

Fig. 16.16. UV absorption spectra of leaf-blade parenchyma bundle sheath of: —— Coastal Bermuda grass, ----- cocksfoot grass and ········ Kentucky-31 tall fescue.

the vascular bundle were imaged at 1509, 1550 and 1469 cm^{-1}, respectively. The greatest concentration of phenolics was found in the sclerenchyma and parenchyma-bundle sheath (which included the mestome sheath). Lipids appeared to be located in the epidermal tissue and protein nearby. It is apparent that MIR microspectroscopy is capable of locating more chemical constituents than UV microspectroscopy, but it is limited by its spatial resolution in discriminating between anatomical structures.

Raman microspectroscopy has been around as long as MIR microspectroscopy but has been more difficult to routinely access. Raman imaging was introduced as the Raman microprobe in 1975 (Delhaye and Dhamelincourt, 1975). Like MIR imaging, it has seen a resurgence for many of the same reasons (Turrell and Corset, 1996). However, the development of the FT-Raman confocal systems (Brenan and Hunter, 1995) or systems based on acousto-optic tunable filter or liquid-crystal tunable filter sources (Morris *et al.*, 1994) and charge-coupled device (CCD) detectors have made this a more viable method. Spatial resolution can be achieved on the order of 1 μm and spectral resolution is nearly equivalent to that in MIR. This technique is not very applicable to the study of forage materials, which contain highly fluorescing compounds, such as chlorophyll. However, it is applicable to the study of residues. It would be especially useful in mapping the location of lignin in digested residues, although it has not yet been used for this purpose.

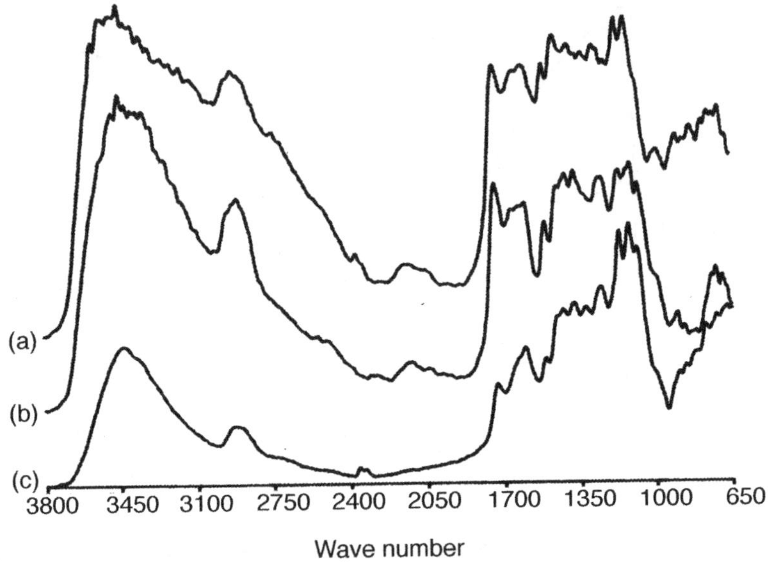

Fig. 16.17. MIR microspectroscopy of cell walls in pearl millet stem: (a) normal rind (sclerenchyma and vascular tissue); (b) vascular bundle within the pith of *bmr* mutant; (c) normal pith.

NIR imaging is just starting to be explored extensively and most MIR imaging systems can be converted to work in the NIR spectral region. Some commercial NIR imaging systems, which are optimized for the NIR, are just becoming available (Treado, 1995) and will make this technique more accessible. Its main advantage is that it can be used to analyse samples that are totally absorbing, which is a situation that can be expected to be encountered in studies on forages. NIR imaging represents a compromise between MIR and Raman imaging with regard to spatial and spectral resolution. It generally has higher spatial resolution but presents difficulties in assigning spectral features, due to severe spectral bands overlapping. Although most of the recent applications of NIR imaging have been in the medical arena, its use is beginning to become more widespread in other biological areas (Dempsey *et al.*, 1996). No applications of NIR imaging to forages could be found, but this technique certainly lends itself to their study, as NIR spectroscopy has already done.

NMR imaging or magnetic resonance imaging (MRI) can also be used to obtain chemical imaging information. This technique has been restricted to solution studies in plants and much of the work is at low spatial resolution, but it can yield information that is inaccessible by other microscopic techniques (Blümich and Kuhn, 1992). The combination of solution 2-D $\{^1H, ^1H\}$ correlation spectroscopy with chemical-shift imaging NMR in an experiment called correlation-peak imaging (CPI) is an example of the state of the art. It has been performed on a plant seedling to reveal the locations of soluble carbohydrates and amino acids (Metzler *et al.*, 1995). The ^1H image obtained at 500 MHz clearly shows the vascular bundles, pith parenchyma and cortex parenchyma. The phloem and xylem within the vascular bundles can also be distinguished, with a spatial resolution of ~24 μm. Separate images of each component can be obtained by selecting individual cross-peaks in the 2-D correlation spectrum. Sucrose shows its highest intensity in the vascular bundles and lowest in the cortex parenchyma. The anomers of glucose can be imaged separately and show their highest intensity in the pith parenchyma. Glutamine/glutamate were located in a ring covering the cortex parenchyma. Lysine and arginine were observed in the vascular bundles. Although this technique only works in solution and has lower spatial resolution than the UV and MIR microspectroscopic imaging techniques, it is a step closer to providing complete imaging of plant tissue.

PyMS can also be used for microspectroscopy but is a destructive technique. It has been used for the characterization of cuticle, parenchyma, sclerenchyma and vascular-bundle tissue for the determination of polysaccharide/lignin ratios and detection of *p*-coumaryl acid and truxillic acid groups in forages (Boon, 1989).

Conclusions

The physicochemical techniques that are available to assess forages, just in the form of spectroscopy, are overwhelming. Not all have been touched on here. NMR is exceptionally specific and even sensitive to minute conformational effects but has practical limitations in its use. Each technique discussed has its advantages and disadvantages. Even with all of the advances over recent years, no technique is the panacea for all assessment problems. It is up to investigators to pick and choose those techniques that will best satisfy their particular needs. However, it does appear that, in order to make significant advances in the understanding of the underlying causes of differential digestibility of forages, some form of spectroscopic imaging needs to be utilized. This is necessary in order to embrace the impact of synergistic effects of the morphological, anatomical and chemical factors that influence nutrient availability.

References

Akin, D.E. and Chesson, A. (1989) Lignification as the major factor limiting forage feeding value especially in warm conditions. In: *Proceedings 16th International Grassland Congress*. Association Française pour la Production Fourragère, INRA, Versailles, France, pp. 1753–1760.

Akin, D.E. and Rigsby, L.L. (1992) Scanning electron microscopy and ultraviolet absorption microspectrophotometry of plant cell types for different biodegradabilities. *Food Structure* 11, 259–271.

Akin, D.E., Hanna, W.W., Snook, M.E., Himmelsbach, D.S., Barton, F.E., II and Windham, W.R. (1986) Normal 12 and brown midrib-12 sorghum. II. Chemical variations and digestibility. *Agronomy Journal* 78, 832–837.

Akin, D.E., Himmelsbach, D.S., Carl, R.T., Hanna, W.W. and Barton, F.E. II (1993a) Mid-infrared microspectroscopy to assess lignin in plant tissue related to digestibility. *Agronomy Journal* 85, 171–175.

Akin, D.E., Morrison, W.H., III and Himmelsbach, D.S. (1993b) Characterization of digestion residues of alfalfa and orchardgrass leaves by microscopic, spectroscopic and chemical analysis. *Journal of the Science of Food and Agriculture* 63, 339–347.

AOAC (1997) *Official Methods of Analysis*,16th edn, Vol. I, Ch. 4, March 1996 Supplement. AOAC, Gaithersburg, Maryland.

Aue, W.P., Bartholdi, E. and Ernst, R.R. (1976) Two-dimensional spectroscopy: application to nuclear magnetic resonance. *Journal of Chemical Physics* 64, 2229–2246.

Bax, A. and Davis, D.G. (1985) Practical aspects of two-dimensional transverse NOE spectroscopy. *Journal of Magnetic Resonance* 63, 207–231.

Bax, A. and Marion, D. (1988) Improved resolution and sensitivity in ^{1}H detected heteronuclear multiple-bond correlation spectroscopy. *Journal of Magnetic Resonance* 78, 186–191.

Bax, A. and Subramanian, S. (1986) Sensitivity-enhanced two-dimensional heteronuclear shift correlation NMR spectroscopy. *Journal of Magnetic Resonance* 67, 565–569.

Bax, A. and Summers, M.F. (1986) ^{1}H and ^{13}C assignments from sensitivity-enhanced detection of heteronuclear multiple-bond connectivity by 2D multiple quantum NMR. *Journal of the American Chemical Society* 108, 2093–2094.

Blümich, B. and Kuhn, W. (1992) *Magnetic Resonance Microscopy*. VCH Publishers, New York.

Boon, J.J. (1989) An introduction to pyrolysis mass spectrometry of lignocellulosic material: case studies on barley straw, corn stem and *Agropyron*. In: Chesson, A. and Ørskov, E.R. (eds) *Physico-Chemical Characterisation of Plant Residues for Industrial and Feed Use*. Elsevier Applied Science, London, pp. 25–49.

Brenan, C.J.H. and Hunter, I.W. (1995) Confocal image properties of a confocal scanning laser visible light FT-Raman microscope. *Applied Spectroscopy* 49(7), 971–976.

Bro, R., Workman, J.J., Mobley, P.R. and Kowalski, B.R. (1997) Review of chemometrics applied to spectroscopy 1985–95. 3. Multiway analysis. *Applied Spectroscopy Reviews* 32(3), 237–261.

Bruch, M.D. (1996) *NMR Spectroscopy Technique*, 2nd edn, revised and expanded. Marcel Dekker, New York.

Colthup, N.B., Daly, L.H. and Wiberly, S.E. (1990) *Introduction to Infrared and Raman Spectroscopy*. Academic Press, New York.

Davis, D.G. and Bax, A. (1985) Simplification of ^{1}H NMR spectra by selective excitation of experimental subspectra. *Journal of the American Chemical Society* 107, 7197–7198.

Delhaye, M. and Dhamelincourt, P. (1975) Raman microprobe and microscope with laser excitation. *Journal of Raman Spectroscopy* 3, 33–43.

Dempsey, R.J., Davis, D.G., Buice, R.G., Jr and Lodder, R.A. (1996) Biological and medical applications of near-infrared spectroscopy. *Applied Spectroscopy* 50(2), 18A–34A.

Fernandez, N. (1990) Carbon-13 NMR study on lignin from bagasse. *Holtzforschung* 44, 35–38.

Gil, A.M., Alberti, E., Tatham, A.S., Belton, P.S., Humpfer, E. and Spraul, M. (1997) Magic angle spinning NMR study of the hydration of wheat seed storage protein omega-gliadins. *Magnetic Resonance in Chemistry* 35(SI), S101–S111.

Harris, P.J., Hartley, R.D. and Lowry, K.H. (1980) Phenolic constituents of mesophyll and non-mesophyll cell walls from leaf laminae of *Lolium perenne*. *Journal of the Science of Food and Agriculture* 33, 516–520.

Hartley, R.D., Morrison, W.H., Himmelsbach, D.S. and Borneman, W.S. (1990) Cross-linking of cell wall phenolic arabinoxylans in graminaceous plants. *Phytochemistry* 27, 349–351.

Himmelsbach, D.S. and Barton, F.E., II (1980) ^{13}C nuclear magnetic resonance of grass lignins. *Journal of Agricultural and Food Chemistry* 28, 1203–1208.

Himmelsbach, D.S. and Hartley, R.D. (1993) Structure of a feruloyl tetra- and penta-

arabinoxylans, that contain the β-D-Xyl*p*-(1 → 2)-α-L-Ara*f* element by one- and two-dimensional homonuclear and heteronuclear proton and carbon-13 spectroscopy. In: *Proceedings of 205th ACS National Meeting, Denver, Colorado.*

Himmelsbach, D.S., Boer, H., Akin, D.E. and Barton, F.E., II (1988) Solid-state carbon-13, FTIR, and NIRS spectroscopic studies of ruminant silage digestion. In: Creaser, C.S. and Davies, A.M.C. (eds) *Analytical Applications of Spectroscopy*. Royal Society of Chemistry, London, pp. 410–413.

Himmelsbach, D.S., Hartley, R.D., Borneman, W.S., Poppe, L. and van Halbeek, H. (1994) Structure of a feruloylated arabinoxylan tetrasaccharide that contains the β-D-Xyl*p*-(1→2)-α-L-Ara*f* element by ^{1}H and ^{13}C NMR spectroscopy. *Magnetic Resonance in Chemistry* 32, 158–165.

Hoffmann, R.A., van Wijk, J., Leeflang, B.R., Kamerling, J.P., Altona, C. and Vliegenthart, J.F.G. (1992) Conformational analysis of the α-L-arabinofuranosides present in wheat arabinoxylans form proton–proton coupling constants. *Journal of the American Chemical Society* 114, 3710.

Ishii, T. (1996) Isolation and characterization of a *p*-coumaryl arabinoxylan tetrasaccharide from bamboo shoot cell-walls. *Mokuzai Gakkaishi* 42, 515–519.

Ishii, T. and Hiroi, T. (1990a) Isolation and characterization of feruloylated arabinoxylan oligosaccharides from bamboo shoot cell walls. *Carbohydrate Research* 196, 175–183.

Ishii, T. and Hiroi, T. (1990b) Linkage of phenolic acids to cell-wall polysaccharides of bamboo shoot. *Carbohydrate Research* 206, 297–310.

Ishii, T., Hiroi, T. and Thomas, J.R. (1990) Feruloylated xyloglucan and *p*-coumaryl arabinoxylan oligosaccharides from bamboo shoot cell walls. *Phytochemistry* 29, 1999–2003.

Jung, H.J.G. and Ralph, J. (1990) Phenolic-carbohydrate complexes in plant cell walls and their effect on lignocellulose utilization. In: Akin, D.E., Ljungdahl, L.G., Wilson, J.R. and Harris, P.J. (eds) *Microbial and Plant Opportunities to Improve Lignocellulose Utilization by Ruminants*. Elsevier, New York, pp. 173–182.

Kato, A., Azuma, J. and Koshijima, T. (1983) A new feruloylated trisaccharide from bagasse. *Chemical Letters* 137–140.

Kato, A., Azuma, J. and Koshijima, T. (1987) Isolation and identification of a new feruloylated tetrasaccharide from bagasse lignin–carbohydrate complex containing phenolic acid. *Agricultural Biological Chemistry* 51(6), 1691–1693.

Kato, A., Azuma, J. and Koshijima, T. (1990) Arabinosylxylotriose mixedly esterified with acetic and ferulic acid from sugar cane bagasse. *Agricultural Biological Chemistry* 54(8), 2181–2182.

Kato, Y. and Nevins, D.J. (1985) Isolation and identification of O-(5-O-(*E*)-*p*-feruloyl-α-L-arabinofuranosyl)-(1→3)-O-β-D-xylopyranosyl-(1→4)-D-xylose as a component of *Zea* shoot cell-walls. *Carbohydrate Research* 137, 139–150.

Kilpeläinen, I., Ämmälahi, E., Brunow, G. and Robert, D. (1994) Application of three-dimensional HMQC–HOHAHA NMR spectroscopy to wood lignin, a natural polymer. *Tetrahedron Letters* 35(49), 9267–9270.

Kowalski, B.R. (1977) *Chemometrics: Theory and Applications*. American Chemical Society Symposium, Series 52, American Chemical Society, Washington, DC.

Martens, H., Alsberg, B., Foulk, S. and Stark, E. (1991) What other spectroscopic techniques could learn from NIR. In: Davies, A.M.C. and Creaser, C.S. (eds) *Analytical Applications of Spectroscopy II*. Royal Society of Chemistry, Cambridge, UK, pp. 221–239.

Messerschmidt, R.G. and Harthcock, M.A. (1988) *Infrared Microscopy: Theory and Applications*. Marcel Dekker, New York.

Metzler, A., Izquierdo, M., Ziegler, A., Köckenberger, W., Komor, E., von Kienlin, M., Hasse, A. and Décorps, M. (1995) Plant histochemistry by correlation peak imaging. *Proceedings of the National Academy of Sciences USA* 92, 11912–11915.

Morris, H.R., Hoyt, C.C. and Treado, P.J. (1994) Imaging spectrometers for fluorescence and Raman microscopy: acousto-optic and liquid crystal tunable filters. *Applied Spectroscopy* 48(7), 857–865.

Morrison, W.H., III, Scheijen, M.A. and Boon, J.J. (1991) Pyrolysis mass spectrometry of coastal Bermudagrass (*Cynodon dactylon* (L.) Pers.) and 'Kentucky-31' tall fescue (*Festuca arundinacea* Schreb.) cell walls and their residues after ozonolysis and base hydrolysis. *Animal Feed Science and Technology* 32, 17–26.

Morrison, W.H., III, Hartley, R.D. and Himmelsbach, D.S. (1992) Synthesis of substituted truxillic acids from *p*-coumaric and ferulic acid: simulation of photodimerization in plant cell walls. *Journal of Agricultural and Food Chemistry* 40, 768–771.

Morrison, W.H., III, Akin, D.E., Himmelsbach, D.S. and Gamble, G.R. (1993) Investigation of the ester- and ether-linked phenolic constituents of cell wall types of normal and brown midrib pearl millet using chemical isolation, microspectrophotometry and ^{13}C NMR spectroscopy. *Journal of the Science of Food and Agriculture* 63, 329–337.

Mueller-Harvey, I., Hartley, R., Harris, P.J. and Curzon, E.H. (1986) Linkage of *p*-coumaryl and feruloyl groups to cell wall polysaccharides of barley straw. *Carbohydrate Research* 148, 71–85.

Newman, R.H. (1992) Nuclear magnetic resonance study of spatial relationships between chemical components in wood cell walls. *Holtzforschung* 46, 205–210.

Opella, S.J. and Fry, M.H. (1979) Selection of nonprotonated carbon resonances in solid-state nuclear magnetic resonance. *Journal of the American Chemical Society* 101, 5854–5857.

Preston, C.A., Trofymow, J.A., Sayer, B.G. and Niu, J. (1997) ^{13}C nuclear magnetic resonance spectroscopy with cross-polarization and magic-angle spinning investigation of the proximate-analysis fraction used to assess litter quality in decomposition studies. *Canadian Journal of Botany* 75, 1601–1613.

Ralph, J., Hatfield, R.D., Quideau, S., Helm, R.F., Grabber, J.H. and Jung, H.J.G. (1994) Pathway of *p*-coumaric acid incorporation into maize lignin. *Journal of the American Chemical Society* 116, 9448–9456.

Ralph, J., Grabber, J.H. and Hatfield, R.D. (1995) Lignin-ferulate cross-links in grasses: active incorporation of ferulate polysaccharide esters into ryegrass lignins. *Carbohydrate Research* 275, 167–178.

Reffner, J.A. (1998) Instrumental factors in infrared microspectroscopy. *Cellular and Molecular Biology* 44(1), 1–7.

Schaefer, J. and Stejskal, E.O. (1979) High resolution ^{13}C NMR of solid polymers. In: Levy, G.C. (ed.) *Topics in Carbon-13 NMR Spectroscopy*, Vol. 3, Wiley-Interscience, New York, pp. 283–324.

Treado, P.J. (1995) Chemical imaging reveals more than the microscope. *Laser Focus World* 31(10), 75–104.

Turrell, G. and Corset, J. (1996) *Raman Microscopy Developments and Applications*. Academic Press, London.

Wetzel, D.L., Eilert, A.J., Pietrazk, L.N., Miller, S.S. and Sweat, J.A. (1998) Ultraspatially-resolved synchrotron infrared microspectroscopy of plant tissue *in situ*. *Cellular and Molecular Biology* 44(1), 145–167.

17 Trace-mineral Status of Forages

A. MacPherson

Scottish Agricultural College, Auchincruive, Ayr KA6 5HW, UK

Important trace elements

Of the 16 trace elements currently recognized as being essential for farm livestock (Mertz, 1987), only eight are likely to be of practical significance, as deficiencies of the others occur only rarely, if ever. Those regarded as essential are cobalt (Co), copper (Cu), iron (Fe), zinc (Zn), manganese (Mn), iodine (I), selenium (Se), chromium, molybdenum (Mo), fluorine, lithium, silicon, vanadium, nickel, arsenic and lead (Pb). Future research will probably provide evidence for the extension of this list. Those of most practical significance are Co, Fe, Cu, Zn, Mn, I, Se and Mo, with the last-named being more important for its interaction with Cu and sulphur (S), rather than in its own right. Other elements, eg. calcium (Ca) and cadmium (Cd) can interact with some of the essential trace elements, such as Cu, Zn and Mn and adversely affect their absorption or availability.

Forage as a Supplier of Trace Elements

Many classes of livestock are solely or mostly dependent for all their nutrients on the quality of the forage available to them, either in its natural state or conserved as hay or silage. Hemingway *et al.* (1968) reported that 90% of hays and 62% of silages provided insufficient Cu to meet the needs of dairy cattle. In Latin America, some 43–60% of forage samples were deficient in Co and Cu and 75% deficient in Zn (McDowell and Conrad, 1990). In Latvia, 85% of analysed forages had low concentrations of Cu, Mn, Zn and Co, with both Cu and Co being some 50% below requirements (Strikauska *et al.*, 1994). Forage species vary widely in the range and quantity of trace elements they provide and this in turn is affected by soil type and condition, its acidity or alkalinity and the stage of maturity of the forage itself. Fertilizer application may also affect trace-element concentration.

Factors Affecting the Copper Status of Forage

Species

Early work in the UK showed much higher Cu concentration in herbs and weeds (10.8–16.6 mg kg^{-1} dry matter (DM)) than in grasses (4.0–8.2 mg kg^{-1} DM) grown as mixed swards (Thomas and Thompson, 1948). Legumes were higher in Cu than grasses but lower than herbs and weeds (Thomas *et al.*, 1952). Minson (1990) confirms that temperate legumes generally contain more Cu than temperate grasses (7.8 vs. 4.7 mg kg^{-1} DM), except when soil Cu is low, when there is no difference (Beck, 1962). In contrast, tropical legumes contain less Cu than tropical grass (3.9 vs. 7.8 mg kg^{-1} DM). Significant species differences in herbage Cu concentration have been reported from Kenya, where kikuyu grass had the highest and Rhodes grass the lowest concentration (Jumba *et al.*, 1995). Also in Kenya, Maskall and Thornton (1989) recorded higher levels of Cu in grass species than in browse plants. However, Mo levels in all plants reached relatively high values and its availability appeared to increase in the more acid soils near lakesides, leading to a risk of Cu deficiency in the grazing ruminant livestock. Species differences are not always consistent between experiments, probably because of the presence of other confounding factors.

Genotype

Large differences were recorded between cultivars of *Trifolium subterranean* which were unaffected by soil Cu concentration. In temperate grasses, large differences were found between cultivars of *Lolium multiflorum* (84%), *Festuca arundinacea* (57%) (Montalvo-Hernandez *et al.*, 1984) and *L. perrene* (24 and 25%) (Forbes and Gelman, 1981; Montalvo-Hernandez *et al.*, 1984). Cultivars of *Dactylis glomerata* varied by 100% in one study (Montalvo-Hernandez *et al.*, 1984) but not at all in another (Forbes and Gelman, 1981).

Plant parts

Leaves of temperate grasses have 35% more Cu, on average, than the stems, but this is affected by age, with little difference between immature stem and leaf (Davey and Mitchell, 1968). Grass flower-heads have similar Cu contents to the leaves. Temperate legumes have higher leaflet Cu than stem Cu, but there is no such difference in tropical legumes (*Lablab purpurea*) (Hendricksen and Minson, 1980). Sheep tend to be more selective grazers than cattle and should therefore benefit from the higher Cu content of the leaves. Occasionally, this can prove to be a hazard rather than a benefit, particularly with breeds susceptible to Cu poisoning (see p. 358).

Stage of growth

Maturity leads to a decrease in the Cu content of forage because of a decline in the proportion of leaf present and a lowering of the Cu content of the stem. Young oat plants (*Avena sativa*) contained 9.4 mg Cu kg^{-1} DM, but fell to 3.2 mg kg^{-1} DM at the milk-ripe stage (Piper, 1942; Burridge, 1970; MacDonald and Wilson, 1980). Similar findings have been reported by many other workers for other grasses (Minson, 1990) and legumes. Moorland plants, including flying bent, deer hair, white bent and draw moss, fell from a mean of 11.1 to 5.7 mg kg^{-1} DM between May and September. On the other hand, Thomas *et al.* (1945) found little seasonal change in Cu concentration of heather samples, which ranged between 11.9 and 14.2 mg kg^{-1} DM. Increasing age from 2 to 10 years also had no effect. Trace-element deficiencies in livestock tend to be rare where heather is prevalent. MacPherson (1967) recorded a decline in the crude-protein content of maturing herbage from 289 to 88 g kg^{-1} DM which was paralleled by a fall in herbage Cu from 10.6 to 5.5 mg kg^{-1} DM.

Soil fertility

If total Cu content of soil is low, forage Cu levels will be low, but they may also be low on soils of normal Cu content but with a high proportion unavailable (Alloway and Tills, 1984). Reduced availability of Cu can be due to the formation of Cu–organic-matter complexes, e.g. peat soils and mineral soils with > 10% organic matter. Many references relate forage Cu to soil type (Lee, 1951; Cunningham, 1960; Alloway and Tills, 1984; Cornforth, 1984) or parent material (Fraser, 1984; Berrow and Ure, 1985).

Fertilizer nitrogen

Nitrogen (N) fertilizer applied to regularly cut cocksfoot and perennial ryegrass swards reduced forage Cu levels, possibly as a result of soil Cu depletion. If soil-available Cu was high, fertilizer N had no effect on the Cu content of star-grass (Rudert and Oliver, 1978), cocksfoot and tall fescue (Reid *et al.*, 1967a). On soils of low Cu status, phosphatic fertilizers can reduce subterranean clover Cu levels by up to 50%, apparently through a growth-stimulation effect, which consequently dilutes the Cu content (Reddy *et al.*, 1981). Other workers have reported no effect of phosphatic or potassic fertilizers on forage Cu contents (Hemingway, 1962; Reith *et al.*, 1964).

Liming

This has generally been found to have little direct effect on forage Cu concentration (Fig. 17.1), but it is nevertheless important in the aetiology of Cu deficiency in sheep on improved pastures, due to a change effected on the forage Cu:Mo ratio,

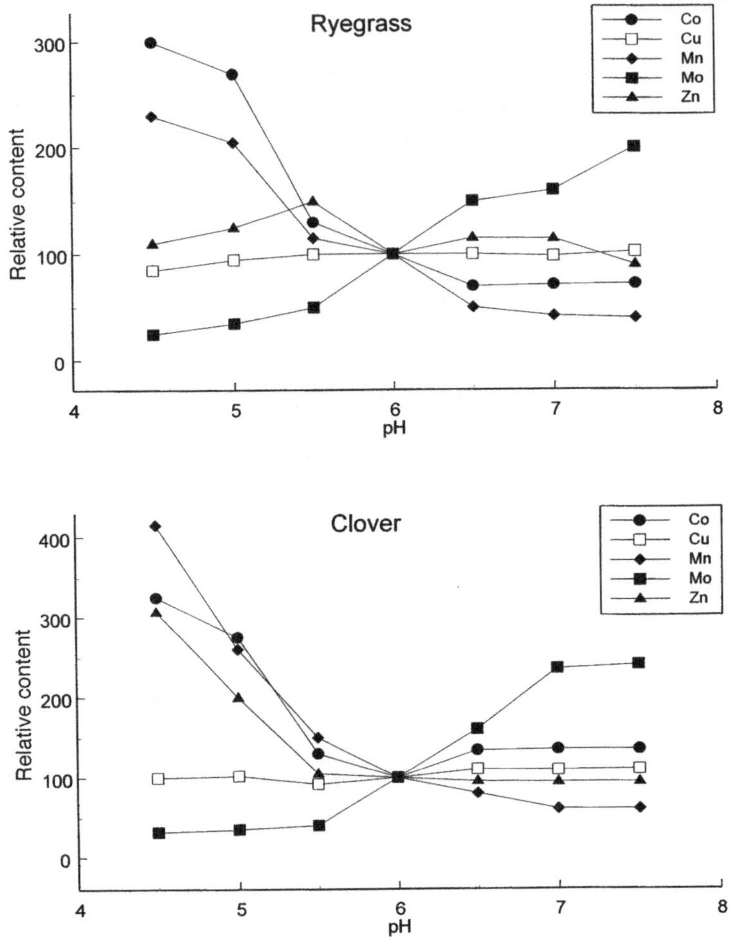

Fig. 17.1. Relative contents of trace elements in ryegrass and clover separated from mixed herbage. Contents in mg kg^{-1} DM at pH 6.0 are equal to 100 and are as follows: Co 0.03; Cu 3.1; Mn 51; Mo 2.7; and Zn 23.

as the more alkaline conditions following liming favour increased Mo uptake (Fig. 17.1). Heather Cu concentrations fell from 12.5 to 11.1 mg kg^{-1} DM following the application of 5 t ground limestone ha^{-1} (Thomas *et al.*, 1945). Stewart (1951) recorded a higher incidence of sway-back on limed pastures, but could not show any effect of treatment on forage Cu. MacPherson (1967) reported that liming at rates up to 4 t ha^{-1}, which gave a range in soil pH from 5.5 to 7.0, had no effect on herbage Cu content.

Climate

In north-eastern Brazil, herbage Cu concentrations ranged from 1.33 to 4.63 mg kg^{-1} DM in the dry season and only from 1.56 to 2.57 mg kg^{-1} DM in the wet season. Herbage Mo concentrations ranged only from 0.51 to 0.54 mg kg^{-1} DM (Sousa *et al.*, 1989). Other Brazilian workers (Pott *et al.*, 1989) considered Cu deficiency a possibility in grazing cattle at the start of the rainy period.

Factors Affecting the Cobalt Status of Forage

Co concentrations in forage range from < 0.01 to 1.26 mg kg^{-1} DM (Beeson, 1950), due mainly to differences in soil Co. Forage legumes generally contain more Co than grasses (Andrews, 1966). Lucerne and clover had from 0.3 to 0.4 mg Co kg^{-1} DM compared with 0.08 mg kg^{-1} DM in grasses grown on the same soil (Latteur, 1962). In Virginia, legumes were reported to contain from 0.06 to 0.48 mg kg^{-1} DM and grasses 0.02 to 0.24 mg kg^{-1} DM (Price and Hardison, 1963). Legumes and grasses both had low and similar concentrations when grown on low-Co soils in Florida (Kretschmer *et al.*, 1954) and New Zealand (Andrews, 1966). Occasionally, lower concentrations have been reported in legumes than in grasses (Winter *et al.*, 1977). On soils with relatively high Co concentrations, little variation in the Co content of different legumes has been reported.

Effect of species

Perennial ryegrass would normally have more Co than timothy. While differences between different grasses and legumes are inconsistent, they mostly follow the order: red clover > white clover > perennial ryegrass > Italian ryegrass > cocksfoot > meadow fescues > timothy (Voss and MacPherson, 1977). These authors state that hill species, such as heather and white bent, contain relatively high Co concentrations. Following upland reseeding and land-improvement schemes, the risk of Co deficiency increases, as relatively Co-poor grasses, such as timothy and ryegrass, replace the natural sward. Indeed, timothy, whether home-grown or imported, has often formed the basis of experimental diets designed to provide low concentrations of the element.

Genotype

There are not many published data available on the effect of genotype on forage Co content. In a study of cultivars of cocksfoot, perennial ryegrass and white clover grown on low-Co soil, no significant differences were found between cultivars of different species or between the forage species (Forbes and Gelman, 1981).

Plant parts

Fleming and Murphy (1968) reported more Co in leaves than in flowering heads and more in flowering heads than in stems of grasses. Handreck and Riceman (1969) found that Co was concentrated at the margins of the leaflets of legumes and at the base and the extreme tips of the leaf blades of grasses.

Stage of maturity

Increasing maturity sometimes, but not always, leads to a decrease in Co content. Fleming and Murphy (1968) and Fleming (1970) recorded lower Co concentrations in perennial ryegrass with maturity, as did Whitehead and Jones (1969) in sainfoin and red clover. These latter, however, reported no consistent changes in lucerne or white clover, confirming similar findings by Rossiter *et al.*(1948) for subterranean clover.

Soil fertility

Low herbage Co is generally associated with soils of low available Co content. Soils derived from Old Red Sandstone and granitic parent material tend to be inherently low in the element, as do soils on limestone parent material. High soil Mn is known to reduce Co uptake by forages (Adams *et al.*, 1969; Norrish, 1975).

Waterlogging of soils allows the release of Co from the soil minerals into soil solution, so that forage on poorly drained soils can have up to seven times more Co than that from well-drained soils. If herbage yield is increased by additional moisture, there may be no effect on Co concentration.

Fertilizer N has been variously reported to have no effect on forage Co concentration (Stewart and Holmes, 1953; Wright and Lawton, 1954; Mudd, 1970), to lower it (Percival *et al.*, 1955; Reith *et al.*, 1964, 1983) and to increase it and the Co status of lambs grazing on it (Voss and MacPherson, 1977). Increased Co concentrations following N fertilization have also been reported for a mineral soil (Klessa *et al.*, 1988) and a peat soil (Reith *et al.*, 1983). Explanations for these varied responses include clover dieback following N application and a dilution effect on herbage Co when growth rate exceeded rate of Co uptake (Reith *et al.*, 1964; West, 1981). Mills and MacPherson (1982) relate the different responses to different soil Co levels. N fertilizer reduces uptake of Co on deficient and borderline status soils and on naturally freely drained soils of whatever status. Klessa *et al.* (1988) suggest the following factors are involved: soil availability; acidification by ammonium fertilizers; and ratio of ammonium (NH_4)-N : nitrate (NO_3)-N taken up by the roots and its effect on the pH of the rhizosphere.

Liming and soil pH

After soil drainage, soil pH is the next most important determinant of Co availability (West, 1981). Decreased Co concentrations following liming have been frequently reported (Ekman *et al.*, 1952; Hill *et al.*, 1953; Wright and Lawton, 1954; Percival *et al.*, 1955; Archer, 1970). Mitchell (1972) reported that an increase in pH of one unit from 5.4 to 6.4 resulted in a halving of the Co concentration of mixed pasture. Others have reported that increasing pH above 6.0 has little further effect (Wright and Lawton, 1954; McLaren *et al.*, 1987; Klessa *et al.*, 1988) (Fig. 17.1). A greater incidence of Co deficiency in grazing livestock occurs on naturally occurring high-pH soils. Finch and Rogers (1979) stated that Co deficiency was more likely on soils formed on limestone and Mitchell (1954) explained the high prevalence of Co deficiency on the machair pastures of the Western Isles by the abundant supply of calcium carbonate ($CaCO_3$) from sea shells. Not only does liming reduce availability by causing greater adsorption on to the various exchange sites, resulting in the precipitation of various Co salts, but higher pH also encourages the establishment of better-quality grasses of lower Co content.

Seasonal changes

No definitive trend can be asserted, but a tendency for concentrations to be higher in spring and autumn has been noted (Voss and MacPherson, 1977), but not in all studies (Andrews, 1966; Fleming, 1970). Silage cut in late spring had higher Co concentrations than hay cut in midsummer (Reith *et al.*, 1983). This can probably be accounted for by a decreased leaf:stem ratio as the herbage matured. Heavy grazing pressure may prevent maturation and maintain pasture at an early physiological growth stage and thus with a higher Co concentration. Repeated cutting maintained Co concentration at almost double that of uncut grass by mid-July, except for clover and timothy. Voss and MacPherson (1977), however, found lower Co concentrations in plots repeatedly cut to simulate grazing. The sporadic nature of seasonal variation is due both to the type of growing season and stocking density. When grass is plentiful, the grazing livestock will not crop it to the same extent, thus allowing maturation to occur. On the other hand, when grass growth is poor or the stocking density is high the grass will remain at a young growth stage throughout the growing season.

Factors Affecting the Selenium Status of Forage

Plants differ markedly in Se concentration ranging from very high concentrations of > 300 mg Se kg^{-1} DM in accumulator plants grown on Se-rich soils to < 0.05 mg Se kg^{-1} DM in normal herbage grown on Se-deficient soils. Although forage species commonly used for grazing livestock will have lower Se levels than accumulator plants, they can still contain toxic concentrations (e.g. *Poa pratense*, 33 mg kg^{-1} DM, and *Medicago sativa*, 40 mg kg^{-1} DM) when grown on seleniferous soils (Hamilton and Beath, 1963). Although knowledge of the toxic effects of Se predates that of its essentiality, it has since been discovered that the detrimental effects of Se deficiency far outweigh those of its toxicity. Areas of Se deficiency have been identified in Australia, New Zealand, China, the USA, many parts of Africa (e.g. Zaïre), much of continental (particularly northern) Europe and the UK

(Gardiner *et al.*, 1962; Gardiner and Gorman, 1963; Long and Marshall, 1973; Ammerman and Miller, 1975; Young *et al.*, 1977).

Effects of species

In New Zealand, grasses contain more Se than white clover at all levels of soil and fertilizer Se (Davies and Watkinson, 1966). Tropical legumes also have lower Se concentrations than associated grasses (Long and Marshall, 1973). No difference was recorded between timothy and white clover on a range of Se-deficient soils in Canada (Gupta and Winter, 1975), between lucerne and grasses in Denmark (Gissel-Nielsen, 1975) or between subterranean clover and associated grasses in Australia (Caple *et al.*, 1980). Ehlig *et al.* (1968) found twice as much Se in lucerne as in timothy, cocksfoot or brome grass. Of the different grass species sweet vernal had the highest concentration followed by Yorkshire fog, bent grass and perennial ryegrass.

Plant parts/stage of maturity

Neither appears to have had many data generated about it. Cocksfoot spikelets had twice as much Se as the leaf fraction, with the leaf sheath and stem intermediate (Davey and Mitchell, 1968). Ehlig *et al.* (1968) recorded that stage of growth had little effect on the Se content of lucerne.

Soil fertility

Forage Se is directly related to the Se content of the soil and the parent material from which it is derived (Robinson and During, 1961; Watkinson, 1962; Kubota *et al.*, 1967; Andrews *et al.*, 1968; Gupta and Winter, 1975; Doyle and Fletcher, 1977). Forage Se tends to be low on sandy soils and lower on mineral upland soils than on organic moorland soils. In Australia and New Zealand, nutritional muscular dystrophy (NMD) is often associated with lush pasture and superphosphate application. It may be due to a higher proportion of legumes, which are lower in Se than grasses, and to a stimulation of growth diluting the Se content (Westerman and Robbins, 1974).

Climate

Annual changes in mean temperature can affect the Se content of forage, e.g. *A. sativa* forage grown at a mean temperature of 14°C contained 0.012 mg Se kg^{-1} DM, compared with 0.042 mg kg^{-1} DM at a mean temperature of 19°C (Lindberg and Lannek, 1970). Areas with higher rainfall may also have a lower Se content in forage – 0.026 compared with 0.036 mg Se kg^{-1} DM, with annual mean rainfall figures of 250–375 and > 750 mm, respectively. Seasonal variations in concentration have been reported from Denmark (Gissel-Nielsen, 1975), where a maximum concentration of 0.20 mg kg^{-1} DM in March fell to a minimum of 0.02 mg kg^{-1} DM in June and September. No such changes were found over the same time period in moorland pasture in Norway (Garmo *et al.*, 1986).

Factors Affecting the Iodine Status of Forage

Low values (0.09 mg kg^{-1} DM) have been recorded for the I concentration of forage grown on sandy soil in the Netherlands (Hartmans, 1974), but much higher values (0.42 mg I kg^{-1} DM) for forage grown in New Zealand (Johnson and Butler, 1957). Forages grown under the same conditions showed species differences (Johnson and Butler, 1957; Alderman and Jones, 1967; Hartmans, 1974), but the ranking order between studies was not always consistent, possibly because different cultivars were used in each comparison. Different cultivars of perennial ryegrass, prairie grass and white clover exhibited large differences in I concentration (Butler and Glenday, 1962; Alderman and Jones, 1967; Rumball *et al.*, 1972). I concentration tends to be higher in leaves (0.24 mg kg^{-1} DM) than in stems (0.10 mg kg^{-1} DM) (Blom, 1934). Concentrations of I tend to be high early in the growing season, but these quickly fall away (Alderman and Jones, 1967).

Soil fertility

I concentrations in forage tend to be lowest on sandy soils and river clays low in organic matter and highest on soils of the younger marine clays and peats. Fertilizer N reduced I concentration in cocksfoot, perennial ryegrass and timothy, but with no difference between species. This effect was greater on younger herbage than on the more mature. N fertilizer also increased the concentration of cyanogenetic glucosides, thereby effectively increasing I requirement, so that it exercised a doubly detrimental effect (Rudert and Oliver, 1978). No consistent seasonal trends have been reported in I concentration.

Factors Affecting the Zinc Status of Forage

Minson (1990) quotes a mean Zn concentration of 36 mg kg^{-1} DM in 719 samples of forages in the world literature; 11% of these had concentrations below 20 mg kg^{-1} DM – a potentially deficient amount. The Pennsylvania State Testing Service analysed 17,000 forage samples over 5 years. The mean Zn concentration was 29 mg kg^{-1} DM (range: 3–300 mg kg^{-1} DM) (Adams, 1975). In Louisana, the mean concentration reported was 27 mg Zn kg^{-1} DM (Kappel *et al.*, 1985), with an even lower mean value of 22 mg kg^{-1} DM recorded for British Columbia (Miltimore *et al.*, 1970). These differences are possibly due to differences in soil Zn, forage species and stage of maturity.

Effect of species

Tropical forages were slightly higher (2 mg kg^{-1} DM) in mean Zn content than temperate ones (Minson, 1981). Although the overall difference was not great, the risk of values below 20 mg Zn kg^{-1} DM was doubled. In general, grasses tend to contain less Zn than legumes. This difference was confirmed when grass and clover were grown together (Metson *et al.*, 1979). The difference increased at higher levels of soil Zn (Gladstones and Loneragan, 1967). Different grass species grown at the same site exhibit large differences in Zn concentration, but the ranking order is not consistent and indeed may be totally reversed (Gomide *et al.*, 1969; Perdomo *et al.*, 1977).

Maximum differences found between genotypes of perennial ryegrass and prairie grass were 21 and 136%, respectively (Butler *et al.*, 1962; Rumball *et al.*, 1972). Minson (1984) recorded only a 10% difference in Zn concentration among five species of *Digitaria* (Pangola and Woolly Finger Grass). Legume species have been found both to vary little in Zn concentration (Loper and Smith, 1961; Whitehead and Jones, 1969) and to vary considerably (Gladstones and Loneragan, 1967).

Plant parts

The most recently formed parts of cocksfoot have the highest Zn concentration, with the leaf blades usually higher than the stems or leaf sheath (Davey and Mitchell, 1968). Fleming (1963) recorded a mean Zn concentration in the leaf blades of cocksfoot, meadow fescue, perennial ryegrass and timothy of 20 mg kg^{-1} DM compared with a mean of 15 mg kg^{-1} DM for their stems. The flowering heads of the grasses were considerably higher at 36 mg Zn kg^{-1} DM. White-clover flowering heads and leaf plus petiole had similar concentrations of 40 and 42 mg Zn kg^{-1} DM, respectively, but stems were much lower, at 12 mg kg^{-1} DM.

Stage of growth

Although it would be anticipated that, as plants mature, their Zn concentration would decline, due to an increasing proportion of stem, results have not always confirmed this. Gladstones and Loneragan (1967) did find that Zn concentration fell with increasing maturity in 24 different forages and was unaffected by level of soil Zn. No such differences were recorded by Whitehead and Jones (1969) or by Gomide *et al.* (1969). Fleming and Murphy (1968) found a high Zn concentration in the first spring of a trial with mixed temperate grasses, but in the following year Zn concentrations remained constant throughout. No change in Zn concentration with maturity was found for Bermuda grass (Miller *et al.*, 1964), tall fescue (Reid *et al.*, 1967a), hyacinth bean (Hendricksen and Minson, 1985) or white clover (Whitehead and Jones, 1969).

Soil fertility

Raising soil Zn effectively increases Zn concentration in forage (Brown *et al.*, 1962; Winter and Jones, 1977). High applications of fertilizer N have increased Zn concentrations in Bermuda grass and perennial ryegrass by around 12 mg kg^{-1} DM (Miller *et al.*, 1964; Hodgson and Spedding, 1966; Large and Spedding, 1966). If ammonium sulphate was used as the source of N the resulting increased acidity could account for the improved Zn absorption. More normal rates of N fertilization (up to 500 kg ha^{-1} $year^{-1}$) had no consistent effect on forage Zn content of giant star-grass (Rudert and Oliver, 1978), tall fescue (Reid *et al.*, 1967a) or cocksfoot (Reid *et al.*, 1967b). Extremely high levels of lime, which raised soil pH from 4.9 to 6.8, resulted in a lowering of the Zn concentration of Bermuda grass from 37 to 28 mg kg^{-1} DM. In Scotland, increasing soil pH had a much greater impact on lowering Zn concentration in clover than in ryegrass grown on the same site (Fig. 17.1). There is little available information on the effect of phosphorus (P) and potassium (K) fertilizers, but they appear to exercise little influence (Reid *et al.*, 1967a).

Factors Affecting the Manganese Status of Forage

Forage manganese concentrations range markedly from 1 to 2670 mg kg^{-1} DM. The mean value on a worldwide basis was 86 mg kg^{-1} DM, with only South Australia (24%) (Piper and Walkley, 1943) and Scotland (8%) (Hemingway *et al.*, 1968) having a reasonable proportion of samples with concentrations of < 20 mg Mn kg^{-1} DM. New Zealand had the highest mean concentrations, at 166 mg kg^{-1} DM (Smith and Edmeades, 1983).

Effect of species

Where Mn concentrations are not above 60 mg kg^{-1} DM, grass and legumes grown on the same site have similar levels. Above 60 mg kg^{-1} DM, however, grasses tend to have considerably higher values than legumes (Hemingway, 1962; McNaught, 1970; Reay and Marsh, 1976: Metson *et al.*, 1979). Of the grass species, cocksfoot and red-top bent tend to have the highest Mn concentrations, with brome grass and tall fescue intermediate and meadow grass and timothy the lowest. Maize silage contains only half as much Mn as grass silage. White clover has higher Mn concentrations than lucerne but some lupins were found to have much higher concentrations, thought to be due to their deep-rooting capability (Gladstones and Loneragan, 1970).

Plant parts

No very consistent pattern emerges. Similar levels have been found in leaf and stem (Fleming, 1963) for cocksfoot, but Mitchell and Davey (1968) reported twice as much Mn in leaf blade as in the sheath and stem. In six species, Mn in leaf was higher than in stem, but in meadow fescue and perennial ryegrass the reverse situation prevailed (Fleming, 1963). Generally, stage of growth has little effect on Mn concentration, but small increases (Beeson and MacDonald, 1951) and decreases (Thomas *et al.*, 1952) have been recorded in a range of legumes. This variability in response may be attributable to differences in availability of soil Mn.

Soil fertility

Mn follows Fe as the second most abundant of the essential elements in the earth's crust. Its availability is inversely related to soil pH. Thus the highest Mn concentrations are found in forages growing on acid soil. Whitehead (1966) reported a decrease in the Mn content of forage from 702 to 127 mg kg^{-1} DM following the liming of acid soil. Large falls in Mn concentrations in both ryegrass and clover with increasing soil pH were also recorded in Scotland (Fig. 17.1). Poor drainage also leads to higher Mn concentrations. The use of ammonium sulphate as a N fertilizer reduced soil pH and led to an increase in mixed-pasture Mn content from 31 to 63 mg kg^{-1} DM (Hemingway, 1962) and in six tropical grasses from 121 to 164 mg kg^{-1} DM (Gomide *et al.*, 1969). Ammonium nitrate applications had little effect on soil pH and had no effect on Mn concentration in cocksfoot (Reid *et al.*, 1967b) or star-grass (Rudert and Oliver, 1978).

No consistent trend in seasonal Mn concentrations has been reported (Minson, 1990).

Factors Affecting the Iron Status of Forage

Iron is naturally well supplied by forages, and deficiencies of the element in grazing livestock are unlikely to occur normally but may result from blood loss due to heavy parasitic infestation or some other cause of haemorrhage. The mean concentration in UK grasses has been reported as 103 mg kg^{-1} DM (range 73–154 mg kg^{-1} DM) (Whitehead, 1966). Legumes would generally contain more Fe than grasses (Kabata-Pendias and Pendias, 1992). Forage Fe concentration is affected by changes in soil conditions, by climate and by stage of growth. Grass grown on soils derived from serpentine had Fe concentrations ranging from 2127 to 3580 mg kg^{-1} DM (Johnson and Proctor, 1977), but values above 300 mg kg^{-1} DM would normally reflect soil contamination of the sample, rather than intrinsic Fe. Such adventitious Fe is important in the context of its impact on Cu availability to the ruminant, and consequently care should be exercised in minimizing the level of any such contamination in grass silage and fodder crops.

What Affects Trace-element Availability

This section deals with those factors that influence the availability of the trace-element content of forages to animals and not those affecting the plant content itself. Table 17.1 details the threshold concentrations for desirable, marginal and

Table 17.1. Threshold concentrations of trace elements in forage for ruminants.

			Cattle	Sheep
Cu	Desirable	(Mo < 1.0)	> 10.0mg kg^{-1} DM	> 5.0mg kg^{-1} DM
	Marginal	(Mo > 3.0)	> 10.0	> 5.0
	Deficient	(Mo > 3.0)	<10.0	< 5.0
Co	Desirable		> 0.11	> 0.11
	Marginal		0.05–0.07	0.07–0.10
	Deficient		< 0.05	< 0.07
Se	Desirable		> 0.10	> 0.10
	Marginal		0.03–0.05	0.03–0.05
	Deficient		< 0.03	< 0.03
I	Desirable (normal diet)		0.50	0.50
	(goitrogenic diet)		2.0	2.0
Zn	Desirable		50	50
	Marginal		20–40	30–50
	Deficient		< 20	< 30
Mn	Desirable		25	25
Fe	Desirable		> 30	> 30
	Calves (< 150 kg)		100	
	Cows in late pregnancy		40	

NB Because of the extreme susceptibility of some sheep breeds to Cu toxicity, it is possible, even with forage Cu levels within the normal range of 8–15 mg kg^{-1} DM, for poisoning to occur if forage Mo is also low (0.2–0.8 mg kg^{-1} DM). Critical Cu:Mo ratio = 20 or greater.

deficient levels of these elements in forages. These perhaps somewhat arbitrary thresholds have been based on the Agricultural Research Council (ARC, 1980), Jumba *et al.* (1996) and the author's own experience and delineate where normal production, some loss of tissue trace element status possibly associated with subclinical deficiency and frank deficiency respectively might be anticipated. Productive responses to supplementation would generally be anticipated in the deficient, occasionally in the marginal and not at all in the desirable ranges. Interactions between elements are frequently critical in the ability of a forage to meet the trace-element requirements of ruminants, and a small change in the concentration of one element can mean a swing from deficiency to excess in the supply of another.

Copper

Its availability to the animal is governed mainly by interactions with Mo, S and Fe. Suttle (1981) produced prediction equations for the availability of Cu in grass, grass silage, hay, cereals and brassicas. Even when within their normal ranges, both Mo and S exercised marked effects and, as the relationships are logarithmic, initial small increases in concentration exercise the greatest inhibitory effects. The equations used to predict Cu availability (ACu) are as follows:

Grass	ACu (%) = $5.72 - 1.297S - 2.785\log_e Mo + 0.227(Mo \times S)$
Silage	ACu (%) = $10.6 - 6.65\log_e S$
Hay	ACu (%) = $8.9 - 0.70\log_e Mo - 2.61\log_e S$
Brassicas	ACu (%) = $0.042 - 0.0096Mo \times S$

Fresh grass has the lowest available Cu, followed by silage, hay, brassicas and cereals. S alone had the greatest influence on silage Cu availability, while both Mo and S were in operation in fresh grass and in hay, but to a much lesser extent. These equations have proved to be helpful in the diagnosis of Cu deficiency in the field over the past two decades and have been adopted by the extension services with considerable success. Some anomalies do, however, occur, as, for example, the development of hypocupraemia in kale-fed cattle, despite its theoretically high available Cu (Bray and Hacker, 1981), and, at the other end of the scale, the occurrence of copper poisoning in sheep while grazing herbage of normal copper concentration but very low Mo content and despite a relatively high S concentration of 4 g kg^{-1}. The presence of a highly susceptible breed, Texel, undoubtedly contributed to the fatal outcome (MacPherson *et al.*, 1997). The Cu from animal-derived feedstuffs is twice as available as that from plant sources (Baker and Ammerman, 1995).

A high Fe content in spring pasture was blamed by Jarvis and Austin (1983) for Cu deficiency in grazing cattle, but, although individual animals showed clinical signs, no overall growth retardation was recorded. Concentrations of Fe > 300 mg kg^{-1} DM are reported to have a deleterious effect on Cu availability; Suttle and Peter (1985) effected a 33% decrease in sheep given 800 mg Fe and fed on a semi-purified diet, while McFarlane *et al.* (1991) recorded Cu deficiency in cattle grazing pasture on rendzina soils high in Fe. Forages are frequently contaminated with soil Fe and this must be taken into account when estimating Cu availability. As swayback occurred on pastures grown on calcareous soils, it was thought that Cu availability might be adversely affected by dietary Ca, but this has not been substantiated. High levels of Zn can inhibit Cu uptake and protect against its toxic effects (Bremner *et al.*, 1976), but such high values are likely to be rare in practice. The presence of high levels of Cd in sewage sludge and in superphosphate fertilizers affects Cu absorption and reduces placental transfer.

Cobalt

The availability of dietary Co has not got the same significance as the other essential trace elements, as the requirement by the animal is not for the element itself but for a preformed Co compound, vitamin B_{12}, which is synthesized by microorganisms in the rumen and then absorbed lower down the gastrointestinal tract. This microbial production of the vitamin is vulnerable to Co deficiency, to a decreasing roughage:concentrate ratio and also to greater or lesser production of inactive vitamin B_{12} analogues at the expense of 'true' vitamin B_{12}. (Gawthorne, 1970; Sutton and Elliot, 1972; Hedrich *et al.*, 1973). Sheep fed lucerne hay apparently retained more Co than those fed timothy hay (Looney *et al.*, 1976). If the Co concentration in the rumen fluid falls below 5 ng ml^{-1}, the rate of vitamin B_{12} synthesis will be insufficient to meet the animal's needs (Smith and Marston, 1970). The production of vitamin B_{12} in the rumen is relatively inefficient, as is its absorption from the small intestine. Opposing findings have been recorded with respect to the efficiency of vitamin B_{12} production in the rumen during deficiency and sufficiency of Co supply. Elliot *et al.* (1971) found 3 and 10%, respectively, while Smith and Marston (1970) found 13 and 3%. Absorptive efficiency of synthesized vitamin B_{12} has been variously quoted at values ranging from 5–20% (Smith and Marston, 1970; Hedrich *et al.*, 1973) to 8–38% (Rickard and Elliot, 1978).

Selenium

In forages, Se tends to occur as protein-bound selenomethionine, with smaller amounts of selenocysteine, and both appear to be well used by ruminants. Cantor *et al.* (1975) reported that Se of plant origin is more readily available than that of animal-derived feeds and this was confirmed by Douglass *et al.* (1981). The bioavailability of Se varies widely, both between and within foods of plant and animal origin (Mutanen, 1986). Organic Se appears to be more readily absorbed from the digestive tract than inorganic (Pehrson, 1993). Pehrson *et al.* (1989) found that selenomethionine was about twice as active as inorganic Se in increasing glutathione peroxidase activity in erythrocytes of heifers. Molnár *et al.* (1997) reported that selenomethionine was more efficiently incorporated into the muscle of sheep than sodium selenite, but that there was no difference in organ tissue. In a study in rats, Whanger and Butler (1988) found increased tissue Se retention from selenomethionine, rather than from selenite, particularly in muscle and brain. Selenomethionine can be incorporated directly into body proteins and stored in this form and can thus inflate bioavailability estimates based on body retention (Henry and Ammerman, 1995). There tends to be a greater excretion in urine of inorganic Se (Aspila, 1991), but the reverse is true of milk, where higher concentrations follow organic supplementation.

Se and vitamin E metabolisms have long been linked, due to their complementary roles as biological antioxidants. Consequently, a good supply of either nutrient was able to make good or at least to mask to some extent a deficiency of the other. However, Se is now known to have several other roles in the body unlinked to vitamin E, so there is a fundamental requirement for a dietary supply of the element that cannot be met by high concentrations of the vitamin. The minimum necessary Se concentration in the presence of high vitamin E levels in forages has not, to my knowledge, been established, but it would be unwise to rely on concentrations below the deficiency threshold of 0.03 mg kg^{-1} DM (Table 17.1) being adequate.

Iodine

Dietary I is apparently well absorbed by ruminants, mostly from the abomasum (Miller *et al.*, 1974; Price, 1989), but its incorporation into the thyroid may be reduced by low dietary Co and Mn concentrations (Underwood, 1977). However, the main inhibitor of I availability/absorption and thyroid hormone synthesis is the presence of goitrogens. Brassicas contain the vinylthio-oxazolidine type of goitrins (Hetzel and Maberly, 1986), while white clovers contain cyanogenetic glycosides (Flux *et al.*, 1956). Grasses, however, appear to be largely free of goitrogenic activity. According to Underwood (1981), in times of a diminished I supply, the thyroid can clear the element from the blood more rapidly and utilize it in hormone synthesis more efficiently. Through its presence in the deiodinase enzymes, Se is required for the conversion of tetraiodothyronine to tri-iodothyronine and may thus influence I bioavailability (Arthur *et al.*, 1988). The processes of ploughing, drainage and liming involved in the improvement of upland pasture have been implicated in a reduction in their capacity to retain I in a bioavailable form (Lidiard, 1995). Miller and Ammerman (1995) conclude a recent review on I availability by saying that little is known about its bioavailability from plant sources.

Zinc

There is a lack of information on the physiological availability of Zn from forages to ruminants. In monogastrics, Zn from animal-derived protein concentrates is more available than that from plant sources, due to its combination with phytate and fibre (Underwood, 1981). High Ca can inhibit Zn absorption, particularly in pig and poultry diets and the ratio of phytate × Ca:Zn is claimed to be a better predictor of Zn availability than the phytate:Zn ratio *per se* (Fordyce *et al.*, 1987). Other elements that adversely affect Zn availability include Cu (although the reverse effect is more potent), Cd and Pb in high concentrations. Some, but not all, chelators can improve Zn bioavailability – their efficacy is related to their stability constants (Vohra and Kratzer, 1964).

Manganese

Absorption of dietary Mn by all domestic livestock appears to be poor and is adversely affected by high concentrations of Ca, P and Fe. Excess P appears to be a greater inhibitor of dietary Mn than Ca and it appears to exercise its effect at gut level (Wedekind and Baker, 1990a, b). Bremner (1970) reported that only 10% of Mn in sheep rumen was in a soluble form, compared with about 50% of dietary Mn being water-soluble. Other workers have shown that no more than 1% dietary Mn is absorbed (Sansom *et al.*, 1976; Vagg, 1976) by adult cattle. Most of the data relating to Mn bioavailability have been obtained from studies with poultry, but Henry (1995) states that results suggest that it is similar for both poultry and sheep. Excess Mn is removed by the liver and excreted in the bile.

Iron

Not much is known about the availability of forage Fe to ruminants other than that it is affected by the solubility of the chemical form, e.g. ferric oxide Fe is significantly less available to calves and sheep than more soluble forms, such as ferrous sulphate and ferric chloride (Morris, 1987). However, as little Fe is lost from the body, the normal content of the grazed herbage is sufficient to supply all that the animal requires.

How to Remedy Low Status

There are a multiplicity of methods by which trace-element deficiency in ruminants grazing forage or consuming conserved products made from it may be remedied. It would require a review in its own right to assess correctly the best options for each situation. All that can be done here is to list the range of methods available and possibly highlight the author's preference or prejudice with respect to some of them.

For animals at grass in non-extensive situations, it is possible to treat the ground with the appropriate trace element(s) and raise herbage concentration(s) sufficiently to meet the need of the grazing livestock. This works well for Co, where a single treatment of 2 kg ha^{-1} can boost herbage concentrations to normal for 3 or 4 years. New Zealand and Finnish experience has shown that Se treatment can be effective in raising herbage and crop Se status sufficiently to correct deficiencies of the element in livestock and humans. Application at a rate of 1 kg ha^{-1} provided protection for 2 years (Aspila, 1991; Watkinson, 1991). Se treatment has not often been attempted in the UK, but there is no reason why it should not operate successfully here. With Cu, however, while there is a big initial boost in concentration, it is not maintained and long-term effects tend to be marginal and generally insufficient on their own to correct a deficiency.

In extensive and rough-terrain grazings, such applications are either uneconomic or impracticable and treatment or prevention of deficiencies has to rely on the provision of free-access minerals. These may be provided loose in a covered trough, to prevent weathering, or in the form of blocks or mineral licks. To encourage optimum usage, they are best positioned close to watering points and one block or trough should be available for every 25–40 cows. The major snag with this method is the variable intake, ranging from nil to an oversupply. This may have to be tolerated if the alternative of individual animal treatment is inconvenient to the nature of the enterprise. Where cattle and sheep are grazing side by side, care needs to be exercised with respect to Cu, as sheep are highly susceptible to excess intakes of the element, whereas cattle are relatively tolerant.

Where handling facilities exist and individual treatments can be administered successfully, there are again a wide variety of possibilities. These include oral treatments of various descriptions and parenteral-injectable preparations, and many of these were reviewed by Allen and Moore (1983) and MacPherson (1983). On the basis that the fewer treatments that are required per annum the better, only such long-efficacy products will be alluded to, unless some other consideration becomes paramount. Various boluses, from the traditional Co one to the more recently developed soluble glass products, provide suitable vehicles for the correction of deficiency in grazing ruminants. These can be either single-element treatments or various combinations of two or more elements and, in some cases, vitamins as well, as in the Agrimin all-trace bolus. These can be effective for 6–8 months at grass and can provide all the necessary trace elements for a complete grazing season, at least in these northern climes, at a single handling. One of the possible drawbacks with the traditional bullets was the risk of their forming an insoluble coating of $CaCO_3$ and being thus rendered ineffective. This was possible because these boluses did not erode from the surface but the active ingredient, which only formed a very small proportion of the total weight, leached out through the interstitial spaces. With the soluble-glass bolus, the whole surface is continually eroding and this is also true of the exposed end of the Agrimin all-trace bolus. Where a traditional bolus is being used in a situation where coating is a problem, it can be overcome, to some extent, by the use of a second bolus or an inert grub screw. Constant abrasion of the two boluses in the reticulum prevents build-up of the coating. Regurgitation is another hazard of this form of treatment but, where care is taken in the administration of the bolus, it should not be too serious. Another treatment that is effective over a whole grazing period is copper oxide wire, which is administered in a gelatin capsule, which dissolves rapidly, allowing the short pieces of wire to become trapped by the intestinal villi, where the acid conditions dissolve some of the Cu which is then absorbed and stored in the liver (MacPherson, 1984). Injectable preparations, which use barium selenate as the source of Se, have also been shown to be long-lasting (MacPherson *et al.*, 1988), being efficacious in cattle for up to a year and in ewes for up to 18 months.

Administration of trace elements via the water-supply is feasible and can be done by means of a plumbed-in metering device or by the addition of trace-element pellets placed in a filter tube at the bottom of water troughs. In order for this method of supplementation to be successful, there must be no access to untreated sources of water, such as streams or springs.

For housed animals, the method of choice is by supplementation of the feed with the required amounts of the appropriate trace elements. This can be achieved by incorporation into a complete diet or into the concentrate portion of it or by dissolving the required elements and spraying them over the forage portion of the ration. These methods ensure that all animals get their requisite intake of these essential nutrients by eliminating the possibility of individuals selecting not to consume the minerals provided.

A matter of some controversy in recent years has been the debate over the relative merits of inorganic versus organic or chelated minerals. Such products include bioplexes, where the minerals are bound to proteins or yeast, and chelates and metalosates, where the mineral is bound to a protein, part of a protein or to a specific amino acid, such as lysine. In theory these organically bound products should be protected against breakdown in the rumen, thus preventing the formation of unavailable chemical compounds. The protected compounds then progress further down the digestive tract and are absorbed. Great claims are made for these new formulations with respect to improved performance and fertility and enhanced resistance to infectious challenge, such as mastitis. To date, however, these claims have not been substantiated by solid scientific evidence; in studies comparing the treatments, when used at the same rate, no productive benefit has been recorded for the complexed minerals and they have the added disadvantage of being considerably more expensive. A situation of high forage Mo would be one where Cu proteinate products would be expected to be beneficial if they could successfully protect the element from forming insoluble compounds with Mo in its passage through the rumen. However, in my own experience, one company would not support a trial in just such a situation because they felt it might present too severe a test of their product's efficacy. Similarly, claims are made for the efficacy of Se yeast products, but many much cheaper alternatives apparently work highly effectively. At present, the best approach to this problem would appear to be to use the considerably cheaper traditional products first and only if these fail to resort to the appropriate bioplex or chelate.

Further Research Needed

As trace-element deficiencies are widespread in forages throughout the world and as some exist now in areas of improved pasture where they did not prior to the improvement, there would appear to be ample scope to improve the trace-element content of the productive forages by selective breeding and to such an extent as to render fertilization or direct animal supplementation unnecessary. In tandem with this could be a selective breeding programme to produce lines of animals that would be less susceptible to the effects of low or deficient intakes. This has been at least begun with the Scottish Blackface breed, where a flock has been separated into 'low'- and 'high'-line animals with respect to their ability to absorb Cu and to resist the development of sway-back in their lambs (Wooliams *et al.*, 1986). Breeding for less susceptibility to Cu poisoning should also be feasible for vulnerable breeds, such as the Texel.

Definitive studies could be undertaken to determine whether chelated minerals do have advantages over inorganic sources and, if they do, to identify the situations where these would be of sufficient economic benefit to justify their use. Better diagnostic criteria would be helpful to identify economically responsive situations more accurately than is currently possible. A reliable indicator for Co deficiency in cattle remains a priority for future research.

References

Adams, R.S. (1975) Variability in mineral and trace element content of dairy cattle feeds. *Journal of Dairy Science* 58, 1538–1548.

Adams, S.N., Honeysett, J.L., Tiller, K.G. and Norrish, K. (1969) Factors controlling the increase of cobalt in plants following the addition of a cobalt fertilizer. *Australian Journal of Soil Research* 7, 29–42.

Alderman, G. and Jones, D.I.H. (1967) The iodine content of pastures. *Journal of the Science of Food and Agriculture* 18, 197–199.

Allen, W.M. and Moore, P.R. (1983) Parenteral methods of trace element supplementation. In: Suttle, N.F., Gunn, R.G., Allen, W.M., Linklater, K.A. and Weiner, G. (eds) *Trace Elements in Animal Production and Veterinary Practice*. Occasional Publication of the British Society of Animal Production, No 7, Edinburgh, pp. 87–92.

Alloway, B.J. and Tills, J.R. (1984) Copper deficiency in world crops. *Outlook on Agriculture* 13, 32–42.

Ammerman, C.B. and Miller, S.M. (1975) Selenium in ruminant nutrition: a review. *Journal of Dairy Science* 58, 1561–1577.

Andrews, E.D. (1966) Cobalt concentrations in some New Zealand fodder plants grown on Co-sufficient and Co-deficient soils. *New Zealand Journal of Agricultural Research* 9, 829–838.

Andrews, E.D., Hartley, W.J. and Grant, A.B. (1968) Selenium responsive diseases of animals in New Zealand. *New Zealand Veterinary Journal* 16, 3–17.

ARC (1980) *Nutrient Requirements of Ruminant Livestock*. Commonwealth Agricultural Bureau, Farnham Royal, Slough, England, 351 pp.

Archer, F.C. (1970) Uptake of magnesium and trace elements by herbage of a re-seeded upland pasture. *Journal of the Science of Food and Agriculture* 21, 279–281.

Arthur, J.R., Morrice, P.C. and Beckett, G.J. (1988) Thyroid hormone concentrations in selenium deficient and selenium sufficient cattle. *Research in Veterinary Science* 45, 122.

Aspila, P. (1991) Metabolism of selenite, selenomethionine, and feed-incorporated selenium in lactating goats and dairy cows. *Finnish Journal of Agriculture* 63, 9–14.

Baker, D.H. and Ammerman, C.B. (1995) Copper bioavailability. In: Ammerman, C.B., Baker, D.H. and Lewis, A.J. (eds) *Bioavailability of Nutrients for Animals*. Academic Press, San Diego, USA, pp. 127–149.

Beck, A.B. (1962) The levels of copper, molybdenum and inorganic sulphate in some Western Australian pastures – a contribution to the study of Cu deficiency diseases in ruminants. *Australian Journal of Experimental Agriculture and Animal Husbandry* 2, 40–45.

Beeson, K.C. (1950) *Cobalt – Occurrence in Soils and Forages in Relation to a Nutritional Disorder in Ruminants*. Agriculture Information Bulletin, Publication 7, United States Department of Agriculture, Washington, DC.

Beeson, K.C. and MacDonald, H.A. (1951) Absorption of mineral elements by forage plants III. The relation of stage of growth to the micronutrient element content of timothy and some legumes. *Agronomy Journal* 43, 589–593.

Berrow, M.L. and Ure, A.M. (1985) Relationships between trace element availability and the soils and geology of the Aberdeen area. In: Mills, C.F., Bremner, I. and Chesters, J.K. (eds) *Trace Elements in Man and Animals – 5*. Commonwealth Agricultural Bureau, Farnham Royal, Slough, England, pp. 841–843.

Blom, I.J.B. (1934) Studies in mineral metabolism 30. Variation in the iodine content of grasses at different stages of growth and a note on the iodine content of milk. *Onderstepoort Journal of Veterinary Science and the Animal Industry* 2, 139–150.

Bray, R.A. and Hacker, J.B. (1981) *Australian Journal of Agricultural Research* 32, 311–323.

Bremner, I. (1970) Zinc, copper and manganese in the alimentary tract of sheep. *British Journal of Nutrition*, 24, 769–783.

Bremner, I., Young, B.W. and Mills, C.F. (1976) Protective effect of zinc supplementation against copper toxicosis in sheep. *British Journal of Nutrition* 36, 551–561.

Brown, A.L., Krantz, B.A. and Martin, P.E. (1962) *Soil Science Society of America Proceedings* 26, 167–170.

Burridge, J.C. (1970) Vegetational factors affecting the trace element content of plants. In: Mills, C.F. (ed.) *Trace Element Metabolism in Animals*. Livingstone, Edinburgh, pp. 412–415.

Butler, G.W. and Glenday, A.C. (1962) *Australian Journal of Biological Science* 15, 183–187.

Butler, G.W., Barclay, P.C. and Glenday, A.C. (1962) Genetic and environmental differences in the mineral composition of ryegrass herbage. *Plant Soil* 16, 214–228.

Cantor, A.H., Scott, M.L. and Noguchi, T. (1975) Biological availability of selenium in feedstuffs and selenium compounds for prevention of exudative diathesis in chicks. *Journal of Nutrition* 105, 96–105.

Caple, I.W., Andrewartha, K.A., Edwards, S.J.A. and Halpin, C.G. (1980) An examination of the selenium nutrition of sheep in Victoria. *Australian Veterinary Journal* 56, 160–167.

Cornforth, I.S. (1984) Mineral nutrients in pasture species. *Proceedings of the New Zealand Society for Aminal Production* 44, 135–137.

Cunningham, I.J. (1960) Molybdate topdressing and animal health. *New Zealand Journal of Agricultural Research* 100, 419–428.

Davey, B.G. and Mitchell, R.L. (1968) The distribution of trace elements in cocksfoot (*Dactylis glomerata*) at flowering. *Journal of the Science of Food and Agriculture* 19, 425–431.

Davies, E.B. and Watkinson, J.H. (1966) Uptake of native and applied selenium by pasture species I. Uptake of selenium by browntop, ryegrass, cocksfoot and white clover from Atiamuri sand. *New Zealand Journal of Agricultural Research* 9, 317–327.

de Sousa, J.C., Nicodemo, N.L.F. and Darsie, G. (1989) Mineral deficiency in cattle in Roraima, Brazil. 5. Copper and molybdenum. *Pesquisa Agropecuriaria Brasiliera* 24, 1547–1554.

Douglass, J.S., Morris, V.C., Soares, J.H., Jr and Levander, O.A. (1981) Nutritional availability in rats of selenium in tuna, beef kidney and wheat. *Journal of Nutrition* 111, 2180.

Doyle, P.J. and Fletcher, W.K. (1977) Influence of soil parent material on the selenium content of wheat from west-central Saskatchewan. *Canadian Journal of Plant Science* 57, 859–864.

Ehlig, C.F., Allaway, W.H., Carey, E.E. and Kubota, J. (1968) Differences among plant species in selenium accumulation from soils low in available selenium. *Agronomy Journal* 60, 43–47.

Ekman, P., Karlsson, N. and Svanberg, O. (1952) Investigations concerning cobalt problems in Swedish animal husbandry. *Acta Agriculturae Scandinavica* 2, 103–130.

Elliot, J.M., Kay, R.N.B. and Goodall, E.D. (1971) Production and absorption of vitamin B_{12} in the sheep – a preliminary study. *Life Science* 10, 647–654.

Finch, T. and Rogers, P.A.M. (1979) Distribution of cobalt and molybdenum in soils in eastern midlands of Ireland. *Irish Journal of Agricultural Research* 17, 107–113.

Fleming, G.A. (1963) Distribution of major and trace elements in some common pasture species. *Journal of the Science of Food and Agriculture* 14, 203–208.

Fleming, G.A. (1970) The influence of stage of maturity and season on trace element levels in perennial ryegrass. *Agricultural Digest* 19, 25–32.

Fleming, G.A. and Murphy, W.E. (1968) The uptake of some major and trace elements by grasses as affected by season and stage of maturity. *Journal of the British Grassland Society* 23, 174–185.

Flux, D.S., Butler, G.W., Johnson, J.M., Glenday, A.C. and Peterson, G.B. (1956) Goitrogenic effect of white clover (*Trifolium repens*). *New Zealand Journal of Science and Technology* A38, 88–102.

Forbes, J.C. and Gelman, A.L. (1981) Copper and other minerals in herbage species and varieties on copper-deficient soils. *Grass and Forage Science* 36, 25–30.

Fordyce, E.J., Forbes, R.M., Robbins, K.R. and Erdman, J.W., Jr (1987) Phytate × calcium/zinc molar ratios: are they predictive of zinc bioavailability? *Journal of Food Science* 52, 440.

Fraser, A.J. (1984) The relationship between New Zealand's geology and soils and trace element deficiencies in grazing animals. *Proceedings of the New Zealand Society for Animal Production* 44, 125–133.

Gardiner, M.R. and Gorman, R.C. (1963) Further observations on plant selenium levels in Western Australia. *Australian Journal of Experimental Agriculture and Animal Husbandry* 3, 284–289.

Gardiner, M.R., Armstrong, J., Fels, H. and Glencross, R.N. (1962) Observation on plant selenium levels in Western Australia. *Australian Journal of Experimental Agriculture and Animal Husbandry* 2, 261–269.

Garmo, T.J., Frøslie, A. and Hoie, R. (1986) Levels of copper, molybdenum, sulphur, zinc, selenium, iron and manganese in native pasture plants from a mountain area in Southern Norway. *Acta Agriculturae Scandinavica* 36, 147–161.

Gawthorne, J.M. (1970) The effect of cobalt intake on the cobalamine and cobinamide composition of rumen contents and blood plasma of sheep. *Australian Journal of Experimental Biology and Medical Science*, 48, 285–292.

Gissel-Nielsen, G. (1975) Selenium concentration in Danish forage crops. *Acta Agriculturae Scandinavica* 25, 216–220.

Gladstones, J.S. and Loneragan, J.F. (1967) Mineral elements in temperate crop and pasture plants I. Zinc. *Australian Journal of Agricultural Research* 18, 427–446.

Gladstones, J.S. and Loneragan, J.F. (1970) Nutrient elements in herbage plants in relation to soil adaptation and animal nutrition. In: *Proceedings of the XI International Grassland Congress*, pp. 350–396.

Gomide, J.A., Noller, C.H., Mott, G.O., Conrad, J.H. and Hill, D.L. (1969) Mineral composition of six tropical grasses as influenced by plant age and nitrogen fertilization. *Agronomy Journal* 61, 120–123.

Gupta, U.C. and Winter, K.A. (1975) Selenium content of soil and crops and the effect of lime and sulfur on plant selenium. *Canadian Journal of Soil Science* 55, 161–166.

Hamilton, J.W. and Beath, O.A. (1963) Uptake of available selenium by certain range plants. *Journal of Range Management* 16, 261–265.

Handreck, K.A. and Riceman, D.S. (1969) Cobalt distribution in several pasture species grown in culture solutions. *Australian Journal of Agricultural Research* 20, 213–226.

Hartmans, J. (1974) Factors affecting the herbage iodine content. *Netherlands Journal of Agricultural Science* 22, 195–206.

Hedrich, M.F., Elliot, J.M. and Lowe, J.E. (1973) Response in vitamin B_{12} production and absorption to increasing cobalt intake in the sheep. *Journal of Nutrition* 103, 1646–1651.

Hemingway, R.G. (1962) Copper, molybdenum, manganese and iron contents of herbage as influenced by fertilizer treatments over a three-year period. *Journal of the British Grassland Society* 17, 182–187.

Hemingway, R.G., MacPherson, A., Duthie, A.K. and Brown, N.A. (1968) The mineral composition of hay and silage grown in Scotland in relation to the ARC standards for the mineral requirements of dairy cattle. *Journal of Agricultural Science* 71, 53–59.

Hendricksen, R.E. and Minson, D.J. (1980) The feed intake and grazing behaviour of cattle grazing a crop of *Lablab purpureus* cv. Rongai. *Journal of Agricultural Science* 95, 547–554.

Hendricksen, R.E. and Minson, D.J. (1985) Growth canopy structure and chemical composition of *Lablab purpureus* cv. Rongai at Samford, S.E. Queensland. *Tropical Grasslands* 19, 81–87.

Henry, P.R. (1995) Manganese bioavailability. In: Ammerman, C.B., Baker, D.H. and Lewis, A.J. (eds) *Bioavailability of Nutrients for Animals*. Academic Press, San Diego, USA, pp. 239–256.

Henry, P.R. and Ammerman, C.B. (1995) Selenium bioavailability. In: Ammerman, C.B., Baker, D.H. and Lewis, A.J. (eds) *Bioavailability of Nutrients for Animals*. Academic Press, San Diego, USA, pp. 295–336.

Hetzel, R.S. and Maberly, G.F. (1986) Iodine. In: Mertz, W. (ed.) *Trace Elements in Human and Animal Nutrition*. Academic Press, London, pp. 139–208.

Hill, A.C., Toth, S.J. and Bear, F.E. (1953) Cobalt status of New Jersey soils and forage plants and factors affecting the cobalt content of plants. Soil Science 76, 273–284.

Hodgson, J. and Spedding, C.R.W. (1966) The health and performance of the grazing animal in relation to fertilizer nitrogen usage. I. Calves. *Journal of Agricultural Science* 67, 155–167.

Jarvis, S.C. and Austin, A.R. (1983) Soil and plant factors limiting the availability of copper to a beef suckler herd. *Journal of Agricultural Science, Cambridge* 101, 39–46.

Johnson, J.M. and Butler, G.W. (1957) Iodine content of pasture plants. 1. Method of determination and preliminary investigation of species and strain differences. *Physiologia Plantarum* 10, 100–111.

Johnson, W.R. and Proctor, J. (1977) Metal concentrations in plants and soils from two British serpentine sites. *Plant Soil* 46, 275.

Jumba, I.O., Suttle, N.F., Hunter, E.A. and Wandiga, S.O. (1995) Effects of soil origin and mineral composition and herbage species on the mineral composition of forages in the Mount Elgon region of Kenya. 2. Trace elements. *Tropical Grasslands* 29, 47–52.

Jumba, I.O., Suttle, N.F. and Wandiga, S.O. (1996) Mineral composition of tropical forages in the Mount Elgon region of Kenya. 2. Trace elements. *Tropical Agriculture (Trinidad)* 73, 113–118.

Kabata-Pendias, A. and Pendias, H. (1992) *Trace Elements in Soils and Plants*, 2nd edn. CRC Press, Boca Raton, Florida, USA, 365 pp.

Kappel, L.C., Morgan, E.B., Kilgore, L., Ingraham, R.H. and Babcock, D.K. (1985) Seasonal changes of mineral content of Southern forages. *Journal of Dairy Science* 68, 1822–1827.

Klessa, D.A., Dixon, J. and Voss, R.C. (1988) Soil and agronomic factors influencing the cobalt content of herbage. *Research and Development in Agriculture* 6, 25–35.

Kretschmer, A.E., Lazar, E.A. and Beeson, K.C. (1954) *Proceedings of the Soil and Crop Society of Florida* 24, 167–176.

Kubota, J., Allaway, W.H., Carter, D.L., Carey, E.E. and Lazar, V.A. (1967) Selenium in crops in the United States in relation to selenium-responsive diseases of animals. *Journal of Agriculture and Food Chemistry* 15, 448–453.

Large, R.V. and Spedding, C.R.W. (1966) The health and performance of the grazing animal in relation to fertilizer nitrogen usage. II. Weaned lambs. *Journal of Agricultural Science* 67, 41–52.

Latteur, J.D. (1962) *Cobalt Deficiencies and Sub-deficiencies in Ruminants*. Cobalt Information Centre, Brussels.

Lee, H.J. (1951) *Journal of the Agriculture of South Australia* 54, 475–490.

Lidiard, H.M. (1995) Iodine in the reclaimed upland soils of a farm in the Exmoor National Park, Devon, UK and its impact on livestock health. *Applied Geochemistry* 10, 85–95.

Lindberg, P.E. and Lannek, N. (1970) Amounts of selenium in Swedish forages, soils and animal tissues. In: Mills, C.F. (ed.) *Trace Element Metabolism in Animals*. Livingstone, Edinburgh, pp. 421–426.

Long, M.I.E. and Marshall, B. (1973) The selenium status of pastures in Uganda. *Tropical Agriculture (Trinidad)* 50, 121–128.

Looney, J.W., Gille, G., Preston, R.L., Graham, E.R. and Pfander, W.H. (1976) Effects of plant species and cobalt intake upon cobalt utilization and ration digestibility by sheep. *Journal of Animal Science* 42, 693–698.

Loper, G.M. and Smith, D. (1961) *Changes in Micronutrient Composition of the Herbage of Alfalfa, Medium Red Clover, Ladino Clover and Brome Grass with Advance in Maturity*. Wisconsin Agricultural Research Station Research Report 8, Wisconsin.

MacDonald, R.C. and Wilson, K.R. (1980) Dry matter yields, digestibilities, mineral levels and cattle growth rates on greenfeed oats at different stages of development. *New Zealand Journal of Experimental Agriculture* 8, 105–109.

McDowell, L.R. and Conrad, J.H. (1990) Mineral imbalances of grazing livestock in tropical countries. *International Journal of Animal Science* 5, 21–32.

McFarlane, J.D., Judson, J.D., Turnbull, R.K. and Kempe, B.R. (1991) An evaluation of copper-containing soluble glass pellets, copper oxide particles and injectable copper as supplements for cattle and sheep. *Australian Journal of Experimental Agriculture* 31, 165–174.

McLaren, R.G., Lawson, D.M. and Swift, R.S. (1987) The availability to pasture plants of native and applied cobalt in relation to extractable soil cobalt and other soil properties. *Journal of Science Food and Agriculture* 39, 101–112.

McNaught, K.J. (1970) Diagnosis of mineral deficiencies in grass–legume pastures by plant analysis. In: *Proceedings of the XI International Grassland Congress*, pp. 334–338.

MacPherson, A. (1967) Studies of copper metabolism in sheep. PhD thesis, University of Glasgow, 403 pp.

MacPherson, A. (1983) Oral treatment of trace element deficiencies in ruminant livestock. In: Suttle, N.F., Gunn, R.G., Allen, W.M., Linklater, K.A. and Weiner, G. (eds) *Trace Elements in Animal Production and Veterinary Practice*. Occasional Publication of the British Society of Animal Production, No. 7, Edinburgh, pp. 93–103.

MacPherson, A. (1984) Long term effectiveness of graded doses of copper oxide needles in suckler cows. *Veterinary Record* 115, 354–355.

MacPherson, A., Kelly, E.F., Chalmers, J.S. and Roberts, D.J. (1988) The effect of selenium deficiency on fertility in heifers. In: Hemphill, D.D. (ed.) *Trace Substances in Environmental Health – XXI*. University of Missouri, Columbia, USA, pp. 551–554.

MacPherson, A., Milne, E.M. and Macpherson, A.J. (1997) Copper poisoning in ewes. *Veterinary Record* 128, 631.

Maskall, J.E. and Thornton, I. (1989) The mineral status of Lake Nakuru National Park, Kenya: a reconnaissance survey. *African Journal of Ecology* 27, 191–200.

Mertz, W. (ed.) (1987) *Trace Elements in Human and Animal Nutrition*, 5th edn. Academic Press, London, UK, 480 pp.

Metson, A.J., Gibson, E.J., Hunt, J.L. and Saunders, W.M.H. (1979) Seasonal variations in the chemical composition of pasture. III. Silicon, aluminium, iron, zinc, copper and manganese. *New Zealand Journal of Agricultural Research* 22, 309–318.

Miller, E.R. and Ammerman, C.B. (1995) Iodine bioavailability. In: Ammerman, C.B., Baker, D.H. and Lewis, A.J. (eds) *Bioavailability of Nutrients for Animals*. Academic Press, San Diego, USA, pp. 157–167.

Miller, J.K., Swanson, E.W., Lyke, W.A., Moss, B.R. and Byrne, W.F. (1974) Effect of thyroid status on digestive tract fill and flow rate of undigested residues in cattle. *Journal of Dairy Science* 57, 193–197.

Miller, W.J., Adams, W.E., Nussbaumer, R., McCreery, R.A. and Perkins, H.F. (1964) Zinc content of coastal Bermuda grass as influenced by frequency and season of harvest, location and level of N and lime. *Agronomy Journal* 56, 198–201.

Mills, C.F. and MacPherson, A. (1982) *Trace Element Deficiency in Ruminants. Report of a Study Group of the Scottish Agricultural Colleges and Research Institutes*. Scottish Agricultural Colleges, Edinburgh, 87 pp.

Miltimore, J.E., Mason, J.L. and Ashby, D.L. (1970) Copper to molybdenum ratio and molybdenum and copper concentrations in ruminant feeds. *Canadian Journal of Animal Science* 50, 293–300.

Minson, D.J. (1981) The effects of feeding protected and unprotected casein on the milk production of cows grazing ryegrass. *Journal of Agricultural Science* 96, 239–242.

Minson, D.J. (1984) Digestibility and voluntary intake by sheep of five *Digitaria* species. *Australian Journal of Experimental Agriculture and Animal Husbandry* 24, 494–500.

Minson, D.J. (1990) *Forage in Ruminant Nutrition*. Academic Press, London, 483 pp.

Mitchell, R.L. (1954) Trace elements and liming. *Scottish Agriculture* 34, 139–143.

Mitchell, R.L. (1972) Trace elements in soils and factors that affect their availability. *Geological Society of America Bulletin* 83, 1069–1076.

Molnár, J., Drusch, S. and MacPherson, A. (1997) The effect of level and form of supplementary selenium on tissue concentrations in lambs and their importance in raising human dietary selenium intakes. In: Fischer, P.W.F., L'Abbé, M.R., Cockell, K.A. and Gibson, R.S. (eds) *Trace Elements in Man and Animals – 9*. NRC Research Press, Ottawa, Canada, pp. 468–470.

Montalvo-Hernandez, M.I., Garcia-Ciudad, A. and Garcia-Criado, B. (1984) In: Riley, H. and Skjelvag, A.O. (eds) *The Impact of Climate on Grass Production and Quality*. Norwegian Agricultural Research Station, As, Norway, pp. 407–411.

Morris, E.R. (1987) Iron. In: Mertz, W. (ed.) *Trace Elements in Human and Animal Nutrition*, 5th edn. Academic Press, London, UK, pp. 79–142.

Mudd, A.J. (1970) Trace mineral composition of heavily fertilised grass in relation to ARC standards for requirements of dairy cows. *Journal of Agricultural Science* 74, 11–21.

Mutanen, M. (1986) Bioavailability of selenium. *Annals of Clinical Research* 18, 48–59.

Norrish, K. (1975) Geochemistry and mineralogy of trace elements. In: Nicholas, D.I.D. and Egan, A.R. (eds) *Trace Elements in Soil–Plant–Animal Systems*. Academic Press, New York, pp. 55–81.

Pehrson, B.G. (1993) Selenium in nutrition with special reference to the biopotency of organic and inorganic selenium compounds. In: *Biotechnology in the Feed Industry. Proceedings of Alltech's Ninth Annual Symposium*. Alltech Technical Publications, Nicholasville, Kentucky, USA, pp. 71–89.

Pehrson, B.G., Knutsson, M. and Gyllensward, M. (1989) Glutathione peroxidase activity in heifers fed diets supplemented with organic and inorganic compounds. *Swedish Journal of Agricultural Research* 19, 53–57.

Percival, G.P., Josselyn, D. and Beeson, K.C. (1955) *Factors Affecting the Micronutrient Element Content of Some Forages in New Hampshire*. New Hampshire Agricultural Experimental Station Technical Bulletin 93, Durham.

Perdomo, J.T., Shirley, R.L. and Chicco, C.F. (1977) Availability of minerals in four tropical forages fed freshly chopped to sheep. *Journal of Animal Science* 45, 1114–1119.

Piper, C.S. (1942) Investigations on copper deficiency in plants. *Journal of Agricultural Science* 21, 762–779.

Piper, C.S. and Walkley, A. (1943) *Journal of the Council of Scientific and Industrial Research (Australia)* 16, 217–234.

Pott, E.B., de Brum, P.A.R., Pott, A., de Almeida, A.L., Comastri Filho, J.A. and Tullio, R.R. (1989) Beef cattle mineral nutrition in the Brazilian Pantenal. 4. Micronutrient survey on the lower Piquiri River. *Pesquisa Agropecuriaria Brasileira* 24, 1369–1380.

Price, J. (1989) The nutritive value of grass in relation to mineral deficiencies and imbalances in the ruminant. Paper read before the Fertilizer Society of London, 14 December.

Price, N.O. and Hardison, W.A. (1963) Minor element content of forage plants from central Piedmont region of Virginia. *Virginia Agricultural Experimental Station Technical Bulletin* 165, 15.

Reay, P.F. and Marsh, B. (1976) Element composition of ryegrass and red clover leaves during a growing season. *New Zealand Journal of Agricultural Research* 19, 469–472.

Reddy, G.D., Alston, A.M. and Tiller, K.G. (1981) Effects of fertilizer on concentrations of copper, molybdenum and sulphur in subterranean clover (*Trifolium subterraneum*). *Australian Journal of Experimental Agriculture and Animal Husbandry* 21, 491–497.

Reid, R.L., Odhuba, E.K. and Jung, G.A. (1967a) Evaluation of tall fescue pasture under different fertilization treatments. *Agronomy Journal* 59, 265–271.

Reid, R.L., Jung, G.A. and Kinsey, C.M. (1967b) Nutritive value of nitrogen-fertilized orchardgrass pasture at different periods of the year. *Agronomy Journal*, 59, 519–525.

Reith, J.W.S., Inkson, R.W.E., Holmes, W., MacLuskey, D.S., Reid, G., Heddle, R.G. and Copeman, D.J.F. (1964) The effects of fertilizers on herbage production. II. The effect of nitrogen, phosphorus and potassium on botanical and chemical composition. *Journal of Agricultural Science* 63, 209–219.

Reith, J.W.S., Burridge, J.C., Berrow, M.L. and Caldwell, K.S. (1983) Effects of the applications of fertilizers and trace elements on the cobalt content of herbage cut for conservation. *Journal of the Science of Food and Agriculture* 34, 1163–1170.

Rickard, T.R. and Elliot, J.M. (1978) Absorption of vitamin B_{12} and factor B from intestine of sheep. *Journal of Animal Science* 46, 304–308.

Robinson, T.G. and During, C. (1961) Comprehensive trials show liveweight responses to selenium in lambs but few responses to cobalt. *New Zealand Journal of Agriculture* 103, 306–310.

Rossiter, R.C., Curnow, D.H. and Underwood, E.J. (1948) The effect of cobalt sulphate on the cobalt content of subterranean clover at three stages of growth. *Journal of the Australian Institute of Agricultural Science* 14, 9–14.

Rudert, C.P. and Oliver, J. (1978) *Rhodesian Journal of Agricultural Research* 16, 23–29.

Rumball, W., Butler, G.W. and Jackman, R.H. (1972) Variation in nitrogen and mineral composition in populations of prairie grass (*Bromus unioloides* H.B.K.). *New Zealand Journal of Agricultural Research* 15, 33–42.

Sansom, B.F., Gibbons, R.A., Dixon, S.M., Russell, A.M. and Symonds, H.W. (1976) Absorption of dietary manganese by dairy cows and the role of plasma proteins and liver in its homeostasis. In: *Nuclear Techniques in Animal Production and Health*. International Atomic Energy Agency, Vienna, pp. 179–189.

Smith, G.S. and Edmeades, D.C. (1983) Manganese status of New Zealand pastures 2. Pasture concentrations. *New Zealand Journal of Agricultural Research* 26, 223–225.

Smith, R.M. and Marston, H.R. (1970) Production, absorption, distribution and excretion of vitamin B_{12} in sheep. *British Journal of Nutrition* 24, 857–877.

Stewart, J. (1951) *Plant and Animal Nutrition in Relation to Soil and Climatic Factors*. HMSO, London, 290 pp.

Stewart, J. and Holmes, W. (1953) Manuring of grassland I. Some effects of heavy dressings of nitrogen on the mineral composition of grassland herbage. *Journal of the Science of Food and Agriculture* 4, 401–408.

Strikauska, S., Latvietis, J. and Degola, L. (1994) *Status of Microelements in Feedstuffs and Mineral Premixes in Latvia*. Verlag Harald Schubert, Leipzig, Germany, pp. 569–572. Abstract in: *Micronutrient News and Information* 1996, 16, 53.

Suttle, N.F. (1981) Predicting the effects of molybdenum and sulphur concentrations on the absorbability of copper in grass and forage crops to ruminants. In: Howell, J.McC., Gawthorne, J.M. and White, C.L. (eds) *Trace Elements in Man and Animals – 4*. Australian Academy of Science, Canberra, Australia, pp. 545–548.

Suttle, N.F. and Peter, D.W. (1985) Rumen sulphide metabolism as a major determinant of the availability of copper to ruminants. In: Mills, C.F., Bremner, I. and Chesters, J.K. (eds) *Trace Elements in Man and Animals – 5*. Commonwealth Agricultural Bureau, Farnham Royal, UK, pp. 367–370.

Sutton, A.L. and Elliot, J.M. (1972) Effect of ratio of roughage to concentrate and level of feed intake on ovine ruminal vitamin B_{12} production. *Journal of Nutrition* 102, 1341–1346.

Thomas, B. and Thompson, A. (1948) The ash-content of some grasses and herbs on the Palace Leas hay plots at Cockle Park. *Empire Journal of Experimental Agriculture* 16, 221.

Thomas, B., Escritt, J.R. and Trinder, N. (1945) The minor elements of common heather (*Calluna vulgaris*). *Empire Journal of Experimental Agriculture* 13, 93.

Thomas, B., Thompson, A., Oyenuga, V.A. and Armstrong, R.H. (1952) The ash constituents of some herbage plants at different stages of maturity. *Empire Journal of Experimental Agriculture* 20, 10–22.

Underwood, E.J. (ed.) (1977) Iodine. In: *Trace Elements in Human and Animal Nutrition*. Academic Press, London, pp. 281–322.

Underwood, E.J. (1981) *The Mineral Nutrition of Livestock*. Commonwealth Agricultural Bureau, Farnham Royal, Slough, England, pp. 77–89.

Vagg, M.J. (1976) Assessment of trace element metabolism in farm animals. *Proceedings of the Royal Society of Medicine* 69, 473–474.

Vohra, P. and Kratzer, F.H. (1964) Influence of various chelating agents on the availability of zinc. *Journal of Nutrition* 82, 249.

Voss, R.C. and MacPherson, A. (1977) *Cobalt and Copper in Soils, Herbage and Ruminant Nutrition*. Technical Note 5, West of Scotland Agricultural College, Ayr, Scotland.

Watkinson, J.H. (1962) Soil selenium in animal health. In: *Transactions of the International Society of Soil Science, Commissions IV and V*, pp. 149–154.

Watkinson, J. (1991) Selenium. *Micronutrient News and Information* 11, 11–12.

Wedekind, K.J. and Baker, D.H. (1990a) Manganese utilization in chicks as affected by excess calcium and phosphorus ingestion. *Poultry Science* 69, 977.

Wedekind, K.J. and Baker, D.H. (1990b) Effect of varying calcium and phosphorus level on manganese utilization. *Poultry Science* 69, 1156.

West, T.S. (1981) Soil as a source of trace elements. *Philosophic Transactions of the Royal Society of London* B294, 19–39.

Westerman, D.T. and Robbins, C.W. (1974) Effect of sulfate sulfur fertilization on the selenium concentration of alfalfa (*Medicago sativa* L.). *Agronomy Journal* 66, 207–208.

Whanger, P.D. and Butler, J.A. (1988) Effects of various dietary levels of selenium as selenite or selenomethionine on tissue selenium levels and glutathione peroxidase activity in rats. *Journal of Nutrition* 118, 846–852.

Whitehead, D.C. (1966) *Nutrient Minerals in Grassland Herbage*. Technical Report 4, Grassland Research Institute, Hurley.

Whitehead, D.C. and Jones, E.C. (1969) Nutrient elements in the herbage of white clover, red clover, lucerne and sainfoin. *Journal of the Science of Food and Agriculture* 20, 584–591.

Winter, W.H. and Jones, R.K. (1977) Nutrient responses on a yellow earth soil in Northern Cape York peninsula. *Tropical Grassland* 11, 247–255.

Winter, W.H., Siebert, B.D. and Kuchel, R.E. (1977) Cobalt deficiency of cattle grazing improved pastures in northern Cape York Peninsula. *Australian Journal of Experimental Agriculture and Animal Husbandry* 17, 10–15.

Wooliams, J.A., Wooliams, C., Suttle, N.F., Jones, D.G. and Weiner, G. (1986) Studies on lambs from lines genetically selected for low or high plasma copper status. 2. Incidence of hypocuprosis on improved hill pasture. *Animal Production* 43, 303–317.

Wright, J.R. and Lawton, K. (1954) Cobalt investigations on some Nova Scotia soils. *Soil Science* 77, 95–105.

Young, L.G., Jenkins, K.J. and Edmeades, D.M. (1977) Selenium content of feedstuffs grown in Ontario. *Canadian Journal of Animal Science* 57, 793–799.

18 Major Minerals in Forages

L.R. McDowell[1] and G. Valle[2]

[1]Department of Animal Science, University of Florida, 125 Animal Science Building, 459 Shealy Drive, Gainesville, FL 32618, USA; [2]Departamento de Producción Animal, CURLA, Apartado 89, La Ceiba, Honduras

Important Major Minerals

Undernutrition is commonly accepted as one of the most important limitations to grazing livestock production. The lack of sufficient energy and protein is often responsible for suboptimum livestock production. However, numerous investigators have observed that ruminants sometimes deteriorate in spite of an abundant feed supply. Mineral imbalances (deficiencies or excesses) in soils and forages have long been held responsible for low production and reproductive problems among grazing ruminants in the tropics. Wasting diseases, loss of hair, depigmented hair, skin disorders, non-infectious abortion, diarrhoea, anaemia, loss of appetite, bone abnormalities, tetany, low fertility and pica are clinical signs often suggestive of mineral deficiencies throughout the world.

The seven major minerals supplied by forages are calcium (Ca), chlorine (Cl), phosphorus (P), magnesium (Mg), potassium (K), sodium (Na) and sulphur (S). Each of these minerals has been found to be deficient for grazing livestock under specific conditions, with the exception of Cl (McDowell, 1985, 1992, 1997). Deficiencies of Cl in animals in general have been unequivocally observed only on specially purified or concentrate diets. There are few forage Cl concentrations reported, and those that are available indicate adequacy of the element. Cl will not be discussed further in this review. The importance of the remaining macrominerals is as follows.

Calcium and phosphorus

Clinical signs of borderline Ca and P deficiencies are not easily distinguishable from those of other deficiencies. An inadequate intake of Ca may cause weakened bones, slow growth, low milk production and tetany (convulsions) in severe deficiencies. Signs of P deficiencies are not easily recognizable, except in severe cases, when fragile bones, general weakness, weight loss, emaciation, stiffness, reduced milk production and chewing of wood, rocks, bones and other objects may be noticed. Abnormal chewing of objects may also occur, however, with other dietary deficiencies. A number of tropical countries (e.g. South Africa, Argentina, Brazil and Senegal) have reported death from botulism as a result of bone-chewing (McDowell, 1992). In areas of Piaui, Brazil, an estimated 2–3% of approximately 100,000 cattle die annually of botulism.

For grazing livestock, the most prevalent mineral-element deficiency throughout the world is lack of P (McDowell, 1976; Underwood, 1981). In a review (McDowell *et al.*, 1984), P deficiency was reported in 46 tropical countries of Latin America, South-East Asia and Africa. P deficiency is more prevalent for tropical than for temperate grazing regions. It is also known that in tropical grasses there is very little or no difference in P concentration between species in a given region (Long *et al.*, 1970; Minson, 1990). Many reports from tropical regions of the world, dating back to the early part of the century, have revealed the beneficial effects of P supplementation on overall performance. Reports of improved weight gains by P-supplemented cattle have been summarized for various world regions (McDowell, 1985). From Brazil, Moraes *et al.* (1982) related supplemental P to daily gains of growing cattle. Treatments of 0, 4.3, 6.7 and 8.4 g of P per day resulted in daily gains of 216, 379, 465 and 564 g, respectively. From Peru, Echevarria *et al.* (1973) obtained weight gains of 0.59 kg day^{-1} in steers supplemented with dicalcium phosphate and 0.27 kg for controls. Bolivian cattle gained 96.4 kg in 338 days when receiving a source of P, versus 79.4 kg for controls (McDowell *et al.*, 1982). Estevez-Cancino (1960) reported increased milk production (up to 24%) with bone-meal supplementation in farms that had P-deficient forages.

The most devastating economic result of P deficiency is reproductive failure, with P supplementation dramatically increasing fertility levels in grazing cattle from many parts of the world. In 2-year observations of 200 South African cattle, calf crops increased from 51% in control cattle to approximately 80% for cattle supplemented with bone-meal or other P sources (Theiler *et al.*, 1928). Latin American countries reporting increased fertility rates resulting from P administration include Bolivia, Brazil, Colombia, Panama, Peru, Uruguay and Venezuela, and from Asia, the Philippines and Thailand (McDowell, 1997). From this review, averaging 18 cattle reports on the effect of supplementation on calving percentage, those receiving P supplement averaged a 73.3% calf crop versus 51.3% for controls (McDowell, 1997; Rojas and McDowell, 1997).

From Zimbabwe (Ward, 1968) and northern Australia (Lamond, 1970), the effects of undernutrition on fertility were apparent. With few exceptions, lactating cows were not calving 2 years in succession. In P-deficient areas, if a calf is produced, cows may not come into a regular oestrus again until body P levels are restored, either by feeding supplementary P or by cessation of lactation.

Compared with P inadequacies, Ca deficiency is rare in grazing cattle, with the exception of cows lactating large quantities of milk or those grazing on acid, sandy or organic soils in humid areas where the herbage consists mainly of quick-growing grasses devoid of legume species (Underwood, 1981). Although Ca deficiency can easily be produced in young, growing animals and lactating dairy cows fed native forages supplemented with concentrates, the deficiency is infrequently reported in grazing beef cattle, even during lactation.

Magnesium

The practical importance of Mg is its relationship to the serious metabolic disorder grass tetany (hypomagnesaemia). Grass tetany is a complex ruminant metabolic disorder that is affected by forage species and mineral composition, soil properties, fertilizer practices, season of the year, temperature, animal species, breed and age.

Clinical signs of hypomagnesaemia in ruminants include reduced appetite, increased excitability, profuse salivation and convulsions. For adult ruminants, milk yield is usually reduced and the animals are nervous. In more severe cases, affected cows may avoid the rest of the herd, walk with a stiff gait and lose their appetite. In the most severe stage, the animal collapses to the ground with continuation of the tetanic muscular spasms. Death occurs after the collapse stage of tetany if the animal does not receive medical treatment.

Grass tetany is a serious problem of grazing ruminants in many parts of the world. Susceptibility to grass tetany is increased in older ruminants because of decreased ability to mobilize skeletal Mg with increasing age. Grass tetany generally occurs during early spring or a particularly wet autumn, among older cattle grazing grass or small-grain forages in cool weather. Clinical tetany is endemic in some countries, affecting only a small proportion of cattle (1–2%). However, individual herds may report tetany as high as 20%. Females are mainly affected, with cattle more susceptible than sheep or goats. This is especially a problem with grazing beef cattle, where a first indication of grass tetany may be finding one or more dead animals. Excellent and extensive reviews concerning hypomagnesaemic tetany are cited by Rendig and Grunes (1979) and Underwood (1981).

Occurrence of tetany is seen where ruminant production is highly developed, high-quality pastures are available and high-yielding, quick-maturing stock is raised. Incidence of hypomagnesaemic tetany occurs in most European countries, North America, Australia, South Africa and New Zealand. Some of the reasons for fewer reports of grass tetany in tropical regions include the prevalence of the condition during cooler temperatures (8–14°C) and, for many tropical countries, the generally low productivity of pastures and of herds and lack of fertilization of pasture with nitrogen (N) and K.

Sodium

The initial sign of Na deficiency is a craving for salt (NaCl), demonstrated by the avid licking of wood, soil and sweat from other animals, and drinking water. Cattle deprived of salt may be so voracious that they often injure each other in attempting to reach salt. A prolonged deficiency causes loss of appetite, decreased growth, unthrifty appearance, reduced milk production and loss of weight. More pronounced signs of Na deficiencies include shivering, incoordination, weakness and cardiac arrhythmia, which can lead to death.

Natural forages low in Na have been reported in numerous tropical countries throughout the world. French (1955) concluded that NaCl is the most needed nutrient for livestock throughout East Africa. According to Sütmoller *et al.* (1966), the insufficiency of Na is the most widespread mineral deficiency in the Amazon

Valley of Brazil (~ 3.5 million square kilometres). From Bolivia, steers receiving salt had significantly higher average weights than unsupplemented controls – 385 versus 370 kg, respectively (McDowell *et al.*, 1984). In Zambia, Walker (1957) reported that the daily feeding of 28 g of NaCl resulted in a marked increase in weight gains for grazing steers. Australian workers (Murphy and Plasto, 1973) noted dramatic responses in beef cows supplemented with salt.

Na deficiency is most likely to occur: (i) during lactation, due to secretion of Na in milk; (ii) in rapidly growing animals; (iii) under tropical or hot semi-arid conditions, where large losses of water and Na occur in the sweat and where pastures are low in Na; and (iv) in animals grazing pastures heavily fertilized with K, which depresses herbage Na levels. Even after prolonged severe deficiency, NaCl levels secreted in milk remain high. Thus, lactating animals suffer most from lack of salt in the diet.

Potassium and sulphur

There are fewer reports of K and S deficiencies under natural grazing conditions than of P, Ca, Na and Mg. K deficiency for ruminants results in non-specific signs, such as slow growth, reduced feed and water intake, lowered feed efficiency, muscular weakness, nervous disorders, stiffness, decreased pliability of hide, emaciation, intracellular acidosis and degeneration of vital organs.

There are very few confirmed reports of K deficiency for ruminants grazing exclusively forages. One report from Nigeria indicated clinical manifestations of K deficiency in 27 cattle raised in the traditional semi-nomadic herding system, which consumed forages analysed as containing 2 g K kg^{-1} dry matter (DM) (Smith *et al.*, 1980). Clinical signs quickly disappeared after administration of K. The main reason for lack of widespread K deficiency, even when forages contain less than the requirement, is probably due to other nutrient deficiencies of forages. Mature tropical forages are often deficient in energy, protein, P, Na and a number of trace elements. It is likely that a K deficiency will not be expressed as long as there are other nutrients that are even more deficient.

Signs of S deficiency have been described as loss of weight, weakness, lacrimation, dullness and death. In S deficiency, microbial protein synthesis is reduced and the animal shows signs of protein malnutrition. S is required for protein formation, and the depressed appetite of animals eating low-S feed is due to a protein deficiency. Low levels of S have been found in young regrowths of *Digitaria decumbens* and feeding an S supplement has increased voluntary feed intake by 28% (Rees *et al.*, 1974). In relation to animal requirements, temperate forages are generally adequate in S compared with tropical plants.

Forage as a Supplier of Major Elements

Forage macromineral concentrations

Frequently, livestock producers, particularly in tropical countries, do not supplement grazing livestock with minerals, with the possible exception of salt. Grazing livestock must therefore depend largely upon forages to supply their mineral requirements. However, often forages cannot completely satisfy each of the mineral requirements for grazing animals. Table 18.1 summarizes the mineral concentrations of 2615 Latin American forages (McDowell *et al.*, 1977). Borderline or deficient levels of macroelements were noted for many entries: Mg, 35%; P, 73%; and Na, 60%. For K, 84.9% of samples were high (> 8 g kg^{-1}) in the mineral and for S there were no analyses.

Most often, tropical forages contain less macrominerals than species grown in temperate regions. A number of publications from different tropical regions of the world have well documented the inadequacy of tropical forages in meeting the mineral requirements of grazing ruminants (French, 1955; McDowell, 1976, 1985, 1997; Underwood, 1981). Table 18.2 evaluates forage macromineral concentrations from one African and six Latin American countries. The majority of forages from the seven countries were low to deficient in Ca, P, Na and Mg; in addition, the majority of forages from two of seven countries were low in K.

Calcium

Reviewing the world literature on forage Ca (1263 samples), Minson (1990) reported that 31% of forages were low in Ca (< 3 g kg^{-1} DM). A similar proportion of 1123 samples of forage collected in Latin America (McDowell *et al.*, 1977) and 103 Caribbean grass samples (Devendra, 1977) contained less than 3 g kg^{-1} DM

Table 18.1. Mineral breakdown and concentrations of 2615 Latin American forages (g kg^{-1}, DM).*

Element	No. of entries	Requirements of ruminants†	Level of nutrients in forages	Low level	Higher level
Calcium	1123	1.8–8.2	Concentrations	0–3	Over 3
			Percentage of total entries	31.1	68.9
Magnesium	290	1–2	Concentrations	0–2	Over 2
			Percentage of total entries	35.2	64.8
Phosphorus	1129	1.8–4.8	Concentrations	0–3	Over 3
			Percentage of total entries	72.8	27.2
Potassium	198	6–8	Concentrations	0–8	Over 8
			Percentage of total entries	15.1	84.9
Sodium	146	0.6–1.8	Concentrations	0–1	Over 1
			Percentage of total entries	59.5	40.5

* Latin American Tables of Feed Composition (McDowell *et al.*, 1977).

† Requirements (DM) based on national research and agricultural research councils for various ruminant species (McDowell, 1992, 1997).

Table 18.2. Percentage of forage minerals below critical concentrations needed by ruminants.*

Mineral level	Critical level based on ruminant needs† ($g\ kg^{-1}$ DM)	Dominican Republic (1)	Bolivia (2)	Colombia (3)	Guatemala (4)	Malawi (5)	Nicaragua (6)	Venezuela (7)
Ca	3	24	57	100	71	13	80	97
P	2.5	83	100	92	57	75	67	98
K	6–8	0	1	15	13	57	20	84
Na	0.6	78	100	100	88	97	80	84
Mg	2	33	64	56	76	31	96	94

* Complete reference citations for each study are given in McDowell (1997).
† Critical levels based on Ruminant National Research and Agricultural Research Councils for various ruminant species (McDowell, 1992, 1997).
1, based on 69 samples (Jerez *et al.*, 1984); 2, based on 84 samples (Peducassé *et al.*, 1983); 3, based on 36 samples (Vargas *et al.*, 1984); 4, based on 84 samples (Tejada *et al.*, 1987a, b); 5, based on 21 samples (Mitimuni *et al.*, 1990); 6, based on 304 samples (Velásquez-Pereira *et al.*, 1997a, b); 7, based on 198 samples (Rojas *et al.*, 1993a, b).

Ca. Temperate forages generally contain more Ca than those grown in the tropics. Hay from Ireland had a mean Ca concentration of 8 g kg^{-1} DM (Wilson *et al.*, 1968), which is similar to that reported by the Pennsylvania State Forage Test Service for over 9500 forage samples grown in 5 years (Adams, 1975). The mean level of Ca was 8.4 g kg^{-1} DM, with mean values of 4 and 11.5 g kg^{-1} DM for grass and legume hays, respectively. Legumes are significantly greater sources of Ca than grasses.

Phosphorus

The mean P concentration of 1823 forage samples from the world literature was 2.9 g kg^{-1} DM (Minson, 1990). This would be higher than the predominantly tropical forages from Latin America, where 73% of 1129 forages contained less than 3 g P kg^{-1} DM (McDowell *et al.*, 1977). From six ranches in northern Mato Grosso, Brazil, forage P averaged 0.8 and 2 g kg^{-1} DM for the dry and wet season, respectively (Sousa *et al.*, 1979). From Costa Rica, 1468 forage samples contained 1.8 g P kg^{-1} DM in the wet season, and 1335 samples averaged 1.1 g P kg^{-1} DM in the dry season (J. Sanchez, personal communication). Table 18.3 illustrates extremely low Ca and P (particularly P) forage concentrations for 11 ranches in Venezuela during the wet and dry seasons. For all farms, the extremely P-deficient farms averaged 0.8 g P kg^{-1} DM in the wet season and 0.6 g P kg^{-1} DM in the dry season.

In the temperate zones, higher levels of P have usually been reported. Forage samples analysed by the Pennsylvania State Forage Testing Service over a 5-year period contained 2.6 g P kg^{-1} DM, with a higher concentration in legumes (3 g kg^{-1} DM) than in grasses (2.2 g kg^{-1} DM) (Adams, 1975).

Magnesium

The Mg content of forage plants is normally higher in legumes than in grasses. Mayland *et al.* (1976) estimated the tetany hazard of several forages and found wheat > oats > barley > rye. Mg levels of wheat were the lowest, whereas those of rye were the highest.

For tropical herbage, Mg concentrations were relatively low, with slightly more than one-third of 288 forages included in the 1974 *Latin American Tables of Feed Composition* containing 2 g Mg kg^{-1} DM or less (McDowell *et al.*, 1977). Forage mineral analysis from a large number of farms in four Latin American tropical countries indicated that a considerable number of forages contained less than 2 g Mg kg^{-1} DM: Bolivia (64%), Colombia (56%), Dominican Republic (33%) and Guatemala (76%) (McDowell, 1985).

Sodium

There are two classes of forage plants, the Na accumulators and the non-accumulators (Youssef, 1988). Na-accumulating forages contain > 2 g Na kg^{-1} DM, provided the soil contains sufficient Na and soil K is not in excess. The non-accumulating forages always contain 2 g Na kg^{-1} DM, even when grown on soils high in Na, and the concentration is almost independent of the level of soil K (Minson, 1990). Forages usually range from 0.07 to 1.2 g Na kg^{-1} DM. The Food and Agricultural Organization (FAO) has published an extensive summary of the composition of tropical feeds (Göhl, 1975). High-Na sources (DM basis) were: salt bush,

Table 18.3. Mean forage calcium and phosphorus concentration (g kg^{-1} DM) from cattle ranches in the states of Monagas and Guarico, Venezuela (adapted from Velásquez, 1979; Faria *et al.*, 1981).*

	Monagas				Guarico			
	Calcium		Phosphorus		Calcium		Phosphorus	
Ranch no.	Dry season	Wet season	Dry season	Wet season	Dry season	Wet season	Dry season	Wet season
1	2.6	3.1	0.5	1.0	1.9	3.0	0.1	1.6
2	1.9	1.8	0.8	1.6	1.9	1.7	0.5	1.6
3	1.4	1.2	1.2	0.5	2.4	3.5	0.2	1.3
4	1.7	2.3	0.7	0.8	1.6	2.9	1.8	0.2
5	1.7	2.2	0.7	0.8	1.5	3.3	20.6	2.2
6	0.6	1.6	0.9	0.7	1.2	1.0	1.0	0.2
7	1.6	1.5	0.8	1.0	1.2	1.8	1.0	0.2
8	–	–	–	–	2.1	1.2	1.0	0.4
9	–	–	–	–	3.2	1.4	4.0	0.4
10	–	–	–	–	4.9	2.8	3.0	0.7
11	–	–	–	–	3.8	3.6	1.0	0.3
Overall mean	1.6	1.9	0.8	0.9	2.3	2.4	0.6	0.8
Standard deviation	0.6	0.6	0.2	0.3	1.2	1.0	0.8	0.7

* Each mean is based on four pastures per farm and ten subsamples per pasture.

63.4 g kg^{-1}; creeping salt bush, 46.6 g kg^{-1}; seaweed meal, 29.3 g kg^{-1}; and sunflower meal, 20.9 g kg^{-1}.

Na deficiency is more likely to occur in cattle grazing tropical pasture species, as these plants generally accumulate less Na than temperate species (Morris, 1980). Natural forages low in Na have been reported in numerous tropical countries throughout the world (McDowell, 1985). Forage Na analysed from cattle ranches in northern Mato Grosso, Brazil, during both the wet and dry seasons, was extremely deficient, meeting only between 14 and 30% of the animals' requirements (Sousa *et al.*, 1982).

Potassium

K in growing forage is usually quite high (10–40 g kg^{-1} DM); thus, it had been assumed that grazing livestock consuming primarily a forage diet would receive adequate K. In tropical regions, it is possible that K deficiencies could arise, in view of the decreasing content of this mineral with increasing forage maturity during the extended dry season and the use of urea, which supplies none of this element. Low K values have been reported for forages in Florida, Brazil, Panama, Nigeria, Swaziland, Uganda and Venezuela (McDowell, 1997). In Brazil, average K content (DM basis) of six grasses at 4 weeks was 14.2 g kg^{-1} DM versus 3 g kg^{-1} DM at 36 weeks of age (Gomide *et al.*, 1969). In Florida, a 1979 forage analysis from four regions indicated adequate K in September–October, but five of seven ranches sampled in February–March contained forage K concentrations considerably less than 6 g kg^{-1} DM (Kiatoko *et al.*, 1982). Kalmbacher and Martin (1981) report creeping bluestem (*Schizachyrium stoloniferum*) affected by seasonal patterns and low K concentrations (DM basis), ranging from 3.3 to 5.4 g kg^{-1} with an average of 4.2 g kg^{-1} DM.

Sulphur

S analysis in biological materials is comparatively difficult. For these reasons, there is a scarcity of S analyses, particularly tropical forages. A limited amount of analysis has indicated that many tropical forages contained less than an optimum S concentration of 2 g kg^{-1} DM. S analyses of 10 forage samples from the llanos rangelands of both Colombia and Venezuela were low, ranging from 0.32 to 0.88 g kg^{-1} DM (Miles and McDowell, 1983). From these regions, severe leaching of soils and frequent burning of grasslands led to the assumption that many, if not most, llanos rangeland forages would be found deficient in S. Ruminants consuming large quantities of maize silage would probably receive inadequate S. More than 80% of the maize silages had less than one-half the estimated 2 g S kg^{-1} needed for lactating cows (Miller, 1979). Forages high in S (e.g. $>$ 4 g kg^{-1} DM) combined with molybdenum will bring about copper deficiencies.

Grazing selectivity and forage mineral intake

Selective grazing patterns of livestock in relation to choosing not only particular species of forage but parts of the plant need consideration. The principles of dietary selection are not clearly understood. Nevertheless, there is recorded evidence that grazing animals tend to select higher protein and more highly digestible fractions from the pasture available (Arnold *et al.*, 1966). In conditions of reduced availability of mineral elements from the soil, the younger leaves of growing plants, which constitute a highly digestible fraction, are often the leaves of lowest content of most elements, which are poorly or only slowly mobilized within the plant. Kincaid and Cronrath (1983) reported that significant portions of Ca and P are associated with the fibrous versus the leaf portion of lucerne (alfalfa) hay and silage. P, Na, K and Mg were all in higher concentration in stems than in leaves of dwarf *Pennisetum purpureum* (Montalvo *et al.*, 1987). Under grazing conditions, however, cattle show a preference for leaves over stems, and leaf consumption can be 46% more than stems of similar digestibility and the animals are capable of selecting a diet of 60% leaves from a sward that has only 20% (Laredo and Minson, 1973; Stobbs, 1976).

What Affects Major-element Status of Forage?

Wide ranges of mineral concentrations in tropical forages have been observed (Reid and Horvath, 1980; McDowell, 1992). The concentration of mineral elements in plants from diverse world regions are dependent upon the interaction of a number of factors, including soil, plant species, stages of maturity, yield, pasture management and climate. As an example, forage mineral content is affected by: (i) plant maturity; (ii) species as well as variety within a species; (iii) management procedures such as grazing or crop removal systems; (iv) fertilization, particularly with K and N; and (v) soil and environmental conditions (McDowell, 1985). Interrelationships with other elements likewise influence forage mineral concentrations. In many tropical countries, high amounts of soil iron (Fe) and aluminium (Al) accentuate P deficiency by forming insoluble phosphate complexes, making P unavailable for plant uptake.

Soils

Soil is the source of all mineral elements found in plants. Most naturally occurring mineral deficiencies in livestock are associated with specific regions and are directly related to both soil mineral concentration and soil characteristics. Of the total mineral concentration in soils, only a fraction is taken up by plants. The 'availability' of minerals in soils depends upon their effective concentration in soil solution (Reid and Horvath, 1980).

Young and alkaline geological formations are more abundant in most elements than are the older, more acid, coarse, sandy formations (Hartmans, 1970). In tropical regions under conditions of heavy rainfall and high temperature, there has been marked leaching and weathering of soils, making them deficient in plant minerals (Pfander, 1971).

Soils derived from basic igneous rocks usually have higher readily soluble contents of the important elements than do corresponding soils from acid igneous or metamorphic parent materials. The amounts of readily soluble elements are generally smaller in coarse-textured than in fine-textured soils.

Soil, drainage and pH

The soil content of an element would seem the most important limitation. However, availability factors, including soil pH, texture, moisture content and organic matter, are probably more often the limiting factors rather than soil mineral content (Williams, 1963). With increasing acidity of soils, there is impaired absorption of Ca, P and Mg (Reid and Horvath, 1980). Elkins *et al.* (1978) found that tall fescue grown in field trials on poorly drained soil had lower concentrations of Mg than fescue grown on well-drained soil.

Forage species and varieties

Mineral intake by animals depends more on the type of plant and level of consumption than on the parent rock from which the soil was derived and on which plants were grown. Large variations in mineral content of different plant species growing on the same soil have been reported (Thompson, 1957; Gomide *et al.*, 1969; Underwood, 1981). *Brachiaria decumbens* nearly doubled its P content when soil P was doubled, but P content of other pasture species, such as *Sporobulus*, did not vary widely (Long *et al.*, 1969). It is a generally accepted view that herbs and legumes are richer in a number of mineral elements than are grasses (Fleming, 1973). From Rhodesia, Jones (1964) reported one variety of *Chloris gayana* (DM basis) to contain 0.3 g Na kg^{-1}, while a second variety grown on the same site had a vastly different Na level of 3.1 g kg^{-1}.

Forage maturity

Variations in mineral content of tropical grasses have been reported to occur with increasing age of the plant. Gomide *et al.* (1969) studied the N, K, P, Ca and Mg content of five tropical grasses of Brazil at different stages of maturity (Table 18.4). It was concluded that, as forages neared maturity, mineral content declined for most minerals, and that K and P could be deficient for grazing cattle. In most circumstances, P, K, Mg, Na and Cl decline as the plant matures (Gomide *et al.*, 1969; Underwood, 1981). Forage Ca concentration is less affected by advancing maturity (Gomide *et al.*, 1969), thereby resulting in a detrimental widening of the ratio of this mineral with other elements (e.g. a wide Ca:P ratio). Stage of maturity is probably the most important factor influencing forage K. Forages harvested at an early stage of maturity among the richer sources of K. Actively growing grass and legumes are usually high in K, containing 10–50 g kg^{-1} DM. However, mature pastures, winter pastures that have weathered or hay that has been exposed to rain and sun can have deficient K levels.

As plants mature, mineral contents decline, due to a natural dilution process (Fleming, 1973) and translocation of nutrients to the root system (Tergas and Blue, 1971). Fleming (1973) concluded that, as photosynthetic areas increase, DM production outstrips mineral uptake, resulting in a decline in mineral concentration. The process of translocation of nutrients is a more serious problem for tropical

Table 18.4. Variation of mineral composition with forage age.

	Forage age (days)	Composition of forage (g kg^{-1} DM)			
		P	K	Ca	Mg
Grass					
Guinea grass (*Panicum maximum*)*	14	1.8	22.4	4.1	2.8
	28	1.4	23.3	3.4	2.3
	42	1.3	28.0	3.4	2.0
	56	1.0	26.4	3.4	1.7
	70	0.8	25.3	3.1	1.4
Molasses grass (*Melinis minutiflora*)*	14	2.0	23.0	2.8	2.5
	28	1.8	23.0	2.7	2.5
	42	1.8	22.0	2.7	2.5
	56	0.6	20.0	2.0	2.0
	70	0.5	17.0	2.0	1.8
Elephant grass (*Pennisetum purpureum*)*	28	3.3	23.8	6.1	4.2
	84	1.5	12.0	3.8	2.8
	140	1.1	3.4	4.3	3.6
Pangola grass (*Digitaria decumbens*)*	28	1.6	13.2	5.6	3.9
	84	1.1	7.4	5.0	3.8
	140	1.2	3.7	6.6	3.9
Jaragua grass (*Hyparrhenia rufa*)*	28	2.8	16.8	4.0	4.6
	56	1.7	6.3	2.0	3.6
	84	1.1	5.7	2.3	5.8
Setaria (*Setaria sphacelata*)†	Hay stage	1.3	–	1.1	1.4
Rhodes grass (*Chloris gayana*)†	Hay stage	1.4	–	1.7	1.5
Legumes:					
Gliricidia (*Gliricidia sepium*)‡	60	1.9	27.5	11.9	4.0
	90	2.1	28.0	17.5	4.0
	120	2.3	25.5	16.9	4.2
Leucaena (*Leucaena leucocephala*)§	21	1.2	25.0	6.3	4.2
	42	1.3	25.0	13.0	2.5
	63	1.0	23.4	12.6	2.5
	84	1.0	22.5	12.1	2.4

* Modified from Gomide *et al.* (1969).
† Modified from Jumba *et al.* (1995).
‡ From Chadhokar (1982).
§ Kabaija and Smith (1989).

versus temperate regions, since freezing conditions in temperate areas will stop translocation, while the movement of nutrients to the root system is continuous throughout the dry season in tropical regions.

Pasture management, forage yield and climate

Pasture management, forage yield and climate influence the species of forage predominating and also change the leaf–stem ratio radically, thereby having a direct bearing on the mineral content of the sward. From India, Whyte (1962) observed that good perennial grasses are 'grazed into the ground' late in the dry season, resulting in disastrous effects on their powers of regeneration and replacement with tougher species of inferior quality. From Africa, uncontrolled, heavy grazing pressure is also causing many palatable genuses, such as *Brachiaria* and *Panicum*, to disappear and to be replaced by highly lignified *Sporobulus* species.

Increased crop yields remove elements from the soil at a fast rate, with the result that deficiencies are frequently found on the most progressive farms. Climate also exerts an effect upon the mineral composition of forages. Reith (1965) notes that the maximum content of major nutrients is obtained at soil temperatures that produce maximum growth, and low temperatures generally produce crops containing lower percentages of N, P, K, Ca and Mg.

Climate limits the potential yield and the magnitude of yield response to mineral additions. Where climate varies markedly between seasons, large between-season differences in nutrient uptake are observed. Digestibility of forages, in general, is reduced due to high temperatures, which stimulate rapid growth and onset of maturity, with decrease of digestibility being 0.8 to 1.0% for each per cent of increase in temperature (Minson and McLeod, 1970; Denium and Dirven, 1975). Under the abundant rainfall and high temperatures of the wet tropics and semi-tropics, soils are often low in soluble minerals (Allman and Hamilton, 1949). Grass tetany predictably occurs primarily in the cool season. Pasture management involving liming and fertilization can usually be extremely beneficial to increasing forage macromineral concentrations. Overuse of N and K fertilizers increases the incidence of grass tetany (Kemp *et al.*, 1961), and K fertilization also dramatically reduces forage Na content (Underwood, 1981).

What Affects Major-element Availability?

There is an extreme lack of data on the biological availability of minerals from forage sources. Availability of Ca and P may vary considerably, according to their chemical combination or physical association with other compounds in feeds. Phytic acid, which is formed from six phosphate molecules combined with myo-inositol, hinders intestinal absorption of P, Ca and other minerals. Phytate P is readily used by ruminants, because they have abundant amounts of microbial phytase in the rumen. Before the rumen is developed, phytase is less utilized. Lucerne meal is quite high in Ca (18 g kg^{-1} DM), but several studies indicate that the Ca is poorly available. The low availability of Ca in lucerne may be owing to the presence of oxalates in lucerne. Ward *et al.* (1979) followed the fate of Ca oxalate. In India, Ca deficiency is reported for cattle fed straw, due to the large amounts of oxalates in this feed. Certain tropical grasses contain high levels of oxalate (Kiatoko *et al.*, 1978).

Biological availability of different sources of Mg for ruminants is considerable.

Peeler (1972) found that the availability of Mg ranged from 10 to 25% in forages and from 30 to 40% in grains and concentrates. Greater Mg availability in concentrates, along with the substantial amounts of grain usually fed to lactating dairy cows, is one of the major reasons why grass tetany is less prevalent among dairy cows than among beef cows in the USA. This would not be the situation in many developing tropical countries, since dairy cows often do not receive concentrates in substantial quantities.

Kemp *et al.* (1961) reported that Mg availability improves with increasing maturity of grasses and may be decreased by heavy K and N fertilization. Likewise, usually Mg in preserved forages is more available than in pastures. From temperate pastures (growing perennial ryegrass and white clover) fed fresh to sheep, apparent availability of Mg, Ca, P, Na and K was found to be 31, 21, 17, 95 and 97%, respectively (Grace, 1972).

Perdomo *et al.* (1977) studied the apparent digestibility in sheep of five minerals at three stages of regrowth for three tropical forages (Table 18.5). Digestibility and retention of Ca, P and Na by sheep tended to decline with maturity, while availability of Mg tended to increase at 56 days of age for the three species. Emanuel and Staples (1990) and Emanuel *et al.* (1991) evaluated mineral availability of six forages cut at 32 to 37 days, using an *in vitro* technique that simulated

Table 18.5. Apparent digestibility of various nutrient elements in three forages harvested at three ages of regrowth and fed fresh to sheep (from Perdomo *et al.*, 1977).

		Age of forage regrowth, days (% digestibility of forage mineral elements)			
Forage	Element	28	42	56	Mean ± standard error
Pangola	Ca	10.9^{b}	-40.0^{a}	23.3^{b}	-1.92 ± 29.4^{e}
	P	43.0^{b}	11.6^{a}	41.4^{b}	$32.00 \pm 10.8^{d,e}$
	Mg	41.4^{a}	32.3^{a}	66.3^{b}	46.70 ± 16.4^{e}
	K	80.6^{a}	80.9^{a}	88.6^{b}	83.4 ± 6.0^{d}
	Na	75.0	68.4	79.8	74.10 ± 7.2^{e}
Guinea grass	Ca	62.6^{c}	8.87^{a}	37.0^{b}	36.20 ± 23.0^{f}
	P	62.4^{b}	20.9^{a}	21.9^{a}	$35.00 \pm 22.3^{d,e}$
	Mg	25.4^{a}	44.5^{b}	56.5^{b}	42.20 ± 15.0^{e}
	K	88.8	87.2	89.2	88.40 ± 2.5^{e}
	Na	69.6	58.5	63.9	64.00 ± 10.1^{d}
Star grass	Ca	58.9^{b}	20.0^{a}	33.1^{b}	37.30 ± 18.2^{f}
	P	63.1^{b}	20.8^{a}	26.6^{a}	36.90 ± 22.1^{e}
	Mg	29.9^{a}	54.3^{b}	57.5^{b}	47.20 ± 16.1^{e}
	K	90.1^{b}	88.6^{b}	18.6^{a}	87.40 ± 3.9^{e}
	Na	62.6	70.0	63.1	65.20 ± 9.1^{d}

Different superscript letters in a row indicate that variations in values are significant for effect of age of regrowth ($P < 0.05$), and different superscript letters in a column for mean ± standard error values for specific elements indicate that variations are significant ($P < 0.05$).

rumen abomasum and intestine mineral release. K release was complete in the rumen. Ranking of minerals based on maximal extent of release was K > Mg > P > Ca. With the exception of K, legumes released more of their minerals than grasses. As was true for Ca, mean P release from legumes averaged across the sites (95%) was greater than the mean P release from grasses (82%) (Emanuel *et al.*, 1991).

Greene *et al.* (1986) performed an experiment to determine if breed or DM intake influenced Mg availability to dry, non-pregnant, mature cows. Hereford, Holstein and Jersey breeds excreted more faecal Mg and absorbed less Mg than Angus and Brahman breeds.

How to Remedy Low Status

Indirect methods of providing minerals for grazing animals include use of mineral-containing fertilizers, altering soil pH and encouraging growth of specific pasture species. Direct administration of minerals to ruminants, through water, mineral licks, mixtures and drenches, heavy pellets and injections, is generally the most economical method of supplementation. Benefits and disadvantages of mineral supplementation methods have been discussed (Underwood, 1981; McDowell, 1992).

Providing mineral-containing fertilizers

Increasing mineral intake of grazing livestock via fertilizers has been reviewed (Reid and Horvath, 1980; Underwood, 1981; Butterworth, 1985; Minson, 1990; McDowell, 1992, 1997; Barnes *et al.*, 1995). Fertilization of grasses and legumes is a practice capable of increasing the mineral level of these plants. However, there are frequently cases in which the plants do not respond to fertilization – i.e. the mineral composition is not increased and, at times, the inverse reaction occurs. As an example, the results obtained for *Panicum maximum*, fertilized or non-fertilized, respectively, cut at 28 days of age, were as follows: 24 and 23.8 g N kg^{-1} DM, 1.4 and 1.5 g P kg^{-1} DM, 23.3 and 24.1 g K kg^{-1} DM, 3.4 and 4.5 g Ca kg^{-1} DM and 2.3 and 2.9 g Mg kg^{-1} DM (Gomide *et al.*, 1969). One frequent response to N fertilization of pastures is a change in botanical composition, with a loss of the legume component and a consequent decline in total herbage concentration of elements such as Ca and Mg. Numerous studies (Reid and Horvath, 1980) have demonstrated a depression of Ca, Mg and Na concentration in herbage with the application of K fertilizers. Fertilizers would not normally be applied to forages for the purpose of increasing plant content of Ca, K, S and Cl. Fertilizer has been used to increase forage P and, to a lesser degree, to elevate plant Na and Mg.

Phosphorus

On sparse P-deficient grazing, the direct-supplementation method is preferred, because the use of phosphate fertilizers involves high transport and application cost and herbage productivity is usually limited by climate or soil problems. The effects of P fertilization on plant mineral uptake are dependent, to a considerable degree, on the properties of the soil. Whitehead (1966) concluded that fertilization often had little effect on the P concentration of herbage, unless the soil was severely deficient.

With infertile soils, P is frequently the major factor limiting pasture growth and, under these conditions, the regular use of fertilizer may not only increase yield substantially but also double the concentration of P in herbage (Reith, 1973). In Australian studies, superphosphate applications at the rate of 125 kg ha^{-1} doubled pasture yields and increased forage P by some 50% (Underwood, 1981). From pastures grown on soils from Florida, USA, Kirk *et al.* (1970) reported increases in the P content of pangola grass (*D. decumbens*) from 0.7–1 g kg^{-1} DM to 2.5–3 g kg^{-1} DM due to P fertilization.

P fertilization to raise forage P would not be effective for very acid soils, as the fertilizer P would be tied up in the soil and forage levels of the element would be increased only slightly (McDowell, 1985). Australian research (Underwood, 1981) has shown that not only does superphosphate fertilizer increase herbage P but it also results in improved palatability and digestibility of the forage. However, unless there are definite forage-yield increases that can be utilized effectively by grazing herbivores, use of mineral-containing fertilizers is economically prohibitive.

Sulphur

Responses to fertilizer S are widespread in the tropics and have been recorded in 40 tropical countries with 23 different crops (International Fertilizer Development Center (IFDC), 1979). However, responses are from crop-yield studies and have not generally been designed to relate fertilizer-S to plant-S concentrations. Rees *et al.* (1974) reported that applying S fertilizer not only raised the S level of the plant but also increased the yield of DM and the leaf percentage. With S fertilizer, the voluntary intake of *D. decumbens* was increased by 44%.

Sodium

The Na concentration in Na-accumulating forages is related to the ratio of Na to K in the soil and can be increased by applying fertilizers containing Na (Henkens, 1965). The Na concentration in forage species that do not accumulate Na is relatively independent of the level of soil Na. With this group of forages, applying Na fertilizers will not increase the Na concentration in the forage or prevent a deficiency in the animal (Minson, 1990).

Magnesium

Mg fertilizers can significantly increase pasture concentrations. This method of control has limitations on many soil types and usually has to be accompanied by other means of supplying additional Mg. Foliar dusting of pastures with calcined magnesite ($MgCO_3$) before or during tetany-prone periods is one such means that has proved effective, provided it is applied at not less than 17 kg ha^{-1} at not more than 10-day intervals (Rogers, 1979).

Free-choice mineral supplementation

For grazing livestock that receive no concentrates, the major method of receiving supplemental macrominerals is to rely on the free-choice feeding of mineral supplements. The main ingredient for the complete free-choice supplement is common salt. The vast majority of minerals are highly unpalatable; however, animals consume a complete mineral mixture because of the favourable taste and their craving for salt. Contrary to popular belief, livestock do not select minerals on the basis of their physiological needs, but rather by taste of the ingredients (McDowell, 1992, 1997). Even though some animals will either overconsume or underconsume mineral supplements, there is no other practical way of supplying certain mineral needs under grazing conditions. There are definite disadvantages to the free-choice mineral-supplementation method, the major concern being the lack of uniform consumption by animals. Observations indicate that some animals may consume double their needs, while others may eat practically none of a supplement.

Factors that affect the consumption of mineral mixtures have been reviewed (McDowell, 1992, 1997) and are as follows: (i) soil fertility and forage type consumed; (ii) available energy–protein supplements; (iii) individual requirements; (iv) salt content of drinking-water; (v) palatability of mineral mixture; (vi) availability of fresh mineral supplies; and (vii) physical form of minerals.

The characteristics of a good free-choice mineral supplement for cattle are listed in Table 18.6 and recommendations for sheep are available (McDowell, 1996). In order to evaluate a mineral supplement for ruminants, it is necessary to have an approximation of: (i) requirements of the target animals for the essential nutrients; this includes the age of the animals involved, stage of current production or reproduction cycle and intended purpose for which the animals are being fed; (ii) relative biological availability of the minerals in the sources from which they will be provided; (iii) approximate daily intake per head of the mineral mixture and of total DM by the target animals; and (iv) concentration of the essential nutrients in the mineral mixture (Ellis *et al.*, 1988).

Problems concerned with mineral-supplementation programmes in diverse regions include: (i) insufficient chemical analyses and biological data to determine which minerals are required and in what quantities; (ii) lack of mineral consumption data needed for formulating supplements; (iii) inaccurate and/or unreliable information on mineral ingredient labels; (iv) supplements that contain inadequate amounts or imbalances; (v) standardized mineral mixtures that are inflexible for diverse ecological regions; (vi) farmers not supplying mixtures as recommended by the manufacturer (e.g. mineral mixtures diluted 10:1 and 100:1 with additional salt); and (vii) difficulties involved with transportation, storage and cost of mineral supplements.

An oral Mg supplement is of value only during seasonal occurrences of grass tetany (Allcroft, 1961). Unfortunately, many commercial Mg-containing free-choice mineral supplements are often of little value because: (i) they contain inadequate amounts of Mg to protect against tetany during susceptible periods; and (ii) provision of such supplements to normal animals during non-susceptible periods is useless as a prophylactic measure, since additional Mg will not provide a depot of readily available Mg for emergency use. Some producers feed Mg supplements

Table 18.6. Characteristics of a 'good' free-choice cattle mineral supplement (from McDowell, 1992).

1. Final mixture containing a minimum of 60–80 g total P kg^{-1}. In areas where forages are consistently lower than 2 g P kg^{-1}, mineral supplements in the 80–100 g P kg^{-1} range are preferred
2. Calcium:phosphorus ratio not substantially over 2:1
3. Provide a significant proportion (e.g. about 50%) of the trace-mineral requirements for Co, Cu, I, Mn and Zn.* In known trace-mineral-deficient regions, 100% of specific trace minerals should be provided
4. Composed of high-quality mineral salts that provide the best biologically available forms of each mineral element, and avoidance or minimal inclusion of mineral salts containing toxic elements. As an example, phosphates containing high F should be either avoided or formulated so that breeding cattle would receive no more than 30–50 mg F kg^{-1} in the total diet. Fertilizer or untreated phosphates could be used to a limited extent for feedlot cattle
5. Formulated to be sufficiently palatable to allow adequate consumption in relation to requirements
6. Backed by a reputable manufacturer with quality-control guarantees as to accuracy of mineral-supplement label
7. An acceptable particle size that will allow adequate mixing without smaller-size particles settling out
8. Formulated for the area involved, the level of animal productivity, the environment (temperature, humidity, etc.) in which it will be fed, and is as economical as possible in providing the mineral elements used

For most regions it would be appropriate to include selenium (Se), unless toxicity problems have been observed. Iron (Fe) should be included in temperate region mixtures but often both Fe and Mn can be eliminated for acid soil regions. In certain areas where parasitism is a problem, Fe supplementation may be beneficial.
Co, cobalt; Cu, copper; I, iodine; Mn, manganese; Zn, zinc; F, fluorine.

about a month before the Mg tetany season to decrease the amount of Mg needed daily during the susceptible period.

Responsible firms that manufacture and sell high-quality mineral supplements provide a great service to individual farmers. However, there are companies that are responsible for exaggerated claims in their advertising. Others produce inferior products of little value or, worse, likely to be a detriment to animal production.

Further Research Needed

Many experiments compare varieties of forages in relation to yield and quality of components (e.g. available energy and protein) or potential for improved forage intakes. A few studies (Ramos-Santana and McDowell, 1993, 1994; Faría-Mármol *et al.*, 1996) have also evaluated mineral concentrations of new tropical forage accessions. However, generally there is little, and usually no, regard for mineral concentrations of new forage varieties taken into account. Forages have even been selected on their ability to grow in environments of low available soil P (e.g.

Stylosanthes in acid, tropical soils), which results in forage exceedingly low in P. In developing new strains and varieties of forages, new research should include mineral concentrations as a component of the 'quality' term.

Other research needs are as follows:

- There is a scarcity of data on the mineral status of vast grazing areas worldwide, particularly in developing countries. Studies are needed which include mineral analyses of forage and animal fluids and tissues to establish which minerals are needed for supplementation.
- Determining the effects of various environmental and production factors upon mineral concentrations of forages, as well as element requirements for different species and classes of grazing livestock under varying conditions.
- It is important to study soil–plant interrelationships as they affect plant mineral concentrations.
- Developing methods to evaluate the mineral status of grazing livestock and biological assay techniques for minerals in both forages and mineral supplements.
- Improved methods of providing supplemental minerals for livestock are needed, particularly for grazing animals, with cost–benefit relationships established.
- To formulate free-choice mineral supplements for various soil types and ecological regions, information on mineral supplement consumption is required.

Cost Implications

Only limited data exist on cost implications of increasing mineral intake by grazing livestock. Providing minerals via fertilizer treatment is economically the method of choice if the treatment improves both yield and mineral composition of herbage. Cost–benefit relationships of fertilizer treatment have generally been restricted as to forage-yield components vs. the value of increased plant minerals.

Under most grazing conditions, increased mineral intake would economically favour direct supplementation. Most studies would conclude that providing direct administration of minerals to livestock as drenches, rumen preparations and injections (practical only for trace elements) is more expensive than providing animals access to a free-choice mineral mixture. Free-choice mineral supplementation is a low-cost insurance with economic returns on mineral investment generally of more than 2:1 in Latin American studies (McDowell *et al.*, 1984) and returns could be as high as 20:1 in some studies (J. de Sousa, personal communication). Berger (1992) calculated that, in a 100-cow herd, preventing the loss from grass tetany of a single cow every 3 years would more than pay the cost of Mg supplementation.

Cost comparisons were made on P supplementation methods for cattle at the King Ranch in Texas, USA, for the following treatments: disodium phosphate in water, bone-meal in self-feeders, fertilized with triple superphosphate and control (Reynolds *et al.*, 1953). The cows that yielded the highest return ha^{-1} after the costs of supplementation had been deducted were those fed disodium phosphate, followed by those on fertilized pasture, those fed bone-meal and the unsupplemented group, the returns being \$50.75, \$49.93, \$42.28 and \$34.78, respectively.

The importance of mineral supplementation on overall production of cattle in the llanos regions of Colombia is presented in Table 18.7. Supplemental minerals

Table 18.7. Four-year Colombian study evaluating supplemental minerals.*

Item	Salt only	Complete minerals†
Abortions (%)	9.3	0.75
Births (%) year^{-1}	50.0	67.0
Death losses, birth to weaning (%)	22.6	10.5
Calves weaned of total herd (%) year^{-1}	38.4	60.0
Weaning weight (9 months) (kg)	117.0	147.0
Gain of growing cattle (572 days) (kg)	86.0	141.0
Average gain per day (g)	150.0	247.0
Weaned calf wt (kg year^{-1} cow^{-1})‡	44.9	88.2

* Evaluation of CIAT (1977) report (Miles and McDowell, 1983).
† Evaluation of the complete mineral supplement indicated adequate concentrations of most minerals but suboptimum levels of Zn and Cu, with no added Se and S.
‡ Weaned calf per cent multiplied by weaning weight.
Zn, zinc; Cu, copper; Se, selenium.

dramatically increased all parameters of production. Multiplication of the weaning percentage by the weaning weight gave 88.2 kg of calf produced per cow with complete minerals, compared with 44.9 kg with salt alone. For this study, free-choice mineral supplementation resulted in a return of 15.6 Colombian pesos for every 1 peso invested in the supplement (Miles and McDowell, 1983).

When there is no information available on mineral status for specific regions, then complete ('shotgun') mineral mixtures are warranted. However, with additional information on likely limiting minerals, more economical mixes can be formulated. In Colombia, ten cattle experiments were designed to evaluate feeding complete commercial supplements as against feeding supplements formulated to contain a minimum number of minerals that had previously been established as deficient on the basis of forage and animal-tissue analyses (Laredo *et al.*, 1989). The specifically tailored mineral supplements produced equal production with a 42% cost saving.

References

Adams, R.S. (1975) Variability in mineral and trace element content of dairy cattle feeds. *Journal of Dairy Science* 58, 1538–1548.

Allcroft, R. (1961) The use and misuse of mineral supplements. *Veterinary Record* 73, 1255–1266.

Allman, R.I. and Hamilton, T.S. (1949) *Nutritional Deficiencies in Livestock*. FAO Agricultural Studies No. 5, Washington, DC.

Arnold, G.E., McManus, W.R. and Bush, I.G. (1966) Studies on the wool production of grazing sheep. 5. Observations on teeth wear and carry over. *Australian Journal of Experimental Agriculture and Animal Husbandry* 6, 101.

Barnes, R.F., Miller, D.A. and Nelson, C.J. (1995) *Forages*, Vol. 1, 5th edn. Iowa State University Press, Ames, Iowa, 516 pp.

Berger, L.L. (1992) *Salt Tracer Minerals* 24, 12. (Salt Institute, Alexandria, Virginia.)

Butterworth, M.H. (1985) *Beef Cattle Nutrition and Tropical Pastures*. Longman, London, 500 pp.

Chadhokar, P.A. (1982) *Gliricidia maculata*: a promising legume fodder plant. *World Animal Review* 44, 36–43.

CIAT (Centro Internacional de Agricultura Tropical) (1977) *Annual Report 1977*. CIAT, Cali, Colombia.

Denium, B. and Dirven, J.P. (1975) Climate, nitrogen and grass. 6. Comparison of yield and chemical composition of some temperate and tropical grass species grown at different temperatures. *Netherlands Journal of Agricultural Science* 23, 69–82.

Devendra, C. (1977) The mineral content of Caribbean feeding stuff. *Tropical Agriculture (Trinidad)* 54, 29–38.

Echevarría, M., Valdivia, R., Barúa, J. and Campos, L. (1973) Efecto de la suplementación mineral sobre la ganancia de peso de toretes nelore de Pucallpa. *3a Reunion de Especialistas e Investigadores Forrajeros del Perú, Pucallpa, Perú* 83, 92–98.

Elkins, C.B., Haaland, R.L., Hoveland, C.S. and Griffey, W.A. (1978) Grass tetany potential of tall fescue as affected by soil O_2. *Agronomy Journal* 70, 309–311.

Ellis, G.L., McDowell, L.R. and Conrad, J.H. (1988) Evaluation of mineral supplements for grazing cattle in Latin America. *Nutrition Reports International* 38, 1137.

Emanuel, S.M. and Staples, C.R. (1990) Ruminal release of minerals from six forage species. *Journal of Animal Science* 68, 2052–2060.

Emanuel, S.M., Staples, C.R. and Wilcox, C.J. (1991) Extent and site of mineral release from six forage species incubated in mobile dacron bags. *Journal of Animal Science* 69, 801–810.

Estevez-Cancino, J.A. (1960) Milk production increases with bonemeal supplementation. *Acta Agronomica, Palmira* 10, 169–183.

Faria, J.A., Velásquez, J. and López, M.G. (1981) *Situacion de la nutricion mineral en fincas de las sabanas orientales del estado Guarico*. Boletin No. 6, Fondo de Investigación Agropecuaria, Guarico, Venezuela.

Faría-Mármol, J., Morillo, D.E. and McDowell, L.R. (1996) *In vitro* digestibility, crude protein, and mineral concentrations of *Leucaena leucocephala* accessions in a wet/dry tropical region of Venezuela. *Communications of Soil Science and Plant Analysis* 27(13–14), 2663–2674.

Fleming, G.A. (1973) Mineral composition of herbage. In: Butler, G.W. and Bailey, R.W. (eds) *Chemistry and Biochemistry of Herbage*. Academic Press, London, pp. 529–563.

French, M.H. (1955) Earth-eating and the mineral needs of livestock. *East African Agricultural Journal* 20, 168–175.

Göhl, B. (1975) *Tropical Feeds*. FAO Feeds Information Centre, FAO, Rome, pp. 1–661.

Gomide, J.A., Noller, C.H., Mott, G.O., Conrad, J.H. and Hill, D.L. (1969) Mineral composition of six tropical grasses as influenced by plant age and nitrogen fertilization. *Agronomy Journal* 61, 120–123.

Grace, N.D. (1972) Studies on the movement of Mg, Ca, P, Na and K across the gut wall of sheep fed fresh pasture. *Proceedings of the New Zealand Society of Animal Production* 32, 77–84.

Greene, L.W, Soils, J.C., Byers, F.M. and Schelling, G.T. (1986) Apparent and true digestibility of magnesium in mature cows of five breeds and their crosses. *Journal of Animal Science* 63, 189–196.

Hartmans, J. (1970) The detection of copper deficiency and other trace element deficiencies under field conditions, In: Mills, C.F. (ed.) *Trace Element Metabolism in Animals: Proceedings of WAAP/IBP International Symposium, Aberdeen, Scotland*. E. & S. Livingstone, Edinburgh and London, pp. 441–445.

Henkens, C.H. (1965) Factors influencing the sodium content of meadow grass. *Netherlands Journal of Agricultural Science* 13, 21–47.

International Fertilizer Development Center (IFDC) (1979) *Sulfur in the Tropics*. Sulfur Institute, Muscle Shoals, Alabama, 69 pp.

Jones, D.I.H. (1964) Mineral content of some cultivated grasses grown in northern Rhodesia. *Rhodesia Journal of Agricultural Science* 2, 57–59.

Jumba, I.O., Suttle, N.F., Hunter, E.A. and Wandiga, S.O. (1995) Effects of soil origin, mineral composition and herbage species on the mineral composition of forages in the Mount Elgon region of Kenya. 1. Calcium, phosphorus, magnesium and sulphur. *Tropical Grasslands* 29, 40–46.

Kabaija, E. and Smith, O.B. (1989) Influence of season and age of regrowth on the mineral profile of *Gliricidia sepium* and *Leucaena leucocephala*. *Tropical Agriculture* 66, 125–128.

Kalmbacher, R.S. and Martin, F.G. (1981) Mineral content in creeping bluestem as affected by time of cutting. *Journal of Range Management* 34, 406–408.

Kemp, A., Dejis, W.B., Hemkes, O.J. and Van Es, A.J.H. (1961) Hypomagnesaemia in milking cows: intake and utilization of magnesium from herbage by lactation. *Netherlands Journal of Agricultural Science* 9, 134–139.

Kiatoko, M., McDowell, L.R., Fick, K.R., Fonseca, H., Camacho, J., Loosli, J.K. and Conrad, J.H. (1978) Mineral status of cattle in the San Carlos region of Costa Rica. *Journal of Dairy Science* 61, 324–330.

Kiatoko, M., McDowell, L.R., Bertrand, J.E., Chapman, H.L., Pate, F.M., Martin, F.G. and Conrad, J.H. (1982) Evaluating the nutritional status of beef cattle herds from four soil order regions of Florida. I. Macroelements, protein, carotene, vitamins A and E, hemoglobin and hematocrit. *Journal of Animal Science* 55, 28–37.

Kincaid, R.L. and Cronrath, J.D. (1983) Amounts and distribution of minerals in Washington forages. *Journal of Dairy Science* 66, 821–824.

Kirk, W.G., Shirley, R.L., Hodges, E.M., Davis, G.K., Peacock, F.M., Easley, J.F. and Martin, F.G. (1970) Production performance and blood and bone composition of cows grazing pangolagrass pastures receiving different phosphate fertilizers. *Bulletin of Florida Agricultural Experiment Station* 735.

Lamond, D.R. (1970) The influence of undernutrition on reproduction in the cow. *Animal Breeding Abstracts* 38, 354–372.

Laredo, M.A. and Minson, D.I. (1973) The voluntary intake, digestibility and retention time by sheep of leaf and stem fractions of five grasses. *Australian Journal of Agricultural Research* 24, 875.

Laredo, M.A., Gonzalez, C.F., Carillo, J.A. and McDowell, L.R. (1989) Mineral supplementation of beef cattle in an Andean region of Colombia. *Nutrition Reports International* 39, 1069–1079.

Long M.I.E., Ndyanabo, W.K., Marshall, B. and Thornton, D.D. (1969) Nutritive value of grasses in Ankole and the Queen Elizabeth National Park, Uganda. IV. Mineral content. *Tropical Agriculture (Trinidad)* 46, 201–209.

Long, M.I.E., Thorton, D.D., Ndyanabo, W.K., Marshall, B. and Szekaalo, H. (1970) The mineral status of dairy farms in parts of Buganda and Busoga bordering Lake Victoria, Uganda. II Nitrogen and mineral content of pastures. *Tropical Agriculture (Trinidad)* 47, 37–50.

McDowell, L.R. (1976) Minerals deficiencies and toxicities and their effect on beef production in developing countries. In: *Proceedings of Beef Cattle Production in Developing Countries, Edinburgh, UK*. Centre for Tropical Veterinary Medicine, Edinburgh, pp. 216–241.

McDowell, L.R. (1985) *Nutrition of Grazing Ruminants in Warm Climates*. Academic Press, New York, 443 pp.

McDowell, L.R. (1992) *Minerals in Animal and Human Nutrition*. Academic Press, San Diego, 524 pp.

McDowell, L.R. (1996) Free-choice mineral supplements for grazing sheep in developing countries. In: Masters, D.G. and White, C.L. (eds) *Detection and Treatment of Mineral Nutrition Problems in Grazing Sheep.* Australian Centre for International Agricultural Research, Canberra, pp. 81–94.

McDowell, L.R. (1997) *Minerals for Grazing Ruminants in Tropical Regions*, 3rd edn. University of Florida, Gainesville, Florida, USA, 81 pp.

McDowell, L.R., Conrad, J.H., Thomas, J.E. and Harris, L.E. (1974) *Latin American Tables of Feed Composition.* University of Florida, Gainesville, 552 pp.

McDowell, L.R., Conrad, J.H., Thomas, Jenny E., Harris, L.E. and Fick, K.R. (1977) Nutrional composition of Latin American forages. *Tropical Animal Production* 2, 273–279.

McDowell, L.R., Bauer, B., Galdo, E., Koger, M., Loosli, J.K. and Conrad, J.H. (1982) Mineral supplementation of beef cattle in the Bolivian tropics. *Journal of Animal Science* 55, 964–970.

McDowell, L.R., Ellis, G.L. and Conrad, J.H. (1984) Mineral supplementation for grazing cattle in tropical regions. *World Animal Review* 52, 1–12.

Mayland, H.F., Grunes, D.L. and Lazar, V.A. (1976) Grass tetany hazard of cereal forages based upon chemical composition. *Agronomy Journal* 68, 665.

Miles, W.H. and McDowell, L.R. (1983) Mineral deficiencies in the llanos rangelands of Colombia. *World Animal Review* 46, 2–10.

Miller, W.J. (1979) *Dairy Cattle Feeding and Nutrition.* Academic Press, New York, 411 pp.

Minson, D.J. (1990) *Forage in Ruminant Nutrition.* Academic Press, San Diego, 483 pp.

Minson, D.J. and McLeod, M.N. (1970) The digestibility of temperate and tropical grasses. In: *Proceedings of the 11th International Grassland Congress*, pp. 719–732.

Montalvo, M.I., Veiga, J.B., McDowell, L.R., Ocumpaugh, W.R. and Mott, G.O. (1987) Mineral content of dwarf *Pennisetum purpureum* under grazing conditions. *Nutritional Reports International* 135, 157–170.

Moraes, E., Italiano, E.C. and Pieniz, C. (1982) Supplemental phosphorus for growing cattle. In: *Proceedings Sociedade Brasileria de Zootecnia*, Piracicaba, *São Paulo, Brazil*, p. 156 (Abstract).

Morris, J.G. (1980) Assessment of sodium requirements of grazing beef cattle: a review. *Journal of Animal Science* 50, 145–152.

Murphy, G.M. and Plasto, A.W. (1973) Liveweight response following sodium chloride supplementation of beef cows and their calves grazing native pasture. *Australian Journal of Experimental Agriculture and Animal Husbandry* 13, 369–374.

Peeler, H.T. (1972) Biological availability of nutrients in feeds: availability of major mineral ions. *Journal of Animal Science* 35, 695–712.

Perdomo, J.T., Shirley, R.L. and Chicco, C.F. (1977) Availability of nutrient minerals in four tropical forages fed freshly chopped to sheep. *Journal of Animal Science* 45, 1114–1119.

Pfander, W.H. (1971) Animal nutrition in the tropics – problems and solutions. *Journal of Animal Science* 33, 843–849.

Rae, S.M. (1963) Discussants paper to topic 2 (feeding). *Proceedings of the 1st World Conference on Animal Production.* European Association of Animal Production, Rome, 2, 109–114.

Ramos-Santana, R. and McDowell, L.R. (1993) *In vitro* digestibility and crude protein and mineral concentrations of Guineagrass accessions in a humid tropical region of Puerto Rico. *Communications of Soil Science and Plant Analysis* 24(3–4), 325–334.

Ramos-Santana, R. and McDowell, L.R. (1994) Yield, *in vitro* digestibility, crude protein and mineral concentration of eight *Digitaria* introductions in Puerto Rico. *Communications of Soil Science Plant Analysis* 25(11–12), 2019–2027.

Rees, M.C., Minson, D.J. and Smith, F.W. (1974) The effect of supplementary and fertilizer sulphur on voluntary intake, digestibility, retention time in the rumen and site of digestion of pangola grass in sheep. *Journal of Agricultural Science* 82, 419–422.

Reid, R.L. and Horvath, D.J. (1980) Soil chemistry and mineral problems in farm livestock: a review. *Journal of Animal Feed Science and Technology* 5, 95–167.
Reith, J.W.S. (1965) Mineral composition of crops. *NAASQ Review* 68, 150–156.
Reith, J.W.S. (1973) Soil conditions and nutrient supplies in hill land. In: Gething, P.A., Newbould, P. and Patterson, J.B.E. (eds) *Hill Pasture Improvement and its Economic Utilization*. Potassium Institute, Henley-on-Thames, pp. 5–12.
Rendig, V.V. and Grunes, D.L. (eds) (1979) *Symposium on Grass Tetany*. American Society of Agronomy, Madison, Wisconsin.
Reynolds, E.B., Jones, J.M., Jones, J.H., Fudge, J.F. and Kleberg, R.J., Jr (1953) *Methods of Supplying Phosphorus to Range Cattle in South Texas*. Texas Agricultural Experiment Station Bulletin 773, Texas A & M University, College Station, 16 pp.
Rogers, P.A.M. (1979) Hypomagnesaemia and its clinical syndromes in cattle: a review. *Irish Veterinary Journal* 33, 115–124.
Rojas, L.X. and McDowell, L.R. (1997) Mineral imbalances in South America and their effects on livestock production. In: *Proceedings of International Conference on Livestock in the Tropics*. University of Florida, Gainesville, pp. 129–134.
Smith, O.B., Kasali, O.B., Adeyanju, S.A. and Adegbola, A.A. (1980) Mineral malnutrition in tropical livestock. I. Potassium deficiency in the bovine. *Journal of Animal Science* 51 (Suppl. 1), 396 (Abstract).
Sousa, J.C., Conrad, J.H., Blue, W.G. and McDowell, L.R. (1979) Interrelacoes entre minerais no solo, forrageiras e tecido animal em fuzendas do norte de Mato Groso. I. Calcío e fósforo. *Pesquisa Agropecuaria Brasileira* 14(4), 387–395.
Sousa, J.C., Conrad, J.H., Mott, G.O., McDowell, L.R., Ammerman, C.B. and Blue, W.G. (1982) Interrelacoes entre minerais no solo, plantas forrageiras e tecido animal – zinco, magnesio, sodio e potassio. *Pesquisa Agropecuaria Brasileira* 17(1), 11–20.
Stobbs, T.H. (1976) Factors limiting milk production from grazed tropical pastures. In: *Memorias Seminario Internacional de Ganaderia Tropical, Acapulco, Mexico, 8–12 de Marzo*. SAG, Banco Mex, FIRA, Acapulco, pp 183–198.
Sütmoller, P., Vahia de Abreu, Ant, Van der Grift, J. and Sombroek, W.G. (1966) *Mineral Imbalances in Cattle in the Amazon Valley*. Department of Agriculture Research, Royal Tropical Institute Communication No. 53, Amsterdam, the Netherlands.
Tergas, L.E. and Blue, W.G. (1971) Nitrogen and phosphorus in Jaraguagrass (*Hyparrhenia rufa* (Nees) Stapf) during the dry season in a tropical savanna as affected by nitrogen fertilization. *Agronomy Journal* 63, 6–9.
Theiler, A., Green, H.H. and Du Toit, P.J. (1928) Studies in mineral metabolism. III. Breeding of cattle on P deficient pasture. *Journal of Agricultural Science* 18, 369–371.
Thompson, A. (1957) The trace-element contents of herbage plants with some reference to their availability to the animal. *Journal of Science and Food Agriculture* 8 (Suppl.) 72–81.
Underwood, E.J. (1981) *The Mineral Nutrition of Livestock*. Commonwealth Agricultural Bureaux, London, 180 pp.
Velásquez, J.A. (1979) Situación de la nutrición mineral del ganado bovino en el estado Monagas. *Agronomía Tropical* 29, 273–287.
Walker, C.A. (1957) Studies of the cattle of northern Rhodesia, the growth of steers under normal veld grazing and supplemented with salt and protein. *Journal of Agricultural Science* 49, 394–400.
Ward, H.K. (1968) Supplementation of beef cows grazing on veld. *Rhodesian Journal of Agricultural Research* 6, 93–101.
Ward, G., Harbers, L.H. and Blaha, J.J. (1979) Calcium-containing crystals in alfalfa – their fate in cattle. *Journal of Dairy Science* 62, 715–722.
Whitehead, D.C. (1966) *Nutrient minerals in Grassland Herbage*. Commonwealth Bureau of

Pastures and Field Crops, Mimeo Publication No. 1/1966, Commonwealth Agricultural Bureau, Farnham Royal, Bucks., UK, 83 pp.
Whyte, R.O. (1962) The myth of tropical grasslands. *Trinidad Tropical Agriculture* 39, 1–11.
Williams, R.D. (1963) *Minor Elements and Their Effects on the Growth and Chemical Composition of Herbage Plants*. Mimeo Publication No. 1/1966, Commonwealth Agricultural Bureau, Farnham Royal, Bucks., UK.
Wilson, R.K., Flynn, A.V. and Conniffe, D. (1968) Digestibility intake and chemical composition of old meadow hay. *Irish Journal of Agricultural Research* 7, 31–36.
Youssef, F.G. (1988) Some factors affecting the mineral profiles of tropical grasses. *Outlook on Agriculture* 17, 104–111.

19 Vitamins in Forages

N. Ballet, J.C. Robert and P.E.V. Williams

Aventis Animal Nutrition, 42 Avenue Aristide Briand, BP 100, 92164 Antony Cedex, France

Vitamins are organic molecules with complex molecular structure and they are essential in minimum quantities for the health, growth and reproduction of ruminants (bovine, ovine and caprine) (Table 19.1). Vitamin deficiency results in a number of metabolic disorders of varying levels of gravity. Forage has the potential to play a significant role in the supply of vitamins to ruminants. However, as will be demonstrated, the vitamin content of forages is highly variable and unpredictable and the production of synthetic vitamins was essential to the advent of intensive livestock production.

Vitamin Requirements of Ruminants

Adult ruminants are different from monogastrics (pigs and poultry) with respect to their dependence on an exogenous supply of vitamins. Synthesis of B-group vitamins (thiamine, riboflavine, niacin ...) and vitamin K occurs during the degradation and fermentation of feed ingredients by ruminal microorganisms. Vitamin D is synthesized by the action of ultraviolet radiation on the sterols present in the skin of ruminants, vitamin C is synthesized from C_6 sugars (glucose and galactose) and niacin from tryptophan (if the amino acid is present in excess). It is therefore mainly with respect to vitamins A and E that ruminants have specific dietary dependence.

However, very young ruminants, which do not possess a fully developed rumen and which lack a fully functional rumen microflora, cannot synthesize adequate B-group vitamins. Calves and lambs can therefore suffer from a B-group vitamin deficiency. In this instance, as in monogastrics, they are dependent on an exogenous supply of B-group vitamins.

Table 19.1. Fat- and water-soluble vitamins with their major functions (adapted from McDowell, 1989).

Names	Functions
Fat-soluble vitamins	
A, retinol	Vision
	Maintenance of epithelial cells
	Reproduction
	Growth
E, α-tocopherol	Biological antioxidant
	Phospholipid membrane stability
	Immunomodulation
D, ergocalciferol D_2, cholecalciferol D_3	Phosphocalcium metabolism
	Growth
	Reproduction
K	Cofactor in coagulation
Water-soluble vitamins	
B_1, thiamine	Coenzyme in oxidative decarboxylation
	Role in neurophysiology
B_2, riboflavine	Intermediary in the transfer of electrons in biological oxidation–reduction reactions
B_3, PP factor, niacin	Constituent of coenzymes NAD and NADP in carbohydrate, lipid and protein metabolism
B_5, pantothenic acid	Coenzyme A precursor
B_6, pyridoxine, pyridoxamine	Amino acid metabolism
	Formation of biogenic amines
B_{12}, cobalamin	Integrity of nervous system
	Haematopoiesis
	Growth
	Gluconeogenesis from propionate
H, B_w, biotin	Coenzyme in carbohydrate, lipid and protein metabolism
M, B_c, folacin	Growth
	Haematopoiesis
	Maintenance of immune system
Choline*	Synthesis of acetylcholine
	Component of phospholipids
	Methyl radical donor
	Lipotrophic factor
C, ascorbic acid	Collagen biosynthesis
	Transfer of electrons
	Oxidation reactions

*Choline is classified as one of the B-complex vitamins.
PP, pellagra-preventing; NAD, nicotinamide adenine dinucleotide; NADP, NAD phosphate.

Thus, in addition to the standard classification of vitamins according to their solubility in either water (C and B group) or in lipid solvent (A, D, E and K), vitamins for adult ruminants can be divided according to:

- self-supply (via either the rumen or endogenous supply) (K, C, D and B group);
- supply from the feed (A and E).

Different rearing methods and changes in ration formulation may alter the endogenous production of vitamins and result in a need to modify this classification. Animals reared indoors or fed rations rich in concentrates cannot produce sufficient vitamin D and thiamine. Furthermore, there is an increasing need for an exogenous supply of niacin, as, following genetic selection, the increase in productivity of ruminants begins to approach the quantitative limits of ruminal vitamin synthesis. The consumer demand for animal products of high quality is also imposing a need for an increased exogenous supply, particularly of niacin and vitamin E. The supply of niacin to high-producing cows at the beginning of lactation has a tendency to increase milk production and also the fat and protein content of milk (NRC, 1989; Hullár and Brand, 1993). In the nutrition of adult ruminants, the five most important vitamins with respect to an exogenous supply are therefore vitamins A, E, D and, in specific conditions, niacin and thiamine.

The supply of vitamins to ruminants other than from the synthetically produced material is totally dependent on a supply from fresh or conserved forage in the ration, since the concentrate portion of the feed is practically devoid of any naturally occurring vitamins A, E, D or their precursors (Brown, 1953; Machlin, 1984; McDowell, 1989).

Vitamin A or retinol exists only in animal products (McDowell, 1989). However, vegetable material can contain provitamins A (carotenoids), of which 80 forms are known, with β-carotene being the most important. β-Carotene is partially absorbed at the intestinal level, where it is transformed into vitamin A. It has been established that the conversion rate in the bovine is of the order of 400 international units (IU) of vitamin A mg^{-1} β-carotene (McDowell, 1989) (where 1 IU is defined as the biological activity of 0.300 μg of retinol). However, the efficiency of this conversion varies according to the isomer. The all-*trans* form of β-carotene has the highest activity, the neo-B and neo-U forms have only 53 and 38% of all-*trans*-form activity, respectively (Aitken and Hankin, 1970).

Vitamin E is found in eight known forms, four tocopherols (α, β, γ and δ) and four tocotrienols (α, β, γ and δ) (McDowell, 1989). The α-tocopherol is the most active and is considered true vitamin E.

Vitamin D forms are derived from the ultraviolet irradiation of sterols in animals (D_3 = cholecalciferol) and in vegetables (D_2 = ergocalciferol), the two forms having the same biological potency in ruminants (McDowell, 1989). One international unit of vitamin D activity is defined as the activity of 0.025 μg of vitamin D_3.

The term niacin is used to cover nicotinic acid and nicotinamide, the two compounds possessing the same vitamin activity in ruminants (McDowell, 1989). Thiamine consists of a molecule of pyrimidine and a molecule of thiazole linked by a methylene bridge (McDowell, 1989).

The physiological requirements for these vitamins are difficult to define and can vary appreciably according to the measures used as reference, particularly whether the criterion is the prevention of deficiency symptoms or the requirement for maximum animal performance. Standard tables of requirements tend to ignore the additional needs that have been identified as a result of different repeated stressors occurring during livestock production (unfavourable rearing conditions, infection, etc.). Present recommendations for ruminants concern vitamins A, E, D and niacin and are greater than the strict physiological needs, since they incorporate a margin of safety in order to ensure that the performance of the animals is not compromised (Table 19.2). The minimal recommendations are given for mean rearing conditions but are increased in accordance with a number of factors, such as the composition of the ration and the performance of the animal (growth, reproduction and milk production). Supply of highly fermentable diets increases the requirements for both thiamine and vitamin A (NRC, 1989). The continued increase in genetic potential of animals as a result of breeding results in a general increase in vitamin requirements.

Forage as a Source of Vitamins

The diet of ruminants is composed essentially of a combination of concentrates (cereals and protein supplements) and forages. Fresh or conserved forages are potential sources of vitamins A, E, D, niacin and thiamine (Aitken and Hankin, 1970; McDowell, 1989). However, although it is relatively easy to list vitamins present in forage, it is more difficult to give precise estimates of the mean values. Levels of vitamins found in forages are highly variable (Table 19.3). The factors that are responsible for this variability, and which include plant origin, climatic conditions, stage of maturity, conservation methods (drying, ensiling, dehydration ...) and storage conditions, are described in a later section. It would appear that the minimal attention given to the vitamin content of forages, other than for the vitamins A, E and, to a lesser extent, D, is the result of the lack of literature data concerning the content of niacin and thiamine, since vitamin analysis is not routinely performed on forages.

At first sight, the majority of forages can make a significant contribution to the dietary supply of β-carotene and α-tocopherol to cover the daily recommendations of ruminants (RPAN, 1998; Fig. 19.1). Indeed, only approximately 15% of the forages sampled supplied less than 50 mg of β-carotene or α-tocopherol kg^{-1} dry matter

Table 19.2. Vitamin recommendations for ruminants (RPAN, 1998).

Ruminants	Vit. A (IU $head^{-1}$ day^{-1})	Vit. E (mg $head^{-1}$ day^{-1})	Vit. D (IU $head^{-1}$ day^{-1})	Niacin (g $head^{-1}$ day^{-1})
Dairy cow, lactating	80,000–120,000	100–1,000	15,000–50,000	1–2
Dairy cow, dry	75,000–125,000	500–900	10,000–20,000	0–1
Finishing cattle	40,000–70,000	200–1,500*	4,000–7,000	1–2

* Improvement of meat quality after slaughter.

Table 19.3. Vitamin content of fresh and conserved forages (adapted from Blaylock *et al.*, 1950; Brown, 1953; Keener, 1954; Wallis *et al.*, 1958; Albonico and Fabris, 1958; Hjarde *et al.*, 1963; Hoffmann and Nehring, 1967; Bunnel *et al.*, 1968; Aitken and Hankin, 1970; NRC, 1989).

	β-Carotene ($mg\ kg^{-1}$ DM)			α-Tocopherol ($mg\ kg^{-1}$ DM)			D ($IU\ kg^{-1}$ DM)			Thiamine ($mg\ kg^{-1}$ DM)			Niacin ($mg\ kg^{-1}$ DM)		
Forages	*n*	M	SD	*n*	M	SD	*n*	M	SD	*n*	M	SD	*n*	M	SD
Green forages (1)	349	196 (15–606)*	108	86	161 (9–400)	91	25	365 (31–1800)	470	18	4.6 (1.9–8.3)	2.4	5	37 (13–56)	17
Dehydrated forages (2)	16	159 (66–271)	73	12	125 (28–238)	57	4	– (176–617)	–	7	4 (3.8–4.5)	0.5	7	53 (39–64)	9
Silages (3)	50	81 (2–276)	68	4	– (0–310)	–	10	440 (80–866)	311	1	0.1 –	–	2	– (1.1–34)	–
Hays (4)	68	36 (1–162)	34	10	61 (10–211)	62	40	1156 (90–5560)	1161	19	2.7 (0.2–4.5)	1.3	14	28 (6–52)	17

(1) Grasses (cocksfoot, timothy, ryegrass, fescue …) and legumes (lucerne, clover …); (2) lucerne; (3) legume silages (lucerne, clover), grass silages (cocksfoot, ryegrass …) and maize silage; (4) legume hays (lucerne, clover) and grass hays (timothy, ryegrass, fescue, cocksfoot …).
n, Number of samples; M, mean; SD, standard deviation.
* Range.

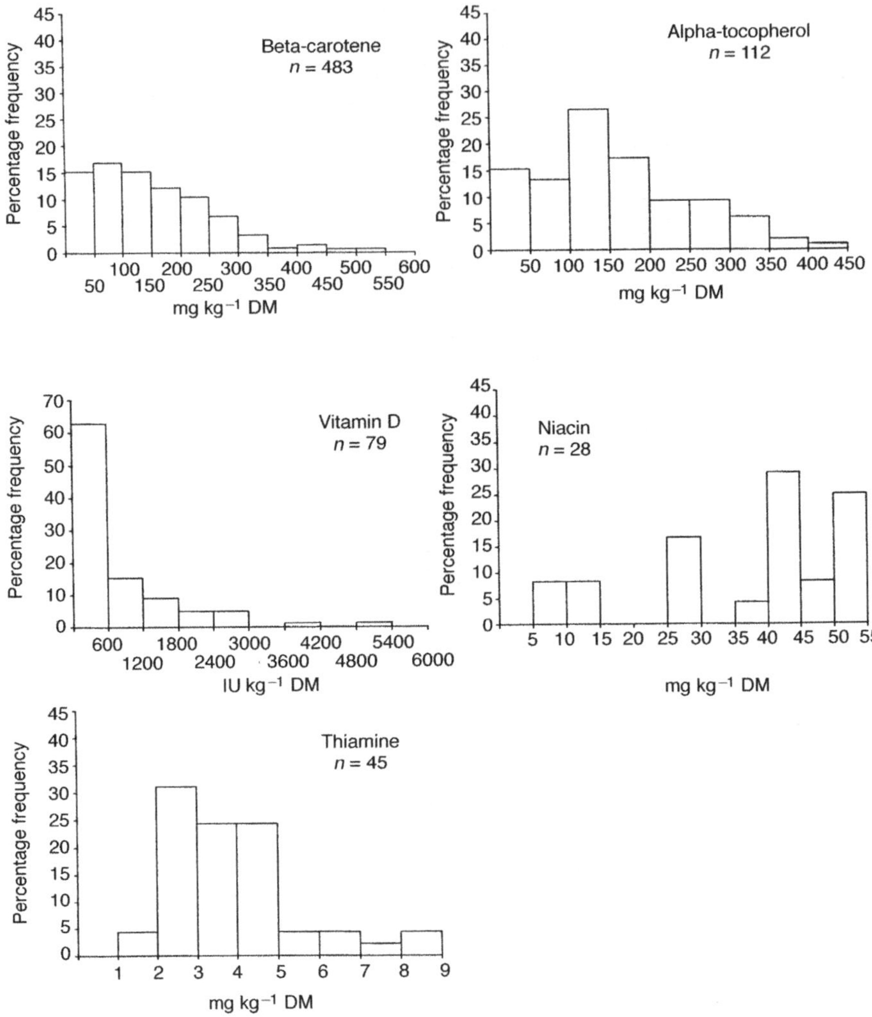

Fig. 19.1. Histograms representing frequency distribution of vitamin content of forages (adapted from Table 19.3). *n* = Number of samples.

(DM), the remaining 85% supplying in excess of 50 mg kg^{-1}. It would appear that forages, in general, are very good sources of the vitamins A and E. However, these data must be viewed with great prudence, due to lack of precision of most analytical methods for vitamins in forages and also to lack of estimates of the bioavailability of β-carotene and α-tocopherol in forages.

Methods of analysis used for feed materials have a tendency to overestimate concentrations of β-carotene and α-tocopherol in a wide range of raw materials (Aitken and Hankin, 1970; Ullrey, 1972; McDowell, 1989). The presence of geometric isomers of all-*trans*-β-carotene that have a lower biological activity results in an overestimation of the biological activity of β-carotene (Table 19.4). This overestimation is dependent on the forage under consideration (+6–11% for fresh lucerne and *Cynadon dactylon*, +13–17% for the same material dried and +19–33% for dehydrated lucerne).

In most cases, even if the concentration of β-carotene and α-tocopherol in the forage is high, the requirements for vitamins A and E of ruminants cannot be adequately covered by their daily forage consumption (Nadai, 1968; Ferrando and Mainguy, 1970; McDowell, 1989). Between the ingestion of β-carotene and α-tocopherol present in forage and the metabolic use of vitamins A and E, two important steps are involved: the passage of the material through the rumen and intestinal absorption, with, in addition, for carotene its conversion into vitamin A (see later). A number of factors can therefore modify the nutritional supply of vitamins A and E from forage. The level of conversion of β-carotene varies (1 mg of β-carotene = 400 IU of vitamin A in a lactating dairy cow) according to the health status and the quantity consumed. The rate of conversion is reduced as the level consumed increases, and it is generally accepted that this conversion factor for ruminants is only valid at levels of intake corresponding to maintenance (Nadai, 1968; Bondi and Sklan, 1984).

Over 60% of the forages analysed (Table 19.3) had a vitamin D content of less than 600 IU kg^{-1} dry matter (Fig. 19.1), resulting in a severe risk of deficiency in housed animals. Niacin and thiamine, although found in forages, are present in quantities insufficient to cover recommendations (RPAN, 1998). The same caution as was given for vitamins A and E must be exercised in estimating the nutritional supply of vitamins D, niacin and thiamine from forages, in particular the

Table 19.4. Stereoisomers of β-carotene and biological efficiency of lucerne (fresh, dehydrated and sun-cured) and *Cynadon dactylon* (fresh and dried at 65°C).

Forages	Stereoisomers of β-carotene (%)			Biological efficiency of total carotene (% of all-*trans* β-carotene)	References
	Neo-B	All-*trans*	Neo-U		
Fresh lucerne	7	84	9	90	Livingston *et al.* (1966)
	4.6	90	5.4	94	Thompson *et al.* (1951)
	6	84	10	91	Kemmerer *et al.* (1944)
Dehydrated lucerne					
T = 104°C*	27	63	10	81	Livingston *et al.* (1966)
T = 160°C	46	36	18	67	
Sun-cured lucerne hay	12	77	11	87	Thompson *et al.* (1951)
Fresh *Cynadon dactylon*	8	81	11	89	Kemmerer *et al.* (1944)
	5	86	9	92	
Cynadon dactylon dried at 65°C	11	70	19	83	Kemmerer *et al.* (1944)

* Outlet temperature of drier.

sensitivity of different analytical techniques and effects on biological availability (McDowell, 1989).

In conclusion, therefore, although forages are potentially good sources of vitamins, the levels which they may contribute in the diet of ruminants are influenced by a number of factors inherent to both the plant and the animal. This range of influencing factors results in a major lack of precision in estimating the quantity of vitamins available to ruminants from forage.

Factors Affecting the Vitamin Content of Forages

The considerable variability found in the vitamin content of forages tends to suggest that a number of factors contribute to the variability encountered. These include:

- the origin of the plant (family, species and variety);
- climatic conditions;
- stage of maturity of the plant;
- conservation methods (drying, ensiling, dehydration …);
- storage conditions.

The origin of the plant

During the early stages of growth, the grasses and legumes have similar mean levels of β-carotene (approximately 300 mg kg^{-1} DM) (Table 19.5). However, at the stage of flowering and at maturity, legumes are richer in β-carotene than the grasses (Moon, 1939). Thus, the stage of growth does not affect the level of β-carotene of legumes to the same extent as in the grasses (Moon, 1939; Seshan and Sen, 1942a). In a comparison of grasses and legumes at the start of the flowering stage (Table 19.6), Livingston *et al.* (1968c) reported that the leaves had similar levels of carotene, but these levels were four to 11 times higher than the levels found in the stems (Seshan and Sen, 1942a; Olsson *et al.*, 1955; Ramanujan and Anantakrishnan, 1958; Park *et al.*, 1983). However, account must be taken of the fact that, at the start of flowering, the proportion of leaves in legumes (mean 40%) is twice that found in the grasses (mean 19%). Thus, at advanced stages of growth, the legumes are the better source of β-carotene, since the proportion of stem is less than in the grasses (Moon, 1939; Olsson *et al.*, 1955).

Results presented by Olsson *et al.* (1955) also indicate the existence of differences between plant species in terms of β-carotene (Moon, 1939; Seshan and Sen, 1942a; Park *et al.*, 1983). At an early growth stage (1st cycle) (Table 19.7) perennial ryegrass has the lowest level of β-carotene of the grasses (146 mg kg^{-1} DM)(Smith and Wang, 1941) and, of the legumes, white clover has the highest level (438 mg kg^{-1} DM). Thus, certain plant species possess an important capacity to synthesize carotene (Park *et al.*, 1983). During the course of a study, Olsson *et al.* (1955) noted that, at a stage of full bud, red clover had a concentration of β-carotene (206 mg kg^{-1} DM) close to that of lucerne (233 mg kg^{-1} DM) and that,

Table 19.5. Influence of plant origin and stage of maturity in β-carotene and α-tocopherol contents of fresh forages (mg kg^{-1} DM) during the first growth cycle (adapted from Brown, 1953; Hjarde *et al.*, 1963; Hoffmann and Nehring, 1967; Aitken and Hankin, 1970; NRC, 1989).

	β-carotene				α-tocopherol			
Forages and stages	*n*	M	Max.	Min.	*n*	M	Max.	Min.
Grasses								
Vegetative to ear	51	278	606	84	29	253	400	121
Early to end flowering	44	133	258	53	17	98	154	40
Mature	30	59	156	4	7	22	30	9
Legumes								
Vegetative to bud	62	309	552	140	19	129	202	79
Early to end flowering	34	192	488	97	6	116	127	109
Mature	7	130	252	80	–	–		

n, Number of samples; M, mean.

Table 19.6. Variation in carotene content of four grasses and three legumes harvested at early flowering during the first growth cycle (from Livingston *et al.*, 1968c).

Forages	Date harvested	Fraction	% (dry wt)	Carotene (mg kg^{-1} DM)
Grasses				
Fescue	18/6	Leaf	12	460
		Stem	88	101
Timothy cv. Verdant	2/7	Leaf	17	387
		Stem	83	95
Cocksfoot cv. Potomac	20/6	Leaf	21	521
		Stem	79	62
Reed canary grass	18/6	Leaf	26	640
		Stem	74	90
Legumes				
Red clover cv. Lakeland	14/6	Leaf	34	724
		Stem	66	66
Lucerne cv. Saranac	9/6	Leaf	39	629
		Stem	61	75
Birdfoot cv. Empire	1/6	Leaf	42	449
		Stem	58	62

at the stage of end ear, timothy (96 mg kg^{-1} DM) had a lower level than that of cocksfoot (175 mg kg^{-1} DM). Table 19.7 also shows that the growth cycle has little effect on the level of β-carotene in lucerne (Bruhn and Oliver, 1978) and in red clover, but strongly influences that in white clover and the grasses. According to these authors, the differences in level of β-carotene are mainly due to the ratio of leaf to stem in the plant and the regrowth.

Table 19.7. β-Carotene content (mg kg^{-1} DM) of legumes and grasses harvested at an early stage of development several times in the vegetation period, in years 1942–1946 (number of samples = 5) (from Olsson *et al.*,1955).

	First cut			Second cut			Third cut			Fourth cut		
Forages	M	Max.	Min.	M	Max.	Min.	M	Max.	Min.	M	Max.	Min.
Legumes												
Lucerne	263	315	235	239	365	172	272	380	225	293	424	244
Red clover	356	462	270	313	362	256	330	391	253	335	411	245
White clover	438	538	387	269	395	184	268	463	177	422	595	342
Grasses												
Timothy	238	369	133	128	222	63	192	233	119	248	356	142
Cocksfoot*	242	267	217	148	191	104	104	–	–	254	331	177
Perennial ryegrass	146	214	95	161	407	68	250	425	123	207	259	160

* Number of samples = 2 (1943 and 1944).
M, mean.

At early stages of growth, grasses contain nearly twice as much α-tocopherol as legumes (Table 19.5). However, at flowering, the legumes and grasses have approximately the same mean levels of α-tocopherol. Similar to the case with carotene, the ratio of leaf to stem is an important factor in the evolution of the level of vitamin E in forages (Brown, 1953), with the leaves being richer in α-tocopherol than the stems (Brown, 1953; Ramanujan and Anantakrishnan, 1958; Booth, 1964). Variability between species is also evident for α-tocopherol. Cocksfoot (313–362 mg kg^{-1} DM) is a better source of α-tocopherol than fescue and timothy (184–249 mg kg^{-1} DM) at the 20–25 cm stage (Brown, 1953). Furthermore, for thiamine, fescue is superior to brome grass and timothy (3.2 mg and 2.6 mg kg^{-1} DM, respectively) and lucerne is superior to clover (7.5 mg and 6.5 mg kg^{-1} DM, respectively) (Robinson *et al.*, 1948).

Brown (1953) reported approximately 22% variation in the level of α-tocopherol when comparing four varieties of lucerne. Thompson (1949) observed a variation of the order of 30% in the level of β-carotene in nine varieties of lucerne. However, comparing three varieties of red clover, Hjarde *et al.* (1963) found no significant difference in the levels of β-carotene and α-ζ-tocopherol. Vail *et al.* (1936), cited by Robinson *et al.* (1948), found no significant difference in the level of thiamine in four varieties of lucerne. Albonico and Fabris (1958) concluded that genetic factors should be considered as having only a weak influence on the vitamin content of forages.

In conclusion, plants that produce little leaf tend to contain low levels of β-carotene and α-tocopherol (Park *et al.*, 1983; McDowell, 1989). Other than the origin of the plant, the ratio of leaf to stem, which is the main factor influencing the levels of β-carotene and α-tocopherol in forage, is also influenced by both climatic conditions and the stage of maturity (Olsson *et al.*, 1955).

Climatic conditions

Bondi and Meyer (1946) measured the level of carotene in lucerne harvested at the budding and flowering stages at different times of the year in Palestine. The level of β-carotene was higher in April and May for the budding stage (266–273 mg kg^{-1} DM) and in November for the flowering stage (170 mg kg^{-1} DM) than in August (110 mg kg^{-1} DM for the budding stage and 72 mg kg^{-1} DM for the flowering stage). Albonico and Fabris (1958) demonstrated that the levels of β-carotene and α-tocopherol of 16 varieties of lucerne sampled in the south of Italy at the beginning of flowering were higher in March/April (349 and 282 mg kg^{-1} DM, respectively) than in July (251 and 110 mg kg^{-1} DM, respectively). Independently of the cycle of growth, the level of β-carotene of legumes and grasses was higher in the summer of 1942 – a fresh and humid summer – than in 1943 and 1944 – two summers that were hot and dry (Table 19.8; Olsson *et al.*, 1955). These results indicate that, for a given stage of growth and independently of the cycle of growth, a forage is richer in β-carotene and α-tocopherol when grown under mild, wet conditions. The beneficial effect of rainy conditions rests in the reduction in the quantity of sunlight received by the plant (Olsson *et al.*, 1955; Hjarde *et al.*, 1963; Park *et al.*, 1983). The ambient temperature and the light contribute more to the variability in the level of β-carotene than the supply of water (Olsson *et al.*, 1955; Repp and Walkins, 1958).

Beck and Redman (1940) showed that heat and intense light were detrimental to the production of carotene in the leaves of clover. However, the positive influence of the reduction in temperature and day length was also related to the increase in the leaf-to-stem ratio (Wilson, 1981). Thus the climatic conditions influence the levels of β-carotene and α-tocopherol of forages because they also influence the leaf-to-stem ratio.

The antirachitic value of forages is also dependent on levels of sunlight (Meissonier, 1981; McDowell, 1989). Green forages have low levels of vitamin D but, when the plant begins to die and the fading leaves are exposed to the ultraviolet light of the sun, high levels of vitamin D are synthesized (Thomas and Moore, 1948; McDowell, 1989).

Table 19.8. β-Carotene content (mg kg^{-1} DM) of grasses and legumes harvested at an early stage of growth during the summer, 1942–1944, in Sweden (from Olsson *et al.*, 1955).

	White clover		Red clover		Timothy		Perennial ryegrass	
Years	First cut	Second cut	First cut	Second cut	First cut	Second cut	First cut	Second cut
1942	538	395	462	362	369	222	214	147
1943	408	210	355	256	209	87	97	100
1944	387	225	270	298	133	63	95	68

The stage of maturity of the plant

The levels of β-carotene and α-tocopherol in the grasses and legumes are very high in the young stages and reduce as the plant matures (Smith and Wang, 1941; Brown, 1953; Clarke, 1953; Bondi and Sklan, 1984; Table 19.5). The levels of β-carotene and tocopherols of lucerne achieve maximum values between the end of the vegetative stage (Park *et al.*, 1983) and the budding stage (Burrows and King, 1968) and then decrease with maturity. The level of carotene drops at the end of the vegetative stage in clover (Olsson *et al.*, 1955) and cocksfoot (Clarke, 1953) and after the start of earing in timothy (Olsson *et al.*, 1955). According to Moon (1939) and Livingston *et al.* (1968c), the level of carotene in the leaves of grasses and legumes is maximal at the start of flowering and then diminishes with varying degrees of intensity (Thompson, 1949). Indeed, the green colour of their leaves is generally a good index of their carotene content (Maynard *et al.*, 1979). At maturity, plants may have 10% (in the case of grasses) to 40% (for legumes) of the value of carotene of immature plants.

As has already been indicated, the principal factor responsible for the variation in levels of β-carotene and α-tocopherol of forages in the course of their maturation is the change in the ratio of leaf to stem, because the leaves are considerably richer in these vitamins than the stems. The formation of stems is accompanied by an increase in the concentration of DM; there is thus a negative correlation between the DM content and the level of β-carotene (Olsson *et al.*, 1955).

Contrary to the situation when vitamin D is produced in fading leaves, the level of thiamine is positively correlated with the verdant nature of the forage leaf (Hunt *et al.*, 1935; Galgan *et al.*, 1950; McDowell, 1989). The vitamin D content of forages is also dependent on the stage of harvest. Indeed, vitamin D level increases even when the plant has reached maturity (Table 19.9; Keener, 1954), because of the increase in dead leaves (Thomas and Moore, 1951; McDowell, 1989). Thus, the levels of β-carotene and vitamin D are negatively correlated. Thomas and Moore (1948) demonstrated that the proportion of dead leaves in lucerne increases from 2.4% at budding to 6.5% at maturity. The dead leaves contain approximately 7064 IU vitamin D kg^{-1} DM, while the green leaves had levels close to zero. Thus, the quantity of dead leaves adhering to the plant during harvest can considerably modify the level of vitamin D, independently of the duration of exposure to sunlight during drying. During the same year, the level of vitamin D of forages at the same stage of development increases with the number of cuts (Keener, 1954; Table 19.9). It was suggested that the difference between a first and second cut was due to differences in the level of ergosterol and the relative importance of the sun's ultraviolet rays.

Methods of harvesting and conservation

The manner in which forage is treated between its harvest and being offered as feed to animals can influence its vitamin content.

Both β-carotene and α-tocopherol are destroyed by oxidation. This reaction is accelerated by ultraviolet light and heat (Seshan and Sen, 1942b; Bauernfeind,

Table 19.9. Influence of stage of maturity and number of cuts in vitamin D content of several forages (from Keener, 1954).

Cutting and date	Red clover (IU kg^{-1} DM)	Timothy (IU kg^{-1} DM)
First cutting		
25 June	31	40
2 August	750	571
15 September	1400	1100
Second cutting		
2 August	130	430
15 September	981	820
Third cutting		
15 September	500	540

1980). However, prolonged heating without oxygen has a minor effect (Seshan and Sen, 1942b; McDowell, 1989).

Before it becomes a photochemical process, the destruction of β-carotene in forages is first an enzymatic process and is due to a lipoxygenase system (Ferrando and Mainguy, 1970; Kalac and McDonald, 1981; Bondi and Sklan, 1984). The lipoxygenases, a group of isoenzymes, are present at varying levels in a wide number of species of plants (Waugh *et al.*, 1944; Larsen *et al.*, 1993). Lucerne is know to be an excellent source of lipoxygenases (Waugh *et al.*, 1944; McDowell, 1989). These enzymes are destroyed at around 80–100°C (Olsson *et al.*, 1955; Ferrando and Mainguy, 1970), wherein lies the interest of dehydration (Blaylock *et al.*, 1950; Bondi and Sklan, 1984).

According to Bauernfeind (1972), the enzymatic destruction of total pigments present in the leaves of lucerne is three to four times greater than photochemical destruction (Mitchell and Hauge, 1946). Enzymatic destruction of carotene begins when the forage is chopped and is more intense in the early stages of growth. Increasing the degree of chopping and grinding also accelerates enzymatic destruction, as does raising ambient temperature and humidity (Mitchell and Hauge, 1946; Ferrando and Mainguy, 1970; McDowell, 1989). During drying, the losses of β-carotene and α-tocopherol can equally be due to isomerization (Livingston *et al.*, 1966, 1970), but this isomerization represents only a small part of the total loss (Kalac and Kyzlink, 1979).

Vitamin D can be destroyed by excessive exposure to ultraviolet light (Ferrando and Mainguy, 1970; McDowell, 1989). When the crop is dry, thiamine is stable at 100°C for several hours, but the presence of humidity accelerates destruction (McDowell, 1989). Thus, thiamine is much less stable to heat in fresh material than in dried material. On the contrary, niacin is highly stable in air, to heat, to light and to the action of alkalis – hence its stability in feedstuffs (McDowell, 1989).

Haymaking

Drying crops either on the ground or in barns reduces the levels of β-carotene and α-tocopherol in forages (King *et al.*, 1967; Burrows and King, 1968; McDowell, 1989). Russell (1929) found that in excess of 80% of carotene from fresh lucerne was lost during the first 24 h of sun-drying. According to Akopyan (1958), cited by Kivimaë and Carpena (1973), the level of tocopherol in leaves of clover and maize was practically zero when the crop was dried for 4–5 days in the sun. With grasses, Ramanujan and Anantakrishnan (1958) obtained carotene and tocopherol losses in the region of 80% after 4 days of sun-drying. Park *et al.* (1983) showed that, as the exposure to the sun was increased, the level of destruction of β-carotene was increased (Table 19.10) and that this occurred irrespective of the stage of harvest of the lucerne. Forages exposed to rain and then dried in the sun have less β-carotene than sun-dried forage. Thus, if the forage rests exposed to the sun for an extended period of time and at the same time is exposed to several showers, the destruction of β-carotene is nearly complete.

According to Akopyan (1958) and Maynard *et al.* (1979), barn-drying has less of a destructive effect on carotene and tocopherol. Galgan *et al.* (1950) reported that barn-dried lucerne hay could contain three times more carotene than sun-dried lucerne hay. However, Ramanujan and Anantakrishnan (1958) reported that, when the drying time of grasses was considerably extended, irrespective of the method, the losses of carotene and tocopherol were practically the same.

During crop-drying on the ground, radiation from the sun can result in synthesis of vitamin D. However, again, it must be emphasized that the degree of radiation must not be excessive (Ferrando and Mainguy, 1970). Thomas and Moore (1948) showed that a sun-dried lucerne contained twice as much vitamin D (971 IU kg^{-1} DM) as a barn-dried lucerne (470 IU kg^{-1} DM).

The level of thiamine is much lower in sun-dried materials than in materials dried in an oven at 62°C (Galgan *et al.*, 1950). The lucerne dried in the sun had high losses of niacin when it was exposed to bad weather (Blaylock *et al.*, 1950; Scott, 1973).

Table 19.10. Effect of time of sun-drying (with or without rain) in β-carotene content of lucerne harvested at three stages of maturity during the second growth cycle (from Park *et al.*, 1983).

Duration of drying (h)	Vegetative (mg kg^{-1} DM)	Bud (mg kg^{-1} DM)	Flowering (mg kg^{-1} DM)
Sun-dry			
48 h	79.0	56.8	53.0
72 h	34.4	24.4	24.2
96 h	29.6	8.1	21.8
Rain damage, sun-dry			
48 h	22.2	22.0	22.6
72 h	13.9	12.0	23.6
96 h	12.7	5.3	10.8

Ensiling

In general, ensiling of grasses and legumes guarantees a better conservation of carotene and tocopherol than haymaking (Bauernfeind, 1980; Wolter, 1988; Hidiroglou *et al.*, 1994). As a result of a series of studies, Watson and Nash (1960) concluded that mean losses of carotene during ensilage were of the order of 30% (cited by Kalac and McDonald, 1981). It should be noted that the losses of β-carotene depend partly on the type of material ensiled, the quality of the ensiling process and the method of silage preparation. Kalac (1983) demonstrated that the losses of β-carotene during the aerobic phase (24 h) were of the order of 17 (cocksfoot) to 30% (lucerne and Italian ryegrass) (Table 19.11) and for maize approximately 19% after 48 h (Kalac and Kyzlink, 1980).

Either wilting prior to ensiling, which often precedes the ensiling of a forage, or the use of additives, such as organic acids (acetic, formic and propionic acids), tends to increase the loss of β-carotene during the preliminary aerobic phase (Kalac and Kyzlink, 1979, 1980; Kalac and McDonald, 1981). Acidification considerably increased the losses of β-carotene in lucerne, clover and maize, but did not have an effect on either Italian ryegrass or cocksfoot (Kalac and Kyzlink, 1980; Kalac, 1983). The destruction of β-carotene under acidic conditions is of enzymatic origin and results from an oxyreductase other than lipoxyreductases (Kalac and Kyzlink, 1980). The losses of β-carotene during fermentation differ according to the forage (Kalac, 1983), and are high for clover and lucerne (Patel *et al.*, 1966) and low for cocksfoot and Italian ryegrass (Table 19.11).

Ferrando and Mainguy (1970) and Kalac (1983) reported that there was no clear relationship between the quality of a silage and its carotene content.

Table 19.11. Changes in β-carotene content and pH of several types of silages (vegetative stage) (from Kalac, 1983).

			β-Carotene (% initial value), pH and quality of silage					
			Delayed sealing (h)			After fermentation and 180 days' storage		
Forages	Silage treatment	β-Carotene in fresh leaves (mg kg^{-1} fresh)	6 (%)	24 (%)	24 (pH)	(%)	(pH)	Quality
Cocksfoot	C	127	97.4	83.1	5.15	81.8	4.70	II
	F	–	92.7	80.8	4.30	76.3	4.40	II
Italian ryegrass	C	196	–	72.2	5.75	70.0	5.15	IV
	F	–	–	77.6	4.10	73.6	4.90	III
Lucerne	C	186	83.6	70.3	5.95	40.2	5.15	V
	F	–	47.0	11.3	4.35	8.4	5.25	IV
White clover	C	237	92.1	75.7	5.95	53.7	5.80	II
	F	–	69.7	48.6	5.10	28.6	5.30	III

C, control; F, formic acid.
Organoleptic assessment of quality: II, good; III, mean; IV, poor; V, very poor.

Bieber-Wlaschny (1988) indicated that grass silage could provide considerable levels of β-carotene (122 mg kg^{-1} DM), but that maize silage is a very poor source (11 mg kg^{-1} DM); lucerne silage (32 mg kg^{-1} DM) and clover silage (19 mg kg^{-1} DM) were intermediate.

Maize silage is practically devoid of vitamin E (Hidiroglou *et al.*, 1977; Bieber-Wlaschny, 1988). On the other hand, the ensiling of lucerne and brome preserves a major proportion of the vitamin E (King *et al.*, 1967).

The process of ensiling is therefore characterized by introducing great variability in the preservation of the β-carotene and α-tocopherol content of forages.

Dehydration

Rapid dehydration allows a major proportion of the β-carotene and α-tocopherol of forage plants to be conserved (Table 19.12). Dehydrated forages contain, on average, seven times the amount of carotene compared with when they are dried on the ground (Meissonier, 1974). The level of residual moisture of the treated material has a direct effect on the destruction of β-carotene and α-tocopherol (Ferrando and Mainguy, 1970; Burdick and Fletcher, 1985). Livingston *et al.* (1968a, b, 1970) demonstrated that the losses of β-carotene and α-tocopherol vary from 0 to 30% during the course of dehydration of lucerne. They observed that the highest losses occurred when the level of residual moisture was equal to or lower than 30 g kg^{-1}. The optimum moisture content appeared to be between 70 and 100 g kg^{-1} (Livingston *et al.*, 1970). Dehydration had no effect on losses of niacin (Blaylock *et al.*, 1950).

Storage conditions

The highest vitamin loss in absolute terms is observed in forages that are well preserved during their preparation, as in dehydrated materials (Park *et al.*, 1983). Bruhn and Oliver (1978) demonstrated that the losses of β-carotene and α-tocopherol in lucerne hay are significantly correlated with the duration of storage.

Table 19.12. β-Carotene and α-tocopherol losses of lucerne during dehydration.

Outlet temperature (°C)	Residual moisture of product (g kg^{-1})	Losses (%)		References
		Carotene	Tocopherol	
153	92	9	5	Knowles *et al.* (1968)
149	92	9	5	Livingston *et al.* (1968a, b)
166	23	18	21	
116–166	15–122	0–33	–	Livingston *et al.* (1970)
–	80	15	–	Ferrer (1972)

Kohler *et al.* (1955) demonstrated that losses in tocopherol during storage also depended on temperature; a ryegrass meal stored for 6 weeks at 3°C lost 8% of its tocopherol but the loss was of the order of 49% at 60°C (Maynard *et al.*, 1979). The conservation of dehydrated lucerne at ambient temperature (32°C) during 3 months results in considerable loss of carotene (42–72%) and tocopherol (35–73%) (Livingston *et al.*, 1968b, 1970). Thus, the storage at low temperature allows the maintenance of reasonable levels of β-carotene and α-tocopherol (Kivimäe and Carpena, 1973; Park *et al.*, 1983). The storage of dehydrated lucerne in a nitrogen atmosphere allows the duration of storage to be extended without any adverse effects on the quality of β-carotene (Melcion and Delort-Laval, 1973; Park *et al.*, 1983). Irrespective of the procedure of conservation, the carotene degrades slowly during storage; the loss is of the order of 50% in the first 6 months, but it can be greater (Fig. 19.2).

A second degradation of the carotene and tocopherol present after the period of conservation is known to occur during the time that forage is held in the feed bunker, prior to its distribution (McDowell, 1989).

Factors Affecting Vitamin Availability in Ruminants

The metabolic utilization of vitamins supplied by forages is not possible without transit through the digestive tract of the animal. The two most important factors limiting vitamin availability in ruminants are passage through the rumen and intestinal absorption. The act of digestion, principally at the level of the rumen,

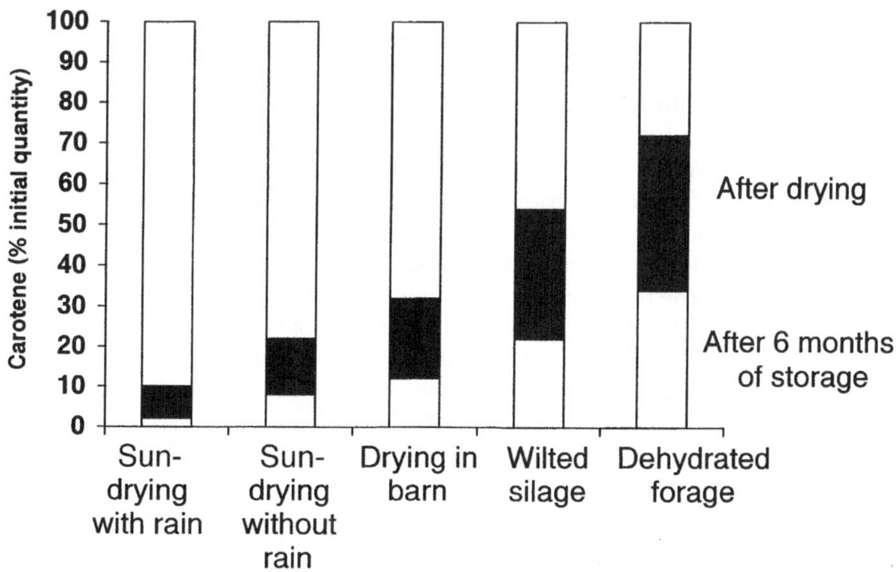

Fig. 19.2. Relative availability of carotene in comparison with its initial content in forages (from Wolter, 1988).

serves to release the vitamins from the combination in which they are found in plant tissue. Thus, the extent of degradation and liberation of vitamins from the forage in the rumen is an essential feature of the bioavailability of vitamins found in forage. According to Nadai (1968), the degree of liberation of carotene in the digestive tract varies according to the type of plant. This may explain why the carotene in green or dehydrated lucerne is more available than that from maize and, in particular, maize silage.

The variability of vitamin supply in forages identified in the foregoing section is one of the reasons why the ruminant-livestock industry has chosen the practice of meeting requirements from synthetic sources. In ration formulation, the supply of vitamins from raw materials is often ignored and the requirements for supplementation with synthetic vitamins are based on meeting 100% of the requirements. Most of the studies concerning the stability in the rumen and intestinal absorption of β-carotene and α-tocopherol have been carried out using synthetic sources. Certain comments on ruminal stability and intestinal absorption of vitamins A, E and D are therefore relevant to this discussion. Factors that need to be considered are the influence of rumen conditions on the availability of vitamins and also intestinal absorption. However, it is beyond the scope of this review to cover the ruminal stability and intestinal absorption of all vitamins and hence comments will be confined to vitamins A, E and D.

Vitamin A

Synthetic vitamin A is produced in the form of the ester, retinyl acetate, in the all-*trans*-isomer form, which has a biological activity of 100% (McDowell, 1989). Vitamin A is sensitive to the action of light, heat and oxidizing agents, as a result of the presence of the unsaturated side-chain. For this reason, synthetic vitamin A is found in encapsulated forms, in order to provide protection against the adverse effects of light and oxidation.

Degradation of vitamin A in the rumen and intestinal absorption

Many biochemical reactions that occur in the rumen are capable of destroying vitamin A, and a large number of studies have provided evidence of major losses of vitamin A in the rumen. These losses may be explained by different phenomena involving the rumen bacteria, including engulfment, oxidation and degradation. To date, no author has demonstrated absorption of vitamin A through the rumen wall, which would appear logical, given the vitamin's large molecular size.

Ruminal losses of β-carotene are, on average, 20%, with a range from 3 to 32% (Table 19.13). Potkanski *et al.* (1974) found that these losses were not affected by diet. Although Keating *et al.* (1964) showed that the addition of an antioxidant (ethoxyquin) or oxidizing agent (potassium nitrate) had no effect on the β-carotene disappearance, Cohen-Fernandez *et al.* (1976) suggested that the destruction of β-carotene could be due to partial oxidation. This contradicts the conclusions of Mitchell *et al.* (1967) and Lichtenwalner *et al.* (1973), who reported that degradation of vitamin A in the rumen is not significantly modified when oxidizing agents, including nitrates, were added to the rumen (Table 19.14). Given that the rumen is

Table 19.13. Ruminal disappearance of β-carotene.

Animal	Diet	Disappearance (%)	Method	References
–	–	31.9	*In vitro* 9 h	King *et al.* (1962)
Dairy heifer	High-hay	25.2 ± 5.6	*In vitro* 7 h	Davison and Seo (1963)
Steer	High-roughage	24.4	*In vitro* 16 h	Keating *et al.* (1964)
Mature wether	High-cellulose	23.1 ± 5.9	*In vivo*	Potkanski *et al.* (1974)
	High-starch	23.3 ± 6.9		
Cow	High-concentrate	4.7 ± 1.7	*In vitro* 24 h	Cohen-Fernandez w*et al.* (1976)*

* β-Carotene from different sources, pure β-carotene or commercial lucerne meal.

Table 19.14. Vitamin A recovery from abomasal fluid of beef cattle (%)* (from Mitchell *et al.*, 1967).

Steers no.	Observation no.	Supplement		Mean
		Control	Nitrate[†]	
1	1	43.1[‡]	62.2	
	2	32.4	46.2	
	3	44.6	55.8	
	4	69.1	32.2	48.2
3	1	41.8	45.1	
	2	35.2	39.5	
	3	42.1	66.4	
	4	35.5	55.7	45.7
Mean		43.0	50.4	

* Calculated from the ratio of vitamin A to chromic oxide in abomasal fluid compared with the ratio administered 24 h later.
[†] 90 mg of potassium nitrate steer^{-1} day^{-1}.
[‡] Each observation is the average result of triplicate determinations.

a highly reducing environment, it is unlikely that oxidation plays a major role. It was therefore concluded that β-carotene disappearance in the rumen was due mainly to the action of bacterial hydrogenases (Ferrando, 1980) and it is probable that vitamin A undergoes reductive degradation. In *in vitro* studies, the highest rate of degradation of vitamin A incubated for 4 h in either rumen fluid or autoclaved rumen fluid was in the presence of an active rumen microflora (Table 19.15; Klatte *et al.*, 1964). It was concluded that the degradation of vitamin A was due to the action of the microflora, although it was accepted that autoclaving could have limited other biological functions. Rode *et al.* (1990) showed that nine different types

Table 19.15. Disappearance of vitamin A after 4 h in distilled water or ruminal fluid (from Klatte *et al.*, 1964).

Incubation fluid	Number of assays	Disappearance (%)
Distilled water	11	16
Rumen liquor	11	36
Autoclaved rumen liquor	2	13

of bacteria were capable of transforming retinol propionate without inducing an accumulation of retinol, which is the only form of vitamin A that can be absorbed (Table 19.16).

Rode *et al.* (1990) measured the disappearance of vitamin A from the rumen of steers given diets based on concentrates, hay or straw and found levels of degradation of 67, 16 and 19% respectively. Weiss *et al.* (1995) confirmed *in vitro* that degradation of retinyl acetate was higher (72%) with rumen fluid obtained from a concentrate-type fermentation, in comparison with fluid obtained when the diet was based on forage (20%).

Using measurements of rumen flow in steers (Mitchell *et al.*, 1967) and sheep (Brongniart, 1987), the degradation of vitamin A *in vivo* was shown to be of the order of 60%. From a review of the literature, a mean rate of degradation of vitamin A in the rumen of 66% was found (Table 19.17). Similar losses of vitamin A from the rumen after intraruminal administration of vitamin A were also reported by Granseigne and Robert (1985) (Fig. 19.3). The rate of degradation of the two esters (acetate and palmitate) was highest in the first 2 h and then tended towards a maximum of 60%. Retinyl palmitate appeared to be less degradable than the acetate form.

Intestinal absorption of fat-soluble vitamins is dependent on the digestion and absorption of fat (Ullrey, 1972; McDowell, 1989). Vitamin A is absorbed in the form of retinol via the same mechanism as lipids, requiring the activity of bile salts,

Table 19.16. Degradation of vitamin A by rumen bacteria (from Rode *et al.*, 1990).

	Retinyl propionate degradation (%)			Retinol accumulation (%)		
Strains	18 h	42 h	66 h	18 h	42 h	66 h
Ruminobacter amylophilus WP109	13	17	30	0	0	0
Ruminobacter amylophilus WP91	67	76	85	0	5	0
Eubacterium ruminantium G1–1(1)	37	45	37	2	3	5
Fusobacterium necrophorum 1A	18	50	69	0	0	0
Fibrobacter succinogenes D	9	26	27	0	0	0
Ruminococcus flavefaciens 4	0	69	63	0	10	5
Propionibacterium acnes ATC6919	48	48	73	0	0	0
Butyrivibrio fibrisolvens XBF	48	95	96	0	0	0
Megasphaera elsdenii AW106	46	59	46	0	0	0

Table 19.17. Measurements of ruminal degradation of vitamin A *in vivo*.

Animal	Diet	Degradation (%)	References
Mature steer	High-roughage	57 (31–67)	Mitchell *et al.* (1967)
Steer	High-hay	73.9 (70–80)	Mitchell *et al.* (1968)
Mature steer	20% concentrate	52	Warner *et al.* (1970)
	40% concentrate	56	
	60% concentrate	70	
	80% concentrate	62	
Sheep	High-concentrate	64 ± 9	Brongniart (1987)

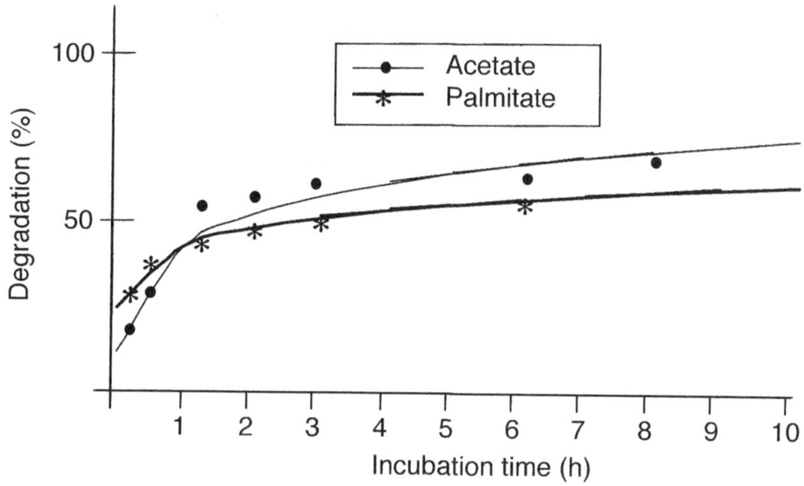

Fig. 19.3. Degradation curve of two esters of vitamin A in the rumen of sheep offered a concentrate ration (from Granseigne and Robert, 1985).

and thus any perturbation of lipid absorption results in a similar perturbation in vitamin A absorption (Bondi and Sklan, 1984; McDowell, 1989). Vitamin A transport across the brush border is not dependent on a membrane transport system but is dependent on the physicochemical characteristics of the surface contact of the molecule with the surface of the enterocyte. Absorption of vitamin A at the level of the intestine was found to be 30–50% of that ingested (Olson, 1984). This figure is in agreement with the level of degradation of 60% in the rumen, on condition that there is no further loss during intestinal absorption.

Ferrando (1980) reported that only 10% of β-carotene from forages would be used by ruminants. The low utilization of β-carotene from forages was due to its low solubilization and hence limited absorption (Bondi and Sklan, 1984), rather than its ruminal degradation. Indeed, Potkanski (1974), cited by Kalac and McDonald (1981), found that postabomasal absorption of carotene was only 5–13% of total carotene intake. Furthermore, intestinal absorption of carotene was reduced as the level consumed increased (Olson, 1984). Following absorption, β-carotene is totally converted into vitamin A in the enterocytes of sheep and goats, but only partially converted in the enterocytes of the bovine. Eaton *et al.* (1959) found that differences existed between breeds in ability to convert carotene into vitamin A. The Holstein breed converted carotene into vitamin A 1.4 times more efficiently than the Guernsey breed.

A particular characteristic of the metabolism of vitamin A is that it is stored in the liver in the A1-all-*trans* esterified form by fatty acids, essentially palmitic acid. The liver may contain 90% of the total vitamin A in the organism and 30–60% of the quantity absorbed each day may be stored in the liver.

Studies in sheep on liver storage of vitamin A as a function of elevated dietary supply showed that the coefficient of liver storage, (vitamin A in the liver/vitamin A ingested) × 100, was of the order of 30% for supplementation with 20 IU g^{-1} of feed (Robert, 1987; Table 19.18). This value is relatively low in comparison with the coefficient of storage of 56% observed in monogastrics (Uzu, 1988), but is in accord with a substantial rumen degradation of vitamin A. In this study, it was also reported that coefficients of liver storage were reduced when levels of supplementation were increased. This could be explained by saturation of either the absorption pathway or the capacity of the liver to store vitamin A or a combination of the two.

During manufacture, synthetic vitamin A is normally encapsulated to guarantee stability against oxidation during fabrication of premixes and compound feed. It has been found that the type of encapsulation can play an important role in the level of ruminal degradation of the vitamin. Robert (1995a) measured the degradation of vitamin A in the rumen of lactating cows, using the nylon-bag technique. The results demonstrated that the encapsulation process could significantly reduce the rate of vitamin degradation in the rumen. It was further demonstrated that the solubilization of the encapsulation was significantly greater with a hay-based ration compared with a ration based on maize silage, probably as a result of the higher pH in the rumen with the hay-based diet compared with maize silage, producing conditions more favourable for the solubilization of the encapsulation material. The observations indicate that, dependent on the diet and the encapsulation process, different levels of vitamin A may bypass the rumen and be absorbed in the intestine.

Table 19.18. Measurements of hepatic storage of vitamin A (from Robert, 1987).

Diet	Number of animals	Planned vitamin A supplementation (IU g^{-1} feed)	Vitamin A level in feed (IU g^{-1} feed)	Total vitamin A hepatic storage (IU × 10^3)	Coefficient of hepatic storage (%)
1	15	0	2.4	11	–
2	15	20	19.8	523	34.1
3	15	40	45.6	961	25.0
4	15	60	61.3	928	18.9
5	15	80	83.1	898	12.1
6	15	100	109.2	989	10.1

Vitamin E

The biological role of tocopherols is related to the antioxidant activity of vitamin E, in particular as a means of protecting polyunsaturated fats in tissue membranes from peroxidation. Tocopherols have a low stability to heat, light and basic pH. Taking into account their antioxidant properties, in the presence of oxygen they themselves are oxidized, with the formation of quinones in the dimer and trimer forms. Acylation of the hydroxyl group present on the ring improves vitamin E stability and, for the most part, synthetic forms of vitamin E are present in the form of esters (acetate, succinate, ...). These synthetic forms have no antioxidant properties *per se* but this property is recovered at the metabolic level, given the fact that absorption at the intestinal level occurs in the form of the alcohol.

Degradation of vitamin E in the rumen and intestinal absorption

Effective supplementation with vitamin E requires that the dietary vitamin supplement arrives intact at the site of absorption in the small intestine, where it can be absorbed and participate in metabolism.

The question of apparent low plasma appearance has been addressed by several authors. Using a preparation in which mature ewes had the pylorus ligatured, Alderson *et al.* (1971) were unable to obtain increased blood levels of vitamin E when 5000 IU of α-tocopherol (1 mg = 1.36 IU) was introduced into the rumen. It was concluded that vitamin E is not absorbed through the rumen wall. The addition of either an oxidizing agent (sodium nitrate) or an antioxidant (ethoxyquin) into the rumen of steers had no influence on the rate of disappearance of supplementary vitamin E added to the rumen at the level of 4190 IU (Tucker *et al.*, 1971; Table 19.19). It is notable that, in these two experiments, quite disparate and unexplainable rates of degradation of vitamin E were reported (27% and 62%). Hidiroglou *et al.* (1970) reported that there was a higher retention of radiolabelled tocopherol in sheep when it was administered intramuscularly compared with oral administration, which could be explained by pre-intestinal destruction of vitamin E. The fact that the rumen microflora contains vitamin E strengthens the hypothesis of a direct action of the rumen bacteria (Hidiroglou and Jenkins, 1974).

The destruction of vitamin E by the rumen microorganisms was studied in rumen-fistulated steers, each supplemented with 20,000 IU of vitamin E day^{-1} and given four diets containing different levels of lucerne hay and maize (Alderson *et al.*, 1971; Table 19.20). Degradation of vitamin E ranged from 8 to 42% as the level of maize incorporation in the ration increased. These results tend to suggest that

Table 19.19. Influence of oxidant or antioxidant agents on the degradation of vitamin E *in vivo* (from Tucker *et al.*, 1971).

	Trial 1		Trial 2	
	Control	Ethoxyquin	Control	Sodium nitrate
% of vitamin E degraded	26.8	24.4	61.5	64.3

Table 19.20. Disappearance of α-tocopherol in the rumen.

Animal	Diet	Disappearance (%)	Method	References
Steer	20% concentrate	8.4 ± 1.2	*In vivo*	Alderson *et al.* (1971)
	40% concentrate	22.2 ± 3.5		
	60% concentrate	25.0 ± 5.3		
	80% concentrate	42.4 ± 7.7		
Adult steer	High-concentrate	39–52	*In vivo*	Shin and Owens (1990)
Sheep	Forage/concentrate 50/50	6 ± 4	*In vitro* 24 h	Astrup *et al.* (1974a)
Beef steer	High-concentrate	4	*In vitro* 24 h	Leedle *et al.* (1993)
Jersey steer	High-forage	7.7 ± 3.5	*In vitro* 24 h	McDiarmid *et al.* (1994)
Dairy cow	High-silage	18	*In vivo*	Robert (1995b)
Dry cow	High-forage	0	*In vitro* 24 h	Weiss *et al.* (1995)
Dairy cow lactating	Forage/concentrate 50/50	0		

vitamin E was partly degraded by the rumen microflora and particularly the amylolytic bacteria. Shin and Owens (1990) also demonstrated that between 39 and 52% of dietary supplemental vitamin E disappeared before reaching the duodenum of steers.

These results are contrary to those obtained by Leedle *et al.* (1993), who, based on assumptions made by Alderson *et al.* (1971), examined the stability of vitamin E in an *in vitro* system, using rumen liquor obtained from steers which had been offered a diet rich in concentrates. After periods of incubation of up to 24 h at 39°C, there was no loss of 1000 IU of DL-α-tocopherol acetate (1 mg = 1 IU). These authors concluded that there was no destruction of vitamin E by the rumen microorganisms and they suggested that the contradictory results obtained by other authors were due to the methods of vitamin E extraction and analysis. Using a diet based on 50% oats and 50% hay and with an *in vitro* rumen system, Astrup *et al.* (1974a) observed only slight degradation of vitamin E (Table 19.20). Similarly, with diets rich in either forage or concentrates, McDiarmid *et al.* (1994) and Weiss *et al.* (1995) observed only slight degradation of DL-α-tocopherol acetate.

Leedle *et al.* (1993) considered that structural differences between vitamins E and A were responsible for the relative differences in stability. Vitamin A possesses four double bonds in the side-chain, which are susceptible to reduction in the anaerobic rumen environment (biohydrogenation), whereas vitamin E possesses only one double bond on the inside of the aromatic ring, conferring greater stability.

It is quite possible that values obtained using *in vitro* fermentation systems may underestimate ruminal degradation, compared with values measured *in vivo*. In an *in vivo* trial with cows cannulated at the duodenum and given diets based on maize silage (ratio of forage to concentrate 80/20), Robert (1995b) reported only moderate degradation of vitamin E (approximately 20%) (Table 19.20). The level of supplementation was 1500 IU animal^{-1} day^{-1}, which was an exceptionally high level but which was used for reasons of experimental methodology. Indeed, it is possible that this very high dose caused a saturation of the microbial degradation of vitamin E.

In ruminants, absorption of vitamin E is slower than in other species, the peak level in the plasma occurring approximately 24–48 h after oral administration. Intestinal absorption is linked to the absorption of lipids and requires the presence of bile salts and pancreatic lipases. Only the alcohol form is directly absorbed from the jejunum. This absorption process can be saturated. Using radiolabelled L-α-tocopherol, Tikriti (1969) obtained levels of absorption of the order of 40–50% of that escaping degradation in the rumen of dairy cows and, in goats, Astrup *et al.* (1974b) obtained levels of absorption of approximately 58% of the vitamin E that was introduced into the abomasum. Similarly to the case with β-carotene, intestinal absorption reduces as the level consumed increases (Machlin, 1984).

Vitamin E is stored both in adipose tissue and in the liver, but all tissues contain detectable amounts. Adipose tissue would appear to be capable of storing an unlimited quantity, whereas storage of vitamin E in the liver occurs for only a limited period. Furthermore, high-level supplementation of vitamin E for a short period does not increase the level of reserves in a similar manner to that following similar supplementation with vitamin A.

A low level of degradation of vitamin E in the rumen, which as yet remains to be confirmed, would be in accordance with responses in terms of animal health and performance obtained in practice. It is also in accordance with the more recent practice of using high-level supplementation with vitamin E to improve carcass quality. Bozzolo *et al.* (1993) (Table 19.21) were able to improve the quality of external carcass fat in sheep by supplementing with 4 g of DL-α-tocopherol acetate for 4 days prior to slaughter. The improvement resulted from the antioxidant properties of tocopherols. It is important to note that these authors obtained a significant increase in the plasma concentration of vitamin E, which was similar to that reported by Hidiroglou and Karpinski (1988). However, the high variability recorded in the concentrations of vitamin E in the treated sheep demonstrates a lack of homogeneity in the response. The authors considered that the variability was a result of the oral administration.

For improvement in meat quality, different authors have recommended supplementation with vitamin E at elevated dose levels (500–1000 IU animal^{-1} day^{-1}) during a period of 90–100 days before slaughter of cattle. Such levels of supplementation have proved beneficial in increasing tissue levels of α-tocopherol and reducing the percentage of metmyoglobin in muscle and the concentration of free radicals in the meat. Mitsumoto *et al.* (1993) orally supplemented steers with 1500 IU vitamin E animal^{-1} day^{-1} for 250 days before slaughter. Proportions of metmyoglobin, which is a brown oxidation product, were, in muscle of the control and treated animals, 19 and 7% on the day of slaughter and 87 and 40% after 9 days of storage, respectively. This reduced oxidation has the potential to dramatically improve the quality and storage life of meat obtained from beef animals and available on supermarket shelves (Faustman *et al.*, 1989a, b; Arnold *et al.*, 1992).

Table 19.21. Vitamin E and external tissue colour of sheep (from Bozzolo *et al.*, 1993).

	Control	Supplemented
Number of animals	120	120
Supplementary supply of DL-α-tocopherol acetate	0	4 g per animal (4 days before slaughter)
Colour of fat (%)		
Creamy white	19	38
Creamy brown	46	27
Yellow	34	34
Plasma level of vitamin E (μg ml^{-1})	0.7 ± 0.18	1.93 ± 0.83

Vitamin D

In high-producing, lactating, dairy cows at the start of lactation and under normal conditions, the presence of vitamin D prevents hypocalcaemia and the occurrence of milk fever. The presence of 1,25-dihydroxycholecalciferol also stimulates mineral uptake and particularly calcium uptake by bone, which explains the occurrence of rickets and bone dystrophy in the absence of vitamin D (rickets found in young animals and osteomalacia in lactating females).

Theoretically, the requirements for supplementary vitamin D in ruminants are low, because animals exposed to sunlight are capable of synthesizing sufficient vitamin D and, as indicated earlier, a supply can be obtained from sun-dried forages. Regular supplementation with vitamin D allows establishment of adequate reserves, which cover requirements during periods when the climate is unfavourable and when diets based on ensiled materials are offered. It is for these reasons that synthetic vitamin D has been developed, which confers the additional advantage of providing a controlled supply.

Studies of degradation of vitamin D incubated for 24 h in rumen liquor obtained from steers fed a diet based on forage showed a loss of approximately 75% of vitamin D (Sommerfeldt *et al.*, 1979b). However, incubation of vitamin D in sterilized rumen fluid resulted in no degradation, indicating that degradation is due to the activity of the rumen microorganisms (Sommerfeldt *et al.*, 1979b). Such activity produces three metabolites of vitamin D (stereoisomers of 10-ceto-19-norvitamin D_3) (Horst and Reinhardt, 1983), which have anti-vitamin D activity but which have the advantage of protecting ruminants from vitamin D toxicity in the event of consumption of high quantities. Although D_2 is comparable to D_3 in antirachitic function (McDowell, 1989), Sommerfeldt *et al.* (1979a, 1983) showed that the two forms of vitamin D (vitamin D_2 and D_3) do not pass through identical metabolic pathways in the rumen, resulting in a reduction of vitamin D_2 activity in comparison with vitamin D_3. Vitamin D is absorbed from the intestine in association with lipids and thus requires the presence of bile salts. Only approximately 50% of the vitamin D ingested is absorbed (Miller and Norman, 1984; McDowell, 1989).

It is probable that a proportion of supplementary vitamin D in the diet is destroyed by the rumen and does not reach the small intestine. This is justification for retaining the use of supplementary vitamin D in the diet as a means of covering requirements during periods of need. However, account must be taken of the ease with which hypervitaminosis and vitamin D toxicity can occur. Severe bone demineralization (osteofibrosis) in conjunction with hypercalcaemia and the occurrence of urinary calculi plus calcification of soft tissues has been observed following chronic doses of between 50 and 100 times requirement. In Europe, legislation limits dietary supplementation for ruminants to a maximum of 4000 IU vitamin D kg^{-1} of diet.

Conclusion

This review has demonstrated that, whilst forages may be considered as a potential source of vitamins, there is great variability in the quantity of key vitamins that remain bioactive in the feed and, perhaps of more importance, it is almost impossible, without costly feed evaluation, to accurately predict the quantity that will be supplied to ruminants. The review has also identified the major influence that rumen fermentation has on the quantity of vitamins reaching the small intestine. The value of correct vitamin supply to ruminants cannot be overstated and there is increasing evidence of the benefits of hypersupplementation with certain vitamins.

In view of the tremendous importance of correct vitamin supply, plus the uncertainty of supply from feed components, it is totally realistic to rely on the precision and reliability that is achieved by supply from synthetic vitamins. Accuracy of vitamin supply to ruminants is assisted by the use of formulation technology, which protects vitamins during feed processing and also helps ensure the delivery of the target dose to the absorption site in the intestine. In this manner, optimum benefit can be gained from these highly valued but relatively inexpensive feed additives. Although forage has the potential to contribute to the vitamin supply of ruminants, it is very difficult to envision how this supply can be reliably incorporated into ration formulation systems.

References

Aitken, F.C. and Hankin, R.G. (1970) *Vitamins in Feeds for Livestock.* Technical communication No. 25, Commonwealth Bureau of Animal Nutrition, Commonwealth Agricultural Bureaux, Farnham Royal, UK.

Akopyan, G.O. (1958) The content of vitamin E in vetches of the Stepanavan region. *Izvvestiâ Akademii Nauk Armânskoj SSR Seriâ Biologiceskih* 11, 95–97.

Albonico, F. and Fabris, A. (1958) Fattori influenzanti il contenuto vitaminico e clorofillico di alcune foraggere leguminose. *Agrochimica* 2, 147–176.

Alderson, N.E., Mitchell, G.E., Jr, Little, C.O., Warner, R.E. and Tucker, R.E. (1971) Pre-intestinal disappearance of vitamin E in ruminants. *Journal of Nutrition* 101, 655–660.

Arnold, R.N., Scheller, K.K., Arp, S.C., Williams, S.M., Buege, D.R. and Schaefer, D.M. (1992) Effect of long- or short-term feeding of α-tocopherol acetate to Holstein and crossbred beef steers on performance, carcass characteristics and beef stability. *Journal of Animal Science* 70, 3055–3065.

Astrup, H.N., Mills, S.C., Cook, L.J. and Scott, T.W. (1974a) Stability of α tocopherol in rumen liquor of sheep. *Acta Veterinaria Scandinavica* 15, 451–453.

Astrup, H.N., Mills, S.C., Cook, L.J. and Scott, T.W. (1974b) Persistence of α tocopherol in the post rumen digestive tract of the ruminant. *Acta Veterinaria Scandinavica* 15, 454–456.

Bauernfeind, J.C. (1972) Carotenoid vitamin A precursors and analogues in foods and feeds. *Journal of Agricultural and Food Chemistry* 20, 456–473.

Bauernfeind, J.C. (1980) Tocopherols in foods. In: Machlin, L.J. (ed.) *Vitamin E: A Comprehensive Treatise*. M. Dekker, New York and Basel, pp. 99–167.

Beck, W.A. and Redman, R. (1940) Seasonal variations in the production of plant pigments. *Plant Physiology* 15, 81–94.

Bieber-Wlaschny, M. (1988) Vitamin requirements of the dairy cow. In: Garnsworthy, P.C. (ed.) *Nutrition and Lactation in the Dairy Cow*. Butterworths, London, pp. 135–156.

Blaylock, L.G., Richardson, L.R. and Pearson, P.B. (1950) The riboflavin, pantothenic acid, niacin and folic acid content of fresh, dehydrated and field-cured alfalfa. *Poultry Science* 29, 692–695.

Bondi, A. and Meyer, H. (1946) Carotene in Palestinian crops. *Journal of Agricultural Science* 36, 1–5.

Bondi, A. and Sklan, D. (1984) Vitamin A and carotene in animal nutrition. *Progress in Food and Nutrition Science* 8, 165–191.

Booth, V.H. (1964) The α tocopherol content of forage crops. *Journal of the Science of Food and Agriculture* 15, 342–344.

Bozzolo, G., Bouillier-Oudo, M. and Phrem, G. (1993) Effet d'une surcharge en vitamine E, administrée oralement avant abattage, sur la coloration du gras de couverture des carcasses d'agneaux de bergerie. *Annales de Zootechnie* 42, 67–73.

Brongniart, I. (1987) *Etude de la dégradabilité de la vitamine A chez le mouton*. AEC/Rhône Poulenc Animal Nutrition Internal Report.

Brown, F. (1953) The tocopherol content of farm feeding-stuffs. *Journal of the Science of Food and Agriculture* 4, 161–165.

Bruhn, J.C. and Oliver, J.C. (1978) Effect of storage on tocopherol and carotene concentrations in alfalfa hay. *Journal of Dairy Science* 61, 980–982.

Bunnell, R.H., Keating, J.P. and Quaresimo, A.J. (1968) Alpha-tocopherol content of feedstuffs. *Journal of Agricultural and Food Chemistry* 16, 659–664.

Burdick, D. and Fletcher, D.L. (1985) Carotene-xanthophyll in field-wilted and dehydrated alfalfa and coastal bermuda grass. *Journal of Agricultural and Food Chemistry* 33, 235–238.

Burrows, F.A. and King, R.L. (1968) Comparative study of tocopherols and beta-carotene in selected forages during a growing season. *Journal of Dairy Science* 51, 973.

Clarke, M.F. (1953) A study of the carotene and crude protein content of orchardgrass (*Dactylis Glomerata* L.). 1. Variation due to stage of growth, cutting management and clones. *Canadian Journal of Agricultural Science* 33, 184–194.

Cohen-Fernandez, S., Budowski, P., Ascarelli, I., Neumark, H. and Bondi, A. (1976) Preintestinal stability of β-carotene in ruminants. *International Journal for Vitamin and Nutrition Research* 46, 439–445.

Davison, K.L. and Seo, J. (1963) Influence of nitrate upon carotene destruction during *in vitro* fermentation with rumen liquor. *Journal of Dairy Science* 46, 862–864.

Eaton, H.D., Myers, G.S., Jr, Dicks, M.W., Dehority, B.A., Grifo, A.P., Jr, Teichman, R., Helmboldt, C.F., Jungherr, E.L. and Gosslee, D.G. (1959) Conversion of carotene from alfalfa to vitamin A by Guernsey and Holstein calves. *Journal of Dairy Science* 42, 642–650.

Faustman, C., Cassens, R.G., Schaefer, D.M., Buege, D.R. and Scheller, K.K. (1989a)

Vitamin E supplementation of Holstein steer diets improves sirloin steak color. *Journal of Food Science* 54, 485–486.

Faustman, C., Cassens, R.G., Schaefer, D.M., Buege, D.R., Williams, S.N. and Scheller, K.K. (1989b) Improvement of pigment and lipid stability in Holstein steer beef by dietary supplementation with vitamin E. *Journal of Food Science* 54, 858–862.

Ferrando, R. (1980) Vitamines. In: Jarrige, R. (ed.) *Alimentation des ruminants*. INRA, Paris, pp. 161–176.

Ferrando, R. and Mainguy, R. (1970) L'évolution des vitamines en fonction des conditions de récolte et de conservation des denrées primaires des animaux destinés à la consommation de l'homme. *Annales de la Nutrition et de l'Alimentation* 24, b445–b476.

Ferrer, A. (1972) Etude des pertes de caroténoïdes de la luzerne de la vallée de l'Ebre par effet de la déshydratation industrielle. *Fourrages* 49, 41–51.

Galgan, M.W., Bell, J.M. and Spencer, E.Y. (1950) Variations in vitamin content with method of cure and stage of growth of alfalfa and oats and their utilization as forages by lambs. *Scientific Agriculture* 30, 316–323.

Granseigne, P. and Robert, J.C. (1985) *Vitamine A et ruminants*. AEC/Rhône Poulenc Animal Nutrition Internal Report.

Hidiroglou, M. and Jenkins, K.J. (1974) Le sort du radiotocophérol administré dans l'appareil gastrique ou dans le duodénum du mouton. *Annales de Biologie Animale, Biochimie, Biophysique* 14, 667–677.

Hidiroglou, M. and Karpinski, K. (1988) Pharmacokinetic disposition in sheep of various vitamin E preparations given orally or intravenously. *British Journal of Nutrition* 59, 509–518.

Hidiroglou, M., Jenkins, K.J., Lessard, J.R. and Carson, R.B. (1970) Metabolism of vitamin E in sheep. *British Journal of Nutrition* 24, 917–928.

Hidiroglou, M., Ivan, M. and Jenkins, K.J. (1977) Influences of barley and oat silages for beef cows on occurrence of myopathy in their calves. *Journal of Dairy Science* 60, 1905–1909.

Hidiroglou, M., Batra, T.R. and Roy, G.L. (1994) Changes in plasma α tocopherol and selenium of gestating cows fed hay or silage. *Journal of Dairy Science* 77, 190–195.

Hjarde, W., Hellström, V. and Åkerberg, E. (1963) The contents of tocopherol and carotene in red clover as dependent on variety, conditions of cultivation and stage of development. *Acta Agriculturae Scandinavica* 13, 3–16.

Hoffmann, M. and Nehring, K. (1967) Carotinuntersuchungen an futterpflanzen. 4. Carotingehalt und -ertrag verschiedener futterstoffe in abhängigkeit vom vegetationsstadium. *Archiv für Tierernährung* 17, 451–461.

Horst, R.L. and Reinhardt, T.A. (1983) Vitamin D metabolism in ruminants and its relevance to the periparturient cow. *Journal of Dairy Science* 66, 661–678.

Hullár, I. and Brand, A. (1993) Nutritional factors affecting milk quality, with especial regard to milk protein: a review. *Acta Veterinaria Hungarica* 41, 11–32.

Hunt, C.H., Record, P.R. and Bethke, R.M. (1935) Effect of the stage of maturity and method of curing upon the vitamin B and vitamin G content of alfalfa, clover and timothy hays. *Journal of Agricultural Research* 51, 251–258.

Kalac, P. (1983) Losses of beta-carotene in unwilted forage crops during silage-making and feeding. *Animal Feed Science and Technology* 9, 63–69.

Kalac, P. and Kyzlink, V. (1979) Losses of beta-carotene in red clover in an acid medium during ensiling. *Animal Feed Science and Technology* 4, 81–89.

Kalac, P. and Kyzlink, V. (1980) The enzymic nature of the degradation of beta-carotene in red clover and in other forage crops during silage-making with acid additives. *Animal Feed Science and Technology* 5, 59–68.

Kalac, P. and McDonald, P. (1981) A review of the changes in carotenes during ensiling of forages. *Journal of the Science of Food and Agriculture* 32, 767–772.

Keating, E.K., Hale, W.H. and Hubbert, F., Jr (1964) *In vitro* degradation of vitamin A and carotene by rumen liquor. *Journal of Animal Science* 23, 111–117.

Keener, H.A. (1954) The effect of various factors on the vitamin D content of several common forages. *Journal of Dairy Science* 37, 1337–1345.

Kemmerer, A.R., Fudge, J.F. and Fraps, G.S. (1944) Constituents of the crude carotene of some forages. *Journal of the American Society of Agronomy* 36, 683–687.

King, R.L., Burrows, F.A., Hemken, R.W. and Bashore, D.L. (1967) Control of oxidized flavor by managed intake of vitamin E from selected forages. *Journal of Dairy Science* 50, 943–944.

King, T.B., Lohman, T.G. and Smith, G.S. (1962) Evidence of rumeno-reticular losses of vitamin A and carotene. *Journal of Animal Science* 21, 1002.

Kivimäe, A. and Carpena, C. (1973) The level of vitamin E content in some conventional feeding stuffs and the effects of genetic variety, harvesting, processing and storage. *Acta Agriculturae Scandinavica* Suppl. 19, 161–168.

Klatte, F.J., Mitchell, G.E., Jr and Little, C.O. (1964) *In vitro* destruction of vitamin A by abomasal and ruminal contents. *Journal of Agricultural and Food Chemistry* 12, 420–421.

Knowles, R.E., Livingston, A.L., Nelson, J.W. and Kohler, G.O. (1968) Xanthophyll and carotene storage stability in commercially dehydrated and freeze-dried alfalfa. *Journal of Agricultural and Food Chemistry* 16, 654–658.

Kohler, G.O., Beier, E. and Bolze, C.C. (1955) The stability of carotene and vitamin E in dehydrated forage crops. *Poultry Science* 34, 468–471.

Larsen, T.W., Yang, A. and Tume, R.K. (1993) The *in vitro* destruction of rumen fluid carotenoids by plant lipoxygenases. *Biochemistry and Molecular Biology International* 30, 197–207.

Leedle, R.A., Leedle, J.A.Z. and Butine, M.D. (1993) Vitamin E is not degraded by ruminal microorganisms: assessment with ruminal contents from a steer fed a high-concentrate diet. *Journal of Animal Science* 71, 3442–3450.

Lichtenwalner, R.E., Fontenot, J.P. and Tucker, R.E. (1973) Effect of source of supplemental nitrogen and level of nitrate on feedlot performance and vitamin A metabolism of fattening beef calves. *Journal of Animal Science* 37, 837–847.

Livingston, A.L., Knowles, R.E., Israelsen, M., Nelson, J.W., Mottola, A.C. and Kohler, G.O. (1966) Xanthophyll and carotene stability during alfalfa dehydration. *Journal of Agricultural and Food Chemistry* 14, 643–644.

Livingston, A.L., Knowles, R.E., Nelson, J.W. and Kohler, G.O. (1968a) Xanthophyll and carotene loss during pilot and industrial scale alfalfa processing. *Journal of Agricultural and Food Chemistry* 16, 84–87.

Livingston, A.L., Nelson, J.W. and Kohler, G.O. (1968b) Stability of α tocopherol during alfalfa dehydration and storage. *Journal of Agricultural and Food Chemistry* 16, 492–495.

Livingston, A.L., Smith, D., Carnahan, H.L., Knowles, R.E., Nelson, J.W. and Kohler, G.O. (1968c) Variation in the xanthophyll and carotene content of lucerne, clovers and grasses. *Journal of the Science of Food and Agriculture* 19, 632–636.

Livingston, A.L., Knowles, R.E. and Kohler, G.O. (1970) Xanthophyll, carotene and α tocopherol stability in alfalfa as affected by pilot- and industrial-scale dehydration. In: *Technical Bulletin No. 1414*. US Department of Agriculture, Washington, DC, pp. 1–13.

McDiarmid, R.E., Majak, W. and Cheng, K.J. (1994) Procedure for analysis of α tocopherol acetate in bovine ruminal fluid. *Canadian Journal of Animal Science* 74, 391–392.

McDowell, L.R. (1989) *Vitamins in Animal Nutrition: Comparative Aspects to Human Nutrition*. Academic Press, San Diego, California.

Machlin, L.J. (1984) Vitamin E. In: Machlin L.J. (ed.) *Handbook of Vitamins: Nutritional, Biochemical and Clinical Aspects*. M. Dekker, New York and Basel, pp. 99–145.

Maynard, L.A., Loosli, J.K., Hintz, H.F. and Warner, R.G. (1979) *Animal Nutrition*, 7th edn. McGraw-Hill, New York, pp. 283–355.

Meissonier, E. (1974) Des vitamines pour les bovins. I. Les vitamines A et E. *Productions agricoles Français* 50(144), 39–42.
Meissonier, E. (1981) *L'Approvisionnement vitaminique des bovins laitiers*. Hoffman-La Roche, Paris, pp. 18–22.
Melcion, J.P. and Delort-Laval, J. (1973) Stabilité en atmosphère contrôlée des caroténoïdes de la luzerne déshydratée. *Annales de Zootechnie* 22, 399–411.
Miller, B.E. and Norman, A.W. (1984) Vitamin D. In: Machlin, L.J. (ed.) *Handbook of Vitamins: Nutritional, Biochemical and Clinical Aspects*. M. Dekker, New York and Basel, pp. 45–97.
Mitchell, G.E., Jr, Little, C.O. and Hayes, B.W. (1967) Pre-intestinal destruction of vitamin A by ruminants fed nitrate. *Journal of Animal Science* 26, 827–829.
Mitchell, G.E., Jr, Little, C.O., Warner, R.L., Klatte, F.J. and Alderson, N.E. (1968) Influence of vitamin E on pre-intestinal disappearance of vitamin A in steers. *Internationale Zeitschrift für Vitaminforschung* 38, 304–307.
Mitchell, H.L. and Hauge, S.M. (1946) Factors affecting the enzymic destruction of carotene in alfalfa. *Journal of Biological Chemistry* 164, 543–550.
Mitsumoto, M., Arnold, R.N., Schaefer, D.M. and Cassens, R.G. (1993) Dietary versus post-mortem supplementation of vitamin E on pigment and lipid stability in ground beef. *Journal of Animal Science* 71, 1812–1816.
Moon, F.E. (1939) The carotene contents of some grass and clover species, with a note on pasture weeds. *Empire Journal of Experimental Agriculture* 7, 235–243.
Nadai, J. (1968) *La Couverture des besoins en vitamine A chez les bovins*. Hoffmann-La Roche, Paris.
NRC (1989) *Nutrient Requirements of Domestic Animals: Nutrient Requirements of Dairy Cattle*, 6th edn. National Research Council, National Academy Press, Washington, DC, pp. 42–49.
Olson, J.A. (1984) Vitamin A. In: Machlin, L.J. (ed.) *Handbook of Vitamins: Nutritional, Biochemical and Clinical Aspects*. M. Dekker, New York and Basel, pp. 1–43.
Olsson, N., Åkerberg, E. and Blixt, B. (1955) Investigations concerning formation, preservation and utilization of carotene. *Acta Agriculturae Scandinavica* 5, 113–184.
Park, Y.M., Anderson, M.J., Walters, J.L. and Mahoney, A.W. (1983) Effects of processing methods and agronomic variables on carotene contents in forages and predicting carotene in alfalfa hay with near-infrared-reflectance spectroscopy. *Journal of Dairy Science* 66, 235–245.
Patel, B.M., Patel, C.A. and Shukla, P.C. (1966) Effect of drying and storage on the carotene content and other constituents in lucerne. *Indian Journal of Veterinary Science* 36, 124–129.
Potkanski, A. (1974) Absorption of carotene from preserved forages in intestinal tract of sheep. *Roczniki Akademii Rolniczej Poznaniu Habilitacyjne* 53, 44.
Potkanski, A.A., Tucker, R.E., Mitchell, G.E., Jr and Schelling, G.T. (1974) Pre-intestinal losses of carotene in sheep fed high-starch or high-cellulose diets. *International Journal for Vitamin and Nutrition Research* 44, 147–150.
Ramanujan, R.A. and Anantakrishnan, C.P. (1958) Tocopherol and carotene contents of green fodder. *Indian Journal of Dairy Science* 11, 101–108.
Repp, W.W. and Watkins, W.E. (1958) Relation of forage carotene to plasma carotene and vitamin A of range cows. *Journal of Animal Science* 17, 398–404.
Robert, J.C. (1987) *Etude du stockage hépatique de la vitamine A chez l'agneau ruminant*. AEC/Rhône Poulenc Animal Nutrition Internal Report.
Robert, J.C. (1995a) Microvit A Supra. Stability in the rumen. Effects of forage composition. In: *Synopsis of Work in Progress at the Centre for Evaluation and Research in Nutrition in 1994–1995*. Rhône Poulenc Animal Nutrition.
Robert, J.C. (1995b) Effects of the rumen environment on vitamin E provided in the diet.

In: *Synopsis of Work in Progress at the Centre for Evaluation and Research in Nutrition in 1994–1995*. Rhône Poulenc Animal Nutrition.

Robinson, A.D., Lynd, L.E. and Miles, B.J. (1948) Thiamine and riboflavin in growing forage crops. *Scientific Agriculture* 28, 162–165.

Rode, L.M., McAllister, T.A. and Cheng, K.J. (1990) Microbial degradation of vitamin A in rumen fluid from steers fed concentrate, hay or straw diets. *Canadian Journal of Animal Science* 70, 227–233.

RPAN (1998) *Microvit Nutrition Guide*. Rhône Poulenc Animal Nutrition.

Russell, W.C. (1929) The effect of the curing process upon the vitamin A and D content of alfalfa. *Journal of Biological Chemistry* 85, 289–297.

Scott, M.L. (1973) Effect of processing on the availability and nutritional value of vitamins. In: Scott, M.L. (ed.) *Effect of Processing on the Nutritional Value of Feeds*. National Academy of Sciences, Washington, DC, pp. 119–130.

Seshan, P.A. and Sen, K.C. (1942a) Studies on carotene in relation to animal nutrition. 2. The development and distribution of carotene in the plant and the carotene content of some common feeding stuffs. *Journal of Agricultural Science* 32, 202–216.

Seshan, P.A. and Sen, K.C. (1942b) Studies on carotene in relation to animal nutrition. 3. Stability of carotene in plant materiel with special reference to hay making and storage. *Journal of Agricultural Science* 32, 275–285.

Shin, I.S. and Owens, F.N. (1990) Ruminal and intestinal disappearance of several sources of vitamin E. *Journal of Animal Science* 68, 544–545.

Smith, A.M. and Wang, T. (1941) The carotene content of certain species of grassland herbage. *Journal of Agricultural Science* 31, 370–378.

Sommerfeldt, J.L., Horst, R.L., Littledike, E.T. and Beitz, D.C. (1979a) Metabolism of orally administred [^{3}H] vitamin D_2 and [^{3}H] vitamin D_3 by dairy calves. *Journal of Dairy Science* 62, 157.

Sommerfeldt, J.L., Horst, R.L., Littledike, E.T. and Beitz, D.C. (1979b) *In vitro* degradation of cholecalciferol in rumen fluid. *Journal of Dairy Science* 62, 192–193.

Sommerfeldt, J.L., Napoli, J.L., Littledike, E.T., Beitz, D.C. and Horst, R.L. (1983) Metabolism of orally administred [^{3}H]ergocalciferol and [^{3}H]cholecalciferol by dairy calves. *Journal of the Nutrition* 113, 2595–2600.

Thomas, J.W. and Moore, L.A. (1948) Vitamin D content of forages as affected by various curing procedures. *Journal of Dairy Science* 31, 689.

Thomas, J.W. and Moore, L.A. (1951) Factors affecting the antirachitic activity of alfalfa and its ability to prevent rickets in young calves. *Journal of Dairy Science* 34, 916–928.

Thompson, C.R. (1949) The carotene content of several varieties and strains of alfalfa. *Agronomy Journal* 41, 294–297.

Thompson, C.R., Bickoff, E.M. and Maclay, W.D. (1951) Formation of stereoisomers of beta-carotene in alfalfa. *Industrial and Engineering Chemistry* 43, 126–129.

Tikriti, H.H. (1969) The metabolism of vitamin E by the lactating dairy cow in relation to oxidized flavour in milk. PhD thesis, University of Maryland, USA, 124 pp.

Tucker, R.E., Mitchell, G.E., Jr, Cannon, W.W. and Alderson, N.E. (1971) Ruminal destruction of vitamin E administered with ethoxyquin or sodium nitrate. *Progress Report* 170, 14–15.

Ullrey, D.E. (1972) Biological availability of fat-soluble vitamins: vitamin A and carotene. *Journal of Animal Science* 35, 648–657.

Uzu, G. (1988) *Mise au point d'un test biologique chez le poulet de chair sur la disponibilité des vitamines A enrobées: Microvit A Supra 500*. Rhône Poulenc Animal Nutrition.

Vail, C.E., Tobiska, J.W. and Douglass, E. (1936) *Further Studies on Vitamins in Alfalfa Hay*. Colorado Experiment Station Technical Bulletin 18.

Wallis, G.C., Kennedy, G.H. and Fishman, R.H. (1958) The vitamin D content of roughages. *Journal of Animal Science* 17, 410–415.

Warner, R.L., Mitchell, G.E., Jr, Little, C.O. and Alderson, N.E. (1970) Pre-intestinal disappearance of vitamin A in steers fed different levels of corn. *International Journal of Vitamin Research* 40, 585–588.

Watson, S.J. and Nash, M.J. (1960) *The Conservation of Grass and Forage Crops*. Oliver and Boyd, Edinburgh and London, 392 pp.

Waugh, R.K., Hauge, S.M. and Hilton, J.H. (1944) Carotene losses in freshly cut plant tissues. *Journal of Dairy Science* 27, 585–590.

Weiss, W.P., Smith, K.L., Hogan, J.S. and Steiner, T.E. (1995) Effect of forage to concentrate ratio on disappearance of vitamins A and E during *in vitro* ruminal fermentation. *Journal of Dairy Science* 78, 1837–1842.

Wilson, J.R. (1981) Environmental and nutritional factors affecting herbage quality. In: Hacker, J.B. (ed.) *Nutritional limits to Animal Production from Pastures*. Commonwealth Agricultural Bureaux, Farnham Royal, England, pp. 111–131.

Wolter, R. (1988) Alimentation vitaminique. In: Jarrige, R. (ed.) *Alimentation des bovins, ovins et caprins*. INRA, Paris, pp. 113–120.

20 Secondary Plant Compounds and Forage Evaluation

J.D. Reed,[1] C. Krueger,[1] G. Rodriguez[1] and J. Hanson[2]

[1]*Department of Animal Sciences, University of Wisconsin-Madison, 1675 University Avenue, Room 256, Madison, WI 57306, USA;* [2]*International Livestock Research Institute, PO Box 5689, Addis Ababa, Ethiopia*

Introduction

Secondary plant compounds (SPC) are a diverse group of molecules that are involved in the adaptation of plants to their environment but are not part of the primary biochemical pathways of cell growth and reproduction. There are over 24,000 structures, including many compounds that have antinutritional and toxic effects on mammals (Harborne, 1993). The SPC that occur in forages include alkaloids, non-protein amino acids, cyanogenic glycosides, volatile terpenoids, saponins, phenolic acids, hydrolysable tannins (HT) and flavonoids, including proanthocyanidins (PA) and oestrogenic isoflavones. Secondary plant compounds are involved in defence against herbivores and pathogens, regulation of symbiosis, control of seed germination and chemical inhibition of competing plant species (allelopathy). Therefore, SPC are an integral part of the interactions of species in plant and animal communities (Harborne, 1993). Research on the SPC in forages has concentrated on their toxic and antinutritional effects on livestock. These effects can be classified into two groups (Reed 1998):

1. Toxic compounds that are present in plants at low concentrations (generally less than 20 g kg^{-1} of the dry matter) and have negative physiological effects when absorbed, such as neurological problems, reproductive failure, goitre, gangrene and death. Examples are alkaloids, cyanogenic glycosides, toxic amino acids, saponins, isoflavonoids and many others.
2. Non-toxic compounds that lower the digestibility and palatability of plants. Higher concentrations (> 20g kg^{-1} of dry matter) of these compounds are required for negative effects and the primary site of activity is in the digestive tract or through sensory organs associated with feeding behaviour. This class includes lignin, tannin, cutin, biogenic silica and volatile terpenoids. Compounds that have a structural role in the plant (i.e. lignin, silica and cutin) lower the extent of microbial degradability in cell-wall polysaccharides. The primary role of tannins and terpenoids may be in plant defence against predators.

The division between these groups of SPC is not well defined. For instance, HT are potentially toxic to ruminants, because microbial 'tannases', which hydrolyse galloyl esters, are present in the rumen (Skene and Brooker, 1995). The gallic acid released is further metabolized to potentially toxic phenols, which are absorbed from the rumen (Murdiati *et al.*, 1992). The major lesions are haemorrhage, gastroenteritis, necrosis of the liver and kidney damage, with proximal-tubular necrosis (Dollahite *et al.*, 1962; Holliman, 1985; Filippich *et al.*, 1991). Excessive consumption of oaks and other tree species that contain more than 200 g kg^{-1} HT results in high mortality and morbidity in cattle and sheep.

However, SPC in forages are also associated with improved nutritive value and may have beneficial effects on animal health. PA, more commonly called condensed tannins in the animal nutrition literature, in forage legumes, such as sainfoin (*Onobrychis viciaefolia*), bird's-foot trefoil (*Lotus corniculatus*) and Lotus *pedunculatus*, are associated with improved protein digestion and metabolism in ruminants and in protecting ruminants against legume bloat (see reviews by Reed, 1995; Waghorn *et al.*, 1998). PA may also protect ruminants against helminthiasis. Undrenched lambs grazing sulla (*Hedysarum coronarium*), a forage that contains PA, had lower faecal egg counts and *Trichostrongylus colubriformis* worm burdens and higher average daily live-weight gains than undrenched lambs grazing lucerne (alfalfa) (*Medicago sativa*), which does not contain PA (Niezen *et al.*, 1995). Growing interest in the potential health-promoting effects of SPC in human foods has prompted research on their potential to prevent or treat cancer, circulatory disease and viral infection. The mechanisms by which SPC have beneficial effects on health are likely to be the same as their toxic effects, and the difference between toxicity and beneficial effects is probably dose-dependent (Shahidi, 1995).

This chapter will not attempt to review the extensive literature on the toxicity and nutritional effects of SPC in forages. Recent reviews are available (D'Mello, 1997; Cheeke, 1998; Reed, 1998) and readers are also referred to a review on mammalian metabolism of SPC by Scheline (1991). In this chapter, we shall discuss results from our research on two forages, *Sesbania seban* and *Trifolium pratense* (red clover), as a way of illustrating the effects of SPC on forage evaluation.

Plant Polyphenols and Protein Digestion in Ruminants

Our research on *Sesbania* and red clover is directed at the manipulation of the plant phenolic chemistry in order to improve protein digestion and metabolism in ruminants. Rapid rates of proteolysis and deamination of amino acids in the rumen are nutritional problems of productive forage legumes, such as lucerne (Broderick, 1985; Broderick and Buxton, 1991). Problems associated with extensive proteolysis and deamination of amino acids limit production in modern feeding systems (Beever *et al.*, 1989). Over 75% of the protein in lucerne is degraded in the rumen (Broderick and Buxton, 1991). Excessive proteolysis and deamination create high levels of rumen ammonia, which is not used by rumen microbes for synthesis of amino acids. This ammonia is absorbed, metabolized to urea in the liver and excreted in the urine and represents a net loss of both energy and protein to ruminants. Lucerne has a higher rate of proteolysis than forage legumes that contain

PA, such as sainfoin and sericea lespedeza (*Lespedeza cuneata* (Dum-Cours) G. Don) (Broderick and Albrecht, 1997). Feeding experiments with forage legumes that contain PA indicate that, at high levels of protein intake, PA have a positive effect on nitrogen (N) retention and increase the flow of essential amino acids to the duodenum (Thomson *et al.*, 1971; Harrison *et al.*, 1973; Egan and Ulyatt, 1980; John and Lancashire, 1981; Barry and Manley, 1984; Beever and Siddons, 1985; Waghorn *et al.*, 1987). Although apparent and true digestibility of protein is depressed, urinary loss of N is sufficiently reduced to compensate for the greater faecal loss. These results suggest that it may be possible to manipulate the PA chemistry of forage legumes in order to improve protein utilization.

The phenolic chemistry of forage legumes also has a large effect on the proteolysis that occurs in silages. Extensive proteolysis during ensiling also creates high levels of soluble non-protein N (SNPN) (Albrecht and Muck, 1991). Plant proteases are responsible for most of the proteolysis that occurs during wilting and the initial phases of the silage fermentation (Muck, 1988). The SNPN in legume and grass silages also contributes to high levels of rumen ammonia and inefficient use of protein. Albrecht and Muck (1991) demonstrated that proteolysis is lower in silage produced from bird's-foot trefoil, sainfoin and red clover in comparison with lucerne. SNPN was negatively correlated with tannin content in silage made from sericea lespedeza, sainfoin, bird's-foot trefoil and lucerne ($r^2 = 0.93$).

However, forage legumes seldom contain a single class of SPC. In addition to PA and related phenolic compounds, forage legumes may contain toxic compounds that interfere with the normal metabolic function of ruminants. As illustrated in the following discussion, the effects of toxic SPC in *Sesbania* and red clover need to be considered in research on the manipulation of phenolic chemistry to improve protein digestion and metabolism.

Secondary Plant Compounds in the Evaluation of Red Clover

Red clover contains a soluble polyphenol oxidase (PPO) and a high level of phenolic substrates (Jones *et al.*, 1995a, b). The action of PPO on phenolic substrates produces quinones and leads to a rapid browning reaction when red clover is harvested and allowed to wilt. The quinones react with amino, sulph-hydryl, thioester, phenolic, indole and imidazole groups of proteins (Matheis and Whitaker, 1984). Subsequent cross-linking of proteins by quinones reduces solubility and enzymatic proteolysis during field wilting and the ensiling process. Although red clover does not contain PA, this legume has lower rates of proteolysis and ammonia production in the rumen in comparison with lucerne (Jones *et al.*, 1995a, b; Yocum, 1995). The effect of the PPO and phenolic substrates on proteolysis in red clover are similar to the effects of condensed tannins in forage legumes that contain PA (Yocum and Reed, 1994). The content of soluble phenolic compounds in red clover is more than twice the content in lucerne and other forage legumes that do not contain PA (Fig. 20.1). The higher levels of soluble phenolics are associated with high amounts of neutral-detergent-insoluble N (NDIN). In a study of six forage legumes, red clover had the highest content of NDIN and this was higher than many species that contain PA (Table 20.1). The NDIN and acid-detergent-insoluble N (ADIN)

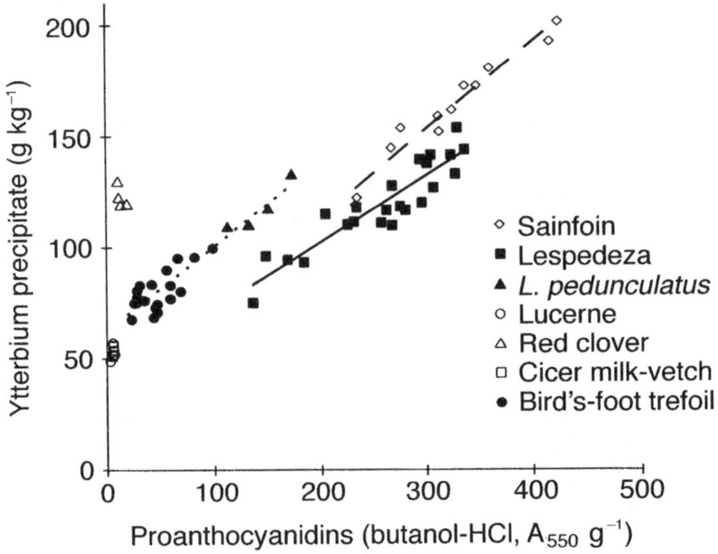

Fig. 20.1. The relationship between soluble phenolic compounds, as measured by ytterbium precipitation, and proanthocyanidins (PA), as measured by butanol-HCl, in seven species of forage legumes. Red clover (*Trifolium repens*) has a similar content of soluble phenolics to sainfoin (*Onobrychis viciaefolia*) and *Lotus pedunculatus* but does not contain PA. Lucerne (*Medicago sativa*) and cicer milk-vetch (*Astragalus cicer*) also do not contain PA, but their concentration of soluble phenolics is less than 40% of the concentration in red clover. The relationship between ytterbium precipitate and PA is highly linear in sainfoin (dashed line, $r^2 = 0.95$), lespedeza (*Lespedeza cuneata*, solid line, $r^2 = 0.84$) and combined *L. pedunculatus* and bird's-foot trefoil (*L. corniculatus*, dotted line, $r^2 = 0.88$). HCl, hydrochloric acid.

in red-clover silage was significantly higher than in lucerne silage and similar to those in sainfoin silage when expressed as a percentage of the total N (Table 20.2) (Yocum, 1995).

Two phenylpropenoid compounds, clovamide and phaselic acid, were isolated and identified in leaves of red clover. Both phenolic compounds are substrates for the PPO in red clover, as indicated by their rapid disappearance during the oxidation process (Fig. 20.2). Red-clover PPO that was isolated by using ammonium sulphate fractionation and chromatography on *Sephadex* G-150 had a pH optimum of 6.5 with catechin and 10.5 with 4-terbutylcatechol. These results indicate that red-clover PPO catalyses the conversion of *ortho*-dihydroxyphenols to quinones (Fig. 20.3). Our laboratory is currently purifying the enzyme, using different chromatographic techniques, in order to determine molecular weight and study enzyme kinetics.

Silages from lucerne, red clover and sainfoin were fed to wethers with rumen cannulae and placed in metabolism crates with feeding mechanisms for continuous feeding in a 3 × 3 Latin-square design (Yocum, 1995). Rumen ammonia and blood urea N was highest and N in the rumen particulate dry matter was lowest in sheep

Table 20.1. Nitrogen, neutral-detergent fibre (NDF) and neutral-detergent-insoluble N (NDIN) in six species of forage legumes. Red clover (*Trifolium pratense*) does not contain proanthocyanidins (PA) but had higher amounts of NDIN than *Lotus corniculatus*, *Lotus pedunculatus*, *Onobrychis viciaefolia* and *Lespedeza cuneata*, which do contain PA. The effect of the polyphenol oxidase and phenolic substrates may decrease the solubility of proteins in detergent.

	N		NDF		NDIN		
Forage	Mean (g kg^{-1} DM)	SD	Mean (g kg^{-1} DM)	SD	Mean (g kg^{-1} DM)	SD	% of N
Medicago sativa	36.0	2.7	406	16	10.7	0.9	29.7
Trifolium repens	28.6	0.8	385	15	18.4	1.9	64.3
Lotus corniculatus:							
var. Norceen	35.4	3.0	350	8	8.7	1.3	24.6
var. Viking	33.6	0.7	349	20	7.9	0.7	23.5
Lotus pedunculatus:							
var. Red Maku	39.2	1.8	345	43	18.4	2.1	46.9
Onobrychis viciaefolia							
var. Remont	35.3	5.6	357	8	15.2	1.6	43.0
var. Eski	39.5	1.8	307	22	16.3	1.7	41.3
Lespedeza cuneata							
var. Au Lotan	28.4	1.6	524	31	14.5	0.9	51.0
var. Serala	27.1	1.2	496	26	12.8	0.3	47.2

DM, dry matter; SD, standard deviation.

Table 20.2. Nitrogen, neutral-detergent fibre (NDF), neutral-detergent-insoluble N (NDIN), acid-detergent fibre (ADF) and acid-detergent-insoluble N (ADIN) in lucerne, red-clover and sainfoin silages. Red-clover silage does not contain proanthocyanidins but has a similar concentration of NDIN and ADIN to sainfoin silage.

Component (g kg^{-1} DM)	Lucerne	Red clover	Sainfoin	LSD
N	26.6^{ab}	25.3^{a}	27.8^{b}	2.3
NDF	478^{ab}	446^{a}	499^{b}	43.1
NDIN	3.2^{a}	7.9^{b}	9.9^{c}	0.04
ADF	358	359	357	21.8
ADIN	1.2^{a}	1.7^{b}	1.8^{b}	0.3
NDIN (% N)	11.80^{a}	31.25^{b}	35.70^{b}	4.65
ADIN (% N)	4.60^{a}	6.70^{b}	6.40^{b}	1.68

Within rows, means with different superscripts differ ($P < 0.05$).
LSD, least significant difference; DM, dry matter.

fed lucerne silage (Table 20.3). Wethers fed sainfoin silage had the highest excretion of total faecal N, faecal NDIN and ADIN, and had the lowest excretion of urinary N (Table 20.3). Wethers fed red-clover and sainfoin silage had higher N retention than wethers fed lucerne silage. The polyphenols in sainfoin and red clover have a large effect on parameters of protein digestion and metabolism, such

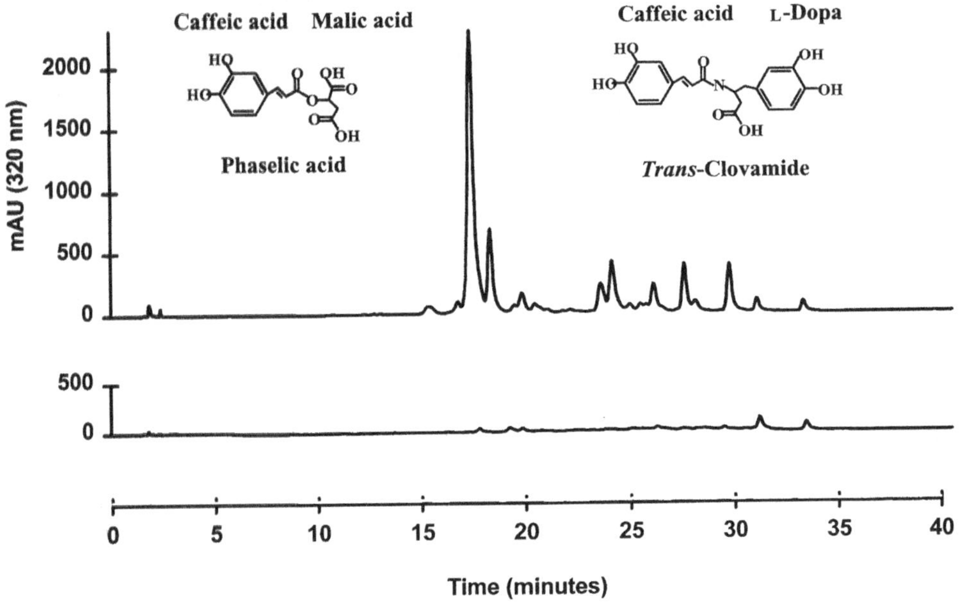

Fig. 20.2. Reverse-phase high-performance liquid chromatography of aqueous methanol extracts of red-clover leaves. Peaks corresponding to phaselic acid and *trans*-clovamide disappear after leaves are allowed to wilt and turn brown, in approximately 30 min.

as lower rumen ammonia and blood urea N and a shift from urinary N excretion to faecal N excretion in comparison with lucerne silages. These effects are accompanied by an increase in the faecal excretion of fibre-bound N and, in the case of sainfoin, a significant depression in the true digestibility of N, but the net result may be a positive effect on N balance. Red-clover silage, on the other hand, had a true N digestibility that was similar to that of lucerne silage, because the NDIN of red clover has a high total-tract digestibility. The NDIN fraction of red-clover silage has an *in vitro* degradability of 74% and digests at a greater rate than the neutral-detergent fibre (NDF) (Yocum, 1995; Fig. 20.4).

Red clover also contains the oestrogenic isoflavones formononetin, daidzein, biochanin-A and genestein (Farnsworth *et al.*, 1975). Formononetin in red clover and subterranean clover (*Trifolium subterraneum*) is the isoflavone that causes reproductive disorders in sheep (Millington *et al.*, 1964; Davies, 1987; Lundh, 1990). Although formononetin is less oestrogenic than the other isoflavones, formononetin is metabolized by rumen microbes to equol, a potent oestrogen (Nilsson *et al.*, 1967; Shutt *et al.*, 1970; Davies and Hill, 1989; Fig. 20.5). Biochanin-A and genestein are degraded to non-oestrogenic compounds by rumen microorganisms (Batterharn *et al.*, 1965; Nilsson *et al.*, 1967). The method of preservation of red clover has an effect on the oestrogenic activity of the forage (Kallela, 1975, 1980). Ensiling may preserve the oestrogenic activity (Kallela, 1980), whereas drying may decrease it.

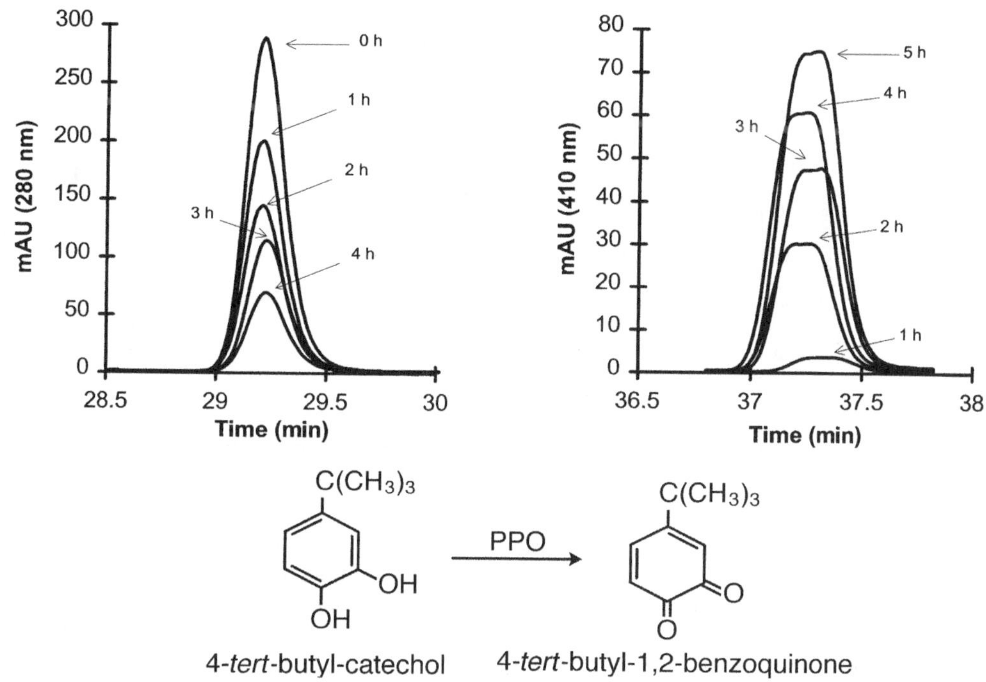

Fig. 20.3. The conversion of dihydroxyphenols to quinones by red-clover polyphenol oxidase (PPO). Reverse-phase high-performance liquid chromatography shows the disappearance of 4-*tert*-butyl-catechol and subsequent appearance of 4-*tert*-butyl-1,2-benzoquinone.

Table 20.3. Parameters of nitrogen metabolism in wethers fed lucerne, red-clover and sainfoin silages. Red clover and sainfoin differ from lucerne because of the effects proanthocyanidins in sainfoin and the effects of polyphenol oxidase and phenolic substrates in red clover on proteins during wilting and ensiling (Yocum, 1995).

	Lucerne	Red clover	Sainfoin
Rumen parameters			
Ammonia N (mg kg^{-1})	415^{a}	166^{b}	24.9^{ab}
Particulate N (mg kg^{-1})	27^{b}	34^{a}	35^{a}
Blood urea N (mg dl^{-1})	18.5^{a}	15.2^{b}	14.1^{b}
Nitrogen excretion (g day^{-1})			
Urinary N (g day^{-1})	14.8^{a}	11.1^{ab}	10.3^{b}
Faecal N (g day^{-1})	6.6^{c}	9.4^{b}	11.1^{a}
Faecal NDIN (g day^{-1})	1.4^{b}	1.7^{b}	6.7^{a}
Digestibility (%)			
True N	93.5^{a}	92.1^{b}	71.7^{c}
Acid-detergent-insoluble N	2.9^{a}	−26.3^{a}	−170.2^{b}
Lignin	1.4^{a}	−19.1^{a}	−73.1^{b}

Within rows, differences between means with different superscripts are significant ($P < 0.05$).

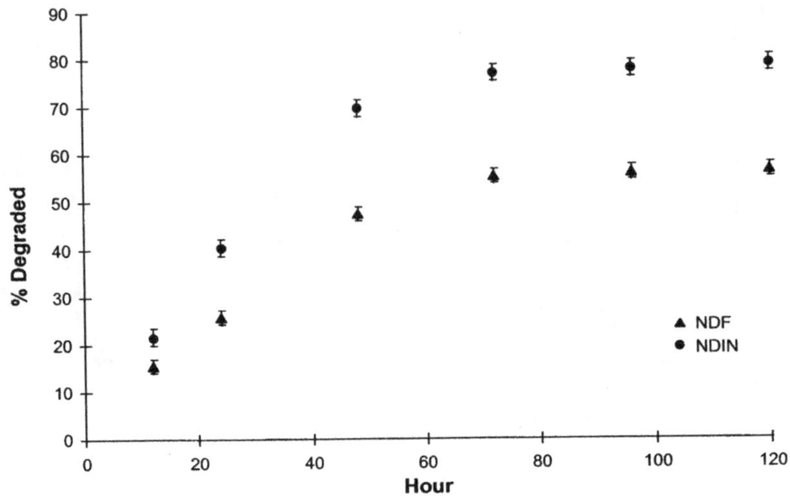

Fig. 20.4. *In vitro* degradability of neutral-detergent fibre (NDF) and neutral-detergent-insoluble N (NDIN) from red-clover silage. The rate of degradation for NDF and NDIN was 0.033 h^{-1} and 0.041 h^{-1}, respectively. The *in vitro* extent of degradation was 74%, a value that is close to the total-tract digestibility obtained in feeding trials with sheep (Yocum, 1995).

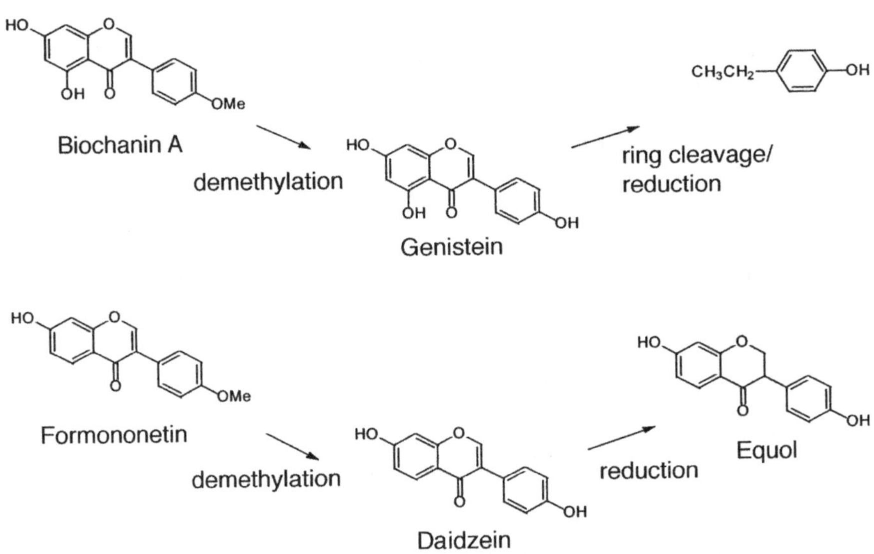

Fig. 20.5. Degradative pathway of red clover isoflavones by ruminal microbes in sheep (from Nilsson *et al.*, 1967).

We studied the effect of feeding ensiled red clover, lucerne and sainfoin to lactating ewes nursing twin lambs during January, February and March in order to determine if differences among silages in protein digestion and metabolism would have an effect on ewe milk yield, as indicated by lamb growth. Ewes were assigned to three silages: 20 ewes to lucerne, 17 ewes to red clover and 15 ewes to sainfoin. During the first 26 days, the average daily gains for lambs from ewes fed lucerne were significantly higher ($P < 0.05$) than for lambs from ewes fed red clover or sainfoin. Average daily gain from birth to 40 days and from birth to weaning were significantly higher for lambs from ewes fed sainfoin ($P < 0.05$) than for lambs from ewes fed lucerne. The higher average daily gains for the first 26 days for lambs from lucerne treatment indicates that milk yield during this period was higher for the ewes fed lucerne than for ewes fed red clover and sainfoin. When lambs consumed solid feed after 26 days, lambs from the sainfoin and red-clover treatments had a higher average daily gain than lambs from the lucerne treatment. These results suggest that the lower protein degradability of red-clover and sainfoin silages may have had positive effects on the average daily gains of lambs but not on ewe milk production. However, none of the 17 ewes from the red-clover treatment lambed when they were bred during the March breeding period. Forty-five per cent of the ewes fed lucerne and 69% of the ewes fed sainfoin lambed when they were bred during this period. The lambing percentages for ewes that were fed lucerne and sainfoin were normal for the breed type for the March breeding (Berger, 1993). These results indicate that formononetin in red clover had a negative effect on conception during the early spring breeding for accelerated lambing. The variety of red clover used in this trial is 'Marathon' and this has a high level of formononetin in comparison with varieties that have been selected for a low content (Fig. 20.6).

Although formononetin is linked to reproductive problems in sheep, there is much interest in the health-promoting effects of the isoflavones in soybeans and soya products in human diets. Soybeans contain the isoflavones genistein and daidzein, which have 0.002–0.001 of the oestrogenic activity of oestradiol (Cassidy *et al.*, 1995). High consumption of soybean products by Asians is linked to decreased incidence of breast cancer compared with Western women and decreased mortality due to prostate cancer compared with Western men (Barnes, 1995; Peterson, 1995; Stephens, 1997). Genistein inhibits the *in vitro* proliferation of tumour cells (Fostis *et al.*, 1995) and the growth of human prostate cancer cells (Peterson and Barnes, 1993).

Secondary Plant Compounds and the Evaluation of *Sesbania*

Sesbania is a large genus of over 50 species from Africa and Asia (Gillet *et al.*, 1971). *Sesbania sesban*, *S. goetzei* and *S. keniensis* from Africa are fast-growing and have potential for cut-and-carry forage production on resource-poor farms. *Sesbania sesban* is the most widely distributed, performs well in tropical highlands and has high potential for development as a forage (Gutteridge and Shelton, 1995). The International Livestock Research Institute (ILRI) maintains the world's largest germplasm collection of this genus and has over 100 accessions of *S. sesban* collected from East and Southern Africa.

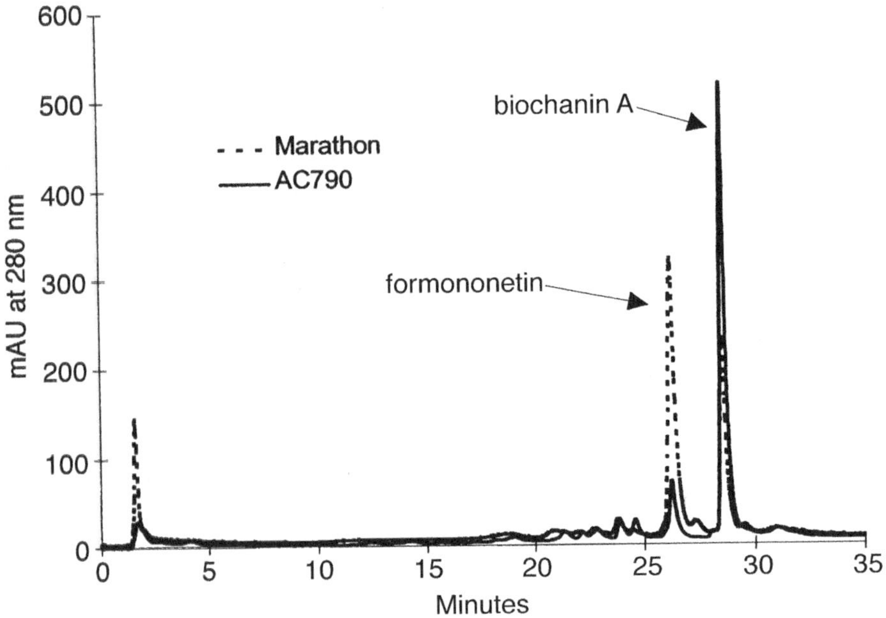

Fig. 20.6. High-performance liquid chromatography of two selections of red clover. 'Marathon' is grown in Wisconsin; AC790 is from New Zealand and has been selected for low formononetin.

However, variation among accessions of *Sesbania* in the quantity and types of secondary compounds affects its nutritive value and utilization as a forage (Tothill *et al.*, 1990; Wiegand *et al.*, 1995). In a sheep metabolism trial at ILRI, accession 15036 (cv. Mount Cotton) had the highest content of PA and the lowest protein digestibility of the three *S. sesban* accessions tested (Wiegand *et al.*, 1995). The presence of other SPC has also raised concerns about the potential toxicity of some *S. sesban* accessions (Brown *et al.*, 1987; Shqueir *et al.*, 1989a, b). The detection and analysis of SPC in the ILRI germplasm collection are required to avoid extensive agronomic evaluation of accessions that are potentially toxic.

Research on the feeding value has concentrated on both the chemical characterization of the entire collection and more detailed nutritional studies, including feeding trials, on some accessions that were identified as promising in agronomic trials. A study of SPC indicated a large amount of variation among accessions (Heering, 1995). High-performance liquid chromatography (HPLC) was used to analyse the flavonoids. A range of distinct fingerprints was found, which showed considerable variation in the content of PA and other types of flavonoids (Heering *et al.*, 1996). Sheep fed accessions that had a high content of PA had the lowest intake of cereal-crop residue and the lowest digestibility of protein, N retention and growth rate (Wiegand *et al.*, 1995). Nevertheless, sheep fed an accession with a moderate content of PA had a higher growth rate and N retention than the sheep fed on the accession with the lowest content. These results suggest that the variation in PA in *Sesbania* could be used to select accessions with a level of PA that does not have a detrimental effect on protein digestion but does improve N metabolism and growth.

However, the results of *Sesbania* toxicity trials with day-old chicks indicate that some accessions contain a toxic SPC (Brown *et al.*, 1987; Shqueir *et al.*, 1989a, b; Reed and Aleemudin, 1995). Researchers in Kenya and Malawi have also reported sporadic toxicity in ruminants (Semenye *et al.*, 1987).

Our research indicated that accessions with a high content of saponin were the most toxic to day-old chicks (Reed and Aleemudin, 1995; Table 20.4). The major saponins in *S. sesban* are glycosides of oleanolic acid (Dorsaz *et al.*, 1988; Fig. 20.7). The biological activity and toxicity of these saponins depend on the pattern of glycosylation (Hostettmann and Marston, 1995). Saponins in *Sesbania* accessions were assayed by thin-layer chromatography (TLC). Six saponins were detected. The total number of spots and their relative areas were larger in *S. sesban* in comparison with *S. goetzii* (Table 20.5). Accessions 10865 and 15019 had the highest relative amounts of saponins and the lowest content of PA.

These results indicate that accessions with a low content of PA are the most toxic to day-old chicks. The toxicity study ranks the accessions in the inverse order to their ranking by sheep growth rate and content of PA (Table 20.6). The chick toxicity contradicts the results from feeding trials, where sheep fed acessions that were high in saponins but low in PA had the highest growth rate. There appears to be an inverse relationship between saponins and PA in *Sesbania*. Toxicity may be an effect of the saponins, which affect chicks more than growing sheep, whereas depressed growth rate in sheep is caused by poor protein digestion in accessions that contain a high level of PA. There is also the possibility that interactions between PA and saponins in accessions with higher levels of PA reduce the toxicity of saponin (Freeland *et al.*, 1985).

Although *Sesbania* saponins may be toxic to day-old chicks, there may be beneficial effects of saponins in the diet of sheep. The molluscicidal activity of the

Table 20.4. Toxicity of *S. sesban* and *S. goetzii* leaves to day-old chicks when included at 30% of the diet. *Sesbania* accessions 10865 and 15019 had the higest content of saponins and the lowest content of proanthocyanidins and were the most toxic to chicks, as indicated by the high mortality rates compared with the lucerne and maize/soybean control diets.

	Per cent mortality	
	Mean	SD
Maize/soybean	0	0
Lucerne	0	0
S. sesban		
10865	37	9.9
15019	35	7.1
15036	5	3.5
S. goetzii		
15007	0	0

SD, standard deviation.

Compound	R^1	R^2
1	β-D-glucopyranosyl	H
2	β-D-glucopyranosyl	β-D-glucopyranosyl
3	β-D-glucopyranosyl	β-D-xylopyronsyl

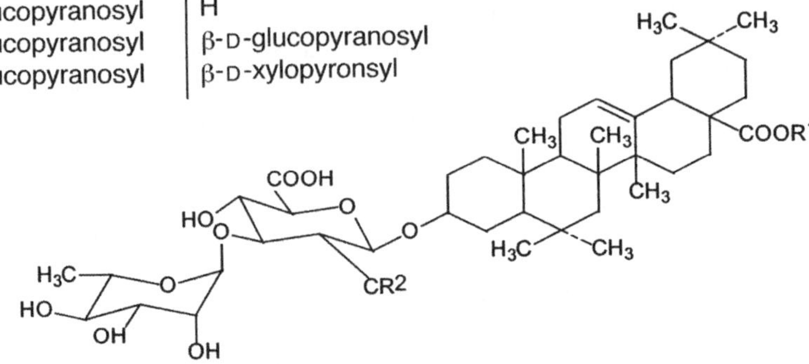

Fig. 20. 7. The oleanane saponins from *Sesbania sesban*. Haemolytic and spermatocidal activity depends on pattern of glycosylation (Dorsaz *et al.*, 1988). Monodesmosidic saponins are not glycosylated at R^1 and have greater haemolytic and molluscicidal activity and may be generated by hydrolytic or enzymatic cleavage of bidesmosidic saponins (Hostettmann and Marston, 1995).

Table 20.5. Thin-layer chromatography of saponins from *Sesbania sesban* and *S. goetzii*. Accessions 10865 and 15019 had the highest total area and area of individual saponins and were the most toxic to chicks.

	Area (mm^2)			
	S. sesban			*S. goetzii*
	10865	15019	15036	15007
TLC spot				
1	22	20	23	23
2	55	40	35	35
3	50	50	38	18
4	52	40	33	ND
5	37	55	33	ND
6	30	30	25	ND
Total area	245	235	186	76

ND, not detected.

oleanane saponins is potentially useful for treating water against the snail vector of schistosomiasis and may have anthelmintic effects. *Sesbania sesban* is used in traditional medicine for protection against mosquito bites, as a remedy for guinea-worms and as a remedy for scorpion stings (Dorsaz *et al.*, 1988). Other plants that contain glycosides of oleanolic acid have anti-inflammatory and antiviral activity (De Tommasi *et al.*, 1991; Liu, 1995).

Table 20.6. Rank, from highest to lowest, of three accessions of *Sesbania sesban* (10865, 15019 and 15036) and one accession of *S. goetzii* (15007) in parameters of nutritive value.

Accession	Sheep growth	Chick mortality	Condensed tannins	Saponins
10865	2	1	4	1
15019	1	2	3	2
15036	3	3	2	3
15007	4	4	1	4

Conclusions

The factors that determine the nutritional value of forages are complex. Laboratory analyses of forages are carried out to determine the relationship of intake and digestibility to content of nutrients (energy, protein, vitamins and minerals). However, many laboratory methods for estimating nutrient content and availability are inaccurate when applied to forages that contain SPC. There are large differences in nutritive value among genotypes of forage species, which may be related to variation in SPC. These differences in nutritive value are not easily predicted by content of nutrients (Wilson, 1977; Wilson and Harrington, 1980) or the analysis of individual SPC, such as phenolics, because plant species seldom contain a single type of SPC. This problem is often greatest in the nutritional evaluation of tropical legumes that have been selected as new potential forage species, as in the case of *S. sesban*. Some species of tropical herbaceous and woody legumes may be agronomically successful but unsuitable for feeding livestock because of their content of SPC. However, it is also apparent that SPC interfere with the nutritional evaluation and utilization of well-known temperate legumes, such as red clover. The variation in the amount and types of SPC within a forage species needs to be investigated from the standpoint of both potential toxic and antinutritional effects and potential beneficial effects.

References

Albrecht, K.A. and Muck, R.E. (1991) Proteolysis in ensiled forage legumes that vary in tannin concentration. *Crop Science* 31, 464–469.

Barnes, S. (1995) Effect of genestein on *in vitro* and *in vivo* models of cancer. *Journal of Nutrition* 125, 777S–783S.

Barry, T.N. and Manley, T.R. (1984) The role of condensed tannins in the nutritional value of *Lotus pedunculatus* for sheep 2. Quantitative digestion of carbohydrates and protein. *British Journal of Nutrition* 51, 493–504.

Batterharn, T.J., Hart, N.K., Lamberton, J.A. and Braden, A.W.H. (1965) Metabolism of oestrogenic isoflavones in sheep. *Nature* 206, 509.

Beever, D.E. and Siddons, R.C. (1985) Digestion and metabolism in the grazing ruminant. In: Milligan, L.P., Grovurn, W.L. and Dobson, A. (eds) *Control of Digestion and Metabolism in Ruminants*. Prentice-Hall, Englewood Cliffs, New Jersey.

Beever, D.E., Gill, M. and Sutton, J.D. (1989) Limits to animal production with high forage diets. *Journal of Animal Science* 67 (Suppl. 1), 298.

Berger, Y.M. (1993) Performance of Romanov crossbred and Finnsheep crossbred ewes in an accelerated lambing system. In: *Proceedings 41st Annual Spooner Sheep Day*. Department of Meat and Animal Science, University of Wisconsin-Madison, Madison, pp. 16–20.

Broderick, G.A. (1985) Alfalfa silage or hay versus corn silage as the sole forage for lactating cows. *Journal of Dairy Science* 68, 3262–3271.

Broderick, G.A. and Albrecht, K.A. (1997) Ruminal *in vitro* degradation of protein in tannin-free and tannin-containing forage legume species. *Crop Science* 37, 1884–1891.

Broderick, G.A. and Buxton, D.R. (1991) Genetic variation in alfalfa for ruminal protein degradability. *Canadian Journal of Plant Science* 71, 755–760.

Brown, D.L., Barnes, D.A., Rezende, S.A. and Klasing, K.C. (1987) Yield, composition and feeding value of irrigated *Sesbania sesban* var. *nubica* leaves harvested at latitude 38°N during a Mediterranean summer. *Animal Feed Science and Technology* 18, 247–255.

Cassidy, A., Bingham, S. and Setchell, K. (1995) Biological effects of isoflavones in young women: importance of the chemical composition of soyabean products. *British Journal of Nutrition* 74, 587–601.

Cheeke, P.R. (1998) *Natural Toxicants in Feeds, Forages, and Poisonous Plants*. Interstate, Danville, Illinois, 479 pp.

Davies, H.L. (1987) Limitations to livestock production associated with phytoestrogens and bloat. In: Wheeler, J.L., Pearson, C.J. and Robards, G.E. (eds) *Temperate Pastures: Their Production, Use and Management*. CSIRO, Melbourne, Australia, pp. 446–456.

Davies, H.L. and Hill, J.L. (1989) The effect of diet on the metabolism in sheep of the tritiated isoflavones formononetin and biochanin A. *Australian Journal of Agricultural Research* 40, 157–163.

De Tommasi, N., Conti, C., Stein, M.L. and Pizza, C. (1991) Structure and antiviral activity of triterpenoid saponins from *Calendula arvensis*. *Planta Medica* 57, 250–253.

D'Mello, J.P.F. (1997) *Handbook of Plant and Fungal Toxicants*. CRC Press, Boca Raton, 356 pp.

Dollahite, J.W., Pigeon, R.F. and Camp, B.J. (1962) The toxicity of gallic acid, pyrogallol, tannic acid, and *Quercus havardi* in the rabbit. *American Journal of Veterinary Research* 23, 1264.

Dorsaz, A., Hostettmann, M. and Hostettmann, K. (1988) Molluscicidal saponins from *Sesbania sesban*. *Planta Medica* 54, 225–227.

Egan, A.R. and Ulyatt, M.J. (1980) Quantitative digestion of fresh herbage by sheep VI. Utilization of nitrogen in five herbages. *Journal of Agricultural Science, Cambridge* 94, 47–56.

Farnsworth, N.R., Bingel, A.S., Cordell, G.A., Crane, F.A. and Fong, H.H.S. (1975) Potential value of plants as sources of new anti-fertility agents II. *Journal of Pharmaceutical Sciences* 64, 213–236.

Filippich, L.J., Zhu, J. and Alsalmi, M.T. (1991) Hepatotoxic and nephrotoxic principles in *Terminalia oblongata*. *Research in Veterinary Science* 50, 170.

Fostis, T., Pepper, M., Adlercreutz, H., Hase, T., Montesano, R. and Schweigerer, L. (1995) Genestein, a dietary ingested isoflavonoid, inhibits cell proliferation and *in vitro* angiogenesis. *Journal of Nutrition* 125, 790S–797S.

Freeland, W.J., Calcott, P.H. and Anderson, L.R. (1985) Tannins and saponin: interaction in herbivore diets. *Biochemical Systematics and Ecology* 13, 189–193.

Gillet, J.B., Polhill, R.M. and Verdcourt, B. (1971) Leguminosae. In: Milne-Redhead, E. and Polhill, R.M. (eds) *Flora of Tropical East Africa*. Crown Agents for Overseas Governments and Administrations, Royal Botanic Gardens, Kew, UK, 1109 pp.

Gutteridge, R.C. and Shelton, H.M. (1995) New herbage plant cultivar, b. Legumes, 24. sesban (a) *Sesbania sesban* (L.) Merill (sesban) cv. Mount Cotton. *Tropical Grasslands* 29, 188–189.

Harborne, J.B. (1993) *Introduction to Ecological Biochemistry*. Academic Press, London, p. 318.

Harrison, D.G., Beever, D.E., Thomson, D.J. and Osbourn, D.F. (1973) The influence of diet upon quantity and types of amino acids entering and leaving the small intestine of sheep. *Journal of Agricultural Science, Cambridge* 81, 391–401.

Heering, H., Reed, J.D. and Hanson, J. (1996) Differences in *Sesbania sesban* accessions in relation to their phenolic concentration and HPLC fingerprints. *Journal of the Science of Food and Agriculture* 71, 92–98.

Heering, J.H. (1995) Botanical and agronomic evaluation of a collection of *Sesbania sesban* and related perennial species. Thesis, Landbouw Universiteit, Wageningen, 127 pp.

Holliman, A. (1985) Acorn poisoning in ruminants. *Veterinary Record* 116, 546.

Hostettmann, K. and Marston, A. (1995) *Saponins*. Cambridge University Press, Cambridge, 548 pp.

John, A. and Lancashire, J.A. (1981) Aspects of feeding and nutritive value of *Lotus* species. *Proceedings of the New Zealand Grassland Association* 42, 152–159.

Jones, B.A., Hatfield, R.D., Muck, R.E. (1995a) Screening legume forages for soluble phenols, polyphenol oxidase and extract browning. *Journal of the Science of Food and Agriculture* 67, 109–112.

Jones, B.A., Muck, R.E. and Hatfield, R.D. (1995b) Red clover extracts inhibit legume proteolysis. *Journal of the Science of Food and Agriculture* 67, 329–333.

Kallela, K. (1975) The effect of storage on estrogenic effect of red clover silage. *Nord. Vet. Med.* 27, 562–569.

Kallela, K. (1980) The estrogenic effect of silage fodder. *Nord. Vet. Med.* 32, 180–186.

Liu, J. (1995) Pharmacology of oleanolic acid and ursolic acid. *Journal of Ethnopharmacology* 49, 57–68.

Lundh, T. (1990) Uptake, metabolism and biological effects of plant estrogens in sheep and cattle. Dissertation, Swedish University of Agricultural Sciences, Uppsala.

Matheis, G. and Whitaker, J.R. (1984) Modification of proteins by polyphenol oxidase and peroxidase and their products. *Journal of Food Biochemistry* 8, 137–162.

Millington, A.J., Francis, C.M. and McKeown, N.R. (1964) Wether bioassay of annual pasture legumes: II. The oestrogenic activity of nine strains of *Trifolium subterraneum* L. *Australian Journal of Agricultural Research* 15, 527–536.

Muck, R.E. (1988) Factors influencing silage quality and their implications for management. *Journal of Dairy Science* 71, 2992–3002.

Murdiati, T.B., McSweeney, C.S. and Lowry, J.B. (1992) Metabolism in sheep of gallic acid, tannic acid and hydrolysable tannin from *Terminalia oblongata*. *Australian Journal of Agricultural Research* 43, 1307.

Niezen, J.H., Waghorn, T.S., Charleston, W.A.G. and Waghorn, G.C. (1995) Growth and gastrointestinal nematode parasitism in lambs grazing either lucerne (*Medicago sativa*) or sulla (*Hedysarum coronarium*) which contains condensed tannins. *Journal of Agricultural Science, Cambridge* 125, 281–289.

Nilsson, A., Hill, J.L. and Davies, L.H. (1967) An *in vitro* study of formononetin and biochanin A metabolism in rumen fluid from sheep. *Biochimica Biophysica Acta* 148, 92–98.

Peterson, G. (1995) Evaluation of the biochemical targets of genestein in tumor cells. *Journal of Nutrition* 125, 784S–789S.

Peterson, G. and Barnes, S. (1993) Genestein and biochanin A inhibit the growth of human prostate-cancer cells but not epidermal growth-factor receptor tyrosine autophosphorylation. *Prostate* 22, 335–345.

Reed, J.D. (1995) Nutritional toxicology of tannins and related polyphenols in forage legumes. Invited paper, Pharmacology/Toxicology Symposium on Toxic Legumes. *Journal of Animal Science* 73, 1516–1528.

Reed, J.D. (1998) Ecological biochemistry of secondary plant compounds in herbivore nutrition. Invited paper, XVIII International Grassland Congress, 8–19 June, Winnepeg, Manitoba, Canada.

Reed, J.D. and Aleemudin, Y. (1995) *Toxicity Effects of* Sesbania sesban *and* Sesbania goetzei *Feed on Broiler Chicks*. Internal Project Report, Department of Meat and Animal Science, University of Wisconsin-Madison, Madison, Wisconsin.

Scheline, R.R. (1991) *Handbook of Mammalian Metabolism of Plant Compounds*. CRC Press, Boca Raton, 514 pp.

Shqueir, A.A., Brown, D.L., Taylor, J.S., Rivkin, I. and Klasing, K.C. (1989a) Effect of solvent extractions, heat treatments and added colesterol on *Sesbania sesban* toxicity in growing chicks. *Animal Feed Science and Technology* 25, 127–135.

Shqueir, A.A., Brown, D.L. and Klassing, K.C. (1989b) Canavanine content and toxicity of sesbania leaf meal for growing chicks. *Animal Feed Science and Technology* 25, 137–147.

Semenye, P.P., Musalia, L., Onim, J.F.M. and Fitzhugh, H.A. (1987) Toxicity of *Leucaena leucocephala* and *Sesbania sesban* sun dried leaf hays. In: *Proceedings of 6th KVA/SR–CRSP Kenya Workshop, Nairobi, Kenya*, pp. 93–101.

Shahidi, F. (1995) Beneficial health effects and drawbacks of antinutrients and phytochemicals in foods: an overview. In: ACS *Symposium Series 662: Antinutrients and Phytochemicals in Food*. American Chemical Society, Washington, pp. 1–9.

Shutt, D.A., Weston, R.H. and Hogan, J.P. (1970) Quantitative aspects of phyto-oestrogen metabolism in sheep fed on subterranean clover (*Trifolium subterraneum* cultivar clare) or red clover (*Trifolium pratense*). *Australian Journal of Agricultural Science* 21, 713–722.

Skene, I.K. and Brooker, J.D. (1995) Characterization of tannin acylhydrolase activity in the ruminal bacterium *Selenomonas ruminantium*. *Anaerobe* 1, 321–327.

Stephens, F.O. (1997) Phytoestrogens and prostate cancer: possible preventive role. *Medical Journal of Australia* 167, 138–140.

Thomson, D.J., Beever, D.E., Harrison, D.G., Hill, I.W. and Osbourn, D.F. (1971) The digestion of dried lucerne and sainfoin by sheep. *Proceedings of the Nutrition Society* 3, 14A.

Tothill, J.C., Reed, J.D. and Asres Tsehay (1990) Genetic resources and fodder quality in *Sesbania*. In: Macklin, B. and Evans, D.O. (eds) *Perennial* Sesbania *Species in Agroforestry Systems*. Nitrogen Fixing Tree Association, Waimanalo, Hawaii, pp. 89–91.

Waghorn, G.C., Ulyatt, M.J., John, A. and Fisher, M.T. (1987) The effect of condensed tannins on the site of digestion of amino acids and other nutrients in sheep fed on *Lotus corniculatus* L. *British Journal of Nutrition* 57, 115–126.

Waghorn, G.C., Reed, J.D. and Ndlovu, L.R. (1998) Condensed tannins and herbivore nutrition. Invited paper, XVIII International Grassland Congrees, 8–19 June, Winnepeg, Manitoba, Canada.

Wiegand, R.O., Reed, J.D., Said, A.N. and Ummuna, V.N. (1995) Proanthocyanidins and the use of *Sesbania sesban* and *Sesbana goetzei* as protein supplements. *Animal Feed Science and Technology* 56, 207–216.

Wilson, A.D. (1977) The digestibility and voluntary intake of the leaves of trees and shrubs by sheep and goats. *Australian Journal of Agricultural Research* 28, 501–508.

Wilson, A.D. and Harrington, G.N. (1980) Nutritive value of Australian browse plants. In: Le Houerou, H.N. (ed.) *Browse in Africa*. International Livestock Centre for Africa, Addis Ababa, pp. 291–297.

Yocum, F.D. (1995) Proanthocyanidins and polyphenols in ensiled forage legumes and nitrogen digestion by sheep. MSc thesis, University of Wisconsin-Madison, Madison, 105 pp.

Yocum, F.D. and Reed, J.D. (1994) Proanthocyanidins and polyphenols in ensiled forage legumes and nitrogen digestion by sheep. ADSA/ASAS Joint Annual Meeting, July 11–15, 1994, Minneapolis, MN. *Journal of Dairy Science* 77 (Suppl.), 385.

21 Current Procedures, Future Requirements and the Need for Standardization

D.I. Givens,[1,2] E. Owen[2] and A.T. Adesogan[3]

[1]*ADAS Nutritional Sciences Research Unit, Alcester Road, Stratford-upon-Avon CV37 9RQ, UK;* [2]*Department of Agriculture, University of Reading, Earley Gate, PO Box 236, Reading RG6 6AT, UK;* [3]*Welsh Institute of Rural Studies, University of Wales, Aberystwyth SY23 3AL, UK*

How Well Do Current Evaluation Techniques Assess Nutrients Available from Forage?

Most past and many current approaches to forage characterization are designed to meet the needs of nutrition/feeding models that essentially ask: how much energy and protein does the animal need to meet the requirements of maintenance and a given output of milk, meat, etc.? Thus, the energy value of forages is commonly estimated from the apparent digestibility of dry matter (DM) or organic matter through the entire digestive tract. This approach has led to the development of *in vitro* and near-infrared reflectance spectroscopy (NIRS) approaches, which predict *in vivo* digestibility, as described in Chapters 6, 7, 8 and 15.

Whilst these approaches have provided a valuable insight into forage quality, ruminant-animal production is facing new challenges, including the need to design diets that not only offer the maximum differential between feed costs and product sales but also reduce negative environmental impact. In addition, for high-genetic-merit dairy cows, at least, it seems essential that the dietary influence on the partition of dietary energy between body-tissue deposition and milk secretion is predicted (Agnew *et al.*, 1998). There is also an increasing need to predict the output of milk constituents (fat, protein, lactose), rather than simply energy and protein. A major consequence of these developments is the industry's desire for approaches to nutrition based on multiple responses of the animal to the nutrients supplied by the diet. Implicit in this is the need to predict voluntary feed intake, microbial transformations in the rumen and the end-products of digestion. This is dealt with in detail in Chapter 4.

For example, one problem that current approaches to herbage evaluation have failed to explain is the poorer productivity of autumn-grown compared with spring-grown grass. Early research (e.g. Corbett *et al.*, 1966) showed that this was likely to be the result of increased partition of metabolizable energy (ME) into heat production, although the reasons for this were unclear. More recently, MacRae *et al.* (1985) showed that, per unit of ME intake, spring grass led to increased incremental intestinal absorption of α-amino nitrogen compared with autumn grass. Recent data (van Vuuren *et al.*, 1992) suggest that most of the difference is likely to be due to increased microbial protein synthesis from the spring grass. MacRae *et al.* (1985) suggested that the additional glucogenic amino acids absorbed would provide additional nicotinamide adenine dinucleotide phosphate (NADPH) and glycerol phosphate necessary for conversion of acetate into long-chain fatty acids and thus improve the flow of energy into animal production. The development of a nutrient-based description of forages, along with suitable models of metabolism, would allow such situations to be revealed and allow corrective action to be taken. However, it is vital not only to know what the nutrient-supply potential of any forage is, but also to understand what factors are responsible for the value of the forage (e.g. molecular structure) and how it could be favourably altered (e.g. by plant breeding).

Volatile fatty acids

In most ruminant-production systems, rumen volatile fatty acids (VFA) represent some 0.6–0.7 of the carbon available for metabolism. It is clear that the various VFA produced during rumen fermentation have different metabolic properties and, for example, do not elicit the same insulin response (Trenkle, 1970). Most rationing systems have not explicitly represented VFA production, although there have been a number of attempts to use stoichiometry-based rumen mechanistic models to predict VFA production and profile (for review, see Sauvant *et al.*, 1995; see also Chapter 4). Many of these approaches have appeared to be unsatisfactory and, whilst improvements to these models are ongoing, the role of *in vitro* systems needs examination.

Early work with a simple gas-production apparatus (O'Hara and Ohki, 1973) showed a strong correlation between gas volume and total VFA *in vitro*. More recently, Blümmel and Bullerdieck (1997) proposed that gas production is, indeed, only a reflection of VFA production, while Rymer and Givens (1999) found that some of the gas-production parameters estimated from the *in vitro* fermentation of various diets based on hay plus micronized maize were closely correlated with at least some of the rumen VFA and pH values estimated *in vivo*. These findings raise the possibility that VFA measurements within *in vitro* gas-production systems may provide a reflection of the amounts and proportions of VFA *in vivo*, although more data are needed. However, this approach may only give the broad picture, since the VFA profile *in vivo* is dynamic, can change substantially after a meal and is influenced by factors such as meal frequency and rate of eating (Sutton *et al.*, 1986).

It is particularly noteworthy that, in the case of grass silage-based diets, there is considerable evidence that the proportions of fermentation acids in the silage can greatly influence the proportions of rumen VFA. In particular, there is a strong relationship between the lactic acid content of the silage and the proportion of propionate in rumen fluid in both cattle (van Vuuren *et al.*, 1995) and sheep (Martin *et al.*, 1994) and there is an effect on the lipogenic : glucogenic VFA ratio (Fig. 21.1). These relationships also appear to hold good for diets containing up to at least

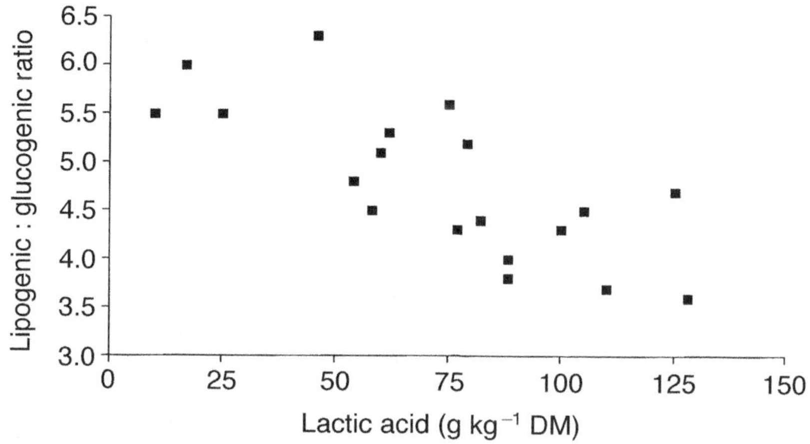

Fig. 21.1. Effect of silage lactic acid on rumen VFA lipogenic : glucogenic ratio (from van Vuuren *et al.*, 1995).

400 g kg^{-1} of concentrates, and Martin *et al.* (1994) suggested that the relationship between the lactic acid content of silage and nutrient supply is likely to be an important factor in the regulation of nutrient utilization for milk production from high-silage diets. This, in part at least, may be a result of its conversion to propionate in the rumen and the impact that the latter may have on gluconeogenesis and the insulin response.

Therefore, it seems vital that, if nutrient supply from silage-based diets is to be predicted, knowledge about the fermentation acids in the silage is required. Whilst this can readily be achieved by gas chromatography, it is slow and expensive. Recent developments in NIRS (Gordon *et al.*, 1998) which allow the obtaining and interpretation of spectral data from fresh (undried) silage have much to offer in this regard. As described in Chapter 15, NIRS can be used to predict lactic acid and VFA.

Microbial protein supply

With most forages, the rumen is the main site of digestion, with some 0.9 of total organic matter being degraded therein (Beever *et al.*, 1986; see also Chapters 2 and 4). A major function of the rumen is the degradation and subsequent fermentation of both soluble and cell-wall carbohydrates to provide energy in the form of adenosine triphosphate (ATP) for microbial maintenance and the synthesis of microbial protein. Microbial protein often provides up to about 0.7 of the absorbed α-amino nitrogen (Sauvant *et al.*, 1995) and, in the case of diets based on grazed herbage, this figure may exceed 0.9. It is now recognized that differences in the amount and proportion of amino acids and other nutrients absorbed can substantially determine the yield and composition of milk (Thomas and Martin, 1988).

It is also noteworthy that microbial growth has a specific demand for hexose in the synthesis of amino acids and nucleic acids from ammonia (Baldwin *et al.*, 1970) and the partition of hexose between synthesis and fermentation can significantly affect the efficiency of microbial protein synthesis (Beever, 1993). Thus, probably the major priority in the development of an improved characterization of forages is a means of better assessing the quality of its complex carbohydrate make-up, how this is likely to be utilized in the rumen and hence the potential for microbial protein synthesis.

There has been considerable interest in the use of the gas-production technique (see Chapter 10) for assessing microbial protein production, both directly and indirectly. Blümmel *et al.* (1994) reported, for cereal straws, an inverse relationship ($r = 0.64–0.77$; $P < 0.001$) between *in vitro* microbial biomass yield, estimated as the difference between the apparent and true degraded residue, and gas production per unit of substrate truly fermented. Although Blümmel *et al.* (1994) proposed a relationship between gas production and *in vitro* microbial biomass yield, it seems unlikely that such a relationship would hold true for microbial protein yield under all rumen conditions. It may be that the combination of estimates of hexose available (e.g. by the *in situ* approach, as described in Chapter 9) and ATP production (i.e. by gas production), together with a suitable model of rumen function, would provide substantially improved estimates of microbial protein production, compared with current relationships based solely on organic matter apparently degraded in the rumen. Further research in this area is needed.

Outflow rates

All current *in situ* and *in vitro* techniques for estimating dynamic aspects of forage degradation and fermentation in the rumen use models to calculate the effective amount of a substrate degraded or fermented, which include assumptions about fractional outflow rates of small particles (see Chapter 9). In many cases, the outcome is very sensitive to the assumptions made, which are still largely those presented by the Agricultural Research Council (ARC, 1984) based on studies with chromium-mordanted protein particles. The ARC (1984) recommended that further data on the outflow of larger forage particles were required and that is still the case today. Of note is the recent finding of Romney and Gill (1998) that the short-term intake rate (STIR) of poor-quality forages at least has considerable potential to predict passage rate. If STIR relates to rumen outflow, it may provide a relatively rapid means of obtaining new data.

Impact of physical structure and sample preparation

Most *in vitro* techniques for forages rely on the use of feeds that have been previously dried and milled. This is derived from the need to use small samples that are homogeneous and representative. However, sample preparation needs considerable care, since, under some circumstances, the destruction of the physical structure can alter the kinetics of the response *in vitro* in a manner different from what occurs in the animal. For example, Sanderson *et al.* (1997) compared the gas-production data from grass silages incubated either as freeze-dried and milled, chopped fresh or unchopped fresh and observed that the freeze-dried and milled material produced substantially higher gas production than the other treatments (Table 21.1). In the same experiment, the authors concluded that wilted silage fermented more rapidly than unwilted silage. This is an important finding which requires further research, since the rate of release of nutrients *in vivo* may also be modified by the likelihood of increased rumination time on the wetter silage.

On the other hand, Deaville and Givens (1998) argued that there may be some merit in pre-drying silages before assessment with the gas-production technique, since the drying largely removes the fermentation acids. If these remain, interpretation of the gas-production data may be flawed, since they lead to gas release from the buffer, which is not associated with fermentation and hence ATP production.

Therefore, it appears vital that *in vitro* techniques are used in a way that prevent destruction of important aspects of physical structure, since failure to account for this may give very misleading results relative to the response *in vivo*.

How Well Do Current Evaluation Techniques Predict Voluntary Intake?

Impact of physical/cellular structure

There is much literature to show that reducing the particle size of forages by a sufficient amount, e.g. by milling (not chopping), increases intake and that the effect is greater in sheep than cattle and is greater the lower the digestibility of the forage (Owen, 1979; Walker, 1984; Fahey *et al.*, 1993). Predicting the response in intake due to physical treatment *per se* has received little recent research. Interest in milling and pelleting high-temperature dried forages declined from the mid-1970s, because of the high cost of fossil fuel. However, as the increase in intake with

Table 21.1. Gas-production data from grass silages incubated either as freeze-dried and milled, chopped fresh or unchopped fresh (after Sanderson *et al.*, 1997).

Silage preparation	Gas asymptote (ml g^{-1} DM)	Lag phase (h)	Fractional rate at 50% of total gas (h^{-1})
Silage 1			
Fresh, unchopped	254[a]	1.44	0.0600[a]
Fresh, chopped to 3 mm	261[a]	1.71	0.0693[a]
Freeze-dried, milled 1 mm screen	284[b]	1.30	0.0995[b]
Silage 2			
Fresh, unchopped	285[c]	0.56	0.0571[a]
Fresh, chopped to 3 mm	303[d]	0.77	0.0670[a]
Freeze-dried, milled 1 mm screen	297[d]	0.87	0.1071[b]

Within columns, different superscripts denote significant differences ($P > 0.05$).

milling is considered to be largely due to increased rate of passage, being able to predict the latter would presumably indicate the response in intake. As noted earlier (see 'Outflow rates' above) and in Chapter 3, STIR (Romney and Gill, 1998) appears to hold promise as a measure of the physical structure of forage and its impact on intake. Future research may well show that intake of forage is better predicted by a combined measure of STIR and rate of digestion in the rumen, estimated by *in vitro* gas production (e.g. Mauricio *et al.*, 1999a). It is likely that the regression relationships between intake and a combination of rate of digestion and STIR will need to be established for forages of similar type (e.g. grasses or legumes) and animals of a given species and level of productivity. Sileshi *et al.* (1996), working in Ethiopia, showed that the correlation between parameters of *in situ* degradation and *in vitro* gas production varied with forage type.

Selectivity

Prediction of intake of non-grazed forage is normally based on *ad libitum* intake, measured indoors, with the amounts offered being restricted to 15% more than the animal will consume, in order to minimize selectivity. However, indoor studies with tropical forages (Zemmelink, 1980) and crop residues, such as barley straw (Wahed *et al.*, 1990) and stovers of sorghum (Osafo *et al.*, 1997) and maize (Methu *et al.*, 1997), show that intake increases with increasing levels offered in excess of *ad libitum*. The intake response is due to animals selecting for the more nutritious leaf and leaf-sheath fractions. Similar selection occurs with grazed forage, provided the allowance rate is sufficient (Gibb and Treacher, 1976). NIRS appears to hold promise as a method of predicting the leaf content, at least in samples of straw (Islam, 1997). A recent workshop on intake in grazing animals (Gibb, 1998) concluded that use of *n*-alkanes as markers was useful for estimating both intake and composition of grazed herbage. On the other hand, Minson and Wilson (1994) concluded that it was probably simpler and more accurate to estimate intake under grazing from equations relating intake of DM and live-weight and production level than to use an equation based on feed composition.

Interactions between forages when fed together

When different forages are fed together, the nutrients they supply often interact and modify the ensuing metabolic processes, to the extent that the digestibility and intake of the combination often differs from the average values for the component forages. Such associative effects reflect the lack of additivity between the nutrients in the forages (Sauvant and Giger, 1989), a process largely ignored by current forage evaluation systems. Interactions may be positive, negative or absent, depending on the nature, physical form and substitution rate of the components, their influence on the rumen ecosystem, the level of feeding and outflow rates and the adequacy of essential nutrients. For instance, synergism was only observed between grass and legume components of a mixed ration when dietary crude protein and mineral levels were deficient in one of the forages and adequate in the other

(Minson, 1990). Likewise, when straw or poor-to-medium-quality hay is supplemented with high-protein feeds, positive effects on intake and digestion are often observed (Butterworth and Mosi, 1986; Reid *et al.*, 1987; Bowman and Asplund, 1988). This is because the additional nitrogen enhances the utilization of the straw.

Associative effects often lead to a higher intake of mixed-forage diets than of the components fed in isolation (Phipps *et al.*, 1995). The reasons for this are unclear, though an increase in palatability upon mixing forages may be involved. Positive associative effects also result from the enhancement of rumen digestion, as seen by decreases in methane production, rumination and heat production (Van Soest, 1994). Simultaneously offered cereal and legume forages, which, respectively, have high readily fermentable energy and protein contents, may synchronously release energy and protein within the rumen and thereby enhance microbial protein production and hence animal performance. Vogel *et al.* (1989) reported that stocking densities could be doubled without adverse effects on live-weight gain in steers fed wheat or Bermuda-grass pastures supplemented with silage, because of positive associative effects between the forages.

Negative associative effects also occur frequently. A common example is when a forage with a high fermentable energy content is fed with one with a high cell-wall content. This is because the fermentation of the carbohydrates in the former produces acidic conditions, which depress the digestibility of the cell walls in the latter. This type of negative associative effect was thought to partly explain the low intraruminal digestion of whole-crop wheat in dairy cows, since the forage has high cell-wall and starch contents (Adesogan *et al.*, 1998). However, there are several accounts of mixed-forage feeding without associative effects (Milne and Bagley, 1976; Eldridge and Kat, 1980; Odoi, 1985), and more work is needed to elucidate the underlying mechanisms and digestive interactions between forages fed together.

As a result of associative effects, the nutritive value of mixed-forage diets is often over- or underestimated when estimated conventionally, because the ingredients are over- or undervalued (Van Soest, 1994). McDonald *et al.* (1995) noted that associative effects pose a serious objection to the determination of the digestibility of components of a mixed ration by difference. They also cited experiments that showed that, for certain ingredients, the efficiency of ME utilization for fattening depends on the nature of the basal ration to which it was added and concluded that values for the efficiency of ME utilization for individual foods are of limited significance. Similarly, Van Soest (1994) noted the continuing problem of oversimplification in relation to prevailing concepts of net energy (NE) values in feed tables and emphasized that it is the composition and character of the total diet that determines the animal response and therefore the NE.

Associative effects between feeds are often ignored when the nutritional potential of diets is assessed. Such interactions can be tested by feeding supplements at different ratios to a basal diet (Van Soest, 1994). Continuous analysis can be applied to evaluate the associative effects of two-component diets, provided the components are, at most, complementary (Franci and Acciaioli, 1998). For more complex diets with several components, the response surface methodology, associated with multiple regression analysis, can be employed (Franci and Acciaioli, 1998). This model was shown by Franci *et al.* (1997) to be a useful and precise means for studying the associative effects of complete diets.

Problems Specific to Silages

Volatile fermentation products in silages

One main difference between silages and the forages from which they are made is the presence of volatile constituents in the silages, which are formed from protein and carbohydrate fermentation during ensiling. Since the volatile compounds typically have higher energy contents than their precursor compounds, silages are usually higher in gross energy (GE) than the forages from which they are formed. Under ideal conditions, the GE of silages would be determined on fresh, unprocessed silage samples. However, the need for uniform samples that are representative of the silage fed necessitates some form of sample processing. Many of the processing methods currently used result in an underestimation of GE content, because they increase the losses of the energy-dense volatiles. The best method appears to be to chop the forage samples in liquid nitrogen. Porter (1992) showed that this method gave more accurate GE results than freezing and cutting with a band saw, shredding, freezing and mincing or chopping in a commercial food processor.

Losses of volatile fermentation products also occur during oven-drying of silages, leading to an underestimation of the DM content. Such underestimations are higher in wetter silages with higher volatile constituents and DM losses of up to 240 g kg^{-1} have been reported (Aerts *et al.*, 1974). If the losses are not corrected, the energy value and chemical constituents in the silage will be overestimated, while the digestibility, DM intake and feed-conversion efficiency (kg feed kg^{-1} gain) will be underestimated (Kaiser *et al.*, 1995). This has resulted in the advancement of several methods for determining the true DM content of silage. One of the most widely used is the toluene distillation technique (Dewar and McDonald, 1961), which quantifies the volatile compounds lost and then uses this information to correct the apparent DM value. However, the method does not always account for all the volatiles lost, especially in silages with high alcohol contents (Kaiser *et al.*, 1995; Porter and Barton, 1996). It is labour-intensive and hazardous and therefore not routinely practicable.

An alternative approach involves quantifying the fermentation products lost during oven-drying and using correction equations, such as that of Porter *et al.* (1984), to account for such losses. However, such equations are forage species-specific and only exist for the more common silages. Many of the equations are also flawed, because they assume that the volatility of the fermentation products is constant across several silage types. Although the development of different equations for grass silage, maize silage and lucerne by Dulphy and Demarquilly (1981) overcomes the interaction between volatility and silage type, the approach does not cover all silage types and requires separate quantification of fermentation products. Such determinations are not required in the method advocated by Givens (1986). This entails using a mean correction factor of 1.9 percentage units, derived from the analyses of 300 silage samples, to account for volatiles lost during oven-drying. While this approach involves no further analysis, it is not sufficiently accurate for some purposes.

Other methods that have been used include low-temperature (70–85°C) oven-drying, freeze-drying, microwave-drying, saponification (Hood *et al.*, 1971), Karl Fischer titration (Galleti and Piccaglia, 1988), gas chromatography and NIRS. Kaiser *et al.* (1995) compared all but the last two of these methods for analysing the DM content of a diverse set of grass silages. They found that only the saponification and Karl Fischer methods appeared to account for all of the volatiles present. They also found that the saponification technique required more time and gave higher standard errors and therefore concluded that the Karl Fischer approach was best for routinely analysing the DM of silage, because it was rapid, accurate and simple. Porter and Barton (1996) also found more accurate DM determinations with the Porter *et al.* (1984) correction equation, gas chromatography or Karl Fischer titration than with the alcohol-corrected toluene distillation and low-temperature oven-drying techniques. However, gas chromatography requires specialist equipment, which may not be available in many laboratories. The availability of odourless, low-health-risk, stable end-point, long shelf-life, pyridine-free reagents overcomes some of the traditional problems of the Karl Fischer titration. However, anhydrous methanol and exclusion of atmospheric moisture are still required and the Dulphy and Demarquilly (1981) correction equation gives comparable results (Galleti and Piccaglia, 1988).

NIRS scanning of undried forages can also provide accurate determinations of silage DM content. Although the cost of purchasing and calibrating the NIRS machine are substantial, once these are done, the technique can be conveniently used for routine DM determinations. Kaiser *et al.* (1995) suggested that NIRS machines should be calibrated with data derived from Karl Fischer titrations.

In conclusion, it is essential to account for volatiles lost during oven-drying of silage. The widely used toluene distillation technique does not account for all the volatiles lost during oven-drying, even when corrected for volatile losses. Therefore, the Karl Fischer titration, gas chromatography, NIRS or corrected oven DM method should be used when accurate DM determinations are paramount. Future work should develop NIRS calibration equations for grass silages and oven DM correction equations for the less common silages.

Volatile compounds can also be rapidly lost from exposed silo faces, leading to differences between the nutritive value of the silage sampled at silo opening and that fed later. Such losses are minimized when good silo management is practised. Nevertheless, future work should examine the extent to which such losses reduce the nutritive value of silages.

High concentration of soluble nitrogen

Soluble nitrogenous compounds in silages include short-chain peptides, ammonia, non-volatile amines and amino acids. In addition to nitrate, these compounds make up the non-protein nitrogen (NPN) fraction, which is about 200–300 g kg^{-1} total nitrogen (N) in fresh forages and 600–700 g kg^{-1} total N in silages (Van Soest, 1994). This difference arises from the proteolytic degradation, which increases NPN at the expense of true protein during ensiling. Deamination of the resulting amino acids leads to increases in concentrations of ammonia, diamines and isoacids (Van Soest, 1994).

Soluble nitrogen is potentially available for ruminal degradation, but its nutritive value is not usually realized, because it is readily converted into ammonia, absorbed and excreted as urea. Some of the soluble N escapes ruminal degradation and is available for peptic digestion in the abomasum. The value of such escape soluble N depends on its amino acid profile, while the extent of rumen escape is affected by the enzyme-to-substrate ratio, the outflow rate and the presence of a readily fermentable energy supply. Brookes and McGuire (1973) found that, in spite of being completely digestible, soluble N was poorly utilized in sheep diets unless the energy-to-protein ratio was improved.

Several authors have found that animal performance improved as nitrogen solubility in the diet decreased. Table 21.2 summarizes some of these findings and reveals the importance of measuring N solubility, so as to regulate its inclusion in ruminant diets. However, there is no standard method for measuring N solubility in silages. Some of the traditional methods include solubility in detergents, cold or hot water, saliva-simulating buffers and salt solutions. The water and buffer solubility methods do not always account for all the soluble true protein and do not differentiate adequately between NPN and true protein in the soluble fraction (Van Soest, 1994). Trichloroacetic acid and tungstic acid methods are superior in these respects.

The UK metabolizable-protein system (AFRC, 1993) determines the quickly degradable protein (QDP) fraction in feeds from the immediately soluble N frac-

Table 21.2. Effect of N solubility (N sol) level in the diet on animal performance.

Author	Feed	Species	Observation
Majdoub *et al.* (1978)	Sorghum silage and concentrates	Dairy cows	Low N sol improved milk yields
Dingley *et al.* (1975)	Maize silage and concentrates	Dairy cows	Low N sol decreased essential and total amino acid supply to the udder
Zook (1982)	Maize silage, lucerne hay and concentrates	Dairy cows	Low N sol improved milk yield and milk fat and protein contents
Beever *et al.* (1976)	Grass	Sheep	Low N sol diets improved efficiency of ruminal VFA synthesis
Mader and Britton (1986)	Lucerne	Steers	Better utilization of lucerne hay than lucerne silage
Knowlton *et al.* (1992)	Lucerne silage and concentrates	Dairy cows	Low N sol diets improved the yield of bacteria per unit silage DM and the metabolizable protein per unit of silage crude protein

tion, which is determined as the N lost when feeds in polyester bags are washed with cold water. This method overestimates true N solubility, due to particulate losses through the pores of the bags, which may include degraded, insoluble N. Such losses can be accounted for by using correction equations (Dhanoa, 1988; Lopez *et al.*, 1994), providing the rate of degradation of the particulate matter lost is similar to that of the material remaining in the bag. Although it is more accurate to determine N solubility by cold-water extraction of feeds suspended in filter-paper (Cockburn *et al.*, 1993), such extractions do not completely account for the solubilization of true protein (Pichard and Van Soest, 1977).

The Agricultural and Food Research Council (AFRC, 1993) assumes that the QDP fraction is utilized in the rumen with an efficiency of 0.8 and recommends a maximum QDP inclusion level of 0.4 of the effective rumen-degradable protein to prevent ammonia toxicity. Although this is a useful guide for preventing excess intake of soluble N, the model is premised on an assumed efficiency of utilization and overestimates the soluble N content of rations, because of the insoluble, readily degradable proteins in the QDP fraction.

The model of Pichard and Van Soest (1977) overcomes some of the shortcomings of the UK metabolizable-protein system by using different solvents to distinguish the water-soluble NPN fraction, containing primarily nitrate, ammonia (NH_3) and free amino acids, from three water-insoluble true-protein fractions that differ in their degradability. This model can improve the accuracy of ration formulation and enhance the efficiency of protein utilization. However, in order to use this model, further work is required to define optimum levels of inclusion of soluble true protein and soluble NPN in livestock diets. Additionally, there is a need for formulae that can be used for estimating the economic and nutritional value of protein supplements from solubility estimates.

Since silage forms the basal diet of most ruminants in the UK, improving the utilization of silage N by reducing silage intake is not always feasible. Alternative methods that have been examined include thermal treatment (Murphy and McNiven, 1994), dehydration (Beever *et al.*, 1976), pelleting (Alloui *et al.*, 1994), wilting (Kowalski *et al.*, 1993) and treatment with iron alum (Antoniewicz, 1993), formic acid, formalin (Beever *et al.*, 1976) and condensed tannins (Ulyatt *et al.*, 1980). Effective methods of minimizing the soluble N content of silages include reducing N fertilization at pasture and employing management practices that ensure good-quality fermentation.

Prediction of silage intake by using near-infrared reflectance spectroscopy

As discussed in Chapter 3 and earlier in this chapter, many approaches have been examined for predicting the voluntary intake of forages. Prediction equations have poor performance when applied to ensiled forages. It was proposed some time ago (AFRC, 1991) that NIRS could be one of the most promising techniques for predicting the voluntary intake of grass silage, since the spectra contain information on the entire chemistry of the feed. In fact, NIRS has been successfully used to predict the intake of mainly non-fermented forages. This is shown by the work of Norris *et al.* (1976) and the more recent studies of Ward *et al.* (1982) and Flinn *et al.* (1992).

However, it has also been shown recently (Steen *et al.*, 1995) that NIRS can accurately predict the voluntary intake of grass silage by cattle. Steen *et al.* (1995) measured the voluntary intakes of 136 grass silages, using 192 individually fed beef cattle, and compared NIRS with some 30 other laboratory measurements on the silage for their ability to predict voluntary intake. The best overall relationship was between NIRS spectra and intake, with R^2 values of 0.86 and 0.71 for calibration and blind validation, respectively, and a standard error of prediction of 5.5 g DM kg^{-1} live-weight $(LW)^{0.75}$ daily. Of the other measurements, the next best relationships with intake were obtained with electrometric titration (R^2 0.53), digestibility *in vivo* (R^2 0.30) and DM degradability (R^2 0.28).

The fact that NIRS seems to offer considerable opportunities for predicting intake suggests that such studies should be extended to include the impact of feeding concentrates and interactions between forages fed together.

Applicability of Techniques

Appropriateness and cost

The discussion in previous chapters has centred on reviewing the literature on given techniques. Whilst there is a need to standardize techniques (see below), there is also a need to consider the appropriateness of techniques for given circumstances. Choice of technique will be a function of many factors. For example, in developing-country laboratories, often the only technique available for a given analysis is one donated by a developed country or development agency. Unfortunately, such donations are often inappropriate, because they do not sufficiently consider local needs and circumstances. Techniques often require chemicals and backup services not available locally, which have to be imported at large expense.

Increasingly, in developed countries, there is concern to devise non-invasive techniques, because of public aversion to vivisection. Thus, for evaluating the kinetics of organic-matter degradation in the rumen, the *in situ* technique (Chapter 9) is likely to be superseded by *in vitro* gas-production methods (Chapter 10). The *in situ* technique has been much promoted and used in developing-country laboratories, because of its low cost, robustness and lack of dependence on a reliable electricity supply. However, animal-welfare issues apart, the *in situ* technique is only able to accommodate a few samples at a given time and is therefore inappropriate when large numbers of samples have to be screened, e.g. as in forage breeding.

The question of cost and numbers also applies to gas-production methods. Automated systems (Chapter 11) are attractive, in that they obviate the need to use labour for night-time recording, but are expensive to purchase and are only able to evaluate a few samples at a given time (e.g. a 12-unit Cone *et al.* (1996) system costs around £10,000). On the other hand, Mauricio *et al.* (1999a) have recently developed a semi-automated version (Reading pressure technique (RPT)) of the manual technique of Theodorou *et al.* (1994). RPT has a capacity to evaluate concurrently 70 forages in triplicate, and its capital cost is low (less than £1000). Importantly, Mauricio *et al.* (1999b) have also shown faeces to have potential for replacing rumen liquor as an inoculum in RPT. Use of faeces in the Tilley and Terry (1963) *in vitro* digestion technique (e.g. Akhter *et al.*, 1999) was discussed in Chapter 7. Use of faeces instead of rumen liquor has much potential application, particularly in developing countries, but its use requires much more research and development.

Increasingly in the future, choice of analysis and technique will be influenced by the use to be made of the information generated. Thus issues such as accuracy, speed and cost of analysis will receive different emphases, depending on whether the customer is a researcher, commercial operator (e.g. feedstuff manufacturer), extensionist or farmer. As mentioned previously, the location will also affect the choice. In developing countries, extensionists are being encouraged to make greater use of simple assessments of nutritive value such as leaf-to-stem ratio, colour and smell (Thorne, 1998). NIRS combines many advantages of accuracy, speed and low recurrent costs and will be used increasingly in developed countries. However, use of NIRS in developing countries will increase only when there is a major reduction in capital cost.

Need for standardization of techniques

Although forage characterization is most accurate when conducted *in vivo*, public concerns and high costs associated with such procedures have necessitated the development of *in vitro* techniques. Such methods are constantly being modified, adapted and ultimately replaced by faster, cheaper or more accurate methods by the commercial firms, consulting analytical chemists and official advisory services that use them. Thus a plethora of methods is currently available for analysing the chemical constituents in forages. Nevertheless, outdated, inaccurate methods are still being used to predict nutritive value. For instance, modified acid-detergent fibre (MADF) is currently used in some laboratories for predicting the ME value of grass silages. Yet the error associated with the prediction is much higher (standard error of prediction = 1.30; $r^2 = 0.14$ (Barber *et al.*, 1990)) than that recommended as the maximum permissible (0.5 MJ kg^{-1} corrected DM) to prevent inaccurate ME predictions and attendant wastes in feed outlay or lost production (Offer, 1993). Also, equations that are accurate for predicting nutritive value in certain feeds are often applied, without validation, to other feeds that were not represented in the population for which the equation was developed. The inaccuracy of this approach has been widely demonstrated in temperate (McLeod and Minson, 1972; Barber *et al.*, 1989; Givens *et al.*, 1995; Adesogan *et al.*, 1998) and tropical (Acamovic *et al.*, 1997) forages.

An added complication is the intra- and interlaboratory variation in expertise and analytical procedures used, such as sample-preparation methods, reagents, equipment, computation of results and technique. This can lead to considerably different results for the same sample. Givens *et al.* (1990) found substantial differences between the organic-matter digestibility of straws measured at different laboratories. After sending maize silage samples to nine different laboratories, Beever *et al.* (1996) concluded that the interlaboratory variation, especially in starch, crude-protein and NH_3-N content, was too great to be acceptable for use in farming practice. Considerable interlaboratory or between-method differences have also been shown in the estimation of ME (Van der Meer, 1983; AFRC, 1993), *in situ* rumen degradability (Huntington and Givens, 1995), gas-production kinetics (Rymer and Givens, 1997), *in vitro* digestibility (Aufrère and Michalet-Doreau, 1988; Narasimhalu, 1985), GE (Porter, 1992), neutral-detergent fibre and ether extract (Lanari *et al.*, 1991). In addition, it is disconcerting that some techniques said to have the least predictive ability in some reports were found to have the highest in others. For instance, De Boever *et al.* (1994) reported that rumen-fluid digestibility techniques were more accurate than cellulase-based methods for estimating *in vivo* digestibility, whereas contrary results were shown by Aufrère and Michalet-Doreau (1988).

These factors emphasize the need for standardization of forage characterization procedures. Within laboratories, there is an ongoing need to ensure that experiments are precise, reproducible and repeatable. This may be effected by internal and external calibration (Minson, 1998). Internal calibration involves using results of standard samples analysed alongside test samples, to correct for week-to-week deviations in accuracy, while external calibration involves comparing test results with those determined *in vivo* (Minson, 1998).

In addition to the intralaboratory standardization, it is imperative that interlaboratory standardization procedures are developed to facilitate comparison of results from different centres. National and European Union (EU)-wide ring tests have already shown that, when the same procedures are employed, similar estimates of starch (McClear *et al.*, 1994) and mycotoxins (Schuhmacher *et al.*, 1997) can be achieved in different laboratories. There is a need to involve more laboratories in such ring tests and ultimately to ensure that accurate, standardized procedures are used throughout the feed industry. This may be achieved by the introduction of an accreditation scheme, whereby results from laboratories will only be considered acceptable if the laboratory is accredited for the test in question. Unless such a scheme is introduced, laboratories using the less accurate techniques may continue generating misleading results, and the validity of comparing analytical results from laboratories employing different techniques will remain questionable.

Health and safety issues related to techniques

The formal consideration of health and safety aspects related to laboratory techniques is relatively new. However, it is now a matter given increasing attention by laboratory managers and institutions in developed countries, if only because of fear of litigation. In the UK, in 1989, the government introduced legislation to protect health and safety of people at work, known as the Control of Substances Hazardous to Health (COSHH) (Simpson and Simpson, 1991). The COSHH requirements cover prevention or control of exposure to hazardous substances, use of control measures, maintenance of control measures, monitoring of control measures, health surveys and information and training. Employers have a duty under the law to ensure, so far as is reasonably practicable, the health, safety and welfare of their employees at work. The regulations require a COSHH risk assessment to be undertaken on each substance used in a given technique. The assessment includes the occupational exposure limits of the substance, whether it is corrosive, an irritant, a carcinogen, a mutagen, a teratogen, a neurotoxin, an allergen or sensitizer and whether it affects the central nervous system. Furthermore, the assessment identifies precautions required for handling the substance, control measures to be used (special ones or 'good laboratory practices'), first-aid procedures, spillage response and action in case of fire. There is an increasing volume of literature on health and safety issues in laboratories (Bretherick, 1986; Web addresses). Health and safety aspects of forage evaluation techniques will need to be considered increasingly in the future, particularly in developing-country laboratories, where this aspect has tended to be neglected to date.

Role of Forages in Animal and Human Health

The effect of toxigenic fungi in forages on human and animal health and performance

There is little published information on the role of forage moulds and their associated mycotoxins in animal and human health. This is partly because the delay between mould contamination and disease incidence makes it difficult to inculpate mycotoxins in the aetiology of animal or human diseases. Although some forage mycotoxins have been implicated in the development of livestock diseases, toxigenic fungi are not perceived as a major threat to animal production in the developed world. Yet they adversely affect the efficiency and profitability of livestock production through subclinical toxicoses, such as immunosuppression, reduced feed intake and decreased feed-conversion efficiency. Moulds also reduce the nutritional value of forages by degrading water-soluble carbohydrates, lactic acid, cellulose and other cell-wall fractions (Oldenburg, 1991). In addition to the threat from mycotoxin contamination, the inhalation of fungal propagules can lead to kidney damage, respiratory-tract problems and irritation of the eye and skin (Scudamore and Livesey, 1998).

Mycotoxin-synthesizing moulds have been isolated from grass, grass silage, maize silage, whole-crop wheat straw and hay samples. The major toxigenic fungi in forages belong to the *Fusarium*, *Penicillium* or *Aspergillus* species. *Fusarium* species are field moulds that produce toxins such as trichothecenes, fumonisins and zearalenone. The moulds are commonly found in grazed or harvested forages and could also be endemic in inadequately dried (> 150 g moisture kg^{-1}) hay and straw bales, where the aerobic conditions enhance their stability. *Fusarium* species pose a lower threat in silages, because the fungi and most of their associated mycotoxins are degraded during anaerobic ensiling (Scudamore and Livesey, 1998). A notable exception is zearalenone, which has been shown to persist in maize silage after 12 weeks of ensilage (Oldenburg, 1991). Although zearalenone toxicity has been linked to infertility, reduced milk production and hyperoestrogenism in ruminants and pigs (Cheeke, 1995; D'Mello and MacDonald, 1996), it is not a major problem in the UK (Scudamore and Livesey, 1998).

Penicillium and *Aspergillus* species are storage moulds that produce ochratoxin and aflatoxin, respectively. Ochratoxin A causes kidney damage in pigs, but ruminants are less susceptible. Ochratoxin A also has carcinogenic, teratogenic, nephrotoxic and immunosuppressive properties (Rosenthaler *et al.*, 1984; Oldenburg, 1991) and is transmissible from grains to edible animal products (D'Mello and MacDonald, 1996).

Aflatoxin is produced from *Aspergillus* species, which thrive between 25 and 30°C (D'Mello and MacDonald, 1996). Hence aflatoxin contamination is mainly associated with tropical feed ingredients, such as groundnut cake and cottonseed cake, which are banned for use as livestock feed ingredients in the UK. Previous surveys have not found aflatoxin to be a major contaminant of forage in the UK (Scudamore and Livesey, 1998). However, formic acid treatment of forage increases the risk of *Aspergillus* infestation and aflatoxin production. In Sweden, an increased incidence of aflatoxin M1 (the metabolic derivative of aflatoxin B1) in milk produced on farms using formic acid culminated in a ban on formic acid treatment of moist grain (Peterson *et al.*, 1989). Nevertheless, formic acid-treated silages may be safe for animals, since aflatoxin B1 degrades during ensiling (Scudamore and Livesey, 1998). Ruminants are less susceptible to aflatoxicosis than pigs, but clinical signs of the acute condition have been observed in cattle (Diekman and Green, 1992). Chronic exposure to aflatoxin lowers feed intake and milk production. This may be due to the fact that, in the rumen, aflatoxin decreases cellulose digestion, VFA production, proteolysis and motility (Scudamore and Livesey, 1998).

Mycotoxin production and mould growth in forages are inhibited by ensuring that the following conditions are not exceeded during storage: temperature 2.2°C, relative humidity 70%, moisture content of grains 150 g moisture kg^{-1} and oxygen 5g kg^{-1} (Diekman and Green, 1992). Mycotoxin activity in grains has also been inhibited by treatment with NH_3 and aluminosilicates (Diekman and Green, 1992). Previous work has also shown that increasing the protein or amino acid content of the diet beyond normal levels could alleviate poor performance in pigs that were fed on mycotoxin-contaminated grain. However, the foregoing relates to inhibition of grain mycotoxins. Further work is required to ascertain the effectiveness of these techniques in forage mycotoxins.

Although the disease risk from toxigenic fungi is relatively small, their role in animal and human health requires further study, because of the reduction in animal performance due to subclinical toxicoses and the possibility of transfer of carcinogenic metabolites from animal tissues to humans. Future studies are required to identify forage mycotoxins that produce residues of toxic metabolites in edible animal products. This requires the development of rapid detection techniques, as the methods currently used for grain mycotoxins (mass spectrometry and thin-layer, gas and high-performance chromatography) have not been fully tested and validated on forage mycotoxins (Scudamore and Livesey, 1998). More studies like those being undertaken at Aberystwyth (A.T. Adesogan, 1999, unpublished) are required to ascertain the effect of mycotoxins on digestive processes in livestock and the extent to which the mycotoxins are degraded during the digestion process. Coker (1999), in reviewing mycotoxins, calls for much further research, particularly to develop rapid and inexpensive mycotoxin analysis methods, which can be used in unsophisticated laboratories, e.g. in developing countries.

In conclusion, although toxigenic fungi have not posed a major threat to livestock production in the UK in the recent past, more attention should be focused on them because of the increasing use of alternatives to grass silage for winter feeding of livestock. Forages such as maize silage, whole-crop wheat, baled-grass silages and legume silages have higher DM contents than clamped-grass silage. They are therefore more susceptible to microfloral contamination and aerobic spoilage.

Forages as sources of drugs

The possible beneficial role of secondary plant compounds was alluded to in Chapter 20. Butter *et al.* (1998), at Nottingham University, reported dietary quebracho tannin and/or high protein being effective in reducing parasite eggs in lambs infected with *Trichostrongylus colubriformis*. The same group is currently investigating the effectiveness of tanniniferous forages in Tanzania for reducing parasitic infection in goats. In view of the complexity of tannin chemistry and the large variation within and between species, there will be a need to develop methods for screening forages for their beneficial or antinutritional secondary plant properties.

Because of the high cost of veterinary drugs in developing countries, there is growing interest in the use of ethnoveterinary medicine based on local plants (ITDG and IIRR, 1996). Current research at the Institute of Grassland and Environmental Research, Aberystwyth, in partnership with Xenova Discovery Limited, is characterizing natural products from temperate forages and plants which may have beneficial drug properties (Nash, 1998). In the same programme, the use of modern analytical techniques, such as gas chromatography and high-performance liquid chromatography coupled to mass spectrometry (GC-MS and HPLC-MS), is enabling the active ingredients in plants used in folk medicines and toxic plants to be identified.

Forages as sources of n-3 fatty acids and antioxidants

Whilst, traditionally, forages have been regarded primarily as sources of energy and protein, the data reviewed in Chapter 19 clearly show that, under certain circumstances, forages can be an important source of antioxidants, which may ultimately make a useful contribution to the human diet. Also in relation to the human diet, there is increasing evidence that an increased intake of *n*-3 fatty acids is desirable in relation to cardiovascular health (see review by Givens *et al.*, 2000). The most important *n*-3 fatty acids are the long-chain polyunsaturates eicosapentaenoic acid (EPA, C20:5 *n*-3) and docosahexaenoic acid (DHA, C22:6 *n*-3), but there is evidence (Scollan *et al.*, 1997) that increased amounts of these acids can occur in the meat and milk of ruminants fed diets enhanced with the shorter-chain linolenic acid (C18:3 *n*-3). Although the total lipid concentration in the DM of most forages is relatively low, because they are often consumed in large quantities they can make a sizeable contribution to total lipid consumed.

In this context, Dewhurst (1997) confirmed earlier reports that lipids in grasses are a rich source of linolenic acid. However, Dewhurst (1997) also showed that this was influenced markedly by grass genotype (Table 21.3), plant maturity and methods of conservation. These results highlight the very important role grass-based forages could play in introducing increased amounts of *n*-3 fatty acids into the human

Table 21.3. The major fatty acids in various species of grasses (after Dewhurst, 1997).

Fatty acid* ($g\ kg^{-1}$ DM)	*Lolium perenne*	*Lolium multiflorum*	*Lolium* hybrids	*Lolium*–fescue hybrids	Fescue	Cocksfoot	Timothy	Effect of species
Total fatty acids	22.5	20.8	23.2	23.0	21.9	19.1	21.8	$P < 0.001$
C16:0	4.7	4.5	4.9	4.6	4.4	4.0	4.1	$P < 0.001$
C18:2 (*n*-6)	3.0	2.7	3.2	3.1	2.6	2.9	3.6	$P < 0.001$
C18:3 (*n*-3)	11.6	10.5	11.9	12.3	12.1	10.3	11.3	$P < 0.05$

* Mean of three sampling dates.

diet. It also highlights the future need to regard forages as more than simply sources of energy and protein for the animal. Much more work in this area is needed.

Environmental factors

The need to minimize environmental pollution from livestock production systems will mean that future evaluation of forages must consider aspects such as methane emission (Moss and Givens, 1997) and efficiency of nitrogen (e.g. Castillo *et al.*, 1999) and phosphorus (e.g. Shah, 1999) utilization when ingested by ruminants.

Conclusions

The results presented and reviewed in all chapters of this book highlight not only the immense importance of forages worldwide as part of the diet of ruminants but also that forages can no longer be regarded as simply sources of energy and protein. They must be characterized in terms of the nutrients they supply to the animal and, more particularly, to the rumen microbes. Their wider potential in relation to their important role as indirect suppliers of important biologically active compounds to the human diet needs considerable further exploitation and development. All of these factors highlight the challenge that faces those developing means of characterizing forages. It also seriously calls into question the value of traditional books and databases on the nutritive value of forages, not only because of the inherent variability of forages but also because the data they contain will rapidly become less relevant to the needs of the industry.

References

Acamovic, T., Stewart, C.S. and Topps, J.H. (1997) Appropriate methods for the evaluation of tropical feeds for ruminants. *Animal Feed Science and Technology* 69, 53–66.

Adesogan, A.T., Owen, E. and Givens, D.I. (1998) Prediction of the *in vivo* digestibility of whole crop wheat from *in vitro* digestibility, chemical composition, *in situ* degradability, *in vitro* fermentation and near infrared reflectance spectroscopy. *Animal Feed Science and Technology* 74, 259–272.

Aerts, J.V., De Brabander, D.L., Cottyn, B.G., Buysse, F.X. and Mooremans, R.J. (1974) Comparison of methods of dry matter determination of high moisture roughages and faeces. *Journal of the Science of Food and Agriculture* 25, 619–627.

AFRC (1991) AFRC Technical Committee on Responses to Nutrients, Report Number 8. Voluntary intake of cattle. *Nutrition Abstracts and Reviews (Series B)* 61, 815–823.

AFRC (1993) *Energy and Protein Requirements of Ruminants. An Advisory Manual Prepared by the Technical Committee on Responses to Nutrients.* CAB International, Wallingford, 159 pp.

Agnew, R.E., Yan, T. and Gordon, F.J. (1998) Nutrition of the high genetic merit dairy cow – energy metabolism studies. In: Garnsworthy, P.C. and Wiseman, J. (eds) *Recent Advances in Animal Nutrition 1998.* University of Nottingham Press, Nottingham, pp. 179–207.

Akhter, S., Owen, E., Theodorou, M.K., Butler, E.A. and Minson, D.J. (1999) Bovine faeces as a source of microorganisms for the *in vitro* digestibility assay of forages. *Grass and Forage Science* 54, 219–226.

Alloui, O., Smulikowska, S., Chibowska, M. and Pastuszewska, B. (1994) The nutritive value of lupin seeds for broiler chickens as affected by variety and enzyme supplementation. *Journal of Animal and Feed Sciences* 3, 215–227.

Antoniewicz, A. (1993) The influence of trivalent cations and thermal treatment on ruminal degradability of field bean (*Vicia faba*) and rape seed (*Brassica napus*) protein. *Journal of Animal and Feed Sciences* 1, 263–278.

ARC (1984) *The Nutrient Requirements of Ruminant Livestock, Supplement No 1*. Commonwealth Agricultural Bureaux, Farnham Royal, 45 pp.

Aufrère, J. and Michalet-Doreau, B. (1988) Comparison of methods for predicting digestibility of feeds. *Animal Feed Science and Technology* 20, 203–218.

Baldwin, R.L., Lucas, H.L. and Cabrera, R. (1970) Energetic relationships in the formation and utilisation of fermentation end-products. In: Philipson, A.T. (ed.) *Physiology of Digestion and Metabolism in the Ruminant*. Oriel Press, Newcastle upon Tyne, pp. 319–334.

Barber, G.D., Offer, N.W. and Givens, D.I. (1989) Predicting the nutritive value of silage. In: Haresign, W. and Cole, D.J.A. (eds) *Recent Advances in Animal Nutrition – 1989*. Butterworths, London, pp. 141–158.

Barber, G.D., Givens, D.I., Kridis, M.S., Offer, N.W. and Murray, I. (1990) Prediction of the organic matter digestibility of grass silage. *Animal Feed Science and Technology* 28, 115–128.

Beever, D.E. (1993) Rumen function. In: Forbes, J.M. and France, J. (eds) *Quantitative Aspects of Ruminant Digestion and Metabolism*. CAB International, Wallingford, pp. 187–215.

Beever, D.E., Thomson, D.J. and Cammell, S.B. (1976) The digestion of frozen and dried grass by sheep. *Journal of Agricultural Science* 86, 443–452.

Beever, D.E., Dhanoa, M.S., Losada, H.R., Evans, R.T., Cammell, S.B. and France, J. (1986) The effect of forage species and stage of harvest on the processes of digestion occurring in the rumen of cattle. *British Journal of Nutrition* 56, 439–454.

Beever, D.E., Cammell, S.B. and Edmonds, S. (1996) Inter-laboratory variation in feedstuff evaluation of two contrasting maize silage samples. *Animal Science* 62, 685–686.

Blümmel, M. and Bullerdieck, P. (1997) The need to complement *in vitro* gas production measurements with residue determinations from *in sacco* degradabilities to improve the prediction of voluntary intake of hays. *Animal Science* 64, 71–75.

Blümmel, M., Steingaß, H. and Becker, K. (1994) The partitioning of *in vitro* fermentation products and its bearing for voluntary feed intake. *Proceedings of the Society for Nutrition and Physiology* 3, 123 (Abstract).

Bowman, J.G.P. and Asplund, J.M., (1988) Evaluation of mixed lucerne and caucasian bluestem hay diets fed to sheep. *Animal Feed Science and Technology* 20, 19–31.

Bretherick, L. (1986) *Hazards in the Chemical Laboratory*, 4th edn. Royal Society of Chemistry, London.

Brookes, I.M. and Maguire, M.F. (1973) Utilisation of water soluble nitrogen in silage by sheep. *Irish Journal of Agricultural Research*, 12, 47–52.

Butter, N.L., Dawson, J.M., Wakelin, D. and Buttery, P.J. (1998) Effect of dietary tannin protein level on the susceptibility of sheep to parasitic infection. *Proceedings of the British Society of Animal Science* 1998, 97.

Butterworth, M.H. and Mosi, A.K. (1986) The intake and digestibility by sheep of oat straw and maize stover offered with different levels of noug (*Guizotia abyssinica* meal). *Animal Feed Science and Technology* 16, 99–107.

Castillo, A.R., Barbi, J.H., Beever, D.E., Sutton, J.D. and Humphries, D.J. (1999) Influence of different energy supplements on nitrogen utilisation by dairy cows fed grass silage diets. *Proceedings of the British Society of Animal Science* 1999, 73.

Cheeke, P.R. (1995) Endogenous toxins and mycotoxins in forage grasses and their effects on livestock. *Journal of Animal Science* 73, 909–918.

Cockburn, J.E., Dhanoa, M.S., France, J. and Lopez, S. (1993) Overestimation of solubility when using dacron bag methodology. *Animal Production* 55, 188.

Coker, R. (1999) *Mycotoxins – the Silent Threat to Human and Animal Health.* Inaugural Lecture Series, University of Greenwich, Chatham, 52 pp.

Cone, J.W., van Gelder, A.H., Visscher, J.G.W. and Oudshoorn, L. (1996) Influence of rumen fluid and substrate concentration on fermentation kinetics measured with a fully automated time related gas production apparatus. *Animal Feed Science and Technology* 61, 113–128.

Corbett, J.L., Langlands, J.P., McDonald, I. and Pullar, J.D. (1966) Composition by direct animal calorimetry of the net energy values of an early and a late season growth of herbage. *Animal Production* 8, 13–27.

Deaville, E.R. and Givens, D.I. (1998) Investigation of direct gas production from silage fermentation acids. *Proceedings of the British Society of Animal Science* 1998, 64.

De Boever, J.L., Cottyn, B.G., Vanacker, J.M. and Boucque, C.V. (1994) An improved enzymatic method by adding gammanase to determine digestibility and predict energy value of compound feeds and raw materials for cattle. *Animal Feed Science and Technology* 47, 1–18.

Dewar, W.A. and McDonald, P. (1961) Determination of dry matter in silage by distillation with toluene. *Journal of the Science of Food and Agriculture* 12, 790–795.

Dewhurst, R. (1997) *Effects of Wilting and Chemical Additives on Fatty Acids in Grass Silage.* Report to MAFF on Project OCS967E, Part 2. IGER, Aberystwyth, 7 pp.

Dhanoa, M.S. (1988) The analysis of dacron bag data for low degradability feeds. *Grass and Forage Science* 43, 441–444.

Diekman, M.A. and Green, M.L. (1992) Mycotoxins and reproduction in domestic livestock. *Journal of Animal Science* 70, 1615–1627.

Dingley, P.Y., Sniffen, C.J., Johnson, L.L., Hoover, W.H. and Walker, C.K. (1975) Protein solubility and amino acid supply to the udder. *Journal of Dairy Science* 58, 1240.

D'Mello, F. and MacDonald, A. (1996) Mycotoxins in grain: an emerging issue. *Feed Compounder* May, 34–36.

Dulphy, J.P. and Demarquilly, C. (1981) Problèmes particuliers aux ensilages. In: Dermarquilly, C. (ed.) *Prévision de la valeur nutritive des ruminants.* INRA, Versailles, pp. 81A–104A.

Eldridge, G.A. and Kat, C. (1980) The influence of hay on the digestibility of pasture and the recovery of faecal marker. *Australian Society of Animal Production* 13, 471.

Fahey, G.C., Bourquin, L.D., Titgemeyer, E.C. and Atwell, D.G. (1993) Postharvest treatment of fibrous feedstuffs to improve their nutritive value. In: Jung, H.G., Buxton, D.R. Hatfield, R.D. and Ralph, J. (eds) *Forage Cell Wall Structure and Digestibility.* American Society of Agronomy, Crop Science Society of America, Soil Science Society of America, Madison, Wisconsin, USA, pp. 715–766.

Flinn, P.C., Windham, W.R. and Dove, H. (1992) Pasture intake by grazed sheep estimated using natural and dosed *n*-alkanes – a place for NIR? In: Hildrum, K.L., Isaksson, T., Naes, T. and Tandberg, A. (eds) *Near Infrared Spectroscopy.* Ellis Horwood, Chichester, pp. 175–178.

Franci, O. and Acciaioli, A. (1998) Mathematical models for the innovative interpretation of the associative effect. *Options Méditerranéennes, Serie B, Etudes et Recherches* 17, 73–78.

Franci, O., Antongiovanni, M., Acciaioli, A., Bruni, R. and Martini, A. (1997) Response surface analyses of the associative effects of lucerne hay, wheat straw and maize gluten feed on growing lambs. *Animal Feed Science and Technology* 67, 279–290.

Galleti, G.C. and Piccaglia, R. (1988) Water determination in silages by Karl Fischer titration. *Journal of the Science of Food and Agriculture* 43, 1–7.

Gibb, M.J. (ed.) (1998) *Techniques for Investigating Intake and Ingestive Behaviour by Farm Animals. Proceedings of the IXth European Intake Workshop held at the Institute of Grassland and Environmental Research, North Wyke, UK, 18–20 November 1998.* Institute of Grassland and Environmental Research, Aberystwyth, 92 pp.

Gibb, M.J. and Treacher, T.T. (1976) The effect of herbage allowance on herbage intake and performance of lambs grazing perennial ryegrass and red clover swards. *Journal of Agricultural Science, Cambridge* 86, 355–365.

Givens, D.I. (1986) New methods for predicting the nutritive value of silage. In: Stark, B. and Wilkinson, M. (eds) *Developments in Silage – 1986*. Chalcombe, Marlow, pp. 66–75.

Givens, D.I., Dewey, P.J.S., Donaldson, E., Jones, D.I.H. and Adamson, A.H. (1990) Within- and between-centre variability in the measurement of organic matter digestibility *in vivo* of cereal straws. *Animal Feed Science and Technology* 28, 103–108.

Givens, D.I., Cottyn, B.G., Dewey, P.J.S. and Steg, A. (1995) A comparison of the neutral detergent-cellulase method with other laboratory methods for predicting the digestibility *in vivo* of maize silages from three European countries. *Animal Feed Science and Technology* 54, 55–64.

Givens, D.I., Cottrill, B.R., Davies, M., Lee, P., Mansbridge, R.M. and Moss, A.R. (2000) Sources of *n*-3 polyunsaturated fatty acids additional to fish oil for livestock diets. *Nutrition Abstracts and Reviews* (Series B) 70, 1–19.

Gordon, F.J., Cooper, K.M., Park, R.S., and Steen, R.W.J. (1998) The prediction of intake potential and organic matter digestibility of grass silages by near infrared spectroscopy of undried samples. *Animal Feed Science and Technology* 70, 339–351.

Hood, R.L., Allen, C.E., Goodrich, R.D. and Meiske, J.C. (1971) A rapid method for the direct chemical determination of water in fermented feeds. *Journal of Animal Science* 33, 1310–1314.

Huntington, J.A. and Givens, D.I. (1995) The *in situ* technique for studying the rumen degradation of feeds: a review of the procedure. *Nutrition Abstracts and Reviews (Series B)* 65, 65–93.

Islam, R. (1997) Evaluation of oat straw for ruminants: effects of genotype, climate and agronomic factors on nutritive value and prediction of nutritive value. PhD thesis, University of Reading, 405 pp.

ITDG and IIRR (1996) *Ethnoveterinary Medicine in Kenya: A Field Manual of Traditional Animal Health Care Practices*. Intermediate Technology Development Group and International Institute of Rural Reconstruction, Nairobi, Kenya, 226 pp.

Kaiser, A.G., Mailer, R.J. and Vonarx, M.M. (1995) A comparison of Karl Fischer titration with alternative methods for the analysis of silage dry matter content. *Journal of the Science of Food and Agriculture* 69, 51–59.

Knowlton, K.F., Pitt, R.E. and Fox, D.G. (1992) Dynamic model prediction of the value of reduced solubility of alfalfa silage protein for lactating dairy cows. *Journal of Dairy Science* 75, 1507–1516.

Kowalski, Z.M., Pys, J.B. and Kanski, J. (1993) The effect of cut, stage of growth and wilting on nitrogenous compounds and protein solubility in grass silages. *Journal of Animal and Feed Sciences* 1, 213–221.

Lanari, D., Pinosa, M., Tibaldi, E. and D'Agaro, E. (1991) Results of two ring test on the chemical composition of some feedstuffs and on digestibility *in vivo*. *Zootecnica e Nutrizione Animale* 17, 285–295.

Lopez, S., France, J. and Dhanoa, M.S. (1994) A correction for particulate matter loss when applying the polyester-bag method. *British Journal of Nutrition* 71, 135–137.
McClear, B.V., Gibson, T.S., Solah, V. and Mugford, D.C. (1994) Total starch measurement in cereal products: interlaboratory evaluation of a rapid enzymic test procedure. *Cereal Chemistry* 71, 501–505.
McDonald, P., Edwards, R.A., Greenhalgh, J.F. and Morgan, C.A. (1995) *Animal Nutrition*, 5th edn. Longman Scientific and Technical, Harlow, pp. 232–233.
McLeod, M.N. and Minson, D.J. (1972) The effect of method of determination of acid detergent fibre on its relationship with the digestibility of grasses. *Journal of the British Grassland Society* 27, 23–27.
MacRae, J.C., Smith, J.S., Dewey, P.J.S., Brewer, A.C., Brown, D.S. and Walker, A. (1985) The efficiency of utilisation of metabolisable energy and apparent absorption of amino acids in sheep given spring- and autumn-harvested dried grass. *British Journal of Nutrition* 54, 197–209.
Mader, T.L. and Britton, R.A. (1986) Utilisation of alfalfa hay and alfalfa silage: protein sources in ensiled corn stover diets. *Journal of Dairy Science* 69, 2342–2347.
Majdoub, A., Lane, G.T. and Aitchison, T.E. (1978) Milk production response to nitrogen solubility in dairy rations. *Journal of Dairy Science* 61, 59–65.
Martin, P.A., Chamberlain, D.G., Robertson, S. and Hirst, D. (1994) Rumen fermentation patterns in sheep receiving silages of different chemical composition supplemented with concentrates rich in starch or in digestible fibre. *Journal of Agricultural Science, Cambridge* 122, 145–150.
Mauricio, R.M., Mould, F.L., Dhanoa, M.S., Owen, E., Channa, K.S. and Theodorou, M.K. (1999a) A semi-automated *in vitro* gas production technique for ruminant feedstuff evaluation. *Animal Feed Science and Technology* 79, 321–330.
Mauricio, R.M., Owen, E., Abdalla, A.L., Bueno, I.C.S., Mould, F.L., Gilmour, S. and Theodorou, M.K. (1999b) Comparison of rumen liquor and faeces, in UK and Brazil, as sources of microorganisms for *in vitro* gas production for assessing twelve forages. *Proceedings of the British Society of Animal Science* 1999, 147.
Methu, J.N., Owen, E., Abate, A., Scarr, M. and Tanner, J.C. (1997) Effect of offering three amounts of maize stover to dairy cows on intake and selection. *Proceedings of the British Society of Animal Science* 1997, 85.
Milne, J. and Bagley, L. (1976) The nutritive value of diets containing different proportions of grass and heather (*Calluna vulgaris* L. Hull) to sheep. *Journal of Agricultural Science, Cambridge* 87, 599–604.
Minson, D.J. (1990) *Forage in Ruminant Nutrition*. Academic Press, London, 483 pp.
Minson, D.J. (1998) *A History of* In Vitro *Techniques*. Occasional Publication No. 22, British Society of Animal Science, Penicuik, pp. 13–20.
Minson, D.J. and Wilson, J.R. (1994) Prediction of intake as an element of forage quality. In: Fahey, G.C., Collins, M., Mertens, D.R. and Moser, L.E. (eds) *Forage Quality, Evaluation, and Utilisation*. American Society of Agronomy, Crop Science Society of America, Soil Science Society of America, Madison, Wisconsin, USA, pp. 553–563.
Moss, A.R. and Givens, D.I. (1997) Methane production from a range of feedstuffs as determined *in vitro* using the cumulative gas production technique and compared with that measured *in vivo*. *Proceedings of the British Society of Animal Science* 1997, 194.
Murphy, S.R. and McNiven, M.A. (1994) Raw or roasted lupin supplementation of grass silage diets for beef steers. *Animal Feed Science and Technology* 46, 23–35.
Narasimhalu, P. (1985) A comparison of laboratory methods involving single or two-stage cellulase and two-stage rumen inoculum-pepsin treatment for predicting digestibility and intake of hays in sheep. *Animal Feed Science and Technology* 12, 101–107.
Nash, R. (1998) *Natural Products from Temperate Plants: the Way Ahead for Drug Discovery.*

IGER Innovations, Number 2, Institute of Grassland and Environmental Research, Aberystwyth, pp. 42–45.
Norris, K.H., Barnes R.F., Moore, J.E. and Shenk, J.S. (1976) Predicting forage quality by infrared reflectance spectroscopy. *Journal of Animal Science* 43, 889–897.
Odoi, F.N.A. (1985) The nutritional value of mixed diets in cattle. In: *Proceedings of the New Zealand Grassland Association*. University of Canterbury, Canterbury, New Zealand.
Offer, N. (1993) Background to the new *in vivo* ME silage prediction equation. In: *Proceedings of the Society of Feed Technologists Conference, Reading*, Coventry; 14–15 July 1993, pp. E1–E7.
O'Hara, M. and Ohki, K. (1973) Studies on the mode of gas production in an artificial rumen and its application to the evaluation of feedstuffs. III The mode of volatile fatty acid production and its relation to the gas production rate. *Japanese Journal of Zootechnical Science* 44, 432–439.
Oldenburg, E. (1991) Mycotoxins in conserved forages. *Landbouforshcung Volkenrode* 23, 191–205.
Osafo, E.L.K., Owen, E., Said, A.N., Gill, M. and Sherington, J. (1997) Effects of amount offered and chopping on intake and selection of sorghum stover by Ethiopian sheep and cattle. *Animal Science* 65, 55–62.
Owen, E. (1979) Processing of roughages. In: Haresign, W. and Lewis, D. (eds) *Recent Advances in Animal Nutrition – 1978*. Butterworths, London, pp. 127–148.
Peterson, H., Holmberg, T., Larson, K. and Kaspersson, A. (1989) Aflatoxins in acid-treated grain in Sweden and occurrence of aflatoxin M1 in milk. *Journal of the Science of Food and Agriculture* 48, 411–420.
Phipps, R.H., Sutton, J.D. and Jones, B.A. (1995) Forage mixtures for dairy cows: the effect on dry matter intake and milk production of incorporating either fermented or urea treated whole crop wheat, brewers grains, fodder beet or maize silage into diets based on grass silage. *Animal Science* 61, 491–496.
Pichard, G.R. and Van Soest, P.J. (1977) Protein solubility of ruminant feeds. In: *Proceedings of the Cornell Nutrition Conference, Ithaca, New York*. Cornell University, Ithaca, New York, pp. 91–98.
Porter, M.G. (1992) Comparison of sample preparation methods for the determination of the gross energy concentration of fresh silage. *Animal Feed Science and Technology* 37, 201–208.
Porter, M.G. and Barton, D. (1996) A comparison of six methods for the determination of dry matter concentration in grass silage. In: Jones, D.I.H., Jones, R., Dewhurst, R., Merry, R. and Haigh, P.M. (eds) *Proceedings of the XIth International Silage Conference*. Institute for Grassland and Environmental Research, Aberystwyth, pp. 186–187.
Porter, M.G., Patterson, D.G., Steen, R.W.J. and Gordon, F.J. (1984) Determination of dry matter and gross energy of grass silage. In: *Proceedings of the Seventh Silage Conference*. Queens University, Belfast, pp. 89–90.
Reid, R.L., Templeton, W.C., Jr, Ranney, T.S. and Thane, W.V. (1987) Digestibility, intake and mineral utilisation of combination of grasses and legumes by lambs. *Journal of Animal Science* 64, 1725–1734.
Romney, D.L. and Gill, M. (1998) Measurement of short term intake rate (STIR) to predict *in vivo* parameters in sheep. *Proceedings of the British Society of Animal Science* 1998, 98.
Rosenthaler, R., Creppy, E.E. and Dirheimer, G. (1984) Ochratoxin A: on the mode of action of an ubiquitous mycotoxin. *Journal of Toxicology–Toxin Reviews* 3, 53–86.
Rymer, C. and Givens, D.I. (1997) A comparison of different types of apparatus for measuring gas production *in vitro*. In: *Proceedings of Winter Meeting of the British Society of Animal Science, Scarborough*, p. 193.
Rymer, C. and Givens, D.I. (1998) Relationship between the gas production profiles of mix-

tures of hay and maize and the patterns of rumen fermentation observed in sheep fed those mixtures. *Proceedings of the British Society of Animal Science* 1998, 63.

Sanderson, R., Lister, S., Sargent, A. and Dhanoa, M. (1997) Effect of particle size on *in vitro* fermentation of silages differing in dry matter content. *Proceedings of the British Society of Animal Science* 1997, 197.

Sauvant, D. and Giger, S. (1989) Straw digestibility calculation and digestive interactions. In: Chenost, M. and Reiniger, P. (eds) *Evaluation of Straws in Ruminant Feeding.* Elsevier Applied Science, London, pp. 47–61.

Sauvant, D., Dijkstra, J. and Mertens, D. (1995) Optimisation of ruminal digestion: a modelling approach. In: Journet, M., Grenet, E., Farce, M.-H., Theriez, M. and Demarquilly, C. (eds) *Recent Developments in the Nutrition of Herbivores.* INRA Editions, Paris, pp. 161–166.

Schuhmacher, R., Krska, R., Weingaertner, J. and Grasserbauer, M. (1997) Interlaboratory comparison study for the determination of the *Fusarium* mycotoxins in deoxynivalenol in wheat and zearalenone in maize using different methods. *Fresenius Journal of Analytical Chemistry* 359, 510–515.

Scollan, N.D., Fisher, W.J., Davies, D.W.R., Fisher, A.V., Enser, M. and Wood, J.D. (1997) Manipulating the fatty acid composition of muscle in beef cattle. *Proceedings of the British Society of Animal Science* 1997, 20.

Scudamore, K.A. and Livesey, C.T. (1998) Occurrence and significance of mycotoxins in forage crops and silage: a review. *Journal of the Science of Food and Agriculture* 77, 1–17.

Shah, M.H. (1999) The effect of forage and concentrate type on phosphorus utilisation in lactating cows. PhD thesis, University of Reading, 286 pp.

Sileshi, Z., Owen, E., Dhanoa, M.S. and Theodorou, M.K. (1996) Prediction of *in situ* rumen-dry matter disappearance of Ethiopian forages from an *in vitro* gas production technique using a pressure transducer, chemical analyses or *in vitro* digestibility. *Animal Feed Science and Technology* 61, 73–87.

Simpson, D. and Simpson, W.G. (1991) *The COSHH Regulations: a Practical Guide.* The Royal Society of Chemistry.

Steen, R.W.J., Gordon, F.J., Mayne, C.S., Agnew, R.E., Kilpatrick, D.J., Unsworth, E.F., Barnes, R.J., Porter, M.G. and Pippard, C.J. (1995) Prediction of silage intake by cattle. *Animal Science* 60, 514 (Abstract).

Sutton, J.D., Hart, I.C., Broster, W.H., Elliot, R.J. and Schuller, E.L. (1986) Feeding frequency for dairy cows: effect on rumen fermentation and blood metabolites and hormones. *British Journal of Nutrition* 56, 181–192.

Theodorou, M.K., Williams, B.A., Dhanoa, M.S., McAllan, A.B. and Frand, J. (1994) A simple gas production method using a pressure transducer to determine the fermentation kinetics of ruminant feeds. *Animal Feed Science and Technology* 48, 185–197.

Thomas, P.C. and Martin, P.A. (1988) The influence of nutrient balance on milk yield and composition. In: Garnsworthy, P.C. (ed.) *Nutrition and Lactation in the Dairy Cow.* Butterworths, London, pp. 97–118.

Thorne, P.J. (1998) *DRASIC: A Dairy Rationing System for the Tropics. Evaluation Version for Windows 3.1.* Natural Resources Institute, Chatham, UK, 40 pp.

Tilley, J.M.A. and Terry, R.A. (1963) A two-stage technique for *in vitro* digestion of forage crops. *Journal of the British Grassland Society* 18, 104–111.

Trenkle, A. (1970) Effects of short chain fatty acids, feeding, fasting and type of diet on plasma insulin levels in sheep. *Journal of Nutrition* 100, 1323–1330.

Ulyatt, J.A., Jones, M.J. and Shelton, I.D. (1980) Factors influencing nitrogen flow from the rumen. *Proceedings of the New Zealand Society of Animal Production* 40, 226.

Van der Meer, J.M. (1983) *European* In Vitro *Ring Test 1983.* Report No. 155, Institute for Livestock Feeding and Nutrition Research, Lelystad, the Netherlands.

Van Soest, P. (1994) *Nutritional Ecology of the Ruminant*, 2nd edn. Cornell University Press, Ithaca, 476 pp.

van Vuuren, A.M., Krol-Kramer, F., van der Lee, R.A. and Corbijin, H. (1992) Protein digestion and intestinal amino acids in dairy cows fed fresh *Lolium perenne* with different nitrogen contents. *Journal of Dairy Science* 75, 2215–2225.

van Vuuren, A.M., Huhtanen, P. and Dulphy, J.P. (1995) Improving the feeding and health values of ensiled forages. In: Journet, M., Grenet, E., Farce, M.-H., Theriez, M. and Demarquilly, C. (eds) *Recent Developments in the Nutrition of Herbivores*. INRA Editions, Paris, pp. 279–307.

Vogel, G.J., Phillips, W.A., Horn, G.W., Ford, M.J. and McNew, R.W. (1989) Effects of supplemental silage on forage intake and utilisation by steers grazing wheat pasture or bermudagrass. *Journal of Animal Science* 67, 232–240.

Wahed, R.A., Owen, E., Naate, M. and Hosking, B.J. (1990) Feeding straw to small ruminants: effect of amount offered on intake and selection of barley straw by goats and sheep. *Animal Production* 51, 283–289.

Walker, H.G. (1984) Physical treatment. In: Sundstol, F. and Owen, E. (eds) *Straw and Other Fibrous By-products as Feed*. Elsevier, Amsterdam, pp. 79–105.

Ward, R.G., Smith, G.S., Wallace, J.D., Urquhart, N.S. and Shenk, J.S. (1982) Estimates of intake and quality of grazed range forage by near infrared reflectance spectroscopy. *Journal of Animal Science* 54, 399–402.

Web addresses on health and safety:
http://physchem.ox.ac.uk/MSDS/
http://www.uwp.edu/admin/safety/
http://www/enviro-net.com/technical/msds/
http://hazard.com/msds.links.html
http://www.chem.utah.edu/chemistry/classes/labs/safety/safety.htm#contents
http://www.sigma.sial.com

Zemmelink, G. (1980) Effect of selective consumption on voluntary intake and digestibility of tropical forages. Doctoral thesis, Centre for Agricultural Publishing and Documentation, Wageningen University, Wageningen, 100 pp.

Zook, A.B. (1982) The effects of two levels of dietary protein solubility on lactating cows exposed to heat stress and in thermal neutral environments. *Dissertation Abstracts International*, B 43, 2091–2098.

Index

Page numbers in bold indicate start of the main section on the subject

Printed and bound by CPI Group (UK) Ltd, Croydon, CR0 4YY

16/09/2026

14963433-0005

Stamp Taxes 2016/17

Stamp Taxes 2016/17

Ken Wright BA, CA, CTA, Senior SDLT and LBTT specialist

Bloomsbury Professional

Bloomsbury Professional Ltd, Maxwelton House, 41–43 Boltro Road, Haywards Heath, West Sussex, RH16 1BJ

Bloomsbury Professional, an imprint of Bloomsbury Publishing Plc

A CIP Catalogue record for this book is available from the British Library.

ISBN: 978 1 78451 320 7

Typeset by Compuscript Ltd, Shannon
Printed and bound by CPI Group (UK) Ltd, Croydon CR0 4YY

Preface

This book provides comprehensive coverage for the non-specialist of what are commonly referred to as the 'stamp taxes', comprising stamp duty land tax (SDLT), stamp duty and stamp duty reserve tax. The complexity and volume of legislation, particularly in relation to SDLT, continues to increase at an alarming rate. Further fundamental changes have been made to the structure of SDLT during 2016. Transactions over commercial and mixed use property ceased to be taxed under the so called 'slab' tax structure and, with effect from 17 March 2016, moved to a progressive tax structure (like income tax) with new tax rates. With effect from 1 April 2016 a 3% surcharge is levied on the acquisition of most residential properties unless an individual is replacing his or her only or main residence or the individual does not own any other property at the time of the acquisition. At the same time the government's tax take from stamp taxes is forecast to increase substantially, with the Office for Budget Responsibility forecasting in its 'Economic and fiscal outlook' report dated March 2016 that SDLT receipts will rise from £10.9 billion in 2014/15 to £17.4 billion by 2020/21. This does however represent a reduction of £1.5 billion compared to the 2020/21 SDLT receipts being forecast back in July 2015. The report suggests that this reflects (a) an expected reduction in the transactions over residential properties worth £2 million or more as a result of the increased SDLT charge on high value properties, (b) increased receipts from commercial property transactions on the basis of rising prices and tax thresholds that are fixed in cash terms, and (c) an increase of £0.9 billion as a consequence of the new 3% surcharge on second homes and buy-to-let properties. This report was of course issued before the EU referendum result and it remains to be seen what effect the decision to leave the EU will have on SDLT receipts. The same report suggests that the receipts from stamp duty will rise from £2.9 billion in 2014/15 to £3.5 billion by 2020/21. This means that for those dealing with stamp taxes, in whatever capacity, the need to have an understanding of those taxes has never been greater.

Non-specialists sometimes find their minds go blank when faced with a stamp tax question. It is true that stamp taxes are different from other taxes, and indeed from each other. However, once the fundamentals have been understood, they are like other taxes in that they comprise a mixture of the relatively simple, the tediously lengthy and the horribly complicated.

The aims of this, the seventh edition of this book, remain to help the tax-literate non-specialist:

- understand the basic ideas and concepts,
- deal confidently with straightforward matters, and
- recognise when specialist input is needed.

To this end, examples, checklists and tables have been included, together with cross-references to legislation, tax cases and HMRC guidance wherever they are expected to be helpful. HMRC material is generally available on their website, and web addresses are given where possible.

As with other taxes, law and practice are fluid. The previous Chancellor of the Exchequer has, also in the past, made it clear he would change the law with immediate or even retrospective effect to close perceived loopholes. Therefore it will always be important to maintain an awareness of recent developments and pronouncements.

Whatever the reader's purpose in consulting this volume, it is suggested that he or she read Chapter 1 first – it is mercifully brief – as a kind of orientation exercise before using the contents, index or table of statutes to research particular points.

Chapter 1A provides an outline of land and buildings transaction tax (LBTT), which applies to most transactions over land in Scotland from 1 April 2015, sets out the principal differences between SDLT and LBTT, and considers the transitional provisions governing the move between the taxes. However, anyone dealing with LBTT on a regular basis may need to consult a more detailed exposition which can be found in *Land and Buildings Transaction Tax* also published by Bloomsbury as part of its new Scottish Tax Series.

As ever I am grateful to all at Bloomsbury Professional for their help as I updated the book, and to my wife, Amanda, for her continued support. I have attempted to translate the legislation and practice into words which are concise, accurate and user-friendly. I hope I have had at least partial success. Any errors or omissions are entirely down to me.

Ken Wright
August 2016

Contents

Table of Statutes

[*References are to paragraph number and appendices*]

Table of Statutory Instruments etc

[*References are to paragraph number*]

Table of EC Materials

[*References are to paragraph number*]

Table of Cases

[*References are to paragraph number*]

List of Abbreviations

ARTL	Automated Registration of Title to Land
ATED	annual tax on enveloped dwellings
CFD	contract for differences
CTA 2010	Corporation Tax Act 2010
DOTAS	disclosure of tax avoidance schemes
FA [year]	Finance Act [year]
GAAR	general anti-abuse rule
HMRC	Her Majesty's Revenue and Customs
ICTA 1988	Income and Corporation Taxes Act 1988
LBTT	Land and Buildings Transactions Tax
LBTT(S)A 2013	Land and Buildings Transaction Tax (Scotland) Act 2013
LLP	limited liability partnership
LTR	land transaction return
MiFID	Markets in Financial Instruments Directive
NPV	net present value
OEIC	open ended investment company
PIK	payment in kind
PIP	property investment partnership
PFI	private finance initiative
RI	recognised intermediary
RSL	registered social landlord
RSTPA 2014	Revenue Scotland Taxation and Powers Act 2014
SA 1891	Stamp Act 1891
SDLT	Stamp Duty Land Tax
SDLTM	Stamp Duty Land Tax Manual
SDRT	Stamp Duty Reserve Tax
SP	Statement of Practice
SPV	special purpose vehicle/company
SRN	scheme reference number
STM	Stamp Taxes Manual
TNPV	total net present value
UTRN	nique transaction reference number

Chapter 1

Introduction to Stamp Taxes

THREE TAXES

1.1 This book deals with stamp duty land tax (SDLT), stamp duty and stamp duty reserve tax (SDRT), which together are often referred to as 'stamp taxes'. However, almost all tax which people call 'stamp duty' is really one of the two more modern of these taxes, being SDRT and SDLT.

With effect from 1 April 2015 SDLT ceased to apply to transactions over land situated in Scotland being replaced, from that date, by Land and Buildings Transaction Tax (LBTT). Chapter 1A of this book briefly considers the key differences between SDLT and LBTT and sets out the transitional provisions which apply on the transition from SDLT to LBTT. For those who require more detail regarding LBTT, the tax is comprehensively covered in *Land and Buildings Transaction Tax* also published by Bloomsbury as one of its Scottish Tax Series.

The *Wales Act 2014*, which received Royal Assent on 17 December 2014, devolved a transaction tax on land situated in Wales to the Welsh Government from a date to be agreed. On 10 February 2015 this led to the Welsh Government issuing a consultation document entitled 'Tax Devolution in Wales – Land Transaction Tax' seeking views on the format of a new Welsh land tax to replace SDLT from 1 April 2018. The *Tax Collection and Management (Wales) Act 2016* became law in Wales on 25 April 2016 and put in place the legal framework for the collection and management of devolved taxes in Wales, including the new land transaction tax. The act establishes the Welsh Revenue Authority ('WRA') with the powers to identify and collect the devolved taxes, gives the WRA comprehensive civil investigation powers, confers criminal enforcement powers on the WRA, creates duties on taxpayers to pay penalties and interest in certain circumstances, and gives taxpayers rights of review and appeal against certain WRA decisions. Consequently, from 1 April 2018 there are likely to be three land taxes in the UK, with SDLT applying to land in England and Northern Ireland, LBTT applying to land in Scotland and a land transaction tax applying to land in Wales.

SDLT accounts for the overwhelming majority of revenue raised by the stamp taxes. It is also the tax most likely to impact ordinary commercial and private transactions. Its application to most ordinary house purchases is fairly straightforward. In relation to commercial transactions and even some house purchases, it can be an inordinately complex tax. Therefore, the greater part of this book is devoted to SDLT.

SDRT is mostly administered and paid by professionals in the financial services industry, often operating on the basis of rules and arrangements negotiated directly with HMRC. These rules and arrangements are beyond the scope of this book, which contains only a fairly brief outline of SDRT, concentrating on the impact of the tax on transactions in unquoted securities.

1.2 Each of the taxes has its own legislative code, with relatively little interaction between SDLT and the other taxes. There is more interaction between stamp duty and SDRT because, for many transactions, they are alternatives. The following table indicates where to find the main legislation governing each tax. The principal charging provisions are in bold, while other important provisions are shown in normal type. Especially in the case of stamp duty, there are numerous other provisions in various *Finance Acts* and other legislation. These are referred to in the text of this book where appropriate. For SDRT and SDLT, most changes since original enactment of the main provisions have proceeded by amendment; if an up–to-date copy of the main law is available, it is not often necessary to consult subsequent *Finance Acts*, unless investigating the effective date of a change, dealing with a transaction governed by transitional provisions or concerned about the application of general anti-avoidance provisions such as the general anti-abuse rule or GAAR.

Tax	Legislation
Stamp duty	***FA 1999, Sch 13*** *FA 1999, ss 109–123, Schs 12, 14–19* *FA 1986, ss 64–85* *Stamp Act 1891*; *FA 1930, s 42*; *FA 1967, s 27* *FA 2003, s 125*
SDRT	***FA 1986, ss 86–99*** *Stamp Duty Reserve Tax Regulations 1986, SI 1986/1711* *FA 1999, ss 116–122, Sch 19*
SDLT	***FA 2003, ss 42–124, Schs 3–19*** *Stamp Duty Land Tax (Administration) Regulations 2003, SI 2003/2837* *Stamp Duty Land Tax (Electronic Communications) Regulations 2005, SI 2005/844* *FA 2004, ss 306–319*; *SI 2004/1864*; *SI 2004/1865*; *SI 2005/1868* *FA 2013, Pt 5*

1.3 There are three reasons why practitioners and taxpayers need some knowledge of the rules relating to stamp duty itself:

(1) because non-stock market share transactions are still generally dealt with under stamp duty, failure to deal with this tax may leave unrecognised SDRT liabilities outstanding (see **10.11**);

(2) some other transactions finalised on or after 1 December 2003 may still give rise to stamp duty charges because of actions taken before that date (see **2.71** and **11.29**); and

(3) many of the principles and ideas in the other two taxes are derived from equivalent provisions under stamp duty and are most easily understood in that context.

It is therefore suggested that at least a brief study of the chapters on stamp duty would be worthwhile, even if the reader is dealing exclusively with transactions which may be subject only to one of the other taxes.

1.4 The names given to SDRT and SDLT are potentially misleading. They have 'stamp' in their name only because they collect tax which used to be collected through stamp duty. Unlike stamp duty, SDRT and SDLT are chargeable on transactions or agreements no matter how, where or even whether documented; no stamps are involved. They are compulsory, self-assessment taxes, with enforcement provisions similar to those which apply to corporation tax and income tax. SDLT, in particular, has draconian anti-avoidance rules and is within the scope of the General Anti-Abuse Rule introduced in 2013.

1.5 In contrast, stamp duty is often referred to as a voluntary tax because, with few exceptions (such as the charge on issue of bearer instruments, **8.55**), there is no direct obligation to pay. The only payment enforcement mechanism comes from the fact that a document which should be stamped cannot be used for UK official purposes, such as registration of title, evidence in a civil court, or for cancelling an SDRT charge, until it has been stamped. Stamping after the due date will generally lead to charging of interest and penalties, so if a document is to be stamped, it is best to deal with this promptly. However, where a document, despite being within the charge to stamp duty, is unlikely to be needed for any relevant UK purpose, the decision may validly be taken not to pay the stamp duty and to leave the document unstamped. Under current rules, this is only likely for some transfers of interests in partnerships which hold stock or marketable securities, or where the transfer relates to non-UK securities – it is highly unlikely that a stampable transfer of UK registered securities may safely be left unstamped. Historically, many other documents were left unstamped, as explained at **11.29**.

1.6 All three taxes are administered by HMRC, who publish two manuals on their website, setting out their views and practices on the taxes:

- The Stamp Taxes on Shares Manual, which is available at www.hmrc.gov.uk/so/manuals/stsmanual/index.htm, deals with stamp duty and SDRT. In this book, references to this manual are in the form 'STSM123456'. This new interactive manual was published on 23 July 2014, replacing the old Stamp Duty Manual which had been available only as a downloadable PDF file with limited interactive functionality. However, the old Stamp Duty Manual remains an essential reference when dealing with the current completion or unwinding of 'old' transactions, dealt with in **Chapter 11**.

- The second manual is the Stamp Duty Land Tax Manual, available at www.hmrc.gov.uk/manuals/sdltmanual/index.htm. This is also published as an online interactive document. It is widely referred to in this book, references being in the form 'SDLTM12345'.

1.7 HMRC provide online tools for calculation of SDLT, especially useful in relation to leases, and a facility for online completion and submission of SDLT returns. Further details of these are set out below in the chapters on SDLT. Use of the online submission facility is not yet compulsory, but it may become so in due course. In Scotland the online completion of SDLT returns was integrated with the Registers of Scotland's Automated Registration of Title to Land system. This was seen as a first step towards a long-term government goal of having all routine SDLT collection and administration dealt with by the Land Registries, leaving HMRC to deal with technical, policy and policing issues. It is not yet known whether this process will now begin in England, with LBTT having replaced SDLT in Scotland with effect from 1 April 2015 and SDLT to be replaced by a devolved land transaction tax in Wales with effect from 1 April 2018.

HMRC also provide a wide array of other useful information and guidance in the stamp tax section of their website (www.hmrc.gov.uk/so). The volume is so great that it is not always easy to find specific items, but references are given in this book where appropriate. A slight worry is that HMRC have taken to 'tidying up' this information, including removal of 'out of date' items. As a result, information which is visible today may disappear tomorrow.

A BRIEF HISTORY OF STAMP DUTY

1.8 Stamp duty was introduced over 300 years ago, copying a Dutch tax. Like income tax, it was intended to be a temporary measure to finance a war. Unlike income tax, it did not require annual renewal. A hundred years later, Pitt the Younger described stamp duty as a tax 'easily raised, widely diffused, pressing little on any particular class'. In other words, the thought was that those who paid the tax would scarcely notice – so it may, perhaps, be thought of as the original stealth tax. Unfortunately, the American colonists had noticed a few years previously – the 'Boston Tea Party', an event widely regarded as marking the start of the American War of Independence, was in fact a protest against the imposition of UK stamp duty on documents in the American Colonies. Pitt also commented that stamp duty was 'safely and expeditiously collected at a small expense'. Government statistics show that stamp duty is by far the cheapest tax to collect, although this may be because the taxpayer and his advisers have to do most of the work in calculating and administering the tax.

1.9 Stamp duty has always been a tax on documents. As the name indicates, payment is evidenced by fixing or embossing stamps on the relevant document; in general, if no document is created, there is no stamp duty. Its range has varied enormously over three centuries. Those whose memories stretch back to the mid-twentieth century may remember the practice of signing receipts across a 2d (0.8p) stamp in order to ensure their legal validity – an example of a postage stamp being used for fiscal purposes. Until 1971, bank cheques were also subject to a 2d stamp duty, and the issue of a new chequebook by the bank led to an immediate deduction from one's bank account.

During the latter half of the twentieth century, the scope of stamp duty was steadily reduced until it applied only to transactions in shares and securities, land and buildings, and intangible assets such as partnership interests, intellectual property and goodwill. In 1986, SDRT was introduced in response to market developments which allowed transfer of beneficial ownership of securities without a document. As explained at **10.1**, it was at one stage proposed to abolish stamp taxes on shares and securities. However, that proposal was abandoned and the new government of 1997 began to increase stamp duty rates on assets other than shares and securities. The yield to the Exchequer did not increase as expected, because avoidance of the tax on larger transactions became the norm. To counter this, stamp duty on property transactions was replaced by SDLT from 1 December 2003.

1.10 The original intention seems to have been for documents to be stamped no later than the time they were finally executed and, to this end, the government sold stamps and pre-stamped paper to 'authorised dealers' such as law firms, who would then use them on documents or sell them on to the end user. The government sold the stamps at a discount to face value, allowing the dealer to make a profit upon onward sale. This was, therefore, an early example of the privatised collection of tax. William Wordsworth was perhaps the most famous individual involved in this process – he was Distributor of Stamps for the county of Westmoreland. It was possible until very recently to buy blank stock transfer forms pre-stamped with a £5 stamp (the rate then applicable to many transfers otherwise than on sale), although sadly no discount was given!

Adhesive stamps were open to abuse (*Stamp Act 1891, s 9* still imposes penalties for fraudulent removal of adhesive stamps from a document with intent to re-use), and eventually the predecessors of HMRC became directly responsible for all stamping of documents at a number of Stamp Offices situated in major cities. The section of HMRC responsible for stamp taxes is still generally referred to as 'the Stamp Office'. Stamp Offices had stamping presses, allowing them to emboss appropriate value stamps on documents whilst keeping track of the value embossed, for audit purposes. The Stamp Offices provided an 'over the counter' service for routine stamping and even, by arrangement, for 'adjudication' (see **8.47**). Stamp Office public counters have now been closed and most of the machines decommissioned. Documents requiring physical stamping are sent to the Birmingham Stamp Office, and specific arrangements have to be agreed in advance for those transactions where, for commercial or perhaps overseas tax or regulatory reasons, stamping of documents on the day of execution is required.

1.11 Stamp duty is generally chargeable on the document which formally transfers the asset in question – the conveyance. In relation to shares and securities, a mere contract to transfer will not normally be liable, as this will not itself transfer the assets. When stamp duty was chargeable on other assets such as goodwill, the contract was usually the document liable to stamp duty, because no separate transfer document is needed to complete the transaction. This led to a practice of completing transactions without a written contract (for example, a written offer to sell, completed by the purchaser paying the consideration). In the absence of a written contract, there was no document on which stamp duty could be charged (see **11.30**).

1.12 As mentioned above, in most cases there is no direct obligation to pay stamp duty or have a document stamped, but an unstamped document cannot be used for certain purposes. Where documents were executed outside the UK, it was the case for a long time that no interest and penalties accrued, provided the document was stamped within 30 days of first being brought into the UK. In the case of large transactions, if there was no current need for the document to be stamped, it became standard practice to execute and retain it offshore (often in the Channel Islands or the Isle of Man) and many such unstamped documents remain there to this day. In an attempt to reduce the attractiveness of this practice, the rules were changed in 1999 to charge interest from 30 days after execution, no matter where this took place. Another common mitigation technique was simply to avoid executing a conveyance (generally referred to as resting on, or in, contract). For example, legal title to the asset might be held by a special purpose vehicle/company (SPV) for the benefit of the beneficial owner. The current beneficial owner would contract to sell to a purchaser. The purchaser would make payment, thus gaining beneficial ownership, and the owner would transfer the SPV to the purchaser. In the absence of a conveyance, no stamp duty arose. There were many disputes between taxpayers and the Stamp Office over whether other documents created in this process could be treated as conveyances. To discourage the practice, the rules were changed in 2002 to treat contracts for transfer of land for more than £10 million as conveyances.

1.13 Numerous other ingenious methods were devised to mitigate stamp duty, and the government began consultations on modernising the tax, concentrating on property transactions. However, it quickly became clear that it would be very difficult to create a tax which was relatively simple and certain, yet dealt fairly with the wide range of modern property transactions. Consultations ended abruptly, and SDLT was introduced in *Finance Act 2003*, replacing stamp duty on transactions in UK land and buildings with effect from 1 December 2003. At the same time, the scope of stamp duty itself was restricted to transactions in stock and marketable securities and interests in certain partnerships which hold stock or marketable securities. The inadequacy of the consultation process is reflected in the complexity of SDLT, the uncertainty over its application to many transactions, and the amendments made to the legislation, by regulations even before the tax came into effect and then in almost every subsequent *Finance Act*. The new rules appear to have had the desired effect, raising the stamp tax take on property transactions from £5 billion in 2003–04 to £10 billion in 2007–08. However, economic factors then intervened and by 2009–10 the total had fallen below the 2003–04 level, largely as a result of the reduced numbers and values of domestic transactions, before recovering to about £6 billion in 2011–12, almost £7 billion in 2012–13, £9.3 billion in 2013–14 and £10.9 billion in 2014–15. These increases in tax receipts appear to reflect a degree of recovery in property prices, and the Office for Budget Responsibility is forecasting in its 'Economic and Fiscal Outlook' report dated March 2016 that SDLT receipts will rise to £17.4 billion by 2020–21. This is however a reduction of £1.5 billion to the forecast SDLT receipts for 2020–21, as compared to that at July 2015, reflecting:

(a) an expected reduction in transactions among residential properties worth at least £2 million, apparently as a result of the increased SDLT charge on high value properties;

(b) increased receipts from commercial property transactions on the basis of rising prices and thresholds that are fixed in cash terms; and

(c) an increase of £0.9 billion as a result of the new 3% SDLT surcharge on second homes and buy-to-let properties (see **2.39**).

Over the five years to 2012–13 there was a slow decline in the total raised by stamp taxes on shares, reaching £2.2 billion for 2012–13. Given that share prices had recovered significantly from their low point of 2009, this probably reflected a reduction in volumes, possibly because of a shift toward trading derivatives, which are not subject to stamp taxes. Since then the tax raised by stamp taxes on shares has begun to increase again with £3.1 billion collected in 2013–14 and £2.9 billion for 2014–15, possibly reflecting the increase in share prices. In relation to stamp taxes on shares, the Office for Budget Responsibility is forecasting (March 2016) that by 2020–21 receipts will have increased to £3.5 billion.

The annual tax on enveloped dwellings (see **2.72** et seq) raised £100m in 2013–14 and £116m in 2014–15 and, with the reduction of the thresholds after which the tax applies, from properties valued at more than £2m down to properties valued at more than £1m from 1 April 2015, and properties valued at more than £0.5m from 1 April 2016, it is likely that the tax raised will continue to increase. However, this will, to some extent, depend on whether residential properties continue to be owned by companies, probably for reasons other than the future avoidance of SDLT, or whether properties are transferred out of companies to avoid the imposition of ATED.

Chapter 1A

Scotland – Land and Buildings Transaction Tax

REPLACEMENT FOR SDLT

1A.1 From 1 April 2015, land and buildings transaction tax (LBTT) replaced SDLT in relation to transactions over land and buildings situated in Scotland. However, there are transitional provisions (see **1A.20**) covering the move from SDLT to LBTT. Under these transitional provisions some transactions which commenced before 1 April 2015 remain within the ambit of SDLT or potentially fall within both regimes. For example, consider a situation where missives are concluded after 1 May 2012 (the date the *Scotland Act 2012,* which devolved the tax over land transactions in Scotland to the Scottish Parliament, received Royal Assent). The contract is 'substantially performed' (see **2.9**) before 1 April 2015 but settlement (completion) takes place after 31 March 2015. In these circumstances the substantial performance of the contract will be chargeable to SDLT whereas the settlement of the contract will be chargeable to LBTT. Under the transitional provisions, credit is given for any SDLT paid on substantial performance in assessing what, if any, LBTT is payable on settlement.

LBTT was introduced by the *Land and Buildings Transaction Tax (Scotland) Act 2013* (*LBTT(S)A 2013*) and references in this chapter are to that Act unless otherwise stated. The administration, management and compliance aspects of the taxes devolved to the Scottish Parliament, including LBTT, are governed by the *Revenue Scotland and Tax Powers Act 2014* (*RSTPA 2014*), which received Royal Assent on 24 September 2014.

What follows is merely an outline of LBTT; for those who want to know more, the tax is comprehensively covered in *Land and Buildings Transaction Tax*, published by Bloomsbury Professional as part of its Scottish Tax Series.

LBTT is similar to SDLT in many respects and some of the legislation borrows heavily from the corresponding SDLT law. In particular LBTT is a self-assessment tax charged on the consideration given for transactions over land in Scotland. It is payable by the buyer (purchaser), generally within 30 days of the 'effective date' of the transaction. The rules to determine the occasions of charge and the amount of consideration closely parallel those for SDLT. As with SDLT, any VAT payable will be included in the measure of consideration, thus increasing the effective rate of tax for a business which is able to recover VAT as input tax (see **2.23**). Most SDLT reliefs are imported into the new tax, generally with similar rules and conditions.

From 1 April 2015, if a single transaction involves land on both sides of the border it will be necessary to complete separate returns under the two taxes for the respective parcels of land, the overall consideration being apportioned on a just and reasonable basis. The rate of SDLT for the part situated in England and Wales should then be determined by reference to the consideration apportioned to that part, not by reference to the overall consideration, because the two parts will not be regarded as linked with each other.

HMRC, in conjunction with Revenue Scotland (the new tax authority with responsibility for the collection and management of the taxes devolved to the Scottish Parliament, including LBTT) have published detailed guidance on the transitional provisions at www.hmrc.gov.uk/government/publications/transitional-land-and-building-transaction-tax-guidance.

KEY DIFFERENCES BETWEEN LBTT AND SDLT

There are some significant differences between LBTT and SDLT and these are covered below.

Higher nil tax rate threshold for residential property

1A.2 The nil rate tax band for transactions in relation to residential property, at or below which no tax will be charged, is £20,000 higher under LBTT (£145,000) than under SDLT (£125,000).

A progressive tax for non-residential property transactions.

1A.3 Under SDLT, until 16 March 2016, the entire consideration for a non-residential property transaction, apart from rent under a lease, was charged at a single rate, which was determined by the amount of the chargeable consideration and the highest tax band into which that consideration fell – the so called 'slab' system. However, for transactions with an 'effective date' on or after 17 March 2016, SDLT is now charged on the same progressive basis as LBTT, with different bands of consideration charged at different rates (see **2.37**). The LBTT tax bands and rates for non-residential properties are as follows:

Purchase price	LBTT rate
Up to £150,000	0%
£150,001–£350,000	3%
Above £350,000	4.5%

RESIDENTIAL PROPERTY

1A.4 Up until midnight on 3 December 2014, SDLT on residential property was also calculated using the 'slab' system described above in relation to non-residential properties. However, from 4 December 2014, the UK government introduced a progressive tax structure for residential property transactions subject to SDLT – see **2.37.**

DIFFERENTIAL TAX CHARGES UNDER LBTT COMPARED TO SDLT

1A.5 It is understood that the Scottish Government's plan was to collect broadly the same amount of tax under LBTT as the UK Government would have collected, in Scotland, under SDLT. However, the proposed tax rates and bands announced by the Scottish Government on 9 October 2014 were set at such a level that there would be winners and losers with purchasers at the lower end of the market paying less tax, and those at the higher end of the market paying more tax, compared to SDLT. Subsequent to the release of the proposed LBTT tax rates and bands, the Chancellor of the Exchequer announced that, with effect from 4 December 2014, a progressive tax structure would apply for the purposes of SDLT in relation to the acquisition of residential property. The new SDLT tax rates and bands would have resulted in an SDLT rate of 5% applying to that part of the purchase price above £250,000 compared to an LBTT rate of 10%. In response to the change in the SDLT tax structure and tax rates and bands, the Scottish Government announced on 21 January 2015 that the proposed LBTT tax rates and bands, in relation to residential property, would be revised. The revised tax rates and bands are shown in the table below.

At the Autumn Statement 2015 the UK Chancellor of the Exchequer announced that on or after 1 April 2016 higher rates of SDLT would apply to purchases of additional residential properties, such as second homes and buy-to-let properties. In fact the legislation is much wider than the policy description and, for example, applies to first purchases of residential properties by companies and other non-individual purchasers. The application of the higher rates of SDLT to acquisition of additional residential property is considered in detail at **2.39** et seq. In response to the Chancellor's announcement, and ostensibly to avoid distortions in the property market between Scotland and the rest of the UK, the Scottish Government announced that it would also introduce a higher rate of LBTT for purchases of additional residential properties. Like its SDLT equivalent, it is also much wider than the policy description and it applies to first purchases of residential properties by companies and other non-individual purchasers. There are however some important differences between the two regimes. Under SDLT the higher rate does not apply to the acquisition of mixed use properties whereas under LBTT, the higher rate would apply, with a just and reasonable apportionment of the consideration being made between the residential and non-residential elements of the property. Under SDLT an individual has 36 months to sell his old main residence and claim a refund of the additional rate of SDLT paid, whereas under LBTT the period is 18 months. Where the LBTT additional rate applies it adds 3% to each of the LBTT rates, including the 0% rate. The LBTT additional rate however does not apply if the chargeable consideration for the transaction is less than £40,000. The LBTT rates, if the additional rate applies, are shown in the table below.

Purchase price	LBTT rate	Additional rate
Up to £145,000	0%	3%
£145,001–£250,000	2%	5%
£250,001–£325,000	5%	8%
£325,001–£750,000	10%	13%
Above £750,000	12%	15%

Under LBTT, residential property (irrespective of whether the additional rates apply) acquired for a consideration of up to £333,000 will incur a smaller or similar tax liability compared to that under SDLT. However, above those purchase price levels the charge under LBTT will be greater as compared to that under SDLT.

As regards non-residential property, under SDLT, an acquisition of a property for a consideration of up to £300,000 will incur a smaller or similar tax liability compared to that under LBTT. However, above those purchase levels the charge under LBTT will be less as compared to that under SDLT.

Set out below are examples comparing the LBTT and SDLT liabilities, for both residential and non-residential property purchases, at different levels of purchase price. These examples demonstrate that at the higher end of the market the LBTT charge in relation to residential property is significantly higher than the equivalent SDLT charge, reflecting the Scottish Government's view that purchases at that level should bear a higher tax burden.

Residential property

House purchase costing £250,000 – assumes that an individual is replacing their main residence so that the additional rate does not apply for either LBTT or SDLT purposes.

LBTT

Purchase Price	Rate	LBTT
£	%	£
145,000	0	NIL
105,000	2	2,100
250,000		2,100

SDLT

Purchase Price	Rate	SDLT
£	%	£
125,000	0	NIL
125,000	2	2,500
250,000		2,500

The LBTT charge is 16% lower compared to the equivalent SDLT charge.

House purchase costing £750,000 – this time the assumption is that both the LBTT and SDLT additional rate for residential property applies.

LBTT

Purchase Price	Rate	LBTT
£	%	£
145,000	3	4,350
105,000	5	5,250
75,000	8	6,000
425,000	13	55,250
750,000		70,850

SDLT

Purchase Price	Rate	SDLT
£	%	£
125,000	3	3,750
125,000	5	6,250
500,000	8	40,000
750,000		50,000

The LBTT charge is 42% higher compared to the equivalent SDLT charge.

Non-residential property

Commercial property purchase costing £250,000

LBTT

Purchase Price	Rate	LBTT
£	%	£
150,000	0	NIL
100,000	3	3,000
250,000		3,000

SDLT

Purchase Price	Rate	SDLT
£	%	£
150,000	0	NIL
100,000	2	2,000
250,000		2,000

The LBTT charge is 50% higher compared to the equivalent SDLT charge.

Commercial property purchase costing £5,000,000

LBTT

Purchase Price	Rate	LBTT
£	%	£
150,000	0	NIL
200,000	3	6,000
4,650,000	4.5	209,250
5,000,000		215,250

SDLT

Purchase Price	Rate	SDLT
£	%	£
150,000	0	NIL
100,000	2	2,000
4,750,000	5	237,500
5,000,000		239,500

The SDLT charge is 10.125% higher compared to the equivalent LBTT charge.

Under SDLT, a rate of 15% applies to the whole of the chargeable consideration where a company or a partnership with one or more corporate members or a collective investment scheme acquires a dwelling for chargeable consideration of more than £500,000 (*FA 2003, Sch 4A*). There are, however, a number of exceptions from that penal rate where, for example, the dwellings are held by property traders and developers or for the purposes of a property development business. There is no similar regime under LBTT and the LBTT rates for residential property, including the top rate of 15% (assuming the additional LBTT rate applies), apply irrespective of the category of buyer.

REGISTRATION OF LAND TRANSACTIONS AND PAYMENT OF LBTT

1A.6 Whilst the collection and administration of LBTT is the responsibility of Revenue Scotland, the Keeper of the Registers of Scotland need not register title in relation to a notifiable transaction until the tax obligations (ie the land transaction return has been filed and any LBTT payable has been paid) have been dealt with by the buyer (*LBTT(S)A 2013, s 43(1)* and the *Land and Buildings Transaction Tax (Ancillary Provision) (Scotland) Order 2014 (SSI 2014/376)*). Differences in land law and practice between Scotland and the rest of the UK have historically meant that there was usually an impetus to register a change of owner of land in Scotland as soon as possible, frequently on the same day as the transaction. This should now be less of an issue following the *Land Registration etc (Scotland) Act 2012* coming into force on 8 December 2014, which, inter alia, introduced the concept of an 'advance notice' to provide protection, for a period of 35 days, for a deed that is intended to be registered in the Land Register. Despite this protection it is still likely to be normal practice for the land transaction return to be filed and LBTT to be paid immediately after, or very soon after, the 'effective date' of a transaction, (even though the return need only be filed, and the tax need only be paid within 30 days of the 'effective date' of the transaction) although in relation to payment this only requires that 'arrangements satisfactory' to Revenue Scotland are in place at that date. The term 'arrangements satisfactory' broadly means that where an electronic return is being filed, arrangements have been made to make payment by direct debit, BACS, or CHAPS, or payment is made by credit card or debit card at the time of filing the electronic return, or payment is made by cheque which should reach Revenue Scotland by the third working day after the date on which the electronic return was filed. If a paper return is being filed the only method of payment is by cheque.

In general under SDLT, if the transaction is notifiable the purchaser must file a land transaction return and obtain an SDLT5 certificate from HMRC, before the effect of the transaction can be registered in the Land Registry. However, there is no requirement to pay the tax at the time the land transaction return is filed, as the tax need only be paid within the 30-day period from the 'effective date' of the transaction.

NO GENERAL SUB-SALE RELIEF BUT THERE IS A 'SUB-SALE DEVELOPMENT RELIEF'

1A.7 Although most of the reliefs available under SDLT have been imported into LBTT, the one major omission is the relief which has probably been most widely used for unacceptable tax mitigation – so called 'sub-sale relief' under *FA 2003, s 45* and *Sch 2A*. By omitting the relief from the LBTT rules, the Scottish government and Revenue Scotland have doubtless saved themselves a lot of trouble. However, sub-sale relief is not only relied upon for unacceptable tax mitigation. Under SDLT it has also been an important factor in arrangements to finance speculative development, which includes most private residential building, and in so-called forward funding transactions where a developer identifies a site but then sells it on to an institutional investor or owner-occupier prior to development. It remains to be seen whether the omission of a general sub-sale relief will have a detrimental effect on this part of the Scottish economy.

Whilst there is no general sub-sale relief within LBTT the Scottish government has introduced a much narrower relief known as 'sub-sale development relief'. The relief is only available where, at the time of the filing of the land transaction return in relation to the substantial performance (see **2.9**) or completion of the first contract, the first buyer has entered into a 'qualifying sub-sale' and a 'significant development' is to take place on the land within five years of the date of the 'qualifying-sub-sale' transaction. The relief will be withdrawn, or partially withdrawn, if the significant development is not actually completed, or only part of the significant development is actually completed, within the requisite five-year period.

Assignment of a contract before substantial performance/completion

1A.8 It should however be noted that there is a form of more general sub-sale relief within the LBTT legislation due to the fact that the assignation of a contract (which is to be completed by a conveyance), before its substantial performance and completion, triggers 'substantial performance' of that contract (*LBTT(S)A 2013, s 14(1)(c)*). As a consequence, the contract is treated as if it were the transaction provided for in the contract, and the 'effective date' of the deemed land transaction is the date of substantial performance (*LBTT(S)A 2013, s 10(1)* and *(2)*). Any LBTT payable in relation to the deemed land transaction will be payable by the original buyer within 30 days of the 'effective date' of the deemed land transaction (*LBTT(S)A 2013, ss 29(3)* and *40(2)*). Following the assignation it will be the assignee who will complete the contract, and completion of the contract will also be a notifiable transaction. However, LBTT will only be payable by the assignee to the extent (if any) that the LBTT payable on completion is greater than that which was paid by the original buyer on substantial performance of the contract (*LBTT(S)A 2013, s 10(3)*). Therefore, if the consideration payable on completion is the same as that payable under the contract at the time of its assignation, and no LBTT reliefs applied at substantial performance but not at completion, and assuming no change in LBTT bands and rates, no further LBTT should be payable by the assignee. In such circumstances only one charge to LBTT will have arisen, although it will have been paid by the assignor (ie the original buyer) rather than the assignee. This is similar to the position had sub-sale relief been available, but here it is the assignee which has not incurred an LBTT liability, whereas under sub-sale relief it would have been the original buyer who would have been relieved from paying LBTT.

Whilst it may not always be commercially possible or acceptable to assign a contract, consideration could be given to such a strategy if a person other than the original buyer may look to get the benefit of the contract. In these circumstances it would be necessary to ensure that under the terms of the contract it is possible for the original buyer to assign the contract. If the prospective assignee is unhappy with some of the terms of the existing contract it may be possible to vary the terms of the contract, without giving rise to an LBTT liability, either before or after the assignment.

LEASES

1A.9 The special rules for leases are contained in *LBTT(S)A 2013, Sch 19*. In principle, the manner of charging LBTT on the grant of a new lease will be similar to that under SDLT. Tax will be calculated separately on the rent and on any premium, and the two amounts added to give the initial tax bill for the lease. The rules relating to leases of indefinite term, leases for a fixed period that continue beyond that period, and renewals are similar to those under SDLT (see **4.64** et seq).

1A.10 However, under SDLT the tax liability on rents payable under a lease is finally determined by reference to the rent payable during the first five years of the term of the lease. Even if it is known that rents will increase after year five, this is not taken into account. There is no such rule in relation to LBTT – any **known** increases throughout the life of the lease will have to be taken into account in the initial assessment of tax, apart from increases in line with the retail price index, consumer price index or similar index of inflation. But even then, the position will not be finally determined. In addition, assuming the grant of the lease was notifiable or becomes notifiable, the tenant will be required to review the position every three years during the period of the lease, and again when either the lease is assigned or the lease comes to an end, to determine whether further tax is due. The requirement for submission of new returns every three years (whether or not there is any further tax to pay) will be onerous for some lessees, especially those businesses that have numerous leased premises.

It is understood that the intention under LBTT is to adjust the tax payable at these three yearly intervals to take account of increases in rent or other changes in the lease. This is not entirely clear from the legislation. *LBTT(S)A 2013, Sch 19, para 10(2)* requires a further return to be made at each third anniversary of the effective date of the grant of the lease, if the grant of the lease was notifiable, or at each third anniversary of the event triggering notification of the lease assuming the lease has not been assigned or terminated. The further return must be made irrespective of whether there have been any changes affecting the amount of the tax payable. *Paragraph 10(4)* states that the return must include an assessment of the tax due on the basis of the information in the return. However, it is not stated in the legislation that such information should include any changes such as rent increases. It is assumed that the intention is to rely on the power of the tax authority to specify what information is to be included in returns (*LBTT(S)A 2013, s 35(1)(b)*) for this purpose.

LEASES OF RESIDENTIAL PROPERTY

1A.11 Generally in Scotland it is now not possible to grant a lease over residential property for more than 20 years (*Land Tenure Reform (Scotland) Act 1974, ss 8–9*), although it should be noted that there are now exceptions to this general rule following the *Housing (Scotland) Act 2010, ss 138* and *139* and the *Private Rented Housing (Scotland) Act 2011, ss 36* and *37*. Also, unlike in England, it is not common for leases of residential property to be granted for long periods. The Scottish government therefore took the view that it was unlikely that the rent payable on such leases would be sufficient to breach the nil rate tax band for the NPV of the rent (£150,000) such that an LBTT liability arose on the grant of the lease. As a consequence the grant, assignation, or renunciation of a lease of residential property is exempt from LBTT unless it is an 'ultra-long lease'. An 'ultra-long lease' is broadly a lease which is registered, was granted for a period of more than 175 years, and which on 28 November 2015 had more than 100 years to run. Such 'ultra-long leases' were converted into ownership on 28 November 2015 hence their exclusion from the exemption from LBTT on leases over residential property.

There is no such exemption under SDLT. Currently a 20-year residential lease at a fixed rent greater than about £8,700 per annum would give rise to an SDLT liability, and at lower levels of annual rent there may be an obligation to submit an SDLT return even if no tax is due.

ANTI-AVOIDANCE PROVISIONS

SDLT specific anti-avoidance rule

1A.12 The SDLT legislation contains a specific anti-avoidance rule (*FA 2003, s 75A*), which is extremely wide in its scope. Whilst its objective is clearly to counter-act SDLT avoidance, there is no requirement that an SDLT avoidance motive exists to fall within the rule. As a result, wholly commercial transactions, perhaps structured in an SDLT efficient manner, potentially fall within the rule, which can lead to significant uncertainty for taxpayers.

The Scottish government has very sensibly decided not to incorporate a similar type of rule into the LBTT legislation, meaning that taxpayers are spared the uncertainty that can arise under SDLT. Within the LBTT legislation the Scottish government places reliance on targeted anti-avoidance provisions and the general anti-avoidance rule (see below), which applies to the devolved taxes, including LBTT.

General anti-avoidance rule

1A.13 The Scottish Government has introduced a general anti-avoidance rule (the Scottish 'GAAR') (*Revenue Scotland and Tax Powers Act 2014, ss 62–72*) which is specifically designed to be broader in scope than the UK general anti-abuse rule found in *FA 2013, ss 206–215*. The objective of the Scottish GAAR is to counteract 'tax advantages' arising from a 'tax avoidance arrangement' which is 'artificial'. The Scottish GAAR does not contain the safeguards of an advisory review panel, and the so-called double reasonableness test, which apply to the UK general anti-abuse rule.

Residential property holding companies

1A.14 The Scottish Ministers can, by regulation, provide that 'qualifying transfers' of interests in 'residential property holding companies' be treated as land transactions chargeable to LBTT (*LBTT(S)A 2013, s 47*). A 'residential property holding company' is a company whose shares are not listed on a recognised stock exchange, whose sole or main activity is holding interests in land (including interests in land located outside of Scotland) and whose assets include interests in residential property. A 'qualifying transfer' is a transfer of an interest in a 'residential property holding company' that results in the transferee (ie the buyer) acquiring the right to occupy some or all of the company's residential property. The Scottish ministers have not yet made any such regulations and therefore this provision is not currently in force.

There is no similar provision under SDLT for transfers of shares in 'land-rich' companies to be treated as land transactions chargeable to tax.

MULTIPLE DWELLINGS RELIEF

1A.15 Under SDLT a claim to multiple dwellings relief cannot reduce the SDLT charge to less than 1% of the 'total dwellings consideration' (*FA 2003, Sch 6B, para 5(2)*). However, following the introduction of the additional rate of SDLT on residential properties with effect from 1 April 2016, it would seem that the 1% minimum charge provision is now effectively redundant, as the additional rate will impose a 3% charge on the first £125,000 of consideration. If a claim for multiple dwellings relief is made the additional rate of SDLT will apply and it is not possible to apply the normal rates of SDLT to the acquisition of, say, one property on the basis it is a replacement of a main residence, and the higher rates to the other properties acquired. This is possible under LBTT.

Multiple dwellings relief in relation to LBTT is similar in most respects; however, it does not specify a lower rate of charge, but states that where the relief is available the tax payable cannot be less than a proportion (25%) of that due had the relief not been available. This is stated by the Scottish government to be a consequence of LBTT being a progressive tax (see **1A.3**) however, following the change to a progressive structure for SDLT on residential property from 4 December 2014, the UK government has retained the minimum 1% charge on the total chargeable consideration given for dwellings. The higher rate of LBTT on the acquisition of additional residential property, and residential property acquired by non-individual purchasers, will apply when a claim is made for multiple dwellings relief (see Example 1A.7) if the relevant conditions for the additional rate to apply are satisfied. However, under LBTT, it is possible for some of the properties to be charged at the additional rate whilst one of the properties is charged at the standard rates, if for example, one of the residential properties being acquired is a replacement of an individual purchaser's main residence. This means that if a claim for multiple dwellings relief is made, and the additional rate of LBTT applies to the purchase of any of the properties, the tax chargeable on each property must be calculated, applying either the normal or additional rate LBTT rates as applicable. The LBTT chargeable on each property is then aggregated to give the total amount of LBTT payable.

ACQUISITION RELIEF

1A.16 Under SDLT a successful claim to acquisition relief reduces the rate of SDLT chargeable to 0.5% of the chargeable consideration for the transaction (*FA 2003 Sch 7, para 8(1)*).

A successful claim to acquisition relief in relation to LBTT reduces the tax payable to a proportion (12.5%) of that due had the relief not been available. Again the fact that there is no single reduced rate of LBTT applied to the chargeable consideration is stated by the Scottish government to be a consequence of LBTT being a progressive tax (see **1A.3**).

CROFTING COMMUNITY RIGHT TO BUY RELIEF

1A.17 Under SDLT, relief (*FA 2003, s 75*) for a transaction entered into under the crofting community right to buy was provided by dividing the total chargeable consideration being given for the acquisition of the crofts by the number of crofts being acquired, calculating the SDLT chargeable on the amount so determined, and multiplying the amount of the SDLT charge by the number of crofts being acquired to get the total SDLT payable. If that average consideration fell within the nil rate band no SDLT was payable as there was no minimum rate of tax payable under this relief as there is for multiple dwellings relief (see **1A.15**).

In contrast, under LBTT (*LBTT(S)A 2013, Sch 9*), the acquisition of two or more crofts under a transaction, which is entered into pursuant to a crofting community right to buy, is wholly exempt from LBTT irrespective of the average consideration paid per croft (the *Land and Buildings Transaction Tax (Addition and Modification of Reliefs) (Scotland) Order 2015, art 2(4) (SSI 2015/93)*).

LICENCES TO OCCUPY PROPERTY

1A.18 Under LBTT the grant, assignation or renunciation of a licence to use or occupy property is exempt from charge unless it is a 'prescribed non-residential licence', which is a licence of a description prescribed under regulations made by the Scottish Ministers (*LBTT(S)A 2013, Sch 1, paras 3(1)(b), 3(4)* and *s 53(1)*). Currently no such licences have been prescribed by the Scottish ministers, meaning that the grant, assignation or renunciation of a licence to use or occupy non-residential property, irrespective of the use to which the property is to be put, will be an exempt transaction for the purposes of LBTT.

In relation to SDLT, a licence to use or occupy land is an exempt interest (*FA 2003, s 48(2)(b)*), so that any transaction in relation to such a licence will not be chargeable to SDLT.

The only practical impact of this difference in treatment should be that under LBTT the variation of a licence to use or occupy property may give rise to LBTT obligations to file a land transaction return and pay tax (*LBTT(S)A 2013, s 6(1)(c), 3(c)* and *(4)(c)*) if the chargeable consideration given for the variation exceeds the £150,000 nil rate tax band (*LBTT(S)A 2013, s 30(1)(c)*). In contrast, under SDLT, the variation of a licence to use or occupy land would be exempt.

OTHER SDLT RELIEFS NOT AVAILABLE UNDER LBTT

1A.19 The following SDLT reliefs are not available under LBTT:

- relief for acquisition by an employer in a case of relocation of employment of an employee (*FA 2003, Sch 6A, para 5*); and
- relief for acquisition by a property trader from the personal representatives of a deceased person (*FA 2003, Sch 6A, para 3*).

TRANSITIONAL PROVISIONS

Disapplication of SDLT

1A.20 The *Scotland Act 2012* contains provisions at *s 29* and *Sch 3* for the disapplication of SDLT, in relation to transactions in land situated in Scotland from 1 April 2015.

Scotland Act 2012, s 29(2) provides that a chargeable interest, for the purposes of SDLT (*FA 2003, s 48*) comprises an interest in land in England and Wales or Northern Ireland, but not in Scotland. Consequently, subject to the transitional provisions described below, a transaction in relation to an interest in land situated in Scotland, with an effective date (*FA 2003, s 119*) which is on or after 1 April 2015, will not be a land transaction for SDLT purposes but will, instead, potentially be within the charge to LBTT (*Scotland Act 2012, s 29(4)*).

Linked transactions

1A.21 Based on the *Scotland Act 2012, s 29(2)* and *(4)*, and on the technical guidance issued by HMRC on 15 February 2015 (see **1A.1**) transactions in land situated in Scotland with an effective date (*FA 2003, s 119*) on or after 1 April 2015, which would be linked (see *FA 2003, s 108*) for SDLT purposes, will no longer be linked for SDLT purposes with either:

- a transaction in land which is situated in England and Wales or Northern Ireland; or
- a transaction in land which is situated in Scotland but which is subject to SDLT.

Example 1A.1 – A transaction chargeable to SDLT is not linked to a transaction chargeable to LBTT

An Cala Ltd, as part of a single deal, agrees to acquire two commercial buildings, one located in Inverness and the other in Derby, from the same company. The seller is based in Inverness and the effective date of the acquisition of both buildings is 28 April 2015.

The acquisition of the building located in Derby is chargeable to SDLT whereas the acquisition of the building in Inverness is chargeable to LBTT. Neither of the transactions is linked for the purposes of SDLT or LBTT.

As a separate deal An Cala Ltd also agrees to acquire two office blocks from a company based in London, with one of the office blocks being situated in Glasgow and one in Dundee. The effective date of the acquisition of the

Glasgow property was 31 March 2005 whereas the property in Dundee is not acquired until later, the transaction having an effective date of 10 July 2015.

The acquisition of the Glasgow property is chargeable to SDLT whereas the acquisition of the Dundee property is chargeable to LBTT. Neither transaction is linked for the purposes of SDLT or LBTT.

Acquisition of land located in both Scotland and the rest of the UK

1A.22 Where a transaction in land with an effective date on or after 1 April 2015 includes both land in Scotland and land in England and Wales or Northern Ireland, the acquisition of the land in Scotland is within the charge to LBTT, whereas the acquisition of the land in England and Wales or Northern Ireland is within the charge to SDLT, and the consideration given for the acquisitions should be apportioned on a just and reasonable basis (*LBTT(S)A 2013, Sch 2, para 4* and *FA 2003, Sch 4, para 4*).

Example 1A.2—Apportionment of consideration – acquisition of properties in Scotland and in the rest of the UK

Super Retailer Ltd acquires a number of shops from Discount Retailer Ltd as a single transaction. Some of the shops are situated in England and Wales and some in Scotland. The effective date of the transaction (the date of settlement) is 31 October 2015.

The acquisition of the shops in Scotland is not chargeable to SDLT, but will be within the charge to LBTT. The acquisition of the shops in England and Wales will be within the charge to SDLT. The consideration given for the respective acquisitions should be apportioned on a just and reasonable basis.

Transitional provisions (Scotland Act 2012, s 29(5) and (6))

Land situated in Scotland – transactions to which SDLT continues to apply

1A.23 SDLT continues to apply to any transaction in relation to land situated in Scotland, which has an effective date (*FA 2003, s 119*) on or after 1 April 2015, and which is:

(1) effected under a contract entered into and substantially performed (*FA 2003, s 44(5)*) on or before 1 May 2012 (the date the *Scotland Act 2012* received Royal Assent); or

(2) effected under a contract entered into on or before 1 May 2012 (but not substantially performed on or before 1 May 2012) and not excluded by reason of falling into (a), (b) or (c) below:

(a) the contract is varied, or there is an assignation of rights under the contract, after 1 May 2012;

(b) the transaction is effected as a result of the exercise after 1 May 2012 of an option, right of pre-emption or similar right; or

(c) after 1 May 2012 there is an assignation, subsale or other transaction relating to the whole or part of the land which is the subject of the contract, as a consequence of which a person other than the purchaser (buyer) under the contract becomes entitled to call for a conveyance.

The technical guidance issued by HMRC on 15 February 2015 (see **1A.1**) includes a useful table setting out the effect of the transitional provisions. The table is reproduced below.

Table to demonstrate the effect of the transitional provisions for land transactions in Scotland

On or before 1 May 2012	2 May 2012 – 31 March 2015	1 April 2015 and after	Tax charge
Conclusion of Missives		Settlement	SDLT on Settlement
Conclusion of Missives + Substantial Performance		Settlement	SDLT on Substantial Performance + SDLT on Settlement*
Conclusion of Missives + Substantial Performance	Variation	Settlement	SDLT on Substantial Performance + SDLT on Settlement*
Conclusion of Missives	Substantial Performance	Settlement	SDLT on Substantial Performance + SDLT on Settlement*
Conclusion of Missives	Variation	Settlement	LBTT on Settlement
Conclusion of Missives	Substantial Performance + Variation	Settlement	SDLT on Substantial Performance + LBTT on Settlement*
Conclusion of Missives		Substantial Performance + Settlement	SDLT on Substantial Performance + SDLT on Settlement*

On or before 1 May 2012	**2 May 2012 – 31 March 2015**	**1 April 2015 and after**	**Tax charge**
Conclusion of Missives		Substantial Performance + Variation + Settlement	SDLT on Substantial Performance + LBTT on Settlement*
	Conclusion of Missives	Settlement	LBTT on Settlement
	Conclusion of Missives + Substantial Performance	Settlement	SDLT on Substantial Performance + LBTT on Settlement*
	Conclusion of Missives + Substantial Performance	Variation + Settlement	SDLT on Substantial Performance + LBTT on Settlement*
	Conclusion of Missives	Substantial Performance + Settlement	LBTT on Settlement
	Conclusion of Missives	Substantial Performance + Variation + Settlement	LBTT on Settlement

1A.24 HMRC have indicated (see technical guidance issued by HMRC on 15 February 2015 – **1A.1**) that they will not regard the following post-1 May 2012 events as excluding a transaction from the SDLT regime under *Scotland Act 2012, s 29(6)* (see point (2) in **1A.23**):

- a variation of the completion date specified in the contract; or
- a transfer to the nominee or bare trustee of the purchaser (buyer) under the contract.

1A.25 Each of the scenarios set out in **1A.23** are now considered in turn.

Contracts entered into on or before 1 May 2012

Scenario 1

The missives (exchange of contracts) are concluded on or before 1 May 2012, substantial performance and settlement (completion) take place at the same time but after 1 April 2015. The settlement transaction is within the SDLT regime (*Scotland Act 2012, s 29(5)(b)* and *(6)*).

Scenario 2

The missives (exchange of contracts) and substantial performance are concluded/takes place on or before 1 May 2012 and settlement (completion) takes place on or after 1 April 2015. The settlement transaction is within the SDLT regime (*Scotland Act 2012, s 29(5)(a)*).

Scenario 3

The missives (exchange of contracts) and substantial performance are concluded/takes place on or before 1 May 2012; however, this time the contract is varied before 1 April 2015. The variation does not bring the settlement (completion) transaction into the LBTT regime, as the contract was substantially performed on or before 1 May 2012 (*Scotland Act 2012, s 29(5)(a)*). The settlement (completion) transaction is therefore within the SDLT regime.

Scenario 4

The missives (exchange of contracts) are concluded on or before 1 May 2012, substantial performance takes place after 1 May 2012 but before 1 April 2015, and settlement (completion) takes place after 1 April 2015. As the contract was entered into on or before 1 May 2012, and has not been varied, the settlement (completion) transaction on or after 1 April 2015 is within the SDLT regime (*Scotland Act 2012, s 29(5)(b)*). As substantial performance took place before 1 April 2015 it will also be within the SDLT regime.

Scenario 5

The missives (exchange of contracts) are concluded on or before 1 May 2012, the contract is varied after 1 May 2012 but before 1 April 2015, and settlement (completion) and substantial performance take place at the same time, but on or after 1 April 2015. As the contract has been varied after 1 May 2012 the settlement (completion) transaction is within the LBTT regime (*Scotland Act 2012, s 29(5)(b) and (6)(a)*).

Scenario 6

The missives (exchange of contracts) are concluded on or before 1 May 2012, the contract is substantially performed after 1 May 2012 but before 1 April 2015, the contract is then varied before 1 April 2015. Settlement (completion) of the contract takes place on or after 1 April 2015. The substantial performance transaction will fall within the SDLT regime, but the settlement (completion) transaction will be within the LBTT regime (*Scotland Act 2012, s 29(5)(b)*). However, LBTT will only be payable on the settlement (completion) transaction if, and to the extent that, the LBTT chargeable on that transaction exceeds the amount of the charge to SDLT, which was paid to HMRC, on the substantial performance of the contract (the *Land and Buildings Transaction Tax (Transitional Provisions) (Scotland) Order 2014, art 3(4) (SSI 2014/377)*).

Scenario 7

The missives (exchange of contracts) are concluded on or before 1 May 2012, there is no variation of the contract, and both substantial performance of the contract and settlement (completion) take place on or after 1 April 2015. Both the substantial performance transaction and the settlement (completion) transaction are within the SDLT regime (*Scotland Act 2012, s 29(5)(b)*).

Scenario 8

The missives (exchange of contracts) are concluded on or before 1 May 2012, the contract is substantially performed on or after 1 April 2015, but is subsequently varied, after which settlement (completion) takes place. The substantial performance transaction is within the SDLT regime as the missives were concluded on or before 1 May 2012 and there was no variation prior to substantial performance taking place (*Scotland Act 2012, s 29(5)(b)*). As the contract is then varied, the subsequent settlement (completion) transaction is within the LBTT regime. However, LBTT will only be payable on the settlement (completion) transaction if, and to the extent that, the LBTT chargeable on that transaction exceeds the amount of the charge to SDLT, which was paid to HMRC, on the substantial performance of the contract (the *Land and Buildings Transaction Tax (Transitional Provisions) (Scotland) Order 2014, art 3(4) (SSI 2014/377)*).

1A.26

Contracts entered into after 1 May 2012

Scenario 9

The missives (exchange of contracts) are concluded after 1 May 2012 but before 1 April 2015, and settlement (completion) and substantial performance take place at the same time after 1 April 2015. The settlement (completion) transaction is within the LBTT regime (*Scotland Act 2012, s 29(2)* and *(4)*).

Scenario 10

The missives (exchange of contracts) are concluded after 1 May 2012 but before 1 April 2015, the contract is then substantially performed before 1 April 2015, but settlement (completion) takes place after 1 April 2015. As substantial performance takes place before 1 April 2015 it is within the SDLT regime (*Scotland Act 2012, s 28(2)*). However, as the missives were concluded after 1 May 2012 the post-31 March 2015 settlement (completion) transaction is within the LBTT regime (*Scotland Act 2012, s 29(2) and (4)*). In these circumstances, LBTT is only payable to the extent that the amount of the LBTT chargeable on the settlement (completion) transaction exceeds the amount of the SDLT chargeable, and paid to HMRC, on substantial performance of the contract (the *Land and Buildings Transaction Tax (Transitional Provisions) (Scotland) Order 2014, art 4(4) (SSI 2014/377)*).

Scenario 11

The missives (exchange of contracts) are concluded after 1 May 2012 but before 1 April 2015, the contract is then substantially performed before 1 April 2015, however on or after 1 April 2015 the contract is varied. Some time after the variation settlement (completion) of the contract takes place. As substantial performance takes place before 1 April 2015, and prior to the variation of the contract, it is within the SDLT regime (*Scotland Act 2012, s 28(2)*). As the contract was entered into after 1 May 2012, the variation of the contract has no impact, since the settlement (completion) is in any event within the LBTT regime (*Scotland Act 2012, s 29(2)* and *(4)*). However, LBTT is only payable to the extent that the amount of the LBTT chargeable on the settlement (completion) transaction exceeds the amount of the SDLT chargeable, and paid to HMRC, on substantial performance of the contract (the *Land and Buildings Transaction Tax (Transitional Provisions) (Scotland) Order 2014, art 4(4) (SSI 2014/377)*).

1A.27

Scenario 12

The missives (exchange of contracts) are concluded after 1 May 2012 but before 1 April 2015, and the contract is then substantially performed on or after 1 April 2015, after which settlement (completion) takes place.

It seems that in these circumstances the substantial performance of the contract will not give rise to either SDLT or LBTT obligations. The technical analysis appears to be that the substantial performance of the contract does not fall within the SDLT regime as the contract was entered into after 1 May 2012, and therefore on or after 1 April 2015 an interest in land in Scotland is no longer a 'chargeable interest' for the purposes of SDLT (*Scotland Act 2012, s 29(2)*). Equally, from an LBTT perspective, *LBTT(S)A 2013, s 10* (Substantial performance without completion) only applies on the substantial performance of a contract for a 'land transaction' which is to be completed by a conveyance. As the contract was entered into prior to 1 April 2015 it cannot be a contract for a 'land transaction' for LBTT purposes, because when it was entered into the land in Scotland, which is the subject of the contract, was still a 'chargeable interest' for the purposes of SDLT, and had not become a 'chargeable interest' for LBTT purposes (*Scotland Act 2012, s 28(2)*). The subsequent completion of the contract would however give rise to LBTT obligations, and clearly there would be 'no credit' for any SDLT or LBTT arising from the earlier substantial performance of the contract, as no tax would have been paid.

Scenario 13

This scenario is the same as Scenario (12) except that following the substantial performance of the contract a variation of the contract takes place, after which settlement (completion) occurs. As the contract was entered into after 1 May 2012, the variation has no impact, since the settlement (completion) is in any event within the LBTT regime (*Scotland Act 2012, s 29(2)* and *(4)*). The substantial performance of the contract should not give rise to any SDLT or LBTT obligations as discussed above in relation to Scenario 12.

1A.28 It should be noted that, in relation to the scenarios set out in **1A.25–1A.27**, a 'variation' means a disqualifying event prescribed by *Scotland Act 2012, s 29(6)*(see **1A.23**) and includes *inter alia* an assignation of rights under the contract or the exercise of an option.

Example 1A.3—Contract entered into post-1 May 2012 but before 1 April 2015 and completed post-31 March 2015

Caroline concludes missives for the purchase of a house in Prestwick on 21 March 2015. The agreement is not substantially performed before the date of settlement and the effective date of the transaction (the settlement date) is on 30 May 2015.

The purchase of the house is within the LBTT regime, and is not chargeable to SDLT, despite the fact that the missives were concluded prior to 1 April 2015. The transitional rules in the *Scotland Act 2012, s 29(5)* and *(6)* do not apply, as the missives were not concluded on or before 1 May 2012.

Example 1A.4—Contract entered into before 1 May 2012 but completed post-31 March 2015

On 1 April 2012 Pamela concludes missives for the purchase of a completed office block, located in Scotland, before construction begins on 1 May 2012. Completion of the construction work is delayed and, by agreement between the parties, the date of settlement is changed to 1 May 2015. The agreement is not substantially performed before the date of settlement and the effective date (*FA 2003, s 119*) of the acquisition of the office block (the date of settlement) is 1 May 2015.

The acquisition is subject to SDLT as it falls within the transitional provisions (*Scotland Act 2012, s 29(5)(b)* and *(6)*), as the missives were concluded before 1 May 2012, and none of the exclusions in *Scotland Act 2012, s 29(6)* should apply. In particular the change in the date of settlement is not treated as a variation of the contract within *Scotland Act 2012, s 29(6)* (see HMRC technical guidance issued on 15 February 2015 – **1A.1**).

Example 1A.5—Substantial performance of a contract before 1 April 2015 with completion post-1 April 2015

Andrew concludes missives for the purchase of a retail unit in Dumfries on 28 February 2015, for a consideration of £300,000 and with the date of settlement agreed for 1 May 2015. The agreement is substantially performed (*FA 2003, s 44(5)(a)*) on 14 March 2015 when Andrew goes into the retail unit to undertake the fit-out.

The substantial performance of the agreement will be within the charge to SDLT as the contract is treated as if it were the transaction governed by the contract and the effective date (*FA 2003, s 119*) of the transaction is when the contract is substantially performed, namely 14 March 2015.

However, as the date of settlement takes place on 1 May 2015, that transaction will have an effective date of 1 May 2015 (*LBTT(S)A 2013, s 63(1)(a)*) which means it will be within the LBTT regime, and will not be chargeable to SDLT. The transitional rules in the *Scotland Act 2012, s 29(5)* and *(6)* do not apply as the missives were concluded after 1 May 2012. However, Andrew is only required to pay LBTT in relation to the settlement transaction if, and to the extent that, the LBTT chargeable on the completion transaction exceeds the amount of the SDLT chargeable, and paid to HMRC, on substantial performance of the contract (the *Land and Buildings Transaction Tax (Transitional Provisions) (Scotland) Order 2014, art 4(4) (SSI 2014/377)*).

However, even if no LBTT is payable, Andrew will have an obligation to file an LBTT land transaction return by 31 May 2015 (*LBTT(S)A 2013, ss 10(3)(a)* and *30(1)* and *6(a)*).

Specific provisions

Contract providing for conveyance to a third party (FA 2003, s 44A and LBTT(S)A 2013, s 11)

1A.29 Under SDLT, *FA 2003, s 44A* applies where a contract is entered into pursuant to which a chargeable interest in land is to be conveyed by (A) at the request of (B), either to (B) or (C) in circumstances where only (A) and (B) are party to the contract.

If such a contract is substantially performed (*FA 2003, s 44(5)*) on or before 1 May 2012 (the date on which the *Scotland Act 2012* came into force) and, at the request of (B), (A) becomes obliged to convey the land to (B), or a conveyance is made under a contract between (B) and (C), such transactions will be subject to SDLT even if the effective date (*FA 2003, s 119*) of the transaction is on or after 1 April 2015 (see para 6.2 of the HMRC guidance on the transition from SDLT to LBTT and *Scotland Act 2012, s 29(5)(a)*).

Where such a contract is entered into on or before 1 May 2012, but not substantially performed by that date, a conveyance from (A) to (B), or a conveyance under a contract between (B) and (C) will be subject to SDLT if the effective date (*FA 2003, s 119*) of the transaction is on or after 1 April 2015, and none of the exclusions set out at **1A.23** (*Scotland Act 2012, s 29(6)*) applies to the contract. However, if any of the exclusions set out at **1A.23** do apply, and the effective date of the transaction is on or after 1 April 2015, a conveyance from (A) to (B), or a conveyance under a contract between (B) and (C), will be subject to LBTT (see para 6.3 of the HMRC guidance on the transition from SDLT to LBTT and *Scotland Act 2012, s 29(5)(b)*). If the contract was substantially performed before 1 April 2015, and tax has been paid to HMRC in respect of that substantial performance, LBTT is only payable on the conveyance to the extent (if any) that the LBTT chargeable is greater than the SDLT chargeable, and paid to HMRC, on the substantial performance transaction (the *Land and Buildings Transaction Tax (Transitional Provisions) (Scotland) Order 2014, article 3(4) (SSI 2014/377)*).

Finally, if such a contract is entered into after 1 May 2012, a conveyance from (A) to (B), or a conveyance under a contract between (B) and (C), will not be subject to SDLT if the effective date (*FA 2003, s 119*) of the transaction is on or after 1 April 2015. Such transactions will be within the LBTT regime (see para 6.4 of the HMRC guidance on the transition from SDLT to LBTT). If the contract was substantially performed before 1 April 2015, and tax has been paid to HMRC in respect of that substantial performance, LBTT is only payable on the conveyance to the extent (if any) that the LBTT chargeable is greater than the SDLT chargeable, and paid to HMRC, on the substantial performance transaction (the *Land and Buildings Transaction Tax (Transitional Provisions) (Scotland) Order 2014, art 4(4) (SSI 2015/377)*).

Pre-completion transactions (FA 2003, s 45 and Sch 2A)

1A.30 *FA 2003, Sch 2A* (see **3.11** et seq) applies where a person (the original purchaser) enters into a contract (the original contract) for the acquisition of an interest in UK land under which the acquisition is to be completed by a conveyance and, before that contract is substantially performed (see **2.9**) or completed, there is 'a pre-completion transaction'. For these purposes a 'pre-completion transaction' is a transaction which results in a person other than the original purchaser (the transferee) becoming entitled to call for a conveyance to him of the whole or part of the UK land interest to be acquired under the original contract. A pre-completion transaction can either be an 'assignment of rights' or a 'free standing transfer'. A pre-completion transaction will be 'an assignment of rights' if the entitlement of the transferee to call for the conveyance to be made to him is an entitlement to exercise rights under the original contract, ie the original contract has been assigned in whole or in part to the transferee. A pre-completion transaction which is not an assignment of rights is a 'free standing transfer', and this term would therefore include a subsale or a novation of the original contract.

The transitional rules in relation to such pre-completion transactions then operate as follows. If the assignment or free standing transfer transaction is completed or substantially performed by the 'transferee' (ie the person other than the original buyer who is now entitled to call for a conveyance) on or after 1 April 2015 neither the completion or the substantial performance of the assignment or free standing transfer transaction, nor a transaction (subsale) or notional transaction (for the purposes of *FA 2013, Sch 2A* on an assignment of rights) entered into by 'the transferor', will be subject to SDLT. For these purposes the transferor in relation to a pre-completion transaction is the party to the pre-completion transaction who immediately before the pre-completion transaction took place was entitled to call for a conveyance of the interest in UK land which was the subject matter of the pre-completion transaction, ie in most cases the original purchaser (see para 7.2 of the HMRC guidance on the transition from SDLT to LBTT).

Unless the original contract was substantially performed on or before 1 May 2012, the assignment of rights or free standing transfer transaction will mean that the substantial performance and completion of the original contract on or after 1 April 2015 will be excluded from SDLT by virtue of the *Scotland Act 2012, s 29(6)(c)* (see **1A.23**).

Instead, if the assignment of rights or free standing transfer transaction is entered into on or after 1 April 2015, LBTT will be chargeable both on the original buyer in relation to the substantial performance of the original contract, as that original contract will be treated as having been substantially performed under *LBTT(S)A 2013, s 14(1)(c)*, and on the substantial performance and completion of the further agreement, ie the assignment, subsale or novation. This is subject to the availability of sub-sale development relief.

If the pre-completion transaction (ie the assignment of rights or free standing transfer transaction) is entered into before 1 April 2015 it is considered that *LBTT(S)A 2013, s 14(1)(c)* should not apply, but the subsequent completion of both the original contract and the assignment or free-standing transfer transaction, on or after 1 April 2015, will be chargeable to LBTT. This is subject to the availability of sub-sale development relief.

Options (FA 2003, s 46 and LBTT(S)A 2013, s 12)

1A.31 If an option or pre-emption right over land in Scotland is acquired before 1 April 2015, the exercise of the option or pre-emption right on or after

1 April 2015 will not be within the SDLT regime, regardless of whether or not the grant and exercise of the option would have been linked transactions (*FA 2003, s 108*) for SDLT purposes, but for the fact that land in Scotland is not a chargeable interest for the purposes of SDLT on or after 1 April 2015 (see **1A.23**). The exercise of the option will potentially be chargeable to LBTT (see para 8.2 of the HMRC guidance on the transition from SDLT to LBTT).

If the option or pre-emption right is acquired on or after 1 April 2015 both the acquisition of that right, and any subsequent exercise, will be chargeable to LBTT and may be linked transactions.

Exchanges (FA 2003, s 47 and LBTT(S)A 2013, s 57)

1A.32 Where, on or after 1 April 2015, land in England and Wales or Northern Ireland is exchanged for land in Scotland, the acquisition of the land in England and Wales or Northern Ireland will be within the SDLT regime, whilst the acquisition of the land in Scotland will be within the LBTT regime. As the acquisition of land in Scotland is not a land transaction for SDLT purposes, the SDLT exchange provisions (*FA 2003, s 47* and *Sch 4, para 5*) will not apply, and SDLT will be charged on the consideration given for the land in England and Wales or Northern Ireland under the normal rules (*FA 2003, Sch 4, paras 1* and *7*) (see para 9.2 of the HMRC guidance on the transition from SDLT to LBTT).

Equally, as the acquisition of the land in England and Wales or Northern Ireland will not be a land transaction for LBTT purposes, the LBTT exchange provisions (*LBTT(S)A 2013, s 13* and *Sch 2, paras 5* and *6*) will not apply. LBTT will be charged on the acquisition of the land in Scotland by reference to the market value of the land in England and Wales or Northern Ireland given as consideration (*LBTT(S)A 2013, Sch 2, para 7*), plus the amount of any equality money or other consideration given by the buyer of the land in Scotland (see para 9.5 of the HMRC guidance on the transition from SDLT to LBTT).

1A.33 Where the transfer of land in England and Wales or Northern Ireland takes place before 1 April 2015, and the transfer of the land in Scotland takes place on or after 1 April 2015, neither the SDLT nor the LBTT exchange provisions will apply (see para 9.3 of the HMRC guidance on the transition from SDLT to LBTT).

1A.34 In contrast, where the transfer of land in Scotland takes place before 1 April 2015 and the transfer of land in England and Wales or Northern Ireland takes place on or after 1 April 2015, both transactions will be within the SDLT regime, and the SDLT exchange provisions apply (*FA 2003, s 47* and *Sch 4, para 5*) (see para 9.8 of the HMRC guidance on the transition from SDLT to LBTT).

Example 1A.6—Exchange of a property in England for one located in Scotland

A company (A Ltd) decides to relocate its head office from England to larger premises in Scotland, and enters into an agreement on 1 March 2015 with another unrelated company (B Ltd) to exchange its offices in England for an office block in Scotland, and to pay £750,000 equality money. The agreement is not substantially performed (see **2.9**) before the date of settlement, which takes place on 30 June 2015.

A Ltd's acquisition of the premises in Scotland is within the LBTT regime. However, the LBTT exchange provisions do not apply as the sale of the English property is not a land transaction for LBTT purposes, as the right over the English property is not a chargeable interest for the purposes of *LBTT(S)A 2013, s 4(2)*. The consideration chargeable to LBTT will be the market value of the property in England plus £750,000.

B Ltd's acquisition of the offices in England will be within the SDLT regime, but the SDLT exchange provisions will not apply as the sale of the Scottish property is not a land transaction for SDLT purposes (*Scotland Act 2012, s 29(2)* and *(4)*). The consideration chargeable to SDLT will be the market value of the property in Scotland given as consideration (*FA 2003, Sch 4, para 7*).

Contingent, uncertain or unascertained consideration (FA 2003, s 51 and LBTT(S)A 2013, ss 18 and 19)

1A.35 If a transaction is within the SDLT regime, the obligation to adjust the chargeable consideration when, on or after 1 April 2015, a contingency ceases or consideration is ascertained (*FA 2003, s 80*) remains in place and is not affected by the introduction of LBTT. Similarly the provisions relating to the deferment of SDLT (*FA 2003, s 90*) remain unchanged. The contingency ceasing or the consideration becoming ascertained, in relation to such a transaction, will not be subject to LBTT (see para 10.1 of the HMRC guidance on the transition from SDLT to LBTT).

Sale and leaseback transaction (FA 2003, s 57A and LBTT(S) A 2013, Sch 3)

1A.36 Where the effective date (*FA 2003, s 119*) of a sale (or lease) of land situated in Scotland is before 1 April 2015 but the effective date of the grant back of the lease (or sub-lease) is on or after 1 April 2015, the sale (or lease) transaction will be within the SDLT regime, but the lease (or sub-lease) transaction will be within the LBTT regime. This means that neither the SDLT nor the LBTT exchange provisions will apply (see **1A.32**).

For SDLT purposes the chargeable consideration for the sale (or lease) transaction will be the consideration given in money or money's worth for the sale (or lease), including the market value, if any, of the lease (or sub-lease) granted back.

For LBTT purposes the grant of the lease (or sub-lease) back on or after 1 April 2015 will potentially be within the LBTT sale and leaseback relief provisions in *LBTT(S)A 2013, Sch 3* and therefore, as long as all of the requisite conditions are satisfied, the grant of the lease back would be exempt from the charge to LBTT, despite the fact that the sale (or lease) transaction took place before 1 April 2015 (see para 11.2 of the HMRC guidance on the transition from SDLT to LBTT).

Multiple dwellings relief (FA 2003, Sch 6B and LBTT(S) A 2013, Sch 5)

1A.37 Where a transaction (or linked transactions) comprise multiple dwellings situated in both Scotland and England and Wales or Northern Ireland, the rate of SDLT payable on a claim for multiple dwellings relief is calculated according to the number of dwellings the acquisition of which is subject to SDLT (see para 12.1 of the HMRC guidance on the transition from SDLT to LBTT).

Similarly, the LBTT payable on a claim for multiple dwellings relief is calculated based on the number of dwellings the acquisition of which is subject to LBTT.

Example 1A.7—Claim for multiple dwellings relief where properties are acquired in both Scotland and in the rest of the UK

Property Investments Ltd acquires eight new houses from the same developer on 30 September 2016. It plans to let the eight houses out to a third party. Four of the houses are in Kendal with two costing £200,000 each and two costing £400,000 each. The other four houses are in Aberdeen with two costing £600,000 each and two costing £500,000 each.

Only the four houses in Kendal are chargeable to SDLT and, if a claim for multiple dwellings relief is made, the SDLT chargeable is determined by dividing the aggregate consideration for the Kendal houses of £1,200,000 by the number of dwellings chargeable to SDLT of four, giving an average price per dwelling of £300,000. As the company is only acquiring four properties chargeable to SDLT (the other four properties are chargeable to LBTT) it is considered that the rule which states that where six or more dwellings are acquired as part of a single transaction the transaction is treated as the acquisition of non-residential property (*FA 2003, s 116(7)*) should not apply. Consequently the SDLT additional rates on residential property will apply such that the SDLT charge per dwelling is calculated as follows:

Purchase Price	Rate	SDLT
£	%	£
125,000	3	3,750
125,000	5	6,250
50,000	8	4,000
300,000		14,000

Based on four dwellings the SDLT payable is therefore £56,000 (4 × £14,000). However, if this amount is less than 1% of the total dwellings consideration then the SDLT payable is that 1% amount.

Total dwellings consideration of £1,200,000 × 1% = £12,000.

Consequently the SDLT payable is £56,000.

Note that as none of the properties are acquired for chargeable consideration of more than £500,000, the 15% rate of SDLT in *FA 2003, Sch 4A* is not in point.

In the absence of a claim for multiple dwellings relief the SDLT liability would have been £99,750.

Similarly, if a claim for multiple dwellings relief is made, the LBTT chargeable on the purchase of the four houses in Aberdeen is determined by dividing the aggregate consideration for the Aberdeen houses of £2,200,000, by the number of dwellings chargeable to LBTT of four, such that the LBTT is determined based on an average consideration of £550,000. As the company is only acquiring four properties chargeable to LBTT (the other four properties are chargeable to SDLT) it is considered that the rule which states that where six or more dwellings are acquired as part of a single transaction the transaction is treated as the acquisition of non-residential property (*LBTT(S)A 2013, s 59(8)*) should not apply. Consequently, the LBTT additional rates on residential property will apply and the LBTT charge per dwelling is therefore calculated as follows:

Purchase Price	Rate	LBTT
£	%	£
145,000	3	4,350
105,000	5	5,250
75,000	8	6,000
225,000	13	29,250
550,000		44,850

Based on four dwellings, the LBTT charge would be £179,400 (4 × £44,850).

This is subject to the 'minimum prescribed amount' being 25% of the LBTT which would have been payable if multiple dwellings relief had not been available (see **1A.13**).

If multiple dwellings relief had not been available the LBTT charge would have been as follows:

Purchase price	Rate	LBTT
£	%	£
145,000	3	4,350
105,000	5	5,250
75,000	8	6,000
425,000	13	55,250
1,450,000	15	217,500
2,200,000		288,350

The minimum prescribed is therefore 25% × £288,350 = £72,087.

In this instance therefore the minimum prescribed amount does not apply and the LBTT payable is £179,400.

1A.38 If a claim for SDLT multiple dwellings relief was made in relation to a transaction which included land in Scotland, the 'adjustment for change of circumstances' provisions in *FA 2003, Sch 6B, para 6* continue to apply to the transaction even if the event which triggers the adjustment occurs on or after 1 April 2015 (see para 12.2 of the HMRC guidance on the transition from SDLT to LBTT).

Group relief, reconstruction and acquisition reliefs (FA 2003, s 62 and Sch 7)

1A.39 Where group relief, reconstruction relief or acquisition relief have been claimed in relation to a transaction within the SDLT regime, and the transaction included land situated in Scotland, the provisions relating to the withdrawal of the relevant relief will continue to apply even if the event giving rise to the withdrawal takes place on or after 1 April 2015 (see para 13.1 of the HMRC guidance on the transition from SDLT to LBTT).

Charities relief (FA 2003, s 68 and Sch 8)

1A.40 Where charities relief has been claimed in relation to a transaction within the SDLT regime, and the transaction included land situated in Scotland, the provisions relating to the withdrawal of the relief will continue to apply even if the event giving rise to the withdrawal takes place on or after 1 April 2015 (see para 14.1 of the HMRC guidance on the transition from SDLT to LBTT).

Shared ownership leases (FA 2003, s 70 and Sch 9)

1A.41 Where the grant of a shared ownership lease in relation to a dwelling located in Scotland was subject to SDLT (and a market value election has not been made), any 'staircasing transaction' following which the lessee does not hold a total share of the dwelling which exceeds 80% is exempt from the charge to SDLT (see para 15.1 of the HMRC guidance on the transition from SDLT to LBTT).

Any 'staircasing transaction' in relation to a dwelling situated in Scotland which takes the lessee's total share of the dwelling over 80%, or the acquisition of the ownership of the land, will be exempt from SDLT if the effective date of the transaction is on or after 1 April 2015 (see paragraph 15.2 of the HMRC guidance on the transition from SDLT to LBTT).

Alternative property finance (FA 2003, s 72 and s 72A and LBTT(S)A 2013, Sch 7)

1A.42 The *Scotland Act 2012, Sch 3, para 10* omits from the SDLT legislation *FA 2003, s 72* (alternative property finance in Scotland: land sold to financial institution and leased to person) and *s 72A* (alternative property finance in Scotland: land sold to financial institution and person in common) with effect from 1 April 2015. As a consequence, if the effective date (*FA 2003, s 119*) of the third transaction (*s 72*) or further transaction (*s 72A*) under arrangements falling within those sections is on or after 1 April 2015, those transactions will be neither chargeable to SDLT nor notifiable for the purposes of SDLT (see para 16.1 of the HMRC guidance on the transition from SDLT to LBTT).

Alternative property finance relief is available under LBTT. The Scottish government has ensured that a party entering into a land transaction in connection with an alternative property finance arrangement prior to 1 April 2015, in relation to land situated in Scotland, is not disadvantaged by the move from SDLT to LBTT, by making subordinate legislation providing that neither a third transaction under *FA 2003, s 72* nor a further transaction under *FA 2003, s 72A*, with an effective date on or after 1 April 2015, will be chargeable to LBTT, provided that the conditions mentioned respectively in *FA 2003 s 72(4)* and *s 72A(4)* have been met (the *Land and Buildings Transaction Tax (Transitional Provisions) (Scotland) Order 2014, art 5 (SSI 2014/377)*).

Alternative finance investment bonds (FA 2009, Sch 61 and FA 2003, s 73C and LBTT(S)A 2013, Sch 8)

1A.43 A land transaction in relation to land located in Scotland and which is a 'first transaction' within the meaning of *FA 2009, Sch 61, para 6* will be exempt from SDLT if the effective date (*FA 2003, s 119*) of the transaction is before 1 April 2015 and the conditions for the relief are satisfied. In such circumstances SDLT will become chargeable if the relief is withdrawn on or after 1 April 2015 under *FA 2009, Sch 61, para 7(3)* (see para 17.1 of the HMRC guidance on the transition from SDLT to LBTT).

A transaction which is a 'second transaction' for the purposes of *FA 2009, Sch 61, para 8* will be exempt from SDLT if the 'effective date' of the transaction is before 1 April 2015 and the conditions in para 8 are satisfied (see para17.2 of the HMRC guidance on the transition from SDLT to LBTT).

In either case the transaction will not be subject to SDLT if the effective date (see **2.74**) is on or after 1 April 2015 (see para 17.3 of the HMRC guidance on the transition from SDLT to LBTT).

Relief for alternative finance investment bonds is also available under LBTT (*LBTT(S)A 2013, Sch 8*). The Scottish government has ensured that a party entering into a land transaction in connection with an alternative finance investment bond prior to 1 April 2015, in relation to land situated in Scotland, is not disadvantaged by the move from SDLT to LBTT, by making subordinate legislation providing that the second transaction for the purposes of *FA 2009, Sch 61, para 8* with an effective date on or after 1 April 2015 will not be chargeable to LBTT provided that the conditions in *FA 2009, Sch 61, para 8(1)(a)* and *(b)* have been met (the *Land and Buildings Transaction Tax (Transitional Provisions) (Scotland) Order 2014, art 6, (SSI 2014/377)*).

Anti-avoidance (FA 2003, s 75A)

1A.44 The general anti-avoidance provision within the SDLT legislation will only apply to a disposal and acquisition of land in Scotland if the effective date of the notional transaction (*FA 2003, s 75A(6)*) is before 1 April 2015 (see para 18.1 of the HMRC guidance on the transition from SDLT to LBTT).

The effective date of the 'notional transaction' is the earlier of:

(a) the last date of completion for the 'scheme transactions' (broadly the transactions which are involved in connection with the disposal and acquisition); and

(b) the last date on which a contract in respect of the 'scheme transactions' is substantially performed.

Later linked transactions (FA 2003, s 81A and s 108)

1A.45 Where a land transaction in relation to land situated in Scotland has an effective date (*FA 2003, s 119*) before 1 April 2015 but it is not notifiable for SDLT purposes (*FA 2003, s 77*) and the transaction becomes notifiable on or after 1 April 2015 by reason of a later linked transaction (*FA 2003, s 108*), which does not relate to land in Scotland, the pre-1 April 2015 transaction remains chargeable to SDLT and a land transaction return must be filed in accordance with *FA 2003, s 81A* (see para 19.1 of the HMRC guidance on the transition from SDLT to LBTT). In these circumstances a paper SDLT1 form must be submitted to Birmingham Stamp Taxes notifying the pre-1 April 2015 transaction, with a covering letter (see SDLTM 50350, para 3), and the effective date of the transaction is the effective date of the later linked transaction. Code 9998 should be entered in Box 29 of the SDLT1 form, and not a Scottish local authority code (see para 19.2 of the HMRC guidance on the transition from SDLT to LBTT).

If the pre-1 April 2015 transaction in relation to land situated in Scotland has been notified but SDLT, or more SDLT, becomes payable as a result of the post-1 April 2015 linked transaction in relation to land in England and Wales or Northern Ireland, then a further SDLT1 form must be submitted by letter to Birmingham Stamp Taxes in accordance with SDLTM50350, paras 1 and 2 (see para 19.3 of the HMRC guidance on the transition from SDLT to LBTT).

Partnerships (FA 2003, s 104 and Sch 15 and LBTT(S)A 2013, Sch 17)

1A.46 *FA 2003, Sch 15, para 17* is an anti-avoidance provision which applies where land in the UK is contributed to a partnership by a partner, or a person who becomes a partner as a consequence of the contribution, or a person connected with a partner or a person who becomes a partner as a consequence of the contribution (which is not a 'property-investment partnership') and there is subsequently a transfer of an interest in that partnership under arrangements which were in place at the time the land was transferred to the partnership. Para 17 treats the transfer of the interest in the partnership as a land transaction chargeable to SDLT, whereas a transfer of an interest in a partnership, which is not a property-investment partnership, would not normally be chargeable to SDLT.

If, in these circumstances, the land transferred to the partnership includes land situated in Scotland, then to that extent the transfer of the interest in the partnership will not be subject to SDLT if the effective date of the transfer of the interest in the partnership is on or after 1 April 2015, even if the transfer of the land to the partnership took place before 1 April 2015 (see para 20.1 of the HMRC guidance on the transition from SDLT to LBTT).

However, the LBTT legislation includes a similar anti-avoidance provision, and the Scottish government have made subordinate legislation to ensure that the transfer of the interest in the partnership will be chargeable to LBTT, even if the contribution of the land in Scotland took place prior to 1 April, 2015. Consequently, where the effective date of the land transfer referred to in *FA 2003, Sch 15, para 17(1)(a)* is before 1 April 2015 and the transfer of the interest in the partnership referred to in *FA 2003, Sch 15, para 17(1)(c)* is on or after 1 April 2015, the earlier land transfer will be treated for the purposes of *LBTT(S)A 2013, Sch 17, para 17* as though it were the land transfer referred to in *LBTT(S)A 2013, Sch 17, para 17(1)(a)*, such that the provisions of *LBTT(S) A 2013, Sch 17, para 17* will apply to the transfer of the partnership interest (the *Land and Buildings Transaction Tax (Transitional Provisions) (Scotland) Order 2014, art 7 (SSI 2014/377)*).

1A.47 In certain circumstances, where land is transferred by a partner or a person who becomes a partner as a consequence of the contribution, or a person connected with a partner or a person who becomes a partner as a consequence of a contribution, to a partnership, and within three years of the land transfer the transferring partner withdraws money or money's worth from the partnership which does not represent income profit, or the partner (or a person connected with a partner) has made a loan to the partnership and that loan is to any extent repaid, or the partner withdraws money or money's worth which does not represent income profit, then a charge to SDLT can arise (*FA 2003, Sch 15, para 17A*).

Where the 'effective date' (*FA 2003, s 119*) of the later deemed land transaction (ie the date the money or money's worth is withdrawn or the loan is repaid) would be on or after 1 April 2015, the deemed land transaction will not be chargeable to SDLT, even if the transfer of land to the partnership took place before 1 April 2015. However, the Scottish government has made subordinate legislation to ensure that such a deemed land transaction will be chargeable to LBTT.

The land transfer to the partnership which took place prior to 1 April 2015 will be treated for the purposes of *LBTT(S)A 2013, Sch 17, para 18* as if it were the land transfer mentioned in *LBTT(S)A 2013, Sch 17, para 18(1)* and the withdrawal of money or money's worth which does not represent income profit, or the repayment of the loan on or after 1 April 2015, will be treated as though it were a 'qualifying event' for the purposes of *LBTT(S)A 2013, Sch 17, para 18(2)* (the *Land and Buildings Transaction Tax (Transitional Provisions) (Scotland) Order 2014, art 8 (SSI 2014/377)*). As a consequence the provisions of *LBTT(S)A 2013, Sch 17, para 18* will apply.

LBTT(S)A 2013, Sch 17, paras 25 and *26* determine the partnership share attributable to a partner where an interest in land situated in Scotland is transferred from a partnership. It is provided that *paras 25* and *26* will have effect even though the relevant date for the purposes of para 26(1) is prior to 1 April 2015. However, any references to obligations to pay tax under *LBTT(S)A 2013*, shall have no effect (the *Land and Buildings Transaction Tax (Transitional Provisions) (Scotland) Order 2014, art 9 (SSI 2014/377)*).

Leases (FA 2003, s 120 and Sch 17A)

Leases that continue after a fixed term (FA 2003, Sch 17A, para 3 and LBTT(S)A 2013, Sch 19, para 20)

1A.48 *FA 2003, Sch 17A, para 3* provides that where a lease continues beyond its fixed term, it is treated as a lease for the original fixed term plus one year longer than the original fixed term and so on. Where, as a consequence of the deemed extension of the lease, SDLT, or more SDLT, becomes payable, an SDLT land transaction return or further return must be filed within 30 days of the last day of the one-year period concerned (*FA 2003, Sch 17A, para 3(4)*).

Where the lease which is treated as continuing beyond its fixed term as described above was granted before 1 April 2015, and is a lease over land situated in Scotland, the lease remains subject to SDLT, even if the period for which the lease is treated as having been extended includes a period on or after 1 April 2015 (see para 21.1 of the HMRC guidance on the transition from SDLT to LBTT).

If the deemed extension of the lease on or after 1 April 2015 causes the lease to become notifiable for the first time, an SDLT1 form must be filed in the normal way (SDLTM12050, para 1), the effective date of the transaction is the last day of the one-year period giving rise to the notification requirement and code 9998, and not a Scottish local authority code, must be entered at Box 29.

If the grant of the lease has already been notified but the deemed extension of the lease causes additional SDLT to become payable, notification should be made by way of a letter to Birmingham Stamp Taxes (see SDLTM12050, para 2ff).

Leases for an indefinite term (FA 2003, Sch 17A, para 4)

1A.49 A lease for an indefinite term (for example a lease for life – see SDLTM 18715 or a lease the duration of which is subject only to a period of notice) is treated in the first instance as though it were a lease for a fixed term of one year and, if the lease continues beyond that first year it is treated as a lease for a fixed term of two years, and so on.

Where a lease of land in Scotland is for an indefinite term, and is granted before 1 April 2015, it will continue to be liable to SDLT even if its deemed term continues beyond 1 April 2015 (see para 21.2 of the HMRC guidance on the transition from SDLT to LBTT). In these circumstances the deemed extension of the lease will not be subject to LBTT.

Example 1A.8—Lease for an indefinite term treated as granted before 1 May 2012

George enters into a verbal contract for a lease of an office in Glasgow at a market rate rent. The verbal contract is 'substantially performed' (see **2.9**) by the first payment of rent on 1 April 2012. *FA 2003, Sch 17A, para 19(3)* treats the verbal contract as the grant of a lease, beginning on the date of substantial performance.

The grant of the lease is within the SDLT regime and it remains within that regime after 1 April 2015 so long as it continues from year to year, and an SDLT land transaction return must be filed, or notification given, if SDLT becomes payable or additional tax becomes due (*FA 2003, Sch 17A, para 4*).

Successive linked leases (FA 2003, Sch 17A, para 5 and LBTT(S)A 2013, Sch 19, para 23)

1A.50 Where successive leases are granted of the same premises, and the grants of the leases are linked for SDLT purposes, SDLT applies as though there was a single lease granted at the time the first lease in the series was granted, for a term equal to the aggregate of the terms of all of the leases in the series, and in consideration of the rent payable under all of the leases in the series (*FA 2003, Sch 17A, para 5*).

Any lease over land in Scotland, where the effective date of the grant of the lease is on or after 1 April 2015 cannot be treated as successive to a lease over the same premises granted before 1 April 2015, as the later lease is not a land transaction for SDLT purposes (*Scotland Act 2012, s 29(2) and (4)*). However, the grant of the lease on or after 1 April 2015 will be within the LBTT regime (see para 21.3 of the HMRC guidance on the transition from SDLT to LBTT).

Adjustment where rent ceases to be uncertain (FA 2003, Sch 17A, para 8)

1A.51 Where the rent payable under a lease subject to SDLT is contingent, uncertain or unascertained *FA 2003, Sch 17A, para 8* provides for a reassessment of the SDLT chargeable on the lease to be made either:

(a) at the end of the fifth year of the term of the lease; or

(b) when the rent payable in respect of the first five years of the term of the lease ceases to be uncertain at an earlier date.

Where such a lease over land situated in Scotland is granted before 1 April 2015 and under the provision of *para 8*, SDLT, or more SDLT, becomes payable on or after 1 April 2015, the lease remains subject to SDLT. An SDLT1 form or further return, must be filed within 30 days of the end of the fifth year of the term of the lease or, if earlier, the date on which the amount of rent payable in the first five years ceases to be uncertain (see para 21.4 of the HMRC guidance on the transition from SDLT to LBTT).

If as a result the grant of the lease becomes notifiable for the first time, an SDLT1 form must be submitted in the usual way. The effective date of the transaction is the date prescribed by *FA 2003, Sch 17A, para 8(1)* and *(3)* and code 9998, and not a Scottish local authority code, must be entered at Box 29.

If the grant of the lease has already been notified but the effect of the reassessment of the SDLT payable is that additional tax becomes payable, notification should be made by letter to Birmingham Stamp Taxes – see SDLTM18525.

Overlap relief (FA 2003, Sch 17A, para 9 and LBTT(S) A 2013, Sch 19, para 24)

1A.52 Both the SDLT and LBTT legislation give overlap relief (see **4.83**) where A surrenders (renounces) an existing lease (the old lease) to B and in consideration of that surrender (renunciation) B grants a lease to A (the new lease) of the same or substantially the same premises.

Where a lease of land in Scotland, which was granted before 1 April 2015, is renounced on or after that date and, as consideration for that renunciation, a new lease of the same or substantially the same premises is granted, the grant of the new lease will be within the LBTT regime, and will not be chargeable to SDLT (see para 21.5 of the HMRC guidance on the transition from SDLT to LBTT). The new lease will, however, be chargeable to LBTT.

To ensure that the party entering into the new lease is not disadvantaged by the move from SDLT to LBTT, the Scottish government has made subordinate legislation which provides that where the old lease was granted before 1 April 2015, and the new lease is granted on or after 1 April 2015, the rent chargeable to LBTT on the grant of the new lease will be reduced by the rent which was charged to SDLT for the period of the overlap of the two leases (the *Land and Buildings Transaction Tax (Transitional Provisions) (Scotland) Order 2014, art 10 (SSI 2014/377)*).

Backdated leases granted to tenant holding over (FA 2003, Sch 17A, para 9A and LBTT(S)A 2013, Sch 19, para 21)

1A.53 For SDLT purposes, where the tenant under a lease continues in occupation beyond the fixed term of the lease, the tenant is granted a new lease of the same premises, and the term of the lease is stated to begin immediately after the contractual termination date of the previous lease, the term of the new lease is treated as beginning on the stated date ie immediately following the termination of the previous lease.

Where the effective date of the grant of a new lease over land in Scotland, in circumstances where an earlier lease has continued beyond the fixed term, is on or after 1 April 2015, but the term of the new lease is stated to begin from the termination date of the previous lease, which date is before 1 April 2015, the new lease will not be subject to SDLT. However, the grant of the new lease over land in Scotland on or after 1 April 2015 will be within the LBTT regime (see para 21.6 of the HMRC guidance on the transition from SDLT to LBTT).

Assignment of a lease on or after 1 April 2015 where the lease was granted before 1 April 2015

1A.54 Where a lease of land in Scotland was granted before 1 April 2015 and is assigned on or after 1 April 2015, the assignation is not chargeable to SDLT, but may be subject to LBTT if a premium is actually paid, or deemed to be paid for the assignation (see para 21.7 of the HMRC guidance on the transition from SDLT to LBTT).

Assignment of lease treated as the grant of a new lease (FA 2003, Sch 17A, para 11 and LBTT(S)A 2013, Sch 19, para 27)

1A.55 Both the SDLT and LBTT legislation provide that where the grant of a lease is exempt from the charge to the respective taxes because one of a number of specified reliefs is claimed, the first assignation of that lease which is not so exempt is taxed as though it were the grant of a new lease (see **4.17** et seq).

Where the lease of land in Scotland is granted before 1 April 2015, an assignation of the lease on or after 1 April 2015 will not be subject to SDLT and the provisions of *FA 2003, Sch 17A, para 11* will not apply to tax the assignation as though it were the grant of a new lease. However, the Scottish government has made subordinate legislation to provide that where a lease of land in Scotland was granted before 1 April 2015 and it was relieved or exempt from SDLT as one of the reliefs specified in *FA 2003, Sch 17A, para 11* applied, the first assignment of that lease, if it takes place on or after 1 April 2015, which is not exempt from LBTT as none of the reliefs specified in *LBTT(S)*

A 2013, Sch 19, para 27(3) apply, will be charged to LBTT as though it were the grant of a new lease (the *Land and Buildings Transaction Tax (Transitional Provisions) (Scotland) Order 2014, art 11 (SSI 2014/377)*).

Increases of rent treated as grant of a new lease (FA 2003, Sch 17A, para 13)

1A.56 Under SDLT, where a lease is varied so as to increase the amount of the rent payable as from a date before the end of the fifth year of the lease, the variation is treated for SDLT purposes as if it were the grant of a new lease in consideration of the additional rent now payable (*FA 2003, Sch 17A, para 13*). This rule does not apply to any increase in the rent payable as a consequence of a provision within the lease.

A variation of a lease of land in Scotland falling within the above provision will not be treated as the grant of a new lease for SDLT purposes where the effective date of the grant of the deemed new lease would be on or after 1 April 2015 (see para 21.9 of the HMRC guidance on the transition from SDLT to LBTT).

However, the Scottish government has made subordinate legislation providing that where the effective date of the grant of the deemed new lease would have been on or after 1 April 2015, had the provisions of *FA 2003, Sch 17A, para 13* continued to apply to the lease, that the increase of rent will be charged to LBTT as though it were the grant of a new lease and the chargeable consideration for LBTT purposes (other than in respect of any premium) will be the additional rent payable (the *Land and Buildings Transaction Tax (Transitional Provisions) (Scotland) Order 2014, art 12 (SSI 2014/377)*).

Missives of let (FA 2003, Sch 17A, para 19(2) and LBTT(S)A 2013, Sch 19, para 26)

1A.57 *FA 2003, Sch 17A, para 19(2)* deals with the situation where a lease is initially constituted by concluded missives of let (the first lease) and at some later time a lease is executed (the second lease). The equivalent legislation within the LBTT regime is *LBTT(S)A 2013, Sch 19, para 26.*

Where the first lease is constituted by concluded missives of let before 1 April 2015, and the second lease is executed on or after that date, the provisions of *FA 2003, Sch 17A, para 19(2)* do not apply as the second lease is not subject to SDLT and cannot be linked with the first lease. However, the second lease will be subject to LBTT and *LBTT(S)A 2013, Sch 19, para 26* will apply to the second lease despite the fact that the first lease was constituted by concluded missives of let prior to 1 April 2015 (see para 21.10 of the HMRC guidance on the transition from SDLT to LBTT).

Extension of lease treated as grant of a new lease (FA 2003, s 43(3)(d)(i))

1A.58 Where a lease of land in Scotland is granted prior to 1 April 2015, there is a variation of the lease on or after 1 April 2015 so as to increase the term of the lease or to extend the premises let (which under English property law is likely to be treated as the grant of a new lease) and, if that variation had taken place before 1 April 2015, it would have been treated as the grant of a new lease for SDLT purposes, then the variation of the lease is treated as the grant of a new lease for LBTT purposes. The consideration chargeable to LBTT, other than in respect of a premium, is the amount of the additional rent payable over the increased period of the lease or the additional rent payable in relation to the extended premises now let (the *Land and Buildings Transaction Tax (Transitional Provisions) (Scotland) Order 2014, art 13 (SSI 2014/377)).*

General anti-avoidance rule (RSTPA 2014, s 62)

1A.59 The general anti-avoidance rule (GAAR – see **1A.12**) applies in relation to any tax avoidance arrangement which is 'artificial' entered into on or after 1 April 2015 being the date on which the GAAR provisions came into effect. Currently the GAAR only applies to the devolved taxes being LBTT and Scottish landfill tax. Where the tax avoidance arrangement forms part of any other arrangements entered into before that date, those other arrangements are to be ignored unless taking account of those arrangements would mean that the arrangement would not be 'artificial' (*RSTPA 2014, s72(2)* and *(3)*).

The UK general anti-abuse rule found in *FA 2013, ss 206–215* will continue to apply to any land transaction within the SDLT regime but will not apply to any land transaction within the LBTT regime.

Chapter 2

SDLT – General Rules

INTRODUCTION

2.1 Stamp Duty Land Tax (SDLT) replaced stamp duty on transactions involving UK land and buildings from 1 December 2003, although it ceased to apply to transactions involving land and buildings situated in Scotland with effect from 1 April 2015. Therefore, with effect from that date, and subject to the transitional provisions (HMRC published guidance in relation to the transitional provisions on 13 February 2015 – see www.gov.uk/government/publications/transitional-land-and-building-transaction-tax-guidance), SDLT only applies to transactions involving land and buildings in England, Wales and Northern Ireland referred to below as 'RUK', ie the rest of the UK. It should also be noted that with effect from 1 April 2018 it is anticipated that SDLT will cease to apply to transactions involving land and buildings situated in Wales, being replaced by a devolved land transaction tax.

Even though SDLT replaced stamp duty on transactions involving UK land and buildings, it would be a mistake to assume that SDLT is merely an updated version of stamp duty. Many transactions which would not have been within the stamp duty net – including informal arrangements and some unilateral actions which might not conventionally be described as transactions – do give rise to SDLT liabilities and obligations. Even where there is no liability, there may be reporting obligations, with penalties for failure to comply. For this reason, this book often refers to 'SDLT obligations', meaning both liability to pay tax and obligation to notify a transaction or make a return. So wide ranging are these obligations, it is essential to consider the possible impact of SDLT on almost any change relating to the occupation of, or ownership of rights over, RUK land and buildings.

2.2 Almost all transactions in UK land and buildings completed from 1 December 2003, even if commenced earlier, are within the scope of SDLT. Some 'old' transactions remain within the ambit of stamp duty, or perhaps potentially fall within both regimes (see **2.45**). In stark contrast with stamp duty, SDLT is a true 'self-assessment' tax. The obligation to report liability and make payment falls on the purchaser. The tax is triggered by entering into or performing 'land transactions' (as defined in the legislation), no matter how or even whether documented. Some, but not all, of the reliefs previously available under stamp duty are available under SDLT. There are enforcement powers and specific anti-avoidance rules similar to those that apply to other taxes, plus far-reaching general anti-avoidance provisions. SDLT is within the *FA 2004* regime for disclosure of tax avoidance schemes, but it has its own set of criteria to determine whether disclosure is required.

2.3 The initial rates of SDLT for purchase of freeholds and existing leases were the same as they were for stamp duty, so the typical home-buyer saw little or no practical change when the tax was introduced. Higher rates have been introduced by subsequent *Finance Acts* – from 6 April 2011, 5% for residential property costing more than £1,000,000 (*FA 2010, s 7)* and from 22 March 2012, 7% for residential property costing more than £2,000,000 *(FA 2012, s 211)*. A higher rate of 15% was introduced, with effect from 21 March 2012, for the purchase of a single dwelling for more than £2,000,000 by certain non-natural persons (see **2.39**). The threshold for application of this punitive rate was lowered to cover a purchase for consideration exceeding £500,000 from 20 March 2014.

The structure of SDLT changed with effect from 4 December 2014 when the tax payable on a residential property transaction (see **2.38**) ceased to be calculated using the so-called 'slab system' and instead a 'progressive system' (like personal income tax) was adopted, with a top rate of 12% for chargeable consideration exceeding £1,500,000 (see **2.37**). Further changes were made to the tax with effect from 1 April 2016 in relation to residential properties. From that date a 3% additional rate of SDLT applies to purchases of additional dwellings and dwellings purchased by companies (and other non-individual purchasers) such that, where the additional rate applies, the top rate of tax is increased to 15% for chargeable consideration exceeding £1,500,000 (see **2.39**).

Finally, until 16 March 2016 the tax payable on the acquisition of non-residential or mixed use property was calculated using the so-called 'slab system' (see **2.37**). However, with effect from 17 March 2016, the charge on such acquisitions is calculated using a 'progressive system' (like personal income tax), with a top rate of 5% for chargeable consideration exceeding £250,000 (see **2.37**). This has substantially increased the SDLT charge on high value commercial property transactions.

The increase in the rates of SDLT, anti-avoidance provisions and the absence of some reliefs mean that many commercial transactions suffer a greater tax burden than they would have incurred under stamp duty. The regime for the grant, variation or surrender of leases is sufficiently complex to warrant a separate chapter in this book (**Chapter 4**). In most cases involving a lease at rent, the SDLT charge is substantially higher than would have arisen under stamp duty, up to ten times as much in extreme cases.

2.4 Obligations in relation to SDLT may extend beyond the time of the original transaction. In some circumstances, reliefs may be withdrawn or liabilities increased as a result of subsequent events as innocuous as an early rent review or the refinancing of a group company. Constant monitoring of potential liabilities is therefore necessary. The tax applies irrespective of country of establishment or residence of the taxpayer, and it is possible that a non-UK entity, unaware of the rules, may unknowingly trigger an SDLT charge. An example would be where RUK land is transferred in the course of a merger of overseas companies by operation of (overseas) law – normally free of stamp duty under the old regime, but subject to SDLT under current rules.

2.5 With the introduction of SDLT, stamp duty was abolished on transfers of all other assets apart from stock and marketable securities and, in some circumstances, interests in partnerships which hold such stock or securities.

GENERAL SCHEME OF SDLT

Importance of definitions

2.6 The SDLT rules rely on a number of concepts and expressions which are defined in *FA 2003*. It is important to apply these definitions and not to rely on the general meaning of the relevant words. The table below details statutory and HMRC SDLT Manual references for definitions of key terms; there is also a useful list of statutory references in *FA 2003, s 122*. Some of the definitions are interdependent, so it is necessary to see the whole picture before it makes sense; the definitions and the following paragraphs may require several readings! These paragraphs give a general outline of the tax and the defined expressions are shown, in the next few paragraphs only, in **bold**. Subsequent paragraphs and chapters examine the interpretation and application of the expressions in more detail.

Term	***FA 2003 section/Schedule***	**SDLTM para**
Acquisition	*s 43(3)*	00270
Chargeable interest	*s 48*	00280
Consideration	*s 52; Sch 4*	03600 et seq
Conveyance	*s 44(10)(b)*	08100
Effective date	*s 119; s 44(4)*	07600
Exempt interest	*s 48(2)–(3A)*	00320
Land transaction	*s 43(1)*	00260
Linked transactions	*s 108*	30100
Major interest	*s 117*	04130

Purchaser	*s 43(3)–(4); Sch 17A, para 15A*	07200
Substantial amount/ substantially all	*s 44(7)*	07950
Substantial performance	*s 44(5)–(7)*	07700 et seq

The charge

2.7 SDLT is charged on **land transactions**, whether effected by a document or not. Neither the places of residence of the parties, nor the place where any documents are executed, are relevant to the charge. The tax is under the control of HM Revenue & Customs (HMRC), and that departmental name is used here despite the fact that much of the legislation still refers to the Commissioners of Inland Revenue (*FA 2003*, *s 42*). A **land transaction** for these purposes means the **acquisition** of a **chargeable interest**. **Acquisition** may include not only a normal land sale and purchase, but also the creation, surrender, release or variation of a **chargeable interest**. The **acquisition** may occur under the terms of a court order, a statutory provision or by operation of law. A complex and indirect definition makes clear the intention that any party whose interest in land benefits under the transaction is to be regarded as the **purchaser**. Poor wording of the legislation originally led to uncertainty when leases were varied, so supplementary rules were introduced to deal specifically with this (see **2.18**). A person will only be treated as a **purchaser** if he has either given **consideration** for, or is a party to, the transaction. It can thus be seen that the new charge catches a wider range of transactions than the old stamp duty regime.

Timing

2.8 In general, SDLT obligations arise 30 days after the **effective date** of the transaction (*FA 2003*, *ss 76(1), 86(1)*). At the latest, the **effective date** is the date of completion. For these purposes, the word 'completion' is not defined and, therefore, bears its normal legal meaning, broadly summarised as the final action required to give the transaction full legal effect. However, in many cases, the **effective date** is triggered by **substantial performance** of the transaction before completion.

Substantial performance

2.9 **Substantial performance** is defined using a combination of terms such as 'possession' and 'occupation', which are themselves not clearly defined, together with measurement of the proportion of any consideration actually paid. In any dispute as to whether an agreement has been **substantially performed**, expert legal advice may be required as to the meaning of terms.

However, in general, a contract is **substantially performed** where either the purchaser takes possession of **substantially all** of the property passing under the contract, or **substantially all** of the consideration is paid or provided. HMRC have indicated that they will regard 90% or more as **substantially all**, but SDLTM07950 indicates they will not consider themselves tied to this figure if they detect an attempt to exploit it for avoidance. *Any* payment of rent under the terms of a lease will be regarded as **substantial performance** of the lease, as will the purchaser becoming entitled to receipt of *any* rent etc from a tenant.

This gives rise to a potential issue – Standard Terms & Conditions for land transactions may entitle the purchaser to rent from a tenant or sub-tenant earlier than the time of payment for the property, which may trigger early substantial performance. It is worthwhile to review the terms and conditions carefully, to avoid such problems.

Formal completion

2.10 If a contract for a **land transaction** is entered into and is completed by a formal **conveyance** not later than the time of **substantial performance**, the contract and **conveyance** are treated as a single **land transaction**, taking effect at the date of completion (*FA 2003, s 44(3)*). Submission of a return and payment of the tax will be due 30 days from that date. Where, however, the contract is entered into and **substantially performed** before completion, the contract itself is treated as a **land transaction**, the effective date being that of **substantial performance** (*s 44(4)*). The return and tax payment are due 30 days from that date. If such a transaction is later completed by a formal **conveyance**, that **conveyance** will be a separate **land transaction** (in some circumstances, but not always, requiring submission of a further return), but SDLT will only be payable to the extent that the tax payable on the **conveyance** is greater than the tax paid on **substantial performance** of the contract – there should be no overall 'double charge' (*s 44(8)*). However, care is needed where reduced or zero SDLT is paid at the time of **substantial performance** because of a relief or exemption – in some circumstances, the later completion can trigger further liability because conditions for the relief or exemption are not satisfied at that time. The circumstances in which two or more returns may be required for the same transaction are explored in more detail at **6.15** et seq.

What is chargeable

2.11 The terms **conveyance** and transfer are used interchangeably, and include any instrument. Accordingly, they cover the grant of a lease as well as a transfer of land. A transaction is chargeable unless it is specifically covered by an exemption. **Chargeable interests** include estates, interests (including beneficial interests), rights (including for example a right to light) and powers over RUK land, as well as the benefit of any obligation, restriction or other condition that would affect the value of any such estate, interest, right or power. However, licences to occupy, mortgages and similar security interests are excluded. The extension of the definition to include obligations and restrictions will mean, for example, that a payment to the owner of a neighbouring plot of land, for a restrictive covenant or wayleave, may give rise to a charge. For some purposes, it matters whether an interest is a **major interest**. This term covers interests generally referred to as freeholds, leaseholds and other tenancies. However, for the purposes of *FA 2003, Sch 4ZA (SDLT: Higher Rates for Additional Dwellings and Dwellings Purchased by Companies) (see* **2.39**), a leasehold interest is not a *major interest* if the term of the lease does not exceed seven years on the date of its grant (*FA 2003, Sch 4ZA, para 2(4)*).

2.12 The grant of an option to acquire an interest in land is specifically identified as a **chargeable transaction**. The transfer or exercise of an option is a separate transaction from the grant, but the two may be **linked transactions** for determining the amount of tax due. This is intended to counter attempts to take advantage of lower rates of SDLT by splitting a purchase into two steps. An option is not, itself, a **major interest**, although this fact is of little practical significance unless it is a component in an exchange.

2.13 If interests in land (at least one being a **major interest**) are exchanged, each transfer is separately chargeable; see **2.32**.

OCCASIONS OF CHARGE

Chargeable transactions

2.14 SDLT is chargeable on a chargeable land transaction – that is, any acquisition of a chargeable interest unless the transaction is itself exempt from charge. Any interest in RUK land and buildings is chargeable unless it is specifically identified as exempt (see **2.16**). Examples of chargeable interests include freeholds, leases, sporting rights and the benefits of restrictive covenants – but this list is not exhaustive. The following give rise to the acquisition of a chargeable interest:

- transfer of all or part of a freehold or leasehold interest in RUK property;
- grant of a lease out of a freehold or superior leasehold interest in RUK property;
- surrender of a lease over RUK property to the landlord;

- variation of a lease over RUK property where that variation is treated as a surrender and re-grant in law, or where the variation includes a reduction in rent or a reduction in term; and
- grant, transfer, variation or surrender of an interest in RUK property other than a freehold or lease – for example, sporting or mineral rights, right of light, right of passage, etc.

2.15 In relation to leases (and arrangements treated as leases for SDLT purposes), certain events, which are not in law any of the above, are nonetheless deemed to be one of this list and therefore to give rise to SDLT obligations. These matters are considered in **Chapter 4**.

Exempt interests

2.16 The following limited list of interests is regarded as exempt, so that no SDLT liability or obligation arises on their creation, transfer, variation or surrender (*FA 2003*, *s 48(2)*). All other interests are chargeable:

- a security interest – that is, an interest or right (other than a rentcharge, feu duty or similar) held for the purpose of securing payment of money or performance of other obligations. This has been expanded to include the interest held by a financial institution as a result of an 'alternative property finance' transaction (see *FA 2003*, *s 73B*);
- a licence to use or occupy land. HMRC have made clear their view that this only applies to an interest which is a licence in both name and substance. They will resist attempts to 'dress up' as a licence an interest which is in substance greater than that, and care is required in relying on this exemption;
- a tenancy at will. Informal occupation arrangements sometimes begin as a tenancy at will but over time may develop into some other form of tenancy which is not exempt and which therefore may give rise to SDLT obligations, see **4.10**; and
- certain somewhat archaic rights which might otherwise be regarded as including interests in land – an advowson (the right to appoint a clergyman to a benefice), a franchise (a right granted by the crown such as the right to hold a fair or market, or the right to take tolls) and a manor (a right of certain kinds of jurisdiction over a specified area or estate).

Exempt transactions

2.17 Even if the land interest is not exempt, the transaction may be so – see **3.7** for details.

THE PURCHASER IS CHARGEABLE

2.18 SDLT obligations fall on the purchaser, identified in the legislation as the person who acquires the chargeable interest, or whose existing interest is enlarged or enhanced. The identity of the purchaser is obvious in straightforward transfers and grants of leases, but less obvious in some other cases. In particular:

- if a chargeable interest is transferred or granted to the trustees of a settlement, the trustees are jointly purchasers (*FA 2003, Sch 16, para 4*);
- if an existing chargeable interest is transferred to bare trustees or nominees, the beneficial owner is the purchaser (*Sch 16, para 3(1)*);
- if a lease is granted to bare trustees, those trustees are (jointly, if more than one) purchasers of the whole interest (*Sch 16, para 3(3)*);
- if a lease is surrendered or the term shortened, the landlord is the purchaser (*FA 2003, s 43(3)(b), Sch 17A, para 15A(2)*);
- if the rent under a lease is reduced, the tenant is the purchaser (*Sch 17A, para 15A(1)*); and
- if any chargeable interest other than a lease (eg an easement) is varied, whoever benefits is the purchaser (*FA 2003, s 43(3)(c)*). This is an enigmatic definition, since it is normally the essence of a commercial transaction that both parties benefit, but it should probably be read as the person whose interest in land is enhanced.

However, a person cannot be a purchaser unless he gives consideration or is a party to the transaction.

Example 2.1 – identifying the purchaser

Peter owns the freehold of property F, and Rachel owns the freehold of property G which abuts F.

Peter pays the HCA* £100,000 for removal of a restrictive covenant preventing use of F for carrying on a manufacturing business. This is a land transaction and Peter is the purchaser.

Rachel agrees to pay Peter £500 per annum for the right to bring maintenance vehicles through the car park of Peter's property F to service the rear of Rachel's building on her property G. This is a land transaction and Rachel is the purchaser.

Peter and Rachel agree to exchange small parts of their respective properties to create better shaped plots, allowing construction of better buildings on both. Peter is purchaser in respect of the parcel of land he gains, and Rachel is purchaser in respect of the parcel she gains – that is, there are two chargeable transactions, even though no money changes hands. Furthermore this is an exchange, see **2.32**.

* The Homes and Communities Agency is a UK government body which has inherited the benefit of restrictive covenants originally imposed on land by other, now defunct, agencies, such as the Commission for the New Towns.

CONSIDERATION

Measuring consideration – valuation

2.19 The basic definition is 'any consideration in money or money's worth, given directly or indirectly by the purchaser or a person connected with him' (*FA 2003*, *Sch 4*, *para 1(1)*). The full face value of consideration must be taken into account – no discount is given for the fact that the consideration may be postponed, payable in instalments etc (*Sch 4*, *para 3*). However, where payment of consideration is postponed because it is contingent or the amount is uncertain, it is sometimes possible to apply for deferral of the corresponding SDLT (see **6.31**). In some circumstances where instalments are payable over many years, the consideration is taken to be the amount payable over 12 years (see **2.28**). Non-sterling consideration is valued at the London closing exchange rate for the effective date of the transaction; consideration which is not in the form of money or debt must be valued at its market value at the effective date. One example of this is where a 'roofspace lease' for the installation of solar panels gives the property owner the right to take electricity generated by the panels at reduced or zero price (see **4.14**). However some forms of consideration are ignored in some transactions (see **2.24**, **2.34** and **3.7**); in some cases, actual consideration is replaced or supplemented by deemed consideration (see **2.29**). When a lease at rent is in point, there are specific computational rules (see **4.43–4.56**).

Contingency and uncertainty

2.20 If payment of any part of the consideration is contingent on some future event, it is assumed that the amount will be payable and SDLT must be paid accordingly (*FA 2003*, *s 51*).

If the amount of consideration is uncertain or has not been finally ascertained by the time payment and submission of the SDLT return are due, a reasonable estimate must be used.

Once the figures are finally known, if the final SDLT amount is higher than originally entered on the return, an amended return must then be submitted and the further SDLT paid. If the final figures show that too much SDLT has been paid, the surplus SDLT may be reclaimed (*FA 2003*, *s 80*). The stamp duty concepts of unascertainable and contingent consideration (see **8.36, 8.39**) do not apply to SDLT. There is one exception to the treatment of contingent consideration set out above, and that relates to 'right to buy' transactions (see **3.43**).

Where contingent and/or uncertain consideration may not be payable until some time after the effective date of the transaction, it may be possible to apply to postpone payment of the corresponding SDLT (see **6.31**). However the tight deadlines for applying for this postponement are rigidly enforced, so prompt application is essential.

In general, when SDLT is paid after the normal due date (that is, more than 30 days after the effective date of the transaction), interest will be charged from the normal due date to the date of payment. This may mean that interest will be charged for a period before the date when the final amount of tax is known (*FA 2003*, *s 87(3)(c)*). However, if HMRC have agreed to postpone the tax as explained in **6.31**, interest will not be charged provided tax is paid in accordance with the postponement agreement (*FA 2003*, *s 87(3)(b)*).

Example 2.2 – measuring consideration

Lissa agrees to transfer a plot of land M to Norman. In return, Norman agrees to pay Lissa £150,000 cash, payable in ten annual instalments of £15,000, commencing immediately, and to have his family company Quickbuild Ltd build Lissa a new house on a different plot of land P, which Lissa also owns. Norman is subject to SDLT on the acquisition of M, and the consideration is £150,000 plus the market value of the building work to be undertaken by Quickbuild Ltd. Although Lissa gets a new house out of the arrangement, this is built on land she already owns, so she has no SDLT obligations.

Example 2.3 – contingent consideration

Norman cannot afford to fund the development of the land purchased in the previous example so he sells it to Investor Ltd for £ 300,000 cash upfront plus 50% of the net profit made on eventual sale of the completed development, up to a further £700,000. As a condition of the sale, Quickbuild Ltd is engaged to carry out the development, being paid on a normal commercial basis by Investor Ltd. Investor Ltd is liable to pay SDLT on the total consideration, consisting of £300,000 cash payable at the outset plus £700,000 contingent cash payable However, provided some part of the contingent amount may fall to be paid more than six months after the effective date of the purchase, Investor Ltd may apply to defer payment of SDLT on the whole £700,000 contingent cash payable. The payments to Quickbuild are being paid for the development works and are not consideration for the transfer of the land. It is however necessary to consider the case of *Prudential Assurance Co Ltd v IRC* [1992] STC 863. HMRC set out their view of the impact of this case at SDLTM04015. HMRC consider that the same principles apply for the purposes of SDLT as those set out in SP 8/93 in relation to the stamp duty in that if the land purchase and development contracts are genuinely independent of each other, SDLT will only be charged on the consideration given for the land. However, if there is, in reality, only one transaction, being the acquisition of the completed development then HMRC are likely to seek SDLT on the consideration paid for the development works. If the amount payable under the development contract is chargeable to SDLT and cannot be finally determined at the effective date of the purchase (for example it is a 'cost plus' contract rather than a fixed price) and again assuming some part may fall to be payable more than six months later, Investor Ltd may also apply to defer SDLT on this amount. However, specific rules apply in relation to works as consideration and Investor Ltd will be required to make SDLT payments on an agreed schedule at not more than six-monthly intervals (see SDLTM50920).

If no application is made to defer payment of SDLT, Investor Ltd must pay tax on the full cash price, including the maximum contingent amount of £700,000, so £1 million in total, plus, if the payments under the development contract are chargeable to SDLT, a justifiable estimate of the amount payable under the development contract, no later than 30 days after the effective date of the purchase.

Attribution

2.21 When two or more related transactions are entered into, or there is a composite transaction for acquisition of various assets, the overall consideration must be apportioned between the assets on a just and reasonable basis (*FA 2003, Sch 4, para 4*). Apportionment between chargeable and non-chargeable assets has in the past often given rise to difficulties. Until 4 December 2014 SDLT was charged on the acquisition of residential property under the so-called 'slab system' under which a single rate of SDLT applied determined by the highest band into which the chargeable consideration fell. Consequently, where the amount apportioned to chargeable interests was close to the upper limit for a lower SDLT rate, it was likely that HMRC would have taken a close interest in whether the apportionment was justified. This should be much less of an issue now that the SDLT charge in relation to residential property is determined using a 'progressive system' with effect from 4 December 2014 (see **2.37**). In the Tax Tribunal (First-tier) case of *Orsman v CIR* [2012] UKFTT 227 (TC), released on 28 March 2012 (see www.financeandtaxtribunals.gov.uk/judgmentfiles/j6342/TC01921.pdf), the taxpayer attributed £250,000 to the fabric of a house and £8,000 to non-liable chattels. On this basis SDLT was due at 1%, £2,500. However HMRC investigated and concluded that some of the 'chattels' were, in fact, fixtures forming part of the house. The Tribunal agreed that a further £800 should be attributed to the house. This took the consideration into the 3% SDLT band, increasing the SDLT due to £7,524. The case contains interesting comments on the general basis for attributing consideration between assets and is worth reading where such attribution is required.

Inherent goodwill

2.22 The purchase and sale of a business often involves consideration attributed to 'goodwill'. Where the assets transferred include interests in land, the question arises as to whether any of that goodwill should be regarded as inherent in or attached to the interests in land, such that it should be subject to SDLT. HMRC set out their view on the point at www.hmrc.gov.uk/svd/goodwill.htm, which includes a link to a fuller practice note. In general terms, if the 'goodwill' element of consideration can be shown to relate to matters such as contracts with customers and suppliers, the specific customer base or a trade name/reputation, it may be accepted that this is true goodwill, separable from the property. However if the goodwill arises from the location of the property or the fact that the property is constructed or adapted in a way which particularly suits the business, HMRC are likely to argue that the amount attributed to goodwill should be added to the amount paid for the property. Many disputes arise in this area and specialist expert help may be required, if the amounts at stake warrant it.

VAT

2.23 If the consideration is subject to VAT, this must be included in the amount on which SDLT is charged. If the purchaser is able to recover VAT suffered as input tax, this means that SDLT is charged on a greater amount than the cost. At current VAT rates, the maximum marginal rate of SDLT on non-residential property is 6%. If the land is not subject to VAT at the effective date of the transaction, but may become so through exercise of the 'option to tax', VAT is not taken into account, provided the option to tax is not exercised until after the effective date (*FA 2003*, *Sch 4*, *para 2*).

Example 2.4 – SDLT on VAT

Big Manufacturer Ltd (B) buys a plot of bare land from Land Trader Ltd (L) for £1,600,000 (plus VAT if chargeable), intending to build a new factory. L has exercised the VAT 'option to tax' the land and charges VAT, but B is able to recover this fully through its own VAT returns. B also negotiates to pay the VAT to L on the same day it recovers it from HMRC. So, commercially, the real cost of the land is simply £1,600,000. SDLT is chargeable at rates of 0%, 2% and 5% on the VAT inclusive price of £1,920,000. This gives a liability of £85,500, which is 5.3% of £1,600,000.

Debt as consideration

2.24 If property is transferred in satisfaction (or in return for the release) of a debt due to the purchaser or owed by the vendor, or in return for the purchaser assuming responsibility for a debt, the amount of debt (including any interest due but unpaid) counts as consideration. However, if the amount of debt exceeds the market value of the property, the consideration is limited to that market value (*FA 2003, Sch 4, para 8*).

2.25 The position may be less clear where there is a debt secured on the property before the transfer, which remains in place after. In principle, it could be possible for the property to be transferred without the purchaser taking on responsibility for payment of the debt, in which case the debt should not be taken into account as consideration for the transfer.

Example 2.5 – debt as consideration?

Parent has purchased a house with the aid of a loan secured on the house. The house is occupied by Student. Parent transfers the house to Student for no consideration and retains full and sole responsibility for the loan. The amount of the loan should not be taken into account as consideration for SDLT purposes.

With a view to prevention of avoidance, it is provided that the amount of debt secured on the property *will* be taken into account as consideration if the rights or liabilities of any party to the transaction in relation to the debt are changed in connection with the transaction (*FA 2003, Sch 4, para 8(1A)*). In the example above, arrangements between Parent and the lender would require very careful drafting to ensure that Parent's rights and obligations in relation to the loan remain unchanged when the property is transferred – and it may prove difficult to persuade a lender to accept the necessary drafting.

Carrying out works on land

2.26 This is a widely misunderstood and, in practice, very limited exclusion from chargeable consideration. The level of understanding is not improved by HMRC's commentary at SDLTM04060 which is incorrect, possibly the result of careless wording. If consideration given for a chargeable transaction includes works of construction, repair or improvement, their value is excluded from chargeable consideration if the following conditions are met (*FA 2003, Sch 4, para 10*):

- the works are carried out after the effective date of the transaction;
- the works are carried out on the land being acquired or on other land already held by the purchaser or someone connected with him; and
- it is not a condition that the works are carried out by the vendor or someone connected with him.

Example 2.6 – works which do not count as consideration

Chemical Co Ltd (C) agrees to sell land to Housebuilder Ltd (H). The land is contaminated and C is under an obligation to carry out decontamination works within one year. In consideration for the transfer, H pays £500,000 and agrees to engage an unrelated company to carry out the required decontamination works at an expected cost of £1 million. The works do not commence until after completion of the transfer. The value of the decontamination works does not count as consideration for the transfer of the land, and H pays SDLT only on the £500,000 cash consideration.

Example 2.7 – works which are consideration

Beryl (B) agrees to buy a plot of land from Landbank Ltd (L) for £200,000 (its true market value) and, on the same day, engages Quickbuild Ltd (Q) to construct a house on the plot at a cost of £300,000. L and Q are associated with each other but unconnected with B. The provision relating to carrying out of works has no application in this scenario. The £300,000 is being paid for construction of the house and is not consideration for the transfer of the plot of land. Instead, it is necessary to consider the case of *Prudential Assurance Co Ltd v IRC* [1992] STC 863. HMRC set out their view of the impact of this case at SDLTM04015. HMRC consider that the same principles apply as those set out in SP 8/93 in relation to stamp duty liability in this kind of transaction. Provided the land purchase and building contracts are genuinely independent of each other, SDLT will be charged only on the price paid for the land. However, if there is, in reality, only one transaction – for example, if in truth L will only sell the land if B also agrees to engage Q to build the house – HMRC will seek SDLT based on the overall price for the land and completed building. In the present example, HMRC are likely to be difficult to convince that this is anything other than a single transaction, with SDLT due on £500,000.

Services as consideration

2.27 Where services are provided as consideration, they are valued at the amount which would have been paid in the open market for such services. In general, this is likely to be more than the cost to the purchaser of providing the services, since an open market provider would expect to make a profit (*FA 2003*, *Sch 4*, *para 11*).

Payment in instalments

2.28 If consideration is payable in instalments, the effective date of the transaction is likely to be triggered by something other than payment of consideration (for example, occupying the property or even formal completion). Once the effective date arrives, SDLT is payable on the whole consideration, no matter when that consideration is payable. In certain circumstances, it is possible to apply to defer payment of some of the SDLT, as mentioned at **2.19** (see **6.31** for details).

However, where the consideration comprises or includes an annuity capable of lasting for at least 12 years, the consideration is taken to be 12 years' instalments (*FA 2003*, *s 52*). For these purposes, any consideration payable periodically (other than rent) is an annuity. If the yearly amounts vary, the highest 12 separate years are taken into account (although increases in line with the Retail Price Index are ignored). If this treatment applies, it is not possible to apply for deferral of any of the SDLT, and there is no adjustment to reflect actual consideration if the total payable proves to be more or less than the 12 years' payments initially assessed.

Example 2.8 – annuities

On 1 July 2013, Pru sells her house to ER plc in return for:

(1) a lump sum of £100,000, plus

(2) a rent-free tenancy of the house for life (valued at £40,000), plus

(3) an annuity for life, initially at £15,000 pa, increasing to £20,000 pa after four years and increasing in line with RPI thereafter.

The consideration on which ER must pay SDLT is the sum of (1) and (2) plus 12 times the £20,000 annuity – £380,000 in total. For these purposes, since the annuity is for life, it is assumed Pru will live at least until 30 June 2029 ,so that at least 12 instalments at £20,000 will be payable, but the subsequent RPI increases are ignored. The SDLT is due by 31 July 2013, with no facility to defer and no adjustment if Pru dies before or after 30 June 2029.

Substitution of market value

2.29 SDLT is generally chargeable by reference to the value of the actual consideration, but in some circumstances this is substituted by the market value of the property acquired. One example is where property is transferred in payment of a debt exceeding the value of the property (see **2.24**); another is where properties are exchanged (see **2.32**) and actual consideration is less than the market value of the property acquired.

Transfer to connected company

2.30 Perhaps the main relevance of market value is when a property is transferred to a company which is connected with the transferor. When this happens, the consideration is to be taken as not less than the market value of the interest transferred (plus any rent where the transaction is the grant of a lease) (*FA 2003, s 53*). For these purposes, 'connected' is as defined in *Corporation Tax Act 2010 (CTA 2010), s 1122*. If the actual consideration, including any VAT chargeable on the transaction, is greater than market value, SDLT is chargeable by reference to the actual consideration, unless a relief or exemption is available.

The market value rule does not apply in the case of some transfers to trustees. Nor does it apply where the transfer is a distribution out of the assets of a company. However, an anti-avoidance rule re-applies the market value rule if group relief has been claimed on the property (or on a superior interest, if the property is a lease) within the previous three years (*FA 2003, s 54*).

2.31 There are two noteworthy points which often cause confusion. First, there is no general relief available where the transfer to a connected company occurs in the course of incorporation of a business by a sole trader or unincorporated partnership. Other tax reliefs are available to ease the process of incorporation but, where the business has substantial property assets, the SDLT cost may act as a significant disincentive. This was the cause of heated parliamentary debate when SDLT was introduced, but the government of the day was not prepared to introduce a relief.

There is a very specific relief on transfer from an unincorporated partnership to a limited liability partnership; see **3.52**. Additionally, application of the special partnership rules may reduce the SDLT on transfers between partners, partnerships and other connected persons, including companies in some cases. For detail on this see **Chapter 5.**

Secondly, where a lease at rent is granted to a connected company, there is no requirement to substitute a market rent. Rather, if the rent is less than a market rate, the lease itself will have a capital value (ie at arm's length, a lease at that rent would also attract a premium). SDLT is chargeable on the actual rent plus the capital value as if paid as a premium.

Exchanges and partitions

2.32 Where interests in property are exchanged, this is treated as two land transactions. Assuming at least one of the interests transferred is a major interest (see **2.6** and **2.11**), the consideration for each acquisition is the greater of (a) the value of the property acquired and (b) the value of the property given in consideration together with any other consideration given (*FA 2003, s 47, Sch 4 paras 5, 6*). Originally the value given in consideration was irrelevant, SDLT was based only on the value of the property acquired. However this was thought to provide an opportunity for avoidance by artificially depressing the value of the property just before it was transferred. The rules were changed to their current form by *FA 2011 Sch 21 para 4* with effect from 24 March 2011 but subject to transitional rules for transactions agreed before that date. How to determine the chargeable consideration on an exchange is covered by HMRC at SDLTM04020. The change will lead to increased uncertainty and administration for some 'innocent' commercial transactions.

Particular issues arise, in relation to both amount and timing of tax liability, when interests in the same property are exchanged – for example when A sells to B and B grants A a lease out of the interest just purchased. See Example 2.23 and **3.40** et seq for further details.

Example 2.9 – exchanges

Erica agrees to transfer a plot of land, X, to Hosta, in exchange for Hosta transferring plot of land Z to Erica. It is agreed that the open market value of X is £200,000 and that of Z is £180,000. The parties agree to ignore the difference in value for commercial purposes. Erica will pay SDLT on the acquisition of Z, based on £200,000, being the greater of the market value of Z, and the value given in consideration. Hosta will pay SDLT on the acquisition of X, based on £200,000, again being the greater of the market value of X and the value given in consideration. The overall SDLT bill would have been lower if each had sold to the other for cash – but there may have been more pressing commercial reasons for proceeding by way of exchange.

In a subsequent transaction, Hosta transfers plot of land W, valued at £1.2 million, to Erica. In return, Erica transfers plot of land Z back to Hosta, and agrees to make cash payments to Hosta based on a formula which takes account of the eventual sales proceeds of houses which Erica will construct on plot of land W. Erica will have to pay SDLT based on a reasonable estimate of the greater of the market value of plot of land W and the total consideration which will be given. Once that total consideration is finally determined – presumably after all of the houses have been constructed and sold – Erica will have to consider whether a further return and payment of SDLT are due (for example if the total consideration exceeds both the original estimate and the original market value of plot of land W). Prior to the 2011 change, the SDLT position would have been finally settled as soon as the value of plot of land W had been agreed.

2.33 If none of the interests is a major interest (that is, the interests are limited to options or matters such as sporting rights, wayleaves, etc), the consideration for each acquisition is simply the amount of any further consideration given apart from the interest in land itself.

Example 2.10 – exchanges, not major interests

Juniper Ltd owns fishing rights over a stretch of Minnow River. Unrelated company Gorse Ltd owns shooting rights over Sparrow Moor. The companies agree to exchange their rights, with Juniper paying Gorse £300,000 cash, being the agreed difference in values. Juniper will pay SDLT based on consideration of £300,000. Gorse will not pay SDLT, as the deemed consideration for its acquisition of the fishing rights will be nil.

2.34 Where two or more persons hold a property jointly, whether as joint tenants or as tenants in common, they are generally treated for SDLT purposes as owning separate and distinct (but undefined) parts of the property. A property held in this way may be partitioned, so that each person takes a specific part.

If this happens, the interest which each gives up is not treated as consideration for the part of which he becomes sole owner. The only consideration for SDLT purposes will be any payment or other valuable consideration which one may give to another, perhaps to compensate for differing values of the parts taken by each (*FA 2003*, *Sch 4*, *para 6*).

Example 2.11 – partition of jointly held property

Brothers Bill and Ben own a plot of land worth £1 million, divided into two paddocks, as joint tenants. They agree to partition it so that Bill owns paddock A and Ben owns paddock B. They agree that paddock A is worth £100,000 more than paddock B, but Bill is short of cash and they agree he will give Ben only £20,000 in compensation for his loss of value. For SDLT purposes, Bill is treated as giving consideration of £20,000 (and, since this is below the relevant threshold, Bill has no SDLT liability or notification obligations); Ben is treated as giving no consideration. Neither the market value of the overall plot nor the difference in values of the divided plots is relevant.

Market values could be relevant where one of the owners is a company which is connected to other owners (see **2.29**).

PFI transactions

2.35 Transactions in which a private sector company performs services, often including provision of premises, for a public sector organisation, are widely referred to as Private Finance Initiative (PFI) transactions. Such transactions often involve transfers and leases of property. There is a specific provision allowing many forms of consideration to be ignored in qualifying transactions. This is dealt with at **3.17**.

Sundry liabilities not treated as consideration

2.36 The following liabilities and obligations taken on by a purchaser do not count as consideration:

- an indemnity given to the vendor in respect of liability to a third party arising from breach of an obligation relating to the land (*FA 2003*, *Sch 4*, *para 16*);
- an agreement to bear any inheritance tax liability arising from the transfer (*para 16A*);

- an agreement to bear any capital gains tax liability arising from the transfer if it is not (or is treated as not being) a bargain at arm's length (*para 16B*); and
- certain specified costs borne by a leaseholder in connection with enfranchisement (*para 16C*).

RATES OF SDLT

2.37 The rate at which SDLT is charged on consideration other than rent depends on the amount of the chargeable consideration, and whether the transaction is a residential property transaction or a non-residential property transaction. Until 4 December 2014 the amount of the tax payable for both a residential property transaction and a non-residential property transaction was calculated using the so-called 'slab system'. Under that system the rates were not progressive; rather, the whole consideration was chargeable at a single rate determined by the highest tax band into which the total chargeable consideration fell. However, the tax payable in relation to a residential property transaction, with an 'effective date' (see **2.8**) on or after 4 December 2014, and in relation to a non-residential or mixed-use property transaction, with an 'effective date' on or after 17 March 2016, is now calculated using a 'progressive system' (like personal income tax). This means that it is only that part of the chargeable consideration falling within each tax band that is taxed at the rate for that tax band.

In relation to residential property transactions there are transitional rules, which provide that where contracts were exchanged to acquire the residential property before 4 December 2014, but the contract was not completed until on or after 4 December 2014, the purchaser can choose to pay SDLT based on the old rules. Similarly, where the contract was exchanged and substantially performed (see **2.9**) but not completed (see **2.8**) before 4 December 2014 then the purchaser can choose to apply the old rules to the completion transaction taking place on or after 4 December 2014. To choose to pay SDLT based on the old rules the purchaser just enters the appropriate amount of SDLT in box 14 of the land transaction return.

Similarly, in relation to non-residential or mixed use property transactions, there are transitional rules which provide that where contracts were exchanged to acquire the non-residential or mixed use property before 17 March 2016, but the contract was not completed until on or after 17 March 2016, then provided that there is no event of the kind listed in *FA 2016, s 127(15)* which results in the effect of the contract being different from the effect of the contract when first entered into, the purchaser can elect to pay SDLT based on the old rules

(*FA 2016, s 127(13)(b)* and *(15)*). Similarly, where the contract was exchanged and substantially performed (see **2.9**) but not completed (see **2.8**) before 17 March 2016 then the purchaser can elect to apply the old rules to the completion transaction taking place on or after 17 March 2016 (*FA 2016, s 127(13)(a)*). To choose to pay SDLT based on the old rules the purchaser simply enters the appropriate amount of SDLT in box 14 of the land transaction return.

A transaction is a residential property transaction if the chargeable interest(s) acquired under the transaction consists **entirely** of an interest in residential property (see **2.38**) or, if the transaction is one of a number of linked transactions (see **2.67**), the chargeable interests acquired under each linked transaction consists **entirely** of residential property (*FA 2003, s 55(1B),(1C),(3)* and *(4)*).

A transaction is a non-residential property transaction if the chargeable interest acquired under the transaction either consists of, or includes, a chargeable interest which is not residential property. If the transaction is one of a number of linked transactions (see **2.67**), it will be a non-residential property transaction if the chargeable interest acquired under any of the linked transactions consists of, or includes, an interest in non-residential property (*FA 2003, s 55(2),(3)* and *(4)*).

A transaction is a mixed use property transaction if the chargeable interest(s) acquired under the transaction either consists of, or includes, a chargeable interest(s) which comprises both residential and non-residential property. If the transaction is one of a number of linked transactions (see **2.67**), it will be a mixed use property transaction if the chargeable interest(s) acquired under any of the linked transactions consists of, or includes, an interest in both residential and non-residential property (*FA 2003, s 55(2), (3)* and *(4)*).

Thus the tax rates for residential property will only be applicable if the transaction(s) consist(s) **entirely** of residential property.

The current rates and tax bands, plus those which applied to residential property transactions with an 'effective date' (see **2.8**) on or before 3 December 2014, and those which applied to non-residential or mixed use property with an 'effective date' (see **2.8**) on or before 16 March 2016, are listed below (*FA 2003, s 55*).

An additional 3% rate of SDLT applies to purchases of additional dwellings (such as second homes and buy-to-let properties) and first purchases of dwellings by companies and other non-individual purchasers (see **2.39** et seq) with an 'effective date' (see **2.8**) on or after 1 April 2016.

On grant (or deemed grant) of a lease at rent, special rules apply for calculation of the SDLT on the rent (see **4.41** et seq).

Consideration (£)	**Residential property[a] – with an 'effective date' on or before 3 December 2014**	**Non-residential[a] or mixed property- with an 'effective date' on or before 16 March 2016**
Not exceeding £125,000	0%	–
Not exceeding £150,000	0%, only for property in a designated Disadvantaged Area[b] Otherwise 1%	0%[c]
Exceeding £125,000/£150,000 but not exceeding £250,000	1%[d]	1%
£250,001 to £500,000	3%	3%
£500,001 to £1,000,000	4%[e]	4%
£1,000,001 to £2,000,000	5%[e]	4%
Exceeding £2,000,000	7%[e]	4%

(a) See **2.38**.
(b) Disadvantaged Areas Relief was abolished with effect from 6 April 2013 (see **3.54**).
(c) 1% in the case of grant of a lease of non-residential or mixed property with a yearly rent of £1,000 or more attributable to the non-residential part (see **4.46**).
(d) 0% for first-time buyers if effective date was before 25 March 2012.
(e) Or 15% in the case of a purchase of an interest in a single dwelling by certain non-natural persons for chargeable consideration of more than £500,000 (see **2.65**).

Consideration (£)	**Residential property[a] – with an 'effective date' on or after 4 December 2014 – additional 3% rate does not apply (see 2.39 et seq)**
Not exceeding £125,000	0%
£125,001 to £250,000	2%
£250,001 to £925,000	5%[b]
£925,001 to £1,500,000	10%[b]
Exceeding £1,500,000	12%[b]

(a) See **2.38**.
(b) Or 15% in the case of a purchase of an interest in a single dwelling by certain non-natural persons for chargeable consideration of more than £500,000 (see **2.65**).

Consideration (£)	Non-residential property[a] – with an 'effective date' on or after 17 March 2016
Not exceeding £150,000	0%
£150,001 to £250,000 Exceeding £250,000	2% 5%

(a) See **2.38**

Example 2.12 – Calculation of SDLT chargeable under the rules which apply for residential property transactions with an effective date on or after 4 December 2014

On 28 April 2016 John (who does not have a spouse or civil partner) exchanges contracts for the purchase of a residential property situated in England for a cash consideration of £2,500,000. This is the first property purchased by John and therefore the 3% additional rate charge (see **2.39**) does not apply. Equally, as John acquires the property in his own right, and not through a company, a partnership with a corporate partner or a collective investment scheme, the single 15% rate of SDLT does not apply (see **2.65**). The purchase of the property completes on 30 June 2016 without any prior event having taken place which would have 'substantially performed' (see **2.9**) the contract prior to that date.

The SDLT chargeable on the purchase of the property is calculated as follows:

Purchase Price (£)	Rate (0%)	SDLT (£)
125,000	0	Nil
125,000	2	2,500
675,000	5	33,750
575,000	10	57,500
1,000,000	12	120,000
2,500,000		213,750

John must therefore file a land transaction return, and pay the SDLT due of £213,750, by 30 July 2016.

Example 2.13 – Calculation of SDLT chargeable under the new rules which apply for non-residential property transactions with an effective date on or after 17 March 2016

On 31 May 2016 Reid and Wright Ltd exchange contracts for the purchase of an office block situated in England for a cash consideration of £5,250,000. The purchase of the property completes on 30 June 2016 without any prior event having taken place which would have 'substantially performed' (see **2.9**) the contract prior to that date.

The SDLT chargeable on the purchase of the property is calculated as follows:

Purchase Price (£)	**Rate (0%)**	**SDLT (£)**
150,000	0	Nil
100,000	2	2,000
5,000,000	5	250,000
5,250,000		252,000

Reid and Wright Ltd *must* therefore file a land transaction return, and pay the SDLT due of £252,000, by 30 July 2016.

Residential property

2.38 Residential property means property used or suitable for use as a dwelling (or in the process of being built or adapted for this purpose) and includes associated gardens, grounds, outbuildings and other rights which subsist for the benefit of the property – for example, access rights across adjacent property (*FA 2003, s 116(1)*). Non-residential property means any property that is not residential property (*FA 2003, s 116(1)*). The status of a house or flat is usually obvious. The legislation provides a list of other categories of building which are treated as residential property, but then removes many examples from the list, as set out below (*FA 2003, s 116(2)* and *(3)*). The distinction between residential and non-residential property is important given the higher rates of tax that apply to residential property.

Where six or more dwellings are the subject of 'a single transaction', they are treated as wholly non-residential property (*s 116(7)*), which effectively limits the highest SDLT rate to 5% subject to a claim for multiple dwellings relief (see **3.48**). The legislation does not define what is meant by 'a single transaction' but it is the author's view that it means one contract between the same vendor and purchaser (and not including parties connected with the vendor and purchaser) for the acquisition of the agreed number of properties. The term 'a single transaction' would not cover, for example, a situation where there are multiple independent contracts between the same vendor and purchaser, since in such a situation there would be a number of transactions, although they would be 'linked' (see **2.67**) for the purposes of *FA 2003, s 108*. The treatment of the purchase of six or more dwellings as the acquisition of non-residential property now has even greater importance given the higher rates of SDLT which generally apply to the acquisition of residential property, and the additional 3% rate which may apply to the purchase of dwellings (see **2.39**). Care will therefore need to be taken to ensure that where six or more dwellings are being acquired the deal is structured as a single transaction, and not a number of separate transactions, if non-residential property treatment is to be available.

Residential	**Non-residential**
Residential accommodation for school pupils	Home or other institution providing residential accommodation for children
Residential accommodation for students	Hall of residence for students in higher or further education
Residential accommodation for members of the armed forces	–
An institution which is the sole or main residence for at least 90% of its residents	Home or other institution providing residential accommodation with personal care to those in need of such care by reason of age, disablement, drug/alcohol dependence or mental illness
–	Hospital, hospice, prison or similar, hotel/inn or similar

Higher SDLT rates for additional dwellings and dwellings purchased by companies and other non-individual purchasers

2.39 At the Autumn Statement on 25 November 2015 the Chancellor of the Exchequer announced that from 1 April 2016 higher rates of SDLT would apply to purchases of additional residential properties (referred to in the legislation as 'dwellings'), such as second homes and buy-to-let properties. In fact the legislation is much wider than the original policy description and, for example, applies to first purchases of residential properties by companies and other non-individual purchasers. The legislation can be found in the new *FA 2003, Sch 4ZA* ('Stamp Duty Land Tax: Higher Rates for Additional Dwellings and Dwellings Purchased by Companies') and it imposes an additional 3% SDLT charge on chargeable transactions falling within its provisions.

A transaction which falls within the legislation is referred to as a 'higher rates transaction' (*FA 2003, Sch 4ZA, para 2(1)*).

The higher rates of SDLT are as follows:

Consideration (£)	**Rate of SDLT on additional dwellings – 'effective date' on or after 1 April 2016**
Not exceeding £125,000	3%
£125,001 to £250,000	5%
£250,001 to £925,000	8%
£925,001 to £1,500,000	13%
Exceeding £1,500,000	15%

The legislation which imposes the higher rate of SDLT (*FA 2003, Sch 4ZA*) is complex and therefore reference should also be made to the HMRC Guidance Note (16 March 2016), which is headed 'Stamp Duty Land Tax: higher rates for purchases of additional residential properties'. The guidance includes some helpful examples and a series of 'Questions and Answers'.

Meaning of 'higher rates transaction'

2.40 Where there is only one purchaser a transaction will be a 'higher rates transaction' if it falls within any of the five scenarios detailed below (see **2.51** – **2.58**) by reference to that single purchaser (*FA 2003, Sch 4ZA, para 2(2)*). Where there are two or more purchasers you must take each of the purchasers in turn and determine whether, by reference to any of those purchasers, the transaction falls within any of the five scenarios detailed below (see **2.51** – **2.58**). If it does then it is a 'higher rates transaction' (*FA 2003, Sch 4ZA, para 2(3)*).

Which of the five scenarios a transaction falls within is generally determined by three factors as follows:

a) whether or not the purchaser is an individual;

b) whether a single 'dwelling' or multiple 'dwellings' are being acquired; and

c) where multiple 'dwellings' are being acquired whether the 'chargeable consideration' (see **2.19**) for only one of the dwellings, or for more than one of the 'dwellings', amounts to £40,000 or more.

Meaning of 'dwelling'

2.41 For the purposes of the higher rates legislation a dwelling is defined (*FA 2003, Sch 4ZA, para 17(2)*) as a building or part of a building that is:

- used or suitable for use as a single dwelling; or
- in the process of being constructed or adapted for use as a single dwelling.

Land that is to be enjoyed with a dwelling as a garden or grounds of the dwelling, including any building or structure on that land (for example a detached garage), and any land which subsists for the benefit of a dwelling, is treated as being part of the dwelling (*FA 2003, Sch 4ZA, para 17(3)* and *(4))*.

There is no definition of the term 'dwelling' in the legislation which must therefore take its ordinary meaning, that is a building which provides those who use it with the facilities required for everyday private domestic existence. The term will cover holiday homes, including those which cannot be used all year round, and furnished lettings.

An 'off-plan purchase' will also count as the acquisition of a 'dwelling' if the following conditions are satisfied (*FA 2003, Sch 4ZA, para 17(5)*):

- a contract has been exchanged for the acquisition of a building which is to be constructed or adapted for use as a single dwelling;
- the contract is substantially performed (see **2.9**); and
- at the time of substantial performance the construction or adaptation of the building has not yet started.

It will be a question of fact as to whether a single dwelling, or more than one dwelling, is being acquired. Generally a self-contained building is likely to be a single dwelling if anyone living in that building can live independently of the residents of the rest of the building. This is likely to include having independent access to the building as well as cooking, toilet and washing facilities.

A building, or part of a building, used for any of the following purposes is not used as a 'dwelling' for the purposes of the higher rates legislation (*FA 2003, Sch 4ZA, para 17(7)*):

- residential accommodation for school pupils;
- residential accommodation for students;
- residential accommodation for members of the armed forces;
- an institution that is the sole or main residence of at least 90% of its residents and does not fall within any of the other types of institution in the remainder of this list;
- a home or institution providing residential accommodation for children;
- a hall of residence for students in further or higher education;

- a home or other institution providing residential accommodation with personal care for persons in need of personal care by reason of old age, disablement, past or present dependence on alchol or drugs or past or present mental disorder;
- a hospital or hospice;
- a prison or similar establishment; and
- a hotel or inn or similar establishment.

It is likely that a bed and breakfast or guest house which has bathing facilities, telephone lines, etc installed in each bedroom will not be a dwelling for the purposes of the higher rates legislation.

Transactions to which the additional rate of SDLT does not apply

2.42 The SDLT higher rates will not apply to the acquisition of:

- **non-residential or mixed use properties** (For example if, as a single transaction, an individual is acquiring a house with a garden and grounds together with some agricultural land close by but clearly not part of the garden and grounds of the house, and not required for the proper enjoyment of the house and its gardens and grounds, then that transaction should be treated as the acquisition of mixed use property (ie both residential and non-residential property is being acquired) and the SDLT higher rates should not apply to any part of the transaction);
- **transactions for which the 'chargeable consideration'** (see **2.14**) **for the acquisition of the single dwelling is less than £40,000** (It should, however, be noted that the £40,000 amount is not a 0% band and, if the transaction is chargeable to the SDLT additional rate, it is chargeable on the whole of the consideration and not only the amount in excess of £40,000);
- **caravans, houseboats and mobile homes** – this is on the basis that these items are moveable and payments in relation to such items are not generally chargeable to SDLT. However, if the moveable asset became sufficiently fixed to the land so as to become a fixture (ie in law the item has become part of the land) SDLT may be chargeable and the SDLT higher rates may apply if the relevant conditions described below are satisfied; and
- **Enveloped residential properties –** the higher rates of SDLT will not apply to the acquisition of a single dwelling for more than £500,000 if the single rate of SDLT of 15% applies to that transaction by virtue of *FA 2003, Sch 4A* (see **2.65**).

Joint purchasers

2.43 If any of scenarios 1–5 detailed below (see **2.51–2.58**) apply by reference to any purchaser under a 'land transaction' (see **2.7**), the 3% additional rate will apply to the transaction. In circumstances where there are two or more individuals acquiring the dwelling as joint purchasers (either as 'joint tenants' or as 'tenants in common'), and any of scenarios 1–5 below (see **2.51–2.58**) apply to any of those purchasers, the transaction will be charged at the higher rates. It does not matter how small the interest of any purchaser is in the dwelling (*FA 2003, Sch 4ZA, para 2(3)*).

A partner in a partnership is treated as a joint purchaser of any 'chargeable interest' (see **2.14**) acquired by or on behalf of the partnership (*FA 2003, Sch 15, para 2(b)*). This means that when a 'chargeable interest' is acquired by or on behalf of a partnership each of scenarios 1–5 detailed below (see **2.51–2.58**) must be tested by reference to each of the partners and if any one partner would be liable to the 3% additional rate then the whole of the acquisition will be liable to the higher rates.

Spouse or civil partner

2.44 If an individual, who is purchasing a dwelling, has a spouse or civil partner who they are living with on the 'effective date' (see **2.8**) of the transaction, and who is not a joint purchaser, the spouse or civil partner will be treated as though they were a joint purchaser in respect of the transaction. Consequently, scenarios 1 and 4 below (see **2.51** and **2.57**) must also be considered in relation to the spouse or civil partner (*FA 2003, Sch 4ZA, para 9 (1)* and *(2)*). Thus if the conditions in scenario 1 and 4 (see **2.51** and **2.57**) below are met in relation to the spouse or civil partner, the higher rates of SDLT will apply. This treatment does not apply if the married couple are either legally separated (by court order or deed of separation) or they are in fact separated in circumstances in which the separation is likely to be permanent (*FA 2003, Sch 4ZA, para 9(3)* and *Income Tax Act 2007, s 1011*).

Individuals – Interests in dwellings treated as owned by an individual

2.45 It will be seen that for either scenarios 1 or 4 below (see **2.51** and **2.57**) to apply one of the conditions is that the individual purchaser owns a 'major interest' (see **2.11**) in another dwelling at the end of the day that is the 'effective day' of the acquisition of the new dwelling.

If an individual is the legal and beneficial owner of a 'major interest' (see **2.11**) in a dwelling they will own that interest for the purposes of the higher rate legislation. Equally, where an individual only has the beneficial ownership of a 'major interest' (see **2.11**) in a dwelling, with legal ownership held within a bare trust or by a nominee, the individual with the beneficial ownership is treated as owning the interest in the dwelling (*FA 2003, Sch 16, para 3* and *Sch 4ZA, para 11(2)* and *(3)*). Equally, if the bare trustee or nominee disposes of the interest in the dwelling it is the beneficiary who is treated as making the disposal for the purposes of the higher rates legislation (*FA 2003, Sch 4ZA, para 11(3)(b)*).

Dwellings held by trusts

2.46 Where a 'major interest' (see **2.11**) in a dwelling is held in a 'settlement' (ie a trust which is not a bare trust: *FA 2003, Sch 16, para 1(1)*), and under the terms of the trust the beneficiary is entitled to (a) occupy the dwelling for life, or (b) the income earned in respect of the dwelling, the beneficiary of the trust is treated as holding the interest in the dwelling for the purposes of the higher rates legislation (*FA 2003, Sch 4ZA, para 11(1)* and *(3)(a)*). Equally, if the trustees of such a trust dispose of the interest in the dwelling it is the beneficiary who is treated as making the disposal for the purposes of the higher rates legislation (*FA 2003, Sch 4ZA, para 11(3)(b)*).

If the 'major interest' (see **2.11**) in the dwelling is held by the trustees of a trust and the beneficiaries are not entitled as described above in this **2.46**, and it is not a bare trust (*FA 2003, Sch 16, para 1(1)*), then the trustees are treated as holding the dwelling for the purposes of the higher rates legislation.

Dwellings held by a partnership

2.47 The general rule is that in deciding, for the purposes of the higher rates legislation, whether an individual purchaser, who is a partner in a partnership, holds a 'major interest' (see **2.11**) in another dwelling, that individual will be treated as owning any 'major interest' held by or on behalf of the partnership (*FA 2003, sch 4ZA, para 14(3)* and *sch 15, para 2(1)(a)*). However, a 'major interest' in a dwelling held by a partnership can be ignored for the purposes of the higher rates legislation if the individual purchaser is not acquiring the new 'major interest' for the purposes of the partnership, the partnership is carrying on a trade and the existing dwelling is being used for the purposes of the partnership's trade (*FA 2003, sch 4ZA, para 14(1)* and *(2)*). A property letting business carried on by a partnership is not a trade, and therefore, for the purposes of the higher rates legislation, an individual who is a partner in a partnership carrying on such a business, will be treated as holding all of the 'major interests' (see **2.11**) in dwellings held by the partnership.

Dwellings situated outside of England, Wales and Northern Ireland

2.48 In determining whether an individual purchaser holds a 'major interest' (see **2.11**) in another dwelling for the purposes of the higher rates legislation, an interest in a dwelling situated outside of England, Wales and Northern Ireland is to be taken into account (*FA 2003, Sch 4ZA, para16(1)*). Other countries may have different land law concepts and it will be a question of fact whether an interest owned by an individual is equivalent to a 'major interest' (ie a freehold interest or a lease granted for a term of more than seven years – see **2.11**), and whether it was originally granted for a term of more than seven years and is not subject to a lease with more than 21 years to run (*FA 2003, Sch 4ZA, para 16(2)(a)*).

Dwellings acquired by inheritance

2.49 Where an individual becomes jointly entitled to a 'major interest' (see **2.11**) in a dwelling in the three-year period before the 'effective date' (see **2.8**) of the acquisition of a new dwelling that interest can be ignored, for the purposes of scenarios 1 and 4 below (see **2.51** and **2.57**), provided the following conditions are satisfied (*FA 2003, Sch 4ZA, para 15(1)* and *(2)*):

- the individual became the joint owner of the interest in the dwelling by inheritance; and
- the individual and any spouse or civil partner's combined interest in the dwelling has not exceeded 50% in the three-year period before the 'effective date' (see **2.8**) of the acquisition of the new dwelling. Where the interest in the dwelling is held as a tenant in common, the declared interest held by the individual and any spouse or civil partner must be 50% or less of the whole interest. Alternatively, if the interest in the dwelling is held as joint tenants, the individual and any spouse or civil partner must make up exactly half, or a minority of the joint tenants

However, if the interest in the dwelling was inherited more than three years before the acquisition of the new dwelling the individual will be treated as holding an interest in that inherited dwelling at the end of the day which is the 'effective date' (see **2.8**) of the acquisition of the new dwelling.

The date of inheritance is the date the individual becomes entitled to the 'major interest' (see **2.11**) which will normally be the date it is transferred to them. In countries where property passes directly to heirs the date of inheritance will be the date of death.

Determining what is a higher rates transaction

2.50 There are five scenarios which need to be considered in determining whether a transaction is liable to the additional rate of SDLT as follows:

- *Scenario 1* (see **2.51**) – the transaction is the acquisition of a single dwelling and the purchaser is an individual (*FA 2003, Sch 4ZA, para 3*).
- *Scenario 2* (see **2.55**) – the transaction is the acquisition of a single dwelling and the purchaser is not an individual (*FA 2003, Sch 4ZA, para 4*).
- *Scenario 3* (see **2.56**) – the transaction is the acquisition of more than one dwelling, at least two of the dwellings are acquired for £40,000 or more and the purchaser is an individual (*FA 2003, Sch 4ZA, para 5*).
- *Scenario 4* (see **2.57**) – the transaction is the acquisition of more than one dwelling, but only one of the dwellings is acquired for £40,000 or more and the purchaser is an individual (*FA 2003, Sch 4ZA, para 6*).
- *Scenario 5* (see **2.58**) – the transaction is the acquisition of more than one dwelling, at least one of the dwellings is acquired for £40,000 or more and the purchaser is not an individual (*FA 2003, Sch 4ZA, para 7*).

Each of these scenarios is now considered in detail.

Scenario 1 – single dwelling transaction (FA 2003, Sch 4ZA, para 3)

2.51 A 'chargeable transaction' (see **2.14**) is a 'higher rates transaction' if each of the following conditions are satisfied:

- the purchaser is an individual;
- the 'chargeable interest' (see **2.11**) acquired consists of a 'major interest' (see **2.11**) in a single dwelling, referred to as 'the purchased dwelling' (for these purposes a 'major interest' does not include a lease originally granted for a term not exceeding seven years);
- the 'chargeable consideration' (see **2.19**) given for the transaction is £40,000 or more;
- on the 'effective date' (see **2.8**) of the transaction, 'the purchased dwelling' is not subject to a lease which at that date has an unexpired term of more than 21 years (for example, where a freehold interest is being acquired that interest is not subject to a lease which has more than 21 years to run at the 'effective date' (see **2.8**) of the transaction);
- at the end of the day which is the 'effective date' (see **2.8**) of the transaction the purchaser (or any of the joint purchasers if there is more than one purchaser) has, or is treated as having (see **2.45–2.49**), a 'major

interest' (see **2.11**) in a dwelling, which has a market value of £40,000 or more, *other* than the 'purchased dwelling', and the interest which the individual holds in that other dwelling is not subject to a lease with an unexpired term of more than 21 years (note that this condition will be satisfied if the individual is a joint owner of another dwelling and the interest which the individual holds has a market value of £40,000 or more); and

- the 'purchased dwelling' is not a replacement for the purchaser's only or main residence (see **2.52** as to what is meant by a purchaser replacing their only or main residence).

In summary, the additional rate of SDLT will apply when an individual acquires a freehold interest, or a leasehold interest which was originally granted for a term exceeding 7 years, for a consideration of £40,000 or more, if at the end of the day the individual holds a freehold interest, or a leasehold interest which was originally granted for a term exceeding 7 years, in another dwelling (see **2.45–2.49** to determine whether an individual is treated as holding another dwelling for the purposes of the higher rates legislation), and that individual is not replacing his only or main residence. This is the case unless either the interest in the dwelling being acquired, or the interest in the other dwelling held, is subject to a lease with more than 21 years to run.

Replacement of main residence test

2.52 The replacement of main residence test can be satisfied by either (a) a disposal of an individual's main residence before, or on the day that, the new main residence is acquired, or (b) the new main residence is acquired first and the disposal of the previous main residence takes place at a later date.

Disposal of the current main residence on or before the date of acquisition of the new main residence

2.52A For these purposes an individual is replacing his only or main residence if the following conditions are satisfied (*FA 2003, Sch 4ZA, para 3(6)*):

- on the 'effective date' (see **2.8**) of the transaction the purchaser intends to use 'the purchased dwelling' as his, or her, only or main residence,
- in the 3 year period prior to the 'effective date' (see **2.8**) of the transaction the purchaser (or the purchaser's spouse or civil partner) disposed of a 'major interest' (see **2.11**) in another dwelling which was the purchaser's only or main residence at any time during that 3 year period,
- at no time during the period between the disposal of the other dwelling and the acquisition of 'the purchased dwelling' has the purchaser (or the

purchaser's spouse or civil partner) acquired a 'major interest' (see **2.11**) in any other dwelling with the intention of it being the purchaser's only or main residence.

It is important to note that the three-year look back period does not apply for purchases of dwellings on or before 26 November 2018, ie the date which is three years after the introduction of the additional SDLT rate was announced by the Chancellor (*FA 2016, s 128(8)* and *(9)*). This is a transitional provision, which ensures that individuals who disposed of their main residence before the announcement of the higher rates on 25 November 2015 are not disadvantaged. Thus an individual who disposed of their main residence say, five years ago, will be treated as replacing their main residence where the 'purchased dwelling' is acquired on or before 26 November 2018. The three-year look back period will only apply to 'purchased dwellings' acquired on or after 27 November 2018.

It is only the first acquisition of a new main residence which counts as a replacement. Consequently, if two or more main residences are acquired within three years of a disposal, or on or before 26 November 2018, it will be the first main residence acquired which will be a replacement.

If an individual rents a dwelling between the disposal of their main residence and the purchase of the replacement this will not prevent the purchase being a replacement provided the rental period is not more than seven years (*FA 2003, Sch 4ZA, para 2(4)*).

Example 2.14

Isobel buys a dwelling which she intends to use as her main residence. The 'effective date' (see **2.8**) of the transaction is 31 January 2019. If Isobel (or her spouse or civil partner) had sold a dwelling at any time on or since 1 February 2016 then the acquisition may be a replacement of her main residence. The dwelling would have to have been Isobel's only or main residence at some time during the three-year period 1 February 2016 to 31 January 2019. She must not have acquired another new main residence after the disposal and before the acquisition on 31 January 2019.

If Isobel had acquired her new dwelling on 30 September 2016 then the three-year look back period would not apply and the requirement would simply be that she had disposed of a previous main residence at some time on or before 30 September 2016. She must still not have acquired another main residence after the disposal and before the purchase of the new main residence on 30 September 2016.

Disposal of the current main residence after the date of acquisition of the new main residence

2.53 Where the individual has not disposed of their current only or main residence by the end of the day on which they acquire 'the purchased dwelling', which they intend to be their only or main residence, 'the purchased dwelling' may become a replacement for their only or main residence if the following conditions are satisfied (*FA 2003, Sch 4ZA, para 3(7)*):

- the purchaser intends 'the purchased dwelling' to be his, or her, only or main residence; and
- in the three-year period starting on the day after the 'effective date' (see **2.8**) of the acquisition of 'the purchased dwelling' the purchaser (or the purchaser's spouse or civil partner) disposes of a 'major interest' (see **2.11**) in another dwelling which at any time during the three-year period ending with the 'effective date' (see **2.8**) of the acquisition of 'the purchased dwelling' was the purchaser's only or main residence.

Where a new dwelling does become a replacement for the purchaser's only or main residence because the purchaser's previous only or main residence is disposed of within the three-year period following the acquisition of 'the purchased dwelling', the land transaction return in respect of the acquisition of the 'purchased dwelling' may be amended to show that the additional rate of SDLT no longer applies, and the amount of the additional rate SDLT may be reclaimed (*FA 2003, Sch 4ZA, para 8(3)*). The amendment to the land transaction return must be made within the following periods, whichever expires later:

- the period of three months starting with the 'effective date' (see **2.8**) of the disposal of the property which was previously the purchaser's only or main residence; and
- the period of 12 months starting with the date by which the land transaction return, in relation to the acquisition of the 'purchased dwelling', had to be filed.

Where the new main residence is acquired first, but the disposal of the previous main residence takes place before the land transaction return is filed notifying the acquisition of the new main residence, that land transaction return should be filed on the basis that the higher rates do not apply, on the basis that the individual is replacing their only or main residence.

Example 2.15

Diana acquires a freehold interest in a new main residence on 30 June 2016 for a consideration of £450,000 but retains her freehold ownership of her previous main residence at the end of that day, and that property is worth £300,000. Neither the freehold interest acquired, nor the freehold interest retained, are subject to a lease with more than 21 years to run. The conditions set out in scenario 1 above are satisfied and the acquisition of the new freehold interest is taxed at the higher rates. The SDLT payable is as follows:

Purchase Price (£)	**Rate (%)**	**SDLT (£)**
125,000	3	3,750
125,000	5	6,250
200,000	8	16,000
------------		----------
450,000		26,000
------------		----------

If Diana sells her previous main residence at any time during the 3 year period to 30 June 2019 then the purchase of the new main residence will cease to be a 'higher rates transaction' and Diana can amend the SDLT return on the original transaction and claim back the additional rate SDLT paid. The SDLT payable if the additional rates do not apply is as follows:

Purchase Price (£)	**Rate (%)**	**SDLT (£)**
125,000	0	Nil
125,000	2	2,500
200,000	5	10,000
------------		----------
450,000		12,500
------------		----------

Diana can therefore claim a refund of £13,500 (£26,000–£12,500).

If a disposal of a previous main residence following the purchase of a new main residence results in the purchase of the new main residence not being chargeable to the SDLT higher rates, as described above, that disposal cannot then be used in deciding if a later purchase of a dwelling is a replacement of a main residence (*FA 2003, Sch 4ZA, para 8(2)*).

What is an individual's main residence?

2.54 There is no definition of only or main residence in the legislation. If the individual lives at only one dwelling that will be their only or main residence.

The HMRC Guidance Note (16 March 2016) indicates at 3.32 that where an individual lives at more than one dwelling all of the facts and circumstances of the case in question must be considered to determine which dwelling is the main residence. The main residence is not necessarily the dwelling where the individual spends the majority of their time although frequently it will be. At 3.34 of the HMRC Guidance Note (16 March 2016) there is a list of points to be considered in determining which dwelling is an individual's main residence:

- If the individual is married or in a civil partnership, where does the family spend its time?
- If the individual has children, where do they go to school?
- At which residence is the individual registered to vote?
- Where is the individual's place of work?
- How is each residence furnished?
- Which address is used for correspondence?
- Where is the individual registered with a doctor/dentist?
- At which address is the individual's car registered and insured?
- Which address is the main residence for council tax?

There are two parts which need to be tested in determining whether or not an individual is replacing their main residence. The first part relates to the dwelling which has been sold in the three-year period either side of the acquisition and it will be a question of objective fact as to whether or not that dwelling was the individual's only or main residence. The second part relates to 'the purchased dwelling'. In this case it will be a question of the intention of the purchaser at the time of the acquisition. At 3.36 of the HMRC Guidance Note (16 March 2016) it is stated:

> 'The intention test will not only be met if there is an intent to immediately occupy, if some works are to be undertaken before occupation commences then this does not prevent the test from being met. If the dwelling is intended to be put to other uses, for example as a source of income, then the intention test will not be met. There may be rare cases of the purchaser's genuine intention at the time of purchase being frustrated by events.'

Therefore the fact that the individual does not move into 'the purchased dwelling' because work is to be carried out on the property will not prevent a property from being an individual's main residence. However, if the initial intention is to let the property out the main residence test will not be satisfied.

Scenario 2 – single dwelling transaction (FA 2003, Sch 4ZA, para 4)

2.55 A 'chargeable transaction' (see **2.14**) is a 'higher rates transaction' if each of the following conditions are satisfied:

- the purchaser is not an individual (eg the purchaser is a company, the trustees of certain trusts, for example a discretionary trust (see **2.59**), etc);
- the 'chargeable interest' (see **2.11**) acquired consists of a 'major interest' (see **2.11**) in a single dwelling (which excludes any lease originally granted for a term of seven years or less), referred to as 'the purchased dwelling';
- the 'chargeable consideration' (see **2.19**) given for the transaction is £40,000 or more; and
- on the 'effective date' (see **2.8**) of the transaction 'the purchased dwelling' is not subject to a lease which at that date has an unexpired term of more than 21 years (for example, where a freehold interest is being acquired that interest is not subject to a lease which has more than 21 years to run at the 'effective date' (see **2.8**) of the transaction),

This means that the additional rate of SDLT will apply where, for example, on or after 1 April 2016, a company acquires a 'major interest' (see **2.11**) in a single dwelling for chargeable consideration of £40,000 or more irrespective of whether or not it holds any other dwellings, ie the additional rate of SDLT applies to first purchases, and all subsequent purchases, of dwellings by companies and other non-individuals. This assumes that the interest in the dwelling acquired is not subject to a lease with more than 21 years to run.

Scenario 3 – multiple dwelling transactions (FA 2003, Sch 4ZA, para 5)

2.56 A 'chargeable transaction' (see **2.14**) is a 'higher rates transaction' if each of the following conditions are satisfied:

- the purchaser is an individual;
- the 'chargeable interest' (see **2.11**) acquired consists of a 'major interest' (see **2.11**) in two or more dwellings, referred to as 'the purchased dwellings';
- the portion of the 'chargeable consideration' (see **2.19**) which is attributable on a just and reasonable basis to each of at *least two* of 'the purchased dwellings' is £40,000 or more;

- on the 'effective date' (see **2.8**) of the transaction at **least two** of 'the purchased dwellings' are not subject to a lease which at that date has an unexpired term of more than 21 years (for example, where a freehold interest is being acquired that interest is not subject to a lease which has more than 21 years to run at the 'effective date' of the transaction); and
- at least two of the purchased dwellings are not 'subsidiary' to any of the other purchased dwellings. For these purposes dwelling 'A' is subsidiary to dwelling 'B' if (a) dwelling 'A' is located within the grounds of, or within the same building as, dwelling 'B', and (b) the amount of the 'chargeable consideration' (see **2.19**) attributed to dwelling 'B' on a just and reasonable basis is greater than or equal to two thirds of the amount of the 'chargeable consideration' attributed, on a just and reasonable basis, and on a combined basis, to the acquisition of dwelling 'A' and 'B' together with any other purchased dwellings located within the grounds of, or within the same building as, dwelling 'B'. This is being referred to as the 'granny flat exception' and is designed to take self-contained dwellings, such as granny flats, etc out of the higher rate charge in appropriate circumstances.

Where two or more dwellings are acquired under a *single transaction* the legislation does not allow there to be a combination of higher rate and 'normal' residential rate transactions, and the transaction will either be a 'higher rates transaction' or it won't. Thus if two or more dwellings are purchased in the same transaction, and at least two of those dwellings satisfy the conditions detailed above, the transaction will be a 'higher rates transaction', and the higher rates will apply to all of the dwellings acquired under the transaction. This is irrespective of whether the individual owns an interest in another dwelling at the end of the day on which the acquisition is made, or whether one of the purchased dwellings replaces a main residence.

The legislation does not define what is meant by 'a single transaction' but it is considered that it means one contract between the same vendor and purchaser (and not including parties connected with the vendor and purchaser) for the acquisition of the agreed number of properties. Thus if two or more dwellings are to be acquired from the same vendor, and those dwellings satisfy the conditions above, and one of the dwellings to be acquired is replacing, or may become a replacement for, the individual's only or main residence, that dwelling should be acquired under a separate contract, which is legally independent from any of the other contracts, such that the higher rate should not apply to its acquisition (see **2.52**), or that in due course, on a disposal of the previous main residence, a subsequent claim can be made that the higher rates do not apply (see **2.53**).

It should be noted that Scenario 4 below may apply if only one of the dwellings acquired in a single transaction are acquired for a consideration of £40,000 or more.

Scenario 4 – multiple dwelling transactions (FA 2003, Sch 4ZA, para 6)

2.57 A 'chargeable transaction' (see **2.14**) is a 'higher rates transaction' if each of the following conditions are satisfied:

- the purchaser is an individual;
- the 'chargeable interests' (see **2.11**) acquired consists of a 'major interest' (see **2.11**) in two or more dwellings, referred to as 'the purchased dwellings';
- there is only one 'purchased dwelling' to which 'chargeable consideration' (see **2.19**) of £40,000 or more is attributed on a just and reasonable basis;
- on the 'effective date' (see **2.8**) of the transaction 'the purchased dwelling' acquired for £40,000 or more is not subject to a lease which at that date has an unexpired term of more than 21 years (for example, where a freehold interest is being acquired that interest is not subject to a lease which has more than 21 years to run at the 'effective date' of the transaction);
- the purchased dwelling in respect of which the conditions in the two bullet points immediately above are satisfied is not a replacement for the purchaser's only or main residence (see **2.52**);
- at least one of the 'purchased dwellings' must not be 'subsidiary' to any of the other purchased dwellings. For these purposes dwelling 'A' is subsidiary to dwelling 'B' if (a) dwelling 'A' is located within the grounds of, or within the same building as, dwelling 'B', and (b) the amount of the 'chargeable consideration' (see **2.19**) attributed to dwelling 'B' on a just and reasonable basis is greater than or equal to two thirds of the amount of the 'chargeable consideration' attributed, on a just and reasonable basis, and on a combined basis, to the acquisition of dwelling 'A' and 'B' together with any other purchased dwellings located within the grounds of, or within the same building as, dwelling 'B'. This is being referred to as the 'granny flat exception' and is designed to take self-contained dwellings, such as granny flats etc, out of the higher rate charge in appropriate circumstances; and
- at the end of the day which is the 'effective date' of the transaction the purchaser has, or is treated as having (see **2.45–2.49**), a 'major interest' (see **2.11**) in a dwelling, other than one of 'the purchased dwellings', which has a market value of £40,000 or more, and the interest in that other dwelling is not subject to a lease with an unexpired term of more than 21 years.

This scenario applies where the purchaser is an individual acquiring more than one dwelling in a single transaction but only one of the dwellings is acquired for chargeable consideration of £40,000 or more and that dwelling is not subject to a lease which has an unexpired term of more than 21 years. In these circumstances the transaction will be a 'higher rates transaction' only if the dwelling acquired for £40,000 or more is not a replacement for the purchaser's only or main residence and the purchaser already holds, or is treated as holding (see **2.45–2.49**) an unexpired term of more than 21 years.

It should be noted that Scenario 3 above may apply if more than one of the dwellings acquired in a single transaction are acquired for a consideration of £40,000 or more.

Scenario 5 – multiple dwelling transactions (FA 2003, Sch 4ZA, para 7)

2.58 A 'chargeable transaction' (see **2.14**) is a 'higher rates transaction' if each of the following conditions are satisfied:

- the purchaser is not an individual (e.g. the purchaser is a company, the trustees of certain trusts, for example, a discretionary trust (see **2.59**), etc),
- the 'chargeable interest' (see **2.11**) acquired consists of a 'major interest' (see **2.11**) in two or more dwellings, referred to as 'the purchased dwellings',
- the 'chargeable consideration' (see **2.19**) which is attributable on a just and reasonable basis to at least one of 'the purchased dwellings' is £40,000 or more,
- on the 'effective date' (see **2.8**) of the transaction at least one of 'the purchased dwellings' is not subject to a lease which at that date has an unexpired term of more than 21 years (for example, where a freehold interest is being acquired that interest is not subject to a lease which has more than 21 years to run at the 'effective date' of the transaction),

Under this scenario, the additional rate of SDLT will apply if, for example, in a single transaction, on or after 1 April 2016, a company acquires a 'major interest' (see **2.11**) in two or more dwellings and the 'chargeable consideration' attributed, on a just and reasonable basis, to at least one of those dwellings is £40,000 or more and at least one of the dwellings is not subject to a lease which has an unexpired term of more than 21 years. The higher rate will apply irrespective of whether or not the company holds any other dwellings.

The differences between Scenarios 2 and 5 are that in Scenario 5 more than one dwelling is being acquired and the 'chargeable consideration' is attributed to the dwellings on a just and reasonable basis, whereas in Scenario 2 there is no need for an attribution of the 'chargeable consideration' as only a single dwelling is acquired.

Trustees purchasing dwellings

2.59 The trustee(s) of a 'bare trust' (*FA 2003, Sch 16, para 1(2)*, ie a trust under which the beneficiary is absolutely entitled to the interest in the dwelling as against the trustee or would be absolutely entitled but for disability or age which prevents the beneficiary from holding legal title) is ignored and the beneficiary is treated as the purchaser for the purposes of the additional rates legislation (*FA 2003, Sch 16, para 3(1)*, and *Sch 4ZA, para 10(2)* and *(3)*).

The trustee(s) of a trust under which the beneficiary is entitled to (a) occupy the dwelling for life, or (b) income earned in respect of the dwelling (referred to below as a 'trust for life or income'), is also ignored, meaning that, for the purposes of the higher rates legislation, the beneficiary with the relevant interest in the trust is treated as the purchaser, and the trustee is treated as not being the purchaser (*FA 2003, Sch 4ZA, para 10(1)* and *(3)*). For these purposes it does not matter whether one, or more than one, beneficiary is entitled as described in (a) and (b) above. Consequently, under Scenario 1 considered at **2.51**, if the beneficiary of a 'trust for life or income' owns another interest in a dwelling at the end of the day on which the new dwelling is acquired by the trustees, and that interest is worth £40,000 or more, and is not subject to a lease with more than 21 years to run, then the trustee's acquisition of the dwelling will be a 'higher rates transaction' unless the trustee's purchase is replacing the beneficiary's main residence (**2.52** et seq).

The trustee(s) of a trust which is not a 'bare trust' or 'a trust for life or income' is treated as the purchaser of the 'major interest' (see **2.11**) in the dwelling, and the transaction will be a 'higher rates transaction', if the conditions in scenario 2 or 5 above are satisfied (*FA 2003, Sch 4ZA, para 13*) (note that if the trustee(s) is an individual the condition that the purchaser is not an individual is ignored in determining whether the conditions in scenario 2 or 5 above are satisfied (*FA 2003, Sch 4ZA, para 13(2)*). For example, for the purposes of the higher rates legislation, the trustees of a discretionary trust will be treated as the purchaser of any 'major interest' in a dwelling acquired by them, in their capacity as trustees of that trust, and the transaction will be a higher rates transaction if the conditions in either scenario 2 or 5 (see **2.55** and **2.58**) are satisfied.

Example 2.16

Cameron and Fraser are trustees of a trust and on 1 September 2016 they use the trust funds to purchase a dwelling for £300,000. The terms of the trust entitle an individual beneficiary, Amanda, to occupy the dwelling for life. Cameron and Fraser must consider whether the higher rates of SDLT would apply as if Amanda were purchasing the dwelling. Consequently, if Amanda has a 'major interest' in another dwelling on the 'effective date' (see **2.8**) of the acquisition of the new dwelling by the trustees, and the new dwelling is not a replacement of Amanda's main residence, then the higher rates will apply to the purchase of the new dwelling by the trustees.

The trustee(s) of a trust which is not a 'bare trust' or 'a trust for life or income' is treated as the purchaser of the 'major interest' (see **2.11**) in the dwelling, and the transaction will be a 'higher rates transaction', if the conditions in scenario 2 or 5 above are satisfied (*FA 2003, Sch 4ZA, para 13*) (note that if the trustee(s) is an individual the condition that the purchaser is not an individual is ignored in determining whether the conditions in scenario 2 or 5 above are satisfied (*FA 2003, Sch 4ZA, para 13(2)*).

Example 2.17

Blair and Nicola are trustees of a trust which enables them to apply the income from the trust's assets at their discretion amongst a class of beneficiaries. Using trust funds, Blair and Nicola acquire a freehold interest in a dwelling on 25 November 2016 for a consideration of £850,000. The trustees are liable to pay the higher rates of SDLT if the conditions outlined in Scenario 2 (see **2.55**) (*FA 2003, Sch 4ZA, para 4*) are satisfied. Those conditions are as follows:

- the purchaser is not an individual – Blair and Nicola are purchasing the dwelling as trustees of the trust and, under the terms of the trust no beneficiary is entitled to (a) occupy the dwelling for life, or (b) the income earned in respect of the dwelling, and the trust is not a 'bare trust'. Consequently, *FA 2003, Sch 4ZA, para13(1)* and *(2)* requires that the condition that the purchaser is not an individual be ignored in determining whether the transaction is liable to the higher rates of SDLT;
- the 'chargeable interest' (see **2.11**) acquired consists of a 'major interest' (see **2.11**) in a single dwelling – this condition is satisfied as a freehold interest in the dwelling is being acquired;
- the 'chargeable consideration' (see **2.19**) given for the dwelling is £40,000 or more – this condition is satisfied as the consideration given for the dwelling is £850,000; and
- on the 'effective date' (see **2.8**) of the acquisition of the dwelling it is not subject to a lease with more than 21 years to run – this condition is satisfied as the trustees are acquiring the unencumbered freehold interest in the property.

All of the conditions in *FA 2003, Sch 4ZA, para 4* are therefore satisfied and Blair and Nicola, acting as trustees of the trust, must pay SDLT at the higher rates and the liability is calculated as follows:

Purchase Price (£)	Rate (%)	SDLT (£)
125,000	3	3,750
125,000	5	6,250
600,000	8	48,000
------------		----------
850,000		58,000
------------		----------

The trustees must therefore file a land transaction return, and pay the SDLT liability, by 25 December 2016, entering code '04' in question 1 of the land transaction return denoting that the higher rate of SDLT applies to the transaction.

Interaction with multiple dwellings relief

2.60 Where two or more dwellings are acquired as part of a single transaction, or under a number of 'linked transactions' (see **2.64**), it may be possible to reduce the aggregate SDLT liability by making a claim for multiple dwellings relief (see **3.48**). The higher rates of SDLT will apply to claims for multiple dwellings relief (see **3.48**) if the relevant conditions in any of scenarios 1–5 (see **2.51–2.58**) apply. If, as part of a single transaction, six or more dwellings are acquired, the SDLT rates for non-residential or mixed use transactions will apply (see **2.38**). In such circumstances the purchaser can therefore choose whether to apply the non-residential rates by not making a claim for multiple dwellings relief. However, see Example 2.20 where a claim for multiple dwellings relief is made in relation to linked transactions.

Example 2.18

Ewan purchases a block of eight flats for £750,000. If a claim for multiple dwellings relief is made, the average consideration per flat would be £93,750 and the SDLT payable would be £22,500 (average price of £93,750 x 3% x 8 flats – a 3% SDLT rate applies to the purchase of a dwelling for £93,750 as the additional SDLT rate applies). If no claim is made for multiple dwellings relief then the non-residential SDLT rates will apply (six or more dwellings are being acquired as part of a single transaction) as follows:

Purchase Price (£)	Rate (%)	SDLT (£)
150,000	0	Nil
100,000	2	2,000

Purchase Price (£)	**Rate (%)**	**SDLT (£)**
500,000	5	25,000
------------		----------
750,000		27,000
------------		----------

In this case it would be beneficial to make a claim for multiple dwellings relief as this would reduce the SDLT liability from £27,000 to £22,500.

Filing a return to which the higher SDLT rate applies

2.61 As for any other 'chargeable transaction' (see **2.14**) a return for a 'higher rates transaction' must be filed within 30 days of the 'effective date' (see **2.8**) of the transaction (see **6.6**). Question 1 on form SDLT 1, on 'type of property', should be completed as code '04' which means that the higher rate of SDLT applies. Payment of the SDLT is made in exactly the same way as that for any other 'chargeable transaction' (see **6.29**).

Claiming a repayment of higher rate SDLT

2.62 If a dwelling previously used by a purchaser as their main residence is sold within three years of the 'effective date' (see **2.8**) of the acquisition of a new main residence a refund of the amount of the additional rate of SDLT can be claimed (see **2.53**). The repayment needs to be reclaimed by the later of (a) three months from the date of sale of the previous main residence, and (b) one year from the date on which the land transaction return notifying the acquisition of the new main residence had to be filed (*FA 2003, Sch 4ZA, para 8(3)*). The online SDLT calculator at www.gov.uk can be used to calculate the amount of the refund.

The refund can be claimed by completing an SDLT repayment request form available on www.gov.uk. Currently, the repayment request form needs to be completed online and the summary printed and posted to HMRC Birmingham Stamp Office (see **Appendix A**). It is understood that in due course a form will be made available which can be submitted online.

The following information will be required to complete the SDLT repayment request form:

- the SDLT UTRN for the land transaction return notifying the purchase of the new main residence, and
- the name of the purchaser of the property which has now been sold.

If the SDLT repayment request form is completed by an agent, and the refund is to be paid to that agent, then a signed letter of consent from the purchaser should be attached to the form.

HMRC aims to process all refunds within 15 working days of receiving the repayment request form. HMRC will issue the refund by payable order, or will issue a letter to explain why the refund request has not been successful.

Transitional rules

2.63 If the relevant conditions are satisfied the higher rates of SDLT will apply to all 'land transactions' (see **2.7**) with an 'effective date' (see **2.8**) on or after 1 April 2016 (*FA 2016, s 117(5))*.

However, the higher rates of SDLT do not apply where the contract for the 'land transaction' (see **2.7**) was (a) entered into and 'substantially performed' (see **2.9**) before 26 November 2015, or (b) entered into before 26 November 2015 and not excluded as described below (*FA 2016, s 117(6)*).

For these purposes a contract entered into before 26 November 2015 is excluded, and therefore the higher rates of SDLT apply, if the following conditions are satisfied:

- there is a variation of the contract, or assignment of rights under the contract, on or after 26 November 2015;
- the transaction is effected in consequence of the exercise, on or after 26 November 2015, of any option, right of pre-emption or similar right; or
- on or after 26 November 2015 there is an assignment, sub-sale or other transaction relating to the whole or part of the land interest acquired under the contract as a result of which a person other than the purchaser under the original contract becomes entitled to call for a conveyance.

For these purposes a variation to a contract would include a change to:

- the land being purchased;
- the parties to the contract;
- the contractual consideration; or
- the term length in an agreement for a lease.

However, some changes, for example a change to the contractual completion date, should be considered too insignificant to be treated as a variation to the contract.

Example 2.19

Hamish exchanged contracts with a construction company on 11 June 2015 for the purchase of four flats, for a consideration of £1,000,000, which he intends to let to third parties. He also jointly owns a property with his wife, which is his only residence. On 11 August 2016 Hamish varies the contract to add his wife as a joint purchaser of the flats. Despite the fact that the contract was entered into before 26 November 2015 Hamish will not be able to benefit from the transitional rules as the contract was varied after 25 November 2015 (*FA 2016, s 117(6)(b)* and *(7)(a)*). The higher rates of SDLT will therefore apply to the transaction.

Linked transactions

2.64 Transactions are linked if they form part of a single scheme, arrangement or series of transactions between the same vendor and purchaser or, in either case, persons connected with them (*FA 2003, s 108:* see **2.67**). Connected persons are as defined in *CTA 2010, s 1122*. The HMRC Guidance Note (16 March 2016) makes it clear that if an individual acquires two or more dwellings in a *single transaction* the legislation does not permit the acquisition of one dwelling to be charged at the normal rates and the acquisition of another dwelling to be charged at the higher rates. However, neither the legislation nor the HMRC Guidance (16 March 2016) makes it clear how linked transactions are treated if, for example, one transaction is chargeable at normal rates and the second transaction at higher rates. The calculation is further complicated if a claim for multiple dwellings relief is made in relation to the linked transactions. It is understood however that the SDLT liability should be calculated as set out below in Example 2.20.

Example 2.20

On 11 March 2017, under separate contracts, Campbell agrees to acquire two houses from a developer one of which he intends to use as his main residence and the other he will let out to third parties. On 14 August 2017 Campbell sells his previous main residence and completes the contract for the purchase of his new main residence for a consideration of £450,000. On 24 November 2017 Campbell completes the purchase of the buy-to-let property for a consideration of £550,000.

Campbell must file a land transaction return, and pay the SDLT, on the acquisition of his new main residence by 13 September 2017. As Campbell is replacing his main residence the normal SDLT rates apply and the liability is calculated as follows:

Purchase Price (£)	Rate (%)	SDLT (£)
125,000	0	Nil
125,000	2	2,500
200,000	5	10,000
------------		----------
450,000		12,500
------------		----------

Following completion of the acquisition of the buy-to-let property on 24 November 2017 Campbell must file a land transaction in relation to that transaction, by 24 December 2017, and pay SDLT at the higher rates as the dwelling is being acquired to let and Campbell is not replacing his main residence. As the purchase of the main residence and the buy-to-let properties are 'linked transactions' (see **2.41**) the SDLT liability is calculated as follows:

Aggregated Purchase Price (£)	Rate (%)	SDLT (£)
125,000	3	3,750
125,000	5	6,250
675,000	8	54,000
75,000	13	9,750
------------		----------
1,000,000		73,750
------------		----------

The SDLT payable in relation to the buy-to-let property would therefore be £40,562 (£550,000/£1,000,000 × £73,750).

The revised calculation of the SDLT liability in relation to the new main residence, using the normal rates of SDLT, would be as follows:

Aggregated Purchase Price (£)	Rate (%)	SDLT (£)
125,000	0	Nil
125,000	2	2,500
675,000	5	33,750
75,000	10	7,500

Aggregated Purchase Price (£)	Rate (%)	SDLT (£)
------------		----------
1,000,000		43,750

The amended SDLT liability in relation to the acquisition of the new main residence would therefore be £19,687 (£450,000/£1,000,000 × £43,750). The total SDLT payable would therefore be £60,249 (£40,562 + £19,687).

However if Campbell makes a claim for multiple dwellings relief it would seem that the SDLT liability should be calculated as set out below.

The average consideration per house would be £500,000 and the SDLT payable at the higher rates would be as follows:

Purchase Price (£)	Rate (%)	SDLT (£)
125,000	3	3,750
125,000	5	6,250
250,000	8	20,000
-----------		----------
500,000		30,000
-----------		----------

The total SDLT payable would therefore be £60,000 (£30,000 × 2 houses) and the amount payable on the buy-to-let property would be £33,000 (£550,000/£1,000,000 × £60,000).

The SDLT payable at the normal rates based on an average consideration per house would be as follows:

Purchase Price (£)	Rate (%)	SDLT (£)
125,000	0	NIL
125,000	2	2,500
250,000	5	12,500
-----------		----------
500,000		15,000
-----------		----------

The total SDLT payable would therefore be £30,000 (£15,000 × 2 houses) and the amount payable on the replacement main residence would be £13,500 (£450,000/£1,000,000 × £30,000).

The total SDLT payable if a multiple dwellings relief claim is made would therefore be £46,500 (£33,000 + 13,500), which compares to a liability of £60,249 if no claim for multiple dwellings relief is made. It is therefore beneficial for Campbell to make a multiple dwellings relief claim.

Campbell must also file an amended land transaction return by 24 December 2017, in relation to the purchase of his new main residence and pay the additional SDLT due of £1,000 (£13,500 – £12,500).

He must also file a land transaction return by 24 December 2017 in relation to the purchase of the buy-to-let property and pay the SDLT due of £33,000.

The above is the author's understanding of how the calculation should be carried out in relation to linked transactions; however, until the matter is clarified by HMRC, it is suggested that confirmation of the calculation methodology be obtained from HMRC.

Enveloped residential properties

2.65 A 15% rate applies where an interest in a single dwelling is acquired, for actual or deemed chargeable consideration of more than the threshold, by a company, by a partnership whose partners include one or more companies, or on behalf of a collective investment scheme. Initially the threshold was £2 million for acquisitions with an effective date of 21 March 2012 or later. *FA 2014, s 111* lowered the threshold to £500,000 for acquisitions with an effective date of 20 March 2014 or later. In both cases transitional provisions protect transactions under contracts entered into before the relevant date, subject to stringent anti-avoidance conditions.

'Dwelling' is defined so as to exclude buildings such as student halls of residence, care homes and the like. There are provisions for apportionment where a single transaction includes interests in dwellings and other properties. It is not possible to take advantage of the relief for transfers of multiple dwellings to reduce the rate charged (see **3.48**). Anti-avoidance provisions do seek to prevent avoidance of the higher rate by fragmenting the purchase (*s 55A* and *Sch 4A*, inserted into *FA 2003* by *FA 2012, Sch 34*).

It seems that *FA 2003, s 116(7)* – reclassification of residential property as non-residential where six or more properties are transferred in a single transaction – has no application in determining whether the 15% rate applies to a transaction. This is because the 15% rate applies to the acquisition of an interest in a single dwelling for a chargeable consideration of more than £500,000, and it is irrelevant whether the acquisition of that single dwelling is part of a single transaction under which five or more other separate dwellings are acquired, such that s116(7) would reclassify the transaction as an acquisition of non-residential property.

The stated purpose of this provision is to discourage the 'enveloping' of high value residential properties within corporate shells, a practice which allows future transfers free of SDLT. As such, it may be regarded as an anti-avoidance provision. The provision had no direct effect on the large number of properties already held in this way at 21 March 2012. However, in addition to the one-off charge on purchase of the property, an annual charge on such properties was introduced from 1 April 2013 – the Annual Tax on Enveloped Properties (ATED). This applies to properties already in corporate ownership before the provisions were enacted as well as future acquisitions. A brief outline of this new tax is set out at **2.72** – more detail can be found in *Property Taxes 2015/16* (Bloomsbury Professional).

2.66 When the 15% rate was first introduced, the only specific exemptions were for a purchase by a company or partnership in the course of a bona fide property development business which it (or a group company) had carried on for at least two years prior to the purchase and for a purchase by a company acting as trustee of a settlement. Bodies already qualifying for relief from SDLT – for example charities, provided the property is acquired for charitable use – automatically escaped the charge. However concern was expressed that this essentially anti-avoidance measure penalised legitimate investment in residential property. In the 2012 Finance Bill debates, the government said that the Treasury was in conversations with 'a number of affected interested parties' and that these conversations would inform the coverage of the annual charge and, possibly, of the 15% rate itself.

Finance Act 2013, s 196 and *Sch 40* amended the relief for property developers and introduced a range of further reliefs. With effect from 17 July 2013 relief is provided for acquisitions by the following (references are to paragraphs of *FA 2003, Sch 4A*, amended or inserted by *FA 2013, Sch 40*):

- property traders and developers or property rental business, with no requirement for the business to have been carried on for any particular period *(para 5)*;
- businesses using the property in the course of a trading activity which includes opening the property to the public *(para 5B)*;
- financial institutions provided the dwelling was acquired in the course of a lending activity and is to be resold *(para 5C)*;
- properties occupied by certain employees or partners working in a trade or property rental business (see *FA 2016, s 131(2)*) carried on by the property owner (or a group company), provided the employee/partner has a less than 10% interest in the business *(para 5D)*;

- a farmhouse occupied by someone working on the farm (or a former long-serving farm worker) *(5F)*.

Inevitably there are detailed conditions governing the reliefs, in an attempt to ensure that they cannot be misused. The reliefs parallel those available in relation to the annual tax on residential properties known as annual tax on enveloped dwellings (ATED) (see **2.72**). However the reliefs from ATED apply from the inception of the tax. It was difficult to see why the reliefs from the 15% rate were not made retrospective so that they, too, applied from the introduction of that rate. This matter was queried during the Finance Bill debates but the Exchequer Secretary to the Treasury, David Gauke, effectively sidestepped the question, using the possibility of the reliefs being amended during the passage of the Bill through Parliament as an excuse for delaying their application until the Finance Act received Royal Assent.

LINKED TRANSACTIONS

2.67 If two or more transactions are linked, the consideration for all of the linked transactions is aggregated to determine the rate of SDLT. Transactions are linked if they form part of a single scheme, arrangement or series of transactions between the same vendor and purchaser or, in either case, persons connected with them (*FA 2003, s 108*). Connected persons are as defined in *CTA 2010, s 1122* (previously *ICTA 1988, s 839*). Whether transactions are linked is a question of fact, but transactions between the same vendor and purchaser which occur at similar times are likely to be presumed to be linked unless there is evidence to the contrary. In *Attorney General v Cohen* [1936] 2 KB 246, [1936] 1 All ER 583, several purchases of properties by the same purchaser from the same vendor at auction were held not to be part of a series of transactions, because each property was the subject of a separate bidding process.

In the case of linked acquisitions of residential properties the amount of the SDLT liability may then be reduced if multiple dwellings relief is claimed (see **3.48**).

Example 2.21 – linked transactions

Over coffee, Tanya agrees to buy two houses, located in different towns, from Gerald – the first for £160,000, and the second for £400,000. Different solicitors are engaged to deal with the two purchases, and separate contracts are entered into for each purchase. Tanya already owns a property which is her main residence and is buying the two houses as buy-to-let properties. The acquisition of the first house is completed on 10 January 2017 without prior substantial performance, but the second is delayed, while old Land Registry entries are investigated, and does not complete (again, without prior substantial performance) until 15 June 2017.

Tanya must file a land transaction return in relation to the purchase of the first house by 9 February 2017 and pay SDLT, based on the higher rates as she is not replacing her main residence, calculated as follows:

Purchase Price (£)	Rate (%)	SDLT (£)
125,000	3	3,750
35,000	5	1,750
-----------		----------
160,000		5,500
-------		-------

Following completion of the acquisition of the second house on 15 June 2017, Tanya must file a land transaction in relation to that transaction by 15 July 2017, and pay SDLT at the higher rates as the dwelling is being acquired for the purposes of buy-to-let. As the purchase of the two houses are 'linked transactions' (see **2.67**) the SDLT liability is calculated as follows:

Aggregate Purchase Price (£)	Rate (%)	SDLT (£)
125,000	3	3,750
125,000	5	6,250
310,000	8	24,800
-----------		----------
560,000		34,800
-------		----------

The SDLT payable in relation to the second house is therefore £24,857 (£400,000/£560,000 × £34,800).

Tanya must also file an amended land transaction return by 15 July 2017, in relation to the purchase of the first house with the amended SDLT liability calculated at £9,943 (£160,000/£560,000 × £34,800). Tanya must therefore pay additional SDLT of £4,443 (£9,943–£5,500).

However if Tanya claims multiple dwellings relief (see **3.48**) the aggregate SDLT liability then falls to £24,800 with SDLT of £7,086 payable on the first house and £17,714 payable on the second.

Example 2.22 – transactions not linked

Gerald attends an auction where he successfully bids for two houses in the same street, which are sold as separate lots. Gerald pays £110,000 for one and £115,000 for the other. Gerald already owns another property which he uses as his main residence and he acquires the two houses as buy-to-let properties. After the auction, he discovers the vendor is the same for both houses. The transactions are not linked, because each purchase was agreed totally independently of the other. SDLT is payable on both purchases at a rate of 3% as the higher rates of SDLT apply since Gerald is not replacing his main residence.

TIMING OF OBLIGATIONS

2.68 In general, payment of any SDLT must be made and SDLT returns must be submitted within 30 days of the effective date of the transaction. The processes for making a return and payment, together with details of the circumstances in which further returns may be needed, are set out in **Chapter 6**.

The effective date

2.69 There is more than one statutory attempt to define the effective date, covering separate specific circumstances, but they all seek to achieve the same overall result (*FA 2003, ss 119, 44(4)*). In essence, the effective date for a transaction is the earlier of:

- the date of legal completion – that is, the date of execution of the conveyance, lease or other document giving the transaction final legal effect, and
- the date on which the transaction is 'substantially performed'. This date depends on the nature of the transaction and is the earlier of the following:
  - the date on which a substantial amount of any consideration is paid. If the consideration includes payment of rent, this is the earlier of the date on which the first payment of rent is made and the date on which a substantial amount of any other consideration is paid. For consideration other than rent, a 'substantial amount' means substantially the whole. Although not set out in the statute, HMRC considers that 90% or more is substantially the whole, but reserves the right to argue for a smaller proportion if they believe avoidance of SDLT is involved; and

- the date on which the purchaser takes possession of the whole or substantially the whole of the property. 'Possession' is not comprehensively defined, but is considered further below.

Particular care is needed when non-monetary consideration is given. For example in an exchange, each person's purchase may be substantially performed at the earlier of taking possession of their purchase or completing the sale of the other property.

Example 2.23 – sale and leaseback timing

Mark agrees to sell his freehold workshop to Luke. In return, Luke agrees to redevelop the site and grant Mark a 100-year lease of a new workshop in the new development. The 100-year lease will be granted for a peppercorn rent and no premium. The development will include new offices, which Luke will lease to third parties to pay for the overall redevelopment and (he hopes) give him a profit. The sale to Luke is completed on 1 March 2017 without prior substantial performance and Mark moves his activities into temporary accommodation. The development is completed and Mark's new lease is granted on 1 May 2018. Clearly Luke must make an SDLT return and pay tax on his purchase of the freehold by 31 March 2017. Mark does not receive the benefit of his lease until 1 May 2018. However he has paid the full consideration for this lease on 1 March 2017, by transferring the freehold to Luke. Therefore his agreement for lease was substantially performed on that date and he too must make an SDLT return and pay tax by 31 March 2017. As noted at **3.41,** HMRC do not consider that sale and leaseback relief is available because of the redevelopment between the sale and the leaseback. The requirement to complete a return and pay tax so early is not entirely reasonable and the author has experience of HMRC agreeing in this situation to have regard to the 'possession' test and accept that the effective date should be the date of grant of the leaseback. However that would appear to be a concession and should not be relied on without specific agreement from HMRC. It would be safer to submit the return and pay the tax within 30 days after the transfer of the freehold.

Possession

2.70 The term 'possession' has a long legal history and cannot readily be comprehensively defined. The following explanation sets out the view which HMRC are understood to take. The purchaser takes possession of the property when he first occupies or uses it, or becomes entitled to occupy or use it. This is not restricted to the time when the purchaser finally moves in, but may include, for example, having the right to enter the property for the purpose of fitting out. However, a person also has possession if he receives rents or other profits from the land, or has the right to receive them. It does not matter whether any of these occurs under the terms of the contract governing the transaction, or under some other arrangement such as a temporary licence or lease. However, if the purchaser already has possession of the property under one interest before entering into a land transaction to acquire a different interest in the same property, the prior possession will not itself cause the later transaction to be substantially performed (SDLTM07900).

Example 2.24 – occupation under previous interest

William occupies an office building under a lease which has 25 years to run. The current rent is £100,000 per year, payable quarterly in advance. His landlord is Vera, who owns the freehold. William agrees to buy the freehold from Vera for £750,000, subject to satisfactory structural survey etc. The agreement provides that William will pay a deposit of £75,000 on signing, a further instalment of £300,000 within 10 days of delivery of the structural survey, and the balance of £375,000 on final completion. The agreement also provides that the final instalment will be reduced by any rent which William pays and which relates to the period after delivery of the structural survey. William occupies and therefore has possession of the property immediately after the agreement is signed. However, he occupies under the terms of the existing lease and not as a result of entering into the agreement to buy the freehold. Therefore, the agreement is not regarded as substantially performed by reason of William's occupation. 50% of the purchase price is paid prior to completion, or possibly slightly more because of the adjustment of the final instalment to allow for rent paid; since this is less than 90%, the agreement will not be regarded as substantially performed until the remaining 50% is paid, on completion.

Example 2.25 – receipt of rents as substantial performance

The same fact pattern applies as in the previous example, except that William's property-holding company Uppercase Ltd, rather than William himself, agrees to buy the freehold from Vera. As part of the agreement, Uppercase Ltd is entitled to receive all rent payable on the property from the date the structural survey is delivered and the £300,000 is paid. The agreement will be regarded as substantially performed no later than the date the structural survey is delivered and the £300,000 is paid, because at that time Uppercase Ltd will become entitled to receive rents.

Transitional provisions

2.71 SDLT was introduced on 1 December 2003, and transactions commenced from that date are wholly within the SDLT rules. Where the contract was entered into on or before 30 November 2003, the transaction may be subject to stamp duty, SDLT or both, as set out below (*FA 2003*, *Sch 19*). *FA 2003* received Royal Assent on 10 July 2003:

(1) **Contract entered into on or before 10 July 2003**: Subject to stamp duty only, no matter when substantially performed or completed. However, if the contract is varied or assigned after 10 July 2003 or given effect as a result of the exercise of an option or similar right after that date, it is effectively treated as entered into on the date of variation/ assignment/ exercise.

(2) **Contract entered into after 10 July 2003**:

  (a) **Contract completed on or before 30 November 2003**: subject to stamp duty only;

  (b) **Contract completed after 30 November 2003**: subject to SDLT, but the contract may also have been subject to stamp duty under *FA 2002*, *s 115*, in which case credit is given for any stamp duty paid.

For these transactions liability to stamp duty or SDLT is not affected by when the transaction was 'substantially performed', as this phrase has no meaning in the context of stamp duty. However, this does affect the 'effective date' and therefore the date on which SDLT is payable. If the transaction is subject to SDLT but the contract was substantially performed before 1 December 2003, the effective date is the date of completion. If the contract is substantially performed after 30 November 2003, the effective date is, as normal, the earlier of substantial performance and completion.

Although SDLT is almost 13 years old, there may still be uncompleted transactions to which these rules may apply. Particular care may be needed in unwinding old structures which may have been set up for stamp duty planning purposes, see **11.26** et seq.

Example 2.26 – application of transitional provisions

On 1 June 2001, Propco Ltd contracted to buy a commercial property Q which was under construction, for £9.5 million. The vendor had previously bought the site using the 'resting on contract' planning (see **11.29**), so title was held by two nominee companies. On 1 August 2003, Propco assigned the purchase contract to its parent company, Holdco. Continuing the planning, Holdco paid the purchase price to the vendor on 30 September 2003 and took a transfer of the shares in the nominee companies. As a result, the transaction had been substantially performed within the SDLT meaning of that term, but not completed and no stamp duty had been paid.

On 1 March 2016, as part of an exercise to 'tidy up' the Holdco group before a stock market listing, title to site Q is transferred from the nominee companies to Holdco. This effectively completes the original purchase contract entered into in 2001 and, but for the SDLT transitional provisions, would give rise to a stamp duty charge.

Although the contract was entered into before 11 July 2003, it was assigned after that date and so is treated as having been entered into on the date of assignment (1 August 2003), which is after 10 July 2003. It was completed after 30 November 2003, and so is within (2)(b) in **2.71**. The effective date is 1 March 2016 (date of completion) and SDLT is payable, and a return due, on or before 31 March 2016.

ANNUAL TAX ON ENVELOPED DWELLINGS (ATED)

2.72 *Finance Act 2013, Pt 3* (*ss 94–174* and *Schs 33–35*) introduced a new tax known as annual tax on enveloped dwellings, or ATED. A brief account of this tax is included here because:

- the tax was introduced partly in response to the alleged avoidance of SDLT by transferring shares of companies which held residential property;
- the rules for administration of the tax have been derived from those relating to SDLT; and
- administration of the tax has become the responsibility of the section of HMRC which also deals with SDLT.

The following description is no more than an outline, primarily to allow readers to determine whether there is a risk that the tax may apply to a particular situation. A more detailed, in-depth explanation may be found in *Property Taxes 2016/17* (Bloomsbury Professional). Section and schedule references in this part are to *Finance Act 2013* unless otherwise indicated.

2.73 The key attributes of ATED are as follows:

1. the tax applies when an interest in a single dwelling, with a value of more than the threshold, is held by a company, partnership or collective

investment scheme at any time in the 'chargeable period'. It applies to a partnership only if the partners include one or more companies;

2. the tax applies to 'chargeable periods', being years from 1 April to 31 March. The first chargeable period for properties with a value of more than £2 million was that commencing 1 April 2013. The first chargeable period for lower value properties commenced on 1 April 2015 in relation to properties with a value of more than £1 million but not more than £2 million and from 1 April 2016 to properties with a value of more than £500,000 but not more than £1 million;
3. the amount of tax payable for each chargeable period depends on:
   a. the value of the interest in the dwelling; and
   b. whether the interest is held for the whole chargeable period;
4. it is a self-assessment tax;
5. there are reliefs for properties held by certain entities;
6. there are anti-avoidance provisions; and
7. interest, penalty and enforcement rules apply and are similar to those applying to SDLT.

Single dwelling

2.74 There is a detailed definition of 'single dwelling interest' (*s 108*), designed to prevent mitigation of the tax. If more than one dwelling is held, each must be evaluated separately. Specific rules attempt to deal with single or linked buildings which may be regarded as a single dwelling or more than one dwelling (*s 112, s 116, s 117*). There is no provision equivalent to SDLT multiple dwellings relief (see **3.48**) to average the values of the properties.

Amount of tax

2.75 The tax is based on the market value of the interest held, initially on 1 April 2012 or on the first day on which the tax applies to the company, etc if later. The market value must then be reappraised every fifth anniversary of 1 April 2012, although the reappraised valuation does not apply in determining the tax payable for the chargeable period beginning with that five-yearly valuation date (*FA 2003, s 102(2A),* inserted by *FA 2015, s 71).* This is to ensure that there is sufficient time to obtain the five-yearly valuation before the ATED return has to be filed, and applies for chargeable periods beginning on or after 1 April 2015. However if there is an acquisition or part-disposal with consideration of £40,000 or more this also gives rise to the need to reappraise the market value (*s 102, s 103*).

The amounts of tax shown in the table below are those which apply for the years commencing 1 April 2015 (*FA 2015, s 70*) and 2016. It is specifically provided that the ATED charge will be increased each year in line with increases in the Consumer Price Index to the previous September, rounded down to the next £50 (*s 101*). However, the government raised the ATED charges for the year commencing on 1 April 2015, for properties with a value of more than £2 million, by 50% above inflation. Consequently, the annual increase of the ATED charge, in line with any increase in the Consumer Price Index under *s 101*, did not apply in relation to the chargeable period beginning on 1 April 2015. The change in the Consumer Price Index to September 2015 was negative and therefore the ATED charges for 2016/17 are unchanged from the 2015/16 charges.

Value	**Tax**	**Tax**
More than £500,000 but not more than £1 million[a]	–	£3,500
More than £1 million but not more than £2 million[b]	£7,000	£7,000
More than £2 million but not more than £5 million	£23,350	£23,350
More than £5 million but not more than £10 million	£54,450	£54,450
More than £10 million but not more than £20 million	£109,050	£109,050
More than £20 million	£218,200	£218,200

(a) Charge first applied for periods commencing 1 April 2016 or later, see *FA 2014, s 110*

(b) Charge first applied for periods commencing 1 April 2015 or later, see *FA 2014, s 109*

Where the tax applies for only part of the chargeable period, for example because the property is acquired or disposed of partway through the year, the amount payable is pro-rated on a daily basis. Tax is initially payable on the assumption that the property will be held from the start of the year or later date of acquisition until the end of the year.

If the property is disposed of before the end of the year (or if its value falls so as to reduce the tax, or if a relief begins to apply), it is possible to apply for repayment of the tax overpaid. That application may be made during or up to one year after the end of, the chargeable period by amending the original return. This means there is effectively a one-year time limit on making such a claim (*s 100*).

Self-assessment

2.76 An obligation to submit a return and pay tax generally arises within 30 days of the start of the chargeable period. If the tax begins to apply to a company, partnership or collective investment scheme part way through a chargeable period (for example because a property is acquired), the return and payment obligations generally arise 30 days after the date of the event which causes the tax to apply (*s 159, s 163*). However the deadline for making the first return in relation to properties with a value of more than £2 million was 1 October 2013 and the deadline for the first tax payment was 31 October 2013, or in each case 30 days after the date of the event which first caused the tax to apply if that was later. (*Sch 35, Pt 2*). Similarly for properties with a value of more than £1 million but less than £2 million the first return was not due until 1 October 2015 nor payment before 31 October 2015. However this concession was not extended to properties with a value of more than £500,000 but not more than £1 million. First return and payment obligations for such properties caught by the rules on 1 April 2016 were due by 30 April 2016.

Where a company, partnership or collective investment scheme is claiming a relief (see **2.77**) from ATED, it must do so by submitting a return (*FA 2013, s 100*). With effect from 1 April 2015 it is possible to file a simplified return called a 'relief declaration return' which must specify the type of relief being claimed but does not need to contain information identifying the particular property (or properties) in relation to which the relief is being claimed or their valuation (*FA 2013, s 159A* inserted by *FA 2015, s 73(3)*. A 'relief declaration return' can be made in respect of more than one property but can only be in relation to one particular relief, ie separate returns must be filed if different reliefs are being claimed for different properties. The return will be treated as made in respect of any property held by the company, partnership or collective investment scheme which is within the charge to ATED, is relievable from the charge to ATED by virtue of the relief specified in the return and there is no day in the chargeable period, prior to the submission of the 'relief declaration return', in relation to which the property is not relieved from the charge to ATED.

Reliefs

2.77 Reliefs are provided for:

- dwellings held by property traders and developers or property rental business (*s 133, s 134, s 138, s 139, s 141*);
- those opened to the public in the course of a trading activity (*s 137*);
- financial institutions, provided the dwelling was acquired in the course of a lending activity and is to be disposed of without undue delay (*s 143*);
- providers of social housing (*s 150*);
- charities (*s 151*);

- properties occupied by certain employees or partners working in a trade or qualifying property rental business (see *FA 2016, s 135(3))* carried on by the property owner (or a group company), provided the employee/partner has a less than 10% interest in the business (*s 145*); and
- farmhouses occupied by someone working on the farm (or a former long-serving farm worker) (*s 148*).

Anti-avoidance provisions

2.78 The rules have been carefully framed throughout in an attempt to provide complete certainty and to prevent avoidance. This accounts for a high proportion of the 80 pages of legislation. In addition to the general attempt to define closely how and when the tax applies, there are specific provisions to prevent reduction of the value of a dwelling by fragmenting ownership between different people (*s 109, s 110*) or by physically dividing the dwelling into separate parts (*s 115*).

The tax is within the DOTAS rules (see **7.26** for application to SDLT), and its application to the tax is set out in the *Annual Tax on Enveloped Dwellings Avoidance Schemes (Prescribed Descriptions of Arrangements) Regulations 2013 (SI 2013/2571)*.

The main effect of the regulations is to add ATED to the taxes covered by the DOTAS regulations. As a result the general requirements for promoters and users to provide information to HMRC, often within very short time limits, would apply to arrangements expected to provide an advantage in relation to ATED. There is a very wide description of the kind of arrangement which is caught, being any arrangement under which:

- a company, partnership or collective investment scheme ceases to meet the ownership condition in respect to the chargeable interest;
- the taxable value of the chargeable interest is reduced to £500,000 or less (prior to 1 April 2016 the limit was £1 million); or
- the taxable value of the chargeable interest is reduced with the consequence that the chargeable interest falls within a lower tax band than it otherwise would.

Certain arrangements are then excluded from the obligation to disclose, including an arrangement under which the transfer constitutes a settlement. It is not clear why such arrangements should be excluded, unless HMRC feel that avoidance arrangements involving such transfers are already blocked.

A separate version of form AAG4 (AAG4 (ATED)) is provided for notification of use of a scheme (see www.hmrc.gov.uk/aiu/form-aag4-kana.htm).

The tax is also subject to the general anti-abuse rule in *FA 2013, ss 206–215* (see **7.3**).

Administration and compliance

2.79 As with other taxes, there are requirements for the taxpayer to keep records (*Sch 33, pt 2*), interest is charged on late paid tax and penalties are chargeable if returns are late or incorrect (*Sch 34, pt 2*). HMRC have power to carry out enquiries (*Sch 33, pt 3*), correct errors in returns and to make determinations of tax payable or assessments as necessary (*Sch 33, pts 4* and *5*). The provisions governing enquiries into returns and settlement of disputes are similar to those for SDLT (for which see **6.36** et seq). Routine assessments cannot normally be made more than four years after the end of the chargeable period, or six years if to recover tax lost as a result of carelessness. However this is extended to 20 years in cases attributable to the deliberate action of the taxpayer, or where there has been a failure to make a return or provide information required under the DOTAS rules (*Sch 33, para 25(3)*).

The SDLT rules on collection and recovery of the tax (*FA 2003*, *Sch 12*) are directly applied to this tax (see **6.35**). HMRC provided some guidance on administrative aspects of ATED in Stamp Taxes Bulletin 1/2014, available at www.hmrc.gov.uk/so/bulletin01-2014.pdf.

2.80 As with SDLT, the tax is charged because the property is located in the UK. The location of the company, etc which owns the property is irrelevant. In the case of SDLT there is a practical need to deal with any liability because it should not be possible to register ownership of the property at the land registries until this has been done. There is no equivalent requirement attached to the payment of ATED. There must therefore be some risk that non-UK companies with no UK presence other than ownership of a dwelling will fail to comply with ATED obligations.

2.81 It may be decided to transfer the property out of the company to prevent future liability to ATED. Care is needed if this is to be done without creating a liability to SDLT. Provided the company is debt-free, the property can be transferred as a distribution in the winding up of the company and no SDLT charge will normally arise. However if the company has debt – typically in the form of a mortgage secured on the property – transfer in return for the shareholder taking on the debt will be a transfer for consideration and liable to SDLT. HMRC have confirmed that, where the debt is owed to the shareholder so that it simply disappears when the company is wound up and the property transferred to the shareholder, this will not be regarded as giving consideration. However if there is a third party debt and the shareholder provides funds for it to be paid off (whether by loan, gift or by subscription for further shares), HMRC will consider whether *FA 2003*, *s 75A* applies (see **7.4** et seq). If they consider that the provision of funds to pay off third party debt is 'involved in connection with the disposal' of the property to the shareholder, they will seek to apply *s 75A* and charge SDLT as if the third party debt had been assumed by the shareholder. See SDLTM04042 for details of HMRC's views on this matter. It is understood that HMRC may consult during 2016 on the introduction of a relief to allow mortgaged residential properties to be transferred out of a company without a charge to SDLT arising.

Chapter 3

SDLT Reliefs

INTRODUCTION

3.1 Some transactions are automatically exempt from an obligation to notify HMRC and do not give rise to any SDLT charge. The legislation labels some (but not all) of these as 'exempt transactions'. In this book, the term 'exemption' is applied to all such cases. Even where an exemption applies, so that there is no liability to pay SDLT or submit a return, the taxpayer is still required to keep appropriate records and documents for up to six years (see **6.62**). However, it should be noted that the requirement to 'self certify' transactions as exempt, using form SDLT60, was abolished by *FA 2008*, *Sch 30* with effect from 12 March 2008. Land Registries should now agree to register exempt transactions on the basis of a statement by the purchaser that the transaction is exempt from SDLT.

3.2 In contrast, there are many reliefs reducing or eliminating SDLT liabilities that would otherwise arise. Such reliefs do not apply automatically (despite the fact that, in many cases, the legislation describes qualifying transactions as 'exempt from charge'), but must be claimed by submission of an SDLT return (see **6.2**, *FA 2003*, *s 62(3)* and SDLTM60050). The relief is claimed in the return by putting appropriate codes in boxes; the form provides no space for disclosure of further information explaining circumstances surrounding the claim. Many reliefs are hedged around with complex anti-avoidance rules, which create genuine uncertainty as to their availability.

3.3 It is often desirable to provide further information in order to ensure full disclosure and reduce the risk that HMRC may reject the claim and make a 'discovery' assessment up to six years later. As a result, it has become commonplace to submit further information by way of letter when the SDLT return is completed (see **6.47**). HMRC have acknowledged this process in informal discussion, but do not appear to have issued any formal guidance on the matter.

3.4 There are several reliefs which are subject to retrospective withdrawal ('clawback') if certain events occur, as explained below in the descriptions of the reliefs. In these cases, it is necessary to keep track of both the property and the owner(s), in order to be able to make necessary returns and payments should clawback apply. Special rules apply where the interest disposed of is a lease, on grant of which group relief (or one of a list of other reliefs) was claimed. If nothing has already happened to cause clawback of that relief, the first transfer of the lease which does not itself qualify for one of the specified list of reliefs is treated as the grant of a new lease for the remaining term of the actual lease and at the rent payable by the transferee. SDLT is chargeable on the transferee accordingly (*FA 2003*, *Sch 17A*, *para 11*). See **4.17** et seq for an explanation of this and an example.

3.5 Finally, to add to the confusion, there is a provision which is generally regarded as a relief but strictly is simply a computational rule. This is the so-called 'PFI relief' in *FA 2003*, *Sch 4*, *para 17* (see **3.17**). This rule exempts certain consideration from charge, applies automatically (like an exemption) but still requires submission of a return (like a relief).

Originally 'sub-sale relief', was also a computational provision which applied automatically to disregard all except the final step in certain series of transfers. However *FA 2013*, *s 195* replaced this with a true relief, requiring a claim in a land transaction return (see **3.8**).

3.6 In this chapter, reliefs with a wide application are dealt with in some detail, but those with relatively narrow or specialised application are dealt with only briefly, the most specialised being summarised in table form. Reliefs which have been abolished or virtually abolished are mentioned at the end of the chapter. A few reliefs which apply specifically to leases are dealt with in **Chapter 4** (see **4.81** et seq).

EXEMPTIONS

3.7 The main exemptions are listed in *FA 2003*, *s 77A* and *Sch 3*, as follows:

(1) A transaction for which there is no chargeable consideration is exempt (*FA 2003*, *Sch 3*, *para 1*). Chargeable consideration is defined in *FA 2003*, *Sch 4* (see **2.19**). If there is no actual consideration in money or money's worth, there will be no chargeable consideration unless *FA 2003*, *s 53* (purchaser is connected company; see **2.29**) or *FA 2003*, *Sch 15* (transfers involving partnerships; see **Chapter 5**) or *FA 2003, Sch 4, para 12(1)(c)* (land transaction entered into by reason of employment), apply. Additionally, *FA 2003*, *Sch 4* provides for certain amounts to be ignored where they might otherwise be regarded as consideration, see **2.36.**

(2) The grant of a lease to a tenant by a relevant housing provider (RHP) for an indefinite term or in circumstances where the lease is terminable

by notice of a month or less is exempt provided certain other conditions are satisfied (*FA 2003, Sch 3, para 2*). For these purposes a 'relevant housing provider' is a non-profit registered provider of social housing or a registered social landlord. Detailed review of the rules affecting RHPs is beyond the scope of this book, but it seems likely in practice that any transaction which might qualify for this exemption may also be exempt under the 'de minimis consideration' rules; see (7) below.

(3) A transfer between parties to a marriage or civil partnership in connection with dissolution of the marriage or partnership or separation of the parties (whether in pursuance of a court order or by agreement) is exempt (*FA 2003, Sch 3, paras 3* and *'second 3A'*, an apparent glitch in numbering of the statute).

(4) A transfer of property to a beneficiary under a will or intestacy is exempt, provided the transferee gives no consideration other than the assumption of any debt secured on the property. If other consideration *is* given, the assumption of any secured debt does not count as consideration in calculating the SDLT charge (*FA 2003, Sch 3, first para 3A*). Similarly, a variation of a will within two years of death (for example, by way of a family deed of arrangement) is exempt, provided no consideration is given other than agreement to other variations (*FA 2003, Sch 3, para 4*).

(5) A transfer not involving the grant of a lease is exempt if the chargeable consideration (together with the consideration for any linked transactions; see **2.67**) is less than £40,000 (*FA 2003, s 77A(1.2), (1.4)*). The figure of £40,000 seems somewhat arbitrary – it is difficult to see why it should not be increased to match the threshold below which no SDLT is chargeable. Note that the consideration must be less than £40,000, not '£40,000 or less'.

(6) The grant of a lease for a term of seven years or more is exempt if the chargeable consideration other than rent is less than £40,000 and the rent is less than £1,000 per annum (*FA 2003, s 77A(1.3)*). This exemption does not specifically require aggregation of chargeable consideration of linked transactions. It may be that HMRC would argue that the need to aggregate is implied by the general provisions of *FA 2003, s 55(4)* but it is by no means clear that this would be correct.

(7) The grant of a lease for a term of less than seven years is exempt, provided the chargeable consideration does not exceed the zero rate threshold – that is, provided no tax is chargeable and no tax would be chargeable but for availability of a relief (*FA 2003, s 77A(1.5), (2)*). In relation to rent no tax will be chargeable if the aggregate of the net present value of the rent (see **4.47** et seq) is £125,000 or less (residential property) or £150,000 or less (non-residential property). For any consideration other than rent (ie a premium), no tax will be chargeable if the amount of the consideration other than rent is £125,000 or less (residential property) or £150,000 or less (non-residential property) (see **2.37**). The higher SDLT

rate for additional dwellings and dwellings purchased by companies and other non-individual purchasers (see **2.39**) does not apply to the grant of a lease for less than seven years, as it only applies where a 'major interest' in a dwelling is acquired, and a lease granted for a term of less than seven years is not a 'major interest' (see **2.11**) for the purposes of the higher SDLT rates legislation (*FA 2003, Sch 4ZA, para 2(4)*).

(8) The assignment or surrender of a lease which was originally granted for a term of less than seven years, where the chargeable consideration for the assignment or surrender does not exceed the zero rate threshold.

SUB-SALE RELIEF

3.8 As originally enacted, *FA 2003, s 45* was not a true relief; it was a computational provision which applied automatically if the conditions were satisfied. However, the original provisions were widely exploited, using structures which gave economic ownership of property while qualifying for sub-sale treatment. *FA 2013, s 195* replaced these provisions with a true relief which has to be claimed in a land transaction return. The relief is designed to prevent avoidance while retaining the same overall tax effect for 'genuine' cases. At the same time *FA 2013, s 194* amended the original provisions with retrospective effect to counter some specific avoidance arrangements (see **3.16**). An attempt to challenge the legitimacy of the retrospective change, by way of judicial review, was blocked by the court (*R (on the application of St Matthews (West) Ltd and others) v HMRC,* High Court, judgment 6 June 2014).

3.9 Sub-sale relief can apply where:

- A agrees to transfer property to B but,
- before that contract is completed or substantially performed, B agrees to transfer all or part of the same property to C, and either
  - B drops out so that substantial performance and completion occurs directly between A and C, or
  - the A–B contract is substantially performed and completed at the same time as, and in connection with, respective substantial performance and completion of the B–C contract.

The relief is often relied on in arrangements to finance speculative developments. A builder may find it difficult to borrow the full cost of a development without certainty of selling a proportion of the apartments or units. If apartments or units are sold 'off plan' to investors, the builder will be able to borrow more easily. When the development is completed, the investors may sell the apartments or units on, without taking possession as illustrated in Example 3.1. A form of sub-sale relief had long been available under stamp duty. When SDLT was introduced in 2003, the initial proposals did not include sub sale-relief. However the importance of this to the construction industry was sufficient to persuade the government to include the relief.

3.10 Under the old rules, the A to B agreement was ignored for SDLT purposes, to the extent that the property passed to C. In respect of the property transferred to C, SDLT obligations arose only in relation to the A to C or B to C transfer (*FA 2003, s 45,* before amendment by *FA 2013*). Because this was an automatic treatment if the transaction qualified, there was neither need nor facility for B to make a return or claim relief or exemption. If the whole property was transferred to C, SDLT was charged only on the overall total amount which C paid to A and/or B. If only part of the property was transferred to C, B also paid SDLT in respect of the part retained.

The new pre-completion transaction rules

3.11 The new rules, which are substantially longer than the old, are in *FA 2003, Sch 2A*. The greater length and complexity attracted criticism from business and advisers but also from MPs during the Finance Bill debates, but no simplification was offered by the government. HMRC provides guidance at http://www.hmrc.gov.uk/so/guidance-sdlt-pre-completion.pdf. This replaces draft guidance issued at the time of the 2013 Budget.

The new rules apply to transactions entered into on or after 17 July 2013. Paragraph numbers in this explanation refer to *Sch 2A* unless otherwise stated. The technical operation of the rules has been completely recast in order to combat avoidance. However in straightforward cases the overall effect should be unchanged, save that B will now normally have to make a land transaction return and claim relief.

3.12 The rules distinguish between two forms of sub-sale. An 'assignment of rights' occurs when B drops out of the transaction so that C gains the rights and obligations of B in relation to the original contract. Performance and completion of the transaction then takes place directly between A and C. A 'free standing transfer' takes place when the A to B contract remains in place and is performed and completed between A and B, but the property is transferred on to C under a separate agreement.

- In the case of an assignment of rights, the A to B agreement is regarded as substantially performed/completed by C and liability to tax in respect of that agreement now falls on C. C therefore is responsible for notifying the transaction and paying the liability. In addition, B is regarded as entering into a notional land transaction on the same terms as provided in the original A to B agreement (*para 5*). Unless excepted under other provisions (eg if the transaction is exempt, see **3.7**) B must notify HMRC of the notional transaction by submission of a land transaction return. See **6.6** in relation to notifying transactions.

- In the case of a freestanding transaction, both the A to B transaction and the B to C transaction are subject to SDLT. *Sch 2A* does not specifically state that the A to B transaction is subject to SDLT. However (in contrast with the old rules) there is no longer any provision disregarding this transaction and since it is substantially performed/completed, it remains subject to SDLT on general principles. Therefore both B and C must notify the transaction by submission of land transaction returns, unless excepted under other provisions such as application of an exemption (see **3.7**).

In both cases, provided the appropriate conditions are satisfied, B can claim relief from SDLT in the return, in respect of the part sub-sold to C (*paras 15, 16*). Apart from actually being a step in a sub-sale, the main condition to be satisfied is that the transaction does not form part of any tax avoidance arrangements (*para 18*). In this context 'tax' is not separately defined and so must be taken to mean only SDLT (*FA 2003, s 121*). Relief is also denied if the second (or final) transfer qualifies for relief under *FA 2003, ss 71A–73* (alternative property finance, see **3.37** onwards). In contrast with the old version of sub-sale relief, there is no specific denial of relief if the B to C transfer qualifies for group, reconstruction or acquisition relief (see **3.20**). It is assumed that HMRC are content to rely on the specific rules for those reliefs and/or general anti-avoidance rules to prevent such a combination of reliefs.

As before the rules allow for successive sub-sales, with each intermediate purchaser being responsible for notifying the transaction and for each, other than the final purchaser, potentially able to claim relief. In the case of an assignment of rights, if only part of the land is sub-sold, the original A to B agreement is regarded as divided into two, covering the part sub-sold and the part retained (*para 7*). The consideration must be apportioned between the two (or more) parts on a just and reasonable basis (*FA 2003, Sch 4, para 4*).

3.13 Most of *Sch 2A* effectively consists of anti-avoidance measures, designed to ensure that SDLT is payable on the full economic consideration given. Terms and expressions are closely defined in an attempt to remove any doubt. A new rule imposes a minimum consideration which C is deemed to give if B and C are connected (as defined in *Corporation Tax Act 2010, s 1122*) or not acting at arms length. Under this rule the consideration given by C is deemed to be the greatest of:

- the actual consideration given by C;
- the consideration to be given under the original A to B agreement; and
- the total net amount of consideration given by B and C.

In relation to the final point, the net amount of consideration is the amount given minus any amount received from someone else. Where only part of the original land is sub-sold, amounts are apportioned accordingly. The rules are modified where there are successive sub-sales. The final point becomes the total net amount given by C and all intermediate purchasers who are connected with (or not acting at arms length with) C. The second point becomes the consideration to be given by the first intermediate purchaser in the chain who is connected with, or not acting at arms length with C (*paras 12–14*).

3.14 Under *para 19*, the Treasury can make regulations to amend the legislation, in order to exempt certain purchasers (B in the narrative above) from making returns or from needing to claim relief. As at 30 June 2016 no such regulations had been made.

3.15 Early in the SDLT era, a mechanism exploited sub-sale treatment to avoid or defer SDLT on substantial performance, using 'alternative contracts'. These provided for the land to be transferred to the original purchaser or a third party, the main idea being that the original purchaser would never gain an interest in land and so would not have to pay SDLT. It is not clear that the planning worked, but *FA 2003, ss 44A* and *45A* were enacted to put the matter beyond doubt and ensure the SDLT liability arose as soon as the original contract was substantially performed.

Retrospective changes to old rules

3.16 Despite the provisions referred to in **3.15**, the old version of sub-sale relief was widely relied on in various attempts to mitigate SDLT. Pending amendment of the rules for sub-sale relief by *FA 2013*, it was announced at the time of the 2012 Budget that any remaining such arrangements would be countered by legislation with retrospective effect back to 21 March 2012. *FA 2012, s 213* amended the DOTAS rules to require re-disclosure of schemes using sub-sale relief. This was to ensure that the users of such schemes were also required to notify HMRC (see **7.38**). From 21 March 2012, the grant or assignment of an option was prevented from qualifying as a sub-sale or assignment (*FA 2012, s 210).*

FA 2013, *s 194* amended the old version of *FA 2003*, *s 45* to deny relief for the A to B transfer where the B to C transfer was substantially performed but not completed at the same time as the A to B transfer and B remained in possession of the property, and it was reasonable to conclude that a main purpose of the arrangements was the obtaining of a tax advantage for B. This was to counter a specific arrangement under which B, the real purchaser, agreed to sub-sell the property to another party, C, for a very small sum but with completion delayed for a century or more. Because the legislation was retrospective, those who had used this arrangement were given until 30 September 2013 to submit a land transaction return disclosing the liability and to pay the tax. If this deadline was met, no penalty would be charged (although interest would be payable on the tax). Anyone who has missed the deadline may expect to suffer a penalty as well.

In an amendment introduced by the government during the Committee stage of the Finance Bill, *FA 2003, s 45* was further amended to counter a recently identified scheme using options, which claimed to escape the restrictions imposed by *FA 2012, s 210* mentioned above.

Example 3.1 – sub-sales

Buildquick Ltd proposes to build a block of apartments. Richman, who has an impeccable credit record and is not connected with Buildquick, agrees to buy six apartments for a total price of £1.5 million (ie, £250,000 each). Each apartment will be acquired under a separate, and legally independent, agreement to facilitate the onward sale, by Richman, to third parties. It is therefore considered that the acquisition of each apartment will be a separate transaction (as there are separate agreements) and consequently *FA 2003, s 116(7)* (which applies where six or more dwellings are acquired as part of a single transaction and treats such a transaction as the acquisition of non-residential property) will not apply to treat the acquisition of the apartments as an acquisition of non-residential property (see **2.38**).

The agreements allow Buildquick to borrow from its usual bank to fund construction. Richman pays the customary 10% deposit. The agreements have not yet been completed or substantially performed, so no SDLT obligations have yet arisen.

As completion of construction approaches, Richman's agent begins marketing the apartments. Sales of the first two are agreed to unconnected individuals at £270,000 each. In each case, the purchaser pays Richman's agent a deposit of 10%. At this stage, no SDLT obligations have arisen.

The apartments are finished and the time for completion of the sales arrives. At the last minute, an unconnected buyer is found for a third apartment, at a price of £240,000. Sales of the first three apartments to Richman are substantially performed and completed at the same time as the onward sales to the unconnected purchasers. The purchases by Richman give rise to the obligation to complete a land transaction return but he should be able to claim relief under *FA 2003, Sch 2A, para 16* so that no tax arises. SDLT is chargeable only on the ultimate purchasers. Since these ultimate purchasers are not connected with each other or with Richman, SDLT is chargeable on each at the rates appropriate to the amount paid – that is, 0%, 2% and 5% on each of two purchases at £270,000 and 0% and 2% on the third purchase at £240,000. This assumes that the ultimate purchasers are not liable to pay the 3% additional rate of SDLT on their acquisitions of the flats (see **2.39** et seq). The amount paid by Richman is irrelevant in calculating the SDLT liability of the third party purchasers.

Richman completes the purchase of the fourth and fifth apartments in his own name, with the intention of letting them out at rent. He assigns the benefit of the contract to purchase the sixth apartment to his sister, who pays £10,000 for the benefit of the contract, and pays Buildquick the balance of the original purchase price, £225,000, to complete the purchase. The arrangements relating to the fourth, fifth and sixth apartments give rise to SDLT issues as follows:

- Although the market value of the sixth apartment is probably at least the £240,000 paid by the last-minute purchaser of the third, the total amount paid by the sister is only £235,000 (£10,000 to Richman for transfer of the contract and £225,000 to Buildquick to complete). However, Richman has paid a £25,000 deposit. Since Richman is connected with his sister, the minimum consideration rule comes into play (*para 12*). The three figures to consider are:

  1. the total amount of consideration given by the sister (£235,000);
  2. the amount payable under the original agreement between Richman and Buildquick (£250,000); and
  3. the total net amount paid by Richman (£25,000 deposit less £10,000 received from sister = £15,000) and his sister (£235,000).

  The second and third amounts are the same, £250,000 and are larger than the first amount. So the sister pays SDLT on £250,000. The new rules remove one ambiguity; under the old version of *s 45*, it could have been argued that SDLT was due on the gross amount paid by Richman and his sister, totalling £260,000 – although this would have been greater than the real net cost of the apartment.

- The purchases of apartments four and five by Richman and the purchase of the sixth by his sister are linked transactions (see **2.67**). The consideration must therefore be aggregated to determine the SDLT rate. The aggregate consideration is £750,000 so prima facie the SDLT rates which apply should be 3%, 5% and 8%. This is on the basis that both Richman and his sister already own separate properties which they use as their main residence, and the apartments are being acquired as buy-to-let properties, such that the higher rates of SDLT apply (see **2.39**). However, if Richman and his sister choose to make a claim, the purchases should qualify for relief under *FA 2003, Sch 6B* (see **3.48**). In this case the rates of SDLT are determined by the average cost of the apartments – £250,000 – and the SDLT higher rates that will apply will therefore be 3% and 5%.

PFI RELIEF

3.17 This is the calculation provision which looks like a hybrid exemption/relief, referred to at **3.5**. Although popularly referred to as PFI relief, *FA 2003, Sch 4, para 17* does not necessarily apply to all PFI-type transactions, nor does it necessarily remove all liability. It merely excludes certain amounts from the calculation of chargeable consideration where:

(1) there is a transfer of land (which may include the grant of a lease) from a 'qualifying body' (A) to a non-qualifying body (B), and

(2) a lease back of that land to A, and

(3) an agreement for B to provide services to, or carry out works for A, to be paid for at least in part in money.

3.18 'Qualifying bodies' include public bodies within *FA 2003, s 66* (broadly, local and central government organisations, health authorities and other statutory bodies, but the list in *s 66* should be checked in any individual case). They also include entities operating Higher and Further Education establishments and Academies – the list in *FA 2003, Sch 4, para 17(2)* should be checked where appropriate.

3.19 Where the relief applies, the chargeable consideration for the initial transfer of land does not include the value of the lease-back or the carrying out of works or provision of services by B. The chargeable consideration for the lease-back does not include the initial transfer of land, any transfer of other land from A to B, or the payment of money by A to B. This may mean that there is no chargeable consideration for any of the land transfers; however, PFI-type transactions are often more complex than the relief appears to envisage, and each transaction must be analysed carefully to establish whether the precise conditions have been met for all consideration to be excluded from charge.

Example 3.2 – PFI relief

Anytown Unitary Authority (A) needs a new building for council offices and a public library. This is to be built on a site mostly owned by A. Buildit plc (B) owns a small piece of land adjacent to this site, which is needed for access. It is agreed that:

(1) B will transfer the freehold of the small piece of land to A;

(2) A will grant B a 25-year lease of the whole site at a peppercorn rent;

(3) B will construct the building to A's specification;

(4) B will lease the building back to A for 25 years less one day, on the basis that B will manage the building (cleaning, maintenance, security etc);

(5) A will pay B a 'unitary charge' each year for provision of the building and services; and

(6) in addition, A will transfer to B the freehold of an old library site, on which B hopes to develop luxury apartments for sale.

Items 3 and 4 do not count as chargeable consideration for the grant of the lease at 2. Items 2, 5 and 6 do not count as chargeable consideration for the grant of the lease at 4. However, the transfer at 1 is not within the relieving provisions and is potentially subject to SDLT. If A is not able to claim any other relief or exemption, it will be necessary to establish how much of the value of 2, 5 and 6 should be attributed as consideration for the transfer at 1, and SDLT will be charged on this. The natural assumption will be that the consideration is the same as the market value of the small piece of land, unless analysis of the figures suggests otherwise. Central government bodies and some others, such as strategic health authorities, have the benefit of blanket SDLT reliefs, but reliefs available to the various local authorities are more limited and probably do not apply here.

GROUP, RECONSTRUCTION AND ACQUISITION RELIEFS

3.20 These reliefs are dealt with together because they appear together in *FA 2003, Sch 7* and many of the conditions and anti-avoidance provisions are similar. The basic operation of each relief is outlined separately below, with an account of the conditions which apply to the particular relief. This is followed by discussion of an anti-avoidance provision which is common to all three reliefs. Finally, there is a review of the circumstances in which each relief may be retrospectively withdrawn (clawed back) as a result of events after the initial transaction.

3.21 The reliefs apply to transactions involving 'companies' (*FA 2003, Sch 7, para 1*). There are no restrictions as to residence or place of incorporation of the companies, providing they can be recognised as bodies corporate. In all cases, with the possible exception of the top company of a group, they must have 'issued ordinary share capital' or its equivalent. This can cause difficulties in relation to some overseas entities, where their 'corporate' nature may be open to question or which may have no concept of shares or share capital. In any case of doubt, it is possible to ask the opinion of HMRC. However, if it is not a type of body on which they have previously opined, HMRC are likely to demand an appropriate legal opinion from a lawyer in the relevant country (paid for, inevitably, by the taxpayer).

3.22 Relief may not be due, or in some cases may be withdrawn, if there are 'arrangements' for certain events to happen – the specific events are mentioned in relation to each relief below. The legislation makes it clear that 'arrangements' may exist even though there is no legal commitment by the parties (*FA 2003, Sch 7, para 2(5)*); in relation to stamp duty, HMRC gave their view that arrangements exist where it is intended that the event will happen and there is no likelihood in practice that it will not (see SP 3/98, para 6); however, in Inland Revenue Tax Bulletin Issue 70 (April 2004 – *'Stamp Duty Land Tax: Group Relief'*) HMRC make the point that the term 'arrangements' was not defined for the purposes of stamp duty and 'Under this new definition [*FA 2003, Sch 7, para 2(5)*] "arrangements" will include schemes, arrangements or understandings not in writing and the existence of arrangements in a particular claim will depend on a consideration of all of the facts relating to the claim and the surrounding circumstances.' (para 19). The guidance goes on to state at para 34 that '... the practical likelihood of the scheme being carried through is not in itself relevant.' This seems to be an attempt by HMRC to suggest that the term 'arrangements' may have a wider meaning for the purposes of SDLT than it does for starhp duty purposes. However, in the view of the author, this is probably not the case and similar criteria to that used in determining whether an 'arrangement' exists for stamp duty should apply in relation to SDLT. The legislation which denies the availability of the relief if a disqualifying arrangement is in existence should, like any other statutory provision, be interpreted purposively, with the relief only being denied if the arrangement is designed to avoid SDLT.

Group relief

3.23 Group relief removes any SDLT charge on land transactions between companies (ie bodies corporate) which are members of the same group (*FA 2003, Sch 7, para 1*). To be members of the same group, one company must be the 75% subsidiary (direct or indirect) of the other, or each must be the 75% subsidiary of a third company. For A to be the 75% subsidiary of B, B must be:

(a) the beneficial owner of not less than 75% of the issued ordinary share capital of A;

(b) beneficially entitled to at least 75% of any profits of A available for distribution to equity holders; and

(c) beneficially entitled to at least 75% of any assets of A available for distribution to equity holders on a winding up.

Where ownership is indirect, proportions are to be calculated in the normal algebraic manner under *Corporation Tax Act 2010 (CTA 2010), ss 1155–1157*. Rights of equity holders are measured in accordance with *CTA 2010, Pt 5, Chapter 6* (omitting references to arrangements which might affect future rights if put into effect). If shares in a particular company are not identical as to par value and rights, or if creditors have greater rights than payment of the relevant debts, it will be necessary to consider the impact of this chapter.

Example 3.3

Fred, an individual, owns 100% of the shares of Gill Ltd and 100% of the shares of Irene Ltd. Gill Ltd owns 90% of the shares of Jack Ltd and 75% of the shares of Kerri Ltd. Irene Ltd owns the other 10% of Jack Ltd and 20% of Harry Ltd. Jack Ltd owns the remaining 80% of Harry Ltd. Fred's business partner Leo owns the remaining 25% of shares in Kerri Ltd. All shares are ordinary, same class and denomination, and there are no share options or other instruments which might change relative entitlement to dividends or asset distribution in a winding up. The structure is therefore as follows:

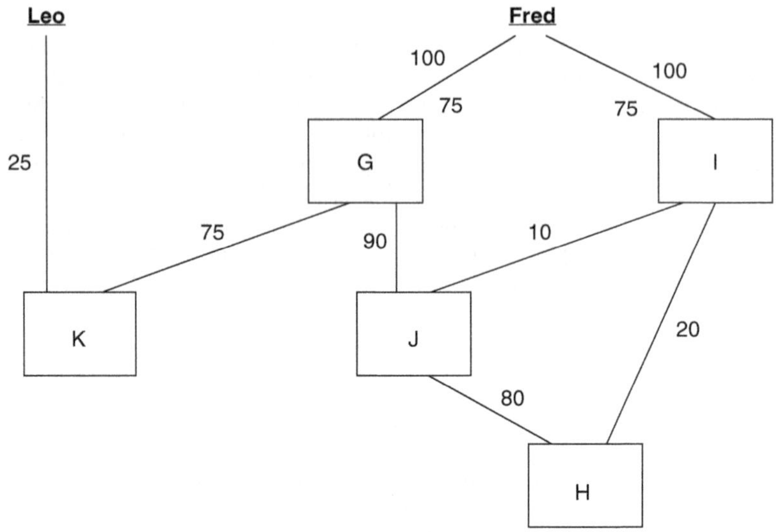

Clearly, Fred ultimately controls all of the companies, so they are associated. This has implications for the measurement of consideration for any land transactions between them (see **2.29**). However, they are not all members of the same group. J and H form a group, so transfers between them are capable of qualifying for group relief. Equally G, J and K form a group, and transfers between them may qualify for relief. However, H is not grouped with G or K. This is because G has (indirect) beneficial ownership of only 72% (90% × 80%) of H, the other 28% being owned directly and indirectly by I (and I is not grouped with any of the other companies). Note that J is separately grouped with both H, and G and K. In contrast with some direct tax provisions, it is acceptable for SDLT purposes for a company to be a member of more than one group in this way. Although H and G (or K) are not directly grouped, it appears at first glance that it should be possible to transfer land between them free of SDLT, for example by G making the transfer first to J, and then J transferring onwards to H. However, group relief is surrounded by anti avoidance provisions (see **3.28**), and it is highly likely that these would apply to such a sequence, potentially denying relief on both transfers and leading to a higher charge than would have applied to a direct transfer.

Treatment of LLP – company or partnership?

3.24 There is a conflict between possible treatments of a limited liability partnership (LLP). For most purposes, an LLP is regarded as a body corporate. On this basis, it can be the top company of a group for group relief purposes, and transfers between companies owned by the LLP may qualify for SDLT group relief. An LLP cannot be a subsidiary within a group, because it does not have issued ordinary share capital; and as it cannot be looked through as it is a body corporate, it therefore breaks any group structure, isolating companies above it in the structure from those below. Therefore, group relief cannot be available for a transfer to (or from) a subsidiary company, from (or to) a partner or other company above the LLP in the structure, **even if all companies are substantively in the same corporate group.**

However, the special rules governing the application of SDLT to partnerships state that any chargeable interest held by a partnership (which includes an LLP) is regarded as held not by the partnership but by the partners (*FA 2003, Sch 15, para 2(1)(a)*). Therefore, group relief is not available for a transfer of property from (or to) an LLP to (or from) a corporate partner, because this is regarded as a transfer from (or to) all of the partners, unless the partners are all members of the same group. A transfer of property from a subsidiary of the LLP to the LLP itself cannot qualify for group relief even if the partners in the LLP are all members of the same group. This is because such a transfer is regarded as a transfer to the partners, and the partners cannot be grouped with the subsidiary of the LLP because the presence of the LLP breaks the group relationship. The provisions at *FA 2003, Sch 15, para 2* which treat partnership property as owned by the partners, do not apply to treat shares owned by the LLP as owned by the partners. This is an unreasonable and potentially unfair consequence of treating the LLP differently in relation to ownership of property and of shares.

After some uncertainty, HMRC eventually confirmed this interpretation in a news release in October 2010 (see www.hmrc.gov.uk/so/group-rel-sdlt-sd.htm). HMRC give some simple examples of possible group relief claims involving an LLP at SDLTM34450 et seq.

Example 3.4

Companies B and C are 100% owned by company D. B and C are equal partners in an LLP. The LLP owns 100% of the shares in company E. The presence of the LLP means that, for SDLT purposes, E is not grouped with B, C or D. A transfer of property from E to the LLP will not qualify for group relief because it will be treated as a transfer to B and C. Similarly a transfer from E to any of B, C or D will not qualify for group relief. However a transfer between any pair of B, C, D or the LLP may qualify for relief.

Arrangements

3.25 In accordance with *FA 2003, Sch 7, para 2*, group relief is not available if, at the time of the transfer, there are 'arrangements' under which:

- at that or a later time, a person could obtain control of the purchaser but not the vendor. This condition may be breached if a third party has options to acquire or subscribe for shares of the purchaser, for example. However, if the arrangements are part of a scheme of reconstruction which itself will qualify for stamp duty relief under *FA 1986, s 75*, relief will not be denied (see SDLTM23035); or

- any part of the consideration is to be provided or received by someone other than a group member. In principle, this could lead to denial of relief where, for example, the purchaser borrows funds from a bank. In Inland Revenue Tax Bulletin Issue 70 (April 2004) HMRC state that borrowing on ordinary commercial terms to fund an intra-group transfer would not normally lead to denial of relief (para 29). In addition HMRC indicate that they will consider all of the facts and interpret this provision as not denying relief unless loan finance is provided as part of a scheme to save SDLT when the property leaves the SDLT group (para 27); or
- the companies are to cease to be members of the same group by reason of the purchaser leaving the group. It is not generally a problem if there are arrangements under which the vendor is to leave the group. However, arrangements even for this must not be too far advanced, or beneficial ownership of relevant shares may already be lost and the group broken before the time of the transfer.

Reconstruction relief

3.26 Reconstruction relief removes any SDLT charge on land transactions in the course of a qualifying reconstruction under which one company, A, acquires all or part of the undertaking of another, B (*FA 2003, Sch 7, para 7*). The reconstruction qualifies if:

(1) the consideration includes the issue of non-redeemable shares in A to all of the shareholders of B;

(2) there is no other consideration, except that A is permitted to assume or discharge debt of B;

(3) the shareholders of B before the transaction end up owning identical proportions (or as near as is possible) of company A; and

(4) the transaction is for bona fide commercial purposes, without a main purpose of avoiding tax.

See **3.28** for a discussion of the last of these conditions.

This relief is the SDLT equivalent of stamp duty relief under *FA 1986, s 75* (confusingly headed 'acquisitions: reliefs' in the legislation), and the stamp duty case law relating to the term 'undertaking' applies here too. The HMRC view on this is set out in SDLTM23201. Broadly, there must be some form of active business originally carried on by B and subsequently continued by A – a mere transfer of assets which are passively held or subsequently used by A for a different purpose will not qualify.

Acquisition relief

3.27 Acquisition relief is the SDLT equivalent of stamp duty relief under *FA 1986, s 76.* It reduces the rate of SDLT chargeable to 0.5% where a company, A, acquires all or part of the undertaking of another, B, and relevant other conditions are met (*FA 2003, Sch 7, para 8*). The conditions are

(1) the consideration includes the issue of non-redeemable shares in A to B, or to any or all of the shareholders of B,

(2) the only other consideration (if any) comprises one or both of:

- payment of cash not exceeding 10% of the nominal value of the shares issued, and
- the assumption or discharge by A of liabilities of B,

(3) A is not associated with any company which is party to 'arrangements' with B relating to the shares issued by A, and

(4) the main activity of the undertaking consists of a trade other than property dealing or development.

Condition (3) is intended to prevent relief where the parties agree in advance that B will sell the newly issued shares to a member of A's group or other associated company, so that as far as B is concerned the sale of the undertaking is effectively for cash. However, the condition can cause difficulties in relatively innocent circumstances, where A and B are themselves associated perhaps because they are held by members of the same family, and shareholder agreements give pre-emption rights where one shareholder wishes to exit. All such transfers require careful detailed analysis to establish whether the relief applies; in general, HMRC seem reluctant to provide guidance on whether particular arrangements fall foul of this condition.

Condition (4), which was added in 2005, means that the relief can no longer apply to the transfer of a property letting business (not a trade) or a property development activity. The relief is, however, potentially useful in relation to the transfer of a trade where property forms a large proportion of the fixed assets – for example, retailing or hotel operation.

Anti-avoidance provision

3.28 Each relief is additionally subject to the condition that the transaction must be for bona fide commercial purposes, without the avoidance of tax forming a main purpose. 'Tax' here means stamp duty, SDLT, and income, corporation or capital gains tax. This condition is effectively a general anti-avoidance provision. There is no legislative guidance as to its scope, and doubt over this has led to significant uncertainty as to whether relief is due. In response to widespread disquiet, HMRC produced a 'white list' of transactions which would not normally be regarded as falling foul of this condition. The list can be found at SDLTM23040. However, the list itself is heavily caveated and gives limited comfort in the context of real commercial transactions. After a meeting between HMRC and representative bodies, including the Chartered Institute of Taxation, HMRC confirmed in August 2013 that group relief would not be denied merely because one company obtained a tax advantage by:

(1) acquiring another company which held property; then

(2) transferring the property within the group.

(See http://www.hmrc.gov.uk/so/bulletin02-2013.pdf, item 3.)

Clawback of relief

3.29 Even where a transaction initially qualifies for one of these reliefs, in certain circumstances the relief may be retrospectively withdrawn, leading to crystallisation of the SDLT charge originally relieved. Because SDLT is a self-assessment tax, it is for the taxpayer to detect and disclose any situation in which a clawback arises (*FA 2003*, *s 81* – 'Further return where relief withdrawn'). Where one of these reliefs has been claimed, it is therefore important to put in place arrangements to monitor ownership of the property, and of the companies involved in the original claim, for an appropriate period, which is generally three years.

Clawback of group relief

3.30 Group relief is clawed back if the purchaser, while it or a 'relevant associated company' still owns the property (or an interest derived from it), ceases to be a member of the same group as the vendor, either within three years of the effective date of the property transfer, or later but under arrangements entered into within that three-year period. A 'relevant associated company' is any other member of the group which leaves the group 'in consequence of' the purchaser leaving. This provision prevents an otherwise simple avoidance arrangement, under which the purchaser transfers the property to a wholly-owned subsidiary before both purchaser and subsidiary leave the vendor's group and the purchaser is thus able to claim that it no longer holds any interest in the property. If the interest which the purchaser (or a relevant associated company) still holds is less than the interest originally transferred, a pro rata amount of the relief is withdrawn (*FA 2003*, *Sch 7*, *para 3*).

Example 3.5

Xerxes Ltd owns 100% of the shares of Yolande Ltd and Zeb Ltd. On 2 March 2017, Y transfers to Z a freehold interest in a commercial property worth £800,000, and group relief from SDLT is claimed. In September 2017, Z grants a capital lease of part of the property to a third party, in return for a premium of £600,000 and a peppercorn rent. It is agreed that the value of the part thus disposed of is three-quarters of the value of the whole property. On 24 February 2018, X sells 50% of the shares of Z to another third party, in the course of setting up a joint venture. Z therefore leaves the X group while still holding an interest in the property with a value of one quarter of the value of the whole property. The SDLT payable is, therefore, one quarter of the SDLT which would have been payable on the original transfer on 2 March 2017, if group relief had not been claimed. The SDLT on the original transfer would have been calculated as follows:

Purchase Price (£)	Rate (%)	SDLT (£)
150,000	0	NIL
100,000	2	2,000
550,000	5	27,500
800,000		29,500

Consequently, the amount chargeable under the clawback is one quarter of £29,500 or £7,375.

Note the SDLT is *not* calculated as if there was a transfer of a property with a value of one quarter of the original; in the present case, such a calculation would give a liability of £1,000 – but, sadly, this is not how the clawback operates! However, even if SDLT rates or thresholds have subsequently changed, the clawback is still calculated on the basis of rates and thresholds which applied at the time of the original transfer.

3.31 Group relief is not clawed back where the purchaser leaves the vendor's group (*FA 2003, Sch 7, para 4*):

- as a result of the winding up of the vendor or a company above the vendor in the group;
- in the course of a reconstruction which itself qualifies for relief under *FA 1986, s 75*, as a result of which the purchaser becomes a member of the same group as the 'acquiring company'; or
- in the course of a qualifying demutualisation of an insurance company.

3.32 Group relief is not clawed back if it is the vendor which leaves the group rather than the purchaser. This allows a group to remove property which it wishes to keep from a subsidiary which is to be disposed of. However, if there is subsequently a change of control of the purchaser within three years of the original transfer, the purchaser is at that time treated as leaving the vendor's group, and clawback applies accordingly (*FA 2003, Sch 7, para 4ZA*). See **3.34** for discussion of what amounts to a change of control. There are complex anti-avoidance provisions designed to ensure the 'change of control' rules apply as intended, where successive transfers of the property might otherwise lead to doubt as to their application. Although it is not clear why, the winding up of the purchaser is also treated as a change of control (*FA 2003, Sch 7, paras 4ZA, 4A*).

However, if the purchaser ceases to be a member of the same group as the vendor as a consequence of the vendor being wound up, or a company above the vendor being wound up, so that group relief is not clawed back (see **3.31**), it is understood that HMRC accept that a subsequent change of control of the purchaser will not cause the group relief to be clawed back under *FA 2003, Sch 7, para 4ZA*.

Clawback of reconstruction or acquisition relief

3.33 These reliefs are clawed back if, within three years of the effective date of the property transfer, or later under arrangements entered into within the three years:

- control of the acquiring company changes; and
- that company or a relevant associated company holds the property or an interest derived from it (*FA 2003, Sch 7, para 9*).

A 'relevant associated company' is a company controlled by the acquiring company immediately before the change in control, control of which changes as a consequence of the change in control of the acquiring company. If only part of the property, or a lesser interest derived from it, is held, an amount of relief is withdrawn pro rata to the relative values of the whole property and the part or interest still held. The calculation is analogous with that for withdrawal of group relief (see Example 3.5).

Change of control

3.34 In many cases, a change of control is easily identified – for example, when a majority shareholder sells the shares to a third party. However, a change of control may be very difficult to identify or define in some circumstances and, in the case of a publicly traded company, may be subject to the whim of the market. In the 2003 Finance Bill debates, the government stated that normal market transactions in the shares of traded companies would not be regarded as giving rise to a change in control (see *Hansard*, Standing Committee B, 10 June 2003, col 425). However, this leaves open the possibility that the transfer of a minority shareholding in the case of a non-traded company may still give rise to a change of control, since the company would then be controlled by a different group of people.

3.35 There is no clawback of reconstruction or acquisition relief where control of the acquiring company changes for one of the following reasons:

(1) as a result of transactions in connection with divorce etc or distribution of an estate, as set out in (3) and (4) at **3.7** (*FA 2003, Sch 7, para 10(2)* and *(3)*),

(2) as a result of a transfer of shares which itself qualifies for group relief under *FA 1930, s 42* (*FA 2003, Sch 7, para 10(4)*) or share acquisition relief under *FA 1986, s 77* (*FA 2003, Sch 7, para 10(5)*)(see **9.9** et seq), or

(3) purely as a result of a loan creditor becoming or ceasing to be treated as having control (*FA 2003, Sch 7, para 10(6)*). This may happen, for example, when a loan creditor gains additional rights if the company defaults on payments under the loan.

There are anti-avoidance provisions to ensure that, where (2) above applies to prevent clawback, subsequent transactions leading to the relevant company either leaving the group or suffering a change of control within the original three-year period (or later under arrangements entered into within the period) will lead to clawback (*FA 2003, Sch 7, para 11*).

3.36 In relation to the clawback of all three reliefs, if the liable company fails to pay within six months of the due date, there are provisions (*FA 2003, Sch 7, paras 5, 12*) allowing tax to be recovered from:

- any company which was above the liable company in a group structure at any time between the original transfer and the event leading to clawback,
- a controlling director of the liable company or any other company having control of it, or
- in the case of group relief, the vendor.

Purchasers of companies should be aware of the risk of liability under these provisions and should seek appropriate protection in the form of warranties, indemnities, retentions etc.

OTHER MAJOR RELIEFS

Alternative finance reliefs

3.37 A series of reliefs has been introduced in an attempt to ensure that additional SDLT liabilities do not arise on property transactions where funding is structured in such a way as to avoid payment of interest. The need to avoid interest typically arises from specific beliefs, particularly the acceptance of Shari'ah principles; however, availability of the reliefs depends only on the nature of the transaction and participants, and not on subscription to any particular set of beliefs. The reliefs fall into two groups. The first (*FA 2003*, *ss 71A–73B*) relates to the purchase or ownership of property where, in other circumstances, mortgage finance might be obtained; these are dealt with in more detail below. The second (*FA 2003*, *Sch 61*, *paras 6–9*) relates to property backed investment bonds. Review of the reliefs available to those involved in the issue and management of such bonds is beyond the scope of this book; it is sufficient to note that a portfolio investor in qualifying bonds, although in fact gaining a direct interest in the underlying property, is deemed not to have such an interest for SDLT purposes (*FA 2009, Sch 61, para 2*). This does not apply if the investor, together with connected persons, has the right of management and control of the underlying property (for example, by virtue of the proportion of bonds held). Careful fact gathering and analysis may therefore be required where anything more than an obvious minority of bonds is held, and especially where family members or connected companies may hold investments in the same products. This difficulty is acknowledged in the legislation. If connected persons obtain control inadvertently, the favourable tax treatment is not lost, provided the control is not in fact exercised and holdings are reduced to remove that control once the issue has come to light (*FA 2009, Sch 61, paras 3–4*).

3.38 The reliefs relating to purchase or ownership of property envisage two forms of transaction. In the first (*s 71A*), under arrangements with the ultimate or intended owner ('Owner'), a financial institution ('Bank') purchases a freehold or leasehold property, leases it to Owner and grants Owner a right (the 'option') to require Bank to transfer the original freehold or leasehold interest to Owner. The terms of the lease and option typically ensure that Bank receives the same economic return as if it had provided mortgage finance secured on the property. If the relief applies, the lease to Owner and the final transfer to Owner under the option are relieved from SDLT. Partial transfers to Owner under the option during the course of the arrangement (for example, as a result of Owner making capital payments) are exempt from SDLT. As a result, the only SDLT payable is on the initial purchase by Bank, which is equivalent to what would happen if Owner purchased the property from a third party with the aid of conventional mortgage finance. However, if Bank buys the property from Owner, that purchase is also relieved from SDLT. In this situation, no SDLT is payable on the overall transaction, reflecting the position where a property owner obtains a loan secured on property already held.

The second form of transaction (*s 73*) envisages purchase of the property by Bank and onward sale to Owner, the onward sale typically being on instalment terms designed to replicate the economics of a repayment mortgage. If the relief applies, the onward sale to Owner is relieved from SDLT. Again, if the purchase by Bank is itself from Owner, that transaction is also eligible for relief.

There were provisions to ensure that the reliefs applied with the same overall effect where the property was located in Scotland (*ss 72, 72A*). However, following the introduction of LBTT with effect from 1 April 2015, these provisions have been repealed by *Scotland Act 2012, Sch 3, paras 2* and *10* from that date.

However, relief is denied if, at the time of the transaction, there are arrangements to transfer control of the financial institution. This is an anti-avoidance provision, to counter the possibility of a bank setting up a subsidiary to enter into the transaction, with every intention of then selling the subsidiary to the erstwhile purchaser of the property. Further anti-avoidance provisions (*s 73A*) seek to prevent the combination of these reliefs with group, reconstruction and acquisitions reliefs described above, to obtain an overall tax advantage. Beyond this, the government appears to accept that it may be possible to exploit the reliefs to avoid SDLT and has stated that the reliefs 'will be kept under careful review'. The implied threat is that they may be withdrawn if abuse is perceived.

The legislation originally relied on a very wide definition of Financial Institution set out in *FA 2005, s 46*, which included licensed credit brokers. The potential tax savings in some transactions were such that it was worthwhile to set up such a credit broker to take advantage of the relief. To counter this, a narrower definition of Financial Institution was imported from *ITA 2007, s 564B* by *FA 2011, Sch 21, para 3* with effect from 24 March 2011. It also includes, for transactions with an effective date on or after 25 March 2015, any person with permission under the *Financial Services and Markets Act 2000, Pt 4A* to enter into regulated 'home purchase plans' as 'home purchase provider' (*FA 2003, s 73BA(3)*, inserted by *FA 2015, s 68(2)*).

3.39 The interest held by Bank is an exempt interest (*s 73B*). This means any transfer to another financial institution (for example, if the bank sells its loan book) will not give rise to an SDLT charge.

Sale and leaseback relief

3.40 If A transfers a property to B (or grants a lease to B) (the Sale), and out of the interest transferred B grants a lease back to A (the Leaseback), relief from SDLT is available on the leaseback provided certain conditions (see **3.41**) are satisfied (*FA 2003, s 57A*). The purpose of the relief is to facilitate transfer of properties from occupiers to investors. Absent the relief, it might be possible for the same effect to be achieved in some cases by A first granting a lease to a group company, with SDLT group relief being claimed on the grant of the lease, then transferring the property, subject to this lease, to B. However, the relief allows a simpler process and resolves doubts as to whether anti-avoidance provisions might otherwise apply.

The relief itself does not reduce the SDLT charge on the original transfer from A to B. However, if the original transfer agreement requires the leaseback to be granted and if this requirement reduces the value of the interest originally transferred, SDLT will be charged on that original transfer, only on the reduced value. HMRC's original agreement with this view is made clear in an article at www.hmrc.gov.uk/so/exchanges.htm, referred to in Stamp Taxes Bulletin 2/2011, see www.hmrc.gov.uk/so/bulletin-22011.htm. Their continued agreement is confirmed at SDLTM04020a Example 3.

Example 3.6

Matthew owns freehold premises comprising a shop with apartment over, valued at £500,000. He sells the premises to Cassandra for £200,000 cash. The sale agreement requires Cassandra to grant a 100-year lease of the apartment back to Matthew at a peppercorn rent. No SDLT will arise on the 100-year lease as sale and leaseback relief should be available. Assuming the transaction is at arm's length, it is reasonable to regard £200,000 as representing the current value of the premises subject to the requirement to lease the apartment back, so Cassandra should have to pay SDLT on that amount.

3.41 The relief should still be available even if, for some reason, no SDLT is payable on the original transfer, but A and B must not be members of the same group and the transfer to B must not be a sub-sale or a 'pre-completion transaction' within the meaning of *FA 2003, Sch 2A* (transactions entered into before completion of contract) (*FA 2003, s 57A(3)(c)* and *(d))* (see **3.8** et seq as to the meaning of sub-sale and 'pre-completion transaction'). The transactions must be entered into in consideration of each other, so it would be wise to have this overtly stated in the documents (*FA 2003, s 57A(3)(a)*). The only other consideration permitted for the original transfer is the payment of money or the assumption, satisfaction or release of a debt (or both) (*FA 2003, s 57A(3)(b))*. Thus, the relief is not available if, for example, in consideration of the first transfer, B also agrees to provide A with services in addition to granting the leaseback to A out of the original property. However, subject to the possible application of the general anti-avoidance rules (see **7.3** et seq), it may be possible to agree the provision of services under a separate contract for cash consideration payable by A but left outstanding as a debt. The transfer from A to B could possibly then be partly in satisfaction of the debt, which is within the terms of the relief.

It appears to be HMRC's view that if A transfers a property to B in return for B agreeing to redevelop it and then grant A a leaseback of part of the developed property, this does not come within the terms of the relief. It is understood that they consider that the agreement to redevelop the property (or procure its redevelopment) is non-financial consideration for the original sale. B would be able to take account of the obligation to grant a lease to A in valuing the original transfer, but A would not be entitled to sale and leaseback relief and would pay SDLT on the full market value of the leaseback from B. This view is not made clear in published HMRC guidance.

Social housing, right to buy and shared ownership

3.42 This group of reliefs applies to a limited range of acquisitions by providers of social housing and qualifying leases/sales to their tenants. The *Housing and Regeneration Act 2008* refers to registered providers of social housing and for SDLT purposes a 'relevant housing provider' (RHP) is defined as either a non-profit registered provider of social housing or a registered social landlord (*FA 2003, s 71(1A)*). An acquisition by an RHP qualifies for complete relief from SDLT if:

(1) the majority of its trustees/board members/management committee members are tenants (*FA 2003, s 71(1)(a)* and *(2)*), or

(2) the vendor is an RHP, a housing action trust or a qualifying local or central Government body (see list at SDLTM27500) (*FA 2003, s 71(1)(b)*), or

(3) the transaction is partly or wholly funded with a grant or other financial assistance from the National Lottery or pursuant to one of the *Housing Acts* (*FA 2003, s 71(1)(c)* and *(4)*).

Profit-making registered providers of social housing are also able to benefit from the relief, but only if condition (3) above is satisfied (*FA 2003*, *s 71(1A)*).

3.43 Certain tenants are entitled to buy their homes from their public sector landlord (or from a successor body under a preserved right) at a discount to market value. The sale conditions normally include a contingent liability to repay the discount if the purchaser re-sells within a certain period. Under normal SDLT rules, this contingent further consideration should be charged to SDLT (see **2.20**). *FA 2003*, *Sch 9*, *para 1* provides that the contingent consideration does not count as consideration for SDLT purposes. This is expressed as a computational provision rather than a relief, but it is the view of HMRC that a claim to relief is required on an SDLT return, entering code 22 in Section 9 of the form. The return should be completed showing the actual consideration paid and not including the contingent amount. *FA 2003*, *Sch 9*, *para 1(3)* contains a comprehensive list of relevant 'public sector bodies', which are mainly central and local government entities and social housing bodies.

3.44 *FA 2003, Sch 9, paras 2–12* provide a treatment which may provide partial relief from SDLT for the tenant who buys an interest in his or her home under 'shared ownership' arrangements. The normal 'shared ownership' arrangement is for the tenant to enter into a lease, paying a premium (representing part ownership of the property) and rent (for the part not purchased). The treatment is available where the tenant has a right to purchase further shares by making capital payments leading to a reduction in rent ('staircasing') and/or, where the freehold reversion is available, has the right eventually to purchase it. Where staircasing is allowed, the lease must state the minimum rent which can be payable as a result of maximum staircasing without acquiring the freehold. The rules relating to this relief are complex. The following paragraphs provide a summary, but it will be important to check the terms of the lease and the status of the parties in detail against the legislation to be sure that the relief is available.

In the absence of any special treatment, SDLT would be potentially chargeable on both the premium and the rent, and further SDLT could be chargeable on later 'staircasing' transactions. This could lead to an element of double charge, since the capital payment on the staircasing transaction would relate to the same interest in the property as the initial SDLT payment on the rent. To avoid such problems, the tenant may elect to have SDLT charged on the full value of the property at the time of the initial transaction. 'Full value' here means market value, or the discounted value where the tenant is entitled to buy at a discount. Equivalent treatment is available for 'shared ownership trust' arrangements often used in relation to blocks of apartments where individual ownership of the freehold reversion is not possible.

3.45 If this election is made, SDLT is not chargeable on the rent, nor is it chargeable on subsequent staircasing transactions and/or the eventual purchase of the freehold. Even if no election is made, staircasing transactions are not subject to SDLT, provided the total share of the property owned by the tenant does not exceed 80%. Any subsequent staircasing purchases which take the proportion owned above 80% and any purchase of the freehold reversion remain subject to SDLT. Whether the election is made or not, the normal 'linked transactions' rules are modified so that the consideration for later transactions is not aggregated with that for the original grant in determining the rate of tax on the original grant. This means the later transactions do not lead to the need to reconsider the tax charged on the original grant. However, if no election is made, the consideration for all transactions (including the grant of the original lease) is aggregated in order to determine the rate of SDLT payable on staircasing transactions which leave the tenant with more than 80% of the property and on acquisition of the freehold reversion.

Election to apply the 'full value' treatment must be made in the SDLT return, or by amendment of the return within 12 months of the due date of the return. Guidance on completion of the standard SDLT1 return form in these circumstances is at SDLTM27080. HMRC instruct that the election is made by showing the market value of the property as the consideration in Box 22, although it seems likely that this should say the 'full value' as defined above, since this is the amount by reference to which the SDLT liability will be calculated. If there is thought to be any uncertainty, the election could also be notified by letter to Birmingham Stamp Office (see **Appendix A**), quoting the UTRN of the return. Careful review of the circumstances is required because the election, once made, is irrevocable.

Example 3.7

On 11 March 2017 Stanley and Oliver enter into identical 100-year shared ownership leases with the Holly Wood Housing Trust. Neither Stanley nor Oliver own any other properties and therefore the higher rates of SDLT would not apply to the transactions. They each pay an initial premium of £100,000 (which is considered to give them each a 50% share in their property) and rent of £5,000 per annum which is subject to yearly adjustment in line with the Consumer Price Index. They have the option to increase their ownership at the end of each year in multiples of 10% of the total value of the property. The minimum rent, which would apply if the tenant held 90% of the property, would be £500. Any acquisition which would take ownership above 90% necessarily includes acquisition of the freehold. Stanley elects to apply the 'full value' SDLT treatment; Oliver does not. After four years, they each opt to acquire a further 30% interest, paying £36,000 (based on the then market value) and reducing rent accordingly. At the end of the sixth year, they each buy a further 10% for £13,000. Finally, at the end of year eight, they each buy the remaining 10% including the freehold reversion, for £14,000. Neither qualifies for first-time buyer's relief or any other special treatment. Their SDLT positions are as follows:

	Stanley	**Oliver**
Initial purchase	Full value is £200,000, SDLT payable at 0% and 2% = £1,500. The NPV of the minimum rent is £13,827, which is below the £125,000 threshold, so no SDLT is payable on it.	Lease premium £100,000 and NPV of rent £138,276, so no SDLT on premium and £132 on rent. As this is residential property, the full zero rate band is available for the premium. The indexing of the rent is ignored for SDLT purposes, so this is treated as a flat £5,000 per annum rent.
30% addition, taking total ownership to 80%	No liability; no SDLT return required.	No liability; no SDLT return required.

	Stanley	Oliver
10% addition	No liability; no SDLT return required.	Aggregate of capital sums paid to date (£100,000 + £36,000 + £13,000) = £149,000, so the 0% and 2% rates apply. SDLT payable on £24,000 at 2% = £480. SDLT return required.
Final 10% purchase including freehold	No liability, but SDLT return must be submitted (see SDLTM27080).	Aggregate of capital sums paid to date (£149,000 + £14,000) = £163,000, so 2% rate still applies. SDLT payable on £14,000 at 2% = £280. SDLT return required.

Overall, Stanley pays SDLT of £1,500, Oliver pays £760 but has to complete an extra SDLT return. Economically, it would not appear to be worthwhile to make an election. However, this depends on the precise facts. In particular, Oliver faces the risk of substantial increases in value or changes in SDLT rules, leading to a higher SDLT cost. Note that, if the final purchase had pushed the aggregate consideration over £250,000, the 2% and 5% SDLT rates would have applied to the last purchase as it is fully linked to the penultimate purchase (see **2.67**).

3.46 The final relief in this section is in *FA 2003, Sch 9, paras 13–14* (introduced by *FA 2009, s 82*) and concerns 'rent to mortgage' and 'rent to loan' transactions. These are transactions where individuals initially rent a property under an assured shorthold tenancy, perhaps while saving for a deposit, and subsequently purchase an interest under a shared ownership lease or trust. The relief provides that the shorthold tenancy and the shared ownership lease/trust arrangement are not linked with each other, and that the effective date of the shared ownership transaction is determined without regard to the shorthold tenancy. This ensures that no SDLT disadvantage arises from the fact that the tenant initially occupies under a pure rental lease.

SPECIALIST AND MINOR RELIEFS

House trader relief

3.47 Full relief from SDLT is available for certain acquisitions of residential property from individuals who are buying another property. The acquisition may be made by a house building company from which the individual is buying a new dwelling, or by a property trader whose business includes buying such properties from individuals. Relief is also available for an acquisition by such a company, whether or not the replacement home is new, if arrangements to sell the old home have broken down or if the individual is moving due to relocation of employment. In the latter case, it is also possible for the employer to acquire the old home free of SDLT. Finally, relief is also available where the property is acquired by an appropriate trader from personal representatives of a deceased person.

The aims of the relief are to improve liquidity in the housing market and to assist with mobility of labour. Conditions apply to ensure the property really was used as the individual's residence, and to guard against exploitation by property developers who may buy property with the intention of refurbishing or redeveloping for onward sale (*FA 2003, s 58A, Sch 6A*).

Transfers involving multiple dwellings

3.48 A further relief introduced by *FA 2011* with effect from 19 July 2011 aims to encourage investment in residential property for letting (*FA 2003, Sch 6B*). The relief may be claimed if the main subject matter of a transaction, or of a number of linked transactions, comprises interests in more than one dwelling. Subject to satisfaction of detailed conditions, the relief may reduce the amount of SDLT chargeable on consideration other than rent.

The 3% additional rate of SDLT on the acquisition of additional dwellings and dwellings purchased by companies will apply to a claim for multiple dwellings relief if the relevant conditions are satisfied (see **2.39** et seq). If, as part of a single transaction, six or more dwellings are acquired, the SDLT rates for non-residential or mixed use transactions will apply (see **2.38**). In such circumstances the purchaser can therefore choose whether to apply the non-residential rates by not making a claim for multiple dwellings relief.

Where a claim for multiple dwellings relief is made the first stage in calculating the SDLT chargeable is to divide the aggregate consideration attributable to interests in dwellings by the number of dwellings to give an average consideration per dwelling. The SDLT chargeable on the acquisition of a residential property for the amount of the average consideration per dwelling is then calculated, and the amount so determined is then multiplied by the number of dwellings actually acquired, to give the amount of the SDLT chargeable in relation to dwellings (*FA 2003, Sch 6B, para 5(1)*). The actual values of the individual properties do not matter – only the average is important. However, the relief cannot reduce the amount of the SDLT chargeable below 1% of the consideration attributable to dwellings (*FA 2003, Sch 6B, para 5(2)*). Following the introduction of the 3% additional rate of SDLT on the acquisition of additional dwellings and dwellings purchased by companies (see **2.39**) the circumstances in which this *de minimus* rate of SDLT might apply are significantly reduced. The relief must be claimed in an SDLT return or an amended return (*FA 2003, Sch 6B, para 4(1)*). The relief is not available if relief could be claimed under *FA 2003, Sch 7* (group and reconstruction reliefs) or *FA 2003, Sch 8* (charities relief), or if any of those reliefs has been claimed and withdrawn (*FA 2003, Sch 6B, para 2(4)(b))*. The relief does not apply to the acquisition of a dwelling to which the 15% SDLT rate applies as described in **2.65** (*FA 2003, Sch 6B, para 2(4)(aa))*. In addition the relief is not available for transactions within *FA 2003, s 74*, which deals with collective purchases of freehold by leaseholders (*FA 2003, Sch 6B, para 2(4)(a))*. Finally the relief is denied if the transaction is linked with another with an effective date before 19 July 2011(*FA 2011, Sch 22, para 9)*. If the relief is claimed, no claim may be made to Disadvantaged Areas Relief under *FA 2003, s 57*. This is of limited significance, as that relief was abolished in 2013.

Example 3.8A

On 14 August 2016 Property Holdings plc (P) purchases the freeholds of five new houses from Housebuilder plc (H) for total consideration of £1.1 million. SDLT would be chargeable at the 3% additional rate (see **2.39**) and, in the absence of a claim for multiple dwellings relief, the amount payable would be £86,750 calculated as follows:

Purchase price	Rate	SDLT chargeable
£	%	£
125,000	3	3,750
125,000	5	6,250
675,000	8	54,000
175,000	13	22,750
1,100,000		86,750

If multiple dwellings relief is claimed, the amount of the SDLT chargeable is calculated as follows:

The average consideration per dwelling is £1,100,000 divided by 5 giving an amount of £220,000 and the SDLT chargeable on a purchase of a residential property for a consideration of £220,000 is as follows:-

Purchase price	Rate	SDLT chargeable
£	%	£
125,000	3	3,750
95,000	5	4,750
220,000		8,500

The SDLT chargeable in relation to the consideration for dwellings is therefore initially calculated as £42,500, being £8,500 multiplied by the number of dwellings of five.

The SDLT chargeable in relation to the consideration for dwellings can never be less than 1% of the consideration attributable to dwellings, which in this case is £11,000 (1% of £1,100,000).

Consequently, the SDLT chargeable in relation to the consideration for dwellings is £42,500, on the basis that a claim for multiple dwellings relief is made.

If there is a change of circumstances within three years of the effective date of the transaction, (or before an onward disposal to an unrelated party if sooner), and if the new circumstances would have led to a higher tax charge at the time of the original transaction, in relation to the five dwellings qualifying for multiple dwellings relief, a further SDLT return must be made and the additional tax paid within 30 days of the change of circumstances (*FA 2003, Sch 6B, para 6:* see **Example 3.9**).

The relief may also apply if the transaction includes property which does not count as a dwelling. Consideration must then be apportioned on a just and reasonable basis between the dwellings and the other property. The relief is applied to the consideration apportioned to the dwellings. SDLT is charged on the consideration apportioned to the other property as the appropriate fraction of the amount of SDLT which would have been chargeable had no claim been made for multiple dwellings relief.

Example 3.8B

As part of the same transaction (ie under a single contract), P also purchases three shop units from H. The total consideration is £1.55 million, of which £1.1 million is justly apportioned to the houses and £450,000 to the shops. In the absence of a claim for multiple dwellings relief SDLT on the house purchases is not charged at the higher rates as the transaction is the acquisition of non-residential property (ie both residential and non-residential property is being acquired as part of the same transaction) and therefore the SDLT rates for non-residential or mixed use property are used to calculate the SDLT charge (see **2.39** et seq).

As detailed above, on the basis that the transaction does not comprise entirely of residential property, as it includes the acquisition of the shops, the transaction is taxed as the acquisition of non-residential property (*FA 2003, s 55(2)(b)*). The SDLT payable on the total consideration of £1.55 million is calculated as follows:

Purchase Price (£)	Rate (%)	SDLT (£)
150,000	0	Nil
100,000	2	2,000
1,300,000	5	65,000
1,550,000		67,000

The SDLT chargeable, in the absence of a claim for multiple dwellings relief, would therefore be £67,000.

The SDLT chargeable on the shops is then calculated as follows:

£450,000/£1,550,000 × £67,000 = £19,452

The SDLT chargeable on the five houses is then calculated as follows:

£1,100,000/£1,550,000 × £67,000 = £47,548.

However, if a claim for multiple dwellings relief is made in relation to the acquisition of the five houses the SDLT payable on acquisition of those houses can be reduced to £42,500 (see **Example 3.8A**). It should be noted that the higher rates of SDLT will apply where a claim for multiple dwellings relief is made as the purchaser is a company and the liability is calculated on the basis of an acquisition of wholly residential property (*FA 2003, Sch 6B, para(1)*).

The total SDLT on the transaction is therefore £42,500 plus £19,452, or £61,952.

For these purposes 'dwelling' means a building or part of a building used or suitable for use as a dwelling or in the process of being constructed or adapted for such use, together with any garden or other grounds to be enjoyed with it. HMRC guidance indicates that the normal Capital Gains Tax criteria will be applied to assess whether grounds may be regarded as part of the dwelling. This does not appear to be correct however as the test for determining what garden and grounds occupied with a dwelling comprise residential property for SDLT purposes is different to the test for determining what land occupied with a dwelling qualifies for capital gains relief. For example, unlike the capital gains legislation, the SDLT definition of residential property does not provide for a permitted area of 0.5 of a hectare. It may also include property which is to be constructed or adapted as a dwelling but where work has not yet commenced – but only if the SDLT arises on 'substantial performance' of a contract (see **2.9**). However, if the dwelling acquired is subject to a lease the term of which was 21 years or more when originally granted, a superior interest to that lease does not count as a dwelling (*FA 2003, Sch 6B, para 2(6)*). This restriction on the availability of the relief does not apply, for transactions with an effective date on or after 26 March 2015, if all of the following conditions are satisfied (*FA 2003, Sch 6B, para 2(7)*, inserted by *FA 2015, s 69(1)*:

(a) the vendor is a 'qualifying body' (which term includes housing authority, housing association – see *FA 2003, Sch 9, para* 5);

(b) the transaction is a sale under a sale and leaseback arrangement within the meaning of *FA 2003, s 57A(2)*;

(c) the sale transaction is the grant of a new lease; and

(d) the leaseback is exempt under *FA 2003, s 57A* (see **3.40** et seq).

FA *2003, s 116(7)* (six or more residential properties to be treated as non-residential) is specifically stated not to apply to properties in respect of which multiple dwellings relief is claimed (*FA 2003, Sch 6B, para 7(7)*). However in a mixed transaction it will still apply in calculating the SDLT chargeable on those properties in respect of which multiple dwellings relief is not due.

Example 3.8C

On 30 November 2016 P Ltd, a property investment company, agrees to purchase six houses from H as part of a single transaction, in circumstances where H has leased three of them to tenants under 99-year leases. In the absence of a claim for multiple dwellings relief, as six dwellings are being acquired under a single transaction *FA 2003, s 116(7)* would apply and the transaction would be treated as the acquisition of non-residential property. However, P Ltd should consider whether a claim for multiple dwellings relief would be advantageous. The consideration justly attributable to the three unoccupied houses is £1 million, with £100,000 attributable to the three leased houses. Multiple dwellings relief only applies to the three unoccupied houses (as the other properties are subject to leases which were originally granted for a term of 21 years or more – *FA 2003, Sch 6B, para 2(6)*), and the amount of the SDLT chargeable is calculated as follows:

The average consideration per dwelling is £1,000,000 divided by three giving an amount of £333,333, and the SDLT chargeable on a purchase of a residential property for a consideration of £333,333 is as follows:

Purchase Price	Rate	SDLT Chargeable
£	%	£
125,000	3	3,750
125,000	5	6,250
83,333	8	6,667
------------		-----------
333,333		16,667
------------		------------

The SDLT chargeable in relation to the consideration for dwellings is therefore initially calculated as £50,001, being £16,667 multiplied by the number of dwellings of three.

As 1% of the consideration attributable to dwellings would be £10,000 (£1,000,000 × 1%) the SDLT chargeable in relation to dwellings is £50,001, as that amount is greater than £10,000.

The SDLT chargeable in relation to those houses which do not qualify for relief is then calculated as follows.

As indicated above, as the transaction comprises the acquisition of six residential properties as part of a single transaction the houses are treated as being non-residential property (*FA 2003, s 116(7)*). The SDLT payable on consideration of £1,100,000 in a non-residential property transaction is as follows:

Purchase Price (£)	**Rate (%)**	**SDLT (£)**
150,000	0	Nil
100,000	2	2,000
850,000	5	42,500
1,100,000		44,500

The SDLT chargeable in the absence of a claim for multiple dwellings relief would therefore have been £44,500. As this is less than the SDLT liability on the purchase of the three houses, assuming a claim for multiple dwellings relief is made, there is no benefit in making that claim, and the transaction should be taxed using the non-residential rates of SDLT, resulting in a liability of £44,500.

Example 3.9

On 1 April 2017 Cupid Ltd pays Venus £800,000 for a building on which work has started to convert it into four flats, on the basis that Cupid Ltd will finish the conversion work. The 3% additional rate will apply as the transaction involves a company acquiring residential property (see **2.39** et seq). Consequently, Cupid Ltd pays SDLT calculated as follows:

The average consideration per dwelling is £800,000 divided by 4 giving an amount of £200,000, and the SDLT chargeable on a purchase of residential property for a consideration of £200,000 is as follows:

Purchase Price	Rate	SDLT Chargeable
£	%	£
125,000	3	3,750
75,000	5	3,750
-----------		----------
200,000		7,500
-----------		----------

The SDLT chargeable in relation to the acquisition of the building is therefore initially calculated as £30,000, being £7,500 multiplied by the number of proposed dwellings of four.

The minimum rate of SDLT does not apply as 1% of the consideration attributable to dwellings would be £8,000 (£800,000 x 1%) and the SDLT chargeable in relation to the transaction is £30,000 (*FA 2003, Sch 6B, para 5(2)*).

Work progresses slowly and, two years later, Cupid Ltd decides to revise the design and create just two larger flats. If the original plan had been to create only two flats the SDLT chargeable would have been calculated as follows:

The average consideration per dwelling is now £800,000 divided by two, giving an amount of £400,000, and the SDLT chargeable on a purchase of a residential property for a consideration of £400,000 is as follows:

Purchase Price	Rate	SDLT chargeable
£	%	£
125,000	3	3,750
125,000	5	6,250
150,000	8	12,000
------------		------------
400,000		22,000
------------		------------

The revised amount of SDLT chargeable on the acquisition of the building is £44,000, being £22,000 multiplied by two. Within 30 days of Cupid Ltd deciding to create only two flats, he must make a further SDLT return and pay the additional tax of £14,000 (£44,000 - £30,000).

Charities relief

3.49 There are three ways in which an acquisition by a charity or charitable trust may be relieved from charge, as follows:

(1) Relief is available if the whole property is to be held for 'qualifying charitable purposes'. This means either furtherance of the charity's

charitable purposes or as an investment to produce profits to be used for those charitable purposes (*FA 2003, Sch 8, para 1*).

(2) Alternatively, full relief is still available if less than all, but more than 50% by value, of the property is to be held for qualifying charitable purposes, but the clawback rules are then different (*FA 2003, Sch 8, para 3*).

(3) Originally HMRC took the view that, where a charity acquired property jointly with a non-charity, no relief was due. However in *The Pollen Estate Trustee Company Limited and Kings College London v HM Revenue & Customs* (June 2013), the Court of Appeal held that the charity was entitled to relief from its share of the total SDLT charge, provided the other conditions for relief were met. The Court confirmed that SDLT payable in respect of the non-charity's share of the property should be calculated at the rate applicable to the whole consideration. In an attempt to ensure clarity, *FA 2014, s 113* and *Sch 23* introduced *FA 2003, Sch 8, paras 3A* to *3C* with effect from 17 July 2014 . These paragraphs give relief where one or more charities purchase property with one or more non-charities, provided the property is acquired by them as tenants in common. The relief given is the proportion of the total charge equal to the lower of the proportion of the property owned by the charity and the proportion of the purchase consideration provided by the charity. Because the relief is given in terms of tax rather than by disregarding consideration, the rates of SDLT are determined by the total consideration given by all parties. Relief is available whether the whole or more than 50% by value of the charity's interest is to be held for qualifying charitable purposes. However the clawback rules are different in those two cases as noted at **3.50.**

The changes do not directly deal with joint acquisition by two or more charities (without involvement of a non-charity). It is thought HMRC may accept the general legislative principle that the singular includes the plural to grant relief but complications may arise in relation to clawback where one charity qualifies for relief under (1) above and the other under (2).

3.50 In all cases, relief is denied if the transaction was entered into for the purpose of avoiding SDLT (*FA 2003, Sch 8, para 1(3)*). Relief is clawed back if, either within three years of the acquisition, or later under arrangements entered into within the three-year period, a 'disqualifying event' occurs and the charity still holds the property or an interest derived from it.

If relief was obtained under (1) above, 'disqualifying event' means either the charity ceasing to be established for charitable purposes only, or the charity beginning to use the property, or any interest derived from it, for non-charitable purposes. If only part of the property is still held when the disqualifying event occurs, or if only part becomes used for non-charitable purposes, an appropriate proportion of the relief is withdrawn.

Alternatively, if relief was obtained under (2) above, 'disqualifying event' *additionally* includes any transfer of all or part of the land, or the grant of a lease at a premium or at a rent of less than £1,000 per year, unless that transfer or grant is made in furtherance of the charitable purposes of the charity.

If relief is obtained under (3) above, the same definitions of disqualifying event are applied, depending on whether the whole of the charity's share, or only part, was originally to be held for qualifying charitable purposes.

3.51 From 1 April 2012, the provisions of *FA 2010, Sch 10* apply to determine whether an organisation qualifies as a 'charity' for the purposes of stamp taxes, including SDLT. *Sch 10* requires that the charity be:

- established for charitable purposes only;
- subject to the courts in the UK, elsewhere in the EU, Iceland or Norway;
- registered, if the laws of the home territory so require; and
- administered by 'fit and proper persons'.

These rules are designed to prevent abuses of charitable status. HMRC are particularly keen on the last requirement and suggest that Trustees be required to self certify that they qualify (see www.hmrc.gov.uk/charities/tax/recognition.htm). The declaration reflects HMRC's view as to what 'fit and proper' means but its use is not obligatory.

Example 3.10

On 1 December 2016, the Q Charitable Trust buys freehold land for £1 million, with the intention of constructing a new operations centre for its own use on three quarters (by area) of the site and selling the other quarter to an unrelated company, R, for a previously negotiated price of £300,000. Because the property is not to be used wholly for charitable purposes, full SDLT relief of £39,500 is claimed under *FA 2003, Sch 8, para 3*, as set out under (2) in **3.49** above. The sale to R falls through, but on 1 May 2017, a slightly different plot is sold to another unrelated buyer, S, for £350,000. It is agreed with HMRC that the value of the plot sold is one quarter of the value of the whole site. At this point, one quarter of the relief previously granted is clawed back, and Q must complete a new SDLT return disclosing the charge and making payment. The SDLT payable is one quarter of the SDLT of £39,500 in relation to which relief was claimed, ie £9,875. The SDLT clawed back is based on the appropriate proportion of the SDLT which would have been payable in the absence of a claim for charities relief and *not* the SDLT which would be payable based on the value of the plot disposed of alone.

Note that, if Q had intended to use the whole site for its charitable purposes at the time of acquisition and thus claimed relief under *FA 2003, Sch 8, para 1*, as set out under (1) above, but subsequently decided to sell the plot to S, this disposal would not have led to any clawback of relief.

Construction proceeds slowly but, as a result of changes in practice at Q during the development period, the operations centre is larger than needed. On 1 November 2017, Q agrees to grant a 25-year lease of one floor to an unrelated company, T, at a premium of £800,000 and a peppercorn rent. The premium is paid and T occupies on 31 March 2018. It is agreed with HMRC that the value of the floor leased out is 40% of the value of the original site acquired for £1 million. At this point, 40% of the relief originally granted and not yet withdrawn is now clawed back – that is 40% of £39,500, or £15,800. At this point, Q owns less than 50% of the original site. This is not a problem as, when the relief was claimed, Q intended to use more than 50% for its charitable purposes. However, given the amount finally disposed of, there may be some risk that HMRC would challenge the original intention.

Incorporation of limited liability partnership

3.52 Relief is available where partnership property is transferred to a limited liability partnership (LLP) from some other form of partnership or from the partners, on incorporation of the LLP (*FA 2003, s 65)*. The relief only applies to a straightforward change of form from another kind of partnership, with no change in partners or their relative interests in the partnership. To qualify, the transfer must happen within 12 months of the date of incorporation of the LLP. In the absence of the relief, the special partnership rules discussed at **5.13** et seq would apply such that, on a transfer from another form of partnership to the LLP, with no change of partners or their relative interests in the partnership, no charge to SDLT would be expected to arise. However, a claim under *FA 2003, s 65* is generally seen as a simpler method of claiming relief assuming the requisite conditions are satisfied. There is an apparent conflict with the rule in *FA 2003, s 53 (see* **2.30**) that the chargeable consideration for a transfer to a connected company (because an LLP is a body corporate and is therefore a 'company' for the purposes of *FA 2003, s 53*) is an amount not less than the market value of the property. However, HMRC have confirmed at SDLTM 34170 that the special partnership rules take precedence over *FA 2003, s 53.*

Reliefs for other specified bodies and transfers

3.53 Relief is available for acquisitions in a range of transactions by other bodies, specified either individually or by class. These are listed below; most are very specialised. Relief is usually subject to conditions, so it is essential to consult the relevant legislation and/or take expert advice before claiming the relief. In general, the relief is claimed by inserting '28' (other reliefs) at the appropriate point in the SDLT return (see **6.23**). In addition, specific relief is often provided in legislation other than *Finance Acts*, in respect of transfers in the course of government-sponsored projects. It is, therefore, generally worth checking such legislation in these cases or, if involved early enough in the process, lobbying for such relief to be provided. Much of the legislation noted below pre-dates SDLT and originally granted stamp duty relief, but it has been amended, largely by statutory instrument under *FA 2003, s 123(3)*. There is a list of miscellaneous reliefs not in the SDLT legislation at SDLTM29600, accessible at www.hmrc.gov.uk/manuals/sdltmanual/sdltm29600.htm.

Qualifying transfer	**Legislation and notes**
Transfer on compulsory purchase facilitating development	*FA 2003, s 60*; relief applies to planning authority or other person making the compulsory purchase order
Transfer to a public authority to comply with planning obligations, for example under *Town and Country Planning Act 1990, s 106*	*FA 2003, s 61*
Between 'public bodies' (local and national government bodies, National Health Service bodies, their wholly-owned subsidiary companies and other statutory bodies, on reorganisation*	*FA 2003, s 66* (see detailed list in legislation) References to NHS bodies are probably redundant (see below)
Between Local Constituency Associations on reorganisation of parliamentary constituencies	*FA 2003, s 67*
Acquisition by British Museum, Natural History Museum, and other specified similar bodies; certain transfers to 'heritage bodies' of property accepted in payment of tax	*FA 2003, s 69*; *National Heritage Act 1980, s 11A**
On demutualisation of an insurance company or a building society	*FA 2003, ss 63, 64* (provided the demutualisation satisfies conditions; see legislation)
On merger of building societies or transfer of business*	*Building Societies Act 1986, s 109A*
Acquisition by NHS Trusts, Foundation Trusts and similar bodies in the UK*, including new bodies established under current and proposed reforms of the NHS	*Health and Social Care (Community Health & Standards) Act 2003, s 33(2)*; *FA 2003, s 67A*, inserted by *FA 2012, s 216*
Acquisition by designated visiting armed forces or NATO headquarters*	*FA 1960, s 74A*
Transfers in the course of an amalgamation of friendly societies or transfer of friendly society business*	*Friendly Societies Acts 1974* and *1992, s 105A* in each case

Qualifying transfer	**Legislation and notes**
Certain transfers in relation to roads, airports and lighthouses*	*Highways Act 1980, s 281A*; *Airports Act 1986, s 76A*; *Merchant Shipping Act 1995, s 221*
Certain transfers relating to inclosures and common land*	*Inclosure Act 1845, s 163A*; *Metropolitan Commons Act 1866, s 33*

* These reliefs are subject to the modified rules on subsequent disposal of a lease; see **4.17**.

OBSOLETE RELIEFS

Disadvantaged areas relief

3.54 This relief, in *FA 2003*, *s 57* and *Sch 6*, originally allowed complete relief from SDLT on transfer of property in some 2,000 specified areas of the UK which, on the basis of rather old census information, were regarded as disadvantaged. They included areas which had subsequently become desirable and valuable, in particular a number of high-profile shopping centres which were therefore quickly transferred into special purpose companies, allowing future transfers of those companies free of SDLT. The relief was modified by *FA 2005*, *s 96* to apply only to wholly residential property where the consideration did not exceed £150,000. Since there was a general threshold of £125,000 below which SDLT was not chargeable on residential property, and there were various temporary increases in this threshold, in practice disadvantaged areas relief applied to few transactions. The government recognised this fact and *FA 2012* abolished the relief from 6 April 2013. This is subject to retention of the relief under certain circumstances for completion of contracts entered into on or before 16 March 2005.

Unit trust seeding relief

3.55 This relief, originally in *FA 2003*, *s 64A* but now abolished, allowed transfer of property to a new unit trust in return for the issue of units, without SDLT cost. Its purpose was to allow bodies such as investment and pension funds to turn fundamentally illiquid property investments into tradable units, in order to comply with new regulations relating to the structure of their balance sheets. However, the relief was loosely drawn and allowed many owners to transfer properties to essentially privately held, non-UK unit trusts. It was then possible for effective ownership of the properties to be transferred free of stamp tax by transfer of the units. Many properties are still held in this way. The relief was withdrawn by *FA 2006*; unusually, the 2006 Budget announcement specified that the withdrawal took effect from 2pm on Budget day, in an attempt to prevent a last-minute scramble to take advantage of the relief.

First-time buyer's relief

3.56 *FA 2003, s 57AA*, introduced by *FA 2010, s 6*, provided a time-limited relief for the first purchase of a wholly residential property by a person or persons ('first-time buyer/s') who had not previously been a 'purchaser' of a residential property. The relief was available for purchases with an effective date between 25 March 2010 and 24 March 2012 inclusive. It applied if the consideration for the purchase exceeded £125,000 but did not exceed £250,000. It therefore only applied to purchases which would otherwise have been subject to SDLT at 1%. No relief was available if the price exceeded £250,000 – there was no question of partial relief. The relief also applied to appropriate acquisitions under 'alternative finance' arrangements (see **3.37**), removing the single charge to SDLT which would otherwise remain after application of the alternative finance reliefs. Strangely, the relief did not overtly specify that the first-time buyers must be one or more individuals, but this was perhaps implied by the condition that the property had to be for occupation by the buyer(s) as their only or main residence. If the transaction was the grant of a lease or the acquisition of a second-hand lease, the lease had to have at least 21 years to run. The relief was denied if the acquisition was linked with any others. To claim the relief, an SDLT return had to be completed and code 32 entered in Section 9.

Zero carbon homes relief

3.57 This relief in *FA 2003, s 58B* was introduced by *FA 2007*. It applied for a limited period of five years from 1 October 2007 and so expired on 30 September 2012. Full relief from SDLT was available on the first acquisition of a dwelling which was a 'zero carbon home' and with a price not exceeding £500,000. For more expensive homes, the SDLT bill was reduced by £15,000 (being the SDLT on a £500,000 home). Zero carbon was defined by reference to zero net carbon dioxide emissions over the course of a year, and therefore effectively required the dwelling to generate power for return to the national grid to compensate for energy used. 'First acquisition' meant acquisition on or before first occupation. As initially drafted, the relief applied only to houses which were separate buildings, but it was realised that, in the immediate future at least, the only dwellings with any practical hope of being 'zero carbon' would be flats. The relief was therefore modified, with retrospective effect, to include parts of buildings. Detailed regulations provided a mechanism for assessment and certification of qualifying homes – it was considered that this could easily be combined with the normal energy efficiency assessment required for new homes under Building Regulations (*FA 2003, ss 58B, 58C*; *SI 2007/3437*). To claim this relief, code 30 had to be entered in Section 9 of the SDLT return.

The relief only applied to the first acquisition of a new dwelling. It did not, therefore, encourage the modification of existing homes to improve their energy efficiency.

Chapter 4

SDLT and Leases

INTRODUCTION

4.1 Each of the grant, variation, assignment and surrender of a lease is a 'land transaction' and therefore potentially gives rise to SDLT obligations as set out in **Chapter 2**. As with other UK land transactions, there may be obligations to make or amend SDLT returns even if no liability arises (see **6.2**). However, where leases are concerned, the general rules are substantially amended and generally made more complex. There are special rules for calculation of SDLT on rent in *FA 2003*, *Sch 5*, but the main technical provisions are in *Sch 17A*. This schedule was inserted after enactment of the main SDLT rules, and has itself been amended several times in an attempt to make the rules easier to apply whilst guarding against avoidance.

Obligations may arise at various times, such as when a payment is made, on first occupation, when rent amounts are agreed or changed, and even when the lease itself is formally executed. They may also arise in connection with arrangements which are not normally regarded as giving rise to a lease – for example, informal occupation or periodic tenancies – and on less formal changes, even if not documented as a variation etc. It is, therefore, important to establish clearly the nature of the transaction. The four key questions to address are the same as for other transactions; but, in relation to leases, the fourth question is likely to require the most attention:

- What SDLT obligations and liabilities may arise?
- At what time do SDLT obligations and liabilities arise?
- What reliefs are available?
- What special rules may affect this transaction?

It has been recognised that some aspects of the rules are oppressively complex. *FA 2013*, *Sch 41* provided some simplification by abolishing rules relating to abnormal rent increases (see **4.27** for explanation of the old rules), and making the rules relating to substantial performance of a lease before execution (see **4.74**) and leases which continue after a fixed term (see **4.13**) more logical. These changes are welcome but the rules remain complex and there is undoubtedly scope for further simplification.

4.2 In general a lease granted before 1 December 2003 is exclusively within the old stamp duty regime. SDLT will only arise in connection with such a lease if an event occurs which is treated as the grant of a new lease – a renewal or extension of the term, a variation which in law amounts to the surrender and regrant of the lease, or an increase in rent within the first five years of the term (SDLTM 10025). Other changes such as an increase in rent after the first five years or continuation or holding over of the lease (see **4.13**), whether or not under the *Landlord & Tenant Act 1954*, will not give rise to SDLT charges (see SDLTM10025 and SDLTM15010).

4.3 SDLT may be chargeable on both rent and any real or deemed premium.

On the grant of a new lease, SDLT is charged on any real or deemed premium on the same basis as it is charged on consideration given for any other 'land transaction' (see **2.7** and **2.37**).

Like other forms of 'chargeable consideration' (see **2.19** et seq), the rates and bands at which SDLT is charged on consideration consisting of rent depends on the amount of the chargeable consideration, and whether the transaction is a residential property transaction or a non-residential property transaction.

Until 16 March 2016 the bands and rates of SDLT applying to rent, on the grant of a new lease over residential property and over non-residential property were the same and were as follows:

Net present value of rent (£)	**Rate (%)**
Up to 150,000	0
Over 150,000	1

The above bands and rates continue to apply to rent on the grant of a new lease over residential property.

However, for transactions with an 'effective date' on or after 17 March 2016 the following bands and rates apply to rent on the grant of a new lease over non-residential property (*FA 2016, s 116(9)*):

Net present value of rent (£)	**Rate (%)**
Up to 150,000	0
Over 150,001 to £5,000,000	1
Over £5,000,000	2

In relation to the rates and bands applying to rent payable under a lease of non-residential or mixed use property there are transitional rules, which provide that where an agreement for lease was entered into before 17 March 2016, but the agreement was not completed until on or after 17 March 2016, then provided that there is no event of the kind listed in *FA 2016, s 116(15)*, which results in the effect of the agreement being different from the effect of the agreement when first entered into, the purchaser can elect to pay SDLT based on the old rates and bands (*FA 2016, s 116(13)(b)* and *(15)*). Similarly, where the agreement for lease was entered into and substantially performed (see **2.9**) but not completed (see **2.8**) before 17 March 2016 then the purchaser can elect to apply the old rates and bands to the completion transaction taking place on or after 17 March 2016 (*FA 2016, s 116(13)(a)*). To choose to pay SDLT based on the old rates and bands the purchaser just enters the appropriate amount of SDLT in box 14 and 25 of the land transaction return.

The grant of a new lease is a residential property transaction if the property over which the lease is granted consists *entirely* of residential property (see **2.38**) or, if the grant of the lease is one of a number of linked transactions (see **2.67**), the chargeable interests acquired under each linked transaction, and the properties over which any new lease is granted, consist *entirely* of residential property (*FA 2003, s 55(1B), (1C), (3)* and *(4)*).

The grant of a new lease is a non-residential property transaction if the property over which the lease is granted either consists of, or includes, non-residential property. If the grant of the new lease is one of a number of linked transactions (see **2.67**), it will be a non-residential property transaction if the chargeable interests acquired under any of the linked transactions, and the properties over which any new lease is granted, consist of, or include, an interest in non-residential property (*FA 2003, s 55(2), (3)* and *(4)*).

The grant of a new lease is a mixed use property transaction if the property over which the lease is granted consists of, or includes, both residential and non-residential property. If the grant of the new lease is one of a number of linked transactions (see **2.67**), it will be a mixed use property transaction if the chargeable interests acquired under any of the linked transactions, and the properties over which any new leases are granted, consists of, or include both residential and non-residential property (*FA 2003, s 55(2), (3)* and *(4)*).

Thus the tax rates and bands in relation to rent for residential property will only be applicable if the property, or properties, over which the lease is granted consist(s) *entirely* of residential property, and the grant of the lease is not 'linked' (see **2.67**) with any 'chargeable transaction' (see **2.14**) over non-residential or mixed use property.

For transactions with an 'effective date' (see **2.8**) on or before 16 March 2016, where the rent payable under a lease of non-residential property was £1,000 or more per annum there was no zero rate band in respect of any consideration other than rent. Consequently, any premium payable under such a lease, in an amount of up to £150,000, was taxed at a rate of 1% (*FA 2003*, *Sch 5*, *para 9A*). However, this rule was abolished, subject to the transitional provisions discussed above, for any transaction with an effective date on or after 17 March 2016 (*FA 2016*, *s 116(11)*). This therefore means that from that date the £150,000 zero rate band applies when calculating the SDLT chargeable on a premium payable on the grant of a lease of non-residential property.

SDLT obligations fall on the 'purchaser'. In most cases, this will be the tenant/lessee but, where a lease is surrendered or varied to shorten the term, the landlord/lessor is the purchaser (*FA 2003*, *s 43*, *Sch 17A*, *para 15A*).

4.4 The general rules relating to timing and notification of liability, availability of reliefs and disclosure of information apply to lease transactions. Specific rules for calculating the SDLT liability and modifications to other general rules are dealt with in detail by way of examples in the text which follows.

4.5 Sometimes, liabilities will arise at a future time on a lease granted now, or now on a lease granted years ago, especially where the lease is transferred to another tenant. It may, therefore, be important to obtain and retain the full detailed history of a lease, and of previous leases of the same premises if they are linked with the present lease (see **4.58**).

EVENTS GIVING RISE TO SDLT OBLIGATIONS AND LIABILITIES

New lease

4.6 SDLT may arise on entering into a lease (*FA 2003*, *s 43(3)(a)*). It is common for the parties initially to enter into an agreement for lease, the definitive lease being executed later, after all necessary details have been settled. The agreement may be acted on before the definitive lease is executed (for example, by payment of rent or the tenant having access to the premises). The general SDLT concept of 'substantial performance' applies to an agreement for lease (see, eg *FA 2003*, *s 44(7)*). SDLT obligations may therefore arise when the agreement is acted on and when the definitive lease is executed – see **4.74**.

As a general principle an amount payable in respect of a period before a lease commences (or is treated as commencing) cannot be regarded as rent even if it is expressed as such (see SDLTM14015). Provided it really is consideration for the lease, any such amount will be treated as a premium. This may occur, for example, when negotiations between landlord and putative tenant have been delayed and the landlord is able to demand 'rent' for an earlier period before the tenant took occupation or signed the lease (see **Example 4.3** for an illustration of this). More usually, the payment of such an amount, (or earlier taking possession of the property) will trigger the commencement of a notional lease, as explained at **4.74**, and the amount paid will be respected as rent for that notional lease.

The one exception arises on renewal of a lease originally entered into within the SDLT regime; see **4.65** and **Example 4.23C**.

New lease for seven years or more

4.7 For non-lease transactions, there is no requirement to make an SDLT return if the consideration is less than £40,000 (*FA 2003, s 77A(1)*). Where a lease for seven years or more is granted, this exemption from making a return only applies if the annual rent is also less than £1,000. This was logical for a non-residential lease granted on or before 16 March 2016 (see **4.3**) involving both rent and premium because, if the annual rent was £1,000 or more, there was no 'zero rate band' for the premium and any premium, however small, gave rise to an SDLT liability (*FA 2003, Sch 5, para 9A*). However, this rule was abolished for transactions with an 'effective date' (see **2.8**) on or after 17 March 2016. Therefore the continuing requirement that the rent must be less than £1,000 per annum, otherwise the grant of the lease is notifiable, would now seem to have little logic.

Example 4.1 – rent and premium below threshold, notifiable

On 1 January 2017, Methuselah enters into a 250-year lease as tenant on a plot of farmland at a yearly rent of £5,200, subject to revision every five years in line with movement in the Retail Price Index. A premium of £35,000 is payable under the terms of the lease. The tenant has the right to terminate the lease at any time after 1 January 2037. Increases in rent in line with the Retail Price Index are ignored for SDLT purposes (see **4.37**), so this is treated as a lease at a fixed rent of £5,200 per annum. The NPV of the rent is £148,542, which is below the threshold of £150,000 so no SDLT is payable on the rent. Equally no SDLT is payable on the premium as it is below the £150,000 threshold, and below the £40,000 threshold above which non-lease transactions must be notified. Nonetheless, an SDLT return must be submitted within 30 days of execution or earlier substantial performance of the lease, as the rent payable under the lease is not less than £1,000 per annum.

New lease for less than seven years

4.8 The grant of a lease for less than seven years does not have to be notified unless the consideration (rent and/or premium) exceeds the relevant zero rate band. The relevant zero rate band for consideration other than rent is up to £125,000 for residential property and up to £150,000 for non-residential property. For rent the zero rate band applies to an aggregate NPV of up to £125,000 for residential property and up to £150,000 for non-residential property.

Example 4.2 – rent and premium below threshold, not notifiable

On 30 June 2016 Moira grants Nathan a six-year lease of a house (single residence) for a premium of £30,000 and an annual rent of £10,000. The NPV of the rent is clearly below the residential threshold for rent (£125,000), and the premium is within the zero rate band (also £125,000). The higher rates of SDLT (see **2.39**) do not apply as a lease granted for a term of less than seven years is not a 'major interest' (see **2.11**) for the purposes of *FA 2003*, *Sch 4ZA* (*FA 2003*, *Sch 4ZA*, *para 2(4)*). Since the lease is for less than seven years and no tax is payable on either rent or premium, no SDLT return or other notification is required.

Example 4.3 – liability on NPV of rent: rent paid for period prior to grant of lease

On 1 March 2017, Craig agrees to grant a five-year lease of a shop to Laura for rent of £50,000 per annum, payable quarterly in advance. It is agreed that the lease will commence on 1 April 2017. However, a dispute over the detailed terms and conditions means the lease is not executed until 1 May 2017. Laura makes the first rent payment, for the period 1 April to 30 June 2017, on 1 May 2017, and takes possession of the shop on the same day. Since nothing happens to cause the agreement to be substantially performed before the lease is executed, the lease is deemed to commence on the date of execution and not at the (earlier) alleged start date – so, for SDLT purposes this is a 4-year, 11-month lease commencing on 1 May 2017. Although £4,167 (£50,000/12 months) of rent is expressed as being for the period 1 to 30 April 2017, rent cannot in fact relate to a period before the lease begins (see **4.6**) (the one exception to this rule arises when a lease, originally executed under SDLT, is renewed – see **4.65** and **Example 4.23C**). It is HMRC's view that this amount must therefore be regarded as a premium for the lease. The NPV of the aggregate rent payable over the term of the lease is £222,244 and SDLT of £722 is payable on that amount of rent. No SDLT is payable on the deemed premium of £4,167 as it is below the £150,000 threshold. Laura must file a land transaction return, and pay the SDLT due of £722, by 31 May 2017.

Licence to occupy and informal arrangements

4.9 A licence to occupy land is not a 'chargeable interest' (*FA 2003, s 48(2)(b)*), so no SDLT obligations arise on entering into or terminating a licence. However, HMRC warn that the nature of an interest in land is not determined by calling it a 'licence'; in any case of doubt, appropriate legal advice should be sought. A tenancy giving exclusive occupation of a property is more than a licence and is potentially subject to SDLT.

4.10 Normally, a transaction creating a tenancy is reduced to writing, but it is possible for a tenancy to be implied by the actions of the parties. No liability should arise on a tenancy at will (*FA 2003, s 48(2)(c)*, see SDLTM 10050), but over time this may develop into a periodic tenancy which is potentially within the charge. However, it may be that the arrangement commenced before 1 December 2003 (the date on which SDLT was introduced). In that case, no SDLT obligations should arise unless and until the arrangement is varied. It is suggested that appropriate legal advice should be taken to establish the status of any such arrangements.

4.11 HMRC's draft guidance on SDLT and partnerships originally suggested that, if a partnership informally occupied a property owned by a partner, the property should be regarded as belonging to the partnership rather than simply let to it. An article in SDLT Technical News 5 (August 2007) announced the more reasonable view that this would depend on the facts of the individual case (see **5.43**). The final guidance on SDLT and partnerships (SDLTM33000 et seq) is silent on the matter.

Lease for an indefinite term and holding over

4.12 A periodic tenancy, whether or not reduced to writing, is probably the most common example of a 'lease for an indefinite term' (*FA 2003, Sch 17A, para 4*). Such a lease, assuming it comes into existence on or after 1 December 2003 (the date on which SDLT was imposed), is treated in the first instance as a lease for one year. If it continues beyond the end of the first year, it is treated as having been a lease for two years from the outset. At the commencement of each successive year thereafter, it is treated as having been, from the outset, a lease for one year longer than at the commencement of the previous year. Any SDLT obligations then must be dealt with within 30 days of the start of each new one-year period. So if the lease is notifiable and/or SDLT is payable as a result of the increase in the deemed period of the lease, the return and/or SDLT payment are due within 30 days of the beginning of the year. This contrasts with the treatment of leases which continue after a fixed term where, for periods beginning on or after 17 July 2013, the SDLT obligations are postponed until 30 days after the end of the year (see **4.13**). There is no adjustment if the lease for an indefinite term ends before the end of a year in respect of which SDLT has been paid (for example the tenant vacates the premises or alternatively enters into a fixed term lease in place of the periodic tenancy). However if the periodic tenancy is superseded by a fixed term lease, allowance should be available in calculating the SDLT on the fixed term lease, in respect of SDLT paid on the periodic tenancy for the overlap period, see **4.83**.

Commonly, particularly in relation to residential tenancies, the rent will not be high enough to produce any SDLT liability in the early years. However, as time passes and the lease 'grows' in length, SDLT obligations may arise. Although the notification requirements for leases under and over seven years are different, it is specifically provided that a periodic tenancy does not become notifiable purely as a result of passing the seven-year mark.

None of this applies to a tenancy which came into existence before 1 December 2003, unless it is varied in such a way as to bring it within the SDLT regime (see **4.2** but note that continuation from one year to the next in accordance with the written or implied terms of the tenancy does not amount to a renewal or extension of the term).

A tenancy at will is also specifically stated to be a periodic tenancy. This seems a bizarre provision, given that a tenancy at will is outside the charge to SDLT, as noted at **4.10**.

Example 4.4 – lease for indefinite term

On 1 September 2010 Quirinius took on a tenancy of a commercial unit. Under the terms of the tenancy Quirinius paid a quarterly rental of £7,500 including VAT, rent to be increased at each anniversary of the tenancy in line with the Retail Price Index. The tenancy may be terminated by either party on giving one quarter's notice on the due date of a rental payment.

The tenancy was initially treated as a lease for one year at an annual rent of £30,000. No SDLT was due as the NPV of the rent was below £150,000. There was no obligation to make a return because the lease was for less than seven years and no tax was due. On 1 September 2011 the continuing tenancy was treated as having been a lease for two years from the outset, at an annual rent of £30,000. The increase in rent in line with the RPI is ignored (*FA 2003, Sch 17A, para 7(5)*). Again no SDLT liability arose and no return was required. The same applied at 1 September in each of 2012, 2013 and 2014. At 1 September 2015, the lease was deemed to have been for six years from 1 September 2010 to 31 August 2016. The NPV of the rent (at £30,000 per year ignoring RPI increases) on the deemed six-year lease was £159,856. As this was over the £150,000 threshold for an SDLT liability to arise a Land Transaction Return was submitted and SDLT was paid at 1% of £9,856 (see **4.47** et seq for calculation principles). In accordance with changes introduced by *FA 2013*, *Sch 41*, *para 2* the return is not due and the SDLT is not payable until 30 September 2016, that is 30 days after the end of the year which gave rise to the liability.

If the tenancy had then been terminated during the sixth year – say on 28 February 2016, so that in total it had lasted five years and six months – there would have been no refund of the SDLT. This is the case even though a fixed term lease for the same period and rent would not have given rise to any SDLT liability, as the NPV of the rent would have been below the £150,000 threshold.

The tenancy continues. At 1 September 2016 it is regarded as becoming a lease for seven years (from 1 September 2010 to 31 August 2017). The NPV of the rent is £183,436. SDLT liability arises on the increase in NPV of the rent in comparison with the previous year (SDLT is 1% of (£183,436 – £159,856) and a land transaction return is required. The return is not due and the SDLT is not payable until 30 September 2017, that is 30 days after the end of the year which gave rise to the liability.

However in January 2017 Quirinius and the landlord enter into negotiations to replace the tenancy with a new 10-year lease. Terms are agreed and the new lease is executed (with no prior agreement) on 28 February 2017. Rent of £15,000 plus the RPI increase is paid in respect of the six months to 28 February 2017, in accordance with the terms of the tenancy. The tenancy is treated as ending on the date of actual termination, SDLT is calculated by reference to the actual total period of the tenancy and based on actual rent paid, including £15,000 for the final six months. Payment of SDLT is due, together with a land transaction return, by 30 March 2017 (*FA 2003*, *Sch 17A*, *para 4* as amended by *FA 2013*, *Sch 41*, *para 2*). Separately, a Land Transaction Return and payment of any SDLT in respect of the new lease are also due by 30 March 2017.

4.13 A tenant may remain in occupation after the end of a lease ('holding over'), continuing to pay rent, perhaps while a new lease is negotiated. The precise SDLT consequences of holding over will depend on whether the old lease was granted under the stamp duty or SDLT regimes and the terms of any new lease. This is one area in which helpful amendments were made by *FA 2013* – see *FA 2003*, *Sch 17A*, *paras 3*, *3A* as amended.

In principle, periods of holding over are treated in the same way as the second and subsequent years of a lease for indefinite term (see **4.12**). The old lease is effectively deemed to remain in force, growing in length one year at a time:

- If the old lease was subject to SDLT, reporting and payment obligations arise in respect of each one year deemed extension. Prior to 17 July 2013 these obligations had to be dealt with within 30 days of the start of each one-year period of holding over. From 17 July 2013 the obligations are postponed to 30 days after the end of each one-year period (or within 30 days of the end of the period of holding over if sooner);

- If the old lease was subject to stamp duty, the deemed extensions do not give rise to SDLT obligations unless there is an increase in rent which was not provided for in the original lease, or there are other changes which may be deemed to give rise to a new lease or agreement for lease (see **4.22**).

These treatments may be modified if the period of holding over is followed by the execution of a new lease of substantially the same premises to the same tenant. For details of the treatment of lease renewals see **4.64**.

Airspace and roof-space leases – renewables and telecoms

4.14 A lease of roof-space or airspace for installation of solar panels, wind turbines or telecoms masts is potentially chargeable just like any other lease. The installation of the equipment will not normally count as consideration, even though required under the terms of the lease, because it will normally fall within *FA 2003*, *Sch 4*, *para 10* (see **2.26**). Any cash premium or rent payable under the lease will be chargeable in the normal way. In relation to renewable energy, such leases often provide a right for electricity to be provided to the owner or occupier of the building either free or at reduced price. This right, being consideration in money's worth, must be valued at the effective date of the lease and included as a premium. The ongoing supply of electricity under the lease does not count as rent.

In *Stamp Taxes Bulletin 1/2014* (see www.hmrc.gov.uk/so/bulletin01-2014.pdf) HMRC comment that in most cases the value of the right to electricity will fall below the £40,000 notification threshold and no SDLT return will be required. However this may not be so for larger installations, perhaps on commercial premises, and HMRC regard it as the responsibility of the taxpayer to check the point.

In the case of domestic solar panels, the common alternative arrangement is that the homeowner buys the panels and receives the full benefit of electricity generated either through free use or payment for feeding back into the grid. No lease is involved in this situation and no SDLT obligations arise.

Transfer of a lease or agreement for lease

4.15 In a straightforward case, the transfer of an existing lease to a new tenant (with no other change in terms and conditions) will be subject to

SDLT in the same way as the transfer of any other interest in land. The new tenant will be subject to SDLT on any consideration given (a) to the outgoing tenant and (b) to the landlord (apart from ordinary rent) – for example, a lump sum paid to persuade the landlord to consent to the transfer. However, neither the rent payable under the lease nor any premium paid to the landlord by the outgoing tenant (for example, in return for the landlord agreeing to release the tenant from the lease) will figure in the new tenant's SDLT liability. Equally, the assumption of normal tenant's obligations and undertakings (such as to maintain the property) do not count as consideration (*FA 2003*, *Sch 17A*, *para 17*). There are provisions at *FA 2003*, *Sch 17A*, *para 12A* designed to ensure that assignment of an agreement for lease rather than the lease itself produces the same overall SDLT liability.

Liabilities which fall on transferee

4.16 Many cases are not straightforward, and it is always necessary for the new tenant to understand the full history of the lease, in order to determine whether the transfer, or the nature of the lease itself, may give rise to other SDLT liabilities and obligations. Additional obligations may arise in two situations. The first is where the rent was uncertain or variable when the lease was granted. If the lease is transferred to a new tenant before the SDLT liability is resolved, the new tenant inherits the obligation to make any necessary disclosures or further returns and to pay any additional SDLT once the uncertainty is resolved. See **4.37** et seq for an explanation of those obligations.

Relief claimed on original grant of lease

4.17 The second situation where the new tenant may incur unexpected liabilities is where one of a specified list of reliefs (see **4.20**) was claimed when the lease was granted (*FA 2003*, *Sch 17A*, *para 11*). If nothing has already happened to cause clawback of that relief, the first transfer of the lease which does not itself qualify for one of the specified list of reliefs is treated as the grant of a new lease for the remaining term of the actual lease and at the rent payable by the transferee. This means, for example, if there have been any rent increases between the dates of original grant and transfer, the current (higher) rent will be taken into account in calculating any SDLT payable. SDLT is also payable, in the normal way, on any premium which the new tenant pays to the landlord. HMRC consider that any amount paid by the new tenant to the outgoing tenant in respect of the transfer is also within the charge to SDLT, as would be the case for a straightforward transfer of a lease on which no relief had previously been claimed. It is logical that this should be the case, but it is by no means clear that the wording of the legislation supports this view.

4.18 SDLT is chargeable on the transferee, and not on the original lessee who claimed the relief. Therefore, strictly, this is not clawback, it is merely a modification of the normal rules for determining a purchaser's SDLT liability. However, as the example below demonstrates, in practice the rules do lead to effective clawback of part of the relief previously claimed, even though it is the new owner/tenant who has responsibility for the resulting SDLT.

4.19 The rules are not subject to any kind of time limitation: they apply on disposal of the lease at any time in its life. Anyone acquiring a second-hand lease should check whether it will be subject to these special rules and seek protection from unexpected SDLT liabilities by way of warranties, indemnities, retentions or whatever may be thought appropriate.

4.20 The special rules apply where any of the following has been claimed on the grant of a lease:

(1) group, reconstruction or acquisition relief,

(2) sale and leaseback relief,

(3) charities relief,

(4) relief for transfers involving public bodies, or

(5) any relief brought forward from the stamp duty rules by regulations under *FA 2003, s 123(3)*.

Care must be taken in relation to the last of these, as many of the reliefs mentioned in the table at **3.53** fall into this category. However, the rules do not apply if the transfer itself qualifies for any of the reliefs in the list (not necessarily the same relief as that previously claimed), provided the transferee claims the relief in an SDLT return.

Example 4.5 – transfer of lease after relief claimed

On 1 March 2009, V plc granted a 25-year lease over commercial property to its subsidiary W Ltd at a rent of £100,000 per annum, subject to five-yearly upwards-only rent reviews. Group relief was claimed. On 1 March 2014, the scheduled rent review leads to a rent increase, to £110,000. On 1 June 2016, W Ltd transfers the lease to an unrelated company X Ltd. X Ltd pays W Ltd £180,000 for the transfer and takes over all of the other terms and conditions of the lease.

X Ltd is not entitled to any relief in the list mentioned above, so the transfer is treated as the grant of a new lease for the remaining 17 years 9 months of the term, at an initial yearly rent of £110,000. This charge effectively claws back part of the relief claimed by W Ltd when the lease was first granted. In this example, it is worse than a simple clawback because it also takes account of the new, higher rent. In addition, in this example, X Ltd pays a sum to W Ltd for the transfer, and this will be subject to SDLT as if it was a premium. The SDLT chargeable on the premium would be £600 (£150,000 at 0% plus £30,000 at 2%). This particular example is worse than that. When X Ltd takes on the lease, a further rent review is due in less than five years. Therefore, the acquisition is treated as a grant of a lease for uncertain or variable rent. It will be necessary for X Ltd to review the position when the rent review occurs, submitting a further SDLT return and paying further SDLT as appropriate (see Example 4.12).

Shared ownership leases

4.21 In the context of affordable/social housing, arrangements are often made for the purchase of a part share in a dwelling. The purchaser then occupies the rest of the dwelling under a lease or tenancy. There are specific provisions (*FA 2003*, *Sch 9*) designed to remove or minimise any SDLT charge on such arrangements, and these are explained at **3.42** in relation to reliefs.

Variation of existing lease

4.22 The original rules dealing with variation of an existing lease were substantially rewritten in 2006. A variation may give rise to SDLT obligations, depending on:

- the precise nature of the variation, including whether, in law, it amounts to a surrender and re-grant;
- whether either party gives consideration or is deemed to give consideration;
- when the variation occurs relative to the date(s) on which the term of the original lease began or was deemed to have begun (see **4.24–4.26**); and
- whether the original lease was granted under the stamp duty or SDLT regime.

Examples of variations

Increase in rent within first five years

4.23 An increase in rent in accordance with the terms of the lease (for example, as a result of a rent review) does *not* arise from a variation. However, where such a change in the rent is possible within the first five years of the term, the rent will have been regarded as uncertain at the inception of the lease. This will have given rise to particular SDLT reporting obligations (see **4.37**).

4.24 A lease may be varied to increase the rent – that is, other than in accordance with the original terms of the lease. If this happens with effect from a date within five years of the beginning of the term, a new notional lease is normally deemed to be granted, covering the period from the effective date of the rent increase to the end of the original lease. The rent for the new notional lease is equal to the increase. This notional lease is then subject to SDLT (*FA 2003*, *Sch 17A*, *para 13*). The new notional lease may be linked with the original lease but is not necessarily so – this is a question of fact. These rules are disapplied where the variation arises from certain provisions relating to agricultural tenancies and holdings. In such cases, a variation has no immediate SDLT consequences.

Example 4.6 – rent increase in first five years

On 1 July 2012, Giovanni entered into a new ten-year lease of a storage and distribution depot at a yearly rent of £25,000 and no premium, with a market rent review due on 1 July 2017. An SDLT return was submitted accordingly. The lease restricted activities in the premises to storage, repackaging and distribution. Giovanni always intended in due course to carry out repairs of faulty goods returned to the depot (and had mentioned this to the landlord when taking the lease on). The landlord agrees to vary the lease to permit this, in return for an increase in rent to £28,000 from 1 January 2016. An additional lease is deemed to come into existence on that date, for a term equal to the remaining 6.5 years of the original lease and at a yearly rent of £3,000. Giovanni must decide whether any SDLT arises on this deemed additional lease and, if so, must submit a further SDLT return accordingly. Since it was always expected that the lease would be varied in this way, it seems likely that the original lease and the deemed additional lease are linked. On this basis, additional SDLT will arise on the deemed lease; see **4.61** and Example 4.22. HMRC appear to accept that the deemed additional lease should not be regarded as of uncertain or variable rent, even though a market rent review is due in respect of the (real) lease just 18 months later.

Increase in rent after first five years

4.25 If the original lease was granted before 1 December 2003, or after that date but with the benefit of the implementation transition provisions (see **2.71**), it will have been subject to the stamp duty regime. A variation to increase the rent after the first five years of the term will not be subject to SDLT, provided it is, in law, a mere variation and not the grant of a new lease (SDLTM10025).

Example 4.7 – original lease under stamp duty

Lucy is tenant of a 1960s office building under a 20-year lease which was granted on 1 January 2000. Rent is increased every five years in line with inflation and is currently below the market rent for an equivalent modern building. The building requires significant work to bring it up to current standards. Lucy agrees to a variation of the lease to provide for market level rent reviews at 1 January 2014 and 2019. In return, the landlord agrees to arrange and fund refurbishment of the building. Because the lease was granted under the stamp duty rules, this variation has no SDLT or stamp duty consequences.

4.26 An increase in rent after the first five years, whether as a result of a rent review, variation or any other matter short of renewal, extension of term or surrender and regrant, is also ignored if the original lease was subject to SDLT, provided the increase takes effect on or after 17 July 2013 (*FA 2013, Sch 41, para 7*). Prior to that date it was necessary to consider whether the increase was 'abnormal'. The old rules are outlined briefly at **4.27.** If, exceptionally, there is a current need to consider whether SDLT liabilities arise on a rent increase taking effect before 17 July 2013, the previous edition of this book should be consulted for guidance. In the Finance Bill debates the government explained that the old rules were being abolished because it was accepted that SDLT avoidance by artificially depressing rent for the first five years of a lease was unlikely to happen. However they made it clear that the matter would be reconsidered should evidence of such avoidance emerge. In practice, it seems lilely that it would be possible to rely on the GAAR to counter any such avoidance, see **7.3**.

Abnormal rent increases – old rules

4.27 The concept of abnormal rent increases only applied to rent increases on leases granted or deemed granted within the SDLT regime, which took effect after the first five years of the lease and before 17 July 2013. The general principle was that a rent increase was abnormal if the new rent was greater than the 'rent previously taxed' by more than 20% (measured on a straight line basis) for each complete year elapsed since the starting date for the rent previously taxed. The rent previously taxed was the last rent by reference to which SDLT had been charged. A five step calculation was necessary to determine the matter. The HMRC website provided a calculator to assist in the calculation, but the calculator may be removed in the near future.

If there was an abnormal increase in rent, this was deemed to give rise to a new lease, running in parallel with the old, for the amount of rent which had not previously been subjected to SDLT. This notional lease was deemed to have been granted on the day on which the new higher rent first became payable (which may not be the same as the date from which the new rent took effect).

Decrease in rent

4.28 A decrease in rent is specifically treated as an acquisition of a chargeable interest by the tenant (*FA 2003, Sch 17A, para 15A(1)*). If the tenant gives or is deemed to give consideration, this will be within the charge to SDLT. Normal compliance rules apply – if no consideration is given or if the consideration is below £40,000, there will be no need to notify HMRC; otherwise, an SDLT return will be required, even if no tax is actually payable.

Decrease in term

4.29 If a lease is varied to decrease the term, this is treated as the acquisition of a chargeable interest by the landlord (*FA 2003, Sch 17A, para 15A(2)*). If the landlord gives or is deemed to give any consideration, this will be within the charge to SDLT. However, if the tenant gives consideration, for example in order to have the term of an onerous lease reduced, this will not be subject to SDLT.

Other variations

4.30 The most common types of variation of a lease are covered by specific rules as set out in preceding paragraphs. All other variations are treated as the acquisition of a chargeable interest by the tenant only if the tenant gives (or is deemed to give) consideration in money or money's worth (*FA 2003, Sch 17A, para 15A(1A)*). Other variations are specifically excluded from the scope of SDLT (*FA 2003, s 43(3)(d)*).

Surrender

4.31 The surrender of a lease is the acquisition of a chargeable interest by the landlord (*FA 2003, s 43(3)(b)*). Therefore, if the landlord gives or is deemed to give chargeable consideration, this will be within the scope of SDLT.

Surrender and re-grant

4.32 Whether a variation of a lease amounts to a surrender and re-grant is a matter of law and, in any case of doubt, appropriate advice should be sought. However, it is generally true that a variation which does not overtly state that the old lease is surrendered is not likely to amount to a surrender and re-grant unless it varies the term or the extent of the premises demised.

Where a surrender and re-grant occurs, SDLT is chargeable in respect of rent under the new lease in the normal way. Any obligation to notify the transaction and pay SDLT falls on the tenant.

4.33 If the original lease was granted under the stamp duty regime, no credit is given for any stamp duty previously paid. For this reason, it is generally best to avoid surrender and re-grant of a lease at rent where the old lease was granted under the stamp duty regime.

4.34 However, if the original lease was granted under the SDLT regime, and if the premises under the two leases are 'the same or substantially the same', credit may be available in respect of rent for the 'overlap period'; see **4.83**.

4.35 Where rent under the new lease is variable or uncertain, it should be noted that the overlap relief is ignored in measuring the highest 12-month rent in the first five years, in order to determine the deemed rent for Year 6 onwards. The surrender of the old lease does not count as consideration for the grant of the new, nor does the grant of the new count as consideration for the surrender of the old (*FA 2003*, *Sch 17A*, *para 16*). If the landlord gives consideration for the surrender of the old lease other than the grant of the new lease, it may be necessary for the landlord to notify the surrender as a separate transaction and pay any SDLT arising on that.

Example 4.8 – superseding lease, no credit for stamp duty

The facts are as in Example 4.7 above. However, the landlord is only prepared to fund the refurbishment if Lucy agrees to a longer lease. Therefore, on 1 January 2014, the original lease is surrendered in return for a new 15-year lease at a market rent and with five-yearly market rent reviews. SDLT is payable on this in the normal way. Because the original lease was dealt with under the stamp duty rules, no credit is available for the stamp duty already paid in respect of the 'unused' period of the old lease. The SDLT liability would be lower if the parties agreed to amend the old lease, as in the original example, and at the same time enter into a 'reversionary lease' to cover the period from expiry of the old lease (31 December 2019) to the intended end date of the new lease (see **4.36**).

Example 4.9 – superseding lease, credit for SDLT

The facts are as in Example 4.6 above. However, shortly after the 1 July 2017 rent review (which increases the yearly rent to £35,000), Giovanni concludes that use of the premises would be enhanced if a small patch of land to the side were added to the lease. The lease is amended on 1 October 2017 to include this, with no additional rent payable. At the same time, the term of the lease is extended by five years to 30 June 2027. This extension of both term and premises is deemed to be a surrender and re-grant of the lease. It is considered that the additional land is small enough for the old and new leases to be regarded as leases of 'substantially the same' premises. SDLT is payable on the 'new' lease, which is for 9 years 9 months at a rent of £35,000 per annum, but credit is given for the £28,000 per annum rent which would have been payable under the old lease (including the deemed additional lease at rent of £3,000 per annum) for the period to 30 June 2022. The rent (before discounting to NPV) to be taken into account in the SDLT calculation for each year of the new lease is as follows:

Period	**Calculation**	**Deemed rent (£)**
Each of 4 years to 30/9/2021	35,000 – 28,000	7,000
Year to 30/9/2022	35,000 – 21,000	14,000
Each of 4 years to 30/9/2026	–	35,000
Period to 30/6/2027	9 months at 35,000 pa	26,750

The highest deemed rent in any 12-month period in the first five years of the new lease appears to be £14,000; however, the credit for rent under the old lease is ignored in determining this 'highest rent' figure, so the deemed rent for periods after the end of the overlap period reverts to the highest figure before relief, which is £35,000 in this case.

Reversionary leases

4.36 Varying a lease to increase the term amounts, in law, to surrender and re-grant, and is subject to SDLT accordingly, as set out above. Sometimes, an alternative is for the parties to enter into a further lease of the premises, covering the period from expiry of the current lease to the desired later expiry date. When this happens, the lease is granted before its start date. The effective date for SDLT purposes is the date of grant, not the commencement date. However, the commencement date is used in calculating the NPV of rents, so no benefit is gained from the fact that rent will not become payable under the new lease until some future date. The original and reversionary leases may be successive linked leases if the reversionary lease was contemplated when the original lease was implemented (see **4.59**). However, this is a question of fact and the leases are not necessarily linked.

Example 4.10 – reversionary lease

Gwen occupies a shop unit under a ten-year lease which expires on 30 September 2016. She wants to remain in occupation beyond this date and, on 1 December 2014, a reversionary lease is executed for a further ten years from 1 October 2016 at a rent of £20,000 per annum. Gwen must submit an SDLT return no later than 31 December 2014 and pay any SDLT due on the new lease by the same date. The NPV of the rents, calculated using 1 October 2016 as the start date, is £166,332. The lease is not linked with the old lease or any other, so the SDLT payable is 1% of £ (166,332 – 150,000), which is £163.

ONGOING OBLIGATIONS

SDLT liability not finally determined when lease was granted

4.37 Sometimes, when a lease is granted, the rent for the first five years is actually or potentially variable or uncertain – for example, because the rent depends on the tenant's business results or there is a rent review scheduled within the first five years. (Note, however, that rent increases in line with the Retail Price Index are ignored (*FA 2003, Sch 17A, para 7(5)*).) In such cases, it is not possible finally to determine the SDLT liability until the uncertainty is resolved, potentially after the end of the first five years. Once the rent for the first five years is known, the deemed rent for the rest of the lease can be determined (see **4.48**). Any uncertainty or variability after the end of Year 5 is ignored. As with other UK land transactions, the return must initially be completed on the basis of a reasonable estimate. The return must then be amended once the consideration is finally determined (*FA 2003, Sch 17A, paras 7, 8*).

Example 4.11 – variable or uncertain rent

On 23 April 2016, Richard grants Harold a 25-year lease of the site for a theme park. The lease requires Harold to construct and open the theme park within two years of grant. No premium is payable. Rent is set at a base level of £10,000 per year but, from Year 4 onwards, the rent is to be 5% of the turnover of the park, if this is higher than £10,000 (additional rent based on turnover to be estimated and paid quarterly with adjustment at the year-end). Reasonably conservative business projections indicate that 5% of turnover will be £8,000 in Year 4 and £12,000 in Year 5.

By 23 May 2016, Harold must submit an SDLT return on the basis of a reasonable estimate of rent, and it seems that £10,000 for Years 1 to 4 and £12,000 for Year 5 would be justifiable. Once the rent for Years 4 and 5 is finally determined (presumably, fairly early during Year 6), Harold must consider whether a further return is required. If the actual rent exceeds the estimate used in the original return, a further return must be submitted and additional SDLT paid. If the actual rent is less than the estimate, Harold is entitled (but not required) to submit an amended return and claim repayment of any SDLT overpaid. See **6.17** for details of submission of further or amended returns.

4.38 Prior to 17 July 2013, potential difficulties arose when a tenant took possession on the basis of an agreement for lease, the final lease being granted some time later, perhaps once minor details had been settled. Such a lease will often be expressed as commencing from the date on which the tenant takes possession with a rent review five years from that date. Previously, for SDLT purposes, the final lease would have been treated as beginning on the date of grant and the first rent review would have been due less than five years later. Hence, a lease with normal five-yearly rent reviews and no other uncertainty over rent would nonetheless often have been treated as a lease with variable or uncertain rent for SDLT purposes. The rules were amended by *FA 2013, Sch 41, para 6* with effect from 17 July 2013 to provide that the final lease is treated as commencing on the date of substantial performance, which will generally coincide with the commencement date stated in the lease. This should prevent such leases, with normal five-yearly rent reviews, being treated as having variable or uncertain rent.

Example 4.12 – rent review within five years of final execution

Tom enters into an agreement for a 12-year lease of a workshop and the agreement is substantially performed by Tom taking possession on 1 November 2016. The lease is finally executed on 20 February 2017, is expressed as ending on 31 October 2028, and provides for rent reviews at 1 November 2021 and 2026.

Under the old rules the term of the lease would have been treated as commencing when it was executed on 20 February 2017. The first rent review falls within five years of this date, so the lease would have been regarded as being at an uncertain or variable rent. When the first rent review was settled, it would have been necessary to review the original SDLT return and calculation. If (as is likely) this would have led to a greater SDLT liability than originally disclosed, a further return would have had to be made and additional SDLT be paid.

The new rules apply to leases with an effective date on or after 17 July 2013 – that is, which are executed, or substantially performed if earlier, on or after that date. The old rules continue to apply to leases which were substantially performed before 17 July 2013, even if they are finally executed on or after that date.

Under the new rules, the term of the lease is treated as commencing on 1 November 2016. The first rent review falls at least five years after this date so the rent is not regarded as variable or uncertain. SDLT obligations can be dealt with on the basis of the starting rent alone and no further action will be necessary when the rent reviews occur.

4.39 Care is needed when the date of the first rent review falls in the final quarter of the fifth year of the term. This may occur, for example, when rent is payable on traditional quarter days but the lease commences on some other date. In this situation, it may be possible to treat the first rent review as falling five years after commencement of the lease, so that the variable or uncertain rent provisions do not apply (*FA 2003*, *Sch 17A*, *para 7A*). However, in applying this rule, it appears that the precise wording of the lease will be important. The lease must provide that the first rent review is due five years after a specified date which itself falls within the three months before the commencement of the lease. It appears that, if the lease merely states 'the first rent review will be on [date within the last quarter of the first five years]', without specifically referring to a date within the three months before the lease began, the rule will not apply and the lease will be treated as having variable or uncertain rent. There does not appear to be any policy reason for this restriction, so it may simply be a case of poorly worded legislation. It is not known whether HMRC allows any flexibility in applying this rule.

Example 4.13 – rent review in final quarter of Year 5

Sales Ltd leases a shop unit from Retail Mall plc for ten years from 1 January 2016. Sales Ltd takes possession on 1 January 2016 but the lease is not finally executed until 10 February 2016. The lease provides for a rent review after five years and, to fit with Retail Mall's other leases, this is due on 25 December 2020. Sales Ltd must make an SDLT return no later than 31 January 2016. Providing the rent review clause of the lease is appropriately worded (on the lines of 'the rent review shall occur five years after 25 December 2015'), the fact that it falls before the fifth anniversary of both substantial performance of the agreement and grant of the lease itself will not cause the rent to be regarded as variable or uncertain.

As a result of the changes introduced by *FA 2013*, *Sch 41*, *para 6*, it no longer matters exactly when the final lease is executed, provided it is before the first rent review.

MAKING AND AMENDING SDLT RETURNS

4.40 SDLT returns relating to lease transactions are made and amended in just the same ways as for other UK land transactions; see **Chapter 6**. Some forms/sections of the return apply only to transactions involving leases. If the return is completed online, these sections only appear if appropriate entries are made in the early sections to indicate that leases are involved. This is set out more fully in **Chapter 6**. As with other transactions, there is generally no prescribed form for amending a return, and amendments are made by writing to SDLT Compliance at Birmingham Stamp Office (see **Appendix A**). HMRC do say that minor amendments may be notified by telephone, but written communications have the advantage of leaving a record.

CALCULATING THE LIABILITY

What is consideration?

4.41 SDLT is charged on 'chargeable consideration' which is defined for leases as for other transactions as 'any consideration in money or money's worth given … by the purchaser or a person connected with him' (*FA 2003, Sch 4, para 1*). In relation to the grant or variation of a lease, consideration may be in the form of rent and/or premium. Since the tax treatments of the two forms of consideration differ, it is important to distinguish between them (see **4.44**). In relation to both rent and premium, the usual valuation rules apply to consideration other than in cash and in sterling (see **2.19**), as do the limited exemptions where the consideration includes the carrying out of works (see **2.26**) or the transaction qualifies for 'PFI relief' (see **3.17**).

4.42 When a lease is surrendered in consideration for the grant of another lease, the surrender and grant do not count as consideration for each other (*FA 2003, Sch 17A, para 16*). The acceptance of normal tenant's obligations other than payment of rent and premium does not count as consideration when a lease is granted (*FA 2003, Sch 17A, para 10*). The acceptance of such obligations *including* payment of rent does not count as consideration for the transfer of a lease (*FA 2003, Sch 17A, para 17*).

Rent

4.43 There is no comprehensive definition of rent, so it must be taken as having its ordinary commercial meaning of consideration paid by a tenant, usually periodically, for use or occupation of the property. Where rent is payable other than in sterling cash, it must be valued in the same way as other consideration, at the effective date of the lease; see **2.19**. A single sum expressed as payable in respect of rent and other matters, without apportionment, is to be treated as entirely rent (*FA 2003*, *Sch 17A*, *para 6*). To minimise the SDLT payable, it may therefore be important to ensure that 'inclusive' rents are overtly apportioned, in the terms and conditions of the lease, between rent proper and other items such as insurance, cleaning and maintenance. However, in practice, HMRC often seem willing to accept a reasonable apportionment even where this is not explicit in the lease. The apportionment, whether explicit in the lease or negotiated with HMRC, must be on a just and reasonable basis (*FA 2003*, *Sch 4*, *para 4*).

Example 4.14

Albert leases an office suite in a larger building owned by Big Office Co Ltd on a ten-year lease, terminable by either party on one year's notice. The rent is £30,000 per annum plus a proportionate share of an insurance premium paid by Big Office and covering the whole building. The rent includes heating/air conditioning and regular cleaning, but amounts are not allocated to these in the lease. Light and power are separately metered and charged. The insurance premium is clearly identified and, since it merely passes on actual cost, is presumably 'just and reasonable', so it will not form part of the rent for SDLT purposes. Since the remaining rent is not allocated between real rent and matters such as cleaning and heating, strictly it is all rent for SDLT purposes. The parties would be well advised to insert a reasonable allocation into the lease. However, even in the absence of numbers in the lease itself, HMRC may be persuaded to accept an allocation. They may require evidence from Big Office to support amounts allocated to non-chargeable matters.

4.44 An anti-avoidance provision at *FA 2003*, *Sch 17A*, *para 18A* is designed to prevent a premium being disguised as rent, which would normally give rise to a lower SDLT liability. If the tenant pays any loan or deposit in excess of a normal 'rent deposit', the loan or deposit is to be taxed as a premium, even though it may be repayable (typically by offset against rent payments as they fall due).

New leases

4.45 When a new lease is granted or deemed to be granted, SDLT must be calculated separately on any rent and on any premium or deemed premium (*FA 2003*, *Sch 5*). The two components are then simply added to give the SDLT liability. As in other cases, any VAT chargeable must be included in both rent and premium. However, if VAT arises as a result of the landlord exercising the 'option to tax' *after* the effective date of the transaction, this does not have to be taken into account (*FA 2003*, *Sch 4*, *para 2*).

SDLT on a premium

4.46 A premium is subject to SDLT at normal rates up to 12% on a progressive basis for a transaction consisting entirely of residential property (however the higher rates of SDLT of up to 15%, on a progressive basis, for additional dwellings and dwellings purchased by companies, may apply (see **2.39**) or a slab rate of 15% could apply in certain circumstances (see **2.65**)) and up to 5% on a progressive basis for any other lump sum payment for an interest in land which is non-residential property (see **2.37**). For transactions with an 'effective date' (see **2.8**) on or before 16 March 2016, if a lease of non-residential property was subject to a rent of £1,000 or more per annum, the zero tax rate for a premium up to £150,000 was replaced by a 1% rate. However, this rule was abolished by *FA 2016, s 116(11)*. This means that for a lease of non-residential property, granted on or after 17 March 2016, SDLT will only be payable on a premium if the amount of that premium exceeds £150,000.

SDLT on rent

4.47 To calculate the SDLT on the rent, it is first necessary to establish the amount of rent to be taken into account for each year of the lease. If the lease is no longer than five years, the rent for SDLT purposes is simply the actual rent. As noted at **4.37**, if these amounts are not determined at the effective date of the lease, SDLT must be paid based on a reasonable estimate. A further return must be made, with payment of any additional SDLT, once the amounts are finally determined. If the final figures indicate that SDLT has been overpaid, an amendment to the return may be submitted and any overpayment reclaimed, but this is not compulsory.

Rent after fifth year

4.48 If the lease is longer than five years but the rent is fixed and at a constant rate throughout the term, again SDLT is based on the actual rent. However, this is probably unusual since most leases for more than five years are likely to provide for rent reviews. If the lease provides for the rent to change over the term for any reason, whether certainly or contingently, the following rules apply to determine the rent for SDLT purposes (*FA 2003, Sch 17A, para 7*):

(1) For the first five years, the actual rent is used.

(2) For each year after the first five, the rent is assumed to be equal to the highest rent for any consecutive 12-month period in the first five years.

This means that an abnormal 'spike' in the rent in the first five years can lead to a very large increase in SDLT payable, as the highest 12-month rent in those early years will form the basis of the deemed rent for later years (see Examples 4.16 and 4.19).

4.49 The rule applies even if the actual rent is known for periods after the first five years, so known increases after the first five years are ignored (see Example 4.15).

Example 4.15 – actual rent after Year 5 ignored

Amy grants Toni a 20-year lease of a shop at an initial rent of £10,000 per annum. The lease is executed on the start date without prior substantial performance. The lease provides for five-yearly rent reviews, with the rent increasing by at least 50% at each review, or to market rent if higher. It is therefore known that the rent will be at least £15,000, £22,500 and £33,750 per annum from the start of Years 6, 11 and 16 respectively. However, for purposes of the SDLT return at the start of the lease, the rent is treated as £10,000 per annum throughout the term of the lease. This is because the rent is £10,000 per annum for the first five years (rule 1), and the highest 12-month rent in the first five years is also £10,000 (rule 2).

Example 4.16 – effect of brief spike in rent in first five years

Julie granted an eight-year lease of a restaurant to Ken, commencing on 1 November 2014. The lease was executed on the start date without prior substantial performance. No premium was payable. Rent is £10,000 per quarter inc VAT but, for any quarter in which the restaurant turnover exceeds £100,000, the next quarter's rent is increased by one tenth of the excess. Over the first five years, the restaurant turnover exceeds £100,000 only in the quarters ending 31 May 2015 (when the rent for the following quarter is agreed at £11,000) and 28 February 2016 (when the rent for the following quarter is £12,300). These increases are not known when the lease is executed, and an initial return is submitted on the basis that the rent is £10,000 per quarter (£40,000 per annum) throughout. At the end of the first five years, a further return is required, with SDLT based on the following rent figures:

Year 1	£41,000	Three quarters at £10,000 plus one at £11,000
Year 2	£42,300	Three quarters at £10,000 plus one at £12,300
Each of Years 3 to 5	£40,000	Four quarters at £10,000
Each of Years 6 to 8	£43,300	This is the rent for the 12 months from 1 June 2015 to 31 May 2016, being the highest rent for any 12-month period in the first five years.

4.50 Once the rent for each year is determined (or estimated if uncertain at the outset), the net present value (NPV) of that rent over the life of the lease must be calculated, using the formula and discount rate specified in the legislation (*FA 2003*, *Sch 5*, *paras 3*, *8*). Since the inception of SDLT, the discount rate has been 3.5%, but in future this may be varied by regulation. For these purposes, the length of the lease is taken as the term specified in the lease, ignoring any break clauses, other right to terminate early, or any right to extend or renew. However, if any right to extend or renew is subsequently exercised, the resulting further lease or extension may be linked with the original lease, whether successively or not; see **4.58** et seq.

4.51 In simple terms, the rent for the first 12 months is divided by 1.035 to give its discounted value, the rent for the second 12 months is divided by $(1.035)^2$, and so on to the end of the term of the lease. The discounted rent figures for all years are then totalled. If this total does not exceed the relevant threshold (currently £150,000 for non-residential or mixed property, or £125,000 for residential property: *FA 2003*, *Sch 5*, *para 2*), no SDLT is chargeable on the rent. The rates and bands used to calculate the SDLT liability on rent, for both residential property and non-residential or mixed use property, are set out at **4.3**.

4.52 It should not be unduly difficult to set up a spreadsheet to perform SDLT calculations for leases. However, an SDLT calculator for this purpose is provided on the HMRC website at www.hmrc.gov.uk/sdlt/calculate/calculators.htm#2. The HMRC calculator works for most cases, provided the correct information is input, and its use is recommended. However, it is purely a mechanical aid. It will not detect whether the correct information is entered, and HMRC have made it clear that they will hold the taxpayer responsible if the calculator gives incorrect answers because incorrect information has been entered. Therefore, it is important to have a full understanding of the nature of the transaction when entering information into the calculator, to guard against errors. A copy of the output (printed or electronic) should be retained in case of any future query. A minor annoyance is that the date boxes in the calculator are not set up to move the cursor to the next box as each is completed – this must be done manually using the mouse or tab key!

Example 4.17 – calculation of NPV

On 1 May 2016, Jeremy agrees to grant Kate a 10-year lease of an hotel at an annual rent of £25,000 inc VAT and for a premium of £250,000. There is an upwards-only rent review on the fifth anniversary. The lease is granted on 1 July 2016 without previously having been substantially performed. The effective date is therefore 1 July 2016, and the rent review is due on 1 July 2021. The SDLT can be calculated manually as follows:

Year	**Actual rent for SDLT purposes**	**Discounted rent**
1	£25,000	£24,154
2	£25,000	£23,337
3	£25,000	£22,548
4	£25,000	£21,786
5	£25,000	£21,049
6	£25,000	£20,337
7	£25,000	£19,649
8	£25,000	£18,985
9	£25,000	£18,343
10	£25,000	£17,722
Total discounted rent		£207,910
Less zero rate band		(£150,000)
SDLT charged at 1% on		£57,910
SDLT on rent		£579
SDLT on premium: £250,000 (£150,000 at 0% + £100,000 at 2%)		£2,000
Total SDLT (on rent and premium)		£2,579

4.53 Alternatively, the HMRC calculator may be used. This has a series of input screens, as follows:

First screen:

GOV.UK

HM Revenue & Customs

◂ Back

Effective date of transaction

Effective date of your transaction.

This is usually the completion date.

For example, 31 3 2014.

Day	Month	Year
01	07	2016

Continue

Get help with this page.

Insertion of the effective date (see **2.6** et seq) and clicking 'continue' leads to the next three screens, as shown below: the first is headed 'Lease dates', in which the start and end dates of the lease are entered, the second is headed 'Lease premium' in which the amount of any premium is entered; and the third is headed 'Rent' in which the rent for the first five years of the term of the lease is entered.

GOV.UK

HM Revenue & Customs

Back

Lease dates

Start date shown in your lease.

This is the date your term starts.

For example, 31 3 2014.

Day	Month	Year
01	07	2016

End date shown in your lease.

This is the date your term ends.

For example, 31 3 2017.

Day	Month	Year
30	06	2026

Continue

Get help with this page.

GOV.UK

HM Revenue & Customs

◂ Back

Lease premium

Total premium payable.

Find this in your lease.

£ 250000

Is your lease premium less than £40,000?

Continue

Get help with this page.

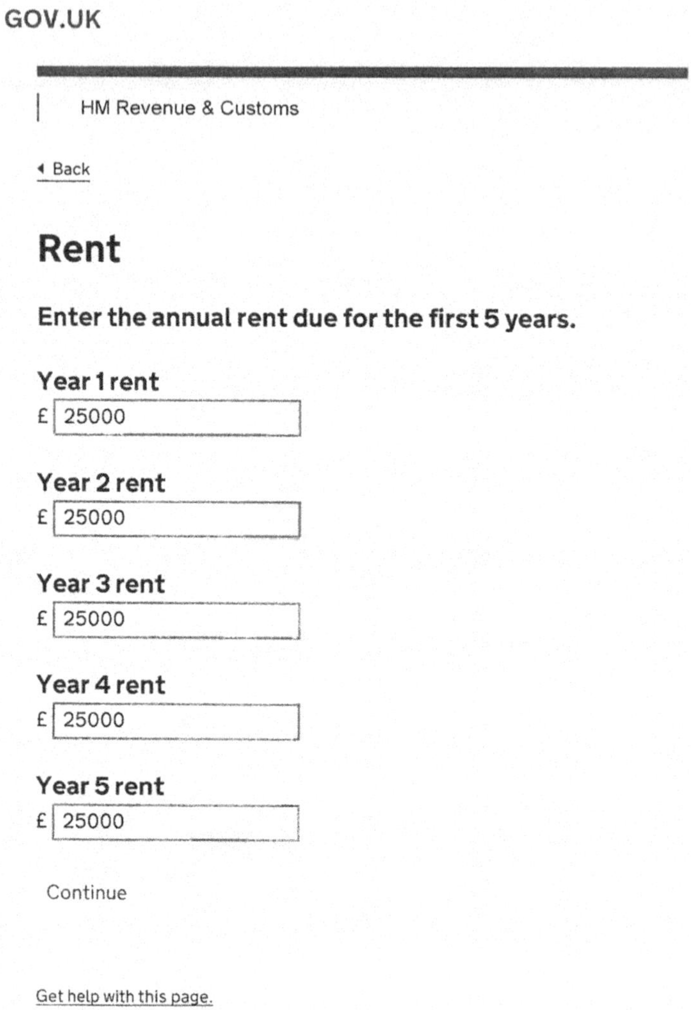
GOV.UK

HM Revenue & Customs

◂ Back

Rent

Enter the annual rent due for the first 5 years.

Year 1 rent
£ 25000

Year 2 rent
£ 25000

Year 3 rent
£ 25000

Year 4 rent
£ 25000

Year 5 rent
£ 25000

Continue

Get help with this page.

4.54 As discussed at **4.46**, where the grant of a lease over non-residential property (see **2.38**) had an 'effective date' (see **2.8**) on or before 16 March 2016, and that lease was subject to a rent of £1,000 or more per annum (referred to as the 'relevant rent' in *FA 2003*, *Sch 5*, *para 9A*), the zero tax rate for a premium of up to £150,000 was replaced by a 1% rate. However, this rule was abolished by *FA 2016*, *s 116(11)* for leases granted on or after 17 March 2016. Consequently, as the lease in Example 4.17 was granted on 1 July 2016 the premium benefits from the zero rate band of £150,000 and the SDLT chargeable on the premium is £2,000 (£150,000 at 0% plus £100,000 at 2%).

Example 4.18 – use of HMRC calculator

Using the numbers from Example 4.17 above, the output from the calculator is:

GOV.UK

HM Revenue & Customs

Back

Check your answers

Start again (https://www.gov.uk/calculate-stamp-duty-land-tax)

Freehold / leasehold	Leasehold	Change
Property type	Non-residential	Change
Effective date of transaction	1 July 2016	Change
Start date as specified in lease	1 July 2016	Change
End date as specified in lease	30 June 2026	Change
Term of lease	10 years 0 days	
Total premium payable	£250,000	Change
Year 1 rent	£25,000	Change
Year 2 rent	£25,000	Change
Year 3 rent	£25,000	Change

Year 4 rent	£25,000	Change
Year 5 rent	£25,000	Change
Highest 12 monthly rent	£25,000	

Calculate

Get help with this page.

GOV.UK

HM Revenue & Customs

◂ Back

Result

Results based on SDLT rules from 17 March 2016

Total amount of tax for this transaction	£2,579	
Net present value	£207,915	
SDLT on rent	£579	Detail
SDLT on premium	£2,000	Detail

Results based on SDLT rules before 17 March 2016

You may be entitled to pay SDLT using the old rules if you exchanged contracts before 17 March 2016.

Total amount of tax for this transaction	£3,079	
Net present value	£207,915	
SDLT on rent	£579	Detail
SDLT on premium (1%)	£2,500	

You can print your calculation.

Or you can start again (https://www.gov.uk/calculate-stamp-duty-land-tax).

Go to the GOV.UK homepage for SDLT.

Get help with this page.

GOV.UK

HM Revenue & Customs

◂ Back

Summary

Freehold / leasehold	Leasehold
Property type	Non-residential
Effective date of transaction	1 July 2016
Start date as specified in lease	1 July 2016
End date as specified in lease	30 June 2026
Term of lease	10 years 0 days
Total premium payable	£250,000
Year 1 rent	£25,000
Year 2 rent	£25,000
Year 3 rent	£25,000
Year 4 rent	£25,000
Year 5 rent	£25,000
Highest 12 monthly rent	£25,000

Results based on SDLT rules from 17 March 2016

Total amount of tax for this transaction	£2,579
Net present value	£207,915

SDLT on rent	£579
SDLT on premium	£2,000

Results based on SDLT rules before 17 March 2016

You may be entitled to pay SDLT using the old rules if you exchanged contracts before 17 March 2016.

Total amount of tax for this transaction	£3,079
Net present value	£207,915
SDLT on rent	£579
SDLT on premium	£2,500

Print

Get help with this page.

4.55 Note that the 'Check your answers' screen shows all of the entries made on the input screens, so it is easy to check there have been no transcription errors. The results based on the SDLT rules before 17 March 2016 have no application as the agreement to enter into the lease was made on 1 May 2016, ie after 16 March 2016 and therefore the transitional rules (see **4.3**) are not in point.

Example 4.19 – HMRC calculator, spike in early rentals

Facts are as in Example 4.17, except that the rent for the period from 1 July 2017 to 30 June 2018 is doubled to £50,000 in recognition of expected better business from a major international sporting tournament. This means that the rent for Year 2 will be £50,000, and that will be the highest rent for any consecutive 12 months in the first five years of the term of the lease. The output from the calculator is as follows:

GOV.UK

HM Revenue & Customs

Back

Check your answers

Start again (https://www.gov.uk/calculate-stamp-duty-land-tax)

Freehold / leasehold	Leasehold	Change
Property type	Non-residential	Change
Effective date of transaction	1 July 2016	Change
Start date as specified in lease	1 July 2016	Change
End date as specified in lease	30 June 2026	Change
Term of lease	10 years 0 days	
Total premium payable	£250,000	Change
Year 1 rent	£25,000	Change
Year 2 rent	£50,000	Change
Year 3 rent	£25,000	Change
Year 4 rent	£25,000	Change
Year 5 rent	£25,000	Change
Highest 12 monthly rent	£50,000	

Calculate

Get help with this page.

GOV.UK

HM Revenue & Customs

Back

Result

Results based on SDLT rules from 17 March 2016

Total amount of tax for this transaction	£3,762	
Net present value	£326,291	
SDLT on rent	£1,762	Detail
SDLT on premium	£2,000	Detail

Results based on SDLT rules before 17 March 2016

You may be entitled to pay SDLT using the old rules if you exchanged contracts before 17 March 2016.

Total amount of tax for this transaction	£4,262	
Net present value	£326,291	
SDLT on rent	£1,762	Detail
SDLT on premium (1%)	£2,500	

You can print your calculation.

Or you can start again (https://www.gov.uk/calculate-stamp-duty-land-tax).

Go to the GOV.UK homepage for SDLT.

Get help with this page.

4.56 The substantial increase in the SDLT charge on the rental element illustrates the effect of a temporary increase in rent during the first five years, as rent for every year after the first five is assumed to be equal to the highest rent for *any* consecutive 12-month period in the first five years. Where rent is variable during the first five years, it is important to establish the precise manner of variation and the correct attribution of rent to each period, as specified by the lease (*FA 2003, Sch 17A, para 7*; see **4.73**). The results based on the SDLT rules before 17 March 2016 are not relevant as the agreement to enter into the lease was made on 1 May 2016, ie after 16 March 2016 and the transitional provisions (see **4.3**) have no application.

Connected company rule

4.57 The normal rule applies, that if the 'purchaser' (the lessee or, in the case of a surrender, the landlord) is a company connected with the other party, the consideration is taken to be at least equal to the market value of the lease in question (*FA 2003, s 53*). Note that it does not matter whether the 'vendor' is a company, individual or other entity. Application of the rule depends only on the 'purchaser' being a company which is connected with the 'vendor'. This is subject to the same exceptions as apply to other transactions (see **2.29**). Where this rule applies to the grant of a lease, it does not lead to any deemed adjustment of the actual rent. If the rent charged under the lease is less than a full market rent, it is assumed that the lease will have a capital value. The amount of this capital value is taxed as a deemed premium.

Example 4.20 – lease to connected company

On 24 June 2016 Maria grants to her family trading company, PPM Ltd, a ten-year lease of a barn on her farm, for use as a storage facility. The lease is at a flat rent of £5,000 per annum with no provision for rent reviews. The NPV of rent over the life of the lease is clearly below the £150,000 threshold. However, it is acknowledged that a current market rent would be £30,000 and a lease at the actual rent of £5,000 would command a market premium of £200,000. The lease is therefore subject to SDLT on this premium and the SDLT chargeable is £1,000 (£150,000 at 0% plus £50,000 at 2%). It is worth noting that, if the rent had been set at a market level of £30,000 so that there was no deemed premium, the NPV of the rent would have been £249,498 and the SDLT payable would have been £994, almost the same as that payable based on the deemed premium. PPM Ltd must file an SDLT return and pay the SDLT due of £1,000, by 24 July 2016.

Linked transactions

4.58 Lease transactions may be 'linked' with other UK land transactions (whether or not involving other leases) if they form part of a single scheme, arrangement or series of transactions between the same vendor and purchaser or persons connected with them (*FA 2003, s 108*; see **2.67**). There are two distinct ways in which lease transactions may be linked, and each has its own implications.

Successive linked leases

4.59 If successive leases of substantially the same premises are granted, and if those leases are linked, the second and subsequent leases are effectively treated as extending the earlier lease (*FA 2003, Sch 17A, para 5*). As each lease in the series after the first is granted (or substantially performed), any SDLT return made in respect of the earlier lease must be amended to show details for what is now deemed to be a longer lease, and any additional SDLT must be paid. Additional SDLT may arise in relation to rent (the NPV of the rent will have increased because the deemed lease is for a longer period) and/or premium (if a further premium is paid, the premiums must be aggregated both to determine the rate of tax and to calculate the amount of tax).

Example 4.21 – successive linked leases

Harriet granted Ian a five-year lease of a warehouse commencing 1 September 2011 for a premium of £130,000 and a yearly rent of £50,000. The lease was executed on the start date without prior substantial performance. The lease agreement included an option for Ian to take a further five-year lease commencing on 1 September 2016 at the same rent, on payment of a further premium of £130,000. Ian exercises the option; the second lease is executed on 1 August 2016 (ie one month before commencement) and the additional premium is paid on 1 September 2016.

During September 2011, Ian submitted an SDLT return for the five-year lease, with premium £130,000 and yearly rent of £50,000. The total SDLT paid was £2,057, comprising £1,300 on the premium (charged at 1% because the rent is £1,000 or more, see **4.46**) and £757 on the rent. A further SDLT return is needed in respect of the second lease. The leases are linked, so there is deemed to be a ten-year lease at a flat rental of £50,000 per annum and a premium of £260,000, commencing 1 September 2011. The SDLT on such a lease is £10,458, comprising £7,800 on the premium (at 3%) and £2,658 on the rent. The SDLT paid on the original lease is deducted, leaving £8,401 payable as a result of the second lease. Although the second lease is not 'substantially performed' early, it is executed one month before commencement, so the further SDLT return and payment are due 30 days after execution, that is no later than 31 August 2016. As the leases are linked, such that there is deemed to be a single lease, for a term of 10 years, commencing on 1 September 2011, it is the SDLT bands and rates which applied on 1 September 2011 which are used to calculate the SDLT on the 'extended' lease, and not those which applied on 1 August 2016 when the second lease was granted.

4.60 Successive leases are not necessarily linked. If, when the first lease was granted, there was no agreement (explicit or implied) that the next lease would be granted, and if the next lease was negotiated independently of the first, they are not likely to be linked. However, if the first lease contained an option for grant of a further lease, they probably are linked. HMRC tend to assume that successive leases are linked unless there is clear evidence to the contrary.

Other linked leases

4.61 Leases of different premises may be linked – for example, a tenant may take on leases of two shops in different towns from the same landlord, negotiated as a single deal. Additionally, any deemed lease arising from an abnormal increase in rent prior to 17 July 2013 (see **4.26**) was automatically linked with the original lease and with any previous such deemed leases (*FA 2003, Sch 17A, para 14(5)(b)*). However, where a deemed lease runs in parallel with the original lease, the two cannot be 'successive'. The deemed lease arising from a variation to increase rent within the first five years of a lease is not automatically linked with the original lease – this is a matter to be determined in the light of the facts. For example, if the rent increase is a result of a statutory rent review, the leases are unlikely to be linked; if the increase relates to improvements of the premises contemplated when the lease was originally executed, the leases are likely to be linked.

4.62 Where leases are linked and the 'successive leases' rules do not apply (for example, leases of two properties agreed between the same parties as part of a single transaction), any premiums are aggregated to determine the correct rate of SDLT, which is then applied separately to each premium. The NPV of the rents under the leases are also aggregated to give a figure for total NPV (TNPV), and the tax on this is calculated in the normal way. The tax applicable to each lease is then calculated by applying the fraction NPV (for the individual lease)/TNPV to the overall tax on rent.

4.63 If the linked leases do not have the same effective date (or if some are and some are not wholly residential), and if, as a result, the zero rate threshold would be different for the separate leases, the tax on the TNPV must be calculated afresh for each lease before applying the NPV/TNPV fraction (*FA 2003, Sch 5, para 2*). This complexity means it is not possible to use the SDLT calculator on the HMRC website to make a direct calculation of the SDLT due on the leases. However, the calculator can still be used to calculate the NPVs for the individual leases, the remainder of the calculation then being performed manually.

Example 4.22 – non-successive linked leases calculation

The facts are as in Examples 4.6 and 4.9 above. The NPV of rent under the original lease is £207,915 and the SDLT paid was £579. The NPV of the deemed additional lease is £21,712, so the aggregate NPV is £229,627. The SDLT payable on a lease with this rental NPV is 1% of £(229,627 – 150,000), which is £796. Thresholds and rates are the same for both actual and deemed lease, so this calculation is appropriate for both leases. The revised SDLT attributable to the original lease is £796 × £207,915/£229,627, which is £720, and the SDLT attributable to the deemed additional lease is £796 × £21,712/£229,627, which is £75 (the £1 overall reduction arises from rounding and seems to be acceptable). As a stand-alone lease, the deemed lease arising from the rent increase would not have given rise to any SDLT charge but, because it is linked with the original lease, a charge arises.

LEASE RENEWALS AND EXTENSIONS

4.64 The interaction between the SDLT rules and commercial practice often leads to complexity when a lease is renewed, or where a lease for indefinite term (see **4.12**) is replaced by a fixed term lease. This complexity has been reduced (but by no means eliminated) by amendments made by *FA 2013, Sch 41*. In some cases the costs of compliance remain greater than the amount of tax payable. Inevitably, the tax payable will depend on the rent (and premium, if any) payable under the new lease and its length. However, the tax may also depend on matters such as:

- when the *old* lease was executed (or is treated as having been executed in the case of a lease for indefinite term);
- whether the old lease had a fixed term and if so,
- whether there has been a period of 'holding over' between expiry of the old lease and finalisation of the new; and
- how the parties agree to treat the holding-over period.

A period of holding over occurs when the tenant remains in occupation after a lease has expired, usually while a new lease is negotiated (see **4.13**)-.

Information required

4.65 The answers to the following questions will be important in determining the SDLT consequences of renewal of a lease:

- Was the old lease subject to stamp duty or SDLT?

- Did the new lease take effect before expiry of the old (is there an overlap in the periods of the lease)?
- If not, did the parties enter into an agreement for lease, which was substantially performed before the new lease was finally executed?
- When is the new lease stated to begin?
- Was there a gap between the end of the old lease and the finalisation of the agreement for the new lease (or execution of the new lease if there was no pre-execution agreement), during which the tenant remained in occupation?
- If there was a gap, was it less than one year, or alternatively a year or more?
- Are the old and new leases linked (see **4.58**)?

The following paragraphs and examples deal with the SDLT consequences for each possible combination of answers to these questions.

Old lease under SDLT

New lease granted on or before expiry of old fixed term lease

4.66 If the new lease is stated to take effect immediately after the end of the old, is executed before or immediately on expiry of the old and is not linked with the old (see **4.58**), SDLT is payable as if the new lease is a simple stand-alone lease for the term and consideration stated in the lease. If the new lease is linked with the old, this affects the calculation of SDLT (see **4.59** and Example 4.21) but not the timing of SDLT obligations.

If the new lease is executed before the end of the old, the new lease is a reversionary lease (see SDLTM17070). It is important then to consider timing – SDLT obligations arise 30 days after the date of execution of the new lease, not 30 days after the start date. See Example 4.23A.

If the new lease takes effect before the end of the old, credit may be available against the SDLT payable, for SDLT paid on the old lease for the overlap period – see Example 4.23B, and Example 4.9 for the calculation of overlap relief.

New lease granted after expiry of old fixed term lease

New lease expressed as beginning immediately on expiry of old

4.67 If the tenant remains in occupation throughout and the new lease is expressed as commencing immediately on expiry of the old, it will be treated as commencing on that date no matter when actually executed. This treatment is in contrast with the SDLT treatment of a completely new lease, where the term cannot be treated as beginning before the lease is executed. See Example 4.23C.

Gap between expiry of old lease and stated start date of new

4.68 For old leases which expire on or after 17 July 2013, provided the term of the new lease is expressed as beginning at some time during the one-year period beginning with the date of expiry of the old, it will still be treated for SDLT purposes as beginning immediately on expiry of the old lease. In other words any period of holding over before the lease is expressed as beginning, amounting to less than one year, will be added to the term of the lease for SDLT purposes. Accordingly, no separate return will be required for the period of holding over.

The most likely scenario is that a new lease will be executed within one year after the date of expiry of the old lease. If the new lease is not executed until more than one year after expiry of the old and if it is not expressed as beginning immediately on expiry of the old, any complete years of holding over will be dealt with as explained in **4.13**.

Once the new lease is finally executed, and provided the latest complete year of holding over commences on or after 17 July 2013, the new lease is treated for SDLT purposes as beginning immediately after the end of the last complete year of holding over. Accordingly no SDLT obligations arise in respect of the final part year of holding over, but that part year is added to the term of the new lease for the purpose of calculating the SDLT on the new lease. The SDLT obligations in respect of the new lease arise 30 days after the date the new lease begins, which will normally be the date of execution.

For leases which expired prior to 17 July 2013, if the new lease was expressed as beginning later than immediately after expiry of the old, the new lease was treated for SDLT purposes as beginning on the date on which it was executed (or the date on which it was stated to begin if later). This then gave rise to a period of holding over between expiry of the old lease and the effective date of the new, which was dealt with for SDLT purposes as explained at **4.13**. On grant of the new lease, overlap relief would normally be available in respect of any part of a deemed one-year extension on which tax had been paid but which had not expired. These fruitless complications have been swept away by the *FA 2013* changes where old leases expire on or after 17 July 2013 or where leases expired before that date but the latest one-year period of holding over commenced on or after that date. However the old rules continued to apply until 17 July 2014, where old leases expired in the year up to 17 July 2013 and new leases were still under negotiation.

New lease superseding old lease for indefinite term

4.69 The new lease will be subject to SDLT as a simple stand-alone lease for a term commencing on the date of execution or later stated start date or, if earlier, the date of substantial performance of any agreement for lease. Note however that occupation of the property under the earlier lease for indefinite term will not cause substantial performance of any agreement for the new lease. If SDLT has been paid on the lease for indefinite term for a period which is partly included in the term of the new lease, it should be possible to claim overlap relief in respect of that part – see **4.83**.

Old lease under stamp duty

4.70 SDLT is payable as if the new lease is a simple stand-alone lease for the term and consideration stated in the lease. The new lease cannot be treated as linked with the old. The main complication arises from the fact that 'rent' for a period before the effective start date of the new lease is treated as a premium as set out at **4.6**.

New lease granted on or before expiry of old fixed term lease

4.71 If the new lease is executed before the end of the old, the new lease is a reversionary lease (see SDLTM17070). It is important then to consider timing – SDLT obligations arise 30 days after the date of execution of the new lease, not 30 days after the start date. See Example 4.23A – the same principles apply even though the old lease was subject to stamp duty.

If the new lease takes effect before the end of the old, no credit is available against the SDLT payable, for stamp duty paid on the old lease for the overlap period.

New lease granted after expiry of old fixed term lease

4.72 If the tenant has simply remained in occupation during the period of holding over, paying rent as if the old lease has continued, this period will be treated as a continuation of the old lease as described at **4.2**. No SDLT obligations will arise in respect of this period. The new lease will be treated as beginning on the date on which it is executed, or 'substantially performed' (see **4.74**) if earlier. A payment of rent under the new lease will be regarded as substantial performance, but the lease will not be regarded as substantially performed merely because the tenant is in occupation/ has possession under the deemed continuation of the old lease (see SDLTM07900).

The lease may be expressed as beginning on a date earlier than the date of execution or substantial performance. However, in this case, because the old lease was not subject to SDLT, the treatment outlined in para **4.67** will not apply. Instead the general rule will apply and the lease will not be treated as beginning earlier than the date of execution or earlier substantial performance. Any rent reserved under the new lease which is expressed as being for a period before the deemed start of the lease, will be treated as a premium. Such amounts are often small and usually below the thresholds for payment of SDLT (£125,000 for residential property, £150,000 for non-residential or mixed property, see **2.37**).

If, exceptionally, any action is taken before the new lease is agreed, which would itself be regarded in property law as a surrender and re-grant (for example a significant change in the premises leased or an explicit agreement to extend the lease for a further fixed term), this is also likely to be regarded as the grant of a new lease for SDLT purposes. In such cases, further advice will be required, preferably taken in advance, based on detailed analysis of the fact pattern. However a mere increase in rent paid in respect of the period of holding over, even if not provided for in the old lease, will not normally give rise to a new deemed lease for the period in question.

Lease renewal examples

4.73 The following examples are referred to in paras **4.65** to **4.72**.

General fact pattern

ServiceCo occupies a commercial unit on lease from unrelated Propco. When the old lease was granted (more than five years ago), there was no agreement between the parties as to whether the lease would be renewed and the old lease does not include any option to renew. The lease expires on 31 December 2016, at which time the annual rent is £100,000 including VAT. A new lease is negotiated, the commercial terms being for seven years from 1 January 2017 (except where noted below) at yearly rent of £120,000 including VAT. No separate premium is payable. There is a rent review due on 1 January 2022.

Example 4.23A

New lease is negotiated in advance and executed on 1 December 2016 to take effect on 1 January 2017. There are no complications, the HMRC calculator gives the NPV of the rent as £733,745, and the SDLT payable as £5,837. The rent review will not occur during the first five years of the lease and so can be ignored. The calculation of SDLT is based on the lease commencing on 1 January 2017. However, because the lease was executed on 1 December

2016, the SDLT return and tax payment are due within 30 days of this date, ie by 31 December 2016, which is before the lease period commences. It does not matter whether the old lease was subject to stamp duty or SDLT (this would matter if the leases could be regarded as linked, see **4.58**).

Example 4.23B

As Example 4.23A, except lease was executed on 1 July 2016 and took immediate effect, displacing the old lease. The new lease is again for seven years (that is, it expires in this case on 30 June 2023). The SDLT return and tax payment were due by 31 July 2016.

(i) If the old lease was a stamp duty lease, the basic calculation is the same as for Example 4.23A; no allowance is given in respect of the unexpired six months of old lease displaced by the new lease.

(ii) If the old lease was an SDLT lease, an allowance is given in the SDLT calculation for the £50,000 rent which would have been paid under the old lease for the period from 1 July to 31 December 2016. See Example 4.9 for the method of calculation. The rent for the first year of the new lease is therefore taken to be £70,000 (that is, £120,000 less the £50,000 'allowance'). The HMRC calculator gives the NPV of the rent as £685,436 and the tax payable as £5,354.

Example 4.23C

New lease is finally executed on 1 July 2017, stated to be for seven years from 1 January 2017 at yearly rent of £120,000 but silent on matter of payment of rent for period before execution. ServiceCo has continued paying rent at yearly rate of £100,000, but pays an extra £10,000 immediately before lease is executed, to bring the yearly rate for the period of holding over (1 January to 30 June 2017) up to the new rent.

(i) Both parties are satisfied that the extra payment is in respect of occupation during the period of holding over and is not an inducement to the landlord to enter into the new lease.

  a. If the old lease was under stamp duty, continued occupation after expiry is treated as an extension of that lease which is not subject to SDLT on either the rent paid at the 'old' rate or the retrospective £10,000 rent increase.

  b. If the old lease was under SDLT, continued occupation is, in the first instance, regarded as an extension of the old lease. Because the old lease expired after 17 July 2013 and the new lease is stated to commence immediately after expiry of the old, the new lease is treated as starting on 1 January 2017. The SDLT is based on an

NPV of rent of £733,754, as in example 4.23A. No further liability or obligations arise in respect of the period of holding over.

c. The same analysis would apply if the new lease was stated to commence on some date after 1 January 2017 – for example on 1 July 2017, the date of execution – such that there was acknowledged to be a gap between the end of the old lease and the start of the new.

(ii) The parties acknowledge that the extra £10,000 payment was made to persuade Propco to grant the new lease.

a. If the old lease was under stamp duty, the continued payment of rent at the old rate during the period of holding over is treated as an extension of the old lease, with no SDLT consequences. However, the additional £10,000 is treated as a premium for the new lease. No SDLT is chargeable on the deemed premium of £10,000 as it falls within the £150,000 zero rate band. The same analysis would apply if the new lease was stated to commence on a date after 1 January 2017 because the lease cannot be regarded as beginning before the date of execution or earlier substantial performance.

b. If the old lease was under SDLT, the new lease is treated as beginning on the stated start date of 1 January 2017 and for a period of seven years. The £60,000 rent paid in respect of occupation in the six months to 30 June 2017 is taken into account as rent under the new lease, irrespective of whether the payment is acknowledged as being in respect of the new lease (*FA 2003, Sch 17A, para 9A*). However SDLT obligations arise on 31 July 2017, 30 days after execution (or earlier substantial performance) of the new lease.

Example 4.23D

New lease is finally executed on 1 July 2018, stated to be for seven years from 1 July 2018, so there is an 18-month gap between expiry of the old lease and stated commencement date of the new. Yearly rent is £120,000 but lease is silent on the matter of the payment of rent for the period before execution. ServiceCo has continued paying rent at yearly rate of £100,000, but pays an extra £30,000 immediately before lease is executed, to bring the yearly rate for the period of holding over (1 January 2017 to 30 June 2018) up to the new rent. If the old lease was under stamp duty, the treatment is covered under Examples 4.23C(i) and 4.23C(ii) above. If the old lease was under SDLT, the treatment of the period of holding over is as follows:

- If, in law, the old lease is treated as continuing (for example, under provisions of the landlord and tenant legislation), the continued occupation is treated initially as a one-year extension of the old lease.

This is likely to lead to the need to recalculate the SDLT payable on that old lease (see **6.17**). An amended return and payment of any additional SDLT will be due by 30 January 2018. However, the additional £20,000 rent for the period will not be taken into account in calculating the additional SDLT.

- The eventual execution of the new lease brings a retrospective end to the deemed continuation of the old lease. This is treated as ending on 31 December 2017 (one year after the actual termination date) and the new lease as beginning on 1 January 2018. As a result the new lease is deemed to have a period of 7 years 6 months, with rent of £120,000 per annum. The SDLT return and tax payment are due by 31 July 2018 despite the deemed start date of 1 January 2018.
- If, exceptionally, the foregoing does not apply (perhaps because the lease disapplied the relevant landlord and tenant provisions), it is likely that an implied new tenancy, a lease for indefinite term, arises for the interim period. This will be separately subject to SDLT as described at **4.12**, but will probably be linked to the new lease, forming a pair of successive leases. SDLT obligations in respect of the first one-year period will have to be dealt with by 30 January 2017 (30 days after the start of the implied tenancy) and those for the second year by 30 January 2018. Credit should then be available for SDLT paid in respect of the six months from 1 July 2018 when calculating the SDLT on the new lease. Overall therefore, apart from bringing deadlines forward, this should have the same net effect as the lease being stated as beginning on 1 January 2017, as in Example 4.23C(ii)(b). In the present case, the landlord's acceptance of continued rent payment suggests this would be the case. However, it would save the need to even consider the matter if the new lease was expressed as beginning immediately after expiry of the old!
- Finally, it may be that the tenant's continued occupation is an act of trespass, or is under a tenancy at will. Such occupation has no SDLT consequences. However, it is likely to be difficult to demonstrate that this is the basis of occupation given that rent is paid and accepted during the period of occupation and a new lease is eventually executed.

WHEN DO SDLT OBLIGATIONS ARISE?

Grant of a lease

4.74 The concepts of 'substantial performance' and 'effective date' (see **2.9** and **2.8**) apply to transactions involving leases, but with some specific 'trigger points'. In the case of a new lease, there are typically several stages which may occur at different times. The first SDLT liability and obligation to submit a return will arise 30 days after the earlier of:

- the lease proper being executed, and
- the agreement for lease being substantially performed.

Substantial performance occurs on the earliest of:

- the first payment of rent being made,
- substantially the whole of any consideration other than rent (eg a premium) being paid, and
- the tenant taking possession of substantially the whole of the premises.

The tenant will be regarded as taking possession if he occupies the premises, including for the purpose of fitting out, or if he becomes entitled to enjoy other benefits of tenancy such as receipt of rent from a sub-tenant (see **2.9**). Where an agreement for lease is substantially performed before the lease is granted, this is deemed to be the grant of a notional lease on the terms provided for in the agreement. The effective date of the grant of the notional lease is the date of substantial performance (*FA 2003*, *para 12A*, *Sch 17A*).

When SDLT obligations have arisen on a notional lease as a result of substantial performance of the agreement, the subsequent grant of the real lease is also potentially subject to SDLT.

Leases with effective date before 17 July 2013

4.75 For leases with an effective date before 17 July 2013, when the real lease is granted, the notional lease is deemed to be surrendered and overlap relief (see **4.82** and Example 4.9) is given for SDLT paid on the notional lease. The practical effect is that, provided the lease is for the same rent and term (and premium, if any) as envisaged in the agreement for lease, no further SDLT is payable. In this case, by virtue of the combined effect of *FA 2003*, *s 77A(3)* and *Sch 17A*, *para 9*, no further notification to HMRC is required. However, difficulty can arise if the rent is uncertain, variable or payable other than in sterling:

> First, if rent is payable in another currency and the exchange rate moves between substantial performance and completion, the sterling equivalent of the rent will differ at the two dates. If the sterling equivalent increases, extra SDLT will be payable. If the sterling equivalent decreases, no refund of SDLT will be due.
>
> Second, for SDLT purposes, the term of the lease will normally be regarded as beginning no earlier than the date of grant. This may result in a rent review falling within the first five years of the term, making the rent 'uncertain' (see **4.37**). It may also be important in determining such matters as when further returns are required if rent is variable or uncertain (see **4.78**). The

exception to this rule occurs when a lease originally executed under SDLT is renewed, see **4.67**.

Leases with effective date on or after 17 July 2013

4.76 The position has been simplified for leases with an effective date on or after 17 July 2013. When the real lease is granted, the notional lease is deemed to be a real lease granted on the date of substantial performance and ending on the end date of the actual lease. The rent and premium are as given in the actual lease. A further return is required only if further tax is payable on grant of the actual lease, for example because the rent, premium or term of the lease are greater than was taken into account in the original return. The lease is regarded as beginning at the effective date of the lease, so the issue relating to uncertain rent, mentioned in **4.75** above, no longer arises.

Deferral of SDLT

4.77 If (perhaps unusually) any amount of a premium is contingent or uncertain at the effective date and will not be payable until more than six months after the effective date, the lessee may apply to HMRC to defer payment of an appropriate amount of SDLT in the same way as for other land transactions (see **6.31**). However, it is not possible to apply to defer payment of SDLT arising in respect of rent, even if the amount of rent is uncertain at the effective date. This means that, if final settlement of the rent leads to a further SDLT liability, say at the end of the first five years of the lease, interest will be chargeable on that SDLT from the original due date (30 days after the effective date of the lease).

Uncertain or variable rent

4.78 The rent for the first five years of the lease may not be finally ascertained at the outset (for example, because the rent is based on results of a business or because there is a rent review within the first five years) – see **4.37** et seq. It remains necessary to submit returns and pay SDLT on normal timescales. SDLT must be calculated on the basis of a 'reasonable estimate' of the rent. Details must then be notified and any additional SDLT paid within 30 days of the earlier of (a) the rent being finally ascertained and (b) the end of the fifth year. The rent for the first five years may not have been finally determined when the five-year deadline arrives (for example, it may depend on profits of a period for which accounts have not yet been finalised). In that case, it may be necessary at the five-year point to submit a further return on the basis of a revised reasonable estimate, and further amend this once the rent is finally determined. In either case for interest purposes, the effective date for the revision remains the original date of the lease, and interest will be charged on any SDLT paid more than 30 days from then.

Example 4.24 – timing of returns and tax payments

The facts are as in Example 4.16 above. Ken must submit an SDLT return and pay SDLT no later than 1 December 2014. At that time, he must make a reasonable estimate of what the rent will be, and it seems likely that his estimate will be £10,000 per quarter. The total rent payable over the first five years (and, therefore, the final SDLT liability) will not be known until sometime after the end of the first five years. Because the actual rent is higher than the estimate, Ken must submit a further return by 30 November 2019 (that is, within 30 days of the end of the first five years), self-assessing the additional SDLT, which must be paid by the same date. If the correct figures are not known by this due date, Ken must make a reasonable estimate, and then correct the return once the actual numbers are known. If any of this leads to additional SDLT being paid after 1 December 2014, interest will be charged on this additional tax from that date.

Informal occupation and holding over

4.79 It can be difficult to establish when SDLT obligations arise in cases of informal occupation. It is essential to establish the precise legal nature of the occupation. A periodic tenancy is initially regarded as a lease for one year. If the annual rent exceeds the relevant threshold, SDLT obligations arise 30 days after commencement. If the tenancy continues beyond one year, it is treated as a lease for two years, then three years and so on. At the commencement of each further year, it is necessary to consider whether the cumulative rent and/or period are sufficient to give rise to SDLT notification or payment obligations. If so, those obligations arise 30 days after the start of the year in question.

The position is similar where a lease does not have a stated end date. In some cases, the position may be similar where a tenant remains in occupation after the end of the lease. However, where the tenant is 'holding over' while a new lease is negotiated, the precise nature and timing of SDLT obligations in respect of the holding-over period depend on the provisions of the new lease (see **4.64** et seq).

Variation or surrender of a lease

4.80 There are no special rules relating to the time for notifying and paying SDLT on the variation or surrender of a lease. The obligations arise 30 days after the earlier of:

- completion of the formalities giving effect to the variation (for example, the parties executing the relevant document), and
- 'substantial performance' of the variation.

Substantial performance has essentially the same meaning as set out in **4.74**, subject to the specific rule in the next paragraph.

RELIEFS

4.81 In general, reliefs and exemptions available for other UK land transactions also apply to transactions involving leases – see **Chapter 3** in respect of reliefs. Some of the reliefs are modified for lease transactions, particularly in relation to possible future clawback of the relief. There are a few additional reliefs which only apply to lease transactions.

Sale (or lease) and leaseback relief

4.82 Where a vendor sells or leases a 'major interest' (see **2.11**) in a property to a purchaser and, in consideration (wholly or partly), the purchaser grants a lease (the 'leaseback') to the vendor out of that major interest, the leaseback may be exempt from SDLT (*FA 2003, s 57A*). The only other consideration permitted for the initial sale/lease (which must not be a sub-sale, see **3.8**) is cash and/or assumption, satisfaction or release of a debt; the vendor and purchaser must not be companies in a group relationship, such that they could qualify for group relief.

A fairly commonplace commercial arrangement is for an owner to transfer a property to a developer in return for the developer redeveloping the site and granting the owner a long lease of part of the new development. The developer recoups his cost and makes a profit by selling the rest of the development to third parties. It is understood that HMRC consider that the lease back to the owner does not qualify for sale and leaseback relief, because the developer's agreement to redevelop the site amounts to non-qualifying consideration. SDLT is therefore due on both the original sale to the developer and the lease back to the owner.

Overlap relief

4.83 This is mentioned at **4.34**, and applies if a lease is granted which supersedes a previous lease of substantially the same premises, which itself was subject to the SDLT regime. In calculating SDLT on the new lease, the rent for the 'overlap period' (that is, the period which would have been covered by the old lease but is now covered by the new) is reduced by the rent which would have been paid under the old lease if that had been left to run its course (*FA 2003*, *Sch 17A*, *para 9*). This is not strictly a relief, it is a statutory treatment. Therefore it is not necessary to claim the 'relief' by answering question 9 of the SDLT1 return form 'yes' and entering a code. It is sufficient, when calculating SDLT on the new lease, to reduce the rent by the rent which would have been paid under the old lease. The relief may also apply where the new lease is granted to someone who acted as guarantor under the original lease, or in certain circumstances where the new lease is granted to someone who was a sub-tenant of the original lessee.

This relief is in addition to the general principle set out in *FA 2003*, *Sch 17A*, *para 16* that, where a lease is surrendered in consideration of the grant of another, neither the surrender nor the grant counts as consideration for the other.

Exercise of collective rights by tenants of flats

4.84 Tenants holding long leases of flats may have the right to gain control of the freehold. This may be achieved, for example, by the tenants jointly forming a company or other entity to buy the freehold. Although the amount paid by any individual tenant may not be large enough to give rise to any SDLT liability, the total amount payable by the company may be very large, possibly attracting SDLT at the top rate. This is clearly unfair in comparison with a leaseholder of an equivalent single house, who may be able to buy the freehold as an individual without triggering an SDLT charge. *FA 2003*, *s 74* provides a relief, under which the rate of SDLT to be applied is determined by dividing the total consideration payable by the company or other entity by the number of flats involved. In its original form, the relief only applied to a specific type of company expected to be established under the *Leasehold Reform*, *Housing and Urban Development Act 1993*. Unfortunately, the provisions of that Act allowing formation of such companies have never been brought into effect, so this relief was ineffective until amended by *FA 2009*, *s 80*. The form of entity buying the freehold is now not specified.

Chapter 5

SDLT and Partnerships

INTRODUCTION

5.1 The main legislation specifically dealing with partnerships is in *FA 2003, Sch 15*. In this chapter, references to paragraph numbers without further identification are references to paragraphs of *Schedule 15*. *Part 1 (paras 1–4)* and *Part 2 (paras 5–8)* set out general provisions, including confirmation of the treatment of an ordinary acquisition of a chargeable interest from an unconnected vendor. *Part 3 (paras 9–40)* was inserted by *FA 2004* and contains the detailed rules applying SDLT to almost all other transactions of partnerships which own, acquire or dispose of chargeable interests. The heading, suggesting that the *Part 3* provisions only apply in 'special' circumstances, is misleading!

5.2 The UK tax system has always had difficulty working out how to deal with partnerships – in particular, whether to treat them as separate entities distinct from their members or as mere aggregations of separately taxed members – and SDLT continues that uncertainty. When SDLT was introduced, many partnership transactions were initially left entirely within the stamp duty rules, while the government tried to work out how best to deal with them under SDLT. Since the enactment of *FA 2004*, SDLT has generally applied to transactions by partnerships or partners, but with modification of and additions to the general rules. Substantial changes were made to the rules by *FA 2006, FA 2007* and *FA 2008*. The last of these amended the *FA 2007* changes with retrospective effect when it was realised that the earlier provisions caused major problems for some very large, widely held property investment vehicles.

5.3 This difficulty and uncertainty is reflected in the fact that HMRC issued guidance in draft, in conjunction with the enactment of *FA 2004*, but only finalised it and included it in the Stamp Duty Land Tax Manual in March 2011.

5.4 Three types of UK-based partnership are recognised:

- a (general) partnership within the *Partnership Act 1890*;

- a limited partnership (LP) registered under the *Limited Partnerships Act 1907*; and
- a limited liability partnership (LLP) formed under the *Limited Liability Partnerships Act 2000* or the Northern Ireland equivalent.

The SDLT rules seek to treat all three types of entity in the same way; they also apply the same treatment to any 'firm or entity of a similar character' formed under the laws of another country. Difficulties arise because the three types of entity listed above and many equivalent overseas entities have quite different legal and commercial characteristics. As a result, the SDLT treatment of a particular transaction may be logical in one case (say, involving a general partnership) but less so in another (say, involving an LLP). These differences have been a factor in the use of partnerships to mitigate SDLT costs, and many of the special rules which now apply were originally intended to combat perceived avoidance. Particular issues arise from the fact that LLPs (together with Scottish LPs and many overseas equivalents) are bodies corporate for the purposes of general law and many other tax purposes and/or have separate legal personality, but are treated as 'transparent' as regards ownership of land for SDLT purposes (see, for example, **3.24**).

GENERAL PRINCIPLES AND ORDINARY TRANSACTIONS

General principles

5.5 Property held by or on behalf of a partnership is treated as held by or on behalf of the partners (and not the partnership); a land transaction entered into for the purposes of a partnership is treated as entered into by or on behalf of the partners (and not the partnership); this is so even if the partnership is regarded as a body corporate or other legal entity for other purposes, such as ability to enter into contracts and to own property (*para 2*). However, the partners at the effective date of the transaction have joint and several liability for payment of any tax (*paras 6, 7*). They also have responsibility, in some circumstances together with any partners who join after the effective date, for payment of any penalty, for example for failure to make a return or notify liability. Although this means all partners have responsibility for compliance, they may nominate 'representative partners' to act on behalf of the partnership. Any such nomination must be notified to HMRC (*para 8*).

5.6 If there is a change in the membership of a partnership, it is nonetheless treated as the same continuing partnership, provided at least one member before the change remains a member afterwards (*para 3*). Thus, despite the assertions in the previous paragraph, there are ways in which a partnership is treated as a separate entity.

Partnership share

5.7 In order to calculate the chargeable consideration for many partnership transactions, or even in some cases to determine whether a chargeable transaction has occurred, it is necessary to know the shares which the different partners have in the partnership. For most SDLT purposes, a partner's share in a partnership is defined as the proportion in which he is entitled to share in the income profits of the partnership at the effective date of the transaction (*para 34(2)*). When property is transferred out of a partnership, a more complicated definition applies, although in many cases it produces the same result (see **5.23**). Entitlement to capital, or to capital profits, does not enter into the calculation. The definition causes practical difficulties – partnerships often decide the precise sharing of profits after the year-end, once total profits and the contribution of each partner to them are known. In such cases, it may be necessary to complete an SDLT return and pay tax based on estimated shares, then amend the return once profit shares are finally settled. Since profit shares do not normally vary from day to day, the reference to the 'effective date' is usually taken to mean the accounting period containing the effective date.

Ordinary transactions

5.8 If a partnership (or a partner on behalf of a partnership) enters into a land transaction with a person who is completely at arm's length, that transaction is treated in the same way as an equivalent transaction not involving a partnership. The only modification is that the partners have joint and several responsibility for any compliance matters. A person is completely at arm's length if he is neither a partner (whether current, joining or departing) nor connected with a partner. Connected for these purposes is as defined by *Corporation Tax Act 2010 (CTA 2010), s 1122*, although the automatic connection of partners with each other is omitted. It follows that, for such an arm's-length transaction, SDLT is generally chargeable by reference to the value of the actual consideration, and it is only necessary to consider the market value of the property itself where that would be required for a non-partnership transaction (eg for exchanges of property interests).

SPECIAL PROVISIONS

5.9 The application of the SDLT rules to arm's-length transactions is relatively straightforward. Complications arise when there is a connection between the partnership and the other party, ie where the other party is a partner (current, departing or joining) or is connected with a partner. It is necessary to consider separately the implications of the partnership acquiring or disposing of property, including modification of an existing interest where that would be treated as an acquisition or disposal under general SDLT rules.

It is also necessary to consider the implications of changes in the membership or relative interests of the members in a partnership which owns property; in certain cases, such changes are deemed to give rise to a corresponding transfer of property interests between partners.

5.10 In general, when partnership transactions are subject to the 'special provisions', SDLT is charged by reference to the market value of the property interest in question and the proportion of the property which is deemed to be transferred. In the original version of the legislation, any actual consideration given was also taken into account. The rules were amended in 2006 to remove this link. Actual consideration (other than any change in partnership share) is now irrelevant in these cases. However, there are many anti-avoidance rules; in some circumstances, amounts which are commercially equivalent to consideration may trigger these (see **5.17**).

5.11 The legislation sets out formulae which must be applied to the market value of the property in question to determine the deemed consideration for the transaction. The formulae appear complex, but are generally seeking to achieve a relatively simple overall effect, which is to ensure that tax is charged on the economic transfer of value which takes place.

Example 5.1 – simple partnership case

Eric and Ernest are in a general partnership, sharing equally. They agree to admit Eddie as an equal partner (so each partner has a one-third share in profits, losses and assets) in return for Eddie agreeing, inter alia, to contribute freehold premises to the partnership. The property is transferred to Eric and Eddie jointly to hold on trust for the partnership (this is a normal arrangement, because a general partnership is not a body capable of holding the title to land). In the absence of any other complicating factors, SDLT is chargeable by reference to two-thirds of the market value of the property. This is the proportion which has economically changed hands by transfer to the other partners; Eddie still has a one-third interest in the property through his membership of the partnership, so no SDLT is charged on that proportion.

5.12 Inevitably, real transactions do include complicating factors, and HMRC are keen to minimise avoidance. Therefore, the basic formulae are supplemented by definitions and rules for special cases which must be taken into account. It would not be safe to assume that the simple apportionment set out above gives the correct SDLT charge, without checking the specific facts of the case against the detailed rules. Many changes to the rules are attempts to combat perceived avoidance, especially attempts to combine the partnership rules with reliefs (for example sub-sale relief, see **3.8**) and other special provisions to mitigate SDLT costs.

Acquisition by a partnership

5.13 Under *para 10*, when a chargeable interest is acquired by a partnership from a partner, or from a person who becomes a partner as a result, or from a person who is connected with either, SDLT is charged on deemed consideration equal to:

$$MV \times (100 - SLP)\%$$

where MV is the market value of the chargeable interest at the effective date of the transfer, and SLP is the 'sum of the lower proportions'. A five-step calculation is required to determine SLP, set out in the legislation (*para 12*) as follows:

(1) Identify the 'relevant owner(s)'. These are persons who have an interest in the property immediately before it is transferred and are partners or connected with partners immediately after the property is transferred.

(2) For each relevant owner, identify the 'corresponding partner(s)'. These are persons who are partners immediately after the transfer and who are relevant owners or are individuals connected with relevant owners. For these purposes, a company may be regarded as an individual connected with a relevant owner *only* if it holds property as trustee and is connected by virtue of *CTA 2010, s 1122(6)* (trustee connected with settlor etc).

(3) For each relevant owner, take the proportion of the property to which he is entitled immediately before the transfer and apportion it between any one or more of his corresponding partners. Where a relevant owner has more than one corresponding partner, the apportionment may be performed in whatever manner gives the best result.

(4) Find the 'lower proportion' for each person who is a corresponding partner in relation to one or more relevant owners. The lower proportion is the lower of:

   (a) the proportion apportioned to him under (3) above, and

   (b) his partnership share immediately after the transfer.

(5) Add together the lower proportions for each partner who is a corresponding partner. This is the SLP.

Example 5.2 – calculating the sum of the lower proportions

Alfred, Betty and Carl are in partnership, sharing 40%, 40% and 20% respectively. Betty and Carl are sister and brother; apart from the partnership, Alfred is not related to or connected with Betty or Carl. The partners agree to buy a building from Danielle, who is Betty and Carl's sister, for £500,000. The calculation steps are as follows:

(1) The only relevant owner is Danielle, because she owns 100% of the property before the transfer.

(2) Betty and Carl are corresponding partners because they are individuals connected with the only relevant owner, their sister Danielle.

(3) Danielle was entitled to 100% of the property before the transfer. We choose to apportion this 50% to each corresponding partner, Betty and Carl.

(4) For Betty, her partnership share immediately after the transaction is 40%, and the proportion apportioned to her under (3) is 50%, so her lower proportion is 40%; for Carl, his partnership share immediately after the transaction is 20%, and the proportion apportioned to him under (3) above is 50%, so his lower proportion is 20%.

(5) The sum of the lower proportions is 40% (Betty) plus 20% (Carl), which is 60%.

The consideration for SDLT purposes is therefore MV × (100 – 60)%, or 40% of the market value, ie £200,000. In this relatively simple case, it can be seen that this represents the proportion of the property effectively transferred to Alfred, the partner who is not connected with the original owner of the property.

5.14 In most cases before the transfer, the property will be owned by one person, so there will be only one relevant owner. If two or more persons own a property as beneficial joint tenants, they are treated as owning as tenants in common, in equal shares.

Connected companies

5.15 The restriction to individuals in step (2) was introduced by *FA 2007, s 72*, and is designed to prevent the use of a partnership arrangement to transfer property to a connected company without suffering SDLT on the full market value of the property. Other anti-avoidance rules are dealt with at **5.17** and **5.27**.

Example 5.3 – partnership including connected company

Derek sets up a company, Fiddler Ltd, in which he is the sole shareholder. Derek and Fiddler Ltd form a limited partnership in which Derek is the limited partner with a 1% share and Fiddler Ltd is the general partner with a 99% share. Derek transfers a valuable property to the partnership. The steps to calculate the proportion of market value on which SDLT is charged are as follows:

(1) Derek is the only relevant owner with 100%.

(2) Derek is also the only corresponding partner. Fiddler Ltd cannot be a corresponding partner, as it is not an individual.

(3) Therefore, Derek's 100% ownership before the transfer must be apportioned 100% to Derek as corresponding partner after the transfer.

(4) Derek's partnership share immediately after the transfer is only 1%, so his lower proportion is 1%; Fiddler Ltd is not a corresponding partner, so it does not figure in the calculation.

(5) The SLP is therefore 1%, and SDLT is chargeable on the 99% of the market value effectively transferred to Fiddler Ltd.

Lease at rent and other transactions

5.16 The same principle applies where the acquisition takes the form of the grant of a lease at rent to the partnership from a partner or person connected with a partner. In that case, the calculation set out above is applied to each year's rent. The resulting apportioned rent figures are then used to calculate the SDLT on the net present value (NPV) of the rent in the normal way (see **4.50**). The same principle also applies where there is a variation of a lease or other chargeable interest which, on normal principles, is regarded as an acquisition by the partnership from a partner or person connected with a partner.

Example 5.4 – lease at rent, partnership shares vary

Twin sisters Gloria and Grace carry on a market research business in partnership with Harry, with whom they are not otherwise connected. Partnership profit-sharing ratios are adjusted at the end of each year in accordance with the partnership deed, using a formula which takes account of matters such as business won, hours worked etc. Gloria grants a 20-year lease to the partnership (strictly, to the partners jointly to hold on behalf of the partnership) at a time when the profit sharing ratios are 25% to Gloria, 40% to Grace and 35% to Harry. The lease is for a rent of £25,000 per annum (thought to be market rent) and no premium, with a market rent review at the end of every fifth year. The partnership has the option to surrender the lease at each rent review date. The calculation steps are as follows:

(1) Gloria is the only relevant owner, owning 100% of the property before the grant of the lease.

(2) Gloria and Grace are corresponding partners, as they are the relevant owner or an individual connected with the relevant owner.

(3) Gloria's 100% ownership of the property before the grant of the lease is apportioned between Gloria and Grace – we choose to apportion 50% to each.

(4) Gloria's partnership share immediately after the grant of the lease is 25%, and Grace's is 40%. In each case, this is less than the apportioned ownership under (3), so in each case the partnership share is the lower proportion.

(5) Therefore the SLP is 25% + 40% = 65%, and the NPV of the remaining 35% of the rent, ie £8,750 per annum, is subject to SDLT. In this particular case, the NPV is less than £150,000, so no SDLT will be payable, but it will be necessary to submit an SDLT return (see **6.2**).

Two years later, the partners agree that the estimate of the market rent was wrong and it should have been £35,000 per annum. They therefore agree to vary the lease to increase the rent to this sum with effect from the start of the third year. At that time, the profit-sharing ratios are Gloria 30%, Grace 30% and Harry 40%. The variation to increase the rent is treated as the grant of a further lease for the remaining 18 years at a rent equal to the increase, which is £10,000 per annum – see **4.24** in this regard. Carrying out the SLP calculation shows that the NPV of 40% of this rent, or £4,000 per annum, is subject to SDLT. On its own, this would be too little to give rise to an SDLT cost. However, in this case, the deemed further lease is probably linked with the original lease (see **4.58** et seq). As a result, SDLT will be payable on both the original lease and the deemed further lease as a result of this rent increase. See Example 4.22 for an illustration of the calculation required.

Withdrawal of money after acquisition of property

5.17 The transfer of property from a partner or person connected with a partner to a partnership marks the start of a three-year period within which certain other events can lead to a further SDLT charge (*para 17A*). If, during that period, the partner concerned withdraws from the partnership money or money's worth which does not represent income profit, but:

- is capital withdrawn from the partners' capital account, or
- arises from the partner reducing his interest or ceasing to be a partner,

the withdrawal is treated as a chargeable land transaction. Where the original transfer was from a person connected with a partner, withdrawal of value by either the partner or any person connected with the partner is treated as a chargeable transaction. However, presumably, the connected person would have also to be a partner to be able legitimately to withdraw capital etc.

If the partner (or, where the original transfer was from a person connected with a partner, the partner or a person connected with the partner) has made a loan to the partnership, the repayment of any of the loan or a withdrawal of any amount by that person which does not represent income profit of the partnership is also treated as a chargeable land transaction.

In either case, the chargeable consideration is the smaller of:

- the amount withdrawn from or repaid by the partnership, and
- the market value of the land at the date of the original land transfer, reduced by any amount which has already been charged to SDLT.

If the withdrawal of money or money's worth is also subject to SDLT as a transfer of an interest in a PIP (see **5.30**), the SDLT charged under the 'withdrawal of money' provision is reduced by the amount of any charge under the 'transfer of interest in PIP' provision (*para 17A(8)*). If a '*paragraph 12A* election' (see **5.38**) has been made in respect of the property transfer, no SDLT arises on a subsequent withdrawal of value.

Example 5.5 – repayment of loan after transfer

Using the facts as in Example 5.2, the partnership does not have enough ready cash to pay the whole price of the building purchased from Danielle, so it is agreed that £350,000 will be left outstanding as a loan from Danielle to the partnership, with interest being charged at 2% above the Bank of England base rate, payable monthly in arrears. The loan arrangement has no effect on the SDLT charge on the initial transfer of the property to the partnership. One year later, after a purge on debtors, the partnership has enough cash and pays off the loan. The loan repayment is taken to be a land transaction, and SDLT is chargeable on the partnership. However, SDLT was paid on £200,000 when the property, with a value of £500,000, was transferred to the partnership. The chargeable consideration for the deemed land transaction cannot exceed the part of that market value which has not yet been taxed. So, SDLT is chargeable on £300,000, and not on the full £350,000 loan repayment. Note that this limitation is assessed by considering the market value of the property when it was transferred, and not the value at the time of the loan repayment or other withdrawal of value.

Had the partnership borrowed from a bank at the outset and paid Danielle in full, the SDLT charge on the original land transfer would still have been based on £200,000, but the subsequent repayment of the bank debt would not have given rise to an SDLT charge. It is difficult to see any policy reason for this discrimination, which therefore looks rather like a poorly directed anti-avoidance provision.

Transfer to a PIP may also involve transfer of partnership interest

5.18 The charge on the partnership when a new or existing partner (or someone connected with him) transfers a chargeable interest to the partnership does not depend on the amount or form of consideration given by the partnership. As suggested in Example 5.1, the consideration could take the form of admitting a new partner or increasing a partner's share. This acquisition of, or increase in, partnership share will be regarded as a transfer of an interest in a partnership. If the partnership is a property investment partnership (PIP), this transfer will be separately subject to SDLT (see **5.30**), chargeable on the new partner or the partner whose share has increased. This is in addition to the charge on the partnership arising from the acquisition of the property. The transaction is an exchange. This fact should not affect the calculation of SDLT because, both in transactions between a partnership and its partners (or persons connected with them) and in an exchange, SDLT is calculated by reference to the market value of the chargeable interests acquired (albeit indirectly, where partnership interest is acquired) rather than the consideration given. As noted below (see **5.34** and **5.36**), the newly transferred property is not taken into account in determining the SDLT charge on transfer of the partnership interest, so there is no double charge.

Example 5.6 – admission of partner, transfer of property

X Ltd and Y Ltd are limited partners in Spendit LP, a limited partnership which owns shopping centres and is clearly a PIP. The general partner is a Jersey-registered company with a negligible partnership share. X Ltd has a 40% partnership share, and Y Ltd has 60%. The shopping centres are worth £100 million. It is agreed that Z Ltd will be admitted as a limited partner with a partnership share of 30%, so that X Ltd's share drops to 28% and Y Ltd's to 42%. By way of payment, Z Ltd will transfer a further shopping centre to Spendit LP, with a value of £45 million. Apart from membership of Spendit LP, there are no other connections between X Ltd, Y Ltd and Z Ltd.

- **Transfer of property to Spendit**: The calculation set out at **5.13** shows that Spendit is liable for SDLT on 70% of the £45 million value of the property brought to the partnership by Z Ltd.
- **Transfer of partnership share to Z Ltd**: Z Ltd is deemed to have acquired an interest in the 'relevant partnership property' owned by Spendit before Z Ltd joined. The consideration is the market value of the property multiplied by the partnership interest acquired, ie 30%. Whether the shopping centres already owned by Spendit are relevant partnership property will depend on their history, and the question of whether the transfer to Z Ltd is 'Type A' or 'Type B' (see **5.33** and **5.35**).

Disposal by a partnership

5.19 Disposal of a chargeable interest should not directly give rise to any SDLT obligations for the partnership; it is for the purchaser to deal with any SDLT. If the disposal is to an unrelated third party, the purchaser will be liable for SDLT in the normal way, based on consideration given. However, if the disposal is to a partner or person connected with a partner, special rules come into play (*para 18*). These are similar to those outlined in the previous paragraph; their purpose is to ensure the purchaser pays SDLT only on the proportion of the property which he and/or connected individual partners did not already own indirectly through membership of the partnership. As in the previous paragraphs, SDLT is charged on deemed consideration equal to:

$$MV \times (100 - SLP)\%$$

where MV is the market value of the chargeable interest at the effective date of the transfer, and SLP is the 'sum of the lower proportions'. Again, a five-step calculation is required to determine SLP, set out in the legislation (*para 20*) as follows:

(1) Identify the 'relevant owner(s)'. These are persons who have an interest in the property immediately after it is transferred and are partners or connected with partners immediately before the property is transferred.

(2) For each relevant owner, identify the 'corresponding partner(s)'. These are persons who are partners immediately before the transfer and who are relevant owners or are individuals connected with relevant owners. For these purposes, a company may be regarded as an individual connected with a relevant owner *only* if it holds property as trustee and is connected by virtue of *CTA 2010, s 1122(6)* (trustee connected with settlor etc).

(3) For each relevant owner, take the proportion of the property to which he is entitled immediately after the transfer and apportion it between any one or more of his corresponding partners. Where a relevant owner has more than one corresponding partner, the apportionment may be performed in whatever manner gives the best result.

(4) Find the 'lower proportion' for each person who is a corresponding partner in relation to one or more relevant owners. The lower proportion is the lower of:

    (a) the proportion apportioned to him under (3) above, and

    (b) his partnership share immediately before the transfer.

(5) Add together the lower proportions for each partner who is a corresponding partner. This is the SLP.

Example 5.7 – transfer of property to partner

Alfred, Betty and Carl are (still) in partnership, sharing 40%, 40% and 20% respectively. Betty and Carl are siblings, Alfred is not related to them. Betty decides to withdraw from the partnership and it is agreed she will take one of the properties in satisfaction of her entitlement to partnership capital. The calculation of Betty's SDLT liability is as follows:

(1) The only relevant owner is Betty, because she owns 100% of the property after the transfer.

(2) Betty and Carl are corresponding partners because they are, or are connected with, the only relevant owner, Betty.

(3) Betty is entitled to 100% of the property after the transfer. We choose to apportion this 50% to each corresponding partner, Betty and Carl.

(4) For Betty, her partnership share immediately before the transaction is 40%, and the proportion apportioned to her under (3) is 50%, so her lower proportion is 40%; for Carl, his partnership share immediately before the transaction is 20%, and the proportion apportioned to him under (3) above is 50%, so his lower proportion is 20%.

(5) The sum of the lower proportions is 40% (Betty) plus 20% (Carl), which is 60%.

The consideration for SDLT purposes is therefore MV × (100 – 60)%, or 40% of the market value. In this simple case, it can be seen that this represents the proportion of the property effectively transferred from the partner who is not connected with the final owner of the property. Again, the same principles apply in relation to rent if the 'transfer' is the grant of a lease at rent.

Example 5.8 – transfer of property to connected company

The initial facts are as in Example 5.7, but it is Alfred who wishes to withdraw from the business. It is decided to wind up the partnership and transfer the properties and other assets to a newly formed company BC Ltd in which Betty and Carl are equal 50% shareholders. It is agreed that BC Ltd will pay for the business and properties by paying Alfred an agreed amount of cash and by issuing shares to Betty and Carl. The calculation of BC Ltd's SDLT liability is analogous to that in Example 5.7:

(1) The only relevant owner is BC Ltd, because it owns 100% of the property after the transfer.

(2) Betty and Carl are corresponding partners because they are connected with the only relevant owner, BC Ltd. This is on the basis that both Betty and Carl will be treated as being connected with BC Ltd by virtue of

Corporation Tax Act 2010, s 1122(3)(b) which states: 'A company is connected with another person ("A") if – ... (b) A together with persons connected with A have control of the company.' As Betty and Carl are both connected with each other, as they are siblings, they are both treated as controlling BC Ltd (*Corporation Tax Act 2010, s 1122(5)(b)*).

The question which then arises is whether Alfred is also treated as having control of BC Ltd, even though he has no equity interest in BC Ltd. *Corporation Tax Act 2010, s 1122(3)(a)* states that: 'A company is connected with another person ("A") if – (a) A has control of the company ...'. Control for these purposes is defined in *Corporation Tax Act 2010, s 450*, which broadly provides that a person has control of a company if he exercises direct or indirect control over the company's affairs. For these purposes the rights of any 'associate' of a person are attributed to that person in determining whether that person has control of a company (*Corporation Tax Act 2010, s 451(4)(c)*). *Corporation Tax Act 2010, s 448(1)(a)* provides that any partner of a person is an 'associate' of that person. Consequently, as Alfred is in partnership with Betty and Carl he will be an 'associate' of both, and attributing the rights and powers of Betty and Carl in relation to BC Ltd to Alfred would mean that Alfred would be treated as possessing the whole of the share capital of BC Ltd, and therefore he would be treated as controlling BC Ltd. Therefore, Alfred is connected to BC Ltd by virtue of *Corporation Tax Act 2010, s 1122(3)(a)*, and is consequently a corresponding partner in relation to BC Ltd.

(3) BC Ltd is entitled to 100% of the property after the transfer. We choose to apportion this 40% to Alfred, 40% to Betty and 20% to Carl.

(4) Alfred's partnership share immediately before the transaction is 40%, and the proportion apportioned to him under (3) above is also 40%, so his lower proportion is 40%. For Betty, her partnership share immediately before the transaction is 40%, and the proportion apportioned to her under (3) is 40%, so her lower proportion is 40%; for Carl, his partnership share immediately before the transaction is 20%, and the proportion apportioned to him under (3) above is 20%, so his lower proportion is 20%.

(5) The sum of the lower proportions is 40% Alfred plus 40% (Betty) plus 20% (Carl), which is 100%.

The consideration for SDLT purposes is therefore MV × (100 – 100)%, which is nil. It can therefore be seen that where all of the partners in a partnership are individuals the incorporation of the partnership should not give rise to a charge to SDLT.

There is an apparent conflict with the rule in *FA 2003, s 53* that a transfer to a connected company is deemed to be at not less than market value. However,

HMRC have confirmed at SDLTM 34170 that the partnership rules override *s 53*.

5.20 In most cases after the transfer, the property will be owned by one person, so there will be only one relevant owner. If the transfer is to two or more persons as beneficial joint tenants, they are treated as owning as tenants in common, in equal shares.

Transfers between partnerships

5.21 The possibility is recognised that a property may be transferred from one partnership to another and that there may be connections or commonality between the partners. In that case, it is necessary to perform calculations of the chargeable consideration twice: once as set out at **5.13** (transfer to a partnership from connected owner); and once as explained at **5.19** (transfer from a partnership). The chargeable consideration is then to be taken as the higher of the two amounts thus calculated (*para 23*).

5.22 However, the following conditions must be satisfied if para 23 is to apply:

(1) There must be a transfer of a chargeable interest from one partnership (the 'transferor partnership') to another partnership (the 'transferee partnership'). This means that the partnerships are treated as opaque for the purposes of the legislation and para 2 (which treats any chargeable interest held by a partnership as being held by the partners, and any land transaction entered into by the partnership as having been entered into by the partners), cannot apply; and

(2) Both para 10 (**see 5.13**) and para 18 (**see 5.19**) must apply to the transfer of the chargeable interest. For para 10 to apply the transfer must be made by a partner of the transferee partnership, or a person who becomes a partner in the transferee partnership in return for the transfer or, in either case, a person connected with that person. For para 18 to apply the transfer must be to a person who is or has been a partner in the transferor partnership or to a person connected with a person who is or has been one of the partners in the transferor partnership. For these requirements to be satisfied the partnerships would have to be treated as transparent either under general law or as a result of the application of para 2 (see above for the impact of para 2). Consequently, if it is accepted that para 2 does not apply to para 23 (it is not known whether HMRC accept this point) then the partnerships would have to treated as transparent under general law, which would mean that neither the transferor partnership nor the transferee partnership could be a partnership with separate legal personality such as an LLP, Scottish partnership or a non-UK partnership which had separate legal personality.

It is considered that the better view is that para 2 does not apply for the purposes of para 23 and therefore that para 23 only applies where the transferor partnership and the transferee partnership are partnerships which under general law do not have separate legal personality such as, an English general partnership, an English limited partnership etc. However, given the complexity of the analysis, it is suggested that consideration should be given to seeking a non-statutory clearance from HMRC if there is doubt as to whether para 23 applies to a transaction.

Partnership share – special rules

5.23 As noted above (see **5.7**), for most SDLT purposes, partnership share is defined as the proportion in which the partner is entitled to share in income profits. A more complex analysis and calculation are required to determine the partnership share of any relevant partner, when this is required for step (4) in the SLP calculation on transfer of a chargeable interest out of a partnership (*para 21*). The complexity arises from an attempt to negate early SDLT mitigation which used partnership structures. In a straightforward case, if stamp duty or SDLT (based on market value or arm's-length consideration) has been paid on all properties acquired after 19 October 2003 and, where necessary, on transfers of partnership interests, the share of any partner should simply be the proportion in which he is entitled to share in income profits. However, to be certain of having the right numbers, it is necessary to work through the following steps:

(1) Check whether the chargeable interest which is the subject of the transfer (or from which any lease is granted, if the transaction is the grant of a lease):

    (a) was transferred to the partnership before 20 October 2003, or

    (b) was transferred to the partnership on or after that date, and either the instrument of transfer has been duly stamped or SDLT has been duly paid on the transfer.

    If neither (a) nor (b) applies, the partnership share of the partner is treated as zero.

(2) Assuming the partnership share is not zero under step (1), find the 'relevant date':

    (a) If (1)(a) applies and the partner was a partner on 19 October 2003, that is the relevant date.

    (b) If (1)(a) applies and the partner became a partner after that date, the date he became a partner is the relevant date.

    (c) If (1)(b) applies and the partner was a partner when the property was transferred to the partnership, the relevant date is the date of that transfer.

(d) If (1)(b) applies and the partner became a partner after the property was transferred to the partnership, the relevant date is the date he became a partner.

(3) Find the partner's actual partnership share on the relevant date.

(4) Add to the share determined under (3) any increases in the partner's share, where:

(a) if the increase resulted from a transfer which occurred on or before 22 July 2004, the instrument of transfer has been duly stamped with *ad valorem* stamp duty, and

(b) if the increase resulted from a transfer which occurred after that date, any SDLT payable in respect of the transfer has been duly paid.

(5) Deduct from the result of (4) any decreases in the partner's partnership share during the period from the day after the relevant date to the day before the date on which the transfer of property out of the partnership occurs.

It can, therefore, be seen that a great deal of historic information may be needed to determine what should be a simple matter of a partner's partnership share.

Transfer from a corporate partnership

5.24 The rules are modified if a property is transferred from a partnership whose members are entirely bodies corporate and the SLP is 75% or more (*para 24*). In this case, no reduction is given for the SLP, and the consideration is deemed to be equal to the market value of the property (and, if the transfer is the grant of a lease at rent, the full NPV of the rent is subject to SDLT). This situation can only occur when the transferee is a partner with a 75% or higher partnership share. In this situation, the only way to reduce the SDLT charge is to claim group relief on the transfer. Group relief is modified for partnership transactions to allow for this (*para 27*; see **5.40**).

Transfer of an interest in a partnership

5.25 Whether a true transfer of an interest in a partnership is possible depends on the law governing the partnership, and the terms of the partnership deed or other agreements governing such matters. In some cases, it may be possible to achieve an effect equivalent to a transfer, only by one partner withdrawing capital or in some other way relinquishing some or all of his entitlement as partner, while another person becomes a partner, introduces new capital or increases his entitlement in some other way. For most transactions, the legislation makes these distinctions irrelevant, defining a transfer of an interest in a partnership as any occasion on which a person acquires or increases a partnership share (*para 36*). This is regarded as a transfer of partnership interest to the person acquiring or increasing partnership share and from the other partners. This is a very wide definition, covering not only situations where a person makes payments to other partners to acquire or increase his partnership share, but also where a person introduces capital on admission to the partnership, such that the actual value of the other partners' interests remains unchanged. The definition of transfer of a partnership interest was introduced from 19 July 2007; prior to that, a definition applied which did refer to changes in the actual value of each partner's interest.

5.26 When the special rules for partnership transactions were first introduced, *any* transfer of a share in any partnership which owned UK property was treated as a transfer of an equivalent share in the UK property. It was eventually accepted that this placed an unnecessary administrative burden on ordinary trading partnerships which happened to own their premises – profit-sharing ratios within large professional partnerships may change several times each year, and each could give rise to complex SDLT calculations as well as the need to value the property held. Therefore, *FA 2006* restricted the scope of the charge on transfers of partnership interests, so that it now only applies to transfers of interests in 'property investment partnerships' (PIPs). As a result, a transfer of an interest in a partnership which is not a PIP will not give rise to SDLT obligations, unless caught by anti-avoidance provisions (see **5.27**).

Transfer pursuant to earlier arrangements

5.27 If:

- a chargeable interest is transferred to a partnership from a partner or person connected with a partner,
- that person or partner then transfers any or all of his partnership interest under arrangements which were in existence at the time of the land transfer, and
- the partnership transfer would not otherwise be a chargeable transaction,

the partnership transfer is a chargeable land transfer (*para 17*). The partners are treated as the purchaser under this deemed transaction, and the chargeable consideration is the market value of the land at the date of the land transfer multiplied by the proportionate share in the partnership transferred under the arrangements. However, this treatment does not apply if an election is made in respect of the land transfer under *para 12A* (see **5.38**). As in other cases (see **3.25**), arrangements are defined as including any scheme, agreement or understanding, whether or not legally enforceable.

5.28 The practical implication is that, if a partner transfers a chargeable interest to the partnership and shortly after transfers his partnership interest, HMRC are likely to question whether the transactions are part of a single arrangement. The deemed land transaction arising as a result of such a transfer of an interest in a partnership is notifiable (by completion of an SDLT return) only if the deemed consideration is sufficient for tax to be payable (para 30). For example, if the initial property transfer to the partnership did not involve the grant of a lease at rent and if there are no linked transactions, the subsequent transfer of partnership interest will not be notifiable unless the deemed consideration exceeds the zero rate limit of £125,000 or £150,000 for residential or non-residential/mixed property respectively. In relation to the £125,000 limit this assumes that the additional rate of SDLT does not apply (see **2.39**) since, if it did apply, a 3% SDLT charge would arise if the property had a value of £40,000 or more.

Property investment partnerships

5.29 A property investment partnership (PIP) is defined as a partnership whose sole or main activity is investing or dealing in chargeable interests, whether or not the activity includes the carrying out of construction activities on the land (*para 14(8)*). So, a partnership which invests in property in order to receive rent, or buys and sells property in order to make a profit (whether as a trade or an investment activity), is a PIP if this is its sole or main activity.

As indicated above the definition of a PIP states that it does not matter whether the activity of the partnership includes the carrying out of 'construction works' on the land that is being held as an investment or being bought and sold. Therefore a partnership whose main activity is investing in or dealing in land will not be excluded from being a PIP merely because it carries out some construction work before any profit is realised. However, as part of the Standing Committee debates on the amendment to the Finance Bill 2003, Sch 7, to restrict the availability of acquisition relief to companies whose main activity is the carrying on of a trade that does not consist wholly or mainly of dealing in chargeable interests the Economic Secretary to the Treasury stated:

> 'I understand that there have been worries that the concept of dealing in chargeable interests is uncertain, a point made by the hon Gentleman. It is unclear, for example, whether a property developer or house builder will be caught. Whilst it is ultimately a question of fact what trade a company is carrying on, I can reassure the Committee that property developers and house builders who derive most of their profits from their work, rather than from buying and selling would not be caught by the measure (*FA 2003, para 8(5A), Sch 7*). Companies that want reassurance about that in relation to their specific trade can approach HMRC to seek clarification.'

On this basis it is understood that HMRC accept that a partnership that generates most of its profits from property development should not be treated as a PIP.

If a partnership carries on more than one activity, some of which do not consist of investing or dealing in chargeable interests, it will be necessary to determine which is the main activity. In many cases, this may be obvious, but it may be difficult in borderline cases, where the partnership perhaps devotes similar effort to two activities, only one of which is caught, and derives similar levels of profit from them. In such a case, it may be necessary to carry out detailed analysis of time spent on the different activities, capital employed in each and profits derived from each, in order to form a defensible view. Unfortunately, there is no room for compromise or apportionment – the partnership must be classified as wholly a PIP or not a PIP. If a partnership is a PIP, all property held must be considered as potential 'relevant partnership property' (as defined below); there is no exemption for any property which may be held for the purposes of a minority 'non-PIP' purpose.

Transfer of an interest in a PIP

5.30 A transfer of an interest in a PIP is treated as a chargeable transaction if the 'relevant partnership property' (see next paragraph) includes a chargeable interest. The consideration for the chargeable transaction is a proportion of the market value of the relevant partnership property; the proportion is the proportion by which the transferee's partnership share increases. Where the transfer of an interest in a PIP is one side of an exchange, the interest in the PIP is deemed to be a major interest in land if any of the relevant partnership property is a major interest – this then confirms the need to calculate SDLT by reference to the market values of the relevant partnership property (see **2.32**).

Example 5.9 – retirement of partner, linked transactions

DEF Partnership is a PIP with three partners D, E and F, each with a one-third partnership share. There is no connection between the partners other than the fact that they are partners in the DEF partnership. F decides to withdraw and D and E each make an agreed payment to F to acquire his share equally; as a result, the shares of D and E each increase to 50%. This is a Type A transfer (see **5.33**). The relevant partnership property (see **5.31**) has a value of £1.2 million. Each of D and E is treated as the purchaser under a chargeable transaction with consideration of one sixth of £1.2 million, or £200,000 (one sixth being the proportion by which their respective partnership shares increase). In the author's view, the two acquisitions are linked (see **2.67**) because they form a series or composite transaction between the same vendor (F) and purchasers D and E, who are connected within the terms of *CTA 2010, s 1122*. Note that, for the purposes of determining whether transactions are linked, the automatic connection of partners applies. The draft HMRC guidance on partnerships appeared to suggest that the transfers would not be linked but the final guidance does not comment on the issue.

Relevant partnership property

5.31 There are two definitions of relevant partnership property, depending on whether the transfer is classified as Type A or Type B (see **5.33–5.35**). This distinction, with its related rules, is the main concept introduced by the *FA 2008* changes to *FA 2003, Sch 15*. The intention of the legislation is to ensure SDLT is paid where a partnership structure is used to transfer economic ownership of property between partners for consideration. It seeks to avoid difficulties for partnership structures used in widely held property investment schemes or where transfers occur for no consideration (for example, normal periodic adjustments of profit-sharing ratios). However, it is necessary to work through the definitions in any particular case. It would not be safe to decide that no SDLT charge arises purely on the basis of the general intention of the legislation. It is important to note that the distinction between Type A and Type B transfers is relevant only for deciding whether the partnership holds relevant partnership property and, if yes, its value. Both types of transfer are potentially subject to SDLT. However, in many Type B transfers, there will be no relevant partnership property and so no SDLT liability.

Market rent leases

5.32 Leases which satisfy certain very stringent conditions are referred to as 'market rent leases' and are always excluded from relevant partnership property. Therefore, if a partnership only holds leases of this kind, no transfer (whether Type A or Type B) will be a chargeable transaction. In practice, such partnerships are likely to be rare. To qualify as a market rent lease, a lease must have been granted for a market rent only (ie no premium and no arrangements for any future premium). Furthermore, either it must be for a term not exceeding five years or, if for a longer term, it must provide for market rent reviews at no longer than five-yearly intervals. Note that the rent review must provide for adjustment to a market rent, which means it must be possible for the rent to fall if market rents have declined. Most rent review clauses provide for upward-only reviews, and such a lease cannot qualify as a market rent lease, even though the inability of the rent to fall may have no practical effect. An anti-avoidance condition requires that there should not have been any amendment to the lease since it was granted which itself leads to the rent being less than a market rent immediately after the transfer of the partnership interest.

Example 5.10 – market rent leases, anti-avoidance rule

Partner X has a 99% share in the XY partnership which is a PIP. X grants a 200-year lease to the partnership at a market rent, with five-yearly market rent reviews. The partnership pays SDLT on the 1% of the NPV of rent, representing the 1% effectively transferred to Partner Y. X then agrees to reduce the rent to a peppercorn. This is a transfer to the partnership (see **4.28**), and the partnership pays SDLT on the 1% of the market value of the transfer (in practice, the value of the transfer is probably close to the market value of the freehold property). The partnership now owns virtually the whole economic value of the freehold property. X then transfers 98% of his partnership share to Y, for consideration. When it was granted, the lease was a market rent lease. But for the prohibition on subsequent amendments, it might still count as a market rent when the partnership share is transferred. The anti-avoidance condition ensures that the lease is not excluded from relevant partnership property. The value of the lease is, therefore, taken into account in determining any SDLT payable on the transfer of the partnership interest.

Type A transfers

5.33 Two situations give rise to a Type A transfer (*para 14(3A), (3B)*). The first occurs when the whole or part of a partner's interest as partner is transferred to another person (whether or not an existing partner) in return for consideration in money or money's worth. For the legislation to make sense, this must mean consideration given to the partner whose interest reduces (or to someone else at that partner's behest). The mere contribution of capital to the partnership by the person who gains partnership share does not amount to the giving of consideration for these purposes. The second occurs when a new partner joins and an existing partner withdraws money or money's worth from the partnership, thereby reducing or extinguishing his partnership share. However, this is not a Type A transfer if the withdrawal of money or money's worth comes from resources available to the partnership prior to the transfer.

Example 5.11 – a Type A transfer

The PQR partnership is a PIP with three partners, P, Q and R, sharing profits equally. The partnership holds assets worth £1 million and cash on deposit of £200,000. P decides to retire. At the same time, the other partners decide to admit S as a new partner with a 20% share. S introduces capital of £240,000 for this share. The partnership uses £200,000 of this cash and the £200,000 previously on deposit to pay out P's capital. This is a Type A transfer.

5.34 If the transfer of an interest in a partnership is Type A, all chargeable interests held by the partnership are regarded as relevant partnership property, apart from market rent leases (see **5.32**) and any property transferred to the partnership in connection with the transfer of partnership interest (*para 5*).

Example 5.12 – Type A transfer, relevant partnership property

The assets held by the PQR partnership (see Example 5.11) consist of: various plots of freehold building land purchased from third parties at various times and currently valued at £600,000; a long leasehold office transferred from P in 2006 and valued at £200,000; a warehouse leased from a third party; and sundry loose furniture and stock-in-trade valued at £150,000. The warehouse lease is at rent, subject to five-yearly market rent reviews and qualifying as a market rent lease (see **5.32**). However, local rents have increased dramatically since the last rent review 18 months ago. Because this lease has the benefit of three and a half years at the current rent, now substantially below current market levels, it is considered to have a value of £50,000.

The freehold building land and long leasehold offices, total value £800,000, are relevant partnership property for a Type A transfer. However, the warehouse lease is not relevant partnership property and may be ignored. The total partnership share transferred is 33.3%, being the share relinquished by the retiring partner: 20% of this goes to the new partner S for deemed consideration of £160,000 (20% of £800,000); and 6.7% goes to each of the continuing partners Q and R for deemed consideration of £53,336 each.

These are three separate transfers, so the question arises as to whether they are 'linked' (see **2.67**). If they are not linked, S will pay SDLT of £200, being 2% of £10,000; Q and R will pay no SDLT, because the consideration paid by each is below the threshold, but each will have to notify the transaction. If they are linked, the aggregate consideration will be £266,672, and the total SDLT liability will be £2,833 divided between the partners based on the consideration each is deemed to give. It is difficult to see how the transfers to Q and R can be anything other than linked, but aggregation of consideration for these transfers does not in itself give rise to a charge. It may be possible to argue that the transfer to S is not linked with the other transfers because, when the transfers were agreed, S was not a partner and so was not connected with Q and R. To strengthen the position, P, Q and R would be well advised to agree the surrender of 13.3% of P's capital *before* it is agreed to admit S on surrender of P's remaining 20%.

Type B transfers

5.35 Any transfer of an interest in a PIP which is not a Type A transfer is a Type B transfer (*para 14(3C)*). This includes:

- a transfer which is the result of an adjustment of profit-sharing ratios between the partners, with no partner paying any other for the transfer and no new partner being introduced; or
- a transfer which is the result of a partner (X) reducing his share or exiting the partnership, where no payments are made by any other partner and either no new partner is admitted, or, if a new partner is admitted, any amount withdrawn from the partnership by X is paid out of resources which were available to the partnership before the transfer. It is considered that 'resources' here includes previously arranged loan facilities.

Example 5.13 – a Type B transfer

The XYZ partnership is a PIP with three partners, Z, Y and Z, sharing profits equally. The partnership holds assets worth £1 million and cash on deposit of £200,000. X decides to retire. It is agreed that X will take £400,000 cash in settlement of his partnership capital and accumulated profits. The partnership borrows £200,000 from its bank to fund this cash payment. Since no new partner has been admitted and no existing partner has paid anything for his increased partnership share, this is not a Type A transfer, so it must be a Type B transfer. Care is needed here. If Y or Z gives any additional security to the bank in relation to the borrowing, this could be interpreted as giving consideration for the additional partnership share acquired on withdrawal of X, making it a Type A transfer. Treatment as Type B will be more secure if the borrowing from the bank is arranged before X decides to retire.

Relevant partnership property for Type B transfers

5.36 In addition to market rent leases, the following property is excluded from relevant partnership property in a Type B transfer (*para 14(5A)*):

- any chargeable interest transferred to the partnership in connection with the transfer of partnership interest. For example, if an incoming partner pays his contribution to capital by transferring a property to the partnership, that property is not relevant partnership property. It is reasonable to exclude this from any SDLT calculation on the transfer of partnership share, since there will be a separate SDLT calculation/charge in respect of this property transfer (see **5.13**);
- any chargeable interest transferred to the partnership on or before 22 July 2004. This is the date from which this kind of partnership transaction was brought within the SDLT regime;
- any chargeable interest not economically attributable to the partnership interest transferred. For example, it may be that other partners have sole rights to profits and proceeds of a property, so that the partner receiving

the partnership share gains no interest in that particular property – although this is an arrangement more commonly seen in relation to the premises of professional and trading partnerships than the assets of PIPs;

- any chargeable interest in respect of which an election has been made under *para 12A* (see **5.38**); and
- any other chargeable interest which was not acquired from a partner (existing or newly admitted) or from someone connected with such a partner.

The net result of these exclusions is that relevant partnership property for a Type B transfer consists only of property transferred to the partnership from a partner or connected person, after 22 July 2004, in respect of which no election has been made under *para 12A*, which is not a market rent lease, and which is not held solely for the benefit of partners other than the recipient of the transfer.

Notifying liability

5.37 A land transaction consisting of the transfer of an interest in a PIP is notifiable just like any other land transaction by completion of an SDLT return, but *only* if the deemed consideration, together with that for any linked transactions, is high enough to give rise to an SDLT liability (*para 30*). This contrasts with many other land transactions, where notification is needed even when no SDLT liability arises.

Paragraph 12A election

5.38 Normally, when a property is transferred to a partnership from one or more partners, or persons connected with partners, SDLT is not charged on the full market value, as explained at **5.13**. However, it is possible for a PIP to make an election under *para 12A* to disapply that rule and have SDLT charged on the full market value of the property. The property is thereafter effectively treated as if it had been acquired from an unconnected third party. The election must be made in the SDLT return for the transaction and, once made, is irrevocable.

If the election is made, it will normally increase the amount of SDLT payable immediately, since there will be no reduction in deemed consideration for the part of the property effectively remaining in the ownership of the transferor or his corresponding partner(s). This is not likely to be an attractive option in many cases, but may be worthwhile where the extra SDLT is not very large and it is anticipated that there will in future be Type B transfers of partnership interests. The election will then exclude the property from the calculation of SDLT on such transfers (see **5.36**).

RELIEFS

5.39 For the avoidance of doubt, *FA 2003*, *Sch 3*, *para 1* (exemption for transfer for no chargeable consideration) is specifically disapplied to transactions involving partnerships (*para 25(1)*). Apart from that, other reliefs and exemption generally do apply to transfers to and from partnerships, to the extent they would apply to transfers to and from the partners (*para 25(2)*). Additionally, a transfer of an interest in a partnership may qualify for relief if a direct transfer of relevant chargeable interests held by the partnership would have qualified.

5.40 Although group relief under *FA 2003*, *Sch 7* may apply to transfers involving partnerships, careful analysis of the various relationships will be required. All partnerships are regarded as 'transparent' as far as ownership of chargeable interests is concerned. However, some kinds of partnership are not treated as 'transparent' as far as ownership of shares is concerned:

(1) English general and limited partnerships do not have separate legal personality and are therefore regarded as transparent in relation to share ownership. As a result if all partners in such a partnership are members of a group, any company at least 75% owned by the partnership will also be a member of the group for SDLT purposes. Transfers of chargeable interests between the partners, the partnership and companies owned by the partnership (satisfying the normal 75% ownership test) will potentially qualify for group relief.

(2) All Scottish partnerships and all UK LLPs have separate legal personality and are therefore not transparent in relation to share ownership. Therefore a company owned by such a partnership cannot be in a group with the partners. Transfer of a chargeable interest between one or more partners and a company owned by the partnership cannot qualify for group relief. Furthermore, because chargeable interests held by the partnership itself are treated as owned by the partners (*para* 2), transfers between the partnership and a company which it owns cannot qualify for group relief.

(3) Scottish general and limited partnerships are not bodies corporate and so cannot function as the parent of a group. Therefore two companies, each directly owned by such a partnership, are not grouped with each other and a transfer of a chargeable interst between them cannot qualify for group relief. In contrast, an LLP is a body corporate and can function as the parent of a group. A transfer of a chargeable interest between companies directly or indirectly owned by an LLP may qualify for group relief, subject to satisfying the normal 75% ownership tests. It remains the case however that a transfer directly between the LLP and a company which it owns cannot qualify as explained at (2) above.

HMRC set out their views on these matters at www.hmrc.gov.uk/so/group-rel-sdlt-sd.htm.

Where group relief does apply, the wording of the provisions is modified to ensure that:

(1) relief is granted by reference to the partnership share of the partner(s) which are in the same group as the transferor/transferee company (*para 27A*), and

(2) in the case of a transfer to a partnership, relief is clawed back if those partners cease to be in the same group as the transferor while the partnership still holds the property or an interest derived from it (*para 27*).

5.41 The wording of charities relief is modified to ensure that it applies only if all of the chargeable interests held by the partnership are held for charitable purposes – that is, there is no partial relief if the partnership holds some properties for charitable purposes and some not (*para 28*).

SUNDRY PROVISIONS

5.42 Stamp duty continues to apply to a document which transfers an interest in a partnership, if that partnership holds stock or marketable securities; see **8.57** (*paras 31–33*).

The acquisition of a partnership interest is only subject to SDLT as set out above (*para 29*). If an interest in a partnership is acquired in some other way, it is not subject to SDLT.

Example 5.14 – non-chargeable acquisition of partnership interest

Cubit LLP is a business advisory partnership with 53 partners, employing 371 staff working out of 11 offices. Cubit has various interests in its office buildings, and a portfolio of business premises which it lets out, but rental income forms only 20 to 30% of its income, the balance coming from advisory and consulting activities. It is therefore considered that Cubit is not a PIP. The partners decide to admit three new partners, two from the existing workforce and one by direct appointment from outside. Each partner contributes £20,000 partnership capital (funded, in the case of the existing staff, by a corresponding loan from the partnership, to be repaid out of profit share) and has an initial profit share of 0.25%.

On the general definition, this amounts to three transfers of partnership interest to the new partners. However, as Cubit is not a PIP, and the transfers are not in connection with the transfer of chargeable interests in UK land, nor are they 'pursuant to earlier arrangements' involving transfers of chargeable interests (para 17), the transfers of partnership interests are not within the charge to SDLT.

Partnership property – guidance

5.43 A chargeable interest is partnership property for SDLT purposes if:

(a) it is held by or on behalf of the partnership, or by the members of a partnership; and

(b) it is held for the purposes of the partnership business (*para 34(1)*).

The HMRC guidance on what constitutes partnership property can be found at SDLTM33390. The guidance states as follows:

> 'In practice, whether a chargeable interest is partnership property or not for SDLT purposes will be decided by reference to whether it is partnership property by virtue of section 20 of the Partnership Act 1890. So a chargeable interest acquired, whether by purchase or otherwise, on account of the firm or for the purposes and in the course of the partnership business, will be partnership property for SDLT purposes.
>
> The mere fact that a business is carried on in property belonging to one or more partners does not make that partner's chargeable interest partnership property. Where a chargeable interest is owned by all the partners whether it is partnership property depends on the basis on which it is held by the co-owners.'

Therefore, for a chargeable interest to be partnership property the partnership must either acquire the chargeable interest or the partners must hold the chargeable interest on behalf of the partnership, and the mere fact that the partnership business is carried on from a property belonging to one of the partners will not be sufficient to make that chargeable interest partnership property for SDLT purposes.

Further guidance on the meaning of partnership property can be found at BIM82058.

Retrospective taxation?

5.44 The *FA 2007* changes contained the usual transitional provisions to ensure that they did not apply retrospectively to transactions already started at the time the changes were announced (2pm on 6 December 2006). The *FA 2008* changes then removed those transitional provisions. As a result, it appears that the rules, as amended by *FA 2008*, apply to all transactions, even those commenced before any of the changes were announced. If currently dealing with the completion of transactions which commenced so long ago, careful analysis will be required to establish whether the current rules lead to greater taxation than would have been the case under the pre-2007 rules and, if so, whether this is retrospective taxation which may be susceptible to challenge.

Chapter 6

SDLT Administration and Compliance

INTRODUCTION

6.1 This chapter deals with questions of what must be notified to HMRC, on whom the responsibility for notification and payment of SDLT falls, and the processes involved. It details deadlines, costs of missing them, and occasions on which payment of SDLT may validly be postponed. It explains how to claim repayment if a payment of SDLT proves excessive. Finally, it reviews the powers which HMRC have to enquire into and amend returns, assess unpaid or underpaid SDLT and seek penalties. The relevant legislation is in *FA 2003*, *ss 76–99* and *Schs 10–14*, and the *SDLT (Administration) Regulations 2003, SI 2003/2837,* as amended by the *SDLT (Administration)(Amendment) Regulations 2011, SI 2011/455*. The *SDLT (Electronic Communications) Regulations 2005, SI 2005/844*, govern the electronic submission of SDLT returns. Precise references are given in appropriate paragraphs below.

SDLT is within the regime for disclosure of tax avoidance schemes (DOTAS), but with its own rules for determining what must be disclosed. Responsibilities in relation to DOTAS are dealt with in **Chapter 7**.

NOTIFICATION

What is notifiable, and by whom?

6.2 In general, (real) land transactions (see **2.7** for definition) are notifiable unless exempt. There is a list of exemptions, which also render the transaction non-notifiable, at **3.7**. If none of the exemptions apply, the transaction is notifiable even if no SDLT is payable. It is especially important to note that a transaction is not excused from being notified merely because a relief is available to reduce or remove the SDLT liability – reliefs must be claimed in a return.

Deemed land transactions also give rise to the obligation to notify. In many cases, a real transaction is deemed to be a land transaction (or to be the acquisition of a chargeable interest, which is automatically a land transaction; see **2.7**). Examples include the substantial performance of a contract before completion (see **2.10**), or variation of a lease to reduce the rent or term (see **4.28** and **4.29**). In other cases, actions or events are deemed to amount to transactions which are specifically stated to give rise to an obligation to submit an SDLT return. One example would be on final determination of uncertain consideration or rent, if that leads to further SDLT becoming payable (see **2.20**).

Three cases involving anti-avoidance provisions are specifically stated to be notifiable (*FA 2003, s 77(1)(c), (d)* and *(e))*. These are the 'notional land transaction' under *FA 2003, s 75A* (see **7.4** et seq), the deemed transaction on substantial performance of a contract providing for conveyance to a third party (see **3.15**), and the notional or additional land transaction under *FA 2003, Sch 2A, para 5* (see **3.11** et seq). In each of these cases, most of the exemptions do not apply; so, for example, a deemed transaction under one of these sections is notifiable even if the consideration is below the normal £40,000 notification threshold. However, the exemptions from notification relating to secondary transactions in 'alternative finance' arrangements (see **3.39**) and transfers of partnership interests (see **5.37**) *do* apply. It is not clear why some exemptions should apply while others do not, and this may simply be the result of poor structuring of the legislation. Making these notional transactions notifiable places responsibility for interpretation and application of the general anti-avoidance rules firmly on the taxpayer.

In relation to timing of obligations, see **2.68** et seq.

Exempt transactions

6.3 If a transaction is exempt, there is no obligation to notify it to HMRC (*FA 2003, s 77A*). However, the taxpayer does still have the obligation to maintain records relating to the transaction for at least six years (see **6.61**). When SDLT was introduced, it was possible for the taxpayer to 'self certify' that the transaction was exempt. The self certificate could then be used as evidence that SDLT obligations had been completed, allowing the land transaction to be registered. Self certificates were abolished by *FA 2008*; when registering an exempt transaction, it is now necessary simply to inform the Land Registry that the transaction is exempt.

6.4 Sometimes, taxpayers choose to complete returns for transactions which appear to be exempt. This is commonly done to guard against future allegations of failure to notify, where exemption is dependent on a treatment of the facts which could be susceptible to challenge. This is considered in more detail at **6.47**.

Responsibility for notifying

6.5 Responsibility for notification and payment of any tax falls on the purchaser (*FA 2003, s 76*; SDLTM00060). The identity of this person is usually self-evident, being the person who gains beneficial ownership of the land interest. However, there are several provisions identifying the purchaser in particular circumstances:

- When property is transferred to a partnership, or where value is withdrawn from a partnership after such a property transfer (see **5.13** and **5.17**), all of the partners are purchasers and have joint and several responsibility for notification. They may nominate one or more responsible partners, but this does not remove the responsibilities of individual partners if the nominated partners fail in their duties (*FA 2003, Sch 15, paras 6–8*).
- When a lease is varied to shorten the term, the landlord is the purchaser. If it is varied to reduce the rent, the tenant is the purchaser (*FA 2003, Sch 17A, para 15A*).
- Where trustees of a settlement other than a bare trust acquire beneficial ownership of a property, they are purchasers (*FA 2003, Sch 16, para 4*). The obligation to make a return may be discharged by any one or more of the trustees, but subsequent obligations, for example in relation to enquiries by HMRC, then fall on the trustees who signed the return. Where trustees (or nominees) acquire a property under a bare trust, the person on behalf of whom the acquisition is made is normally treated as the purchaser (*Sch 16, para 3(1)*). However, if the acquisition by bare trustees takes the form of the grant of a lease to them, they are treated as purchasers of the whole beneficial interest in the lease (*Sch 16, para 3(2)*; see **2.18**). In origin, this is an anti-avoidance provision.
- Where the Official Solicitor is acting for a person under a disability (for example, mental incapacity), the Official Solicitor may notify on behalf of that person (*FA 2003, Sch 10, para 1B*). However, the person for whom the Official Solicitor acts remains the purchaser and therefore is liable for tax, interest and penalties where these arise.

How to notify

6.6 Notification is by submission of a land transaction return (LTR) to HMRC. The LTR is due within 30 days from the effective date of the transaction (see **2.8**). The legislation provides for the 30-day period to be shortened by regulation; in the early days of consultation on modernising stamp duty, a senior member of the Stamp Office suggested that the eventual target period might be five days. For many commercial transactions, this would be an impossibly tight deadline. The government did announce, as part of the 2015 Autumn Statement, that it will consult during 2016 on changes to the

SDLT filing and payment process, including a proposal to reduce the filing and payment window from 30 days to 14 days. The proposal is that these changes would come into effect in 2017 to 2018. At the time of writing (30 June 2016) the consultation document had not yet been issued.

Paper forms

6.7 An LTR must be in a form and containing information prescribed by regulations, or in such other form as has been 'approved by the Board', ie HMRC (*FA 2003, Sch 10, para 1*). The principal regulations are the *SDLT (Administration) Regulations 2003, SI 2003/2837,* as amended by the *SDLT (Administration) (Amendment) Regulations 2011, SI 2011/455*. The 2011 regulations include a facsimile of the general LTR (form SDLT1) and supplementary forms required in some cases (forms SDLT3 and SDLT4). Form SDLT2 was not amended by the 2011 regulations and so is as included in the 2003 regulations. The forms are reproduced at **Appendix B**. Form SDLT 2 should be completed if there are either more than two purchasers or more than two sellers. Form SDLT 3 is filled in if either more than one property is acquired or, there is only one property but it is not possible to fit the full address of the property into the space on the SDLT 1 form. Finally form SDLT 4, as well as form SDLT 1, should be completed in the following circumstances:

- the land or property acquired is mixed use property (see **2.37**) or non-residential property (see **2.37**);
- the purchaser is a company;
- the transaction relates to the grant of a lease over residential property, or the acquisition of an interest in residential property where there is a lease involved, and there is more than one property;
- the transaction is part of a sale of a business;
- the purchaser has asked HMRC for advice on how the law applies to the transaction by requesting a non-statutory clearance or a post-transaction ruling in accordance with CAP 1 has been requested;
- any part of the consideration other than rent is contingent or uncertain;
- arrangements have been made with HMRC to defer SDLT on contingent or uncertain consideration (see **6.31**); and
- there are mineral rights reserved.

The regulations permit use of a different form if approved by HMRC. If paper forms are to be submitted you must use the official HMRC print of the form. Each SDLT1 has a unique reference number (the unique transaction reference number or UTRN), and an original form must be used for each return. Previously, it was acceptable to use photocopies of blank supplementary forms, but HMRC's latest instructions insist on originals of these too. Insistence on a single original paper form causes difficulties in some cases, especially when the required signatories are scattered over several locations or perhaps located overseas; electronic submission may be easier in such cases. Supplies of forms may be obtained from HMRC; see **Appendix A**. Forms must be completed in black ink (*reg 9(1)*) to aid reading by an electronic scanner. Further guidance on the completion of the forms is set out at **6.19** et seq.

6.8 Routine paper returns should be sent to the processing centre at Netherton; see address in **Appendix A**. Apart from the return, any supplementary forms, a cheque in payment of the SDLT (if paying by cheque) and a plan (if one is provided), no correspondence or other documents should be included. This requirement was mentioned in the previous HMRC guidance notes (SDLT6), although it seems to be missing from the latest online guidance. For guidance where it is considered necessary to submit other information or documents to HMRC, see **6.47**. HMRC request that the papers be submitted without folding, in an A4 or larger envelope, as folding may cause problems with machine reading. They also encourage payment by bank transfer, other direct instruction to the bank or other electronic means (including debit cards) where possible, to minimise the risk of payment going astray. Links to appropriate destination bank account details are at www.hmrc.gov.uk/payinghmrc/stamp-land.htm, which should be consulted whenever payments are made in case account details change.

Electronic returns

6.9 The *SDLT (Electronic Communications) Regulations 2005, SI 2005/844*, permit delivery of LTRs electronically and, on this basis, HMRC offer the facility to submit LTRs via the internet once the person making the submission has registered for the purpose. The forms can be completed and submitted using basic software provided free of charge by HMRC, or by using commercial software which has more features. The forms are fundamentally the same as the paper versions, and the guidance set out below (at para **6.19** et seq) applies equally to the electronic forms. Links to appropriate paragraphs of HMRC's own guidance on completion, equivalent to the guidance for paper returns (see **6.20**), should be available as each section of the return is completed. The online forms are 'intelligent', in that they decide which sections to present for completion in accordance with sections already completed; they should, therefore, make it clear when completion of supplementary forms is required. Part-completed forms can be stored for later finalisation and submission. The software should also check for inconsistencies or omissions and prevent submission if these are found. It will not, however, check for technical errors, so the fact that submission is allowed does not guarantee that the return is correct! The electronic forms are just as restrictive and limited as their paper counterparts, so the same difficulties arise in ensuring adequate disclosure (see **6.19** and **6.47**).

6.10 Paper returns have to be signed by the purchaser(s) (*FA 2003, Sch 10, para 1(1)(c)*), but clearly this is not possible for electronic forms. Instead, before an agent is permitted to submit the final return, he must confirm that the purchaser has approved the draft return and that appropriate evidence of this is held. The only fact which is allowed to be missing from the draft approved by the purchaser is the effective date. This allows the forms to be drafted before completion of the transaction, approved by the purchaser, and then finalised and submitted immediately on the effective date (*Sch 10, para 1A*).

6.11 When a return is submitted online, the HMRC certificate (form SDLT5) is issued automatically and immediately. This contrasts with submission of paper returns, when there may be a delay of several days or even weeks before the SDLT5 is received. The SDLT5 certificate must be submitted to the Land Registry to permit registration of the purchaser's interest or the change brought about by the transaction, so there is a clear advantage in receiving this immediately.

6.12 Before using the online submission facility, it is necessary to register with HMRC, following the process which begins with a link under 'Do it online' on the SDLT Index page of the HMRC website, at www.hmrc.gov.uk/sdlt/index.htm. The process includes waiting for postal delivery of a PIN before the service can be activated, so it is not a good idea to wait until a return is due before registering.

Integration of SDLT compliance and registration of land transactions

6.13 It is understood to be a long-held government ambition to integrate collection and routine administration of SDLT with registration of land transactions, leaving HMRC to deal with policy and policing. Prior to 1 April 2015 it was possible for law firms, licensed conveyancers, mortgage lenders, etc for many transactions involving registered land in Scotland, to deliver LTRs via the Registers of Scotland under the ARTL system for Automated Registration of Title to Land. However, in practice, for a variety of reasons, few firms actually used the system. It remains to be seen whether the integration of SDLT compliance and the registration of land in England and Northern Ireland will ever be achieved. As a tax on land transactions is to be devolved to Wales from 1 April 2018 it will presumably be a decision for the Welsh government as to whether to integrate the property registration process and the administration of the Welsh land transaction tax.

Electronic submission of LTRs may eventually become compulsory but, for the time being, paper forms are still acceptable.

On 8 July 2014 HMRC produced updated guidance on submission of LTRs, in relation to land situated in Scotland, for transactions with an 'effective date' (see **2.8**) both before 1 April 2015 and on or after 1 April 2015 (the date from which Land and Buildings Transaction Tax applies to transactions over land in Scotland), including the facility to have paper forms SDLT5 produced very quickly when necessary. This guidance is available at www.hmrc.gov.uk/sdlt/return/scottish.htm#7

Computer-generated paper forms

6.14 When SDLT was introduced, as a first step towards automation, it was possible to complete return forms on screen (using software provided by HMRC or independent suppliers) before printing out, signing and submitting the printed copy. Because these computer-generated forms did not bear a unique number (UTRN, see **6.7**), HMRC provided separate numbered payslips for payment of SDLT. The number from the relevant payslip was added to the computer-generated return form and functioned as its UTRN. Software was then developed to summarise information in the forms in a barcode printed on the forms ('2D Bar-coded Forms') which could be machine read more easily. These methods of submitting returns are no longer available; the choice is only between handwritten paper forms and full online submission.

Further and amended returns

6.15 A number of transactions and events, described at various points in this book, give rise to the requirement to submit a further LTR when one has already been submitted for the original transaction, or to amend the original LTR. It is important to follow the correct procedure for amending or submitting further returns (see **6.17** et seq). HMRC have expressed concern that they too often receive a duplicate return with the request to cancel the original, when an amendment or correction is required (see Stamp Taxes Bulletin 1/2014, www.hmrc.gov.uk/so/bulletin01-2014.pdf). In most cases amendments and further returns do not require submission of a further form SDLT1!

Example 6.1 – further return, variable rent

Dave enters into a seven-year lease of a pub; the lease is executed without prior substantial performance of any agreement. The rent is provisionally fixed at £30,000 per annum, but this is to be adjusted at the end of each year using a formula which takes account of turnover in the year just ended. An LTR is submitted within 30 days of grant of the lease. At the end of the first five years, it is established that the rent has been above £30,000 in all years, so further SDLT is payable. A further return is due within 30 days of the end of the fifth year of the lease.

Example 6.2 – further return, clawback of relief

Dozy Ltd purchases a property from group company Dee Ltd and submits an LTR on which group relief from SDLT is claimed. Two years later, Dozy Ltd, while still owning the property, is sold to a third party. Dozy Ltd must make a further return reporting the withdrawal of group relief within 30 days of whichever step leads to its old parent group ceasing to have beneficial ownership of its shares.

Example 6.3 – further return, final determination of consideration

Beaky Ltd buys a property from connected company Mick Ltd and makes an LTR, paying SDLT based on an estimated market value. Six months later, it is discovered that the market value was under-estimated. Beaky Ltd must amend the original return to show the correct value, within 30 days of discovering the error. Nine months after the original transaction, it is realised that Beaky and Mick are group companies. Beaky may, if it so wishes, further amend the return to claim group relief and seek repayment of the SDLT paid. However, amendment is not compulsory because it would lead to a decrease in the SDLT charge.

6.16 It is important to establish whether the required action is:

(1) submission of a new return for a new transaction, which may or may not be linked with the original transaction;

(2) submission of a further return for the original transaction, perhaps now modified in the light of subsequent events or facts; or

(3) amendment of the return originally submitted in respect of the original transaction.

In some cases, (1) will be necessary as well as either (2) or (3). As noted below, there is no practical difference between (2) and (3). Even where amendment is not *required*, a purchaser may voluntarily amend a return within 12 months of the filing deadline for the original return (for example, in order to claim a relief which was omitted from the original return, or to correct a calculation error which led to payment of the wrong amount of SDLT). If an error is discovered which led to an underpayment of SDLT, it is an offence to fail to amend the return which may give rise to a penalty (see **6.52**). A new return for a new transaction is generally made by completion of a new SDLT1 (whether paper or online), together with supplementary forms if necessary. If the new transaction is linked with a previous transaction, there is space on the return to note this fact. This linking may also lead to the need to amend the return, or to submit a further return, for the original transaction.

Example 6.4 – further return, later linked transaction

The facts are as in Example 2.21. Tanya submits an LTR (using form SDLT1) in respect of the first house purchase, before 9 February 2017. This shows tax chargeable of £5,500, being £125,000 at 3% plus £35,000 at 5%. When the second purchase completes, Tanya submits an LTR (another form SDLT1) in respect of the second purchase. In this return she claims multiple dwellings relief (see **3.48**) which results in tax chargeable on the second purchase of £17,714. She also makes a further return (by letter, not form SDLT1, see **6.17**) in respect of the original purchase, also claiming multiple dwellings relief and showing the tax now increased to £7,086. With this return, Tanya pays the additional tax of £1,586 now due on the first purchase. The new return for the second purchase, the further return for the first purchase and the payments of £17,714 and £1,586 respectively are all due by 15 July 2017. As noted below, the further return in respect of the first purchase is made by letter, not completion of form SDLT1.

How to submit further and amended returns

6.17 In principle, submission of a further return for the original transaction and amendment of the original return are different processes; the legislation specifies which is required in each situation. In practice, HMRC guidance states that each requirement is to be satisfied by submission of a letter to the Birmingham Stamp Office (see **Appendix A**) (www.hmrc.gov.uk/sdlt/return/amend.htm). The letter must quote the UTRN which appeared on the original return and set out details of the changes and/or further information, including any further payment required. The letter should refer to the appropriate numbered boxes on the original return, so it is important that a copy of that return be retained. Where the legislation specifies that a further return must be submitted but, in practice, HMRC require that a letter be sent, they are presumably relying on the fact that the regulations permit use of a form other than SDLT1 'in a form that has been approved by the Board' (see **6.7**). In the case of minor corrections not affecting the substance of the transaction, HMRC may be willing to accept amendment by telephone, but in the author's view it is always safer to write.

Correction of errors in returns

6.18 HMRC may also unilaterally correct minor errors and omissions, in which case notice must be given to the person(s) who submitted the return

(*FA 2003*, *Sch 10*, *para 7*; see comment at the end of SDLTM60100). The person(s) to whom notice is given may reject the amendment:

- by re-amending the return within the permitted 12 months, or
- if notice is received from HMRC later than nine months after the filing date, by giving notice that the amendment is rejected within three months of the date of issue of HMRC's notice.

This facility to correct obvious errors was well used in the early days of SDLT, but it is now much more likely that a return with errors which prevent processing will be rejected and will require re-submission after correction by the taxpayer. A major point in favour of online submission is that many errors are highlighted and must be corrected before the return can be submitted.

Practical completion of return forms

6.19 The return forms are input documents for an automated system, and the paper versions are designed to be machine read, which is why the regulations specify completion in black ink (*SI 2003/2837*, *reg 9*). The forms contain a series of numbered boxes and are very limited and limiting in terms of information which may be included. Some boxes are for matters such as names, addresses, dates and monetary amounts; beyond this, they are completed by inserting 'X' to show that a particular statement applies, or by inserting a letter or number to indicate which statement or item applies from a list.

6.20 Guidance on completion, including lists of the relevant codes, was previously contained in booklet SDLT6 'How to complete your Land Transaction Return', which was available as a paper booklet, a downloadable PDF file or an online interactive document. However this was withdrawn on 2 July 2012 and replaced by separate guidance on the different forms. This is available at http://www.hmrc.gov.uk/sdlt/return/paper.htm. The latest guidance is dated 1 April 2016. Paper copies of the guidance are no longer available, although it is possible to print the online guidance. The new guidance contains links to appropriate sections of the Stamp Duty Land Tax Manual and other HMRC pronouncements where it is thought these will be helpful. If the new guidance is printed out for future use, it will still be safest also to check the current online version for recent changes.

SDLT Return forms were amended in April 2011 to require the input of information allowing HMRC to link together tax records of the purchaser (or lead purchaser if more than one). It is now necessary to disclose the National

Insurance number of an individual or the UK or overseas tax reference number of a company or partnership. If the purchaser is VAT registered, the VAT number must also be shown on the principal SDLT1 form. If none of these identifiers is available, another unique identifier such as a passport or driving licence number is acceptable. If the purchaser has none of these numbers, it is necessary to contact the Stamp Taxes Helpline (see **Appendix A**) to obtain a reference number before submitting the return. Some of the extra information now required on form SDLT1 was previously required on form SDLT4, so this form has also been revised to remove possible duplication. Finally, a very minor change was made to form SDLT3. The revised forms have been in use for online submission since April 2011. It was permitted to use the old versions of paper forms until 3 July 2011, but now only the new forms may be used. Old forms will be rejected so any remaining stocks should by now have been destroyed.

6.21 The LTR has two functions:

(1) to set out the information to establish the amount and timing of liability to SDLT; and

(2) to provide the government with statistical information (which was previously collected under the stamp duty regime from the 'Particulars Delivered' or PD form submitted with many transfer documents) (*FA 2003, Sch 10, para 1(4)*).

In relation to (2), the information on the LTR is significantly more detailed – some would say intrusive – than that on the PD form. In particular, in relation to the purchase of properties leased to multiple tenants, the purchaser is required to collate a lot of detail which may not be readily available. The author is aware of cases where the need to gather this information has delayed submission of returns beyond the due date. It is, therefore, important to be aware of the requirements and to begin gathering information as early as possible to minimise this risk.

6.22 In general, the manner of completion of the boxes is self-evident or adequately explained by the HMRC instructions, although careful close reading of the instructions is essential. This section of the book does not aim to replicate the HMRC guidance, and therefore provides comment only where something useful can be added. There is one important caveat: HMRC's instructions for completion of some of the forms have changed significantly over the last few years. The reader must take care to follow the latest instructions.

Form SDLT1

6.23

Box	Heading	Comment
1	Type of Property	A new code '04' has been added, with effect from 1 April 2016, and is to be used where the purchaser is an individual acquiring a residential property, and the individual owns more than one residential property at the end of the day which is the 'effective date' (see **2.8**). The use of code '04' means that the additional rate of SDLT applies to the transaction (see **2.39**). However, if the residential property being acquired is a replacement for the purchaser's main residence, which was sold in the three-year period prior to the 'effective date' of the transaction then code '04' should not be used and, instead, code '01' should be used. If the new residential property is a replacement for the purchaser's main residence which has not yet been sold, code '04' should still be used however the purchaser may be able to claim a refund of the additional rate of SDLT if the main residence is sold within three years of the purchase of the new main residence (see **2.39** et seq). Code '04' must also be used if the purchaser is a company or non-individual purchasing any residential property even if the property acquired is the only residential property owned. Code '01' should be used for the purchase of residential property where the additional rate of SDLT does not apply e.g. an individual purchaser replacing his main residence. Code '02' is used where the property being acquired is mixed use and code '03' where the property being acquired is non-residential.
2	Description of transaction	Care is needed because the meaning of terms used in the guidance notes differ slightly from normal legal meanings – a transfer of a leasehold interest, or a transfer of a freehold subject to a lease, is A (other, involving lease), not F (conveyance or transfer). Where L or A is entered here, it is very important to complete boxes 16 to 21 (in the case of A or L) and 22 to 25 (in the case of L) – see www.hmrc.gov.uk/so/bulletin01-2014.pdf for further guidance.
3	Interest transferred or created	Code LG is labelled 'long leasehold', but in fact length of lease is irrelevant. Any lease at a ground or nominal rent with vacant possession is LG.

Box	Heading	Comment
5	Restrictions etc	This requires judgement as to what is unusual; the entry in this box has no effect on whether the form will be processed, but omission of facts which could influence how HMRC views the transaction may make it more difficult to resist a discovery assessment later (see **6.47**). The examples of restrictions given by HMRC are (a) the vendor has the right to buy the property back at less than market value, (b) the lease covenants restrict shop use, and (c) agricultural occupancy condition applies.
9	Reliefs	'PFI relief' (see **3.17**) must be claimed as code 28 ('other relief'). The relief may reduce consideration to zero, but a return showing zero consideration may be rejected. If this applies, the normal approach seems to be to insert consideration of £1 and write a 'further information' letter to HMRC (see **6.47**) explaining that this has been done. New codes are introduced from time to time, so the guidance at www.hmrc.gov.uk/news/new-guide-sdlt.htm should be checked where necessary.
10	Consideration	It is a common error to put an entry here when a lease is granted at a premium. If a lease is granted, this box must be left blank – the premium and rent are reported later in boxes 22 and 23 (where zeroes are required if there is no premium/rent). However for a lease it is still necessary to put a figure in Box 14 for the total amount of tax due. The total tax due for the lease will be the sum of the figures in boxes 24 (premium) and 25 (NPV of rent), each of which may be zero if, for example, relief is being claimed.
13	Linked transactions	The 'total consideration' box should show the aggregate capital consideration paid for linked transfers of freeholds and existing leases, plus any premiums paid on grant of new leases or amendment of existing leases where these are also linked. This should include amounts in respect of the transaction(s) reported on this SDLT1. However, it should not include any amount in respect of the NPV of rent on any leases. There is no place on the return to show the aggregate NPV of rents under linked leases. The aggregate NPV must be taken into account in calculating the entry in box 25 (Tax due in respect of NPV) – see **4.58** et seq for an illustration of the significance of this.

Box	**Heading**	**Comment**
15	Total amount enclosed/paid	If application has been made to defer payment of any of the tax (see **6.33**), HMRC will give instructions on completion of this box. If necessary, contact the Stamp Taxes Helpline (see **Appendix A**) for guidance. It is **not** possible to apply for deferral by entry on the return form – application must be made separately and before the return is due.
26	Number of properties	See **6.26** for use of schedules where 100 or more properties are involved.
28	Address	In paper returns, if the address does not fit in the space available, start writing it here and simply put on form SDLT3 the part which does not fit – do not write the whole address on the supplementary form, or you may receive an SDLT5 certificate with a jumbled address for the land, which the Land Registry may reject.
29	Local Authority number	It is now compulsory to insert the Local Authority number, which can be obtained at SDLTM62320. Scottish Local Authority numbers ceased to be valid from 1 April 2015 as Land and Buildings Transaction Tax will apply to most land transactions in relation to land in Scotland (see Chapter **1A**) and use of these codes will lead to rejection of the return. Where a land transaction in relation to land in Scotland has an effective date on or after 1 April 2015, but remains liable to SDLT under the transitional rules (for example an SDLT lease that continues after a fixed term), an SDLT1 form must be filed in the normal way and code 9998, and not a Scottish local authority code, should be entered at box 29.
30, 31	Title number and NLPG UPRN	If this information is not readily available, these boxes can be left blank. This will have no effect on whether the form is processed successfully.
49–51	About the purchaser	See **6.20** for requirement if purchaser does not have an NI number or Unique Tax Reference number.
61	Agent authority	Accidentally leaving this blank is a major cause of agents failing to receive SDLT5 certificates or other communications from HMRC.

Form SDLT4

6.24

Part 2	About leases	Where a property is transferred while subject to more than one lease to tenants, the first lease is reported in questions 16 to 21 of form SDLT1. HMRC *originally* required details of all other leases to tenants to be reported on form SDLT4, or (where multiple leases were involved) as a schedule to SDLT4. This requirement caused confusion, because form SDLT4 is designed (and mainly used) for reporting details of leases which are themselves the subject of the transaction. HMRC have now clarified the matter: if the property transferred is subject to more than one lease to tenants, details of all except the first lease are to be reported on a schedule to form *SDLT1*, not on form SDLT4. A pro-forma schedule is available at www.hmrc.gov.uk/so/sdlt-sublease-guide.htm, but equivalent schedules created for the purpose are acceptable.

SPECIAL CASES

Linked transactions

6.25 Generally, separate transactions require separate returns, and it is generally acceptable to report the acquisition of separate properties on separate forms SDLT1, even if they are acquired under a single transaction or arrangement. Where two or more linked transactions (see **2.67**) have the same effective date, *FA 2003, s 108(2)* provides that the purchaser(s) may opt to make a single land transaction return as if all of the transactions formed a single notifiable transaction. If there are two or more purchasers, they must make the return jointly, which implies they have joint and several liability for payment of the SDLT and any interest and penalties which may arise from late submission/payment or errors. The statute indicates that the 'single return' approach is available for any such linked transactions. However, HMRC impose restrictions because a single LTR form can only cope with two or more transactions if they are sufficiently similar (see HMRC further guidance for question 13 of SDLT1 form). Therefore, it is only permitted to report two or more transactions on a single return if:

- the vendor(s) and purchaser(s) for each are identical and each transaction has the same 'effective date' (see **2.8**);

- the transactions all fall within the same category (that is, one of the following: transfer of freehold; grant of lease; other transaction involving lease; other transaction); and
- either no relief is being claimed, or the same relief is claimed in respect of all transactions.

If this leads to a need to complete two or more returns when the legislation allows a single return and if the returns are late so that a penalty is chargeable under *FA 2003, Sch 10, para 3,* it is the author's view that only a single penalty should be charged. However it is likely that the matter will have to be argued with HMRC!

Multiple properties

6.26 Where linked acquisitions are reported on one form SDLT1, details of one of the properties must be included on form SDLT1, with the others on separate forms SDLT3 (and forms SDLT4 where the transaction includes the grant of leases). If many properties are acquired under a single transaction, it is possible to submit the SDLT3 and any SDLT4 information on schedules rather than on separate forms (SDLTM60215). This instruction originally applied to six or more properties, but now says it applies to 100 or more. It is probably worth asking in any case where the alternative is completion of an oppressively large number of forms SDLT4.

HMRC will provide guidance on acceptable formats for such schedules. The guidance notes suggest calling the SDLT helpline in the first instance but, in the author's experience, direct contact with Birmingham Stamp Office Complex Transactions Unit is best (see **Appendix A** for contact details). It is usually possible to construct a suitable spreadsheet; HMRC require an electronic copy, even if a paper copy is sent with a paper return, to allow them to copy and paste information into their systems. Most of the information entered on these schedules is for the purposes of government statistics rather than determining tax liability and, where information is not readily available, HMRC will often accept schedules with gaps – it is worth asking!

When to notify

6.27 In general, the deadline for submission of a land transaction return is 30 days from the effective date of the transaction (*FA 2003, s 76(1)*). If an amended or further return is required as a result of an event occurring after the effective date, the return is due within 30 days of the date of that event. In the case of a lease with uncertain or variable rent (see **4.37**), the latest date for submission of an amended return is 30 days after the end of the fifth year of the term of the lease. Where the rent depends on the annual results of a business, this may be before the rent for the fifth year has been finally determined. In that case, the return should be submitted using a reasonable estimate of the rent, and then amended as necessary once the rent is finally determined. As noted at **6.17**, if a return is submitted in respect of the original transaction, further and amended returns are to be submitted by way of a letter to Birmingham Stamp Office, not by completion of further forms.

Penalties for late submission

6.28 If a return is submitted up to three months late, a penalty of £100 is chargeable, irrespective of whether SDLT is chargeable or whether the SDLT is paid on time. If the return is more than three months late, the penalty increases to £200. In addition, if a return is filed 12 months or more late, a tax-geared penalty may be charged. This penalty may be up to 100% of the tax chargeable on the relevant transaction (*FA 2003*, *Sch 10*, *paras 3–4*; SDLTM86120).

PAYMENT OF SDLT

Payment methods

6.29 Originally, payment of SDLT was linked to the submission of the return, so that the return was not regarded as duly submitted unless and until any SDLT had been paid. This link was broken by *FA 2007*, so that submission of the return and payment of the SDLT are now regarded as separate actions. The breaking of the link makes it very important to ensure that HMRC can identify sums received and attribute them to the correct return. To this end, the 11-digit UTRN of the related return must be quoted when SDLT is paid. SDLT may be paid by sterling cheque drawn on a UK bank, payable to 'HM Revenue & Customs Only', followed by the UTRN. If a single cheque is used to pay SDLT on several returns, the UTRNs should be listed on the back of the cheque, with the amount of SDLT against each. The cheque may be sent by post to the processing centre at Netherton (see **Appendix A**), or paid in at a bank or Post Office using the payslip(s) from the paper return(s). Alternatively (and, in the eyes of HMRC, preferably), payment may be made by any of the usual electronic or bank transfer methods. Relevant bank account details for these purposes are at www.hmrc.gov.uk/payinghmrc/stamp-land.htm.

Due date and deferral

6.30 SDLT is generally due 30 days from the effective date of the transaction – usually the earlier of substantial performance of the agreement and legal completion of a transfer or execution of a lease (see **2.8**). See **2.68** et seq for the general position in relation to the due date. There is no automatic right to defer payment where there are delays in paying consideration, occupying the property or determining the amount of consideration. If, for any reason, the amount of consideration has not been finally determined when payment of SDLT falls due, payment must be based on a reasonable estimate. Further SDLT may then be due once the consideration is finally determined, and this must generally be paid within 30 days of that final determination. However, interest will be chargeable on this additional SDLT from the original due date of the main payment, unless deferral has been granted in respect of the additional tax.

6.31 If, at the effective date of a transaction:

- the SDLT cannot be finally determined because any amount of consideration is uncertain or contingent, and
- any part of that consideration falls or may fall to be paid more than six months after the effective date,

the taxpayer may apply to defer payment of SDLT on the consideration which falls to be paid later (*FA 2003*, *s 90*; *SI 2003/2837*, *Pt 4*, see SDLTM50900), potentially including any part which may fall to be paid within six months. However, it is not possible to defer payment on any element of the consideration which is certain or ascertainable at the effective date or later date of application and either has already been paid or is known to become payable within six months of the effective date. Deferral is only possible if uncertainty about the consideration arises from future events – if the amount of consideration depends on accounts drawn up to the effective date but is not known at that date because the accounts have not yet been drawn up, it is not dependent on future events and deferral is not possible on these grounds. Deferral is not possible in respect of any SDLT which arises on rent under a lease, but deferral may be possible in respect of SDLT on a lease premium.

6.32 If deferral is granted, the notification of that fact will set out details of when further payments of SDLT must be made. If the specified payments are made by the specified dates, interest will not be charged (*FA 2003*, *s 87(3)(b)*). If payments are late, interest will be charged only from the dates on which they should have been made. The fact that deferral can be interest-free is the principal reason for going to the trouble of making an application. If deferral is not applied for (or refused), interest will be charged on any SDLT paid more than 30 days after the effective date of the transaction, even though the amount of SDLT cannot be determined until some later date. It is important to consider applying for deferral in any case where further consideration may become payable at a later date, even if that further contingent consideration is expected to be negligible – retrospective application is not possible if the final amount is unexpectedly high!

Example 6.5 – deferral

On 1 January 2016, Elvis agrees to buy a freehold block of flats from Cliff for £10 million payable now and up to a further £5 million payable on 1 January 2017. The amount payable on 1 January 2017 will be determined in accordance with a formula where the only variable is the net rents received in the year to 30 June 2016. The agreement is completed without prior substantial performance on 31 March 2016. Cliff must pay SDLT on the £10 million consideration no later than 30 April 2016. He may apply to defer payment on the additional consideration because (a) that consideration falls to be paid more than six months after the effective date of the purchase, and (b) the amount cannot be determined at the effective date (or later date of application) since it depends on rent receipts for a period ending after that date. However, if the contingent amount had depended on the net rents for the year to 31 March 2016, deferral would not have been possible. Although, at the effective date, that amount might not have been finally worked out, it would in principle be possible to know the amount at that date, so the amount is not 'uncertain' as defined by *FA 2003*, *s 51*. Equally, had the amount payable on 1 January 2017 been subject to a minimum, it would not be possible to defer payment of tax on that certain minimum amount because payment of it would not be contingent or uncertain. SDLT on that minimum would be due by 30 April 2016.

Applying for deferral

6.33 An application for deferral must be made no later than the due date for the original return, ie 30 days after the effective date (*FA 2003*, *s 90*). It is understood that this deadline is rigidly enforced, and HMRC encourage early submission of applications, even before the transaction is finalised if possible. Application is made by letter to the Birmingham Stamp Office (see **Appendix A**), marked 'SDLT Deferral Application' (SDLTM50900 et seq). The letter must set out:

- the identity of the purchaser;
- the location of the land involved;
- the nature of the contingency/uncertain payment;
- the amount of consideration for which deferral is sought (or a best estimate where it is uncertain);
- as full details of the times of expected payments as it is possible to give;
- a reasoned opinion as to when this part of the consideration will cease to be contingent or can be ascertained;

- a calculation of the SDLT payable on the total of the actual and the contingent/uncertain consideration (based on best estimates); and
- a calculation of the SDLT in respect of which the application to defer payment refers.

HMRC are entitled to ask for further information, and are likely to do so if the transaction is unusual or if avoidance is suspected. If deferral is denied, it is possible to appeal (see **6.57** et seq), but in practice this will be worthwhile only where very large sums are at stake.

Interest on SDLT

6.34 As with other taxes, interest is charged on SDLT which is paid late (*FA 2003*, *s 87*), at the rates set by regulation under *FA 1989*, *s 178*. In general, the rate is the same as for income tax and NIC (see www.hmrc.gov.uk/rates/interest-late-pay.htm for details and do not be misled by the absence of SDLT from the table heading!). The rate of interest on late paid SDLT is currently 3% per annum. In most cases, tax is late if it is paid after the due date, which is 30 days after the 'relevant date'. The relevant date in various situations is as follows:

(1) for an ordinary acquisition of a chargeable interest, whether by transfer or grant of a lease, the effective date;

(2) where certain events occurring after execution or earlier substantial performance of a lease are treated as the grant of a new lease (eg a deemed surrender and regrant, (see **4.32**), or transfer after a relief was claimed on grant (see **4.17**)), the date of the event which is treated as the grant of a new lease;

(3) where a relief is withdrawn as a result of a later event (eg group relief withdrawn because the purchaser leaves the group within three years), the date of the event giving rise to the withdrawal;

(4) where the tax arises on an earlier transaction as a result of a later linked transaction (eg because aggregation of consideration throws the original transaction into a higher rate band(s)), the effective date of the later linked transaction;

(5) where tax arises on a 'growing' lease (see **4.12**), for periods commencing on or after 17 July 2013, the last day of the year or shorter period giving rise to the liability, For earlier periods, the day on which the lease becomes treated as being for a longer fixed term (ie, the first day of the period giving rise to the liability); and

(6) where deferral has been granted as set out at **6.33**, the date on which the deferred SDLT payment is due.

However, when further tax arises after the normal due date and none of items (2) to (6) applies, interest is chargeable from the original due date, ie 30 days after the effective date of the original transaction. See SDLTM85930 for confirmation of these points. Under the so called 'slab system' (which applied to residential property until 4 December 2014 and to non-residential and mixed use property until 16 March 2016: see **2.37**) if deferral was granted, interest would still arise where contingent consideration proved to be greater than expected and pushed the transaction into a higher SDLT rate. However, this should no longer be the case under the progressive tax structure which now applies to residential, mixed use and non-residential properties. This is because the tax paid on the initial consideration will not be changed when the final consideration is determined, as the total amount of the consideration chargeable does not affect the SDLT paid on the initial consideration, which remains the same, with any additional consideration being added to the initial consideration and taxed at the marginal rate(s).

When SDLT is repaid, interest is payable (*FA 2003, s 89*), again at the same rates as generally apply to direct taxes as set under *FA 1989, s 178*. Such interest is not income for tax purposes. The interest is payable for the period from receipt of the tax by HMRC to the date of repayment. At the date of writing (30 June 2016) the repayment interest rate is 0.5%.

Example 6.6 – relevant date and interest charge

On 31 May 2016, David enters into a five-year lease of a commercial property at a provisional rent of £200,000 per year, but the amount is to be finally determined by reference to the performance of the business carried out in the property during the five-year term of the lease. On 29 June 2016, David pays SDLT on the basis of the £200,000 annual rent. On 15 June 2021, just after the end of the lease, the annual rent is finally determined to be £225,000. The SDLT liability is recomputed, and the additional SDLT is due 30 days after the date on which the rent is finally determined. However, interest is charged on this further SDLT from 30 June 2016, being 30 days after the effective date of the lease (*FA 2003, s 87(3)(c)*). It is not possible to apply to defer tax on rent, so the only way to escape this interest charge would be to over-pay SDLT on 29 June 2016 (*FA 2003, s 90(7)*).

Example 6.7 – deferral and interest charges

On 1 April 2016 Goliath buys a plot of bare land for consideration of £100,000 plus unascertainable consideration equal to 25% of the profit made on development and sale of the site. Goliath estimates that the unascertainable consideration will be £125,000. He successfully applies to defer SDLT on the unascertainable element, and pays tax of £1,500, being the SDLT payable on overall consideration of £225,000 (£150,000 at 0% plus £75,000 at 2%).

When the development is sold on 31 March 2018, the contingent consideration proves to be £175,000. The SDLT payable on total consideration of £275,000 is £3,250 (£150,000 at 0%, £100,000 at 2% and £25,000 at 5%). Goliath pays tax of £1,750 (£3,250 - £1,500) on the £175,000 unascertainable amount by 29 April 2017 and no interest is charged.

Collection and enforcement

6.35 SDLT is due and payable, whether under a self-assessment (that is, as a result of submission of a return), or under a determination or assessment by HMRC, unless deferral has been granted as outlined above or postponement has been agreed/granted (see below). HMRC are empowered to enforce payment of outstanding SDLT by distraint ('diligence' in Scotland) or by civil proceedings in magistrates' courts (for amounts not exceeding £2,000), county courts, the High Court or Scottish equivalents (*FA 2003*, *Sch 12*).

HMRC ENQUIRIES

6.36 The process for HMRC to enquire into an SDLT return is similar to that which applies to other direct tax returns (*FA 2003*, *Sch 10*, *Pt 3*). It begins by HMRC giving notice of enquiry to the purchaser before the end of the enquiry period; it ends when HMRC gives a 'closure notice'. This may be given because HMRC are satisfied that their enquiry is complete or because the Tax Tribunal gives a direction to that effect on application by the purchaser. The enquiry period is potentially relatively short, being nine months from:

- the filing date (which is itself 30 days after the effective date; see **2.8**) if the return was filed on time,
- the actual date of filing if the return was late, or
- the date of amendment if the return is amended by the purchaser.

6.37 However, if the notice of enquiry is given as a result of an amendment to the return, and the normal enquiry period for the original return has ended (or an enquiry into the original return has already been completed), the enquiry is limited to matters to which the amendment relates or are affected by the amendment. Outside these time limits, HMRC may only pursue potential underpayment of tax by making a 'discovery assessment' (see **6.46**).

Example 6.8 – time limits for enquiry into return

On 1 March 2016 (effective date and date of completion), Hattie Ltd buys a business from Eric Ltd. Payment takes the form of Hattie accepting responsibility for debts of the business amounting to £2 million plus an issue of shares in Hattie Ltd to Eric Ltd with a market value of £1 million. The business includes a commercial property provisionally valued at £1.5 million. On 20 March 2016, Hattie Ltd submits an LTR on the basis of this valuation, claiming relief under *FA 2003, Sch 7, para 8* (acquisition relief; see **3.27**). Hattie Ltd pays SDLT at 0.5% accordingly. On 1 February 2017, Hattie Ltd gives notice to amend the return, decreasing the value attributed to the property to £1.2 million. On 15 February 2017, HMRC give notice of enquiry into the return. The normal enquiry period for the original return ended on 31 December 2016 (nine months after the filing date), so the enquiry may only deal with matters affected by the amendment. In this case, it is considered that the enquiry may relate to the valuation of the property but not to the availability of acquisition relief.

6.38 As part of an enquiry, HMRC may demand to see documents and information, giving at least 30 days' notice (*FA 2003*, *Sch 10*, *para 14*). The purchaser may appeal against such a demand (see **6.58**). If no appeal is lodged, or after any appeal is settled, initial and daily penalties may be imposed for failure to produce the outstanding documents and information. In addition to a general right to demand documents and information from the purchaser, HMRC are given powers to seek information and documents from third parties, including tax accountants and other advisers. Exercise of these powers is subject to consent or review by the Tax Tribunal or, in some cases, a judge or sheriff. Situations in which these powers may be invoked are beyond the scope of this book. The relevant parties will normally require specialist advice in these cases.

6.39 While an enquiry is in progress, HMRC may amend the return if they conclude that additional tax is due and amendment is necessary to prevent a loss of the tax (*FA 2003*, *Sch 10*, *para 17*). Written notice of the amendment must be given to the purchaser. This power is normally only used where a prima facie underpayment has been established, but the taxpayer refuses to make an appropriate payment on account, suggesting there may be a risk of eventual default. The purchaser may appeal to the Tax Tribunal against the amendment and may apply to have collection of any additional tax postponed. Postponement may be agreed by HMRC, or the question may be determined by the Tax Tribunal if the taxpayer and HMRC cannot reach agreement.

6.40 The purchaser may also amend the return, if the normal period allowed for amendment has not yet expired (*FA 2003*, *Sch 10*, *para 18*; see www.hmrc.gov.uk/sdlt/return/amend.htm). Voluntary amendment to correct any errors which have come to light may be taken into account in determining any penalty position (see **6.55**). HMRC may, however, reject the amendment if they do not agree with it.

6.41 HMRC and the purchaser may agree to refer matters to the Tax Tribunal during an enquiry (and, if they see fit, the First-tier Tribunal may pass the referral on to the Upper Tribunal). Subject to the normal rights of appeal through the courts, the decision of the Tribunal is binding on all parties.

6.42 Normally, enquiries are closed by HMRC giving notice of closure once they are satisfied no further amendments are required (*FA 2003*, *Sch 10*, *para 23*). The notice must make it clear whether any taxpayer amendments of the return are accepted. The purchaser may appeal to the Tax Tribunal against a closure notice (*Sch 10*, *para 35*). Alternatively, the purchaser may apply to the Tax Tribunal for a direction that the enquiry be closed (*Sch 10*, *para 24*). Details of the procedure are on the Tribunals Service website (see **6.59**).

FAILURE TO DELIVER A RETURN

6.43 If a purchaser fails to deliver a return, or if HMRC believe that a return should have been delivered, HMRC may issue a notice requiring delivery of a return (*FA 2003*, *Sch 10*, *paras 5, 25–27*). At least 30 days' notice must be given and the transaction in question must be specified. If the person to whom the notice is given fails to comply, HMRC may apply to the Tax Tribunal for the imposition of a daily penalty. There does not appear to be any provision for appeal against such a notice, for example on the grounds that the purported transaction to which it relates has not taken place, or is not notifiable, or that the person to whom notice was given is not the purchaser. Any penalty is in addition to those which apply automatically for late submission (see **6.28**). Alternatively or additionally, if no return is forthcoming, HMRC may make a determination of the SDLT which they believe to be unpaid, and this is then treated as a self-assessment (that is, as if a return had been submitted) for the purposes of enforcement, including the charging of interest and penalties. A determination may not be made more than six years after the effective date of the transaction in question. If the taxpayer subsequently makes a self-assessment by delivering a return, provided this occurs not later than 12 months after notice of the determination is given or six years after the original due date for the return (whichever is later), the self-assessment displaces the determination. This, of course, then allows HMRC to launch an enquiry into the return. As an alternative, the taxpayer may appeal to the Tax Tribunal against the assessment, but only on the grounds that the transaction has not taken place or been substantially performed, or that it has but it is not notifiable. Logically, if such an appeal is successful, it should also displace any daily penalty awarded for failure to comply with HMRC's notice to deliver a return, but the legislation does not overtly say that this is so.

6.44 A further bizarre and potentially wholly unfair provision allows HMRC to treat a return (or any other document) as not having been delivered or made available if, whilst in their care, it has been destroyed or so damaged as to become illegible. The author is not aware of any case in which HMRC have in practice relied on this provision.

DISCOVERY ASSESSMENT

6.45 If a return has been made (or is deemed to have been made by virtue of a determination) and the enquiry period (see **6.36**) has not yet ended, any possible loss of tax will normally be investigated by launching an enquiry. Once an enquiry has commenced, if HMRC are concerned that recovery of additional tax is at risk, they may amend the return as described at **6.39**, thereby allowing action to be taken to collect the tax. Once the enquiry period has ended, or an enquiry has been completed, HMRC has no automatic right to reopen matters and seek further tax.

6.46 However, if no return has been made, or in certain circumstances where a return has been made but the time for launching an enquiry has passed, HMRC may assess any tax which they discover to have been underpaid (*FA 2003*, *Sch 10*, *Pt 5*). If a return has been made, a 'discovery' assessment may only be made to recover tax lost:

(1) as a result of fraud or negligence on the part of the purchaser or an agent or partner of the purchaser, or

(2) because, when the normal time for enquiry ended (or any previous enquiry was completed), HMRC could not, on the basis of information then available to them, have reasonably been expected to be aware of the loss of tax.

For the purposes of (2), information is available if it is contained in a land transaction return, has been furnished in the course of an enquiry, could reasonably be expected to be inferred from such information, or is notified to HMRC by the purchaser or a person acting on his behalf (*Sch 10, para 1(5)*).

6.47 In reliance on the last of these points, taxpayers and their agents often submit further information to HMRC at the time of submission of the LTR. HMRC guidance indicates that such further information should not be enclosed with paper returns, but rather should be sent to the Birmingham Stamp Office (see **Appendix A**), quoting the UTRN(s) of the return(s) to which it relates.

6.48 The decision as to what further information, if any, to submit requires careful thought. Submission of further information may increase the likelihood of the transaction being selected for enquiry. If the treatment originally adopted in the return is clearly correct, any such enquiry will normally be fruitless, but will incur administrative costs. Alternatively, if there is genuine doubt over the treatment of the transaction, it will be important to ensure that the information submitted is sufficiently comprehensive. If HMRC cannot see the full facts and circumstances to enable them to make their own judgement of the treatment adopted, they are likely to consider that a discovery assessment is justified, should further information come to their attention later.

6.49 The general time limit for HMRC to make a discovery assessment is four years from the effective date of the transaction (*FA 2003, Sch 10, para 31(1)*). This period is extended to six years where the loss of tax is brought about carelessly by the purchaser or a 'related person' (*FA 2003, Sch 10, para 31(2)*).The period is further extended to 20 years where the loss of tax is:

(a) brought about deliberately by the purchaser or a 'related person'; or

(b) attributable to a person failing to deliver an LTR; or

(c) is attributable to an arrangement in respect of which a person has failed to give notification under the Disclosure of Tax Avoidance (DOTAS) legislation (see **7.26 et seq**) (*FA 2003, sch 10, para 31(2A)*).

If the purchaser has died, the period is shortened to four years from the date of death (or six years from the effective date of the transaction, if that is earlier) (*FA 2003, Sch 10, para 31(4)*). If the assessment is to recover an excessive repayment of SDLT, it may also be made at any time during an enquiry into the relevant return, or up to 12 months after the repayment is made, even if this is outside the normal time limit (*FA 2003, Sch 10, para 31(3)*).

6.50 The purchaser may appeal against a discovery assessment and may apply to have payment of any additional tax postponed (*FA 2003, Sch 10, para 35*). As in other cases, postponement may be agreed by HMRC, or the matter may be determined by the Tax Tribunal.

MISTAKE IN RETURN ETC

6.51 If a mistake is discovered in a return within the period allowed for amendment (usually up to one year after the normal submission deadline – *FA 2003, Sch 10, para 6(3)*), this should be dealt with by amending the return as set out at **6.40**. If this deadline is missed, it may be possible to make a claim for relief from excessive assessment arising from a mistake in a return. Claims made on or after 1 April 2011 may only be accepted if made within four years of the effective date of the transaction (*FA 2003, Sch 10, para 34* and *34B* as substituted by *F(No 3)A 2010, s 28* and *Sch 12*). There is no provision to extend this period where, for example, the return was made after the normal deadline. Prior to 1 April 2011, claims were permitted up to six years after the effective date, and most errors could be corrected. The amendments from

1 April 2011 place severe restrictions on the acceptable grounds for making a claim – for example, if there was a mistake in making or failing to make a claim or election, it is no longer possible to correct this by way of a claim under these provisions. Similarly, if the taxpayer knew or ought reasonably to have known the grounds for the claim before the normal period for making the claim had expired, relief under these provisions will be refused. If the claim is made on behalf of a partnership, all of the partners must nominate someone who was a partner at the time of the relevant transaction to make the claim. This provision may be to prevent a dissident partner seeking to change the treatment of an old transaction.

Making a claim under these provisions will also remove restrictions on HMRC's ability to make a discovery assessment. Therefore careful analysis will be required before deciding to pursue such a claim.

A claim for relief may also be made where a person believes he has been assessed more than once in respect of the same transaction. There does not appear to be any time limit on such a claim (*Sch 10, para 33*).

PENALTIES, FINES AND OTHER SANCTIONS

Penalties

6.52 As noted at **6.34**, it is the view of HMRC that interest charged when SDLT is paid after the date on which it ought to have been paid does no more than compensate the government for loss of use of the money. As a separate matter, penalties are provided for various failures. SDLT is potentially within the new penalty regime in *FA 2009, Sch 55* but at the time of writing (June 2016) this had not yet been activated in respect of this tax. Meanwhile the current penalties are generally in *FA 2003, Sch 10* and *Sch 11, para 6*. Some are fixed amounts and apply automatically; others are for variable amounts up to a ceiling. They apply as follows:

(1) Late submission of return – up to three months after the due date, £100; three months or more after the due date, £200.

(2) Late submission of return – 12 months or more after the due date, an amount up to the amount of tax chargeable on the transaction. This is in addition to the penalty chargeable under (1).

(3) Failure to submit return after HMRC have given a formal notice to deliver – up to £60 per day of continuing failure. This requires a direction from the Tax Tribunal and is in addition to any penalty under (1) and (2).

(4) Careless or deliberate delivery of incorrect return, or failure to correct return after error has come to light – up to the amount of understated tax.

This penalty was brought into the general *FA 2007* regime for penalties for incorrect returns (*FA 2007, Sch 24*) by *SI 2009/571*.

(5) Failure to keep and preserve records as required – up to £3,000.

(6) Failure to produce documents or information after formal notice has been given or obstructing an officer from HMRC in the course of an inspection – an initial amount of £300 and, for continued failure, a daily penalty of £60 and at the discretion of the Upper Tribunal a penalty based on the amount of the SDLT payable. However, these penalties cannot be imposed after the failure has been remedied, so they should be seen as encouragement to comply rather than punishment for not complying.

6.53 In principle, HMRC impose most penalties by making a formal determination (or, where indicated above, by seeking such a determination from the Tax Tribunal) and issuing notice of the determination to the taxpayer (*FA 2003, Sch 14*). The taxpayer may then appeal against the determination, initially to the Tax Tribunal, subsequently to the courts (*FA 2003, Sch 14, para 5*). In the case of fraud, penalty proceedings may be brought in the High Court (or the Court of Session in Scotland) (*FA 2003, Sch 14, para 7*). The general time limit for determining a penalty or commencement of proceedings for a penalty is four years from the date on which the penalty was incurred (*FA 2003, Sch 14, para 8(1)*). However, where the penalty is payable in relation to a case involving a loss of tax brought about carelessly the time limit for determining the penalty is extended to six years (*FA 2003, Sch 14, para 8(4A)*). The period for determining the penalty is further extended to 20 years where the loss of tax is brought about:

(a) deliberately by the person; or

(b) by a person failing to deliver an LTR; or

(c) by an arrangement in respect of which a person has failed to give notification under the Disclosure of Tax Avoidance (DOTAS) legislation (see **7.26 et seq**) (*FA 2003, Sch 14, para 8(4B)*).

For tax-geared penalties, the time limit for determining the penalty is extended to three years after determination of the tax (*FA 2003, Sch 14, para 8(3)*). A penalty for assisting in preparation of an incorrect return may be sought up to 20 years after the penalty was incurred (*FA 2003, Sch 14, para 8(5)*).

6.54 In practice, most fixed and daily penalties are paid without formal determination. These penalties will be waived only in the most exceptional circumstances – for example, if the taxpayer is able to show that the return was completed and posted in good time, but lost in the post, or if the taxpayer's agent suffered a serious illness at just the wrong time. Details are given in leaflet SD7 available from the HMRC publications and stationery service (see **Appendix A**) or online at www.hmrc.gov.uk/leaflets/sd7.pdf.

6.55 Most tax-geared penalties are included in a negotiated settlement, where it is rare for the maximum amount to be imposed. The amount of any such penalty will depend on matters such as size and gravity of adjustment and taxpayer cooperation; some guidance is given in a rather old Tax Bulletin, issue 71, which may be found on the HMRC website at http://webarchive.nationalarchives.gov.uk/20101006151632/http://www.hmrc.gov.uk/bulletins/index.htm. Guidance on HMRC's general approach to tax geared penalties may be gleaned from their Compliance Handbook Manual at CH82000 et seq.

Fines and other sanctions

6.56 Any person who conceals, destroys or falsifies a document, the production of which has been requested or required, or permits or causes such concealment etc, if convicted, is liable to a fine and/or up to two years' imprisonment (*FA 2003, Sch 13, para 53*). A person who fails to comply with a court order to produce documents or information may be held to be in contempt of the relevant court and may be fined and/or imprisoned accordingly. As with other taxes, criminal prosecution is possible for the most serious cases involving fraud, potentially leading to fines and/or imprisonment. Dealing with such matters is not within the scope of this book.

APPEALS

6.57 There are various references in this chapter to appealing against HMRC decisions, determinations and assessments. The normal rules for appeals on direct tax matters apply equally to SDLT (*FA 2003, Sch 10, Pt 7*). These are set out at www.hmrc.gov.uk/complaints-appeals/how-to-appeal/direct-tax.htm on the HMRC website. Initially, appeals are made to HMRC. This may be done either by using the form usually supplied with the notice of decision etc, or by letter. A letter of appeal must set out the details necessary to allow HMRC to identify the tax return or transaction and decision involved (usually the HMRC reference) and the grounds of the appeal.

6.58 In general, the appeal must be made within 30 days of the decision, assessment etc. If the taxpayer and HMRC are unable to reach agreement on the disputed matter, HMRC may offer an 'independent review' (that is, a review by one or more other HMRC officers). Within 30 days of such an offer, the taxpayer must decide whether to accept, or to refer the matter directly to the Tax Tribunal. If a review is accepted, it is still possible for the taxpayer to appeal to the Tax Tribunal if he does not accept the outcome of the review.

6.59 SDLT matters are dealt with by the First-tier Tribunal (Tax) of the Tribunals Service. Guidance on making an appeal or reference to the Tax Tribunal, and the necessary forms, are available on the Tribunals Service website at www.justice.gov.uk/guidance/courts-and-tribunals/tribunals/tax/index.htm, or may be obtained by writing to the Tribunal Service (see **Appendix A**).

RECORD KEEPING

6.60 Purchasers are required to keep and preserve whatever records may be required to enable the delivery of correct and complete returns (*FA 2003*, *Sch 10*, *Pt 2*). Such records include original contracts and transfer instruments, supporting maps, plans and other documents, and records of payments receipts and financial arrangements. The records must be kept for at least six years from the 'effective date' (see **2.8**) of the transaction. They must also be kept until any enquiry into the return is completed, and until HMRC no longer have the right to enquire into the return, if these dates are later than the six-year time limit. It is acceptable to keep the information from the documents (for example, scanned copies) rather than the documents themselves. However, taxpayers may be concerned about the ability to demonstrate that such information is in accordance with the original document. So, in most cases, the original document is likely to be retained, albeit in a document storage facility, even if the information is also stored electronically.

6.61 The requirement to keep and preserve records applies where a return has to be made, but also where no return is required because the transaction is exempt (*FA 2003*, *Sch 11*) (see **6.3**). For most commercial transactions, the requirement to keep records for six years is reasonable and in line with other record-keeping requirements. In the case of routine home purchases, where the property may well be re-sold within the six years, the requirement is not so clearly justified. Many homeowners, unaware of the requirement, are likely to dispose of records in a shorter period.

Chapter 7

SDLT Anti-Avoidance Rules

INTRODUCTION

7.1 Prior to the introduction of SDLT in 2003, it was relatively easy to minimise stamp duty costs on many property transactions. It had proven difficult to introduce effective anti-avoidance rules for this document-based tax. These were major factors in the decision to replace stamp duty with SDLT, a transaction-based tax not reliant on the manner in which the transaction is documented. The initial SDLT rules were not perfectly drafted and, inevitably, transactions were structured to take advantage of this and minimise tax. Initially, HMRC tackled this by introducing targeted anti-avoidance rules intended to prevent exploitation of specific reliefs and treatments. Eventually, it was concluded that these were having only limited success and, on 6 December 2006, a general anti-avoidance rule for SDLT was introduced in the form of *FA 2003, s 75A*. SDLT is also within the scope of the general anti-abuse rule (GAAR) newly introduced by *FA 2013, ss 206–215*. Finally SDLT is within the system for disclosure of tax avoidance schemes (DOTAS), but it has its own rules for determining what must be disclosed, and there are other differences from the system which applies to other taxes. Disclosures under DOTAS have continued to lead to further targeted anti-avoidance provisions. Some changes have been with retrospective effect, as promised by the Chancellor of the Exchequer in his 2012 Budget speech. A recent attempt to challenge these retrospective provisions by way of judicial review failed *(R (on the application of St Matthews (West) Ltd and others) v HMRC*, High Court, judgment 6 June 2014).

7.2 It is arguable that the introduction of the GAAR and *FA 2003, ss 75A–75C* has rendered some of the previous targeted anti-avoidance rules redundant, but they all remain in force. During the 2013 Finance Bill debates the government stated that, before relying on the GAAR, HMRC will first try to combat abuse by relying on existing specific anti-avoidance rules.

This chapter first looks briefly at the GAAR, then at *FA 2003, ss 75A–75C*. It then provides a list of specific targeted rules – these are dealt with more fully in the relevant chapters examining the transactions or reliefs to which they apply. Finally, it explores the SDLT version of DOTAS.

General anti-abuse rule

7.3 The general anti-abuse rule (GAAR) is set out in *FA 2013, ss 206–215* with procedural rules in *FA 2013, Sch 43.* HMRC guidance on the application of the GAAR can be found at www.hmrc.gov.uk/avoidance/gaar.htm. The rule applies to a wide range of taxes including SDLT and so is couched in fairly general terms. It applies to any arrangements, a main purpose of which is the obtaining of a tax advantage. 'Tax advantage' is not comprehensively defined but stated to include relief from or repayment of tax, reduction, avoidance or deferral of a tax liability, or avoidance of an obligation to deduct or account for tax. Where HMRC consider that the rule may apply, they will seek to make adjustments to counter the tax advantage. The taxpayer may then make a claim for 'consequential adjustments', effectively to relieve any multiple taxation caused by the initial adjustments. The making of adjustments is governed by the procedural rules in *Sch 43.* These require HMRC first to give written notice to the taxpayer, setting out what arrangements are considered to be abusive and why. The taxpayer has the opportunity to respond and refute HMRC's view. If HMRC are not satisfied with the response, they must refer the matter to the GAAR Advisory Panel for an opinion. After receiving representations from both the taxpayer and HMRC, the GAAR Advisory Panel must express an opinion on the matter. In the light of this, HMRC may then make adjustments, or instruct the taxpayer as to what adjustments must be made, to counter the tax advantage. Although the GAAR provisions do not specifically cover the matter, it appears to be assumed that adjustments made by way of assessment or refusal of a claim are potentially subject to appeal to the Tax Tribunals and beyond in the normal way.

SDLT differs from other taxes covered by the GAAR in that it is a transaction tax. It is perhaps less clear which arrangements have the saving of tax as a main objective when, in the absence of the transaction there would be no tax. Presumably HMRC would have sought to have applied the GAAR to combat the largely residential property sub-sale schemes which have been retrospectively attacked by specific legislation (see **3.16**).

FINANCE ACT 2003, SECTIONS 75A–75C

Introduction

7.4 These sections really form a general anti-abuse rule specifically aimed at SDLT and introduced more than six years before the GAAR. They potentially apply to all transactions other than those consisting of simple single steps, unless specifically excepted. They operate by comparing the total SDLT actually paid on all steps with the SDLT which would be payable on a 'notional transaction'. If the SDLT actually paid is less, any real land transactions are disregarded and, in their place, the notional transaction becomes chargeable. Credit is given for any SDLT paid on real transactions, and further SDLT is payable to bring the total up to that on the notional transaction.

7.5 *Section 75A* was initially introduced by regulation; by the time it was confirmed by *FA 2007*, it had expanded to three sections, which now appear as *FA 2003, ss 75A–75C*. These sections demonstrate a hardening of attitudes to avoidance of SDLT, which was further illustrated by the June 2010 edition of 'Spotlight', HMRC's web publication about avoidance (see www.hmrc.gov.uk/avoidance/spotlights.htm), which made it clear that HMRC intended to attack schemes which sought to exploit 'sub-sale relief'. In any event the legislation governing 'sub-sale relief' has now been replaced by a relief in relation to pre-completion transactions which has been specifically designed to prevent avoidance (*FA 2003, Sch 2A* – **see 3.8**).The final version of the *ss 75A –75C* provisions is in some ways more stringent than the original version, but is still stated to apply with effect from 6 December 2006. During the parliamentary debates on the 2007 Finance Bill the government was eventually persuaded that this could amount to retrospective taxation, so a clause was introduced (*FA 2007, s 71(3)*) to state that, for transactions entered into before Royal Assent (19 July 2007), if the liability under the original version was less than that under the final version, the lesser liability would apply. This chapter deals with the final version of the rules. If, exceptionally, a transaction potentially within the scope of *s 75A* commenced prior to 19 July 2007 and is only now being dealt with, the transitional rules, which remain in *FA 2007*, should be checked.

7.6 As set out in the legislation, the test is completely objective – that is, there is no need to consider the motives of the parties. This means that a series of transactions which are not entered into with any intention of avoiding tax, but which happen to reduce the SDLT bill below that implied by the notional transaction, could still be caught. For this reason, it is not safe to ignore the rules merely because there is no avoidance motive; their application must be considered in all multi-step transactions.

7.7 HMRC guidance on the application of *ss 75A–75C* is not consistent with the case law on the subject, nor is it clear when HMRC will take the view that the anti-avoidance provision will apply. At SDLTM09175, HMRC state their view that, because *s 75A* is an anti-avoidance provision, it is considered that it only applies where there is avoidance of tax, and HMRC specifically state that they will not seek to apply the provision where they consider the transaction has already been taxed appropriately. The problem is that it can often be difficult to decide whether HMRC are likely to consider that a particular transaction has been appropriately taxed. This therefore appears to be an example of possibly unnecessarily draconian legislation which HMRC will not apply rigorously if they feel happy with the actions of the parties. Although that may be a reasonable approach, it is supported by neither the legislation nor the recent decisions of the First-tier Tribunal, the Upper Tribunal and the Court of Appeal. The First-tier Tribunal decision in *Project Blue Limited v HMRC* [2013] UKFTT 378TC states: 'Whilst it is clear that the purpose of section 75A is to counteract the avoidance of SDLT, the provision contains no requirement that the taxpayer should have a tax avoidance motive or purpose as a precondition or defence to the application of the provision … Parliament obviously intended that the provision should apply regardless of motive'. This is quoted, apparently with approval, in HMRC's note on de-enveloping transactions issued on 20 December 2013 (see www.hmrc/gov.uk/so/stamps-de-enveloping.htm). In the decision of the Upper Tribunal ([2014] UKUT 0564 (TCC)), the presiding judge Mr Justice Morgan held that whilst there was no specific avoidance motive test in s *75A,* a requirement that SDLT has been avoided is implicit as the section only applies if the tax saving test in *FA 2003, s 75A(1)(c)* is satisfied. However, this contention is open to debate as it is possible (and perhaps even common) for there to be a number of ways of structuring a transaction each of which may result in a different amount of SDLT being payable, and for a particular structure to be pursued, for wholly commercial reasons, but in circumstances where an SDLT saving is achieved. In such circumstances it is arguable that there is no SDLT avoidance however *s 75A* could still apply. The decision of the Court of Appeal ([2016] EWCA Civ 485), given on 26 May 2016, supports the view that *s 75A* can apply in the absence of an SDLT avoidance motive. Lord Justice Patten states at 39:

> 'The first is that Mr Thomas was wrong in my view in his submission that *s 75A* has no operation unless it can be shown that the object of the relevant scheme transactions was the avoidance of tax. Although, as the side-note to *s 75A* makes clear, the provisions were clearly introduced to combat the avoidance of SDLT, they operate according to their terms and nowhere in *s 75A* is there any reference to the purpose of the scheme transactions being tax avoidance or any requirement to establish the existence of such a purpose or objective as a pre-condition to the operation of the section. The UT [Upper Tribunal] was right in my view to reject Mr Thomas's construction of *s 75A* and to treat avoidance as spelt out by the conditions for the application of *s 75A* contained in *s 75A(1)*.'

It therefore seems clear that there is no requirement that an SDLT avoidance motive exist before *s 75A* can apply. If the actual liability to SDLT on a transaction is less than the liability which would arise under the notional transaction fiction of *s 75A* (see *FA 2003, s 75A(1)(c)*), then the anti-avoidance provision is in point, assuming that the other conditions in *FA 2003, s 75A(1)* are satisfied (see **7.8** et seq). This makes advising on the potential application of *s 75A* very difficult because on the one hand you have HMRC stating that they will not apply *s 75A* if they consider that a transaction has been appropriately taxed and on the other hand you have the case law telling the adviser (or purchaser) that if there is an SDLT saving, compared to the notional transaction under *s 75A*, then *s 75A* can apply, and a significantly higher SDLT liability may arise.

The SDLTM guidance may therefore no longer be current. For what it is worth, it gives examples of where HMRC considers that the rules do or do not (normally) apply. These are helpful so far as they go but the examples do not amount to a 'white list' since HMRC only say that the rules are 'unlikely' to apply. Most of the examples in the guidance where it is stated the rules will apply involve partnerships and/or the grant, variation or termination of leases. This confirms the nature of the avoidance against which the legislation is primarily directed.

When do the rules apply?

7.8 They apply where one person (V) disposes of a chargeable interest, and another person (P) acquires it or a chargeable interest deriving from it. It does not matter that there may be various steps between the disposal and the acquisition, nor does it matter if the chargeable interest goes through various transformations between the disposal and the acquisition. The legislation was introduced to deal with schemes which involved precisely that kind of step and transformation. If the asset disposed of or acquired is not a chargeable interest, the rules do not apply.

Example 7.1 – no transfer of a chargeable interest

X Ltd owns 100% of the shares of Y Ltd. Y Ltd owns a valuable property. X Ltd makes a loan to Y Ltd, allowing Y Ltd to pay a cash dividend to X Ltd, thus reducing the value of Y Ltd to £1. X Ltd then sells the shares of Y Ltd to unrelated party Z Ltd. Assuming there are no other steps, *s 75A* does not apply to the transaction, because there has been no disposal or acquisition of a chargeable interest. There may have been a saving of stamp duty on transfer of the shares, but stamp duty is not within the ambit of *s 75A*. Shares in a company are not 'chargeable interests', even if the company's only asset is property. However, this may change in future – from time to time, the government has threatened to introduce land-rich company rules, which may deem shares in such a company to be chargeable interests. It is also unlikely that this arrangement would fall foul of the GAAR in respect of stamp taxes, as stamp duty is not within the scope of that rule. However such an arrangement may give rise to advantages in relation to other taxes and it would be important to consider the possible impact of the GAAR in this respect.

7.9 The rules only apply if there are 'a number of transactions' (which may include non-land transactions) involved in the disposal and acquisition. To make sense of this provision, 'a number' must be taken to mean more than one. However, 'transaction' includes an agreement, offer or undertaking not to take a specified action, and any kind of arrangement, even if it would not normally be described as a transaction (*FA 2003, s 75A(2)*). In other words, apparently unilateral acts and failures to act may be regarded as transactions. It does not matter when the transactions occur – they merely have to occur 'in connection with' the disposal and acquisition. The transactions are referred to as 'scheme transactions'(*FA 2003, s 75A(1)(b)*).

7.10 The legislation (*FA 2003, s 75A(3)*) lists examples of 'scheme transactions' as follows, but makes it clear this is not an exhaustive list:

- acquisition of a lease derived from a freehold formerly owned by the vendor;
- sub-sale to a third party;
- grant of a lease to a third party subject to a right to terminate;
- exercise of a right to terminate a lease or take some other action;
- agreement not to exercise a right to terminate a lease or take some other action; and
- variation of a right to terminate a lease or take some other action.

It is clear from this list that transactions involving the grant and termination of leases were foremost in HMRC's collective mind when the rules were written. Although not exhaustive, this list is important when considering 'incidental transactions' (see **7.12**).

Example 7.2 (part 1) – scheme caught by s 75A

Agnes owns a freehold residential property worth £2.1 million. She grants a lease to Bill to hold as nominee for herself. Bill is a friend but not otherwise connected with Agnes. The terms of the lease are as follows:

- yearly rent, one peppercorn, no premium; and
- term one month, but with an option (exercisable before the end of the month) to extend to 100 years in return for payment of £1.

There is no SDLT on the grant of the lease because it is treated as granted beneficially to the nominee (see **2.18**), and there is no reason to substitute the actual consideration with market value or any other figure. However, on the assumption that the option is likely to be exercised, the value of the freehold interest is now minimal. Agnes sells the freehold to Claire for £1. Again, there is no SDLT on this sale as there is no reason to substitute anything else for the actual consideration.

Agnes then undertakes to refrain from exercising the option to extend the lease if Claire pays her £2.15 million. There could possibly be an argument that this is a land transaction subject to SDLT, but in real, slightly more complicated arrangements based on this general structure, it was considered that this was not a land transaction. The overall result then is that Claire has spent £2.15 million to acquire a property but paid no SDLT.

There are a number of transactions, comprising the grant of the lease, sale of the freehold and undertaking not to exercise the option, so this series of transactions is potentially caught by *s 75A*.

The notional transaction

7.11 The rules only apply if the total SDLT paid on all transactions in the scheme is less than the amount which would be payable on a 'notional land transaction' (*FA 2003, s 75A(1)(c)*). This notional transaction is defined as a transaction effecting the transfer from V to P, for consideration (*FA 2003, s 75A(5)*) equal to the largest amount (aggregated if more than one):

- given by or on behalf of any one person, or
- received by or on behalf of V (or a person connected with V),

by way of consideration for the scheme transactions. Reference to a notional land transaction sometimes causes confusion, because the legislation does not specify the nature of the transaction. However, this is unimportant. All that matters in this regard is that there is deemed to be a chargeable transaction for a specified amount of chargeable consideration, which determines the minimum SDLT charge.

Incidental transactions and reliefs

7.12 In measuring the chargeable consideration for this notional land transaction, certain amounts must be left out of account. The amounts are:

- the consideration for any transaction which is merely incidental to the transfer of the chargeable interest (*FA 2003, s 75B(1)*- see **7.16**); and
- any consideration paid in respect of a transaction covered by the following reliefs in *FA 2003* (*FA 2003, s 75C(4)*):
  - *s 60* (compulsory purchase facilitating development);
  - *s 61* (compliance with planning obligations);
  - *ss 63, 64* (demutualisation of insurance company or building society);
  - *s 65* (incorporation of limited liability partnership);

- *s 66* (transfers involving public bodies);
- *s 67* (transfer in consequence of reorganisation of parliamentary constituencies);
- *s 69* (acquisition by bodies established for national purposes);
- *s 71* (acquisition by registered social landlord);
- *s 74* (collective enfranchisement by leaseholders);
- *s 75* (crofting community right to buy);
- *Sch 6A* (relief for acquisition by housing intermediaries etc); and
- *Sch 8* (charities relief).

These various reliefs are considered in more detail in **Chapter 3**. Without these exclusions, many of the reliefs would become relatively useless in real-life transactions which often involve multiple steps for entirely innocent reasons. Perhaps of more importance are the reliefs which are not listed above, the most important of which are the group, reconstruction and acquisition reliefs set out in *FA 2003, Sch 7*. HMRC are concerned that these reliefs may be exploited in ways which were not intended, and are not keen to facilitate this (although this does beg the question of who is to judge the intention of reasonably plain legislation).

Example 7.2 (part 2) – SDLT payable under s 75A

The total SDLT paid on the basis of the scheme transactions is nil. The consideration for the notional transaction is £2.15 million (plus £1), being the aggregate amount given by Claire. Therefore, *s 75A* applies and the SDLT payable is calculated as follows (assuming that the additional rate of SDLT does not apply (see **2.39**)):

Chargeable consideration	Rate	SDLT
£	%	
125,000	0	Nil
125,000	2	2,500
675,000	5	33,750
575,000	10	57,500
650,001	12	78,000
2,150,001		171,750

This scheme is precisely the kind of arrangement that *s 75A* appears to be designed to block.

Effective date

7.13 The effective date of the notional transaction is the last date of completion of a scheme transaction or, if earlier, the last date of substantial performance of a scheme transaction (*FA 2003, s 75A(6)*). In the example above, this is likely to be the date on which Claire pays Agnes £2.15 million. If the transaction was a simple sale for £2.15 million, the effective date would typically be the date on which the consideration was paid, so this appears reasonable.

Exclusions from s 75A

7.14 The potential scope of *s 75A* is very wide. In particular, it could inhibit a legitimate arrangement of transactions in order to come within the terms of a relief or favourable treatment. To guard against this, there are two specific exclusions in *s 75A* itself, and some general restrictions on application set out in *ss 75B* and *75C*.

7.15 *Section 75A* does not apply if the only reason for the reduced SDLT is that *FA 2003, ss 71A–73* (alternative property finance reliefs; see **3.37**) or *Sch 9* (right to buy etc reliefs; see **3.42**) apply (*FA 2003, s 75A(7)*). It is important to remember that the exclusion only applies if these reliefs are the *only* reason for the SDLT saving. A combination of steps which qualify for one of the reliefs with other steps risks losing the protection of the exclusion and bringing the whole transaction within the scope of *s 75A*.

Section 75B – incidental transactions

7.16 Consideration for a real transaction in a series may be ignored in calculating the consideration for the notional transaction, if the real transaction is 'merely incidental' to the transfer of the chargeable interest from V to P. There is no comprehensive definition of 'merely incidental'. The legislation gives three examples of transactions which may be incidental (*FA 2003, s 75B(3)*), as follows, but the list does not claim to be exhaustive:

- a transaction undertaken for the purpose of constructing a building on the land;
- a transaction for the sale of anything other than land; and
- a loan or other provision of finance to enable someone to pay for part of the process by which the land transfer takes place.

Since the legislation only says these transactions 'may' be incidental, the list is of very limited use. It is further limited by the statement that transactions in the list at **7.10** are not incidental (*FA 2003, s 75B(4)*). A transaction cannot be regarded as incidental if it forms part of the process by which the transfer of the chargeable interest is effected, or if the transfer is conditional on completion of the transaction (*FA 2003, s 75B(2)*).

Example 7.3 – incidental transactions

The facts are as in Example 7.2 above, but Claire also pays Agnes £25,000 for the furniture and other chattels in the house. This is a payment for something other than land. Subject to confirmation that this represents a 'just and reasonable' allocation of the overall amounts paid, this will be regarded as an incidental transaction, and the amount paid will not be taken into account in determining the consideration for the notional land transaction.

Section 75C – other exclusions and conditions

7.17 A transfer of shares or securities is ignored for the purposes of *s 75A* if (but only if) it would otherwise be the first of a series of scheme transactions (*FA 2003, s 75C(1)*). This allows a degree of corporate reorganisation in order to permit a claim to relief on a subsequent land transaction.

7.18 If a real transfer equivalent to the notional transaction would have been eligible for a relief, the relief applies to the notional transaction (*FA 2003, s 75C(2)*).

7.19 If the notional transaction is a transfer to a company which is connected with the vendor, *FA 2003, s 53* applies to deem the consideration to be at least equal to market value (*FA 2003, s 75C(6)*) – see **2.29**).

7.20 If there are two notional transactions which, together, amount to an exchange, *FA 2003, Sch 4, para 5* applies to deem the consideration for each notional transaction to be equal to the market value of the interest deemed transferred under that notional transaction (*FA 2003, s 75C(7)*).

Example 7.4 – exclusion of first step, transfer of shares

Henry owns all of the shares in two companies (Anne Ltd and Jane Ltd). Anne Ltd owns the freehold of a property which is let to Jane Ltd at a rent which is below the market rate. The lease held by Jane Ltd therefore has a capital value. Henry wants to put the value of the property into a single company, partly to extract cash and, later, to raise finance. Henry transfers the shares of Jane Ltd to Anne Ltd for payment in cash. He then arranges for Anne Ltd to transfer the freehold property to Jane Ltd as a contribution to capital (claiming group relief from SDLT), at which point the lease held by Jane Ltd collapses into the freehold by operation of law.

This is a series of transactions forming part of a single scheme. There is a transfer of property from Anne Ltd to Jane Ltd. This is for no consideration, but as a transfer to a connected company it is deemed to be for market value consideration. The notional transaction is the transfer from Anne Ltd to Jane Ltd, which is deemed to be for market value consideration by virtue of the application of *s 53* as noted above. The SDLT paid (nil) is less than would be payable on this notional consideration, so *s 75A* potentially applies. However:

(1) As the first step in the series, the transfer of shares of Jane Ltd to Anne Ltd is ignored. There is, therefore, only one step in the transaction (the transfer of the property to Jane Ltd), so *s 75A* does not apply.

(2) If there were other steps such that this was not enough to disapply *s 75A*, the notional transfer (being a transfer from Anne Ltd to Jane Ltd) should qualify for group relief just as the real transfer does. So, even if *s 75A* does apply, the SDLT charge could be removed by a claim to group relief.

7.21 It should be noted that the application of a relief to the notional transaction is subject to the normal conditions and restrictions which would apply in relation to that relief for a real transaction.

7.22 Other provisions which are applied to the notional transaction as they would apply to a real transaction are as follows:

- an interest in a property-investment partnership (see **5.29**) is treated as a chargeable interest (*FA 2003, s 75C(8)(a)*); and
- if any of the scheme transactions is entered into in connection with a transfer of an undertaking (such that *FA 2003, Sch 7, paras 7–8* apply), the notional transaction is also treated as entered into in connection with the same transfer (*FA 2003, s 75C(3)*).

7.23 Where V or P is a partnership, the normal partnership rules were originally applied in relation to a notional transaction which comprises a transfer of land to or from the partnership, from or to a partner or someone connected with a partner (see **5.13** and **5.19**). However, it was concluded that this might allow avoidance. *FA 2010, s 56* therefore amended *s 75C(8)* and inserted *s 75C(8A)* to ensure that the rules in *FA 2003, Sch 15, Pt 3* do not apply to any notional transaction which consists of a transfer to or from a partnership. As a result, any such notional transaction will be subject to SDLT as if it was a transfer between two ordinary persons. This may produce an unfair result where an actual transfer to a partnership involves more than one step, thus bringing *s 75A* into play.

Example 7.5 – potentially unjust result from application of s 75A

RST is a partnership of three otherwise unconnected individuals R, S and T, who share partnership profits equally. The partnership requires further premises. R's wife owns a suitable building which is currently vacant and in need of refurbishment. R's wife leases the building to R for 10 years for no premium and a peppercorn rent. In turn, R sub-lets the building to the partnership for 10 years at a market rent, but with the first year rent-free to reflect the cost of refurbishment which the partnership will incur.

If the transactions are regarded separately, no SDLT arises on the lease from R's wife to R, and SDLT will be chargeable on the partnership in respect of only 67% of the NPV of rents on the lease to the partnership: 33% of the NPV represents the interest effectively retained by R through his membership of the partnership (see **5.11** et seq for an explanation of this).

If the transactions are regarded as a series within *s 75A*, the notional transaction will be the grant of a lease at rent from R or R's wife (it does not matter which, for these purposes) to the partnership. The normal partnership rules will not apply, and the full NPV of rents under the lease to the partnership will be subject to SDLT. The SDLT cost will, therefore, be increased by the application of *s 75A*, even though there was no avoidance in the original series of transactions. In this situation, the taxpayers would hope to rely on HMRC's assertion that *s 75A* will not be applied where there is no avoidance – but it would be far better to have this explicit in the legislation.

7.24 *Section 75C(5)* and *(10)* makes it clear that any apportionment of amounts for the purpose of measuring consideration on the notional transaction must be on a just and reasonable basis, and that any SDLT paid in respect of an actual transaction which is ignored under *s 75A* is to be treated as paid in respect of the notional transaction. This, therefore, should avoid any element of double charge. However, it is possible that *s 75A* could apply to more than one group of steps in a complex series of transactions. This could lead to more than one charge under *s 75A*, even where a 'simple' transfer of the property from the original vendor to the ultimate purchaser would only have led to a single charge.

SPECIFIC ANTI-AVOIDANCE RULES

Introduction

7.25 Many of the conditions surrounding reliefs or treatments could be regarded as anti-avoidance rules. Their purpose is to ensure that the relief or treatment is only available in the circumstances envisaged by the government. Nonetheless, there are some specific rules which may be recognised as having been included to counter particular arrangements and may be properly regarded as anti-avoidance provisions. These specific rules are described at various points in this book, when dealing with the transactions or circumstances in which they may apply. For convenience, they are also listed below, with references. Some seek to attack specific structures which are considered to give rise to a tax advantage; others only apply if there is an avoidance motive. The rules apply whether or not the transaction is caught by *FA 2003 ss 75A–75C*; indeed, by denying a relief or treatment, the specific rules may ensure that the transaction is not within the terms of the general anti-avoidance rule.

Subject	**Description**	**Statute (FA 2003) or other reference**	**Reference in this book**
Sub-sale relief	Alternative contracts No options Motive test	*ss 44A, 45A, Sch 2A, para 18*	**3.15** **3.16** **3.12**
Sale and leaseback relief	Not combined with other reliefs	*s 57A(3)(c), (d)*	**3.40**
Group, reconstruction and acquisition reliefs	Motive test	*Sch 7, paras 2(4A), 7(5), 8(5B)*	**3.28**
	Arrangements	*Sch 7, paras 2, 8(4)*	**3.25**
	Clawback on leaving group	*Sch 7, paras 3, 4A*	**3.30**
	Clawback on change of control	*Sch 7, paras 9, 11*	**3.33**
Alternative finance reliefs	Not combined with group or sub-sale relief	*s 73A* *s 45(3)*	**3.39** **3.16**
	Definition of financial institution and arrangements for change of control	*s 73AB, 73BA*	**3.38**
Charities relief	Change of use from charitable	*Sch 8, para 2*	**3.50**
Partnerships	Arrangements to transfer	*Sch 15, para 17*	**5.27**
	Withdrawal of value after land transfer	*Sch 15, para 17A*	**5.17**
Trusts	Grant of lease to bare trustee	*Sch 16, para 3(4)*	**2.18**
Leases	Assignment of lease after relief claimed on grant	*Sch 17A, para 11*	**4.17**
	Loan or deposit in conjunction with grant	*Sch 17A, para 18A*	**4.44**
Building contracts	Separation of building and land contracts	*Prudential Assurance v CIR* [1992] STC 863	Example 2.6

SDLT AND DISCLOSURE OF TAX AVOIDANCE SCHEMES (DOTAS)

Introduction

7.26 *FA 2004, Pt 7* introduced obligations for the notification of schemes for the avoidance of direct tax (potentially including SDLT). The legislation sets out the nature of the obligations in general terms and states on whom they fall. The detailed application is specified in regulations. SDLT was not initially in the list of taxes covered, but was added in 2005. The rules governing the scope are now consolidated in the *Tax Avoidance Schemes (Information) Regulations 2012, SI 2012/1836* with effect from 1 September 2012. The nature of SDLT schemes brought within the rules was set out in the *SDLT Avoidance Schemes (Prescribed Descriptions of Arrangements) Regulations 2005, SI 2005/1868*. These regulations were amended by the *Stamp Duty Land Tax Avoidance Schemes (Prescribed Descriptions of Arrangements) (Amendment) Regulations 2012, SI 2012/2395*. As explained below, the rules for identifying notifiable SDLT schemes set out in these regulations are very different from the rules for other taxes.

FA 2014, Pt 4 provides for HMRC to demand accelerated payment of the tax which will arise if an avoidance scheme proves ineffective. On 15 July 2014 HMRC issued a list of reference numbers of 1,200 schemes notified under DOTAS, including schemes to mitigate SDLT, saying that demands for payment of tax in respect of these schemes will be made over the following 20 months. The list is available at www.gov.uk/government/publications/tax-avoidance-schemes-on-which-accelerated-payments-may-be-charged-by-hmrc. Any taxpayer who has used a notifiable scheme is advised to check the list to establish whether it includes the reference number of their scheme and, if so, to be prepared to receive a demand for payment of tax. However, at the time of writing (June 2016), it is understood that no accelerated payment notices have been issued in relation to schemes specifically designed to mitigate SDLT.

7.27 It is important not to be misled by the title – the regime is not restricted to complex avoidance schemes, but may apply to the simplest modification to a transaction to reduce the tax cost. The parties to a transaction and anyone advising them may be required to make disclosure to HMRC in a very short timeframe. There are penalties for failure to comply. For many advisers, the original penalties were insignificant, but the stigma of being identified as non-compliant was regarded as worse than any financial penalty. *FA 2010, Sch 17* increased the maximum potential penalty to £1 million, which is likely to be significant to all advisers. Many firms of advisers have instituted systems and procedures to ensure rapid identification of obligations. However, such systems are worthless unless the individuals providing advice are vigilant and recognise occasions when the rules apply.

It is important to remember that notification obligations fall on 'Promoters' and 'Scheme Users' automatically, without any prompting from HMRC. However a person who only acts as an 'Introducer' is only required to give information to HMRC if HMRC specifically demand it.

Application to SDLT

7.28 SDLT was brought within the DOTAS regime from 1 August 2005. From that date, the rules applied to prescribed arrangements where the property in question was not purely residential and had a value of £5 million or more. Residential property with a value of at least £1 million was then brought within the rules from 1 April 2010 by *SI 2010/407*. Note that mixed residential/non-residential property was within the rules if the total value was £5 million or more and/or if the residential element was £1 million or more. However the value thresholds were removed by *the Stamp Duty Land Tax Avoidance Schemes (Prescribed Descriptions of Arrangements) (Amendment) Regulations 2012, SI 2012 2395* for transactions with an effective date of 1 November 2012 or later.

Scheme reference numbers (SRNs)

7.29 As with other taxes, HMRC provide a scheme reference number (SRN) for each scheme disclosed (*FA 2004, s 311*). Prior to 1 April 2010, promoters of SDLT schemes were not obliged to notify users of the SRN, and users had no obligation to notify HMRC of the number of any scheme used. This was because HMRC's primary concern was to understand the nature of SDLT avoidance arrangements rather than to challenge specific schemes. From 1 April 2010, as is the case with other taxes, the SDLT SRN must be provided to users and reported by them to HMRC, along with other information (*FA 2004, ss 312–313*). HMRC have provided a new form for this purpose at www.hmrc.gov.uk/aiu/form-aag4sdlt.pdf (see Appendix B). It is evident that HMRC consider that the obligation for scheme users to report to HMRC applies only in respect of schemes first notified to them by promoters on or after 1 April 2010. The way in which regulations have been sequentially amended does not clearly support this. However taxpayers should be able to rely on the clear statement in HMRC Guidance that the relevant date is 1 April 2010.

The absence of any requirement to report use of schemes first notified before 1 April 2010 (or perhaps 2009) is referred to in HMRC guidance as 'grandfathering'. It was decided that the grandfathering of certain politically sensitive schemes involving residential property and sub-sales is preventing HMRC from detecting their use. Regulations under *FA 2012, s 213* modified the rules to make users notify the use of those schemes (see **7.38**).

HMRC guidance

7.30 HMRC have provided general guidance on the application of the DOTAS rules. The latest version of this may be found online at www.hmrc.gov.uk/aiu/guidance.htm, or paper copies may be obtained from HMRC's publications department (see **Appendix A**). Some sections of this guidance do not apply to SDLT, and other sections apply only to SDLT. This guidance was last updated on 13 January 2016.

Notifiable arrangements and proposals

7.31 A notifiable arrangement is any arrangement which:

(1) falls within the description set out in the regulations (*SI 2005/1868*); and

(2) might be expected to enable any person to obtain an SDLT advantage as a main benefit from the arrangement (*FA 2004, s 306(1)*).

This is a two-pronged test. The mere obtaining of a tax advantage as a main benefit does not make the arrangement notifiable – it must also fall within the description set out in the regulations. Equally, an arrangement which happens to fall within the description set out in the regulations, but which does not have a tax advantage as a main benefit, is not notifiable. However, in the light of the severe penalties for failure to notify under these rules, most advisers 'play safe' and make a disclosure in all uncertain cases.

7.32 A notifiable proposal is a proposal for arrangements which would be notifiable if entered into (*FA 2004, s 306(2)*).

No hallmarks

7.33 In relation to income, corporation and capital gains tax, notifiable arrangements are those which have certain characteristics, referred to as 'hallmarks'. For those taxes, an arrangement is not notifiable unless it has one or more of the hallmarks. The 'hallmarks' have no application in relation to SDLT. Rather, the rules operate the other way round. *All* arrangements relating to the acquisition of chargeable interests, which might have a main benefit of enabling anyone to obtain a tax benefit in relation to SDLT, are notifiable, unless they fall within certain exclusions set out in *SI 2005/1868* as amended by subsequent regulations (see **7.28** and **7.38**), or are 'grandfathered' (see **7.37**). There is a useful flowchart in section 8 of the HMRC Guidance mentioned at **7.30**.

7.34 The exclusions are expressed in the form of six steps which the arrangement might involve, and two rules setting out which combinations of steps are acceptable. The order of the steps is not generally important. The steps are set out in the regulations as follows:

'Step A: Acquisition of a chargeable interest by special purpose vehicle

The acquisition of a chargeable interest in land by a company created for that purpose ("a special purpose vehicle").

Step B: Claims to relief

Making–

(a) a single claim to relief under any of the following provisions of the Finance Act 2003–

(i) section 57A (sale and leaseback arrangements);

(ii) section 58B (relief for new zero carbon homes);

(iii) section 58C (relief for new zero carbon homes: supplemental);

(iv) section 60 (compulsory purchase facilitating development);

(v) section 61 (compliance with planning obligation);

(vi) section 63 (demutualisation of insurance company);

(vii) section 64 (demutualisation of building society);

(viii) section 65 (incorporation of limited liability partnership);

(ix) section 66 (transfers involving public bodies);

(x) section 67 (transfer in consequence of reorganisation of parliamentary constituencies);

(xi) section 69 (acquisition by bodies established for national purposes);

(xii) section 71 (certain acquisitions by registered social landlords);

(xiii) section 74 (collective enfranchisement by leaseholders);

(xiv) section 75 (crofting community right to buy);

(xv) Schedule 6 (disadvantaged areas relief);

(xvi) Schedule 6A (relief for certain acquisitions of residential property);

(xvii) Schedule 6B (transfers involving multiple dwellings);

(xviii) Schedule 7 (group relief and reconstruction acquisition reliefs);

(xix) Schedule 8 (charities relief);

(xx) Schedule 9 (right to buy, shared ownership leases etc.);

(xxi) Schedule 61 to Finance Act 2009 (alternative finance investment bonds), or

(b) one or more claims to relief under any one of the following provisions of the Finance Act 2003–

(i) section 71A (alternative property finance: land sold to financial institution and leased to individual);

(ii) section 72 (alternative property finance in Scotland: land sold to financial institution and leased to individual);

(iii) section 72A (alternative property finance in Scotland: land sold to financial institution and individual in common); or

(iv) section 73 (alternative property finance: land sold to financial institution and resold to individual).

Step C: Sale of shares in special purpose vehicle

The sale of shares in a special purpose vehicle, which holds a chargeable interest in land, to a person with whom neither the special purpose vehicle, nor the vendor, is connected.

Step D: Not exercising election to waive exemption from VAT

No election is made to waive exemption from value added tax contained in paragraph 2 of Schedule 10 to the Value Added Tax Act 1994 (treatment of buildings and land for value added tax purposes).

Step E: Transfer of a business as a going concern

Arranging the transfer of a business, connected with the land which is the subject of the arrangements, in such a way that it is treated for the purposes of value added tax as the transfer of a going concern.

Step F: Undertaking a joint venture

The creation of a partnership (within the meaning of paragraph 1 of Schedule 15 to the Finance Act 2003) to which the property subject to a land transaction is to be transferred.'

7.35 The two rules setting out which combinations are acceptable are as follows:

'Rule 1

Arrangements involving any combination of Steps B, D, E and F, with or without a single instance of one of steps A, C and D, are excluded arrangements unless rule 2 applies. Note however that Step B is in two parts.

It is HMRC's view that if there is more than one claim to any relief listed in part (a), or claims to reliefs in both part (a) and part (b), this combination is no longer within Step B and the arrangement is not excluded.

Rule 2

Arrangements are not excluded arrangements if they–

(a) include all, or at least two of, steps A, C and D; or

(b) involve more than one instance of step A, C or D.'

7.36 It is important to bear in mind that these rules merely make certain transactions exempt. Any multi-step transaction involving any steps other than steps A to F listed above, where the unlisted step(s) are essential to obtain the SDLT advantage, *cannot* be exempt from disclosure, no matter what combination of listed steps it involves.

Example 7.6 – step not in list, notification required

Notax LLP, a firm of tax advisers, develops a scheme for SDLT-free acquisition of leasehold commercial buildings. The scheme involves the client setting up an SPV (not a listed step), the SPV entering into a joint venture with the vendor (Step F), the SPV acquiring an interest in the property (Step A), and a claim to group relief (Step B). Under the rules set out above in **7.35**, this includes only one instance of a step listed in rule 2, so the scheme should be exempt from notification. However, the scheme also involves a step which is not listed but is essential to obtaining the tax advantage – setting up an SPV – so the scheme is not exempt from notification. In practice, it seems that few real schemes will be exempt from notification under these rules.

Arrangements which are 'substantially the same'

7.37 Separate, different schemes must be disclosed separately. However, once a scheme has been disclosed, no further disclosure is needed by the promoter if the same scheme is implemented again, whether for the same or different clients. This is subject to the exception mentioned in **7.38**. Equally, if subsequent implementation involves only minor differences such that the two schemes are substantially the same, no further disclosure is required (*FA 2004, s 308(5)*). HMRC have provided some guidance as to when schemes may or may not be regarded as substantially the same, as follows:

'In our view a scheme is no longer substantially the same if the effect of any change would be to make any previous disclosure misleading in relation to the second (or subsequent) client.

In general provided the tax analysis is substantially the same we will regard schemes as "substantially the same" where the only change is a different client including a different company in the same group.

We will not regard schemes as substantially the same where there are changes to deal with changes in the law or accounting treatment, changes in the tax attributes e.g. schemes creating income losses instead of capital losses or other legal and commercial issues.' (DOTAS guidance, last updated on 13 January 2016 para 14.2.3 at p 100)

7.38 From 1 April 2010, most arrangements are excluded from the need to disclose if they are the same, or substantially the same, as arrangements first made available for implementation before that date. This is generally referred to as 'grandfathering'. Where such arrangements related to non-residential property, they should have been disclosed under the rules then in force; however, this concession means that, in relation to residential property, only novel or significantly changed schemes need be disclosed. However *FA 2012, s 213* amended *FA 2004, s 308* to allow the Treasury to make further amendments to the rules by regulation. *The Stamp Duty Land Tax (Avoidance Schemes) (Specified Proposals or Arrangements) Regulations 2012, SI 2012/2396* came into force on 1 November 2012. These removed the grandfathering concession in respect of schemes which rely on sub-sale relief in combination with a distribution in specie, acquisition by a partnership or settlement, a gift or transfer at an undervalue, the grant of an option or an assignment or novation. Schemes relying on such combinations and used on or after 1 November 2012 have to be disclosed one more time. As a result, these schemes have been brought within the rules which require *users* of the schemes to disclose their use (see **7.53**). The retrospective revision of sub-sale relief in *FA 2013, s 194* (see **3.16**) and the new rules from 17 July 2013 (see **3.11**) were informed by disclosures under these provisions.

Example 7.7 – arrangements which are 'substantially the same'

Tax advisory firm Fiscality has developed an arrangement which allows a company to grant a lease to a fellow subsidiary at low SDLT cost, so there is no need to claim group relief and therefore no clawback if the lessee leaves the group. The arrangement requires the establishment of an orphan company registered overseas to hold an interest in the land as nominee. The arrangement was disclosed under the DOTAS rules before 1 April 2010, and was used by two separate clients of Fiscality. In December 2010, Fiscality concludes that using a nominee company established in a different overseas jurisdiction would make it easier to administer the arrangement without affecting the tax analysis. In the author's view, if the only change is to use a company established in a different jurisdiction purely for convenience, the new arrangement will be 'substantially the same' as the old arrangement and no new disclosure obligation will arise.

Shortly afterwards, Fiscality concludes that the tax analysis will be strengthened if the order of execution of two steps in the arrangement is reversed. This change will be sufficient to prevent the new arrangement from being 'substantially the same' as the old, and a new disclosure obligation will arise when the new version is 'made available for use' or, if earlier, when Fiscality first become aware of a transaction forming part of arrangements implementing the new version of the scheme (see **7.54**).

Promoters and introducers

7.39 *FA 2010, Sch 17* amended the provisions relating to identity and obligations of 'promoters', and introduced a new concept of 'introducers'. The amendments took effect from 1 January 2011 (*SI 2010/3019*).

7.40 From the beginning, the primary obligations have fallen on 'promoters'. A promoter is a person who, in the course of a tax advisory, banking or securities house business and in relation to a notifiable arrangement or proposal:

(1) is to any extent responsible for the design of the proposed arrangements, or

(2) makes a 'firm approach' to another person with a view to making the proposal available for implementation by that person or anyone else, or

(3) makes the proposal available for implementation by any person.

7.41 The second point was inserted by *FA 2010, Sch 17* to ensure the responsibility net is spread as widely as possible. A firm approach is made if, at a time when the arrangements have been 'substantially designed', information about them, including an explanation of the tax advantages, is communicated to anyone with a view to entering into transactions forming part of the proposal. Arrangements are substantially designed if, at that time, the nature of the transactions is sufficiently developed for it to be reasonable to believe that someone would enter into them (or into transactions not substantially different from them) in order to obtain the expected tax advantage.

7.42 Prior to *FA 2010*, the expression 'substantially designed' did not appear in the legislation, but HMRC relied on a similar concept to determine whether a scheme had been 'made available'. In the view of HMRC, a scheme is substantially designed at any time if, at that time, the nature of the transactions to form part of the scheme has been sufficiently developed for it to be reasonable to believe that a person who wishes to obtain the tax advantage communicated might enter into either:

- transactions of the nature developed at the time; or
- transactions not substantially different from those developed at the time.

(DOTAS Guidance last updated on 16 January 2016 para 3.6.2, see **7.30**). The rules are expressed in wide, all-inclusive terms, so that reporting obligations may arise at a very early stage.

7.43 The question of whether an arrangement is substantially designed or developed to such a stage etc is particularly unclear. In any case in which HMRC may allege that a promoter has failed to make timely disclosure, they may be expected to wish to see copies of emails, notes of telephone conversations and other similar evidence, to show the process by which the scheme was developed.

Example 7.8 – timing of obligations

Notax LLP is developing an arrangement which, they believe, will allow purchase of houses at a much reduced SDLT cost. On 1 February 2016, Mark, a partner in Notax, attends a conference with leading tax counsel at which counsel advises that their technical analysis is generally sound. Counsel highlights areas needing some further attention, such as the country of incorporation of a new company required to execute the arrangement. On 2 February 2016, Judith, another partner in Notax who has not been involved in developing the proposal, calls business contact Kim at the local branch of estate agent Houseller plc to tell him about the wonderful tax wheeze her firm has just developed, and to encourage Kim to talk to suitable house purchasers about it with a view to introducing her firm to them. There is an agreement in place between Notax and Houseller under which commissions are paid for successful introductions. Kim is not a tax person, so Judith does not give him any technical details, but she does outline the general steps required and explains that, if successful, the arrangement would reduce the SDLT cost of house purchase to less than one tenth of what it would otherwise be.

It is almost certain that Judith has fallen within paragraph (2) above and has triggered Notax LLP's disclosure obligations on 2 February 2016. It does not matter that Judith has given no technical details, nor that she personally was not involved in developing the arrangement. In the eyes of HMRC, Notax LLP as an entity is carrying on a tax advisory business and is the promoter.

7.44 A person is an introducer if he makes a marketing contact with another person in relation to a notifiable proposal. A marketing contact is made if the first person communicates information about the proposal to the second person, including an explanation of the expected tax advantage, with a view to anyone entering into the proposed arrangements. An introducer, provided he is not also a promoter, has no direct disclosure responsibilities, but may be required by HMRC to pass on to them such information as he has about the scheme.

Example 7.9 – an 'introducer'

In Example 7.8, Kim immediately calls customer Irena who is negotiating to buy a UK house for about £2 million. He passes on the outline information received from Judith, and asks Irena to think about talking to Notax LLP. Kim is not working for or as a tax adviser or a bank or securities house, so neither he nor Houseller can be a promoter. However, since Houseller stands to earn commission from the introduction, this is undoubtedly a 'marketing contact'. Therefore, Houseller is an introducer.

Example 7.10 – no obligations

Following on from the previous example, Irena (who works for an advertising agency) mentions the Notax scheme to a colleague who is also about to buy an expensive house, passing on the contact details for Notax. Neither Irena nor her employer is carrying on a business as a tax adviser or a bank or securities house. Therefore, Irena cannot be acting as a promoter. This is a casual discussion with a colleague – Irena does not stand to make any financial gain from passing the information on to her colleague. Therefore, this is unlikely to be a marketing contact, and neither Irena nor her employer should be regarded as acting as an introducer.

Who is not a promoter?

7.45 A person who is only involved in the design of a scheme, and is not in any way involved in implementation or organising or managing implementation, is not a promoter if:

- his input is benign – that is, the advice he gives does not contribute to the tax-saving aspect of the scheme (for example, advice as to whether a particular shareholding arrangement permitted a claim to group relief would be benign, but advice as to how to change the arrangement in order to qualify for group relief would not be benign), or
- he is not acting as a tax adviser – for example, an unrelated lawyer consulted only on the company law implications of a proposal would not be acting as a tax adviser, or
- he could not reasonably be expected to have sufficient information to know whether there is a disclosable scheme, or to be able to make that disclosure – for example, a newly appointed tax adviser might ask the outgoing tax adviser for an opinion on the implications of a proposed action (in the light of previous transactions) and the new adviser might use that advice in devising a notifiable proposal. However, if the outgoing adviser is not given any details of the proposal, he probably could not reasonably be expected to have sufficient information to know whether there is a disclosable proposal or to make disclosure.

7.46 A person who acts purely as an intermediary or introducer, without being involved in design, implementation or management of implementation, is not a promoter (but HMRC may demand information from such a person now that the changes in *FA 2010, Sch 17* have been activated; see **7.44**).

Co-promoters

7.47 If two or more persons are promoters in relation to the same (or substantially the same) proposal or arrangement, it is possible for only one disclosure to be made, and for this to be made by one of them, provided:

(1) the other promoter holds or is provided with the information disclosed to HMRC,

(2) the disclosing promoter provides HMRC with a note of the name and address of the other promoter, and

(3) (if the co-promoter becomes a co-promoter after initial disclosure has been made) the disclosing promoter provides the other promoter with a note of the SRN,

the co-promoter is exempt from making disclosure. Where the proposal or arrangement made by the co-promoter is not identical with that disclosed, it is for the co-promoter to decide whether it is sufficiently similar for the exemption to apply. Co-promoters are not exempt from other obligations, such as providing their client with SRNs.

Overseas promoters

7.48 In principle, the DOTAS rules apply to a promoter resident outside the UK. However, in practice it may be difficult for HMRC to enforce compliance, especially where the promoter is not carrying on business through a UK permanent establishment. If no promoter is resident in the UK, and the overseas promoter does not comply with DOTAS obligations, the obligations fall instead on any person (the 'client') who enters into any transaction forming part of the arrangements. Note that the obligations fall on 'any person who enters into any transaction …', which therefore includes the vendor, even though any tax advantage will normally accrue to the purchaser. A vendor who does not know about any arrangements which the purchaser may be implementing cannot be faulted for failing to disclose them. However, it is common for a vendor to know at least something of the purchaser's arrangements, so this is a matter over which vendors must take care.

These disclosure obligations are removed from the client(s) if a promoter complies with them (*FA 2004, s 309(2)*).

Arrangements with no promoter

7.49 If a person enters into a transaction forming part of notifiable arrangements in relation to which there is no promoter (for example, a scheme which the taxpayer has developed himself), that person has responsibility for making an appropriate disclosure. This includes the situation where a scheme is devised by one company in a group and used by another in the same group. The group company devising the scheme is not regarded as a promoter but the company using the scheme must make disclosure.

Promoters with the benefit of legal privilege

7.50 In some circumstances, promoters who are lawyers may be able to resist providing some of the information required under DOTAS, on the grounds that the information is subject to legal privilege. If this happens, any user of the scheme is required to make the same disclosure as would be required if there was no promoter (see previous paragraph). However, as noted at **7.60**, the time limit for making disclosure is five days and not the 30 days which applies for 'in-house' schemes.

Duties of promoters

7.51 Promoters (or persons entering into a transaction if there is no promoter, or if there is no promoter in the UK and no overseas promoter complies with DOTAS):

- must provide HMRC with certain prescribed information (see **7.61** et seq), and
- must do this within the prescribed period after the relevant date (see **7.60**).

Promoters must make the disclosure by completing and submitting form AAG1 (see Appendix B), which may be downloaded for manual completion or completed and submitted online, as noted at **7.53**. In response to this, HMRC will normally provide a scheme reference number (SRN). The promoter must notify the SRN to any person to whom he is providing services in relation to the scheme, or any scheme which is substantially the same (see **7.37**). This notification must be made on form AAG6 (this new form should be used from 1 March 2016 – prior to that date it was form AAG6 (SDLT)), which is downloadable from www.hmrc.gov.uk/aiu/aag6_sdlt.pdf. The number must be passed on within 30 days of the later of:

- the promoter first becoming aware of any transaction which forms part of the arrangements, and
- HMRC notifying the promoter of the SRN.

Promoters are also required to keep appropriate records and notify HMRC of details of clients implementing the schemes (see **7.64** below). *FA 2013, s 223* added a requirement that promoters must, if required by notice from HMRC, also supply prescribed details of any other persons whom the promoter might be expected to know may be a party to the arrangements. The details are prescribed in the *Tax Avoidance Schemes (Information) (Amendment,etc) Regulations 2013 (SI 2013/2592), reg 20*, which came into force on 4 November 2013 and comprises the name and address of any person, other than the client, who the promoter might be expected to know may be a party to the arrangements, but only those persons who may either sell the arrangements to another person or achieve the tax advantage themselves by implementing the arrangements, the tax reference numbers of such persons and sufficient information, to allow HMRC to understand the manner in which such a person is involved in the arrangements.

7.52 As noted above, the requirement to notify the SRN for SDLT schemes to the client was new from 1 April 2010. It appears that HMRC consider this only applies to schemes notified to HMRC from 1 April 2010, although the regulations appear to say that it applies to schemes notified from 1 April 2009.

Duties of scheme users

7.53 Also from 1 April 2010, obligations were imposed on users of SDLT schemes and others to whom an SRN is notified. Anyone who is notified of an SRN is now required to report their use of the scheme to HMRC, quoting the SRN. Some SDLT schemes operate by taking the transaction out of the charge so that, if the scheme works, there is no obligation to submit an SDLT return; in any case, the SDLT return has no space to report an SRN. To ensure relevant details are captured, HMRC have provided a new SDLT-specific version of form AAG4 (form AAG4 (SDLT) – see Appendix B) for scheme users to report details. The form can be downloaded from www.hmrc.gov.uk/aiu/form-aag4sdlt.pdf. Alternatively, the form may be completed and submitted online via a link at www.hmrc.gov.uk/aiu/forms-tax-schemes.htm. Use of this form is now compulsory in all cases. *FA 2013, s 223* added the requirement for the user to provide prescribed information to the promoter. The details are prescribed in the *Tax Avoidance Schemes (Information) (Amendment,etc) Regulations 2013 (SI 2013/2592), reg 16*, which came into force on 4 November 2013 and the information to be supplied to the promoter comprises the user's tax reference number and national insurance number or confirmation that the user does not have a tax reference number and/or a national insurance number.

7.54 Anyone to whom an SRN is provided is also required to notify it to anyone else they reasonably expect to be a party to, and to gain a tax advantage from, implementation of the scheme, using form AAG6, as noted at **7.51**. This must be done within 30 days of the later of:

- first becoming aware of any transaction which forms part of the arrangements, and
- receiving notification of the SRN.

Time for disclosure – promoters

7.55 The rules relating to timing of disclosure by promoters are the same as for other taxes. A scheme promoter is required to make disclosure within five days (excluding weekends and bank holidays) of the 'relevant date'. The relevant date is the earlier of the date on which the notifiable proposal is 'made available for use', and the date on which the promoter first becomes aware of any transaction forming part of arrangements implementing the proposal.

7.56 The legislation does not state when a proposal is regarded as 'made available for use'. The HMRC guidance draws a distinction between bespoke and marketed schemes. A marketed scheme – that is, one which is developed for possible use by a range or category of users – is made available when it:

- has been developed to such a stage that the promoter has a high degree of confidence in the tax analysis applying to it; and
- is communicated to the first potential user in sufficient detail that he could be expected to:
  - understand the expected tax advantages; and
  - decide whether or not to enter into it.

HMRC accept that, where a firm of advisers has an internal approval process, a proposal will not normally be regarded as made available before that approval has been given.

7.57 In contrast with this, a bespoke scheme – that is, one which is designed for a specific client's situation – is not generally regarded as being 'made available'. For such a scheme, the relevant date will therefore be the date on which the promoter first becomes aware of any transaction forming part of the arrangements. There is a danger here where more than one individual in a firm of advisers has contact with the client; if the first individual to learn that a transaction has been implemented does not understand the significance of this, the five-day deadline may easily be missed.

7.58 Occasionally, an adviser may outline a partially developed planning idea to a client, in such general terms that no proposal can yet be regarded as made available for use. The client may then do further development work and proceed to implement its own version of the arrangement without further reference to the adviser. It might be thought that, in this situation, the adviser has not made a notifiable proposal and has no disclosure obligations. However, even here the legislation seeks to impose obligations on the adviser. Having provided the basic idea, the adviser is likely to be regarded as 'to any extent responsible for the design of the proposed arrangements' and therefore a promoter. If no disclosure of the proposal has previously been made, the adviser is required to make disclosure within five days of first becoming aware of any transaction forming part of the notifiable arrangements. It is very difficult to be sure of capturing such information – good communications with clients and robust internal systems are essential!

Time for disclosure and other notification – scheme users

7.59 As noted at **7.29**, the general requirement for users to notify use of SDLT schemes was introduced in April 2010. The rules governing timing of this notification are *not* the same as for other taxes. The differences mirror the tight timetable for submission of SDLT returns. The deadlines are as set out below. Guidance on the nature and content of the various forms is also provided below.

Forms for disclosure and notification

7.60

Form	Responsible person	Reason	Deadline for receipt by HMRC or other party
AAG1	Promoter	Disclosing scheme	5 days from making available or becoming aware of first transaction in scheme
AAG2	User	Disclosing scheme from overseas promoter, no other disclosure made	5 days from entering into first transaction in scheme
AAG3	User	Disclosing scheme from lawyer claiming legal professional privilege	5 days from entering into first transaction in scheme
AAG3	User	Disclosing internally developed scheme, no external promoter	30 days from entering into first transaction in scheme
AAG4 (SDLT)	User	Notifying use of scheme with UK promoter (or overseas promoter who complies with disclosure requirement)	30 days from later of receipt of SRN and effective date of first transaction in scheme
AAG5	Anyone	Continuation sheet where information does not fit on another form	As related form
AAG6	Promoter and user	Informing other users of SRN	30 days from later of receiving SRN and becoming aware of first transaction in scheme

Content of disclosure

7.61 The forms noted above prescribe the information to be provided. Use of the appropriate forms (including the continuation sheet AAG5 where necessary) is compulsory. Notification given in any other way will not be regarded as sufficient. The forms are reproduced at **Appendix B**. PDF versions (for manual completion and posting) may be downloaded, or the forms may be completed and submitted online, via links at www.hmrc.gov.uk/aiu/forms-tax-schemes.htm. Most of the forms are the same as those used for disclosure of other direct tax schemes, but there is an SDLT-specific version of form AAG4. Most of the entries required on the forms are self-explanatory, and help on completion is available, if needed, in the HMRC guide mentioned at **7.30**.

7.62 The only sections of the forms which commonly cause difficulty are those in forms AAG1 to AAG3 relating to the description and explanation of the scheme. HMRC guidance (Guidance on DOTAS last updated on 13 January 2016 para 15.3.4) on this subject is as follows:

> 'Sufficient information is required to be provided such that an Officer of the Board of HMRC is able to understand how the expected tax advantage is intended to arise. The explanation should be in straightforward terms and should identify the steps involved and the relevant UK tax law. Common technical or legal terms and concepts need not be explained in depth.
>
> If the scheme is complex then copies of any prospectus or scheme diagrams will help us understand what is proposed. But even where you send such documents you must still use form AAG 1, AAG 2 or AAG 3 as appropriate. Where such documents are supplied there is no objection to these documents excluding information that would identify a client.'

7.63 Each form has a separate section for identification of relevant statutory provisions, but in the author's view it is also likely to be necessary to refer to appropriate sections of legislation in the body of any description and explanation.

Requirement to notify HMRC of users

7.64 Under *FA 2004, s 313ZA*, inserted by *FA 2010, Sch 17*, a promoter who provides services to a client in connection with the arrangements is required to give information about the client to HMRC. The promoter must provide details of the name and address of each client to whom he has provided services in relation to a notifiable scheme (and to whom he would be required to provide the SRN: see **7.51**) together with the SRN of the scheme, details of the promoter's own name and address and the end date of the calendar quarter in which the services were provided. This information must be submitted within 30 days of the end of the relevant calendar quarter.

Example 7.11 – details to be disclosed

The following is an example of the level of detail which may be appropriate in describing/explaining a scheme. There is no suggestion that what follows is a description of a scheme which is currently or has previously been valid or effective – it is purely intended to illustrate the amount of detail which may be appropriate!

Title of arrangement: Partner retirement plan

Explanation:

(1) This arrangement permits a partner to withdraw from a property-investment partnership and a new partner to join the partnership at reduced SDLT cost.

(2) Statutory references are to Finance Act 2003.

(3) ABC is a property-investment partnership as defined in Schedule 15 para 14(8), established in [overseas territory]. The partners are A, B and C who are all individuals and otherwise unconnected and who share equally in the profits and capital of the partnership. Although the majority of ABC's assets are chargeable interests (none of which was acquired from a partner or person connected with a partner), substantial other assets are held including property assets situated overseas.

(4) A wishes to retire and will withdraw value equal to his 33% share in the partnership capital; it is proposed to admit D, another individual who is not connected with A, B or C. D will contribute cash to secure a 33% share in the capital of the partnership.

(5) The partnership deed will be amended to specify that individual partners' shares entitle them to share in the economic value of specific assets. As a result, A's share will entitle him to share in the economic value of the overseas assets and a one-third share in just one chargeable asset; the shares of the other partners will entitle them to share in the economic value of the remaining chargeable assets plus a smaller share in the values of the non-chargeable assets. The values will be such that the overall partnership share of each partner will be unchanged. Appropriate legal advice has been obtained confirming that such subdivision is effective under the law of [overseas territory]. Copies of the current deed and draft amendment are available for inspection if required.

(6) The partnership will use a combination of cash already held and cash borrowed from a bank under an existing overdraft facility to pay A the value of his partnership share.

(7) D will contribute cash on admission to the partnership and will become entitled to the interests in assets relinquished by A.

Analysis:

(8) The division of assets into categories and subdivision of partnership interests will not affect any partner's partnership share as defined by Schedule 15 paragraph 34(2); therefore it will have no SDLT consequences.

(9) The withdrawal of A and admission of D will be a Category A Transfer of partnership interest as defined in Schedule 15 paragraph 14(3B). However the only relevant partnership property as defined in Schedule 15 paragraph 14(5) is the one UK property in which A has an economic interest. SDLT will be chargeable accordingly by reference to 33% of the market value of that property alone, in accordance with Schedule 15 paragraph 14(6) & (7).

Statutory provisions:

Finance Act 2003 Schedule 15 paragraphs 14(3B), (5)(c), (6), (7) and 34(2).

Information powers

7.65 If HMRC suspect that a scheme has not been disclosed, they are entitled to require anyone they suspect of being a promoter to formally state whether they consider a scheme to be disclosable and, if not, why not. The notice from HMRC will give a time by which a reply must be given – potentially, as short as ten days later – and penalties may be imposed for non-compliance. Beyond that, there are powers to require information to be given, but these generally require an order of the Tax Tribunal. These situations are beyond the scope of this book.

Penalties

7.66 The Tax Tribunal may impose penalties for failure to disclose a notifiable proposal or arrangement or for late disclosure, and for failure to comply with orders requiring disclosure or provision of information. The amount of the penalty may be up to £5,000, plus further amounts of up to £600 (or, in some cases, £5,000) for each day during which the failure continues. However, the Tax Tribunal is empowered to award penalties of up to £1 million in exceptional cases if the daily penalties are not considered sufficient to induce compliance. Any person in danger of suffering such penalties will doubtless require specialist advice beyond the scope of this book.

Chapter 8

Stamp Duty – General Rules

INTRODUCTION

8.1 The last comprehensive consolidation of stamp duty legislation was in the *Stamp Act 1891 (SA 1891)* and the *Stamp Duty Management Act 1891*. Unsurprisingly, the law has changed massively since then. The main changes have been to remove particular assets or transactions from the charge, so that *SA 1891, Sch 1*, which originally defined the scope of the tax, is now largely irrelevant. However, the principles set out in the body of *SA 1891* still apply, some sections of that Act having been amended by 21st-century *Finance Acts*.

Subsequent stamp duty legislation is scattered across more than a century of *Finance Acts*. Specific exemptions often appear in other legislation (for example, an exemption for transfers in the privatisation of rail services; see *Railways Act 1993, s 112* and *Sch 9*). For this reason, when dealing with government-sponsored transactions, it is usually worth checking enabling legislation for exemptions. However, for most transactions the key legislation is found in *FA 1986, ss 66–85*, *FA 1999, Sch 13* and *FA 2003, ss 125* and *195*.

The government has embarked on a programme to remove redundant provisions from the statute book. *FA 2012, Sch 38* repealed several pages of stamp duty reliefs which have little or no current application, with effect from 6 April 2013. This small step of simplification is welcomed. However, a better step might be to abolish stamp duty, bringing all UK share transactions into the SDRT net.

8.2 Since 1 December 2003, stamp duty has been chargeable only on transactions in stock and marketable securities, and Transfers of interests in partnerships with stock or marketable securities amongst their assets. (In this book, the expression 'Transfer' is capitalised to make it clear that this is a reference to the potentially stampable document.) The restriction to stock and marketable securities is in *FA 2003, s 125*; however, the charge on Transfers of partnership interests is then reinstated by *FA 2003, Sch 15, paras 31–33*. The latter provision was something of a 'last minute' amendment, when it was realised that it might otherwise be possible to escape stamp duty on Transfers of securities by first contributing them to a partnership, then transferring the interests in the partnership. It is doubtful whether the charge on partnership interests has any significant practical effect. The application to partnerships is considered at **8.57**; the imposition of stamp duty on certain *issues* of securities is dealt with at **8.55**. The following paragraphs deal primarily with the application of stamp duty to Transfers of stock and marketable securities.

8.3 Although, in principle, stamp duty is chargeable on Transfers of a wide range of shares and securities, in practice the transfer of most debts is exempted by either the 'loan capital exemption' or the 'non-marketable debenture' exemption (see **8.16–8.17**). Transfers of units in unit trusts are outside the scope of stamp duty (*FA 1999, Sch 19, para 1*), but are instead subject to a special SDRT regime (see **10.30**).

8.4 There is no direct compulsion to pay stamp duty on Transfers – enforcement is indirect, arising from the fact that a stampable document cannot be used to register title or as evidence in a civil court unless duly stamped (*SA 1891, ss 14, 17*). It is sometimes asserted that a document which ought to be stamped but has not been stamped is not legally valid. This is not true. The problem is simply that the document cannot be used as evidence. So, in many cases, if sufficient evidence of the terms of a contract can be adduced without reference to any (stampable) document, it should be possible to enlist the help of the courts in enforcing the contract, even if no stamp duty has been paid. This principle was widely relied on when stamp duty applied to transfers of assets such as goodwill, when contracts were entered into on the basis of oral agreements and conduct (see **11.30**).

In most cases, the legislation does not specify who should pay stamp duty (the exception is an agreement for sale of an equitable interest – *FA 1999, Sch 13, para 7(1)*); but, in practice, the purchaser/ transferee will normally pay or arrange stamping in order to cancel any SDRT charge (see **10.11**) and to be able to register ownership of the asset.

SCOPE – DOCUMENTS

Stamp duty on Transfers

8.5 A 'Transfer' is a document (the legislation used to refer to 'conveyance or transfer', but it seems to be thought that 'conveyance' should be reserved for land transactions, so references to that word were largely removed from the stamp duty legislation by *FA 2003*). The concept of a Transfer is bound up with the separation in English law between legal and beneficial ownership. Under English law, parties may agree to buy and sell an asset; they may take actions, such as payment of consideration and perhaps physically handing over the asset, to give effect to the agreement. However, for certain assets such as land and shares, this is not sufficient to complete the process. At this point, the purchaser may have some form of beneficial ownership of the asset, but he does not have legal title. He only gains legal title when all steps to perfect the transfer of title, as required by law, have been completed – and, in the case of UK shares held in certificated form, this includes execution of an appropriate Transfer document. (A different process applies to shares held in dematerialised form, such as quoted shares held in CREST – see **10.20** for further details.)

8.6 In the UK, the most common form of share Transfer is a stock transfer form. However, other documents may also function as Transfers. In particular, a document under which the registered owner of legal title acknowledges that he holds as trustee or nominee for someone else – a 'declaration of trust' – can function as a Transfer (see **8.13** and **Example 9.3**). Note that a new version of the standard stock transfer form became available in April 2012. Old versions may now be rejected by company registrars.

Restricted to Transfers on sale

8.7 Stamp duty is currently payable only if the Transfer is 'on sale' (*FA 1999, Sch 13, para 1*) – that is, there is valuable consideration which counts for stamp duty purposes. A Transfer for no consideration (eg a gift) used to be subject to a fixed £5 stamp duty charge, which could sometimes be escaped by completing a certificate on the form. But, in a welcome piece of administrative simplification, *FA 2008, Sch 32* abolished the fixed £5 charge for Transfers completed on or after 13 March 2008. From that date, a Transfer for no consideration is simply not liable, there is no requirement to complete any certificate and the Transfer is deemed to be 'duly stamped' as soon as it is executed.

However, company registrars found it difficult to decide whether unstamped forms presented for registration of a change of ownership were genuinely exempt, or had been left unstamped incorrectly. In April 2012, two new certificates were added to the back of the standard stock transfer form. Certificate 2 (see **8.8** et seq for the circumstances in which Certificate 1 on the reverse of the stock transfer form should be used) is to be completed when the transfer is exempt from stamp duty, for example, transfers in connection with divorce or the dissolution of a civil partnership or because, although there is consideration, there is no consideration which is chargeable stamp duty. The wording of the certificate is as follows:

> 'I/We certify that this instrument is otherwise exempt from ad valorem Stamp Duty without a claim for relief being made or that no chargeable consideration is given for the transfer for the purposes of Stamp Duty.
>
> I/We confirm that I/we have been authorised by the transferor to sign this certificate and that I/we am/are aware of all the facts of the transaction.'

The certificate should be signed either by the transferor(s) (in which case the second sentence should be omitted) or on their behalf.

An internet search should produce the form as a downloadable PDF file.

Small transaction exemption

8.8 The main charging provision in *FA 1999, Sch 13, para 1* was amended by *FA 2008, s 98*, to remove the stamp duty charge where the consideration does not exceed £1,000 (*FA 1999, Sch 13, para 1(3A)*). Certificate 1 on the back of the stock transfer form should be used when this small transaction exemption applies. The wording of the certificate is as follows:

> 'I/We certify that the transaction effected by this instrument does not form part of a larger transaction or series of transactions in respect of which the amount or value, or aggregate amount or value, of the consideration exceeds £1,000.
>
> I/We confirm that I/we have been authorised by the transferor to sign this certificate and that I/we am/are aware of all the facts of the transaction.'

It is therefore only possible to certify if the *total* consideration for all relevant assets transferred in the same transaction or set of linked transactions does not exceed £1,000 (*FA 1999, Sch 13, para 6*).

8.9 In arriving at the total consideration, it is only necessary to take account of types of consideration which 'count' for stamp duty purposes (see **8.25**). It is also possible to leave out of account any consideration attributed on a just and reasonable basis to 'goods, wares or merchandise' (such as stock in trade of a retail business), intellectual property or goodwill, provided the Transfer itself does not also act to transfer those assets (*FA 1999, Sch 13, para 6(2)*). It would be unusual for a share Transfer to also transfer other assets, but this is a point where care may be needed in drafting documents.

However, amounts allocated to some other assets, especially debts and other financial assets, cannot be left out of account, even though no stamp duty will be chargeable in respect of them. Therefore, where there are such assets, a stock transfer form or other document transferring shares with a value below £1,000 may not qualify for certification, simply because the value of other relevant assets transferred under the same agreement take the total over £1,000.

8.10 As indicated above in **8.8**, when using the current version of the stock transfer form, Certificate 1 should be completed if the small transaction exemption applies. In relation to any other form of transfer document it is HMRC's view that:

(1) the certificate should be an integral part of the Transfer, not on a separate piece of paper (STSM01160) (it is not clear what should be done if there is simply no space on the transfer document to insert a certificate), and

(2) if the certificate is added after execution of the Transfer, it should be signed on behalf of all parties to the Transfer. Signature on behalf of the vendor is considered the more important and, in practice, such documents are not always signed by the purchaser.

8.11 Where a pre-printed certificate is not present, the following wording would be appropriate:

> 'The transaction effected by this instrument does not form part of a larger transaction or series of transactions in respect of which the amount or value, or aggregate amount or value, of the consideration exceeds £1,000.'

Example 8.1 – consideration below £1,000

Ian transfers to Joan the assets of his business trading under the name Tidy Gardens and comprising 100 shares in Tidy Gardens (Mudfield) Ltd (a company set up purely for name protection purposes), various loose tools and garden supplies, and a memory stick containing technical information such as details of plants which do well in local areas plus customer and contacts lists. Joan pays consideration of £10,000 with an agreed (and just and reasonable) allocation of £5,500 to tools and garden supplies, £4,400 to the memory stick and £100 to the shares. The tools, garden supplies and memory stick should all be regarded as 'goods, wares and merchandise' and may be ignored in assessing consideration for the overall transaction, which is therefore £100. Provided the stock transfer form transferring the shares contains the certificate of value set out above, no stamp duty will be payable on it and there will be no need to submit it to the Stamp Office.

Part of the amount allocated to the memory stick may be for goodwill and intellectual property, but the provisions which abolished stamp duty on these assets also provided for them to be left out of account in determining the overall level of consideration (*FA 2000, Sch 34, para 4*; *FA 2002, Sch 37, para 3*). However, if the assets had included, say, £1,200 of customer debts, the allocated consideration for which was agreed at £901, this would not be left out of account, even though no stamp duty would be due on the transfer of the debts themselves. The total consideration would then be £1,001 (£901 + £100) and it would not be possible to add a certificate of value. Stamp duty of 0.5% of £100, rounded up to £5, would be due in respect of the consideration correctly allocated to the share Transfer.

Sub-sales and successive transfers

8.12 Sometimes, perhaps most commonly in the course of a reorganisation, there is a need to transfer ownership of shares and securities several times in succession. In the interests of speed and efficiency, the parties may wish to rely on agreements for each transfer except the last, and then to execute a single Transfer of shares from the original owner to the final transferee. This is commonly referred to as a sub-sale. Provided the consideration paid by the final transferee is at least equal to the market value of the shares and securities, the Transfer will be liable to stamp duty only on that consideration. No stamp duty will be payable on the consideration paid by other parties under earlier agreements in the chain (*SA 1891*, *s 58(4)*).

However, if the shares and securities are within the charge to SDRT, any SDRT liability arising on those earlier agreements will remain chargeable, because no Transfer will be stamped in respect of those steps in order to cancel the SDRT (see **10.11** for further details). It will often be the case that the stamp duty charge on a transfer for some or all of the intermediate steps would be less than the SDRT charge, most commonly because a relief such as group relief (see **9.9**) could be claimed. Many reliefs apply only to stamp duty and not to SDRT. *In such a case, the sub-sale route should not be used.* Instead, a Transfer should be executed for each step in the series, and each Transfer should then be stamped.

8.13 Execution of a stock transfer form for each step may not be convenient, perhaps because it is considered that each such form should be stamped and registered before the next is executed. In that case, it may be simpler for the parties to execute a declaration of trust to act as the Transfer for each step, and to submit all of the declarations for stamping together. In this regard, see **Example 9.3**.

SCOPE – ASSETS

8.14 The legislative definitions of 'stock' and 'marketable security' are in *SA 1891, s 122.* They are not wholly enlightening, and the meaning of these terms is partly determined by judicial precedents, some of which are similarly ancient and difficult to interpret. The precise definition is not usually too important in practice, because most debt instruments which may fall within the definition are exempted under the 'loan capital' exemption (see next paragraph).

'Stock' includes shares (equity) of companies (UK and overseas) but also other 'instruments of a capital nature' issued by UK and overseas national and local government and by companies and societies. This includes many debentures and may include instruments issued by incorporated partnerships such as UK LLPs. In many cases, such stock will also be a security.

The term 'security' encompasses a wide range of instruments evidencing indebtedness. 'Marketable' is defined by statute as meaning capable of being sold on any stock market in the UK. This is considered to exclude securities issued by UK private companies, partnerships and individuals and many overseas entities, which are therefore automatically outside the scope of stamp duty unless they also fall within the definition of stock.

Exempt loan capital

8.15 Although many debt instruments issued by companies, governments and other organisations are within the definition of 'stock and marketable securities', they are mostly removed from the scope of stamp duty by the loan capital exemption in *FA 1986, ss 78–79*. For these purposes, loan capital includes:

- all government stock and marketable securities,
- debt instruments or any other borrowed capital of companies,
- capital raised as 'alternative finance investment bonds' within *FA 2005, s 48A*, and
- loan capital raised by certain international organisations.

The last of these is completely exempt from stamp duty on transfer (*FA 1986, s 79(3)*).

8.16 All other loan capital is exempt from stamp duty on transfer (*FA 1986, s 79(4)*) unless it has equity-like characteristics (*FA 1986, s 79(5)* and *(6)*), such as rights to:

(1) conversion into shares or other securities,

(2) receipt of further or other shares or securities,

(3) interest which is dependent on the results of a business or the value of other assets,

(4) interest at a rate greater than a reasonable commercial return, or

(5) repayment at a premium greater than that generally available on similar quoted securities.

Point (3) is disapplied in relation to wholly commercial 'capital market' loans (so-called 'ratchet loans'), where the interest rate increases if business results worsen or decreases if they improve, or where interest reduces if the borrower does not have funds to pay (*FA 1986, s 79(7A)*).

Points (3) and (5) are disapplied if the interest and/or amount repayable vary in accordance with the UK Retail Price Index or Consumer Price Index (*FA 1986, s 79(7)*).

Point (4) has led to dispute with HMRC in relation to so-called 'junk bonds', which carry a high interest rate to reflect greater risk; but HMRC have generally accepted that these are at a commercial rate where comparable publicly issued bonds have been identified.

Point (4) is disapplied in relation to alternative finance investment bonds (often referred to as Islamic bonds), and point (5) is modified to require that the total amount repaid should not exceed the amount originally subscribed plus a reasonable commercial return (*FA 1986,s 79(8A)*).

The risk of loan securities failing to qualify as exempt loan capital is considered in more detail at **11.13**. Care is needed when dealing with debt instruments which are not plain vanilla loans at interest.

Other financial assets outside the scope

8.17 Some debt instruments may not be loan capital, perhaps because they are too short term to be regarded as capital, or perhaps because they arise other than through the borrowing of money. Such instruments are unlikely to be stock and, in many cases, they will not be marketable and so will automatically fall outside the scope of stamp duty. Any which are marketable will often be exempt by virtue of *FA 1999, Sch 13, para 25(a)* because they are debentures. Overall, it is highly unlikely that a straightforward debt instrument bearing a commercial rate of interest will be within the stamp duty charge.

8.18 Transfers to recognised investment exchanges, clearing houses etc, are exempt, if the exchange or clearing house has entered into an appropriate arrangement with HMRC. This exemption was introduced specifically to facilitate reforms to the securities markets, and is not considered further in this book.

8.19 *FA 2014, s 115* and *Sch 24, Pt 2* introduced a new exemption for shares traded on recognised growth markets but not listed on any market. The exemption removes all stamp duty charges in relation to the shares, apart from those applying to bearer securities (for which see **8.55**) for transactions entered into or transfers executed on or after 28 April 2014. For more information see **10.32**).

SCOPE – GEOGRAPHICAL

8.20 In principle there is no territorial limit to the application of the tax. However, the lack of direct enforcement provisions means that this can be ignored for completely non-UK transactions. However, in theory at least, any document with a sufficient connection with the UK can be subject to stamp duty. In this regard, the scope is effectively defined by *SA 1891, s 14* as covering any document executed in the UK, or any document transferring any property located in the UK, or any document of transfer relating to any matter or thing to be done in the UK. For these purposes, shares of UK incorporated companies and shares of foreign companies which are recorded on a share register located in the UK are regarded as located in the UK.

8.21 As a general principle, a document is executed where the last action occurs to make it legally valid/effective – usually addition of a signature but, in the case of documents signed in escrow, it may be satisfaction of the escrow conditions followed by dating and release on behalf of the signatory.

8.22 The idea of any matter or thing to be done in the UK is very wide, and might include the payment of sale proceeds out of a UK bank account, for example (see *Faber v IRC* [1936] 1 All ER 617; *IRC v Maple & Co (Paris) Ltd* [1908] AC 22). So, in theory, a document transferring German shares from a French person to a Dutch person can be subject to UK stamp duty if signed in the UK, or even possibly if signed in Italy but paid for out of a UK bank account. Given the lack of enforceability, in most such cases stamp duty can safely be ignored. However, the existence of a theoretical but unenforceable liability can be difficult to explain to clients, and administrative difficulty can arise in multinational transactions involving both UK and overseas share transfers. HMRC do not appear to understand the point clearly – for example, the author has experience of the Stamp Office trying to reject a composite claim for group relief which included transfers of overseas shares because 'stamp duty is not payable on overseas shares'. This is, therefore, a piece of legislative nonsense well past its sensible abolition date.

CALCULATING THE CHARGE

Rates

8.23 For Transfers of stock and marketable securities, the normal rate of stamp duty is 0.5% of the amount or value of the consideration for the Transfer (*FA 1999, Sch 13, para 3*). A higher rate of 1.5% applies to certain transactions (see **8.54**). If application of the rate does not produce a precise multiple of £5, the charge is rounded up to the next £5 above (*FA 1999, s 112*). Restricting charges to multiples of £5 allows the stamping machines to be simpler than if they had to deal with lower denominations of duty; it also provides a small but useful increase in the tax take. The exemption for small transactions (consideration not exceeding £1,000) is mentioned at **8.8**, but apart from that, there is no 'zero rate band' and no variation in rate with consideration.

8.24 In principle, Transfers of interests in partnerships owning stock and marketable securities are subject to stamp duty at rates of up to 4%, depending on the amount of consideration. However, in practice, it is unlikely that such high rates would apply – see the explanation at **8.57**.

Consideration

8.25 In general, stamp duty is charged on the actual consideration given for a Transfer, irrespective of the value of the asset transferred. This principle is modified in some cases where the transfer is in satisfaction of a debt (see **8.33**). Only cash, securities, debt and the assumption, satisfaction or release of a pecuniary liability count as consideration. This is HMRC's long-established interpretation of words now in *FA 1999, Sch 13, para 2*: 'the amount or value of the consideration for the sale'; see STM1.11 for example. This contrasts with both SDRT and SDLT, where tax is chargeable by reference to any consideration 'in money or money's worth'. However, if relying on the restricted definition of consideration to escape or minimise stamp duty, care is needed in the wording of agreements, as shown in the examples below.

Example 8.2 – consideration which does not count

Pauline agrees to transfer her shares in Big plc, worth £1 million, to Robert in return for his collection of rare gemstones of similar value. The gemstones are not 'chargeable consideration' for stamp duty purposes. A stock transfer form may be completed showing 'nil' consideration. No stamp duty arises, certificate 2 (see **8.7**) on the reverse of the stock transfer form should be completed as there is consideration but it is not consideration chargeable to stamp duty, and there is no need for the document to be submitted to the Stamp Office for stamping.

Example 8.3 – satisfaction of a debt

Meanwhile, in an unrelated transaction, Pauline agrees to buy Sara's 100% shareholding in Tacky Ltd, an online jewellery retailer, for £1 on condition that Pauline also pays off Tacky's bank overdraft of £500,000 (thus releasing Sara from the guarantee she provided to the bank). The consideration for the Transfer of shares in Tacky is £500,001, and stamp duty at 0.5% rounded to the next £5 above is £2,505.

Example 8.4 – specified monetary value

The facts are as in Example 8.2 above, except that the agreement specifies that the consideration for the share transfer is £1 million, to be satisfied by transfer of the collection of gemstones. Although it may be subject to dispute, HMRC are likely to consider that stamp duty is payable on consideration of £1 million. This is on the grounds that the agreement creates a debt of £1 million (so the Transfer is for a debt) which is then satisfied by transfer of the gemstones.

When is a debt assumed?

8.26 The mere fact that a target company has outstanding debts does not mean that these debts are added to the consideration paid for the company. In Example 8.3 above, Pauline's agreement to pay off a debt counts as consideration. However, the debt is owed by Tacky Ltd, and Sara requires it to be paid off only because she needs to be released from her guarantee to the bank. Were it not for this, she might be happy to sell the shares leaving the debt in place, in which case the consideration would be £1 and it would be possible to certify the Transfer as a small transaction with no stamp duty liability.

8.27 Alternatively, if the debt must be paid off, it may be that a similar economic effect could be achieved without stamp duty cost. For example, Pauline could subscribe for further shares in Tacky Ltd, allowing Tacky Ltd itself to pay off the debt. No stamp duty should arise on the subscription. Other alternatives are less clear-cut – for example, Pauline might agree to ensure that Tacky Ltd pays off the debt. It may be open to dispute whether this amounts to the giving of consideration. Great care is needed in the wording of documents in such cases, and further specialist advice should be sought.

8.28 Shares are sometimes issued 'part paid' – that is, only part of the capital which the shares represent has been paid, and further amounts may become payable to the company at a later date. The liability to make such further payment falls on the shareholder who owns the shares when payment becomes due. When part-paid shares are transferred, provided payment of the further capital has not been called for (or if it has, it has been paid) at the date of the Transfer, stamp duty is chargeable only on the actual consideration given (STSM021210).

Distributions

8.29 In general, a distribution from a company to its shareholders is, by definition, for no consideration. The Stamp Office has occasionally claimed that a distribution of assets *in specie*, declared as a final dividend, should necessarily be re-characterised as the declaration of a dividend (creating a money debt) followed by the (stampable) transfer of the asset in satisfaction of the debt. It would doubtless be possible to create and satisfy a debt in this way. However, the idea that a distribution *in specie* should *necessarily* take this form appears to come from an over-enthusiastic application of an idea set out in STM4.404–4.408 of the old Stamp Duty Manual and is probably not correct; STSM021130 is a more accurate statement of the position. To minimise the risk of dispute, it is worthwhile ensuring, where possible, that distributions *in specie* from UK companies are declared as interim dividends, as these do not create a debt between company and shareholder. Distributions from non-UK companies will be subject to the law of the country in which the distributing company is established – it would be worth checking whether declaration of the *in specie* distribution creates a money debt due to the shareholder under the governing law of the company concerned.

Valuation of consideration

8.30 Consideration expressed in a foreign currency must be translated into sterling. If the Transfer specifies the exchange rate, that rate is to be used – although HMRC would seek to set aside an unrealistic exchange rate inserted to reduce the stamp duty. Otherwise, the translation is to be at the appropriate market rate on the day the Transfer is executed; and, for this purpose, HMRC normally seek to use the closing spot exchange rate as listed in the Financial Times (STSM02050).

Stock and securities as consideration

8.31 If consideration includes stock or marketable securities, these are to be valued at market value on the date of execution of the stampable document (which will not necessarily be the same as the date on which the securities given in consideration are transferred) (*SA 1891*, *s 55(1)*).

For quoted securities, HMRC adopt the capital gains tax basis of valuation (STSM021060) – that is, the lower of:

- midway between the highest and lowest price of bargains on the day, and
- the lower closing price for the day plus one quarter of the difference between the lower and higher closing prices.

If securities given in consideration are marketable but not quoted, HMRC's share valuation division is likely to become involved in agreeing the value.

8.32 Where the consideration includes securities which are not marketable, their value is taken to be 'the amount due on the day of the date thereof for principal and interest' (*SA 1891, s 55(2)*). This curious phrase is taken to mean the stated principal plus any interest due but unpaid at the date of execution of the stampable document.

Transfer in satisfaction of debt

8.33 Where assets are transferred to a person in satisfaction of a debt, the consideration is taken to be the amount outstanding on the debt, including any interest due but unpaid at the date of Transfer (*SA 1891, s 57*). However, provided the Transfer is to the creditor, if the amount of debt is greater than the value of the assets transferred, the consideration is restricted to the value of the assets transferred (*FA 1980, s 102*; STSM021080). To take advantage of this relief, the Transfer must be adjudicated by HMRC (see **8.47**).

Consideration payable later or in instalments

8.34 If consideration is payable after the Transfer is executed, or in instalments, stamp duty is chargeable on the total amount, with no discount or postponement to allow for delay in payment. However, where the instalments take the form of an annuity which may last longer than 12 years but ceases on a death, only the first 12 years' instalments are taken into account (*SA 1891, s 56(3)*).

In other cases where the instalments will or may continue for more than 20 years, only the first 20 years are taken into account (*SA 1891, s 56(2)*). The equivalent provision under SDLT only takes account of 12 years' instalments although, if these are variable, the highest 12 individual years are taken into account (see **2.28**).

VAT

8.35 Where consideration for a Transfer is subject to VAT, the VAT must be included in the amount on which stamp duty is chargeable (see *Glenrothes Development Corporation v IRC* [1994] STC 74). This will not normally be an issue in relation to Transfers of shares and securities, which do not normally give rise to VAT charges. It was widely regarded as unreasonable when stamp duty applied to Transfers of other assets, but the same principle has been carried into the SDLT regime (see **2.23**).

Unascertainable consideration

8.36 Stamp duty is only chargeable on consideration which is 'ascertainable' at the date of the Transfer. This principle is not overtly stated in the legislation, but see STSM021100 and 021170 for confirmation of the HMRC view. Consideration is ascertainable if it is possible to deduce, with certainty, one or more specific figures for it from documents or facts extant at the time of execution of the Transfer. It does not matter whether payment of that particular amount is certain or contingent on future events. If more than one figure can be deduced, the consideration is assumed to be the highest of these, in accordance with the 'contingency principle' explained below. Where any amount of consideration is genuinely unascertainable at that date, it will not be subject to stamp duty. Note that the test is whether the amount is capable of being ascertained, not whether it has been ascertained.

Example 8.5 – unascertainable consideration

On 1 January 2016, Conglomerate plc agrees to buy all of the shares in Dan Ltd from Daniel for consideration equal to twice the pre-tax profits of Dan Ltd for the year to 28 February 2016 (with no upper or lower limit). The share Transfer is executed on 1 February 2016. The consideration is entirely unascertainable at the date of the Transfer, because it is based on events which will happen after that date, so there is no consideration on which stamp duty may be charged. There is no requirement to submit the Transfer to HMRC for stamping.

Example 8.6 – an ascertainable sum

The facts are as in Example 8.5, except that the consideration is specified to be at least £1 million. This is an ascertainable amount, and stamp duty will be chargeable on £1 million.

Example 8.7 – an ascertainable sum, variable up or down

The facts are as in Example 8.5, except that the consideration is specified to be £1 million, plus twice the amount by which the profits for the year to 28 February 2016 exceed £500,000, or minus twice the amount by which they fall short of that figure. Stamp duty will again be charged on the £1 million, with no subsequent adjustment to the stamp duty if the consideration proves to be greater or less than £1 million.

Example 8.8 – ascertainable but not yet ascertained

The facts are as in Example 8.5, except that the share transfer is executed on 1 March 2016. The year has now ended, on the profits of which the consideration will be based. The profits are not yet likely to have been ascertained, but they are nonetheless ascertainable, because all events impacting on the profit figure have happened. Stamp duty will be chargeable on the actual consideration.

Since it is unlikely this will be known when the deadline for payment of stamp duty arrives, Conglomerate plc may wish to apply for provisional stamping (see **8.46**).

8.37 The *LM Tenancies* case ([1996] STC 880) involved an attempt to render consideration unascertainable but limit the risk of change, by linking it to the value, a few days later, of a security with a very stable price. It was held that the consideration was not truly unascertainable and should be taken as the figure based on the value of the security at the date of the Transfer.

Contingency principle

8.38 The calculation of consideration is complicated by the application of the so-called 'contingency principle'. This principle, not explicit in the legislation, is based on case law, especially the *Underground Electric Railways* cases (STSM021120). The principle states that any sum stated to be payable by way of consideration on the happening of some contingency is subject to stamp duty as if certainly payable. The principle is subject to the question of whether the consideration is ascertainable, but generally means that the highest possible stamp duty is payable.

Example 8.9 – two or more ascertainable sums

The facts are as in Example 8.5, but the consideration is subject to a minimum of £500,000 and a maximum of £2 million. £500,000 and £2 million are both ascertainable sums, and the contingency principle requires the assumption that the higher amount of £2 million will be paid. Stamp duty is therefore chargeable on this sum, with no adjustment or refund if the actual consideration eventually proves to be less.

8.39 On occasion, when this principle has produced an unforeseen and manifestly unjust result, HMRC have been known to accept stamp duty based on actual consideration – but it would not be safe to rely on such largesse.

Different definitions of consideration for stamp duty and SDRT

8.40 The definition of consideration for SDRT purposes (see **10.4**) is wider than that for stamp duty purposes. As a result, the stamp duty on a Transfer may be less than the SDRT on the related agreement. In Example 8.6, SDRT would be chargeable on the actual consideration once finally ascertained. However, as explained at **10.11**, appropriate stamping of a Transfer within six years of the agreement to transfer the shares cancels the SDRT liability. Since, in that example, the stamp duty on a Transfer would be nil, it would clearly be a good idea to execute a Transfer (which, in this case, would immediately be deemed to be 'duly stamped'), in order to cancel the SDRT liability. This principle is widely relied on in order to mitigate stamp tax costs on share transactions (see Example 8.11 and **10.21**).

ADMINISTRATION

No need to stamp

8.41 If an exemption (as opposed to a relief – **see 9.2**) applies (for example the loan capital exemption in *FA 1986, s 79(4)*) to eliminate any stamp duty charge, the Transfer is effectively deemed to be duly stamped, without any need to submit it to HMRC. As noted at **9.2**, it is possible for the availability of the exemption to be challenged, in which case the document may need to be submitted for adjudication, although in practice this rarely happens.

Routine stamping

8.42 Documents for routine stamping should be sent to the Birmingham Stamp Office (see **Appendix A**) with payment of the duty and, if necessary, a brief note setting out the calculation of the duty. Most importantly, to guard against the stamped document going astray, an envelope should be included, addressed to the person to whom the stamped document must be returned. Long advocated by advisers, HMRC finally made this required practice in June 2014. Until 7 April 2014 there was also a stamping facility at the Edinburgh stamp office and it was possible to present documents there for stamping, on a 'same day' basis when essential. This stamping facility was withdrawn on that date – Birmingham is now the only location for stamping documents.

8.43 In accordance with *SA 1891, s 5*, all facts and circumstances affecting liability must be set out in the Transfer. In practice, important facts may be in a separate agreement rather than on the stock transfer form. Where relevant, a copy of any such agreement should be sent with the stampable document. In the past, when consideration for a transfer included issue of new shares, it was sometimes necessary to have the agreement itself stamped 'not liable to any duty' in order to satisfy *Companies Acts* requirements; those requirements have now changed, so that this is no longer necessary.

8.44 The deadline is 30 days after the date of execution of the transfer, and late submission or payment will lead to penalties and/or interest being charged (see **8.49**). Payment may be by cheque, or by bank transfer to the HMRC stamp duty account (see **Appendix A**). If bank transfer is used, an appropriate reference must be given to the destination bank and repeated in the papers sent to Birmingham – for example, the name of the transferee. Unless there is manifestly an error, the documents will normally be stamped on the basis requested.

8.45 In rare circumstances it may be important to have a transfer stamped on the day of execution, perhaps to enable registration of the transfer prior to a further transaction. Same day stamping is possible, but it is essential that this be agreed with the Birmingham Stamp Office in advance. If same day stamping is required you should write to the Birmingham Stamp Office ('same day' stamping service) giving as much detail as possible including (a) a description of the transaction, (b) the number of instruments to be stamped, (c) the specific reason(s) for the request, and (d) the amount of the stamp duty payable. In exceptional circumstances a request for 'same day' stamping can be made by contacting the stamp taxes helpline on 0300 200 3510. It will be necessary to convince the stamp office staff that same day stamping is really necessary. HMRC make it clear that same day stamping will only be available when there is a business critical requirement to have a document stamped and that the service cannot be used if the urgency could have been avoided by either party or their respective agents. Same day stamping is not available if there is a requirement that the instrument be adjudicated (see **8.47**) either because a relief is being claimed (eg group relief) or for some other reason. HMRC expects the number of occasions on which same day stamping will be required to be minimal.

Provisional stamping and 'wait and see'

8.46 Where the consideration for a transfer is ascertainable at the date of the Transfer, but not yet ascertained when the due date arrives (see Example 8.8), delay in submission of the Transfer for stamping will lead to interest and penalty charges. It will also prevent any change of shareholder from being registered (see **8.4**). Penalties may be avoided by lodging the Transfer with the Stamp Office, either for them to keep until the duty can be determined ('wait and see') or for provisional stamping. Interest will still be charged on late paid stamp duty, but this may be minimised by paying an estimated amount, to be adjusted later. If the taxpayer needs the stamped Transfer, or if the delay in finalising figures will be more than a few weeks, provisional stamping should be requested. The Stamp Office will require a formal undertaking *from the person submitting the document*, confirming that they will return it for adjustment of the duty once figures are finalised. For this reason, advisers may ask clients to submit the documents in such cases, so that the client may give the necessary undertaking.

Adjudication

8.47 Technically, routine stamping is always provisional – that is, if the document is then required for an official purpose such as evidence in court, it would be open to any party to challenge whether it was adequately stamped. In practice, most documents are stamped in this way and no problems arise. To obtain complete certainty, the document must be submitted for HMRC to express their formal opinion as to the amount of duty (and, where appropriate, penalties) due – normally referred to as adjudication. Once the amount of duty has been adjudicated (and assuming no appeal is lodged against this within 30 days), it is no longer open to challenge apart from, possibly, on grounds of fraud.

8.48 There are certain cases where adjudication is compulsory, particularly in relation to claiming reliefs. It is rare for parties to request adjudication when it is not compulsory, but it is possible for all documents. HMRC takes adjudication seriously because of its finality. They will normally need to see all relevant documents and may seek further explanations; rarely, they may request assurances as to motives etc by way of formal statements (Statutory Declarations) from individuals involved in the transaction. Adjudication is evidenced by impressing an 'adjudicated' stamp on the document alongside the duty stamps. As noted at **8.4**, there is no general compulsion to pay stamp duty. However, once the duty on a document has been adjudicated, if it is not paid within 30 days, a further penalty may be added to the duty when eventually paid.

Interest and penalties

8.49 Although there is no general compulsion to pay stamp duty, payment later than 30 days after the date of execution of the Transfer will lead to the addition of interest to the amount of stamp duty (*SA 1891, s 15A*). Interest is rounded down to the nearest £5 and, if the total is less than £25, it is not charged. Apart from this, however, interest charges are never mitigated, and the document will not be stamped unless any interest is paid in addition to the stamp duty itself. As with other taxes, the rates are set by regulation in accordance with *FA 1989, s 178*.

8.50 Penalties may be charged if a document is submitted for stamping after the deadline (*SA 1891, s 15B*). For a document executed in the UK, the deadline is the same as that for payment of duty: 30 days from execution. If the document is executed outside the UK, penalties do not arise unless it is submitted for stamping more than 30 days after first being brought into the UK. In practice, it will be rare for a Transfer of UK shares and securities to be executed and retained offshore, because of the SDRT implications (see **10.11**).

8.51 Provided the Transfer is submitted for stamping within one year of the deadline, the maximum penalty is 10% of the duty payable but capped at £300.

If a Transfer is submitted for stamping more than one year but no more than two years after the deadline, the penalty becomes 20% of the duty payable, and for delays of more than two years the penalty is 30% of the duty payable. It follows that where a relief is successfully claimed, reducing the stamp duty to zero, no penalty will normally be charged. However, see **8.52** below.

8.52 Where the delay is more than one year the penalty may be higher if there is evidence that the failure to submit the document for stamping was deliberate. The more serious the reason for the delay, the greater the penalty is (STSM15020). A table showing the maximum and minimum penalty percentages, dependent upon the type of failure, is shown at STSM15020. It should be noted that if there is a reasonable excuse for the delay no penalty is payable (*SA 1891, s 15B(5)*). Examples of acceptable and unacceptable reasons for a delay in stamping a document are set out at STSM015040.

Repayment of stamp duty

8.53 Occasionally, it may be discovered that too much stamp duty has been paid. This may occur, for example, where assets are subsequently found to be exempt, where an available relief has not been claimed, where consideration was incorrectly allocated to chargeable assets, or where a Transfer was inadvertently duplicated and both were stamped. The Stamp Office is empowered to consider claims for repayment (*Stamp Duties Management Act 1891, s 10*). The original stamped document must be submitted to Birmingham Stamp Office (see **Appendix A**) with a letter claiming repayment and setting out an explanation of the circumstances. In most cases, the claim must be made within two years of the date of execution of the document in question. However, where the claim relates to repayment of duty overpaid in the course of provisional stamping (see **8.46**) the two-year time limit does not apply. Interest may be payable in accordance with regulations under *FA 1999, s 110* and *FA 1989, s 178* on stamp duty repaid and, at the time of writing (June 2016), the official interest rate for these purposes is 0.5%.

Higher rates on stock and marketable securities

8.54 Where stock or marketable securities are transferred to a clearance system or depository, any stamp duty charge is at the higher rate of 1.5% (*FA 1986, ss 67–72A*). In practice, such transfers are usually carried out electronically with no transfer document, so that the relevant tax is SDRT rather than stamp duty. The 1.5% higher rate charge to stamp duty does not apply to the issue of new shares into a clearance system or depository since stamp duty generally only applies to documents transferring existing shares. The relevant tax in relation to the issue of new shares into a clearance system or depository is SDRT (see **10.27**).

Bearer instruments

8.55 A 1.5% charge also arises on the *issue* of certain bearer instruments (*FA 1999, Sch 15, para 1(1)*), but subject to certain exemptions. In contrast with a registered security, a bearer instrument is a security or other financial asset which is transferable by delivery – that is, there is no need for a transfer document, but transfer is effected instead by giving the document which constitutes the asset to the transferee (*FA 1999, Sch 15, para 3*). Where stamp duty is due, it is charged on the market value of the security and is payable by the issuer (*FA 1999, Sch 15, paras 4* and *22*). This is the one situation in which disclosure and payment of the stamp duty liability are compulsory. It should however be noted that, with effect from 26 May 2015, a UK incorporated company has been prohibited from issuing new bearer shares. In addition, from 26 February 2016 all existing bearer shares issued by a UK incorporated company had to have been converted into registered shares or cancelled (the *Small Business, Employment and Enterprise Act 2015*). These changes are part of the UK Government's promotion of corporate transparency as the ownership of unregistered shares can be hard to determine.

8.56 If the stock, in registered form, would be exempt from stamp duty on transfer, no stamp duty is due on issue in bearer form (*FA 1999, Sch 15, para 14*). There are various other exemptions especially for stock expressed in a foreign currency (*FA 1999, Sch 15, para 13*). Issuers of bearer instruments will require specialist advice on compliance with stamp duty obligations. As far as holders of bearer instruments are concerned, a key point to note is that it is *possible* to transfer a bearer instrument by means of a transfer document rather than by delivery of the security itself, but if this is done, a stamp duty charge may arise if stamp duty was not charged on the issue of the bearer instrument (*FA 1999, Sch 15, para 2*).

Partnerships

8.57 With the introduction of SDLT, stamp duty charges were removed from most assets apart from stock and marketable securities. However, stamp duty remains chargeable on a Transfer of an interest in a partnership where the assets of the partnership include any stock or marketable securities (*FA 2003, Sch 15, paras 31–33*). The stamp duty charge is limited to the smaller of:

- the charge at 0.5% which would arise on a transfer of a proportion of the stock or marketable securities for market value consideration, however the consideration is reduced by the proportion of any loan secured solely on the stock or marketable securities; the proportion is the same as the fractional share in the partnership which is transferred; and
- the charge at rates of up to 4% on the consideration actually given for the transfer of the partnership share; for these purposes, the consideration

is reduced by a proportion of the 'net market value' of any UK land and buildings held by the partnership, and that proportion is the same as the fractional share in the partnership which is transferred. The 'net market value' is the actual market value less any debt secured solely on the property. The logic of this is that SDLT will often be charged by reference to the value of the UK land and buildings (see **5.25**). (When this provision was first enacted, it was expected that SDLT would always be charged. The SDLT provisions have subsequently been amended so that no charge will arise on many transfers of interests in property-owning partnerships (**see 5.29** et seq), but the stamp duty provisions have been left unchanged. This possibly indicates that HMRC understand the limited effect of the retention of stamp duty on such transfers.)

HMRC provide a couple of examples of calculation of stamp duty on partnership transfers at STSM091060 and SDLTM34610.

8.58 In practice, it appears unlikely that stamp duty will be paid on many transfers of partnership interests. Most partnerships do not have substantial holdings of securities liable to stamp duty; many transfers will be left unstamped because, unless there is a dispute between the parties, it seems unlikely that a stamped transfer document will be required; the net consideration given is often very low; where necessary, it may be possible to construct a transfer of a partnership interest without creating a Transfer document. Where a Transfer is created, it should be noted that any attempt by the partners to agree contractually that they will not challenge the validity of the Transfer on the grounds it is not stamped is ineffective because of *SA 1891, s 117*. Where a stamped transfer document is required, it must be submitted for adjudication.

Stamp duty mitigation

8.59 There are three legitimate approaches to minimising the stamp duty cost when assets within the scope of the tax are transferred:

(1) Avoid creating a document which requires stamping. This approach was widely used when stamp duty applied to business assets such as goodwill, but is unlikely to be applicable in relation to transfers of stock and marketable securities because of the need to execute and stamp a document to cancel any SDRT charge (see **10.11**).

(2) Ensure the transfer is within the terms of a relief or exemption and claim the relief where appropriate – see **Chapter 9**. Where a relief or exemption may be available, care is needed to ensure the transaction qualifies; reliefs and exemptions are often subject to precise and detailed conditions, and it is often too easy accidentally to fall outside these.

(3) Minimise the amount which will be taken into account as consideration. Stamp duty is charged on the 'amount or value of the (actual)

consideration'. There is no general requirement to substitute the market value of the assets transferred, even where the transaction is between connected persons and at an undervalue.

Example 8.10 – actual consideration below market value

Mustard plc is in the process of selling subsidiary Cress Ltd. Cress Ltd owns 50 shares in UK company Salad Ltd, with an historic cost of £10,000 but a current market value of £1 million. M always intended to keep the shares in S, but the process to sell C has now advanced far enough to create genuine doubt over whether M still has beneficial ownership of C. Therefore, it is not certain that group relief will be available for the transfer of S shares to M at full value. However, if C is able to transfer the S shares to M for payment of historic cost only, the stamp duty will be limited to £10,000 at 0.5%, which is £50, instead of the £5,000 which would apply to a transfer at market value.

Example 8.11 – minimising consideration

Non-UK company Cabbage SA owns UK company Sprout Ltd (current value £9 million) and non-UK company Broccoli Sarl (current value £1 million). Cabbage has agreed to sell both companies to Kale Pty Ltd for £10 million. A direct transfer of Sprout for £9 million would cost Kale £45,000 stamp duty. However, Cabbage first transfers Sprout to Broccoli in return for the issue of two shares by Broccoli. Broccoli already has 1,000 shares in issue, all held by Cabbage. Further (and this is very important), under the company law and constitution governing Broccoli, all shares are identical, giving shareholders identical rights in all respects. Cabbage then transfers Broccoli to Kale for £10 million.

For stamp duty purposes, the consideration given by Broccoli is simply the value of the two shares issued. This will be 2/1002ths of the £10 million value of Broccoli after the transfer of the Sprout shares, or £19,960. Most of the value of Sprout flows into the Broccoli shares already held by Cabbage, but this value shift does not count as consideration for stamp duty purposes. The stamp duty on this transfer will be £100 (after rounding). The SDRT, on the other hand, would probably be £45,000 based on the increase in the value of Cabbage's holding of shares in Broccoli, ie £9 million. It is therefore essential that the Transfer of Sprout shares to Broccoli is submitted for stamping, in order to cancel the SDRT charge which otherwise arises. It is unlikely to be necessary to pay stamp duty on the Transfer of Broccoli to Kale because the shares of Broccoli are not UK registered.

Chapter 9

Stamp Duty Reliefs

INTRODUCTION

9.1 Perhaps the best-known stamp duty relief is that for transfers between group companies, commonly referred to as '*section 42* relief' because it is governed by *FA 1930, s 42*. This is modified by *FA 1967, s 27* and subject to HMRC guidance given in SP 3/98. The other main legislation on stamp duty reliefs and exemptions is in *FA 1986, ss 75–85*, but there are also relevant provisions at various points in other *Finance Acts*. References are given in the text below. Guidance on reliefs may be found at STSM040000 et seq in the HMRC Stamp Taxes on Shares Manual, available at www.hmrc.gov.uk/so/manuals/stsmanual/index.htm. This new interactive manual was published on 23 July 2014 replacing the old Stamp Duty Manual.

9.2 To be consistent with the chapters on SDLT and SDRT, this book uses the term 'exemption' to refer to a measure which applies automatically to remove or reduce a stamp duty charge, and 'relief' to refer to a measure which requires a claim to be made. This is, in some ways, a false distinction. Even where a Transfer is considered to be exempt from stamp duty, it is possible for someone to challenge this. (In this book, the expression 'Transfer' is capitalised to make it clear that this is a reference to the potentially stampable document.) The only way to prove that the exemption applies is to ask HMRC to adjudicate the liability, as explained at **8.47**; this effectively amounts to making a claim for the exemption. However, it is not common for an apparently exempt document to be challenged in this way; so, in practice, it is rarely necessary to have the availability of an exemption settled by adjudication.

In contrast, where a relief applies, the relevant document is usually specifically stated not to be properly stamped unless it has been subject to adjudication (eg *FA 1930, s 42(1)*). If the document is not properly stamped, any SDRT liability which has arisen on the underlying agreement is not cancelled (see **10.11**) – so proper stamping is essential.

9.3 There is generally no theoretical time limit on claiming a relief. However, if stamp duty is paid on a document and it is subsequently realised that a claim to relief could have been made, the claim must be lodged within two years of the date of execution of the document, or the stamp duty will not be repaid (*Stamp Duties Management Act 1891*, *ss 9–10*; STM7.3 et seq). Where an SDRT charge has arisen, the Transfer must be executed and duly stamped within six years of the effective date of the agreement which gave rise to the SDRT charge, if that charge is to be cancelled (see **10.11**).

This has sometimes been a problem when subsidiaries have been moved around within a group on an informal basis. A typical scenario is as follows:

- parent company decides that ownership of a company should be transferred within the group, and instructs the relevant subsidiaries to take the necessary actions;
- the subsidiaries pass the necessary board resolutions, perhaps with consideration for the transfer being left outstanding on inter-company account;
- the resolutions, taken together, amount to an agreement to transfer the shares, so an SDRT charge arises at that point;
- however, no formal share Transfer is ever processed.

After six years have passed, the SDRT charge cannot be cancelled. HMRC are aware of the issue, and have made it clear that, where such situations come to light, they will pursue the SDRT charge.

9.4 Reliefs and exemptions may also apply to Transfers of interests in partnerships, although the absence of any SDRT charge means that, sometimes, documents are left unstamped rather than incurring the administrative costs of making claims (see **8.4**). Most of the reliefs described here also applied to property transactions before the introduction of SDLT in 2003, and most of them have been carried forward to the SDLT legislation (but often with more stringent anti-avoidance rules).

Exemptions

9.5 Exemptions are wide in scope, and are not limited to certain types of transaction or participant. They are listed here, but described in greater detail in **Chapter 8**:

(1) transfers which are not 'on sale';

(2) transfers for consideration not exceeding £1,000, if appropriately certified;

(3) transfers of exempt loan capital and other debt;

(4) transfers of units in a unit trust (see **10.3**);

(5) transfers to recognised investment exchanges, clearing houses etc, where an appropriate arrangement has been entered into between the exchange etc and HMRC; and

(6) transfers of securities traded on recognised growth markets but not listed on any market (*FA 2014*, *Sch 24*, *Pt 2*, see **10.32**).

It must be remembered that the mere fact that a security is registered overseas or issued by a non-UK entity does not make a Transfer of that security exempt from stamp duty. However, no SDRT liability will normally arise on an agreement to transfer such a security, so there would not normally be any practical need to stamp the Transfer (see **10.5** and **10.11**).

Approved share incentive plans etc

9.6 No stamp duty is chargeable on a Transfer of 'partnership shares' or 'divided shares' by the trustees to an employee under an approved share incentive plan (or an old employee share ownership plan) (*FA 2001*, *s 95*).

GROUP, RECONSTRUCTION AND ACQUISITION RELIEFS

9.7 Group, reconstruction and acquisition reliefs apply only to Transfers between companies. 'Company' is defined in the legislation for group relief as meaning body corporate, and so potentially encompasses all UK incorporated entities such as private and public companies, statutory corporations and limited liability partnerships. It also includes overseas entities, provided they have the necessary characteristics. This aspect is not defined in the legislation governing the reconstruction and acquisition reliefs, probably because those reliefs originally applied only if the acquiring company was UK registered. However, it was realised that this could cause difficulties in relation to European law, and *FA 2006*, *s 169* removed the requirement for UK registration. Nevertheless, a body corporate can only function as a subsidiary in a group if it has ordinary share capital, and can only participate in the other reliefs if it has non-redeemable share capital (STSM042200et seq).

9.8 HMRC have examined the characteristics of a range of foreign entities and expressed their view as to whether they are regarded as bodies corporate for the purposes of stamp duty group relief. A fairly comprehensive list of foreign entities which HMRC will accept as being bodies corporate for the purposes of stamp duty group relief is at STSM042260. It is important to note however that this is only a list of those foreign entities which HMRC accept as being bodies corporate and no confirmation is given that the bodies have the equivalent of 'ordinary share capital' which is the other essential requirement if an entity is to be part of a stamp duty group (**see 9.11**), unless it is the parent of that group. Consequently, even if a foreign entity is included on the list it will still be necessary, in many cases, to confirm that the entity has the equivalent of 'ordinary share capital'. If a foreign entity is not included at STSM042260, it is worth asking HMRC whether they have already considered it (see **Appendix A** for contact details). If they have not done so, they will usually require details of the law under which the entity is formed, and a legal opinion (in English, or translated) setting out why the entity should be regarded as a body corporate.

Group relief – general description

9.9 According to the legislation, this should be referred to as 'associated companies relief'; it is widely called '*section 42* relief', after *FA 1930, s 42* which sets out the basic rules. In this book, it is referred to as group relief. Group relief is available to remove the stamp duty charge on a Transfer which transfers a beneficial interest in an asset between two associated companies.

9.10 The reference to beneficial interest sometimes causes confusion, because it is often considered that the *beneficial* interest in an asset passes to the purchaser as a result of payment of the consideration under an agreement, while the Transfer does no more than transfer *legal title*. However, HMRC interpret the legislation as requiring only that the Transfer be a formal step in the process of transferring a beneficial interest. On this understanding, a stock transfer form or other document giving effect to an agreement to transfer an asset on sale is able to qualify for group relief. However, there is a related point which does cause difficulties. If a beneficial interest is to be transferred, the transferor must hold that beneficial interest at the start of the process and the transferee must hold it (even if only briefly) at the end. If the asset is ultimately to be transferred to a third party, care must be taken to ensure that the group companies do nothing in relation to that disposal which may lead to loss of beneficial ownership, until the intra-group Transfer has been completed. Beneficial ownership may be lost, for example, as soon as there is a binding commitment to sell the asset to the third party, even though no payment has been made. See STSM042230 for HMRC's comments on the point.

9.11 Companies are associated if one is the parent of the other, or a third company is the parent of both. A company is the parent of another if it:

(1) is beneficial owner of at least 75% of the ordinary share capital of the other,

(2) is beneficially entitled to at least 75% of any profits available for distribution to equity holders, and

(3) would be beneficially entitled to at least 75% of assets available for distribution in a winding up.

'Ordinary share capital' means all share capital apart from fixed-rate preference shares. The rules in *Corporation Tax Act 2010, Pt 5, chapter 6* (previously *ICTA 1988, Sch 18*) apply in considering (2) and (3), but omitting the paragraphs taking account of potential future changes in rights. Where ownership is indirect, the usual algebraic rule applies to determine the effective percentage ownership – see Example 3.3 (SDLT group relief) for an illustration. The requirement for *beneficial* ownership/entitlement is important. As noted below (**9.16**), an arrangement to break the group relationship by disposal of the transferor does not prevent group relief on a prior intra-group Transfer. However, if that arrangement has progressed too far, there may be a risk that the parent company has lost beneficial ownership of the transferor and that the group is therefore broken. Cases such as *Wood Preservation Ltd v Prior* ([1968] 2 All ER 849) provide guidance on this point. It is important that the necessary group relationship is intact at the time of execution of the Transfer – it is not enough for the companies to be associated only at the time of any agreement to transfer (see STSM042240).

Bodies corporate will not be associated if at the time the Transfer is executed arrangements exist under which at that time or some later time any person has or could obtain control, or any persons together have or could obtain control of the transferee but not of the transferor (*FA 1930, s 42(2)*). For these purposes 'control' is defined in *Corporation Tax Act 2010, s 1124* (*FA 1930, s 42(7)*) and means the power of a person to secure that the affairs of a company are conducted in accordance with that person's wishes through the holding of shares, possession of voting power or as a result of powers granted by the company's articles of association or other documentation regulating the company. The sort of circumstance at which this provision would seem to be aimed would be where a person has, for example, an option to subscribe for shares in the transferee and, if the option were exercised, that person would obtain 'control' of the transferee but not the transferor. In such circumstances it would seem that group relief would not be available.

There are exceptions to this rule for certain arrangements entered into by a joint venture company (*FA 1930, s 42A*) and certain arrangements involving a mortgage secured by way of shares or securities (*FA 1930, s 42B*).

Example 9.1 – group relief, associated companies

K LLP holds all of the shares in two UK companies, J Ltd and H Ltd. J Ltd also holds a portfolio of minority holdings of shares in key suppliers to K LLP. The partners of K LLP decide to transfer the shares of J Ltd into the ownership of H Ltd in return for an issue of H Ltd shares. They also decide J Ltd should transfer the shares of suppliers to H Ltd in settlement of an inter-company debt. K LLP, J Ltd and H Ltd are all associated companies for these purposes, so any Transfers between them are potentially eligible for relief. It should not matter whether the supplier shares are transferred before or after the Transfer of shares of J Ltd.

Group relief – partnerships within the structure

9.12 Because a UK LLP is a body corporate but has no share capital, it breaks the group relationship between companies above and below it in the structure even if, economically, the necessary 75%+ group relationship is there. As a result group relief cannot be claimed for transfers from (or to) a company above the LLP in the structure to (or from) the LLP itself or any company owned by the LLP. HMRC have previously been inconsistent in relation to this matter, sometimes allowing group relief, but the correct interpretation, as set out here, was confirmed in a news release in October 2010, see www.hmrc.gov.uk/so/group-rel-sdlt-sd.htm. This same news release confirmed that English limited and general partnerships are regarded as transparent, so that the partners should be regarded as owning the partnership assets when determining whether the necessary group structure exists. In contrast Scottish limited and general partnerships have legal personality and it is HMRC's view that they cannot be 'looked through' in this way; however as they are not bodies corporate, they cannot be regarded as parents of groups either. This could have a major adverse impact as it means direct 'subsidiaries' of a Scottish partnership will not be grouped with each other, either. Again this is an issue which HMRC has not always treated in a consistent manner, often granting group relief when, on this interpretation, it is not due. The different treatment for English and Scottish partnerships contrasts with the normal tax position, under which if Scots law would produce a tax result which differs from that under English law, the English law result is applied both sides of the border. It is surprising that HMRC's treatment – doubtless technically correct – is politically acceptable.

Group relief – anti-avoidance

9.13 Anti-avoidance rules are set out in *FA 1967, s 27*. Relief is denied if the Transfer is made in connection with any arrangements under which:

(1) any part of the consideration is to be provided or received by anyone other than another group company,

(2) the transferor and transferee are to cease to be associated by the transferee leaving the group, or

(3) the asset being transferred was previously conveyed by anyone other than a group company.

These rules have the potential to deny relief in a wide range of relatively innocent transactions, so HMRC have issued guidance on how they apply them (SP 3/98). The key points of this are as follows:

(1) 'Arrangements' is a very wide term, covering informal understandings as well as formal contracts; there are arrangements if there is an expectation that the event or series of events will happen and no likelihood in practice that it will not.

(2) Nevertheless, relief may be denied if it is found that there are arrangements even if, in the event, those arrangements do not come to fruition.

(3) Relief is not denied merely because the transferee obtains third party debt funding to buy the asset, but relief will be denied where the funding arrangement may lead to transfer of the economic risk and reward of ownership of the asset out of the group – for example, if the transferee borrows at an economically unsustainable level, or the lender has shareholder-type rights which could allow it to take control of the asset.

(4) The rule relating to 'previous conveyance' was intended to prevent a specific type of avoidance device which relied on the stamp duty version of sub-sale relief. This is unlikely to be of interest in relation to Transfers of shares and securities, because of the application of SDRT (see **8.12**). Relief will not be denied if the asset was previously transferred into the group from a third party under a Transfer on which stamp duty was paid.

(5) In any case of doubt, HMRC reserve the right to see relevant documents and to require formal statements of the facts.

9.14 Until March 2000, HMRC routinely required group relief claims to be backed by a Statutory Declaration of the facts by an officer of the claimant company or its parent. It is now rare for this to be required, and claims are generally made by letter (see **9.17** et seq).

9.15 The most common point of doubt is whether there are offending arrangements, particularly where there is a possibility that the transferor and transferee companies will cease to be associated after the Transfer. It is generally accepted that mere uncommitted discussions with a third party potential purchaser do not cause there to be arrangements, unless perhaps it is economically inevitable that the relevant company will be sold. However, signature of heads of terms for the sale, even though not legally obliging the vendor to proceed, is likely to bring arrangements into existence (STSM042270 and 042280; SP 3/98, para 6). To minimise risk, any intra-group Transfers should be done at the earliest possible stage, and certainly before any final decision is made as to whether the transferee company will leave the group.

9.16 The following matters often cause confusion:

(1) There is no question of stamp duty group relief being 'clawed back'. Clawback is a concept which applies only to SDLT. If stamp duty group

relief is validly and successfully claimed on a Transfer, events thereafter have no impact on the claim. However, HMRC are entitled to have regard to subsequent events when considering claims, so it is wise to submit group relief claims quickly after the intra-group Transfer.

(2) Group relief will not usually be denied merely because there are plans for the asset itself to leave the group, provided that onward transfer of the asset is subject to stamp duty as normal – but, as noted above, care is needed in relation to timing of that onward disposal (see SP 3/98, para 9).

(3) No problem is caused by plans to dispose of the transferor, because the legislation specifically refers to the ownership of the transferee, provided those plans have not progressed too far (see **9.11**).

(4) A company may repurchase its own shares from a shareholder. This repurchase may qualify for group relief, provided the companies are in the necessary group relationship, subject to the usual anti-avoidance rules (STSM042250).

Example 9.2 – group relief

Jersey plc owns 100% of Friesian Ltd, and Friesian Ltd in turn owns 100% of Galloway Ltd. It is decided to sell F but retain G. The group markets F, a preferred bidder is identified, and non-binding heads of terms are signed. Before there is any formal commitment to sell to that preferred bidder, G is transferred to J for cash. Assuming there are no other avoidance aspects, the Transfer of G should qualify for stamp duty group relief, even though there are certainly arrangements for F to leave the group. As noted at **9.10**, it is important that the group does not become committed to sell F before the group Transfer is finalised, or there may be a risk that J will have lost beneficial ownership of F, the companies thus ceasing to be associated for group relief purposes.

Example 9.3 – document essential for group relief

A multinational group headed by Dorian BV has a direct UK subsidiary Eve Ltd, which itself has a UK subsidiary Frank Ltd. It is decided that Frank Ltd should be sold. Eve Ltd agrees to transfer Frank Ltd to Dorian BV in satisfaction of a £10 million inter-company debt. In turn, Dorian BV agrees to sell Frank Ltd to a third party purchaser. Dorian BV then instructs Eve Ltd to transfer the shares of Frank Ltd directly to the purchaser.

An SDRT liability of £50,000 (£10 million at 0.5%) arises on the agreement to transfer Frank Ltd to Dorian BV. A further SDRT liability arises on the agreement to sell Frank Ltd to the third party purchaser. It is likely that a stock transfer form will be completed to transfer the shares from Eve Ltd to the third party purchaser. The purchaser will normally pay stamp duty on this, which will cancel the SDRT liability on the second step. However, no document is executed to transfer the shares from Eve Ltd to Dorian BV, so the SDRT liability on that step remains payable. If, instead, a document had been executed to transfer the shares from Eve Ltd to Dorian BV, stamp duty group relief could have been claimed on this under *FA 1930, s 42*, and a successful claim would have cancelled the SDRT charge. A separate document would then have been required to transfer the shares from Dorian BV to the third party purchaser.

A stock transfer form would not necessarily have been required for the transfer from Eve Ltd to Dorian BV; in appropriate circumstances, Eve could have executed a 'declaration of trust', declaring that it now held the shares of Frank Ltd on trust for the benefit of Dorian BV. This declaration of trust could have been submitted for stamping, without the need first to update the share register of Frank Ltd to show Dorian BV as the shareholder. In complex reorganisations, where shares of UK companies may be transferred several times in quick succession within a group, it is commonplace to effect Transfers by way of declarations of trust or similar documents, to avoid delays arising from the need to get documents stamped before share registers can be updated. As noted at **10.21**, where shares are held by a custodian, a 'letter of direction' fulfils a similar function. Appropriate legal advice should be taken where required, to establish what kind of documentation will be effective as a 'Transfer', which can then be submitted for stamping.

Group relief – claims process

9.17 Group relief is claimed by submission of documents to the Birmingham Stamp Office (see **Appendix A**) for adjudication. The outside of the envelope should be clearly marked 'Stamp duty claim under FA 1930, s 42' to ensure it is sent directly to the correct section. The following documents must be sent (STM6.166):

(1) the original, executed Transfer(s) (typically, for shares, a stock transfer form);

(2) certified* copies of the Transfer(s);

(3) certified copies* of any agreements governing the Transfer;

(4) where the parties to the transaction are indirectly associated, a family tree*;

(5) a certified* copy of the register of members of all subsidiary and intermediate companies;

(6) where the consideration was payable in cash, a copy of the latest accounts of the transferee company to demonstrate that the company had sufficient cash to make the payment;

(7) copies of any other documents (eg board minutes) which help explain the transaction and the reasons for it;

(8) a letter of claim; and

(9) a self-addressed envelope for return of the document. This is not a requirement but, in the writer's experience, reduces the risk that the stamped document might be returned to a registered office address as shown on the stock transfer form, for example.

* Items marked thus either are required to be 'certified', or may require certification, so it may be as well to have them certified at the outset. 'Certified' means either the normal certification by a solicitor or other appropriate lawyer that the document is a true copy of an original, or an equivalent statement signed by a director or the company secretary of the company in question. It is not clear why the copy of the Transfer needs to be certified, since HMRC will also have the original in front of them, but this is a long-standing requirement. If any of the companies is non-UK incorporated, it may also be necessary to provide a copy of its incorporation documents or other information, to satisfy HMRC that it is indeed a body corporate with issued ordinary share capital (see **9.11**). Again it is probably safest to have the copy incorporation documents certified.

9.18 The precise format of the letter of claim is not specified in the legislation, but HMRC do require it to make statements about particular matters. As a result, letters of claim tend to follow a fairly standard pattern. Guidance entitled 'Relief from Stamp Duty in respect of documents effecting intra-group transfers of stock or marketable securities' was published by HMRC on 11 December 2014 and can be found at www.gov.uk. The letter must explicitly state (STM6.164):

- that a claim under *FA 1930, s 42*, as amended, is made in respect of the documents (which must be listed at this point);
- details of the dates of all instruments on which relief is claimed;
- the authorised and issued share capitals of all companies in the group structure (or at least that part of it which includes the transferor, transferee and any intermediate companies up to a common parent) at the dates of all the documents. The reference to 'authorised' is probably redundant as UK companies are no longer required to have a specified level of authorised capital. The letter should also state whether any of the shares carry a right to only a fixed dividend and no other share in profits. If a company has more than one class of share in issue the letter should provide details of the profit rights attached to each class of shares and

set out the asset distribution rights attached to each class of shares in any winding up of the company;

- what shares are held by each shareholder in each subsidiary company;
- that it is intended that the existing shareholder relationships are to be maintained (or give details of any change envisaged);
- that the group relationship between the companies meets the requirements of *Corporation Tax Act 2010, s 151(4)(a), (b)* and *Pt 5, chapter 6* (previously *ICTA 1988, s 413(7), Sch 18*);
- that none of the bodies corporate is in liquidation (or give further details if they are);
- that the transferee body corporate intends to keep the beneficial ownership of the asset (or give details of any intended transfer onwards);
- the amount of the consideration and whether it was paid in cash, by issue of shares or securities, through an inter-company loan or from a third party loan; and
- that none of the documents was executed in pursuance of, or in connection with, an arrangement as described in *FA 1967, s 27(3)* or *FA 1930, s 42(2)*, as appropriate.

9.19 Although some of these statements overlap and may seem unnecessary, HMRC does require them all to be made. The letter must be signed by an appropriate person with personal knowledge that the statements are correct: a director or other officer of the ultimate parent company, or of a company which is parent of a sub-group containing all of the other companies; a shareholder of the ultimate parent if that shareholder has a right to attend board meetings which discuss policy; a lawyer (in-house or external) acting for the parent company; or an administrator/receiver appointed to the parent company under insolvency law. For a long time, HMRC also readily accepted letters signed by other advisers (eg auditors and tax advisers) acting for the parent company, but they have been known to challenge whether these are based on personal knowledge, particularly of the intentions of the parties. It may, therefore, be safest for the letter to be signed by an appropriate officer of the parent company.

9.20 In major reorganisations, there may be numerous Transfers on which group relief is to be claimed, sometimes several transfers of the same shares. It is usually acceptable to combine all of the group relief claim letters into one, provided all relevant documents are made available for each Transfer. Where there are successive Transfers of the same shares, it is important that each step is completed by an appropriate Transfer document, in order to cancel any SDRT charge which would otherwise arise. However, there are possible company law difficulties in executing successive stock transfer forms before earlier forms have been stamped and the transfers registered. A solution may be for the original owner of the shares to execute a declaration of trust in respect of each successive transfer agreement, to confirm that it now holds the shares for the benefit of the new owner. Such a declaration acts as a Transfer and can be submitted for stamping.

9.21 Another potential issue, where there are successive transfers, concerns the question of beneficial ownership. If a company is committed to transfer shares onwards, even before it has acquired them, that company may never gain beneficial ownership; yet group relief is only available where beneficial ownership is transferred. Equally, if the transferor or transferee is itself to be transferred at a later stage in the reorganisation, there may be a risk that the necessary beneficial shareholding relationship does not exist when the earlier Transfer is made. Care should be taken in the timing of Transfers and the signature of agreements to minimise such risks. However, when dealing with purely internal group reorganisations, the writer has not known HMRC challenge the availability of group relief on these grounds. A challenge is perhaps more likely if the reorganisation is to be followed by disposal of some entities out of the group.

Reconstruction and acquisition reliefs – general description, common features

9.22 This section covers two reliefs (that is, reconstruction relief, and relief for insertion of a new holding company) which remove the stamp duty charge for certain Transfers which do not involve any change in ultimate ownership but which may not qualify for group relief. There used to be a third relief (*FA 1986, s 76*), which could properly be described as acquisition relief and which limited the stamp duty rate to 0.5%, even where there was a change in economic ownership. Acquisition relief was useful when stamp duty applied to land and other assets but, because the normal stamp duty rate on transfers of stock and marketable securities is 0.5%, it became of very little practical use. The relief was abolished from 6 April 2013 (*FA 2012, Sch 38*). The equivalent relief in relation to SDLT, which remains in force, is described at **3.27**. The common feature of both remaining reliefs is that the consideration must include the issue of non-redeemable shares by the acquiring company, and either no other consideration, or only consideration from a very restricted list.

9.23 Reconstruction relief and relief for insertion of a new holding company are also subject to a general anti-avoidance condition, that the acquisition must be for bona fide commercial purposes and must not form part of a scheme or arrangement which has avoidance of stamp duty, income tax, corporation tax or capital gains tax as a main purpose. In many cases, taxpayers will seek advance clearance from HMRC that the transaction will qualify as a reconstruction for direct tax purposes. Where such clearance is given, HMRC should accept this as proving that this condition is satisfied. If no clearance is obtained, the relief may still be available, but it will be necessary to provide HMRC with further information to satisfy them on this point, as follows:

- a copy of the latest accounts of the 'target company' ('T' in the explanation below);
- full details of any scheme or arrangement of which this transaction forms part;
- confirmation that the 'acquiring company' ('A' in the explanation below) still holds the shares or undertaking of T, as appropriate; and
- a *detailed* (HMRC emphasis) note of the bona fide commercial reasons for the transaction.

Reconstruction relief

9.24 In the heading to the legislation (*FA 1986, s 75*), this is described as 'acquisition relief' but, because it applies only to schemes of reconstruction, it is widely known as 'reconstruction relief'. The relief reduces stamp duty to zero where, in pursuance of a scheme of reconstruction, a company (A) acquires all or part of the undertaking of another company (T). The consideration for the acquisition must consist of or include the issue of non-redeemable shares by A to all of the shareholders of T (*not* to T itself). The only other consideration permitted is that A assumes or discharges debts of T. The general anti-avoidance condition applies, as mentioned above.

9.25 The other stated condition is that, after the acquisition, the shareholders of A must be the same as those of T, and the proportion of shares in A held by each shareholder must be the same as the proportion of shares in T held by that shareholder. Minor variations in proportions, as a result of the inability to allot fractional shares, for example, are permitted. HMRC insist that the shares be properly issued for these purposes – that is, all formalities to record the shareholders in the statutory books of A must be completed – and shares in A may only be issued to a nominee if the equivalent shares in T were held by the same nominee for the benefit of the same beneficial owner. In determining these proportions, any shares held in treasury by T itself are ignored.

9.26 In view of the limited application of stamp duty, clearly the relief is only applicable where the undertaking of T consists of or includes stock or marketable securities, or interests in partnerships which hold stock or marketable securities. There is no statutory definition of 'undertaking', but tax case precedents indicate that the mere passive holding of assets, as investments for example, does not amount to an undertaking. Furthermore, for the undertaking of T to be 'transferred', A must carry it on in much the same form as did T. So, if T holds a portfolio of shares as an investment, or if, after transfer to A, the shares are held as investments, relief is not likely to be available; however, if T carries on a share dealing business and, after the transfer, A also carries on a share dealing business, the Transfer could qualify for relief. HMRC also accept that acting as the holding company of controlling interests in one or more trading companies may amount to an undertaking, providing the holding company does in practice act as a parent company, taking an appropriate interest in the running of the subsidiary/ies. So, if T holds a controlling interest in trading company Z, this may amount to an undertaking and the relief may apply on transfer to A. HMRC's comments on this matter are at STSM042380 in relation to stamp duty. Their comments in relation to SDLT are likely to be relevant here too (see SDLTM23201).

9.27 It is also the view of HMRC, again supported by precedent, that a transaction is not a reconstruction if there is any appreciable change in the ultimate ownership of the assets. The legislation does not explicitly require that the shares issued by A should be of the same class as each other or as the shares in T. However, if shares of different classes are issued and if these give the shareholders different economic rights, the transaction may not be a reconstruction and so may not qualify for relief. The following example is fairly complex, but illustrates many of the difficulties which can arise in relation to this relief.

Example 9.4 – reconstruction relief

Target Ltd is the parent company of a trading group, holding all of the shares of trading companies Yak Ltd and Zebra Ltd. The shareholders of T are two siblings, Rupert and Sandy, each holding 5,000 of the 10,000 £1 ordinary shares (all the same class) in issue. Rupert also holds 2,000 non-voting, non-cumulative 5% £1 redeemable preference shares, reflecting extra cash he put into the group many years ago. It is decided to separate Y and Z, with a view to the possibility that each sibling may wish to take a greater interest in a different company in future. It is proposed to liquidate T, transferring each of Y and Z under a new holding company (YH and ZH respectively) in return for an issue of shares to Rupert and Sandy. The following issues arise:

(1) The preference shares are a potential problem because they are redeemable, but all shares issued by YH and ZH must be non-redeemable. If YH and ZH issue non-redeemable preference shares to Rupert, their economic value may not be the same as redeemable shares. A solution would be for the redeemable shares to be either redeemed or converted to non-redeemable shares before the reconstruction proceeds.

(2) The preference would be to give Rupert a greater interest in Y, and Sandy a greater interest in Z, taking care to ensure that each ended up with the same overall commercial value as at the outset. They consider the following possibilities:

    (a) YH issues 20 £1 preference shares and 70 £1 ordinary shares to Rupert, and 30 £1 ordinary shares to Sandy in return for the shares of Y. ZH issues 70 £1 ordinary shares to Sandy and 30 £1 ordinary shares to Rupert in return for the shares in Z. Neither the transfer of shares in Y to YH nor the transfer of Z to ZH would qualify for *s 75* relief, because at the end of the process Rupert and Sandy would not hold the same proportion of shares in either YH or ZH as they initially held in T.

(b) YH issues 20 £1 preference shares and 50 £1 class A ordinary shares to Rupert, and 50 £1 class B ordinary shares to Sandy in return for the transfer of Y. ZH issues 20 £1 preference shares and 50 £1 class B shares to Rupert, and 50 £1 class A shares to Sandy in return for the transfer of Z. This could qualify for *s 75* relief, as the proportions of shares in each company held by each of Rupert and Sandy at the end are the same as the proportions initially held in T. However, if the relative economic values of the three classes of shares in each company differ from the relative economic values of the shares in T, it is likely that the transaction will not amount to a 'reconstruction' within the terms of the relief, and relief will be denied. To avoid this problem, the rights and entitlements arising from the class A and class B shares in each company must be the same. In short, the siblings cannot skew the relative interests in the two businesses, so that one has a greater share of Y and the other a greater share of Z. If they wish to change this relative ownership, that must be done as a separate matter at a later date. The preference shares remain a particular problem. As shares in T, they give Rupert a reserves-dependent entitlement to a dividend of £100 per year. To achieve the same with 40 £1 shares in YH and XH (20 in each), the coupon rate would have to be 250% – and even that assumes the shares in YH and ZH are of equal worth. It really would be easier to get rid of the preference shares before commencing the reconstruction.

Relief for insertion of new holding company

9.28 In the heading to the legislation (*FA 1986, s 77*), this is also described as relief for acquisitions, but the paragraph heading above is more accurate. No stamp duty is chargeable on a Transfer of shares in one company (T) to another company (A) where the Transfer forms part of an arrangement for A to acquire all of the shares in T, provided other conditions are satisfied. Again, the general anti-avoidance condition applies (as mentioned in **9.23**), and the information listed there will have to be supplied if direct tax clearance has not been obtained for the transaction. The other conditions are that:

(1) the only consideration is the issue of shares in A to the shareholders of T;

(2) every shareholder of T immediately before the acquisition is a shareholder of A after the acquisition, and holds the same proportion of shares in A as previously held in T;

(3) if T has shares of different classes, the classes and relative proportions by number of each class are replicated in A; and

(4) the proportion of shares of A of any particular class, in issue or held by any particular shareholder, is the same as the proportion of shares of that class in T, in issue or held by that shareholder, immediately before the acquisition.

(5) at the time the instrument transferring the shares in T to A is executed there are no 'disqualifying arrangements' (as set out in *FA 1986, s 77A*) in existence (*FA 2016, s 137(5)*). This additional condition only applies to instruments executed on or after 29 June 2016, however the 'disqualifying arrangement' could have been entered into at any time, including before 29 June 2016.

The reference to different share classes relates to the rights and characteristics of the shares, not particularly how they are labelled. So, if shares are labelled class A and class B purely to identify by which member of a consortium they are held and if they have identical rights etc, they will not be regarded as being of different classes. But shares which have different voting or dividend rights or different nominal values will be regarded as being of different classes.

The legislation specifically provides that the references to 'shares' and 'share capital' also include a reference to 'stock' (*FA 1986, s 77(4)*). For these purposes 'stock' takes its meaning from *SA 1891, s 122* by virtue of *FA 1986, s 114(4)*. The *SA 1891, s 122* definition of 'stock' refers not only to shares or share capital but also includes the *funded debt* of any county council, corporation, company, or society in the UK, or of any foreign or colonial corporation or society. Although not clear from this definition, HMRC Stamp Taxes takes the view that for these purposes 'funded debt':

> '… means "debt instruments" *issued* by the target company such as bonds and loan notes (irrespective of whether the note is exempt or chargeable to stamp tax if transferred independently) which can be traded separately in a manner akin to that of equities.'

They go on to state:

> 'We would not expect "funded debt" to include mortgages, bank loans, financing and overdrafts within the meaning because these are not normally issued and tradeable instruments with the characteristics of equity' (STSM042415).

Consequently, for an arrangement to qualify for relief under *FA 1986, s 77*, A must not only acquire all of the shares in T, it must also acquire any 'funded debt' issued by T, and A must then issue 'funded debt' which mirrors that issued by T. Therefore if T, for example, has issued loan notes to a shareholder or a third party, A must acquire those loan notes, along with all of the shares in T, and A must then itself issue identical loan notes to the shareholder or third party. However, mortgages, bank overdrafts, bank term loans, etc can be ignored.

9.29 Prior to 29 June 2016 there was no denial of the relief under *FA 1986, s 77* if the control of A changed shortly after the share for share exchange transaction took place. However, HMRC's view is that the purpose of the *FA 1986, s 77* relief is to ensure that there is no stamp duty charge where there is no real change of ownership of T following its acquisition by A. Consequently, a new condition was introduced in relation to instruments executed on or after 29 June 2016 which requires that, at the time the instrument transferring the shares in T to A is executed, there must be no 'disqualifying arrangements' in existence.

The term 'arrangement' is widely defined as including '... any agreement, understanding, or scheme (whether or not legally enforceable)' (*FA 1986, s 77A(6)*).

An arrangement will be a 'disqualifying arrangement' if it is reasonable to assume that the purpose, or one of the purposes, of the arrangement is to secure that (*FA 1986, s 77A(2)*):

- a particular person obtains control of A; or
- particular persons together obtain control of A.

For these purposes 'control' is defined in *Corporation Tax Act 2010, s 1124* and means the power of a person ('P') to secure by the means of holding shares or the possession of voting power in relation to A (or any other body corporate) so that the affairs of A are conducted in accordance with P's wishes. In addition 'control' can be exercised as a result of any powers conferred by the articles of association or other document regulating the affairs of A (or any other body corporate). Consequently, any arrangement under which a person or persons could acquire the majority of the shares in A, or the majority of the voting power in A, would be a 'disqualifying arrangement'. In the author's view, for such an arrangement to exist, the particular person or persons would have to have been identified at the time the relevant transfer is executed.

9.30 It is specifically provided that the following are not 'disqualifying arrangements' (*FA 1986, s 77A(3)* and *(4)*):

- any issue of shares by A as consideration for the acquisition of T within *FA 1986, s 77(3)*, and
- 'relevant merger arrangements' – such an arrangement will exist where the share for share exchange transaction between the shareholders of T and A is followed by a second share for share exchange between the shareholders in a second company (B) and A. The second share for share

exchange transaction must meet the following conditions (*FA 1986, s77A(4)*):

(1) the only consideration given by A for the acquisition of the whole of the share capital of B must be the issue of shares to the shareholders of B;

(2) the acquisition of B must be for bona fide commercial reasons and must not form part of an arrangement with a main purpose of avoiding a liability to stamp duty, stamp duty reserve tax, income tax, corporation tax or capital gains tax;

(3) after the acquisition of B has been made each person who was a shareholder of B must be a shareholder of A;

(4) after the acquisition of B the shares in A must be of the same class as were the shares of B immediately prior to the acquisition and the relative proportions by number of each class of shares in B must be replicated in A; and

(5) the proportion of shares of A of any particular class held by any particular shareholder, must be the same as the proportion of shares of that class in B held by that shareholder, immediately before the acquisition of B.

In assessing whether conditions (4) and (5) above are satisfied the shares issued by A for the acquisition of T are ignored since, if this were not done, the conditions could never be satisfied. (*FA 1986, s 77A(4)(c)(i)*).

As for *FA 1986, s 77*, the definition of 'shares' and 'share capital' within the meaning of 'relevant merger arrangements' includes 'funded debt' (see **9.28**). This means that A must not only acquire all of the share capital of B but must also acquire any 'funded debt' issued by B, and must then issue identical 'funded debt' to the persons who previously held the 'funded debt' in B (*FA 1986, s 77(4)*).

The exception of 'relevant merger arrangements' from 'disqualifying arrangements' means that *FA 1986, s 77* relief will still be available in relation to A's acquisition of T even if there is an 'arrangement' for A to acquire the whole of the share capital of another company (B), provided that under that 'arrangement' all of the shareholders of B are to become shareholders of A, the shareholdings (class and proportion of shares held) in A are a mirror image of the shareholdings in B immediately before the acquisition of B (ignoring the shares issued by A to acquire T), and the acquisition of B is for bona fide commercial reasons and not part of a scheme to avoid any of the taxes listed in point (2) above.

If the 'relevant merger arrangements' or the share for share exchange within *FA 1986, s 77(3)*, which are excluded from 'disqualifying arrangements', are part of a wider scheme or arrangement which has a purpose of securing that a particular person(s) obtains control of A (see **9.29** for the meaning of control), but those wider arrangements do not fall within *FA 1986, s 77A(3)*, then the 'relevant merger arrangements' or share for share exchange are treated as 'disqualifying arrangements' (*FA 1986, s 77A(5)*).

9.31 The introduction of the additional condition that there must be no 'disqualifying arrangements' in existence at the time the instrument transferring T to A is executed will prevent the insertion of a new holding immediately prior to a sale to a third party (or the introduction of a new controlling shareholder). However, in these circumstances, it should still be possible to claim the relief provided the new holding company is inserted before any third party purchaser (or new controlling shareholder) is identified, and therefore before any 'arrangement' to sell A to a particular person (or to introduce a new controlling shareholder into A) has come into existence.

9.32 The transaction must be part of the acquisition of all of the target company's share capital. If A already holds some shares in T (for example, after stake building in a quoted company) or if any shares of T remain with another shareholder (for example, a dissenting shareholder), the relief will not be available. In practice, therefore, the relief is only of use for insertion of a new holding company by a 100% shareholder or by agreement amongst a group of shareholders with 100% between them. This is most commonly done as a first step in a further reconstruction or reorganisation, or perhaps to put a 'clean' company at the top of a group prior to a stock market initial public offering.

Example 9.5 – insertion of holding company

In Example 9.4, it is possible that the proposed reconstruction would be easier if, as a first step, a new holding company H Ltd was inserted between T and the shareholders, Rupert and Sandy. This might, for example, allow T to be liquidated first, thus removing any difficulty arising from an accumulated deficit on reserves. To this end, Rupert might establish H, taking one £1 ordinary share as founder subscriber. Rupert, Sandy and H would then agree that the shares in T be transferred to H in return for H issuing 2,000 £1 non-redeemable preference shares and 4,999 £1 ordinary shares to Rupert and 5,000 £1 ordinary shares to Sandy. Note the same comments apply as in the original example with regard to the redeemable preference shares. Note also that the key requirement is that numbers proportions of shares held in H after the transaction are the same as those in T before. Since Rupert starts with one share in H, he must agree to be issued with one share less than expected, to ensure the final numbers are correct. In the original form of this relief, precise replication of proportions was vitally important, and real claims failed because the subscriber share was overlooked. However, an amendment inserted by *FA 2006* relaxed this requirement, and now proportional shareholding must be replicated 'as nearly as may be'. This relaxation was intended to allow for the inability to issue fractional shares, and still may not cover overlooking subscriber shares so, to be safe, they should be taken into account as suggested here.

Acquisition reliefs – claims process

9.33 Each of the reliefs just described is claimed by submission of documents to the Birmingham Stamp Office (see **Appendix A**) for adjudication. The outside of the envelope should be clearly marked 'Stamp duty claim under FA 1986, s [75/77, as appropriate]' to ensure it is sent directly to the correct section. The following documents must be sent:

Item
The original, executed Transfer(s) (typically, for shares, a stock transfer form)
Certified* copies of the Transfer(s)
Certified copies* of any agreements governing the Transfer
Certified* copy of the register of members of the target company immediately before the Transfer
Certified* copy of the register of members of the acquiring company after the Transfer
Copies of any other documents (eg board minutes) which help explain the transaction and the reasons for it
A letter of claim
A self-addressed envelope for return of the stamped document[#]

* Items marked thus either are required to be 'certified', or may require certification, so it may be as well to have them certified at the outset. 'Certified' means either the normal certification by a solicitor or other appropriate lawyer that the document is a true copy of an original, or an equivalent statement signed by a director or the company secretary of the company in question.

[#] This became a requirement in June 2014 and reduces the risk that the stamped document might be returned to a registered office address, as shown on the stock transfer form, for example, rather than the postal address of the appropriate individual.

9.34 HMRC provide model letters of claim for these reliefs, most recently reproduced in Stamp Taxes Bulletin 2/2013 available at www.hmrc.gov.uk/so/bulletin02-2013.pdf. The model letters are reproduced in **Appendix B**.

Charities relief

9.35 This is a simple relief – any Transfer to a charity or charitable trust is relieved from stamp duty (*FA 1982, s 129*) – and HMRC are keen to ensure that charities benefit from the relief where appropriate. Provided the Transfer is clearly to a registered charity, the transferee normally only needs to submit the Transfer to HMRC with a note of the charity's registration number and a request that the Transfer be adjudicated exempt under the charities exemption. If the charity is not registered, it may also be necessary to submit the deed or other document establishing the charity, and perhaps engage in correspondence to satisfy HMRC that the entity is indeed established for charitable purposes. It is, however, expected that HMRC will have regard to the updated definition of a charity for tax purposes in *FA 2010, s 30* and *Sch 6*. In future, HMRC may challenge the availability of relief where they do not believe the charity is managed by 'fit and proper' persons.

Relief for transfer to new LLP

9.36 No stamp duty is chargeable on a Transfer to a limited liability partnership within one year of formation of that LLP, provided the Transfer is, effectively, a step in the incorporation of some other form of partnership (such as a UK general or limited partnership) with no change in overall economic ownership (*Limited Liability Partnerships Act 2000, s 12*). The partners in the LLP immediately after the Transfer must be the same as the partners in the old partnership immediately before, and the proportionate interest of each in the assets transferred must remain unchanged. The relief is available whether the assets were previously owned directly by the 'old' partnership or by a partner or other person holding as nominee for some or all of the partners. The Transfer must be submitted for adjudication for the relief to apply.

Recognised intermediary exemption etc

9.37 Any Transfer to a recognised intermediary (RI) is relieved from stamp duty (*FA 1986, s 80A*). An RI is a person who carries on a bona fide business of dealing in stock (or certain other financial instruments) and has been recognised for these purposes. Adjudication is required to take advantage of this relief, but the same recognition also exempts such Transfers from SDRT and, in practice, most such Transfers are electronic and so only within the SDRT net; further details about this exemption are provided at **10.22**. Other reliefs are provided specifically to facilitate financial market activity in the form of repurchases and stock lending, where in practice recognised intermediaries are normally involved (*FA 1986, ss 80C–84*).

Demutualisation of insurance companies

9.38 A Transfer executed in connection with the demutualisation of a mutual insurance company is relieved from stamp duty. This allows the transfer of investments from the mutual insurer to the new company taking over the insurance business, without stamp duty cost, provided the Transfer is submitted for adjudication (*FA 1997*, *s 96*).

Mergers of unit trusts and related transactions

9.39 Transfers executed in connection with the merger of authorised unit trusts, conversion of an authorised unit trust into an open ended investment company (OEIC), or merger of an authorised unit trust into an OEIC, are relieved from stamp duty (*FA 1997*, *s 95*; *SI 1997/1156*).

Other reliefs

9.40 The *Stamp Duty and Stamp Duty Reserve Tax (Collective Investment Schemes) (Exemptions) Regulations 2013* provide an exemption from stamp duty in relation to contractual collective investment schemes – see **10.31**. This will be of interest only to professionals and those investing large sums.

As announced on Budget Day 2013, *Finance Act 2014* contains provisions removing the charge to stamp duty on transfers of shares quoted on 'growth markets' such as the Alternative Investment Market of the London Stock Exchange (AIM) (*FA 2014*, *Sch 24*, *Pt 2*). From 28 April 2014, qualifying shares are removed from the scope of the principal charge to stamp duty and from other potential charges such as those on purchase of own shares and on transfer of shares to a clearance service or depositary. See **10.32** for details of which shares qualify.

There are numerous other reliefs from stamp duty, most of which are no longer relevant to real transactions since the scope of stamp duty was restricted to transfers of stock and marketable securities. In recognition of this, as mentioned at **8.1**, *FA 2012, Sch 38* abolished a number of redundant reliefs from 6 April 2013.

Chapter 10

Stamp Duty Reserve Tax (SDRT)

INTRODUCTION

10.1 SDRT was introduced by *FA 1986, ss 86–99*, in response to market developments under which the paperless transfer of beneficial ownership of shares became more common. Because no transfer document is produced, a paperless transfer does not give rise to stamp duty; SDRT is designed to collect the same amount of tax as stamp duty, even where there is no document. (In this book, the expression 'Transfer' is capitalised to make it clear that this is a reference to the potentially stampable document.) At that time, the London Stock Exchange was trying to develop an electronic trading platform ('Taurus'). The government of the day was eventually persuaded that charging stamp duty or SDRT would inhibit trading on this new platform, and *FA 1990, Pt III* was enacted to abolish both taxes on transfers of securities, from a date to be fixed by statutory instrument (the 'abolition day'). This date was intended to coincide with the launch of Taurus. Unfortunately, technical difficulties meant that Taurus was abandoned before completion. By the time the current electronic trading system was introduced in 1996, the government view had changed. *FA 1990, Pt III*, although still on the statute book, has not been activated, and both SDRT and stamp duty remain potentially chargeable on securities transactions.

10.2 SDRT is potentially within the rules for disclosure of tax avoidance schemes (DOTAS) but, at the time of writing (June 2016), regulations had not yet been made to activate this.

10.3 In practice, virtually all SDRT is paid on electronic stock market transactions (where it is collected automatically). The special SDRT regime, which used to apply to trustees and managers of UK unit trust schemes, was abolished from 30 March 2014. A general explanation of those arrangements is set out below, but detailed consideration of them is beyond the scope of this book. Outside these special areas of commerce, the main need is for taxpayers and advisers to understand how SDRT may arise on ordinary transactions in shares and securities, and how to ensure any such SDRT charge is appropriately dealt with. This chapter aims to provide that appreciation and understanding. In most cases, 'appropriately dealt with' means cancelled by execution of a

Transfer document and ensuring it is 'duly stamped'. The interaction with stamp duty and the actions necessary to ensure cancellation of any SDRT charge are outlined below at **10.11**.

BASIC RULES

What gives rise to an SDRT charge?

10.4 SDRT arises when there is an unconditional agreement to transfer 'chargeable securities' for consideration in money or money's worth (*FA 1986, s 87*). The tax is chargeable whether the agreement is documented or not; in principle, it remains chargeable even if the transaction is subsequently abandoned, unless it can be argued that abandonment is evidence that the original agreement was not in reality unconditional. The normal rate is 0.5% of the consideration (*FA 1986, s 87(6)*), which is the same rate as stamp duty. As with stamp duty, a higher rate of 1.5% is chargeable in certain circumstances, but this is only likely to impact financial market transactions. However, the definition of consideration – anything which is money or money's worth – is far wider than for stamp duty (see **8.25**). As a result, an ordinary transaction may give rise to an SDRT liability which is greater than any stamp duty liability on the same transaction.

Example 10.1 – contrast between SDRT and stamp duty

Abigale agrees to transfer 500 shares in UK company Brian Ltd to Carol, a professional decorator. In consideration, Carol agrees to redecorate Abigale's house (a job for which Carol would normally charge £6,000). The agreement gives rise to an SDRT liability of £30 (£6,000 × 0.5%), subject to possible debate over the 'value' of the decorating work. However, the 'consideration' simply does not count for stamp duty purposes, so no stamp duty is payable on any document transferring the shares. As noted below (**10.11**), if a document of transfer is executed and 'duly stamped' within the period allowed, this will cancel the SDRT charge.

Chargeable securities

10.5 Chargeable securities are defined in *FA 1986, s 99(3)–(12)*. These subsections begin with a blanket definition to include stocks, shares and loan capital, various interests in and options over such assets, and units in unit trust schemes. This definition is then restricted by excluding:

(1) securities which are neither UK registered nor paired with securities which are so registered. (It is uncommon for UK and non-UK securities

to be paired. This particular provision was enacted to deal with the company formed to build the Channel Tunnel, which issued paired UK and French shares.);

(2) securities issued by an SE (a 'European company' formed under EC Council Regulation 2157/2001) unless its registered office is in the UK;

(3) units in a unit trust if all of the trustees are resident, and the registered office is maintained, outside the UK, or if it can only invest in assets which are themselves not within the charge to stamp duty, SDRT or SDLT;

(4) shares in an open ended investment company (OEIC) subject to equivalent requirements to those in (3);

(5) securities traded on a recognised growth market but not listed on that or any other market (see **10.32**);

(6) interests in depositary receipts; and

(7) stock or marketable securities, the transfer of which is exempt from all stamp duties.

In practice, the main categories which remain within the definition of chargeable securities are UK registered shares and UK registered corporate debt with equity-like characteristics (for example, convertible debt). Most debt securities are not chargeable as a result of (7) above, because they are exempt from stamp duty on transfer in accordance with the exemption for 'loan capital' in *FA 1986, s 79* (see **8.15**). However, this exemption may not apply if the debt has equity-like characteristics such as convertibility, a return which depends on business results or asset values, the potential for repayment at a premium, or a coupon which exceeds a reasonable commercial return. The last of these restrictions has caused many disputes in relation to so-called 'junk bonds', which carry a high interest rate in recognition of a high risk of default by the borrower. Where there is a transfer of a UK corporate debt, the terms of the debt should be checked carefully to ensure that they do not contain features which may disapply the *s 79* excmption.

10.6 The scope of stamp duty was restricted in association with the introduction of SDLT in 2003. At that time, the rules excluding stock etc which is exempt from stamp duty were amended, presumably in an attempt to fit in with the new stamp duty rules. The result is a highly convoluted definition, written in such a way as to suggest that there may be securities which are, in practice, outside the stamp duty charge, but which may still be subject to SDRT. There are indeed such securities – units in a UK unit trust; it would have been simpler for the definition to say so.

10.7 Although not specifically excluding the securities from being 'chargeable securities', *FA 1986, s 90* lists a number of cases when no SDRT is chargeable, as follows:

- various agreements relating to unit trust schemes (see **10.30** for the special arrangements for unit trusts);
- agreement to transfer a non-UK bearer instrument, or many UK bearer instruments (but the exemption for UK instruments does not apply in certain cases, mainly where no stamp tax was payable on original issue of the instrument and it would be subject to stamp tax on transfer in registered form);
- agreement to transfer securities to a charity or charitable trust or to certain bodies established for national purposes; and
- agreement to transfer shares held in treasury by the company which issued them.

10.8 No SDRT liability arises on a conditional agreement until such time as it becomes unconditional (*FA 1986, s 87(3)(a)*).

Administration and compliance

10.9 Liability to SDRT should be notified to HMRC and the tax paid by the 'accountable date' (*SDRT Regulations 1986, SI 1986/1711, regs 3, 4*). For financial market transactions, the accountable date is fixed by agreement between HMRC and the operators of the electronic transfer systems used on those markets; it is generally 14 days from the date of the transaction. In the rare cases where SDRT remains payable on non-financial market transactions, the accountable date is the seventh day of the calendar month following that in which the agreement is made or becomes unconditional. So, if an agreement to transfer shares is entered into on the last day of a month, SDRT obligations arise just seven days later.

10.10 Responsibility for notification and payment rests with the 'accountable person'. When a broker or other market professional acts in the transaction, that broker or other professional is the accountable person. When more than one such person is involved, the rules specify which one has the responsibility. The broker then normally adds the SDRT to the amount charged to the client; the client has no direct responsibility to HMRC, but is still liable for the tax, and this fact gives the market professional the right to recover the tax from the client.

In non-financial market transactions, the accountable person is normally the purchaser (*SI 1986/1711, reg 2*). No standard return form is specified; liability must be notified by sending appropriate information to the SDRT Unit at Birmingham Stamp Office (see **Appendix A**). The information required is a full description of the securities (including number/amount), amount of consideration, when the SDRT is due, and the fact that this is a private or off-market transaction. Payment may be made by any of the normal methods but, for any form of direct bank transfer, it will be important to check with HMRC to ensure the correct current destination bank account details are used. These are set out on the HMRC website at www.hmrc.gov.uk/payinghmrc/stamp-reserve.htm.

Interaction between SDRT and stamp duty

10.11 Chargeable securities are usually also 'stock or marketable securities'. Stamp duty is chargeable on Transfers of these types of asset, so clearly both stamp duty and SDRT can apply to the same transaction. To avoid a double charge, any SDRT charge is automatically cancelled (and any tax paid can be reclaimed) if a document giving effect to the transaction (usually a stock transfer form) is executed and 'duly stamped' within six years of the agreement (*FA 1986, s 92*). Private transactions in securities are almost invariably documented and any stamp duty paid. As noted at **10.14**, there are several reasons why it is often best to arrange to fall within the stamp duty regime.

10.12 In some circumstances, Transfers are automatically deemed to be 'duly stamped' (**8.41**), so that mere execution of the Transfer is sufficient to cancel the SDRT charge. In some other cases, especially where a relief is claimed, the Transfer is not duly stamped unless the liability has been adjudicated (see **8.47**). In yet other cases, adjudication is not required, and simple stamping with the correct duty is sufficient to render the Transfer duly stamped unless and until the adequacy of stamping is challenged. Thus, no SDRT is normally paid on such transactions; the main risk is that the parties fail to execute a suitable document and ensure it is 'duly stamped' before the six-year deadline. HMRC have made it clear that, where such cases come to light, they will pursue the outstanding SDRT.

10.13 Often, the execution and stamping of a Transfer will not happen until after the SDRT has technically become due and payable. In recognition of this, HMRC do not in practice seek to collect SDRT on private, non-financial market transactions, or pursue penalties for failure to notify liability, where it is expected that the liability will be cancelled within a few weeks. This delay effectively amounts to an interest-free deferral of payment of the tax. However, should the liability eventually prove payable, interest will be charged from the accountable date, and a penalty may be charged for failure to notify liability.

Issues in non-financial market transactions

10.14 In relation to non-financial market transactions, it is normally best to ensure that the stamp duty rules apply and that any SDRT liability is automatically cancelled for the following reasons:

- Stamp duty reliefs such as group transfer relief (*FA 1930, s 42*) and the reorganisation reliefs (*FA 1986, ss 75* and *77*) do not apply directly to SDRT. An appropriate document must be executed and submitted for stamping in order to claim the relief under the stamp duty rules and cancel any SDRT liability at the same time.
- The definition of 'consideration' for stamp duty purposes is more restricted than that for SDRT purposes. So, the tax bill is often less under stamp duty than under SDRT, but payment of the correct tax under the stamp duty rules is still sufficient to cancel a potentially larger SDRT liability.

10.15 Note that the document of transfer must transfer the securities to the person who was the original transferee under the agreement (or their nominee) – *FA 1986, s 92(1A)*. If A agrees to transfer to B, and B agrees to transfer to C, but there is a single stock transfer form from A to C, this does *not* cancel the SDRT charge on the A to B transfer agreement. In practice, this can be a particular problem where a relief is to be claimed on the A to B transfer. It is usually necessary to ensure that a transfer document is produced for each individual transfer step, to avoid incurring multiple SDRT charges (see Example 9.3).

10.16 In general, shares and securities issued by companies not registered in the UK are outside the scope of SDRT. The exception is where those shares are held on a share register in the UK. Some non-UK registered companies, especially those registered in the Channel Islands and the Isle of Man, maintain copies of their share registers in the UK – often for administrative convenience where, for example, the shareholders are in the UK. In these situations, care must be taken to ensure that the UK copy truly is no more than a copy of the main register kept in the country of registration. It is not unknown for HMRC to allege that the UK 'copy' is in fact a main share register, bringing the shares within the UK SDRT net.

10.17 *FA 1986, s 88* specifies a few situations in which execution (and stamping, where necessary) of a Transfer does not cancel the SDRT charge. These are mostly anti-avoidance provisions to prevent exploitation of exemptions such as those for financial market operators. There are two cases of potential interest to ordinary investors – although, even here, a financial market professional will normally be involved and will deal with the SDRT obligations:

- units in a UK unit trust are exempt from stamp duty on transfer, but may still be liable to SDRT on agreement to transfer to anyone other than the managers of the scheme; execution of a Transfer does not cancel any SDRT charge; and

- when a company issues bonus shares or makes a rights issue, the right to receive the shares may be in 'renounceable letters of allotment' issued to the existing shareholders. Entitlement to the shares may be passed to others by transferring the renounceable letters. A Transfer is normally automatically exempt from stamp duty, but execution of a Transfer does not cancel any SDRT liability which may arise.

SDRT AND FINANCIAL MARKETS

10.18 The vast majority of quoted UK securities are held in electronic systems such as CREST (see below) and traded electronically, without production of any document of transfer. When a client buys quoted chargeable securities in this way, SDRT remains the final tax. The securities are usually purchased through market operators such as stockbrokers and banks; these are normally members of the relevant stock exchange (or 'qualified dealers'), and the SDRT regulations make them responsible for collecting and paying any SDRT which arises (*SI 1986/1711*, *reg 2*). When such an operator buys securities on behalf of a client, the SDRT is normally automatically added to the client bill by the electronic system and passed directly to HMRC. The client has no responsibility for notifying liability or making payment to HMRC.

10.19 Quoted securities are usually purchased for monetary consideration, and the liability under stamp duty would not be any less than under SDRT, so no disadvantage arises to the client from suffering SDRT. In fact, the SDRT liability is often slightly lower, because stamp duty is generally rounded to the next £5 above, whereas SDRT is only rounded to the nearest penny. The one exception is where the total consideration does not exceed £1,000, in which case stamp duty would be nil but SDRT is still charged.

CREST

10.20 Buying or selling securities generally involves two distinct stages: first, an agreement or contract; then a transfer in settlement of the contract. CREST is an electronic settlement system for quoted securities, originally established by a consortium of banks in the mid 1990s but now owned by the Euroclear Group. It formed a key part of the 1996 transition from paper-based transfers to electronic transfers on the London Stock Exchange. This transition was part of the general deregulation or freeing up of the market, colloquially referred to as 'Big Bang'. Stock market transactions are mostly processed within CREST without the need for paper documents. When the sale and purchase of securities within CREST is agreed, the brokers or other market operators acting on behalf of vendor and purchaser enter details of the transaction into the CREST system. Provided the details entered on behalf of vendor and purchaser agree, CREST matches them and processes the transfer. At the same time, CREST transfers the purchase/sale price between the accounts of the parties. Finally, CREST also collects (from the purchaser's broker's account) any SDRT due on the transaction and pays this to HMRC. As part of the compliance arrangement, each evening CREST sends details of that day's trades to HMRC.

Letters of direction

10.21 As explained above (at **10.14**), an SDRT charge may arise on certain agreements to transfer securities when no stamp duty would arise on a paper transfer, because the consideration for the transfer does not count for stamp duty purposes. In the financial sector, this commonly occurs in relation to large holdings of quoted securities in connection with mergers and consolidations – for example, certain mergers of pension funds. In theory, it would be possible to remove all of the quoted securities in question from CREST, carry out a paper transfer, then put them all back into CREST – but this would be an administrative nightmare, where hundreds of different securities might be involved. The securities are typically held on behalf of the pension fund or similar owner by custodians (often special purpose subsidiaries of banks). The transferor can effectively create a transfer document by issuing a letter of direction to the custodian, directing him to hold the shares for the benefit of the transferee. This is known as a letter of direction. It may be submitted for stamping if necessary and, once 'duly stamped', is sufficient to cancel any SDRT charge which would otherwise arise.

The 'recognised intermediary' exemption

10.22 The efficient operation of the securities market depends on the existence of traders who 'make a market' by offering to buy and sell securities from/to all comers. The profit on such trading, as a proportion of price, is often very small, and the imposition of SDRT would inhibit the market. An exemption is, therefore, available for purchase of securities by appropriate market operators, generally referred to as 'recognised intermediaries' (RIs) (*FA 1986, s 88A*).

10.23 Initially, the exemption was restricted to persons (individuals, companies or other entities) which were members of the London Stock Exchange, although it was gradually expanded to include other exchanges. However, in order to comply with the requirements of the EU Markets in

Financial Instruments Directive (MiFID), the scope of the exemption was considerably widened from November 2007 to include:

- 'trading platforms' which would not otherwise be regarded as fully functioning stock exchanges; and
- non-members of stock exchanges carrying on share trading who are authorised by a state in the European Economic Area to provide investment services or activities as set out in the Directive.

10.24 Note the use of the word 'recognised' – the relief does not apply unless the trader and/or trading platform has applied for and gained HMRC recognition for the purpose. HMRC may approve an exchange or platform for these purposes by making regulations under *FA 1991, ss 116, 117*. The exchange then takes on delegated responsibility for approving RIs in relation to transactions through that exchange. There is a steady stream of regulations granting approval of European exchanges or trading platforms – see for example *SI 2011/666*. In contrast, individual approval of authorised traders is obtained by direct application to HMRC.

10.25 An RI must not carry on any 'excluded business'. Excluded businesses encompass those which involve making or managing investments (for example, insurance or running unit trusts) as well as the provision of services to related companies. In practice, financial services groups often also carry on excluded businesses, and so carry on their RI business through a separate company. However, a few have negotiated arrangements with HMRC under which they 'ring fence' their RI business within a greater company, and treat movements of stock between parts of the company as if they were sales and purchases.

Example 10.2 – SDRT on market transactions

Harry instructs his broker, Irena Associates, to buy up to 10,000 shares in Jeremiah plc. Meanwhile, Karen instructs her bank Liam Corporation to sell her holding of 3,000 shares in Jeremiah plc. Irena Associates agrees to buy Karen's shares, initially on its own account. Because Irena Associates is a recognised intermediary, no SDRT arises on the purchase. Once Irena Associates has purchased a further 7,000 shares, it transfers all 10,000 to its client account, for the benefit of Harry. At the moment of transfer, CREST charges Irena Associates SDRT on the price which Harry is paying for the shares. As regards the 3,000 shares originally owned by Karen, one lot of SDRT is paid, just as if Karen had sold them directly to Harry (and the same is true in principle for the other 7,000 shares).

Contracts for differences (CFDs)

10.26 The relief is available to prevent double charges when a trader buys shares and sells them onwards to a customer. However, it applies to all purchases of chargeable securities by the intermediary and, in practice, one of the biggest uses is to allow traders to buy shares free of SDRT to hedge 'contracts for differences' (CFDs). Under a CFD, a customer gains the economic effect of having bought shares without actually buying them.

Example 10.3 – Contracts for differences

Moira believes shares of quoted UK company Nick plc will increase in value over the next six weeks. Broker Olive Ltd agrees to write a CFD for 5,000 shares in Nick plc. Under the CFD, Moira immediately pays Olive Ltd an amount equal to the current price of 5,000 shares in Nick plc (plus a small commission). In return, Olive Ltd agrees that, in six weeks' time, it will pay Moira the value of 5,000 Nick plc shares at that time. Moira, therefore, is in the same economic position as if she had invested in 5,000 Nick plc shares for a fixed six-week period. However, writing the CFD does not involve any agreement to transfer securities, so no liability to SDRT arises for either party. Meanwhile, Olive Ltd does not wish to suffer the risk that Nick plc shares will rise in value, so it goes into the market and buys 5,000 Nick plc shares for its own account. Because Olive Ltd is a recognised intermediary, it does not pay SDRT on this purchase. In six weeks' time, it will sell the Nick plc shares and use the proceeds to pay Moira under the CFD.

It is thought that up to half of all UK stock market transactions relate to use of CFDs. HMRC do not generally seek to impose SDRT on such transactions, provided they are implemented with due care; there would probably be no point in trying to do so, as most such transactions would simply not happen if there was an SDRT cost. However CFDs have gained a negative reputation with the general public because many activities of traders which are thought to cause major fluctuations in prices of shares (and resulting economic instability) involve the purchase and sale of CFDs.

Depositories and clearing systems

10.27 Overseas investors often prefer to buy UK securities through a clearance system operating in their home country or in the form of depository receipts. The rate of SDRT is increased to 1.5% where there is an agreement to transfer securities to a depository receipt issuer or clearance system. The justification for this is that, while the securities remain within the clearance system or depository, subsequent transfers occur without imposition of SDRT.

The higher rate was previously also charged where new shares or securities were *issued* to a clearance system or depository receipt issuer, until this was found unlawful under EU law in two tax cases, both involving HSBC. The first, a 2009 European Court decision (*HSBC & Vidacos v HMRC*), considered the issue of shares to a European clearance system. The second, a UK 2012 First-tier Tax Tribunal decision (*HSBC & Bank of New York Mellon v HMRC*), decided that the EU Capital Duty Directive applies if the issuer of the securities is in the EU, irrespective of the location of the depositary receipt issuer. HMRC have accepted this decision and invited claims for repayment of SDRT paid under the previous incorrect interpretation of the law, subject to the usual four-year time limit (see www.hmrc.gov.uk/so/sdrt-hsbc-holdings.pdf). However, in Stamp Taxes Bulletin 2/2012 (www.hmrc.gov.uk/so/bulletin02-2012.pdf), HMRC make clear their view that the decisions apply only to the specific capital instruments set out in the Capital Duty Directive. In practice the Directive covers all instruments by which companies commonly raise capital – shares, bonds, debentures, etc – but it appears that HMRC may be worried that unusual instruments may be devised to take advantage of the exemption in the course of tax avoidance.

The 1.5% charge is sometimes referred to as a 'season ticket'. There is at least anecdotal evidence that, on average, such shares are transferred fewer than three times within the system before being transferred out of the clearance or depository system, so the 1.5% charge is probably punitive. Clearing systems can avoid the 1.5% charge by agreeing with HMRC to charge SDRT as normal on transfers (*FA 1986, s 97*).

The judgment in the *HSBC/Mellon* case includes comments which imply that the holder of a depositary receipt does not have beneficial ownership of the underlying security. HMRC have stated their view that these comments relate specifically to the position for this specific fact pattern under New York State law and have made it clear that they will continue to treat the holder of a depositary receipt as having beneficial ownership of the underlying security for direct tax purposes.

Public issues

10.28 When there is a public issue of shares, the underwriters typically take temporary ownership of any shares not immediately taken up, and then sell them in the market over the next few weeks. There is normally no SDRT or stamp duty on the issue of shares, but this process would give rise to an SDRT charge on the subsequent sale, which would add to the costs of the share issue. A relief (*FA 1986, s 89A*) therefore applies to stop this charge arising, provided the strict conditions are complied with.

Stocklending

10.29 In origin, stocklending was a process by which institutional investors 'lent' their investments to traders to allow them to sell shares they had not yet bought, all in the interests of maintaining market liquidity. However, this is not a true loan – rather, it is a sale and repurchase agreement which could have given rise to SDRT. So, to maintain liquidity, subject to stringent conditions, there is a relief (*FA 1986*, *s 89AA*). The process gained notoriety in 2008, as speculators borrowed huge numbers of bank shares, flooding the market with them, thus driving the share price down, in turn allowing them to buy back at a reduced price and return the loan – and all helped by an SDRT relief!

Unit trusts and open ended investment companies

10.30 In accordance with the *Stamp Duty and SDRT (Open Ended Investment Companies) Regulations 1997, SI 1997/1156,* references in these paragraphs to unit trusts should be regarded as including references to OEICs. Units in unit trusts are specifically exempted from stamp duty. If units in a UK unit trust are transferred directly between investors, the usual SDRT charge applies, and this cannot be cancelled by execution of a document of transfer. However, in practice, such units in publicly traded unit trusts are not usually transferred, they are surrendered to the manager (who then cancels them or reissues them to a new investor). No charge arises on the investor when this happens. Prior to 30 March 2014, under a special regime in *FA 1999*, *Sch 19*, the manager had to pay a composite SDRT charge which took account of numbers of surrenders and issues, and the composition of the unit trust investment portfolio. The manager recouped the SDRT cost through the margin between the prices at which he issued and repurchased units. There were some adjustments in 2011 to improve the fairness of the charge where one unit trust invested in another and some of the underlying investments were themselves exempt from SDRT. However this did nothing to alter the view of the industry that the regime was burdensome and unduly complex. The government accepted this view and the *Sch 19* charge was abolished for surrenders of units on or after 30 March 2014 by *FA 2014*, *s 114*. This provision led to lively debate in the passage of the Finance Bill through Parliament, the opposition labelling it as a gift to the rich.

Contractual collective investment schemes

10.31 Article 1.3 of Directive 2009/65/EC of the European Parliament and of the Council of 13 July 2009 requires Member States to make provision for the formation of undertakings for collective investment constituted in accordance with contract law. In the UK, the *Collective Investment in Transferable Securities (Contractual Scheme) Regulations 2013* (*SI 2013/ 1388*) modified the *Financial Services and Markets Act 2000* to allow such 'contractual schemes' to be set up and to be regulated by the Financial Conduct Authority. Contractual schemes are specifically excluded from the *FSMA 2000* definition of unit trusts. Broadly speaking, such schemes are intended to be mechanisms for groups of professional or wealthy investors to hold and manage investments in common. To this end, small retail investors are not permitted to participate directly.

The *Stamp Duty and Stamp Duty Reserve Tax (Collective Investment Schemes) (Exemptions) Regulations 2013* (*SI 2013/1401*) amend *FA 1986, s 90* and *FA 1999, Sch 13* to exempt from SDRT and stamp duty transfers of securities to a contractual scheme in return for units in the scheme, with effect from 28 June 2013. Transfers between entities holding the securities on behalf of the scheme are also exempted from both taxes.

Growth markets

10.32 In an attempt to encourage investment in smaller high-growth companies, *FA 2014, s 115* and *Sch 24* amended *FA 1986, Pt 4* to remove certain securities from the scope of both SDRT and stamp duty. From 28 April 2014, shares and other potentially chargeable securities are outside the charge to these taxes if they are admitted to trading on a recognised growth market but not listed on that or any other market. There is a detailed definition of 'recognised growth market', making it clear that the market must deal primarily with lower capitalisation companied with compound annual growth in revenue or employment of at least 20%. Markets must apply to HMRC for recognition to benefit from this change. There is a list of recognised growth markets at STSM041330 and at www.hmrc.gov.uk/fid/rse-recognised-exchanges.htm, which includes the Alternative Investment Market (AIM) in the UK. A company is 'listed' on a market if it is in the Official List maintained by the authority responsible for financial services in the country in question. In the UK, shares traded on the London Stock Exchange are generally 'listed', but those traded on AIM are not.

Chapter 11

Planning, Pitfalls and Legacy Liabilities

INTRODUCTION

11.1 The complexity and illogicality of stamp tax rules have often been exploited through contrived avoidance arrangements. However, increasing volumes of general and specific anti-avoidance legislation including, in the case of SDLT, the application of the general anti-abuse rule and the requirement to disclose novel planning arrangements, have dramatically reduced the scope for tax mitigation. Planning now is often as much a matter of avoiding unintended and unexpected additional liabilities as reducing the headline tax cost of a transaction.

This chapter draws together thoughts on planning, both to minimise costs and to avoid potential pitfalls arising from all three stamp taxes. It also considers situations in which current liabilities may arise or increase as a consequence of disturbing transactions carried out, or structures adopted, long ago.

References to particular arrangements or structures are included in this chapter purely for illustrative purposes. Their appearance should not be taken as advising that they succeed in reducing liabilities. Some of the arrangements described are believed to have been effective under previous legislation but have subsequently been blocked. It is essential that expert advice be taken before entering into any arrangement to mitigate tax, and no responsibility is accepted for the consequences of any persons acting or failing to act on the basis of the comments or examples in this book.

STAMP DUTY LAND TAX

Planning

11.2 SDLT mitigation has generally involved one or more of four general approaches:

(1) Arrange to have the benefit of a relief or exemption, or a combination of two or more reliefs or exemptions.

Example 11.1 – combining relief and exemption

Charles wishes to buy a house at a cost of £600,000 to give to his nephew Dylan. Charles contracts to buy the house, paying the usual deposit. He then agrees to sell it to Dylan for £1,000. The two purchases (by Charles and by Dylan) are substantially performed and completed at the same time. If the arrangement were effective, the purchase by Charles would be ignored and only the onward sale to Dylan would be subject to SDLT. The consideration for this is below the SDLT threshold, so no tax would be payable. Note that, even if this arrangement might once have worked, it is now blocked by the *FA 2013* changes to *FA 2003, s 45,* see **3.8** et seq.

(2) Avoid transferring land.

This may be achieved either by transferring something else (for example, shares in a land-owning company) or by entering into a commercially different transaction.

Example 11.2 – transfer land-owning company

Dale Group plc owns a subsidiary company Hill Ltd which owns a number of valuable commercial properties. It is decided to sell one property. The other properties are transferred out of Hill Ltd to sister company Valley Ltd for cash, and surplus cash is paid out of Hill Ltd to Dale Group plc as a dividend/return of capital. The value of Hill Ltd is now the same as the value of the one remaining property which it holds. Hill Ltd is then sold. The purchaser has to pay stamp duty at 0.5% on the price paid for the shares, rather than SDLT at 4% on a land purchase. Note that versions of this planning have been widely used in the past and, in principle, HMRC do not appear to have regarded it as objectionable. However, the specific circumstances of each transaction must be examined in detail to establish whether a version of the planning could be of use.

Example 11.3 – different transaction

Street Ltd owns land ripe for development. Street Ltd's first thought is to sell the land to a developer for onward sale after redevelopment, with a profit sharing clause in the sale agreement to ensure Street Ltd shares in the benefits of the redevelopment. However, after careful thought, the directors of Street Ltd decide to retain the land for sale directly to the ultimate owner after contracting the development work to the developer. This is, of course, a commercially different transaction, potentially giving Street Ltd greater rewards but exposing the company to greater risks if the redeveloped property cannot be sold at the expected price. However, correctly structured, this could reduce the total SDLT cost by the amount which the developer would otherwise have paid. Care would be needed in structuring and documentation. An attempt to obtain the SDLT benefits of this alternative transaction whilst keeping the commercial effect of a straight sale to the developer is likely to fall foul of anti-avoidance rules.

---

(3) Reduce the value of the land transferred (with most of the consideration paid for something else) – see Example 7.2 for an explanation of a type of planning which may have been used until blocked by *FA 2003, s 75A*.

(4) Arrange to fall within a specific statutory treatment which gives rise to a low or zero SDLT charge.

Example 11.4 – PFI relief

Litebuild plc is negotiating with Marsh Education Authority (MEA) to buy a disused school site for housing development. MEA intends to use the sale proceeds to pay part of the cost of rebuilding a college. It is agreed that MEA will grant to Litebuild a 30-year lease of the college site, and Litebuild will grant an underlease back to MEA for two days short of 30 years. Litebuild will rebuild the college, or procure its rebuilding, and will maintain the grounds and fabric of the building during the lease. In return, MEA will transfer the disused school site to Litebuild in part-payment for the rebuilding of the college, will make a one-off initial payment to cover the balance of the rebuilding cost, and will make yearly payments under the lease for the maintenance services. On the face of it, this arrangement could fall within the terms of *FA 2003, Sch 4, para 17* (PFI relief), so that no SDLT would be payable. The facts outlined here appear readily to fall within the terms of this treatment. The treatment could be available for a transaction starting from a less favourable fact pattern, but it would be important to examine the precise circumstances of the parties and the proposed transactions before there could be any confidence that this would be the case.

11.3 Specific anti-avoidance legislation and *FA 2003, ss 75A–75C* make it difficult to find 'off the shelf' combinations of these approaches which can be applied to a range of transactions to reduce the SDLT liability below the expected level. New versions of old ideas emerge from time to time, but the DOTAS rules allow HMRC to block them quickly with new legislation when necessary. HMRC now also have the benefit of the general anti abuse rule in *FA 2013, ss 206–215* to deploy against any unacceptable arrangements which escape more specific anti-avoidance measures, see **7.3.**

11.4 The best approach is probably to have the fullest possible knowledge of the various statutory reliefs, exemptions and treatments which are available, so that these can be applied to each individual transaction in the most favourable way. In virtually every case, it remains essential to consider carefully whether the combination of steps required to complete the transaction bring it within the scope of *FA 2003*, *ss 75A–75C* (see **7.3** et seq) and whether any disclosure responsibility arises under the DOTAS rules (see **7.26**).

Pitfalls

11.5 At least as important is the need to be aware of the pitfalls waiting to trap the unwary. In relation to SDLT, these arise from current legislation and those listed here are mostly dealt with in more detail elsewhere, as indicated:

- clawback of relief previously claimed, where there is a change of control of a company (**3.29** et seq);
- transfer to a company which is connected with the transferor, but does not fulfil the precise and strict conditions to be treated as in a group with the transferor (**2.29**). Mere common ownership is not enough to qualify for group relief;
- informal arrangements for occupation of a property, which continue for long enough to give rise to a tenancy on which SDLT arises (**4.8** et seq);
- transfer of a lease, when relief was claimed on grant (**4.16**);
- transfer of property by operation of law, especially in non-UK mergers etc, which may still be treated as a transfer for consideration for SDLT purposes (**2.4**);
- withdrawal of cash or other value from a partnership after transfer of property to the partnership from a partner or connected person (**5.17**); and
- timing issues, being unaware of factors which can lead to SDLT becoming payable much earlier than commercial completion of a transaction or can lead to the expiry of time limits to claim relief or deferment of tax earlier than expected (**2.68** et seq; **6.33**).

Prior to 17 July 2013, difficulties could arise where a lease was formally executed after substantial performance of the related agreement, so that the first rent review fell less than five years after the date of execution. *FA 2013* amendments mean this ceased to be an issue in most cases where substantial performance occurs on or after 17 July 2013 (**4.38**);

11.6 In the context of transactions in land as well as other assets, there are also stamp duty pitfalls which arise from legacy structures – that is, structures and arrangements which were put in place in the course of previous tax planning, often many years ago. These may give rise to current liabilities if not dealt with in the correct manner. Examples are described at **11.26** et seq.

STAMP DUTY AND SDRT

11.7 There are no general anti-avoidance rules for stamp duty and SDRT equivalent to *FA 2003, ss 75A–75C* for SDLT. There are specific anti-avoidance rules relating to particular assets and transactions. SDRT is potentially within the DOTAS rules but, at the time of writing (June 2016), no regulations had been issued to activate this. Stamp duty is not within the DOTAS rules, presumably reflecting the fact that it is not strictly speaking a 'tax'.

SDRT planning

11.8 SDRT is only payable on an agreement to transfer chargeable securities. 'Planning' to minimise SDRT usually takes one of three forms, or a combination of them:

(1) minimising the amount of consideration which is subject to tax;

(2) transferring something which is not a chargeable security; or

(3) cancelling the SDRT charge by ensuring a Transfer document is executed and duly stamped (see **10.11**). (In this book, the expression 'Transfer' is capitalised to make it clear that this is a reference to the potentially stampable document.)

11.9 The first approach is mostly of interest for financial market transactions. It commonly involves the use of options and other financial instruments, detailed consideration of which is beyond the scope of this book. However, SDRT is only charged by reference to the value of consideration given in money or money's worth. Therefore, a transfer of valuable securities which is genuinely for zero or low consideration (for example, between family or group members) will attract zero or low SDRT, the market value of the securities themselves being irrelevant.

11.10 The second approach is also mainly of interest to financial and capital markets, although the risk of transferring securities which are *unintentionally* within the charge to SDRT is a pitfall described at **11.13**. Outside financial and capital markets, it may be possible to achieve SDRT-free transfer of chargeable securities by placing them in some kind of 'wrapper'.

Example 11.5 – chargeable securities in a wrapper

Serena owns 100% of the shares in a valuable company T Ltd, which she intends to sell to Vera. Serena arranges to set up a Jersey-registered company U Ltd, and she subscribes £100 for 100 shares in the company. Once all formalities for formation of the company and issue of the shares have been completed, Serena sets up a Jersey-registered limited partnership, W LP, with herself and U Ltd as partners. Serena contributes the shares of T Ltd to W LP as partnership capital. This is a transfer for no consideration for stamp duty purposes, and execution of a stock transfer form by Serena is sufficient to cancel any SDRT charge which might otherwise arise. Serena then sells the shares of U Ltd and her interest in W LP to Vera for full value (the same as the value of T Ltd). Although this effectively transfers the shares of T Ltd to Vera, none of the assets actually transferred is within the charge to SDRT. Theoretically, the transfers may be subject to stamp duty but, as none of the assets is located in the UK, it is unlikely to be necessary to stamp the related documents. The benefit is shared with Serena through a price adjustment.

11.11 Alternatively, the nature of the security itself may be chosen to be outside the scope of SDRT.

Example 11.6 – securities outside the scope of SDRT

Yorrick intends to set up a company to manufacture better mousetraps. He anticipates that, within five years, he will be able to sell the successful company to a major manufacturer for a large sum. He therefore chooses to set up a company registered in the Isle of Man but managed and controlled in the UK. For most other tax purposes, this is treated in the same way as a UK incorporated company. But, provided Yorrick does not make the mistake of keeping the company's principal share register in the UK, its shares will remain outside the SDRT net when the time comes to sell.

11.12 In practice, for non-stock market transactions, the third approach (see **11.8**) is normally what automatically happens, and any planning is directed at minimising the stamp duty (see **11.17**).

Example 11.7 – rely on stamp duty definition of consideration

On 1 April 2016, Raj agrees to sell his minority holding of shares in unquoted UK company Courgette Ltd to Petra in exchange for Petra's collection of rare books, valued at £120,000. This agreement gives rise to an SDRT liability of £600 which is technically payable by 7 May 2016. However, the consideration does not count for stamp duty purposes (see **8.23** et seq) so, provided a stock transfer form is completed no later than 31 March 2022, the SDRT charge will be cancelled. In these circumstances, where there is consideration but it is not consideration chargeable to stamp duty, the stock transfer form is deemed to be 'duly stamped' as soon as it is executed although the certificate on the reverse of the stock transfer form, stating that there is no chargeable consideration, should be completed (see **8.7**). If there is consideration of £1,000 or more which counts for stamp duty purposes such that a liability arises, the form would have to be executed and presented for stamping no later than 31 March

2022 in order to cancel the SDRT charge. It would generally be desirable to do this as soon as possible after 1 April 2016.

SDRT pitfalls

11.13 The main pitfall to be wary of is the risk of a security unintentionally being or becoming a 'chargeable security' and therefore within the SDRT net. The commonest category of chargeable security consists of shares in UK companies. Most loan capital and debt instruments, such as corporate bonds, as well as most shares of non-UK companies are outside the scope of SDRT (see **10.5**). However, bonds and other loan capital may be within the SDRT net if they have certain 'equity-like' characteristics. In particular, any of the four following factors may bring the security into the SDRT net:

(1) A high coupon/interest rate. If the interest rate exceeds 'a reasonable commercial return on the nominal amount of the capital', the security will not qualify for the loan capital exemption from SDRT (although it may still escape under other exemptions). This is often a problem in relation to so-called 'junk bonds', which have a high interest rate to reflect risk. It may be necessary to gather evidence that the interest rate is comparable to that on other publicly traded securities with a similar risk profile, to convince HMRC that the interest rate does not exceed a reasonable commercial return.

Example 11.8 – high coupon bond

Ralph and Simon are brothers. Ralph owns all of the shares of travel agent Timeoff plc, a UK company. Timeoff needs additional funds, and Simon agrees to lend £2 million at an annual interest rate of 18%, reflecting perceived risk. The debt is formalised as a two-year bond. After six months, Ralph agrees to buy the bond from Simon for face value. The high interest rate means that the bond may not qualify for the loan capital exemption from SDRT, and the agreement to transfer may give rise to an SDRT charge of £10,000 (£2 million at 0.5%). The problem may be avoided if Ralph makes a new loan to Timeoff, allowing it to prepay the original bond – assuming the terms of the bond allow this.

---

(2) A variable interest rate, where the issuer of a debt security has the right to vary the interest rate, including the right to cancel or withhold a payment, or the interest varies in the light of business results or asset values. Note, however, that so-called 'ratchet loans', where the interest rate goes up to reflect increased risk when business results or asset values deteriorate (and/or goes down when values or results improve) may still qualify for exemption. Equally, no problem is caused by the fact that the

interest or amount repayable on repayment may vary by reference to the UK Retail Price Index or a similar index.

Example 11.9 – variable interest rate bond

UK company Zebra plc borrows from a private equity firm P (which is also a shareholder) and issues a loan note which pays monthly interest at an annual rate 3% above Bank of England base rate. The loan note provides that the interest rate will fall to base rate for the year following any year in which Zebra makes a loss, but will increase to 6% above base rate for the year following any year in which Zebra makes a pre-tax profit of more than £1 million. No stamp tax consequences arise from the issue of the note. However, two years later, P agrees to sell the loan note to Q, a wealthy individual. The variable interest rate means the loan note will not qualify for the loan capital exemption and is a 'chargeable security', so the sale will be subject to SDRT on the sale price. Assuming it has no features which prevent it from being 'stock or (a) marketable security', a Transfer document will be subject to stamp duty on the sale price.

---

(3) Right of conversion into shares or other securities, or right to receive shares or securities, for example in lieu of interest. These are often called PIK notes, because interest is or may be 'paid in kind' – that is, paid by the issue of further debt instruments. It is understood however that whilst a right to receive securities in lieu of interest would technically preclude the loan capital exemption from applying to a transfer of the underlying securities (*FA 1986, s 79(5)*) HMRC Stamp Taxes take the view that provided the new securities issued in lieu of interest have identical terms to those of the underlying securities, and would themselves qualify for the loan capital exemption if they were transferred, then the loan capital exemption will apply to the transfer of the underlying securities. It should also be noted that if the security bears interest at an acceptable rate, but the *issuer* has a unilateral right to make payment in the form of further securities, and those further securities do not have identical terms of issue as the underlying security, the underlying security is not thereby prevented from qualifying as loan capital until the issuer exercises that right and commits to making an interest payment in kind. This is because, until that time, the security does not carry a right, exercisable by the holder of the security, of conversion into another security.

Example 11.10 – right to further securities, PIK notes

Yak plc issues five-year loan notes to a consortium of investors. The loan notes provide for yearly payment of interest at 6%, which is considered to be a normal commercial rate, and repayment at par at the end of the term.

However, they also give Yak plc the right to make up to two interest payments by issuing further loan notes with a face value equal to the amount of interest due, an interest rate of 10% per year and redeemable at par at the same time as the original notes (a PIK payment). Yak must give three months' notice of intention to make each PIK payment. The 10% notes, if issued, only provide for payment of interest in cash. In all other respects, both kinds of note qualify as exempt loan capital.

No SDRT arises on the issue of the notes. Unless and until Yak gives notice to make a PIK payment, the securities qualify for the loan capital exemption, and transfers in that period should not give rise to any SDRT. However, from the time Yak gives notice until after the PIK payment is actually made, the original securities cease to qualify for the loan capital exemption, and any transfers of those notes will be subject to SDRT. Once the PIK payment is made, the notes again qualify as exempt loan capital, unless or until notice of a further PIK payment is given.

Had the PIK notes been issued on identical terms to those of the original five-year loan notes (ie a 6% rate of interest, a right to make up to two interest payments by issuing further loan notes, etc) then, based on the views of HMRC Stamp Taxes, the loan capital exemption would have continued to be available even after Yak had given notice of the intention to make a PIK payment.

The notes issued in lieu of interest do not carry any PIK option, and so should qualify for the loan capital exemption on any transfers after they are issued, assuming the interest rate is not regarded as excessive.

---

(4) A high premium on repayment – for example, a bond with a face value of £100 is issued at a price of £100 but, at the end of its five-year life, the amount repayable is £160, being the face value plus a premium of £60. Remarkably, the same economic effect may be achieved without risk of falling within the SDRT net, by issuing the bond with a £160 face value but at a £60 discount, so that the amount paid at issue is £100. However, this may affect the treatment of the bond for other purposes.

Any of these characteristics may lead to denial of the loan capital exemption. This is a complex area, and specialist advice is likely to be needed to determine whether SDRT applies to any transfer if these characteristics are present.

11.14 The other main risk is that shares of a non-UK incorporated company, normally outside the SDRT net, will become chargeable securities if they are 'registered in a register kept in the UK' (*FA 1986*, *s 99(4)(a)*). As suggested at Example 11.6, non-UK incorporated companies (often incorporated in the Channel Islands or the Isle of Man) are widely and validly used by UK-based individuals and businesses. The share register of such a company is normally kept in the territory of incorporation, but a copy may be maintained in the UK for administrative convenience. Care must be taken to ensure that the UK document (or electronic record) is no more than a copy, and that the overseas register is the definitive document for determining the identity of the shareholders. HMRC have been known to examine this matter very closely, with a view to showing that the UK record is a definitive register so that the shares become chargeable securities.

Stamp duty planning

11.15 Stamp duty is currently chargeable only on documents effecting transfers of 'stock and marketable securities' and interests in partnerships owning stock and marketable securities.

11.16 In relation to transfers of partnership interests, the simplest planning is to choose not to present any transfer document for stamping. This is effective provided the parties are satisfied that there is no need for a stamped document. A stamped document is normally only required if it is to be produced in evidence or used for other official purposes. Therefore, the most likely occasions on which a stamped document may be needed are in relation to a dispute between the parties or in order to prove to HMRC that the transfer has taken place. In practice, it appears to be rare for transfers of partnership interests to be stamped. An alternative approach is to structure the transaction so that there is no Transfer document which may require to be stamped. For example, an effective transfer of a partnership interest might be achieved by the outgoing partner withdrawing capital from the partnership, and the incoming partner introducing the same amount of capital. This may be regarded as a transfer for some purposes (including SDLT, where relevant – see **5.25** et seq). However, it should not be necessary to execute a Transfer and no stamp duty liability should arise.

11.17 It is usually only necessary to stamp a transfer document for stock and marketable securities, if they are chargeable securities, so that the SDRT charge is cancelled (see **10.11**) and any registration formalities may be completed. The simplest form of planning is, therefore, to transfer securities which are not chargeable – for example, shares in a non-UK incorporated company – and then choose not to present the Transfer for stamping (but beware the risk of making them chargeable by keeping the register in the UK). Alternatively, it may be possible either to arrange to fall within the terms of a relief (see **Chapter 9**) or to minimise the consideration which is taken into account, as set out below.

11.18 See **8.25** et seq for discussion of the measurement of consideration for stamp duty purposes. Some planning has taken the form of attempting to make the consideration wholly unascertainable (see **8.36**), but there are risks. In particular, in the case of *L M Tenancies 1 plc v IRC* ([1996] STC 880), HMRC succeeded in arguing that linking consideration to the price of a publicly traded security, when that price was not likely to change much, did not make the consideration wholly unascertainable. It is safer to make the consideration ascertainable but low (see **8.57** et seq).

Stamp duty pitfalls

11.19 A major risk in relation to stamp duty is linked with the SDRT risk identified at **11.13**. Transfers of many shares and securities may theoretically be within the charge to stamp duty but, if the securities are outside the charge to SDRT, there is normally no imperative to stamp the Transfers. If shares or debt instruments unintentionally become chargeable securities, an SDRT charge will arise on an agreement to transfer for valuable consideration. This may be cancelled by stamping a Transfer, but that in itself will require payment of stamp duty unless the Transfer qualifies for a relief. In either case the transfer must be stamped within six years of the agreement to transfer, or it becomes impossible to cancel the SDRT charge (see **10.12**).

11.20 A related difficulty arises where different assets are transferred by means of the same Transfer. This is unlikely to be a problem where UK shares are transferred by means of a stock transfer form, because that form is normally only valid to transfer UK shares. However, where the Transfer takes the form of a declaration of trust (see Example 9.3), it would be possible for a single document to transfer UK and non-UK shares. This should generally be avoided!

Example 11.11 – mixed assets in a single Transfer

Reptile plc, a UK company, agrees to pay market value to buy the shares of two companies, Lizard Ltd and Gecko Ltd, from Reptile's 51% shareholder, Zoo Inc. Lizard is a UK company worth £100,000, Gecko is a Cayman-registered company worth £1 million. The agreement gives rise to an SDRT liability of £500. Reptile intends immediately to carry out a reorganisation which will involve further transfers of the shares of these companies, so it is agreed that Zoo Inc will execute a declaration of trust in favour of Reptile in relation to the shares. Zoo Inc will then eventually transfer the shares to the final shareholder within Reptile's group. Because Reptile is a UK company and makes payment out of a UK bank account, it is likely that any Transfer will relate to 'any matter or thing to be done' in the UK and so will be liable to stamp duty (see **8.17** et seq). If a single declaration of trust is executed, it will be necessary to pay stamp duty in relation to both the Lizard and the Gecko shares – a total of £5,500 – before the Transfer can be regarded as 'duly stamped' thus cancelling the SDRT charge. To avoid the problem, separate declarations of trust should be completed for the two companies, allowing Reptile to stamp only the one relating to the Lizard shares. Alternatively, the SDRT could be paid on the

Lizard shares and the declaration of trust could be left unstamped, but only if Reptile is happy it will never be necessary to produce the declaration of trust for any UK official purpose (see **8.3**).

11.21 Many other risk areas relate to specific reliefs (especially group and reconstruction reliefs), and these are described in relation to those reliefs in **Chapter 9**. It is important to note that the restrictions which may lead to denial of these reliefs only apply at the time of the transaction in respect of which relief is claimed. There is no question of stamp duty reliefs being clawed back as a result of subsequent events, although relief may be denied if there were arrangements for those subsequent events at the time of the original transaction. This is in contrast to the equivalent reliefs under SDLT, where a change in control too soon after the original transaction, even if unrelated to that original transaction, may lead to retrospective disallowance of the relief.

11.22 The other major pitfall in relation to stamp duty concerns the measurement of consideration and, more specifically, the impact of the contingency principle. As noted at **8.38**, under this principle any ascertainable consideration which may be payable on the happening of a particular contingency is regarded as payable and subject to stamp duty. In contrast with SDLT, there is no relief or adjustment if the consideration eventually proves not to be payable.

Example 11.12 – contingency principle

David buys the shares of a trading company from Goliath for consideration of £1 million payable now, plus further consideration equal to 20% of the company's trading profits for the next two years. The further payment is subject to a ceiling of £2 million. Meanwhile, Saul buys a similar company from Goliath for consideration of £1 million, plus 20% of the company's trading profits for the next two years with no ceiling specified. David will have to pay stamp duty of £15,000 (£1 million certain plus £2 million contingent, at 0.5%) but Saul will only have to pay £5,000, because the further consideration is wholly unascertainable.

Unfortunately, there is often a conflict between the desire to minimise stamp duty and the commercial need to limit the contingent consideration. As a result, David's fact pattern is perhaps seen more often than Saul's.

LEGACY LIABILITIES

11.23 When a company or other entity is bought, there is always a risk that the purchased entity has failed to comply with tax obligations and may harbour undisclosed liabilities, including stamp taxes. The purchaser will doubtless minimise the risk by a combination of due diligence investigation and warranties or indemnities. However, unexpected stamp tax liabilities may also arise on purchase of shares or assets, even though the vendor or purchased entity has fully complied with past obligations. This section seeks to highlight the liabilities which may arise.

SDLT

11.24 The acquisition of a company which has claimed group, reconstruction or acquisition reliefs from SDLT may lead to clawback of that group relief, as described at **3.29** et seq.

11.25 Apart from the normal SDLT charge, unexpected SDLT liabilities may also arise on acquisition of existing leases. The acquisition of an existing lease may be treated as the grant of a new lease if certain reliefs were claimed on original grant of the lease – see **4.17** for details of this. If the SDLT liability was not determined when the lease was granted (normally because the rent was variable or not finalised at that point), a new tenant taking on the lease may inherit ongoing obligations and liabilities – see **4.37** for details.

Stamp duty

11.26 In past years, stamp duty has been payable on Transfers of many different categories of assets, such as goodwill, intellectual property and, of course, land and buildings. Various structures and strategies were adopted to reduce or avoid this charge. These strategies and structures often merely defer the crystallisation of the charge, until such time as another event occurs. Stamp duty has been abolished on Transfers of these categories of assets, but that abolition generally only applies to subsequent transactions. Normally, stamp duty remains potentially payable on documents relating to transactions entered into before the time of abolition. It is, therefore, important to understand the old transactions when dealing with the assets in question, in order to avoid having to pay the old stamp duty charges.

Documents executed and retained offshore

11.27 A Transfer of UK situs assets which are within the charge to stamp duty is liable to stamp duty, no matter where executed. However, if the document is executed and retained offshore, payment of any stamp duty may be deferred. This deferral is potentially permanent if the document is never brought to the UK. Alternatively, if it becomes necessary to stamp the document (for example, in order to produce it in evidence in a civil court case), no late stamping penalty will be incurred, provided the document is presented for stamping within

30 days of first arrival in the UK. For documents executed before 1 October 1999, no late payment interest is incurred, provided the duty is paid within 30 days of the first arrival of the document in the UK.

For this reason, many transfers of certain types of asset (but not generally shares or registered land) have been executed and retained outside the UK (typically in the Channel Islands or the Isle of Man). Although stamp duty no longer applies to Transfers of such assets, those Transfers executed under the old rules remain subject to whatever stamp duty charge applied at the time of execution. To retain the benefit of not incurring interest and penalties should it eventually be necessary to stamp them, it is important that the documents are not brought to the UK in the meantime.

Shares and land

11.28 Offshore execution has not been widely used in relation to transfers of shares and securities. This is because a duly stamped Transfer is required (a) for registration of the new owner in the company's books, and (b) since 1986, to cancel the related SDRT charge. For transfers of registered land prior to the introduction of SDLT, a duly stamped transfer was also required to change the owner at the Land Registry, so offshore execution was not appropriate. The alternative approach adopted for land transfers was generally known as 'resting on (or in) contract' or, less commonly, 'split title planning'. It was widely considered that this planning could not be used in Scotland because of differences in land law, although some lawyers maintained that these could be overcome. In an attempt to curtail the planning, any contracts for purchase of land for consideration exceeding £10 million executed after 24 July 2002 were treated as stampable Transfers.

Resting on contract

11.29 Bare legal title to the land was transferred to a nominee who held the land for the benefit of the beneficial owner (for reasons of land law, generally this would be two 'special purpose companies' or SPVs holding as joint trustees). With care, especially in relation to one anti-avoidance provision, such a transfer was liable only to a fixed stamp duty of £5 (50p prior to *FA 1999*). The vendor then entered into a contract to sell the land to the purchaser. The contract entitled the purchaser to call for a Transfer of the land. The purchaser paid the consideration (which gained him a beneficial interest in the land), and the vendor transferred the shares of the SPVs to the purchaser. The purchaser now had a beneficial interest in the land and control of the legal title through ownership of the SPVs which held title. This was generally accepted as equivalent to full legal and beneficial ownership. However, because no Transfer had been executed to transfer title to the purchaser, the purchase contract remained technically uncompleted and virtually no stamp duty had been paid.

If necessary, the purchaser could subsequently require the SPVs to transfer legal title to him. This Transfer would complete the purchase contract and so would crystallise the stamp duty charge on the original purchase. However, the charge would only crystallise at the date of the Transfer, so interest- and penalty-free deferral would have been achieved. In most cases, title was left in the hands of the SPVs, no Transfer was ever executed, and the stamp duty charge was not crystallised. When the time came for the purchaser to sell the land, the process could be repeated, leaving title in the hands of the same SPVs and transferring the shares in those companies. At that point, it became clear that the first sale and purchase would never be technically completed (because title to the land would never be transferred to the original purchaser), so the deferral of stamp duty on that original purchase would become permanent. This could be repeated many times on successive sales and purchases of the land. However, it would always be the case that the stamp duty on the latest sale and purchase in the sequence was merely deferred. That latest stamp duty charge would become payable if the latest purchase was completed by Transfer of legal title from the SPVs to the current beneficial owner.

Therein lies the danger. There are many properties still held in this 'split title' structure. If the structure is collapsed by transferring title from the SPVs to the beneficial owner, any deferred stamp duty on acquisition by the current beneficial owner will become payable. To avoid this problem, the structure must be maintained at least until the next time ownership of the land is transferred. The structure is not effective for saving SDLT, and so may as well be collapsed when a sale is entered into which is subject to SDLT, provided this can be done without also crystallising any old stamp duty charges. To achieve this, it is important that title is transferred directly from the SPVs to the purchaser under the contract which is subject to SDLT. Title must not first be transferred to the vendor under this latest contract.

Example 11.13 – completion of 'resting on contract' transaction

In 1971, Oldmoney Ltd purchased a prime city centre office building from Armateur Ltd for £2 million. The purchase was effected by the 'resting on contract' scheme – so the parties entered into a sale and purchase contract, Oldmoney paid the purchase price and Armateur transferred to Oldmoney the shares of two Jersey-registered SPVs (S1 and S2) which jointly held legal title to the land. No Transfer was completed in conformity with the contract and no stamp duty was paid. At the time, the rate of stamp duty was 2%, so duty of £40,000 was effectively deferred.

On 1 April 2016, Oldmoney contracts to sell the property to Newrich Ltd for £80 million. Newrich will pay SDLT of £3,989,500 (£150,000 at 0%, £100,000 at 2% and £79,750,000 at 5%) on the purchase. The contract provides that Oldmoney will convey the property to Newrich, and Newrich refuses to accept a conveyance from S1 and S2. The property is therefore first conveyed from S1 and S2 to Oldmoney. This crystallises the stamp duty charge deferred in 1971, and Oldmoney has to pay this before completing the onward conveyance to Newrich. Had Newrich accepted the conveyance direct from S1 and S2, the 1971 deferral of stamp duty would have become permanent and Oldmoney would have been spared the need to pay it. Under either scenario, beneficial and legal ownership of the property are reunited, whether in the hands of Oldmoney or directly in Newrich.

A potentially interesting subsidiary point arises. Stamp duty rates have increased since 1997 to a current maximum of 4% but, at each increase, the old rates have been retained for contracts already signed but uncompleted. So, there should be no risk of the increased rates applying on completion of Oldmoney's original purchase. For a brief period between 1 August 1972 and 30 April 1974, rates were reduced, and the reduction also applied to any contracts signed but uncompleted at the start of the period. This means that it should be possible to argue that if the original Oldmoney purchase is completed as noted above, the stamp duty payable should be reduced to £20,000 (£2 million at 1%, being the rate introduced from 1972). HMRC may not agree.

Avoiding a document; the memorandum rule

11.30 The practice of 'resting on contract' described above was really just a specific example of this general principle: because stamp duty is chargeable on certain documents, no stamp duty is payable if no Transfer document (or document which is deemed to be a Transfer) is created. It was the development of share transfers without a Transfer document which led to the introduction of SDRT. SDLT was introduced because avoidance of stamp duty on property transactions was widespread, much of it historically by resting on contract. It was also often possible to avoid a liability on transfers of other assets such as goodwill, intellectual property and interests in partnerships by avoiding creation of the wrong type of document. For such assets, there is generally no legal requirement for a Transfer document in a particular form; compliance with the terms of the contract itself (eg making payment) usually gives effect to the transfer. This means the contract itself was usually the document liable to stamp duty. To avoid stamp duty, it was necessary to avoid creating a written contract. The normal approach was for the vendor to make a written offer to sell to the purchaser. The purchaser accepted the offer by his action – paying the purchase price to the vendor's solicitors, for example – and not by anything in writing. In some cases, the purchaser accepted orally, the acceptance being video-recorded and given in front of reliable witnesses!

11.31 In relation to transactions of this kind, HMRC (relying on cases such as *Associated British Engineering Ltd v IRC* ([1941] 1 KB 15) considered that any document which recorded the agreement in such a way as to become a primary record of the contract was subject to stamp duty. On this basis, any memorandum which the purchaser prepared to record the terms could be subject to stamp duty, even if prepared purely for internal purposes and a long time after the original transaction. Of course, this was only relevant if there was a need to produce such documents for official purposes – but that included proving the terms of the contract to HMRC for other tax purposes!

As time passes, it becomes less likely that any such memoranda will be created or required for official purposes. However, taxpayers who were purchasers under such undocumented contracts should continue to think carefully before 'tidying up' old arrangements by creating written records for posterity.

Last word

11.32 The idea of countering 'tax avoidance' has great popular appeal. However, in relation to stamp taxes as other taxes, much 'planning' consists of trying to ensure complex rules do not give rise to commercially damaging liabilities. Stamp taxes are very different from taxes on income and profits, and are often encountered only occasionally. As a result, the risk of accidental and unforeseen liabilities is high. It is hoped that this book may help reduce that risk.

Appendix A

Addresses, contact details, etc

SDLT RETURNS

If SDLT returns are not filed electronically, original paper returns and any accompanying cheques should be sent to the HMRC Data Capture Centre at:

HM Revenue & Customs – SDLT
Netherton
Merseyside
L30 4RN
United Kingdom

Alternatively, they can be sent via the DX system to:

HM Revenue & Customs SDLT
Rapid Data Capture Centre
DX 725593
Bootle 9
United Kingdom

Returns should not be folded and therefore require an A4 or larger size envelope. Only forms SDLT1 to 4, plus plans if provided and cheques if that payment method is used, should be included – correspondence should be sent to Birmingham Stamp Office as noted below (see **6.8**).

In case of postal delay the recommendation is that at least three working days are allowed for the return to reach HMRC.

DOTAS FORMS

Paper copies of forms relating to DOTAS (see **7.26**) may be ordered from the Anti-Avoidance Group on 03000 588 993, but they may be more conveniently downloaded from www.gov.uk/guidance/forms-to-disclose-tax-avoidance-schemes, or completed online following links from that webpage. If not completed and submitted online, completed DOTAS forms must be sent to:

HM Revenue and Customs – Counter Avoidance
DOTAS Taskforce S0483
Newcastle
NE98 1ZZ
United Kingdom

OTHER FORMS AND LEAFLETS

Paper copies of other forms and explanatory leaflets may be ordered from the HMRC publications department. The postal address is PO Box 37, St Austell PL25 5YN, but HMRC strongly prefer requests to be made online using the form at www.hmrc.gov.uk/contactus/staustellform.htm or by phone to 0300 200 3511.

SCOTTISH TRANSACTIONS

As SDLT ceased to apply in relation to transactions over land situated in Scotland, with effect from 1 April 2015, an SDLT5 certificate will only be relevant for such a transaction with an effective date on or after that date if, unusually, the transaction remains liable to SDLT under the transitional provisions (see **1A.20** et seq). If the SDLT return is filed online then it should be possible to print an SDLT 5 straightaway. If a paper return is being filed then HMRC will send the SDLT 5 by post. The SDLT 5 should then be sent to the Registers of Scotland to register the transaction. It may be possible to ask for the certificate urgently if, for example, the buyer could lose the property if there is a delay in registration. (www.gov.uk/guidance/sdlt-scottish-transactions#transactions-on-or-after-1-april-2015).

- Where a certificate is needed within two days to avoid a risk of losing the property, appropriate details including scanned copies of documents can be emailed to the stamp office, followed up with original documents sent to the Birmingham Stamp Office at the address noted below.
- Where a paper return has been submitted but the form SDLT5 has not been received after 16 days, Edinburgh Stamp Office should be contacted on 03000 561 627. They will require copies of the SDLT1 and other documents to be faxed to 03000 561 703 and, providing the forms are all correct, will issue a handwritten SDLT5.

Edinburgh Stamp Office is located at:

HM Revenue & Customs
Edinburgh Stamp Office
Elgin House

20 Haymarket Yards
Edinburgh
EH12 5WN

DX ED 543303 Edinburgh 33

OTHER CORRESPONDENCE

All other stamp tax correspondence for HMRC must be sent to the Birmingham Stamp Office. The address is:

HM Revenue and Customs – Birmingham Stamp Office
[Reference]
9th Floor, City Centre House
30 Union Street
Birmingham
B2 4AR
United Kingdom

Fax 03000 570 316

Alternatively the DX system may be used; the address is:

HM Revenue and Customs – Birmingham Stamp Office
[Reference]
DX 15001
Birmingham 1

An appropriate reference or other description should be inserted in place of '[Reference]', as noted in the table below, to ensure that the correspondence is directed to the right section:

Purpose of correspondence	[Reference]	Paragraphs in this book; notes
Submission of further information at same time as SDLT return is submitted*	SDLT (return, further information)	**6.47**
Amendment of SDLT return; submission of further return after ordinary return was submitted for original transaction*	SDLT amended/ further return	**6.17**
Advice on format of schedules listing multiple properties	Complex Transactions Unit	**6.26**
Application for deferment of SDLT	SDLT Deferment Application	**6.33**

Purpose of correspondence	**[Reference]**	**Paragraphs in this book; notes**
Claim for relief from excessive SDLT assessment, after normal period to amend return has expired*	SDLT Excessive Assessment claim	**6.51**
Submission of documents for routine stamping, including provisional stamping or 'wait and see' cases	Stamp Duty section	**8.41–8.44** If same-day stamping required, call 0121 616 4513 first
Submission of documents for adjudication, including claims to relief where adjudication is compulsory	Stamp Duty section (adjudication, *insert reference to the relief, eg Group Relief*)	**8.47**; **9.7**
Claim for repayment of stamp duty	Stamp Duty section (repayment claim)	**8.53**
Payment of bearer instrument duty	Stamp Duty Section (bearer instruments)	**8.55**
Enquiry as to status of foreign entity; enquiry/claim to be recognised as an Intermediary	Stamp Duty Section (Technical)	**9.8**; **9.37**; **10.22**
Notifying and paying SDRT (other than Financial Markets transactions)	SDRT	**10.9**

* UTRN of original return must be quoted in the heading to the letter; see **6.7**.

SAME DAY STAMPING

In exceptional circumstances it may be essential to have documents relating to transfers of shares and securities stamped on the day of execution, or very rapidly afterwards (see **8.45**). HMRC may agree to arrange this but only if they are convinced it is absolutely necessary. It must be arranged in advance by writing to the Birmingham Stamp Office same day stamping service or, in exceptional circumstances by calling 03000 200 3510. Note this is a dedicated line and will not deal with any other matters. In Scotland such matters used to be dealt with by the Edinburgh Stamp Office, but their stamping press was removed on 7 April 2014 and all stamping of documents now takes place in Birmingham.

THE STAMP TAX HELPLINE

HMRC assert that answers to most stamp tax questions can be found on their website, starting from www.hmrc.gov.uk/so. Although most information probably is on the website, it can be frustratingly difficult to find; where possible, this book gives addresses of appropriate pages. Furthermore, the website does not (and realistically cannot) deal with all technical nuances arising from real-life transactions. A telephone enquiry line is provided on 0300 200 3510 and the fax number is 03000 570 316. Unfortunately, like so many 'enquiry' numbers, this is answered by a recorded voice offering a limited range of menu options and offering lengthy homilies on subjects likely to be of no relevance to your query. If you do get through to a human being, he/she is not likely to be a technical expert and is unlikely to provide definitive answers to difficult questions. However, with persistence, it is possible to have questions referred to such experts who, in the author's experience, are generally helpful. If you obtain a name, it is possible to contact the person by email – HMRC email addresses are in the form firstname.lastname@hmrc.gsi.gov.uk.

TAX PAYMENTS

Cheques should be payable to HMRC (*insert name of tax, eg SDLT*). For SDLT, the UTRN of the return to which the payment relates should be added to the payee line. However, for all three taxes, HMRC prefer electronic payment, either by direct bank transfer or by online card payment. HMRC maintain different bank accounts for the different taxes. Bank account details (including SWIFT codes and IBAN numbers for payments from outside the UK) may be found via links at www.hmrc.gov.uk/payinghmrc. The bank accounts have changed in recent years, so it is worth checking to ensure that current details are used (to avoid possibly being out of date, the details are deliberately not reproduced here). If payments are made electronically, it is important to quote an appropriate reference. For SDLT, this is the UTRN which appears on the paper return and payslip (see **6.7**) or on the electronic SDLT5 issued where returns are submitted online (see **6.11**). For stamp duty and SDRT, taxpayers may create their own reference numbers, but these should be quoted on the physical documents submitted for stamping or other correspondence, so that HMRC can easily link the payment to the document.

THE TRIBUNAL SERVICE

It is unlikely to be necessary to contact the Tribunal Service unless a matter is already in dispute with HMRC and under appeal – in which case, HMRC should provide appropriate contact details. The general address for correspondence, or to obtain a paper form for an appeal to the First-tier Tribunal is noted below. Forms and related information are available from the website.

First-tier Tribunal (Tax Chamber)
HM Courts & Tribunals Service

PO Box 16972
Birmingham
B16 6TZ

Telephone 0300 123 1024

Email: taxappeals@hmcts.gsi.gov.uk

Website: www.gov.uk/tax-tribunal

Appendix B

Sample forms etc

PART 1 – SDLT RETURN FORMS

Paying HMRC

Secure and efficient ways to pay

Make sure you pay by the deadline.

Bank details for online or telephone banking, CHAPS, Bacs
You can make a transfer from your bank account by Faster Payments, CHAPS or Bacs. Use sort code 08-32-10, account number 12001020, account name 'HMRC Shipley' followed by the reference on the payslip.

By debit or credit card online
Go to www.gov.uk/pay-tax-debit-credit-card and follow the instructions.

At your bank or building society
Take your payslip to your branch and pay by cash or cheque.

Make cheques payable to 'HM Revenue and Customs only' followed by the reference on the payslip.

At the Post Office
Take your payslip to any participating Post Office and pay by debit card, cash or cheque made payable to 'Post Office Ltd'.

Payment questions?

For further payment information, go to www.gov.uk/dealing-with-hmrc/paying-hmrc

Other ways to pay

By cheque through the post
Make your cheque payable to 'HM Revenue and Customs only' followed by the reference on the payslip. Send your payslip and cheque to:
HM Revenue and Customs SDLT
NETHERTON
Merseyside
L30 4RN

Don't fold either the payslip or the cheque or fasten them together. A stamp for the correct postage is required.

By DX
As above, but send to:
HMRC SDLT
Rapid Data Capture Centre
DX725593
Bootle 9

5014538 HMRC 09/15

Please do not write or mark below this perforation

Santander Corporate & Commercial
Bootle Merseyside L30 4UA

Trans cash

Payslip

HM Revenue & Customs

bank giro credit

ABC

159

24

Reference: 311785749MG

Credit account number: 225 0357

Amount due (no fee payable at PO counter) £

CHEQUE ACCEPTABLE

For official use

Name

Cashier's stamp and initials

Signature

Date

For official use

Barclays Bank Plc
Automated Bulk Credit Clearing
HM Revenue and Customs
Account number 63464695
11-47

CASH

CHEQUE

£

25-03-57

SDLT1/P

HMRC 09/15

Please do not fold this payslip or write or mark below this line

311785749MG &7242250357 000000000 74 X

HM Revenue & Customs

Land Transaction Return

Additional vendor/purchaser details

When to fill in this return

You must fill in this return for each additional vendor and/or purchaser. The guidance notes will help you answer the questions.

If you need help with any part of this return or with anything in the guidance notes, please phone the Stamp Taxes enquiry line on **0300 200 3510**. For our opening hours go to **www.hmrc.gov.uk/contactus**
You can get further copies of this return from the Orderline on **0300 200 3511**.

Reference

Insert the reference number from the payslip on page 7 of the Land Transaction Return, SDLT1, here.

For official use only

Vendor or Purchaser

1 **Please indicate if this return is for an additional vendor or an additional purchaser.** Put 'X' in one box

Vendor Complete section below only

Purchaser Complete section below and the section over the page

Complete in all cases

2 **Title** Enter Mr, Mrs, Miss, Ms or other title
Note: only complete for an individual

3 **Surname or company name**

4 **First name(s)**
Note: only complete for an individual

5 **Address**
Postcode

House or building number

Rest of address, including house name, building name or flat number

Please turn over for additional purchaser details

SDLT2 v3 PG 1

HM Revenue & Customs

Land Transaction Return

Additional details about the land

When to fill in this return

Fill in this return when you cannot fit all the details on the main Land Transaction Return, SDLT1. The guidance notes will help you to answer the questions.

If you need help with any part of this return or with anything in the guidance notes, please phone the Stamp Taxes Helpline on **0300 200 3510**. For our opening hours go to **www.hmrc.gov.uk/contactus**

You can get further copies of this return from the orderline on **0300 200 3511**.

REFERENCE

Insert the reference number from the payslip on page 7 of the Land Transaction Return, SDLT1, here.

For official use only

ABOUT THE LAND

1 **Type of property**

Enter code from the guidance notes

2 **Local authority number**

3 **Title number, if any**

4 **NLPG UPRN**

5 **Address or situation of land**

Postcode

House or building number

Rest of address, including house name, building name, flat number or continuation from the SDLT1

Is this a continuation of the address in box 28 of the SDLT1 return

Yes No

6 **If agricultural or development land, what is the area (if known)?**

Hectares Square metres

Area

7 **If there are any minerals or mineral rights reserved enter the code below**

Enter code from the guidance notes

8 **Is a plan attached?** Put 'X' in one box

Note: the UTRN should be written on the plan

Yes No

9 **Interest transferred or created**

Enter code from the guidance notes

SDLT3 v3 PAGE 1 5014135 HMRC 02/14

HM Revenue & Customs

Land Transaction Return

Additional details about the transaction, including leases

When to fill in this return

You must fill in this return where additional information about the transaction and/or lease can be provided. The guidance notes will help you to answer the questions.

If you need help with any part of this return or with anything in the guidance notes, please phone the Stamp Taxes Helpline on **0300 200 3510**. For our opening hours go to **www.hmrc.gov.uk/contactus**

You can get further copies of this return from the orderline on **0300 200 3511**.

REFERENCE

Insert the reference number from the payslip on page 7 of the Land Transaction Return, SDLT1, here.

For official use only

ABOUT THE TRANSACTION

1 **If this transaction is part of the sale of business, what does the sale include?**
Put 'X' in relevant boxes

Stock   Goodwill

Other   Chattels and moveables

What is the total amount of the consideration for the sale of the business apportioned to these items?

2 **If the property is for commercial use, what is it?**
Put 'X' in the appropriate box(es)

Office

Hotel

Shop

Warehouse

Factory

Other

Other industrial unit

3 **Have you applied for and received a post transaction ruling in accordance with Code of Practice 10, or asked us for advice on the application of the law to this transaction?**
Put 'X' in one box

Yes   No

If 'Yes' have you followed it when completing this return?
Put 'X' in one box

Yes   No   Ruling not received

4 **Is any part of the consideration contingent or dependent on uncertain future events?**

Yes   No

5 **Have you agreed with HM Revenue & Customs that you will pay on a deferred basis?**

Yes   No

6 **If there are any minerals or mineral rights reserved enter the code below**

Enter code from the guidance notes

7 **Give a description of the purchaser(s)**
Enter code from the guidance notes

SDLT4 v4   PAGE 1   SD14146   HMRC 02/14

PART 2 – DOTAS FORMS

HM Revenue & Customs

Disclosure of avoidance scheme
(Notification by scheme promoter)

Scheme reference number

(for HMRC official use only)

Who should use this form?

This form is for use by a scheme promoter notifying under:

- Section 308, Finance Act 2004; and/or
- Regulation 8 of the National Insurance Contributions (Application of Part 7 of the Finance Act 2004) Regulations 2012 (SI 2012, Number 1868).

Guidance on making a disclosure is available on our website, go to **www.hmrc.gov.uk**

Scheme promoter's details

Name of promoter

Address of promoter

Address

Postcode

Email address of promoter (if any)

Phone number of promoter

Internal reference number of promoter (if any)

Name of co-promoter

Address of co-promoter

Address

Postcode

Scheme details

Title of the arrangements (if any)

Specify the provision under which the disclosure is being made.
Tick any box that applies

SDLT Avoidance - acquisition of chargeable interest (SI 2005, Number 1868) ☐

Please indicate which type of property is the subject of the SDLT arrangement (optional)

- Residential property ☐
- Non-residential property ☐
- Use of property not known or mixed ☐

IHT avoidance - transfers into trust (SI 2011, Number 170) ☐

NICs avoidance (SI 2012, Number 1868) ☐

IT, CT and CGT avoidance (Prescribed Descriptions of Arrangements) Regulations (SI 2006, Number 1543) ☐

For IT, CT, CGT and NICs avoidance only, tick the regulation below that applies. Where more than one regulation applies, please specify only the main regulation.

- Regulation 6 Confidentiality ☐
- Regulation 8 Premium fee ☐
- Regulation 9 Off market terms ☐
- Regulation 10 Standardised tax products ☐
- Regulation 12 Loss schemes ☐
- Regulation 13 Leasing arrangements ☐
- Regulation 17A Pensions ☐

AAG1 Page 1 HMRC 10/12

Summary of proposal or arrangements

Explanation of each element in the proposal or arrangements from which the expected tax and/or NICs advantage arises

If you need more space continue on form AAG5 (continuation sheet)

Statutory provisions relevant to those elements of the proposal or arrangements from which the expected tax and/or NICs advantage arises

If you need more space continue on form AAG5 (continuation sheet)

Declaration

The information I have given on this form, and any continuation sheets, is correct and complete to the best of my knowledge and belief.

Signature

Date *DD MM YYYY*

Name of signatory (use capital letters)

Capacity in which signed

Number of forms AAG5 (continuation sheet) used

You should send this form to the:
Anti-Avoidance Group (Intelligence)
HM Revenue & Customs
CTIAA Intelligence S0528
PO Box 194
Bootle
L69 9AA

Page 2

HM Revenue & Customs

Disclosure of avoidance scheme (Notification by scheme user where offshore promoter does not notify)

Scheme reference number

(for HMRC official use only)

Who should use this form?

This form is for use by a scheme user notifying under:

- Section 309, Finance Act 2004; and/or
- Regulation 10 of the National Insurance Contributions (Application of Part 7 of the Finance Act 2004) Regulations 2012 (SI 2012, Number 1868).

Guidance on making a disclosure is available on our website, go to **www.hmrc.gov.uk**

Scheme user's and promoter's details

Name of scheme user making disclosure

Address of scheme user making disclosure

Address

Postcode

Email address of scheme user making disclosure

Phone number of scheme user making disclosure

Internal reference number of scheme user (if any)

Name of promoter

Address of promoter

Address

Postcode

Scheme details

Title of the arrangements (if any)

Specify the provision under which the disclosure is being made.
Tick any box that applies

SDLT Avoidance - acquisition of chargeable interest (SI 2005, Number 1868) ☐

Please indicate which type of property is the subject of the SDLT arrangement (optional)

Residential property ☐

Non-residential property ☐

Use of property not known or mixed ☐

IHT avoidance - transfers into trust (SI 2011, Number 170) ☐

NICs avoidance (SI 2012, Number 1868) ☐

IT, CT and CGT avoidance (Prescribed Descriptions of Arrangements) Regulations (SI 2006, Number 1543) ☐

For IT, CT, CGT and NICs avoidance only, tick the regulation below that applies. Where more than one regulation applies, please specify only the main regulation.

Regulation 6 Confidentiality ☐

Regulation 8 Premium fee ☐

Regulation 9 Off market terms ☐

Regulation 10 Standardised tax products ☐

Regulation 12 Loss schemes ☐

Regulation 13 Leasing arrangements ☐

Regulation 17A Pensions ☐

Summary of proposal or arrangements

Explanation of each element in the proposal or arrangements from which the expected tax and/or NICs advantage arises

If you need more space continue on form AAG5 (continuation sheet)

Statutory provisions relevant to those elements of the proposal or arrangements from which the expected tax and/or NICs advantage arises

If you need more space continue on form AAG5 (continuation sheet)

Declaration

The information I have given on this form, and any continuation sheets, is correct and complete to the best of my knowledge and belief.

Signature

Date *DD MM YYYY*

Name of signatory (use capital letters)

Capacity in which signed

Number of forms AAG5 (continuation sheet) used

You should send this form to the:
Anti-Avoidance Group (Intelligence)
HM Revenue & Customs
CTIAA Intelligence S0528
PO Box 194
Bootle
L69 9AA

Page 2

HM Revenue & Customs

Disclosure of avoidance scheme

(Notification by scheme user where no promoter, or promoted by lawyer unable to make full notification)

Scheme reference number

(for HMRC official use only)

Who should use this form?

This form is for use by a scheme user notifying under:

- Section 310, Finance Act 2004; and/or
- Regulation 11 of the National Insurance Contributions (Application of Part 7 of the Finance Act 2004) Regulations 2012 (SI 2012, Number 1868).

Guidance on making a disclosure is available on our website, go to **www.hmrc.gov.uk**

Scheme user's details

Name of scheme user making disclosure

Address of scheme user making disclosure

Address

Postcode

Unique Taxpayer Reference (UTR)

Email address of scheme user making disclosure

Phone number of scheme user making disclosure

Scheme user's internal reference number (if any)

Scheme details

Title of the arrangements (if any)

Specify the provision under which the disclosure is being made. *Tick any box that applies*

SDLT Avoidance - acquisition of chargeable interest (SI 2005, Number 1868) ☐

Please indicate which type of property is the subject of the SDLT arrangement (optional)

- Residential property ☐
- Non-residential property ☐
- Use of property not known or mixed ☐

IHT avoidance - transfers into trust (SI 2011, Number 170) ☐

NICs avoidance (SI 2012, Number 1868) ☐

IT, CT and CGT avoidance (Prescribed Descriptions of Arrangements) Regulations (SI 2006, Number 1543) ☐

For IT, CT, CGT and NICs avoidance only, tick the regulation below that applies. Where more than one regulation applies, please specify only the main regulation.

- Regulation 6 Confidentiality ☐
- Regulation 8 Premium fee ☐
- Regulation 9 Off market terms ☐
- Regulation 10 Standardised tax products ☐
- Regulation 12 Loss schemes ☐
- Regulation 13 Leasing arrangements ☐
- Regulation 17A Pensions ☐

Summary of proposal or arrangements

Explanation of each element in the proposal or arrangements from which the expected tax and/or NICs advantage arises

If you need more space continue on form AAG5 (continuation sheet)

Statutory provisions relevant to those elements of the proposal or arrangements from which the expected tax and/or NICs advantage arises

If you need more space continue on form AAG5 (continuation sheet)

Declaration

The information I have given on this form, and any continuation sheets, is correct and complete to the best of my knowledge and belief.

Signature

Date *DD MM YYYY*

Name of signatory (use capital letters)

Capacity in which signed

Number of forms AAG5 (continuation sheet) used

You should send this form to the:
Anti-Avoidance Group (Intelligence)
HM Revenue & Customs
CTIAA Intelligence S0528
PO Box 194
Bootle
L69 9AA

Page 2

HM Revenue & Customs

Disclosure of SDLT avoidance scheme – notification of scheme reference number

Scheme details

Name of scheme user

Title number of the relevant property (if allocated)

Address of scheme user

Address

Postcode

Address of the relevant property

Address

Postcode

Phone number of scheme user

Market value of the relevant property

Unique transaction reference number (UTRN) (if allocated)

Date of first land transaction *DD MM YYYY*

Scheme reference number

Who should use this form?

You should use this form if you have received an 8-digit scheme reference number (SRN) from HM Revenue & Customs (HMRC), a scheme promoter, or the client of a scheme promoter, and you expect to obtain a Stamp Duty Land Tax (SDLT) advantage. This will usually be because you are the purchaser of a property.

If you are a partner in a partnership, read the notes about 'How to complete this form' on page 2.

Declaration

The information I have given on this form, and any continuation sheets, is correct and complete to the best of my knowledge and belief.

Signature

Date *DD MM YYYY*

Name of signatory (*use capital letters*)

Capacity in which signed

If the scheme user named above is also notifying as the representative partner on behalf of other partners in a partnership to which he belongs, indicate here the number of additional partners who are liable to notify, and list them separately

For guidance on completing this form please see the notes overleaf or go to **www.hmrc.gov.uk**

AAG4 (SDLT) Page 1 HMRC 10/12

Notes

Background

Details of certain SDLT avoidance schemes must be notified to HMRC, usually by the scheme promoter. The user of the scheme must make the notification where the scheme is devised 'in-house', the promoter is outside the UK and no promoter notifies, or the information is to be covered by legal professional privilege.

When a SDLT scheme is notified, HMRC may issue a SRN to the person making the notification. A promoter to whom a SRN is issued must provide it to each client who implements the scheme. If there are other users of the scheme the client must provide the SRN to them.

If you have received a SRN, whether from HMRC, the scheme promoter, or a client of the promoter, you are required to declare the use of the scheme to HMRC. This is done by entering the SRN and certain other information on this form.

How to complete this form

Partnerships

If you are a partner in a partnership, you should consider any expected SDLT advantages from the perspective of your own liabilities to SDLT, notwithstanding that the partnership may be regarded as a legal person or corporate body under the law of the country or territory under which it is formed.

If you and other members of the partnership have a duty to notify the SRN, a representative partner may make a single notification on behalf of the others provided that the names of all of those with a duty to notify are scheduled and attached to this form.

Title number of the relevant property

Enter the title number of the property forming the subject of the arrangements, if any is allocated.

Address of relevant property

Enter the address of the property forming the subject of the arrangements.

Unique transaction reference number (UTRN)

Enter the UTRN for the land transaction return relating to the transaction on which you expect to obtain an SDLT advantage. The UTRN can be found either on the submission receipt (called the Submission Receipt and Electronic SDLT5 certificate) if you have filed your return online, or at the bottom of page 2 and on the payslip on page 7 of your paper SDLT return.

Open market price of the relevant property

Enter the price which the property might reasonably be expected to fetch if on sale on the open market. If a formal valuation has not been obtained this should be a reasoned estimate.

Scheme reference number (SRN)

Enter in the boxes the 8-digit SRN provided to you.

Date of first land transaction

Enter the effective date of the first land transaction which forms part of the scheme.

When to send in this form

You must send in this form within 30 days of the first land transaction which forms part of the scheme or receipt of the SRN, whichever is later. Submission should not be delayed if for example a formal valuation of the property has not been obtained.

Where to send this form

You should send this form to:

Anti-Avoidance Group (Intelligence)
HM Revenue & Customs
CTIAA Intelligence S0528
PO Box 194
Bootle
L69 9AA

Any SDLT return due to be submitted should not accompany this form and should be sent to HMRC in the normal way.

HM Revenue & Customs

Disclosure of Tax Avoidance Schemes
Notification of Scheme Reference Number

You are involved in a Disclosed Tax Avoidance Scheme

Your scheme is **NOT** HM Revenue and Customs (HMRC) approved

You may not have been aware that you are involved in a Disclosed Tax Avoidance Scheme. However, you need to understand your responsibilities or you could face **significant penalties.** Being involved in tax avoidance means:

- HMRC may subject your tax affairs to close scrutiny and you may be marked out as a high risk taxpayer
- you may find yourself tied up in a dispute, which can last for years
- you may soon receive an Accelerated Payment Notice requiring you to pay any claimed tax saving
- you may end up in front of a tax tribunal with your tax affairs being made public
- you may have to **pay more in tax, interest and penalties** than the scheme claims to save you

Can you still withdraw?

If you are worried about what you have signed up to, want to withdraw from it, or discuss the implications with HMRC, phone **03000 588 993**.

For more detailed guidance, go to **www.gov.uk/government/publications/disclosure-of-tax-avoidance-schemes-guidance**

Failure to comply with any of the requirements below may lead to payment of significant penalties

Receipt of this notice means:

- you must report this Scheme Reference Number (SRN) to HMRC for all affected years on your tax returns or, in specific circumstances, on form AAG4, 'Disclosure of avoidance scheme' (or their IHT, SDLT or ATED equivalent forms, if applicable) - go to **www.gov.uk/forms-to-disclose-tax-avoidance-schemes** for information on how to do this and to download a copy if you are required to pass this form on
- when entering the SRN on a tax return, use **only** the boxes specified for that purpose - **don't enter the SRN in the white space** - the guidance notes for the tax return in question will specify where to put the SRN, which may be on special or additional pages
- you must pass a completed copy of this form to anyone you know may be a party to the scheme and who may obtain a tax or National Insurance contributions advantage - if you pass this form to another person enter the date (DD MM YYYY) below:
- if you are an employer you must complete form AAG7, 'Disclosure of Tax Avoidance Schemes' and provide the SRN to each employee who is a party to the scheme

Important information on page 2.

Details to be provided by the promoter

1 **Promoter's full name**

2 **Promoter's full address**

Postcode

3 **Scheme Reference Number (SRN)**

4 **Brief description or name of the scheme**

5 **Date** DD MM YYYY

Things tax avoiders should consider

The vast majority of people in the UK don't try to avoid paying tax. They pay on time and without intervention from HM Revenue and Customs (HMRC). Some tax avoiders are not aware of the consequences.

Your scheme is not HMRC approved

Getting an avoidance Scheme Reference Number (SRN) doesn't mean HMRC has in any way approved or cleared the scheme. HMRC issues these numbers to identify and track you as a user of an avoidance scheme.

Most schemes don't work

You may be told that avoidance is legal but if the scheme doesn't work you'll have made an incorrect tax return which is not in accordance with the law. You are legally obliged to pay tax that is due and you may be charged interest and penalties if you try to avoid it.

It could cost you more than you bargained for

Avoidance schemes are complex. They can give rise to unintended additional tax consequences, and the fees you pay the promoter don't count as tax paid. So you could end up **paying much more than just the tax** you're trying to avoid.

You may have significant legal fees to pay

If the scheme is taken to litigation, you're likely to have large **legal fees to pay**. Your promoter may ask you to pay into a 'fighting fund' up front.

You could face criminal conviction

If you deliberately mislead or conceal information from HMRC, you could be prosecuted and convicted.

You could face negative publicity as a tax avoider

If you are named in court papers when a scheme is litigated, or your name is in a public register, you could be reported in the media as a tax dodger.

You'll have to pay the tax up front anyway

You won't get a cash flow advantage while HMRC investigates a scheme. New legislation means you'll have to **pay the disputed tax up front**.

PART 3 – PRECEDENT LETTERS FOR STAMP DUTY CLAIMS UNDER FA 1986, SS 75 AND 77

These letters are based on model letters provided by HMRC in Stamp Taxes Bulletin 2/2013, available at (www.hmrc.gov.uk/so/bulletin02-2013.pdf.) Although the letters for the two reliefs are similar, there are sufficient differences to make it worthwhile to reproduce them separately. They are written on the assumption that all companies are UK registered, that certain details are set out in an agreement governing the transaction, and that the letter is submitted by an adviser acting for the claimant. They have been slightly adapted from HMRC's original precedent letters, previously set out in leaflets ADJ473 and 475, to recognise changes in the UK Companies Acts, for example removing the concept of authorised share capital. They may require adapting further to accommodate characteristics of overseas companies or to reflect the precise form of the agreement. Where words may or may not be required, depending on circumstances, they are enclosed in square brackets. Information to be inserted is indicated in [square brackets] and in italics. In places the language used is quaint, but it is as well to use the approved wording so far as it fits the facts. It is also worth reviewing the draft agreement between the parties before execution, to ensure its provisions accord with the statements which must be made in the letter of claim.

(a) Section 75 claim

Dear Sir

[name of claimant company]

1. We act for *[name of claimant company]* ("the Acquiring Company").

2. In connection with the transactions referred to below we hereby apply on behalf of the Acquiring Company for exemption from transfer duty under Section 75 Finance Act 1986.

3. The Acquiring Company, whose registered office is at *[address]* was incorporated in *[country]* on *[date]* under *[name of law]* with number *[registered number]*. A copy of the certificate of incorporation [and of certificate(s) of change of name] is enclosed marked "A".

4. *[Name of target company]*, ("the Target Company"), whose registered office is at *[address]* was incorporated in *[country]* on *[date]* under *[name of law]* with number *[registered number]*. A copy of the certificate of incorporation [and of certificate(s) of change of name] is enclosed marked "B". A list of all the members of the Target Company immediately prior to *[date of transaction]* certified by the [Registrars] [Company Secretary] of the Target Company, is enclosed marked "C".

[Note – a computer print-out of the Register of Members may be used as "C" and need not be certified]

5. The Acquiring Company has acquired [the whole][part] of the undertaking of the Target Company in pursuance of a scheme for the reconstruction of the Target Company in order that *[set out brief reasons for explanation of the transaction]*.

6. By an agreement dated *[date]* and made between the Target Company and the Acquiring Company ("the Agreement"), it was provided (inter alia) that the Target Company should sell and the Acquiring Company should purchase [the whole][part] of the undertaking of the Target Company as described in *[refer to appropriate part of the Agreement]* ("the Business") in pursuance of a scheme for the reconstruction of the Target Company and that as consideration for such sale the Acquiring Company should allot credited as fully paid to all the shareholders of the Target Company the respective numbers of shares of the Acquiring Company as specified in *[refer to schedule or other part of the Agreement setting out the numbers and classes of shares to be allotted to each shareholder of the Target Company]* ("the Consideration Shares"). A copy of the Agreement is enclosed marked "D".

7. The said sale was duly completed on *[date]* when the Target Company transferred the business to the Acquiring Company.

8. At a meeting of [the Directors]*[or specify committee or similar]* of the Acquiring Company held on *[date]* the consideration shares were duly allotted to the shareholders of the Target Company pursuant to the provisions of the Agreement. We enclose marked "E" [and "F" respectively] a certified copy of the resolution of the Directors of the Acquiring Company passed on *[date]* [appointing the said committee and of the Resolution of the said Committee] making such allotment. We also enclose marked "G" a certificate under the hand of [the senior official of the Registrars][the company secretary] of the Acquiring Company, confirming that the names of the respective allottees of the Consideration Shares have been entered in the Register of Members of the Acquiring Company in respect of the Consideration Shares, together with a list of all members of the Acquiring Company immediately following the allotment, certified by the [Registrars][company secretary] of the Acquiring Company and marked ["H"].

[Note – as in (4) above, a computer printout may be substituted and need not be certified]

9. It is confirmed that no part of the consideration for the acquisition consisted of the issue of redeemable shares in the Acquiring Company.

10. It is confirmed that, immediately after the acquisition:

    a. Each shareholder of the Target Company was also a shareholder of the Acquiring Company;

    b. Each shareholder of the Acquiring Company was also a shareholder of the Target Company;

    c. Each shareholder held the same proportion (or as nearly as may be the same proportion) of shares in the Target Company as that shareholder held in the Acquiring Company.

11. [No application for clearance under the relevant direct tax provisions] [An application for clearance under [section numbers]] was made by the Acquiring Company. [A copy of the application together with copies of the correspondence with HMRC is enclosed marked ["I"]].

[Note – if no clearance application was made, the information listed below must be supplied]

12. It is submitted that the acquisition was effected for bona fide commercial reasons and did not form part of a scheme or arrangement of which the main purpose or one of the main purposes is the avoidance of liability to stamp duty, income tax, corporation tax or capital gains tax and that the appropriate conditions of Section 75 of the Finance Act 1986 have been complied with, and accordingly exemption from ad valorem stamp duty under the head "Transfer on Sale" is claimed in respect of the Transfers executed pursuant thereto.

13. We enclose for adjudication [give adequate list or description of the documents enclosed] together with certified copies.

Further information to be provided if no clearance application was made as set out in (11) above:

1. A copy of the latest accounts of the Target Company;
2. Full details of any scheme, arrangement or overall transaction of which this transaction forms part;
3. Confirmation that the Acquiring Company still holds the Business;
4. A <u>detailed</u> note of the bona fide commercial reasons for the acquisition.

(b) Section 77 claim

Dear Sir

[name of claimant company]

We act for *[name of claimant company]* ("the Acquiring Company").

1. In connection with the transactions referred to below we hereby apply on behalf of the Acquiring Company for exemption from transfer duty under Section 77 Finance Act 1986.

2. The Acquiring Company, whose registered office is at *[address]* was incorporated in *[country]* on *[date]* under *[name of law]* with number *[registered number]* and immediately prior to *[date of transaction]* the issued share capital of the Acquiring Company was *[amount]* divided into *[give details of the classes, denominations and numbers of shares]* such shares being fully paid up *[or give details if shares were part paid]*. A copy of the certificate of incorporation [and of certificate(s) of change of name] is enclosed marked "A".

3. *[Name of target company]*, ("the Target Company"), whose registered office is at *[address]* was incorporated in *[country]* on *[date]* under *[name of law]* with number *[registered number]* and immediately prior to *[date of transaction]* the issued share capital of the Target Company was *[amount]* divided into *[give details of the classes, denominations and numbers of shares]* such shares being fully paid up *[or give details if shares were part paid]*. A copy of the certificate of incorporation [and of certificate(s) of change of name] is enclosed marked "B". A list of all the members of the Target Company immediately prior to *[date of transaction]* certified by the [Registrars][Company Secretary] of the Target Company, is enclosed marked "C".

[Note – a computer print-out of the Register of Members may be used as "C" and need not be certified]

4. The transactions referred to below were carried out in order that [set out reasons for the transactions. If no direct tax advance clearance was obtained, give a detailed explanation of the bona fide commercial reasons for the overall transaction].

5. By an agreement dated [date] and made between [names of shareholders of the Target company] ("the Shareholders"), the Target Company and the Acquiring Company ("the Agreement") it was provided (inter alia) that the Shareholders should sell and the Acquiring Company should purchase the respective numbers of shares of the Target Company as described in [refer to appropriate part of the Agreement] (such shares amounting in aggregate to the whole of the issued share

capital of the Target Company and that as consideration for such sale the Acquiring Company should allot credited as fully paid to the Shareholders (being all the shareholders of the Target Company) the respective numbers of shares of the Acquiring Company as specified in [refer to schedule or other part of the Agreement setting out the numbers and classes of shares to be allotted to each shareholder of the Target Company] ("the Consideration Shares"). A copy of the Agreement is enclosed marked "D".

6. The said sale was duly completed on [date] when the Shareholders delivered to and in favour of the Acquiring Company duly executed transfers of the whole of the issued shares, together with relevant Share Certificates, of the Target Company. [Immediately following such delivery one Ordinary Share of the Target Company was transferred to [name] as nominee of the Acquiring Company.]

7. At a meeting of [the Directors][or specify committee or similar] of the Acquiring Company held on [date] the consideration shares were duly allotted to the shareholders of the Target Company pursuant to the provisions of the Agreement. We enclose marked "E" [and "F" respectively] a certified copy of the resolution of the Directors of the Acquiring Company passed on [date] [appointing the said committee and of the Resolution of the said Committee] making such allotment. We also enclose marked "G" a certificate under the hand of [the senior official of the Registrars][the company secretary] of the Acquiring Company, confirming that the names of the respective allottees of the Consideration Shares have been entered in the Register of Members of the Acquiring Company in respect of the Consideration Shares, together with a list of all members of the Acquiring Company immediately following the allotment, certified by the [Registrars][company secretary] of the Acquiring Company and marked ["H"].

[Note – as in (4) above, a computer printout may be substituted and need not be certified]

8. Immediately after the acquisition the issued share capital of the Acquiring Company was *[give details of classes, denominations and numbers of shares]* such shares being fully paid up *[or give details if not]*.

9. It is confirmed that immediately after the acquisition the classes of shares in the Acquiring Company were in the same proportions (or as nearly as may be the same proportions) as they had been in the Target Company.

10. It is confirmed that, immediately after the acquisition the proportion of shares of any particular class in the Acquiring Company held by each respective shareholder was the same (or as nearly as may be the same) as the proportion of shares of that class in the Target Company held by that shareholder immediately before the acquisition was made.

11. [No application for clearance under the relevant direct tax provisions] [An application for clearance under *[section numbers]*] was made by the Acquiring Company. [A copy of the application together with copies of the correspondence with HMRC is enclosed marked ["I"].

[Note – if no clearance application was made, the information listed below must be supplied]

12. It is submitted that the acquisition was effected for bona fide commercial reasons and did not form part of a scheme or arrangement of which the main purpose or one of the main purposes is the avoidance of liability to stamp duty, income tax, corporation tax or capital gains tax and that the appropriate conditions of Section 75 of the Finance Act 1986 have been complied with, and accordingly exemption from ad valorem stamp duty under the head "Transfer on Sale" is claimed in respect of the Transfers executed pursuant thereto.

13. We enclose for adjudication *[give adequate list or description of the documents enclosed]* together with certified copies.

Further information to be provided if no clearance application was made as set out in (11) above:

1. A copy of the latest accounts of the Target Company;

2. Full details of any scheme, arrangement or overall transaction of which this transaction forms part;

3. Confirmation that the Acquiring Company still holds the shares of the Target Company and that there is no intention to dispose of them (or give details if disposal will occur in the course of a greater overall transaction);

4. A <u>detailed</u> note of the bona fide commercial reasons for the acquisition.

Index

[*References are to paragraph number and appendices*]